Plant Biodiversity

Rahmatullah Qureshi
Editor

Plant Biodiversity

A Key to Sustainable Agriculture and Climate Adaptation

Springer

Editor
Rahmatullah Qureshi
Department of Botany
Faculty of Sciences
Pir Mehr Ali Shah Arid Agriculture University
Rawalpindi, Pakistan

ISBN 978-981-92-1470-9 ISBN 978-981-92-1471-6 (eBook)
https://doi.org/10.1007/978-981-92-1471-6

This Springer imprint is published by the registered company Springer Nature Singapore Pte Ltd.
The registered company address is: 152 Beach Road, #21-01/04 Gateway East, Singapore 189721, Singapore

Foreword

Plant biodiversity constitutes the biological capital upon which sustainable agriculture, food security, ecosystem resilience, and human well-being fundamentally depend. In the twenty-first century, accelerating climate change, biodiversity erosion, and nutritional insecurity have emerged as deeply interconnected global challenges, particularly affecting agrarian societies and climate-vulnerable regions of the Global South. In this context, plant biodiversity is no longer merely a conservation concern; it is a strategic asset for climate adaptation, resilient food systems, and sustainable development.

The edited volume *Plant Biodiversity: A Key to Sustainable Agriculture and Climate Adaptation* responds to this global imperative with scientific rigor, interdisciplinary depth, and policy relevance. Bringing together contributions from ecology, agriculture, ethnobotany, food science, biotechnology, and conservation biology, the book provides a comprehensive synthesis of current knowledge on biodiversity-based approaches to sustainable agriculture. The early chapters establish a strong conceptual foundation by examining biodiversity–climate–food linkages and highlighting the critical role of plant genetic diversity in stabilizing agricultural productivity under climate stress.

These conceptual insights are reinforced by thematic chapters addressing traditional knowledge systems, ethnobotany, food as medicine, algae, hemp, and other underutilized plant resources—domains increasingly recognized for their potential to enhance nutritional security, diversify livelihoods, and strengthen ecosystem resilience. The volume is further distinguished by its rich collection of regional case studies spanning arid deserts, mountain ecosystems, forest landscapes, and agroecosystems of Pakistan, Saudi Arabia, Egypt, Morocco, and the Indo-Pak subcontinent. These contributions provide empirical evidence of how biodiversity-driven solutions function under real ecological and socioeconomic constraints.

The inclusion of applied innovations such as nano-fertilizers, climate-smart agricultural practices, and ex situ and in situ conservation strategies—adds significant practical value to the book. By bridging traditional knowledge with modern scientific advances, the volume demonstrates how integrated approaches can support resilient food systems and adaptive land-use strategies in a changing climate.

Importantly, this book aligns closely with international policy frameworks, including the United Nations Sustainable Development Goals, the Convention on Biological Diversity, and global climate adaptation agendas. By linking robust scientific evidence with actionable insights, it offers valuable guidance for researchers, educators, policymakers, and practitioners.

Overall, this volume represents a significant contribution to the growing body of literature advocating biodiversity-centered solutions to climate and food system challenges. It will serve as an authoritative reference and a source of inspiration for advancing sustainable agriculture and climate resilience through the conservation and sustainable use of plant biodiversity.

Institute of Biology, Biotechnology
and Environmental Protection
Faculty of Natural Sciences, University of Silesia
Katowice, Poland 16-03-2026

Adam Rostanski, PhD

Foreword

All living beings are dependent on "biodiversity" as it fulfils our needs and earthly life depends on this diversity. Nearly two million plant taxa are reported to live on our planet, which unfortunately are used in an unsustainable way, thereby suffering a great loss. The last 200 years have faced a great loss as more than 5000 plant taxa are said to have gone extinct and every year nearly 20 plants are lost. This is threatening all earthly living beings. As such, sustainable use of resources is highly important, which should be followed in harmony. Looking at the biodiversity as an irreplaceable resource, authors have attempted to correlate the subject with human existence.

This edited volume covers 22 chapters, all elucidating the foundational principles underpinning their integration into sustainable agricultural practices. The readers are introduced to the intricate molecular mechanisms and nanoscale phenomena that dictate the efficacy of these biostimulants, laying the groundwork for a comprehensive exploration of their potential applications. First chapter discusses the "Plant Biodiversity—A Key to Sustainable Agriculture and Climate Adaptation" prepared by the authors from Pakistan, highlightening the fact that how plant diversity contributes to the efficacy, stability, and resilience of agricultural land systems, especially in addressing climate change breakdown. It shows how to manage plant biodiversity in improving the climate resilience of agriculture and ensuring long-term food security in a practical and low-cost way. It is followed by the title "Plant Diversity and Food Security" compiled by a group of authors again from Pakistan, emphasizing the importance of crop diversity in improving nutritional stability, covering the regional imbalances, and the danger of excessive dependence on a small pool of staple crops. It highlights the significance of the wild food plants, neglected and underutilized species, and traditional types of crops, in increasing the dietary diversity, nutrition, and health, especially among the vulnerable population. It stresses the fact to move to food systems based on biodiversity in order to attain sustainable development concerns and the provision of sustainable food and nutrition security to the growing world population. The title "Climate Change and Biodiversity in Agriculture" covers data by the contributors from China and Pakistan enlightening the facts related to the agricultural sustainability and food security in the twenty-first century, directly impacting agricultural biodiversity as well as exploring the relationship between climate change and agricultural biodiversity and highlighting its importance for ecosystem functioning. The contributors from Pakistan and China have jointly prepared another chapter on "Traditional Knowledge and Ethnobotany" stating the sensitive interface of ethnobotany with traditional knowledge and highlighting the central role they play in conserving biodiversity preserving culture and ensuring sustainable development. The chapter looks into current documentation strategies, challenges with keeping oral knowledge, and the role of community-driven and digital archives. The role of food in medicine prepared by authors from Pakistan has attempted to enlighten the fact that recognizing the healing power of food can help healthcare to move toward a more preventive, patient-focused approach that appreciates the important role of diet in maintaining health. Another title deals with the subject of "From Pond to Plate: Safety Perspectives on Algae in Food and Feed Systems" compiled by the authors from Pakistan. It emphasizes the need for toxicological validation, monitoring of cultivation environments, and harmonized regulatory frameworks to ensure consumer safety and promote algae-based food and feed industries. "Hemp: A Potential Crop for Resilient Food Security" is another title, compiled by the authors from Pakistan, the United States, and Scotland, exploring how hemp, as a vital crop, can enhance food security through its nutritional benefits, environmental sustainability, and economic viability. A compilation from Egypt and Pakistan on Sustainable Use and Biodiversity of Egyptian Desert Plants: Economic Potential under Climate Change has been discussed in another chapter which assesses the current status of Egypt's natural plant resources through representative case studies from three desert regions, the traditional uses of desert plants for fodder, folk and herbal medicine, as well as household applications. It also presents the biogeographical distribution of economically important plant species across Egypt's major ecological regions. "From Forest to Table: Role of Wild Food Plants in Promoting Food Security and Biodiversity Conservation in Kabal Valley, Swat from Pakistan" highlights the urgent need to use indigenous knowledge systems and cultural heritage in ensuring sustainable utilization of the wild plant resources, and gives policy and research suggestions on how WFPs can be integrated into the national plans on food security, rural development, and biodiversity management in Pakistan is another chapter we come across. The title "Nano-Fertilizers and Plant Biodiversity: A Pathway Toward Climate-Smart Agriculture" has been evaluated and prepared by the authors from Pakistan and the United States. It presents the potential and restrictions of nano-based fertilizer technologies for the environment and biologically sustainable agriculture systems, written in a way to assist decision-makers, researchers, teachers, and farmers in implementing nanotechnology-based technologies that boost production, control the functioning of ecosystems, and promote sustainable development while fulfilling current and future climate adaptation targets. The follow-up chapter is titled as "Plant Biodiversity and Food Security in the Wake of Climate Change in AlUla Region, Saudi Arabia" which has been submitted by the authors from Pakistan-Egypt. As per the contributors AlUla region in northwestern Saudi Arabia is a unique ecological and cultural landscape shaped by millennia of natural and human history. The chapter provides an in-depth account of this region's plant biodiversity. It is expected to serve as a reference work for researchers, conservationists, and policymakers working to protect biodiversity while promoting sustainable food systems in hyper-arid regions. Another chapter presents the data on the topic prepared by the contributors from Pakistan and UK titled as "Medicinal Plant Wealth of Pakistan with Potential to Combat Cardiometabolic Disorders." It summarizes an overview of the metabolic syndrome (MetS), and its cardiometabolic complications, including type 2 diabetes mellitus, hypertension, dyslipidemia, and central obesity, have become a significant and growing public health problem in Pakistan. The authors stress that strict pharmacological and clinical studies, phytochemical standardization, sustainable cultivation, and the development of a solid regulatory framework are necessary to translate traditional knowledge into safe, effective, and clinically validated therapeutic solutions for cardiometabolic diseases. A group of researchers from Pakistan have compiled the chaper on "Forage Plants and Palatability Preferences in the Nara Desert: Ethnobotanical, Ecological, and Conservation Perspectives." Authors evaluate the seasonal availability and palatability of native forage plant species in the Nara Desert, Sindh, Pakistan, providing a foundation for further investigation into the nutritional and mineral composition of palatable species to support rangeland management and livestock productivity. A joint attempt has been made in the preparation of the titled "effects of climate change on fruit crops," where authors mention that in recent years, climate change has become a major driving force element affecting the growth and development of fruit crops, particularly in Pakistan, more than other industrialized countries. They have tried to reveal the potential to aid fruit crop growers in adapting novel farming methods that might help in boosting the cultivation of high-quality fruit crops. The title "Ex Situ and In Situ Conservation Strategies for Medicinal Plants of Arid Climates" is a contribution from Pakistan, emphasizing the need for collaborative action, integrating traditional knowledge with modern science, and promoting sustainable use and conservation of medicinally important plants in arid ecosystems. The authors from Pakistan have also submitted the title "Community-Led Plant Biodiversity Conservation in Pakistan: Case Studies and Lesson Learnt." It significantly contributes to deeper understanding of complex relationships between people, plants, and the environment as well as data on enhancing and strengthening community led conservation practices. Another group of workers from Pakistan have contributed the title on Biodiversity in Action: Case Studies of Resilient Solutions for Food Security. It emphasizes the need for NbS to focus on food access and utilization, involving community engagement and strong monitoring systems. While these strategies offer promising opportunities, challenges remain regarding their widespread adoption. Economic feasibility, long-term studies, and the integration of social benefits are areas that need further exploration and development. Additionally, local contexts and community involvement are crucial for the successful implementation of biodiversity-focused solutions aimed at improving food security. A chapter from Saudi Arabia-Pakistan is titled as: Integrating Plant Biodiversity into Climate-Smart Agriculture, emphasizing the use of traditional knowledge in the formulation of biodiversity management strategies, underlining the importance of the international cooperation and policy harmonization for further development of the biodiversity-related CSA practices. It is well established that education and public awareness are important factors to encourage cultural and behavioral shifts toward incorporating biodiversity in agri-systems. This will offer an effective means to connect ecological science and policymaking with implementation and foster sustainable farming and biodiversity conservation. The title "Traditional Food Plant Resources of the Mountain Ranges in the Indo-Pak Subcontinent" submitted from Pakistan summarizes the data on the mountainous range, the Himalaya, Karakoram, Hindu Kush, and the neighboring hill systems as significant hub of agrobiodiversity and traditional ecological knowledge. A collaborative team of researchers from the United States, Canada, and Pakistan contributed to the subsequent chapter entitled "Impact of Climate Change on Food Security and Biodiversity of Pakistan." The authors emphasize that achieving meaningful and lasting outcomes requires a comprehensive action plan based on public–private partnerships, with clearly defined responsibilities, accountability mechanisms, and effective enforcement of environmental laws and regulations. From this zero milestone, environmental protection will guarantee economic prosperity and sustainable development. The last contribution is titled "Integrating Biodiversity into Agri-Food Systems: A Transition Roadmap for Sustainable Agriculture in Morocco" jointly compiled by the authors from Morocco and Pakistan. It concludes with policy-relevant recommendations to operationalize biodiversity mainstreaming across agri-food value chains at regional and national scales.

The effects of plant diversity on ecosystem functioning aboveground have been previously reviewed. However, major knowledge gaps in how belowground effects contribute—and could be managed—to promote sustainable agriculture remain. These should be addressed to reliably predict conditions under which we can reach the desired outcomes. Uncertainties include the efficacy related to the different management practices with respect to belowground processes and the context dependency—including abiotic, biotic, and cultural—that should be accounted for to develop general strategies in sustainable agriculture. Nevertheless, growing evidence emerging from various fields highlights that plant diversity could be purposefully used to guide soil biodiversity, and we should take advantage of this potentially highly effective yet environmentally friendly and cost-effective management strategy now that sustainable agriculture is in greater demand than ever before. Evidence is mounting that promoting plant diversity in agricultural systems can drive soil microbial diversity, and jointly above- and belowground diversity should enhance and maintain the desired ecosystem functions of sustainable cropping systems. However, two major knowledge gaps remain. Building sustainable agroecosystems requires more than ecological and scientific–technological knowledge, as these systems are a combination of nature and society. Implementation of diversity in agroecosystems will end up with benefits of diversification which at the same time depend on other management practices, including conservation tillage, reuse of crop residues, and integrated pest management, which have been more extensively studied. The functional diversity, particularly the distribution of trait abundances, is a strong predictor of agricultural ecosystem multifunctionality, including weed suppression, nitrogen retention, inorganic nitrogen supply, increase in aboveground biomass, and sometimes yield.

Modern agriculture often relies on large fields of uniform crops and thus has large potential for diversification which can occur at varying spatial and temporal scales such as agroforestry systems incorporating trees and shrubs on and between the agricultural fields and the spatial arrangement of fields or rows of different main crops or main crops and service crops can create diverse landscapes. Also within fields, the mixture of different same-crop cultivars or multiple different species—be it multiple main crops or main crops with service crops—can contribute to local diversification. When different crops grow on the same field but are temporally separated, diversification occurs in time. In relay cropping systems, a subsequent crop needs to be planted before the prior crop is harvested. Thus, there is a time period when the two crops grow together, but not throughout their entire life cycles, so diversification occurs both in space and time. In crop rotation, different crops are sown after the harvest of the prior crop and diversification occurs solely in time. The plant diversity impacts multitrophic interactions and microbial community assembly and growth, more-diverse rhizodeposits will promote microbial nutrient cycling, increasing microbial community growth efficiency and the formation of more complex necromass, which will form soil organic matter, more persistent to degradation. High plant diversity results in more positive soil multitrophic interactions, which will impact soil carbon cycling. While evidence suggests that any of these diversification measures helps to provide some ecosystem function, we lack a general framework to maximize multiple ecosystem functions without compromising yields. Increasing biodiversity is often associated with reduced disease risk due to dilution effects, particularly when disturbance-caused biodiversity loss underlies biodiversity gradients. Overall, future studies should focus on how multiple dimensions of ecosystem functions and their underlying mechanisms may be impacted by biodiversity through time.

The contents in this volume enlighten the fact that plants play a role in all walks of our lives. It discusses different approaches toward conservation and evaluation of our ecosystems in light of future climate change scenarios. Hopefully the information provided in this edited volume will open new perspectives and prove useful at the global level. The readers will find important data while analyzing the chapters from different authors. The volume will serve as an important source for policymakers to bring up the general economy of their countries. I strongly advise the readers to make use of this important source in search for innovative and much-needed models to achieve better economic goals as a blessing for humans in their economic as well as food security. Throughout the book, readers will find different scientific inquiries on innovative applications, interdisciplinary collaborations, all redefining the boundaries of sustainability. It will inspire us all toward fostering interdisciplinary dialogue and collective action toward sustainability of our natural resources as a unique source and will be an idyllic source of scientific information for research scholars, faculty, and scientists globally.

Prof. Dr. Munir Ozturk (M.Sc., Ph.D., D.Sc.) has served at the Ege University, İzmir, Turkiye, for over 50 years. He is the Fellow and Ex-Vice President of "The World Islamic Academy of Science"; Fellow of Honour -Russian Academy of Ecology (Moscow); Foreign Fellow Pakistan Academy of Science; Distinguished Fellow (Division of Natural Sciences)-Core Academy, Singapur. He has served as Consultant Fellow, Faculty of Forestry, University Putra Malaysia, Malaysia, and Distinguished Scientist at ICCBS Karachi University, Pakistan. He has received the following fellowships: Alexander von Humboldt (Munich Technical University), Japanese Society for Promotion of Science (Tohoku University-Sendai), and National Science Foundation (USA-University of North Carolina-Chapel Hill). He has served as Ex-Chairman Botany Department and Founder Director Centre for Environmental Studies, Ege University, İzmir, Turkiye. His fields of scientific interest are plant ecophysiology, medicinal plant diversity, biosaline agriculture and crops, pollution, and biomonitoring. He has published 70 books, over 90 book chapters, and 182 scientific manuscripts in journals, and has served as guest editor for more than 14 journals. He is a member of 25 different national and international bodies.

His total citations are: 16163, h-index: 055, i-index: 376.

Department of Botany, Ege University Munir Ozturk
Izmir, Turkey

Preface

Climate change poses an unprecedented challenge to global food systems, biodiversity conservation, and rural livelihoods. Increasing climatic variability, land degradation, and ecosystem fragmentation are undermining agricultural productivity, particularly in climate-vulnerable regions. Against this backdrop, plant biodiversity emerges as a strategic resource for enhancing food security, nutritional diversity, ecosystem resilience, and adaptive capacity. This book was developed to capture and advance this critical discourse.

The volume comprises 22 interlinked chapters that offer conceptual analyses, thematic reviews, and region-specific case studies. The opening chapters focus on plant biodiversity as the foundation of sustainable agriculture and food security, followed by discussions on climate change impacts on agricultural biodiversity. These chapters provide a scientific framework for understanding why biodiversity is central to climate adaptation strategies.

A major strength of the book lies in its emphasis on traditional knowledge, ethnobotany, and food–medicine linkages, recognizing indigenous food systems as repositories of climate-adaptive wisdom. Chapters on algae, hemp, wild food plants, forage species, fruit crops, and medicinal plants highlight the potential of both cultivated and neglected resources to diversify diets, enhance resilience, and support livelihoods.

The book further integrates innovation and conservation, addressing nano-fertilizers, ex situ and in situ conservation, and community-led biodiversity management. The concluding chapters synthesize applied lessons through biodiversity-driven solutions for food security and climate adaptation.

This volume is intended for researchers, postgraduate students, policymakers, conservationists, and development professionals. By bridging science, tradition, and policy, it aims to support biodiversity-centered approaches aligned with global sustainability and climate agendas.

Climate change, biodiversity loss, and food insecurity constitute some of the most pressing and interlinked challenges of the twenty-first century. Rapid environmental change, declining ecosystem services, and increasing pressure on agricultural systems threaten global food security, particularly in arid, semi-arid, and biodiversity-rich regions of the Global South. Within this complex context, plant biodiversity offers a powerful, nature-based pathway for strengthening agricultural resilience, enhancing nutritional security, and supporting sustainable livelihoods. This book was conceived to explore these connections in an integrated and solution-oriented manner.

The volume *Plant Biodiversity: A Key to Sustainable Agriculture and Climate Adaptation* brings together *22 interdisciplinary chapters* that collectively demonstrate how biodiversity-centered approaches can contribute to multiple *United Nations Sustainable Development Goals (SDGs)*, notably *SDG 2 (Zero Hunger), SDG 3 (Good Health and Well-Being), SDG 12 (Responsible Consumption and Production), SDG 13 (Climate Action),* and *SDG 15 (Life on Land)*. Through conceptual analyses, thematic reviews, and region-specific case studies, the book highlights the central role of plant diversity in building climate-resilient food systems.

The opening chapters—"Plant Biodiversity: A Key to Sustainable Agriculture and Climate Adaptation," "Plant Biodiversity and Food Security," and "Climate Change and Biodiversity in Agriculture"—establish the scientific foundation linking plant genetic diversity with ecosystem stability, agricultural productivity, and climate adaptation. These chapters directly address *SDG 2* and *SDG 13* by demonstrating how diverse crop systems and agro-biodiversity reduce vulnerability to climatic extremes and contribute to sustainable food production.

Traditional knowledge and human health dimensions are explored in chapters such as "Traditional Knowledge and Ethnobotany," "Role of Food in Medicine," and "Traditional Wealth of Food Plants of Pakistan with Potential to Combat Cardiometabolic Disorders." These contributions emphasize the value of indigenous food systems and medicinal plants in improving dietary diversity, preventive healthcare, and community resilience, thereby supporting *SDG 3* and *SDG 12* while reinforcing the cultural dimensions of biodiversity conservation.

Several chapters focus on innovative and underutilized biological resources with strong relevance to sustainable production and climate-smart agriculture. "Algae as a Source of Food and Feed: Potential and Prospects" and "Hemp: A Potential Crop for Resilient Food Security" illustrate how alternative crops and bio-resources can diversify food systems, reduce environmental footprints, and generate green economic opportunities, contributing to *SDG 2*, *SDG 8 (Decent Work and Economic Growth)*, and *SDG 13*.

Regional case studies form a central pillar of the book, grounding global sustainability goals in local realities. Chapters such as "Sustainable Use and Biodiversity of Egyptian Desert Plants: Economic Potential under Climate Change," "Plant Biodiversity and Food Security in the Wake of Climate Change in AlUla Region—Saudi Arabia," "From Forest to Table: Role of Wild Food Plants in Promoting Food Security and Biodiversity Conservation in Kabal Valley, Swat, Pakistan," "Forage Plants and Palatability Preferences in the Nara Desert, Pakistan," and "Traditional Food Plant Resources of the Mountain Ranges in the Indo-Pak Subcontinent" provide empirical evidence of how plant biodiversity supports livelihoods, food security, and ecosystem resilience under diverse ecological conditions. These chapters strongly align with *SDG 15* and *SDG 13*.

The book also addresses future-oriented scientific and technological solutions through chapters such as "Nano-Fertilizers and Plant Biodiversity: A Pathway Toward Climate-Smart Agriculture," "Climate Change and Fruit Crop Productivity: Challenges and Adaptation Strategies," and "Harnessing Plant Biodiversity for Resilient Food Systems." These contributions demonstrate how innovation, when guided by ecological principles, can enhance productivity while minimizing environmental degradation, supporting *SDG 9 (Industry, Innovation and Infrastructure)* and *SDG 12*.

Conservation and governance frameworks are examined in chapters including "Ex Situ and In Situ Conservation Strategies for Medicinal Plants of Arid Climates," "Community-Led Plant Biodiversity Conservation in Pakistan: Case Studies and Lessons," and "Biodiversity in Action: Case Studies of Resilient Solutions for Food Security." These chapters highlight participatory conservation, institutional strengthening, and policy integration as essential components of long-term biodiversity stewardship, reinforcing commitments under the *Convention on Biological Diversity* and advancing *SDG 15*.

Thematic chapters addressing biodiversity, climate change, and food systems further strengthen the book's contribution by linking ecological knowledge with global sustainability agendas. Chapters such as "Traditional Food Plant Resources of the Mountain Ranges in the Indo-Pak Subcontinent," "Impact of Climate Change on Food Security and Biodiversity of Pakistan," and "Integrating Biodiversity into Agri-Food Systems: A Transition Roadmap for Sustainable Agriculture in Morocco" collectively highlight the critical role of plant diversity, indigenous knowledge, and adaptive agricultural strategies in ensuring resilient food systems. These contributions demonstrate how traditional plant resources underpin nutritional security, how climate change threatens both biodiversity and agricultural productivity, and how strategic integration of biodiversity into agri-food systems can facilitate sustainable transitions. Together, these chapters provide a comprehensive perspective that aligns strongly with *SDG 2 (Zero Hunger)*, *SDG 13 (Climate Action)*, *SDG 15 (Life on Land)*, and *SDG 12 (Responsible Consumption and Production)*, emphasizing the need for coordinated policy, innovation, and community-driven approaches to achieve sustainable agriculture.

This book is intended for researchers, postgraduate students, policymakers, conservation practitioners, and development professionals. By explicitly linking plant biodiversity with the Sustainable Development Goals and presenting evidence-based, regionally grounded solutions, the volume seeks to inform policy, guide research, and inspire action toward sustainable agriculture and climate adaptation in an increasingly uncertain world.

Rawalpindi, Pakistan Rahmatullah Qureshi

Acknowledgments

It is with great pleasure that I share the editorship of this volume. The true credit for this volume's content lies with the chapter authors. Their insightful contributions underwent a rigorous peer-review and proofreading process, ensuring the highest quality content. Their cooperation, responsiveness, and commitment throughout this journey were truly commendable. Any shortcomings, however, remain the responsibility of the editor. The exceptional quality of this volume is also a testament to the invaluable input and dedication of our reviewers. I, the editor, am deeply grateful for their willingness to share their time and expertise. Their critical analyses and expertise significantly enhanced the readability and overall quality of the chapters. I would like to express our sincere gratitude to the following reviewers:

Prof. Dr. Raza Bhatti, Professor Emeritus & Former Vice Chancellor, Sheikh Ayaz University, Shikarpur, Sindh, Pakistan
Prof. Dr. M. Qaiser, Professor Emeritus & Former Vice Chancellor, University of Karachi, Pakistan
Dr. Khan Bahadar Marwat, Professor Emeritus, Agriculture University, Peshawar, Pakistan
Prof. Dr. Tariq Mukhtar, Dean, Faculty of Agriculture, Pir Mehr Ali Shah Arid Agriculture University, Rawalpindi, Pakistan
Dr. Muhammad Ismail Bhatti, Pakistan Museum of Natural History, Islamabad, Pakistan
Prof. Dr. Mushataq Ahmad, Quaid-e-Azam University, Islamabad
Dr. Nasir Mehmood, IG Forest, Islamabad, Pakistan
Prof. Dr. Sher Wali, Department of Botany, Karakorum International University, Gilgit-Baltistan, Pakistan
Dr. Muzaffar Sirohi, Associate Professor of Botany, Shah Abdul Latif University, Khairpur, Pakistan
Dr. Ibrar Shinwari, Associate Professor, Department of Environmental Sciences, International Islamic University, Islamabad, Pakistan
Prof. Dr. Munir Ozturk, Department of Botany, Ege University, Izmir, Turkey

Prof. Dr. Volkan Okatan, Faculty of Agriculture, Department of Horticulture Eskişehir Osmangazi University, Turkey
Dr. Arshad Mehmood Khan, Assistant professor, Government Hashmat Ali Degree College, Shamsabad, Rawalpindi, Pakistan
Prof. Dr. Ömer KILIÇ, Adıyaman University, Turkey
Dr. Moha Haddouch, UNESCO Biosphere Reserves National Committee, Morocco
Prof. Dr. Humayun Shaheen, Department of Botany, University of Azad Jammu & Kashmir, Muzaffarabad
Prof. Dr. Muhammet Uzun, Marmara University, Turkiye
Dr. Amal M. Fakhry, Professor of Plant Ecology and Biodiversity, Faculty of Science, Alexandria University, Egypt
Prof. Dr. Muhammad Azim Khan, Agriculture University, Peshawar, KPK, Pakistan
Prof. Dr. Monier M. Abd El-Ghani, Faculty of Science, Cairo University, Giza, Egypt
Prof. Dr. Rahmatullah Qureshi, Dean, Department of Botany, Faculty of Sciences, Pir Mehr Ali Shah Arid Agriculture, Rawalpindi, Pakistan
Dr. Mudassar Iqbal, Clinical Assistant Professor, College of Health and Human Services, University of North Carolina Wilmington, USA
Dr. Muhammad Ilyas, Associate Professor, Government Boys College, G-10/4, Islamabad, Pakistan
Dr. Francisco Martín Huerta Martínez, Profesor Investigador Titular C, Departamento de Ecología, CUCBA-Universidad de Guadalajara, México
Dr. Muhammad Maqsood, Pakistan Atomic Energy Commission, Islamabad, Pakistan
Dr. Wasim Ahmad, Government College, Havelian, KPK, Pakistan
Dr. Gul Rahim, Department of Botany, University of Malakand, KPK, Pakistan
Dr. Mehmooda Munazir, GC-Women University, Sialkot, Punjab, Pakistan
Dr. Mubashrah Munir, Chairperson, Department of Botany, University of Veterinary & Animal Sciences, Lahore, Patoki Campus, Pakistan

Contents

Editor and Contributors

About the Editor

Rahmatullah Qureshi is an internationally recognized botanist and leading researcher in plant biodiversity, conservation ecology, and natural products. He serves as Dean, Faculty of Sciences at PMAS Arid Agriculture University, Rawalpindi. With over 30 years of experience, he earned M.Sc. (1996) and Ph.D. (2005) in Botany from Shah Abdul Latif University, Khairpur, Pakistan.

After qualifying as a Seed Certification Officer (2002), he gained expertise in seed science and quality assurance. In 2007, he joined PMAS AAUR, rising to Professor, Dean, and Director of the National Hemp Cultivation Project, introducing new degree programs. His national hemp initiative positioned Pakistan for global medicinal and industrial hemp markets.

He is a foremost expert on Pakistan's plant biodiversity, conducting fieldwork across all ecological zones. He collaborated with WWF Pakistan and other bodies, supporting biodiversity assessments of the Indus Ecoregion, Hingol and Khunjerab National Parks, Lal Suhanra, Taunsa Wildlife Sanctuary, Sindh's Ramsar sites, Nara, Thar, Thal, Cholistan Deserts, and Neelum Valley. He has led 12 national projects and 12 consultancies.

He has published over 300 peer-reviewed articles, 4 books, and 36 book chapters (Springer, Taylor & Francis). His work has >8200 citations, h-index 47, and i10-index 166, and he is ranked among the Top 2% Global Scientists by Stanford Elsevier. He has supervised 19 PhD, 77 M.Phil., and over 200 M.Sc. students.

He serves as Vice President (Punjab) of the Pakistan Botanical Society and on editorial boards of various high-impact journals. He received 5 Research Productivity Awards, 3 consecutive Best Research Professor Awards, organized 25 conferences, and delivered over 50 keynote/invited talks worldwide.

He filed seven patents in herbal therapeutics, nutraceuticals, and hemp technologies, founded university spin-off Himalaya Hemp & Health (Pvt.) Ltd., which launched 20 commercial products. His work shaped environmental policy, biodiversity strategies, and sustainable development in Pakistan. He also established a YouTube channel "Explore Nature with Prof. Dr. Rahmatullah Qureshi" to raise public awareness about plant diversity and ecosystem services.

Through scientific rigor and visionary leadership, Prof. Dr. Rahmatullah Qureshi stands as one of the most influential botanists in the region, inspiring future generations and advancing sustainable use of plant resources.

Contributors

Naila Parveen Abbasi Department of Geography, Shah Abdul Latif University Khairpur Mirs, Khairpur, Sindh, Pakistan

Syed Asad Abbas Department of Forestry, Kohat University of Science and Technology, Kohat, Khyber Pakhtunkhwa, Pakistan

Monier M. Abd El-Ghani Cairo University Herbarium, Department of Botany and Microbiology, Faculty of Science-Cairo University, Giza, Egypt

Sadia Abdullah Department of Botany, Rawalpindi Women University, Rawalpindi, Pakistan

Ozodbek S. Abduraimov Institute of Botany, Academy of Sciences, Republic of Uzbekistan, Tashkent, Uzbekistan

Arusa Aftab Department of Botany, Lahore College for Women University, Lahore, Pakistan

Qadeer Ahmad Department of Pharmacology, Faculty of Pharmaceutical Sciences, Government College University, Faisalabad, Pakistan

Mushtaq Ahmed Department of Plant Sciences, Quaid-e-Azam University, Islamabad, Pakistan

Adiba Akram Department of Biological Sciences, University of Veterinary and Animal Sciences, Lahore, Pakistan

Jameel M. Al-Khayri Department of Agricultural Biotechnology, College of Agriculture and Food Sciences, King Faisal University, Hofuf, Saudi Arabia

Ambar Amin Department of Biotechnology, Lahore College for Women University, Lahore, Pakistan

Maria Amin Center for Interdisciplinary Research in Basic Sciences, Faculty of Sciences, International Islamic University, Islamabad, Pakistan
Department of Botany, University of Mianwali, Mianwali, Pakistan

Muhammad Anas Department of Plant Sciences, Faculty of Biological Sciences, Quaid-I-Azam University, Islamabad, Pakistan

Maliha Asma Center for Interdisciplinary Research in Basic Sciences, Faculty of Sciences, International Islamic University, Islamabad, Pakistan

Mohammad Athar California Department of Food and Agriculture, Sacramento, CA, USA

Yamin Bibi Department of Botany, Rawalpindi Women University, Rawalpindi, Pakistan

Ozair Chaudhry DECAC, Environment and Climate Advisory. Planning & Economic Development. Durham, Regional Municipality, Rossland, Ontario, Canada

Samreen Dawood Department of Botany, University of Mianwali, Mianwali, Pakistan

Ishrat Fatima Department of Biological Sciences, University of Veterinary and Animal Sciences, Lahore, Pakistan

Samreen Fatima Department of Biological Sciences, University of Baltistan, Skardu Gilgit Baltistan, Pakistan

Moha Haddouch Independent Consultant, SECADERD, Hay Ibn Rochd Wifak Temara, Temara, Morocco

Muhammad Noor ul Hasan Department of Entomology, Faculty of Agricultural Sciences and Technology, University of Layyah, Layyah, Punjab, Pakistan

Chaudhry Sadam Hussain Department of Botany, Phir Mehr Ali Shah Arid Agriculture University, Rawalpindi, Pakistan

Muzzammil Hussain Department of Wildlife & Ecology, Faculty of Life Sciences, University of Okara, Okara, Punjab, Pakistan

Aqsa Ilyas Department of Biological Sciences, University of Veterinary and Animal Sciences, Lahore, Pakistan

Mohammad Ilyas Department of Biology, Islamabad Model College for Boys, Islamabad, Pakistan

Muhammad Zafar Iqbal Department of Botany, University of Karachi, Karachi, Pakistan

Sidra Islam Department of Botany, Lahore College for Women University, Lahore, Pakistan

Sumera Javad Department of Botany, Lahore College for Women University, Lahore, Pakistan

Bisma Javed Department of Botany, Rawalpindi Women University, Satellite Town, Rawalpindi, Pakistan

Mehrab Khalil Department of Zoology, Wildlife and Fisheries, University of Agriculture, Faisalabad, Punjab, Pakistan

Naveed Ahmed Khan School of Science, College of Science and Engineering, University of Derby, Derby, UK

Waseem Ahmed Khattak Department of Plant Sciences, Faculty of Biological Sciences, Quaid-I-Azam University, Islamabad, Pakistan

Ömer Kiliç Pharmacy Faculty, Adıyaman Üniversitesi, Eczacılık Fakültesi, Adıyaman, Turkey

Chad A. Kinney Institute of Cannabis Research, Colorado State University Pueblo, Pueblo, CO, USA

Saba Latif Department of Botany, PMAS-Arid Agriculture University, Rawalpindi, Pakistan

Muhammad Majeed Department of Botany, University of Gujrat, Gujrat, Punjab, Pakistan

Abdul Malik Department of Pharmacology, College of Pharmacy, University of Sargodha, Sargodha, Pakistan

Nyla Mansha The Design Build Inc, Sacramento, CA, USA

Mubeena Mansoor Department of Biological Sciences, University of Veterinary and Animal Sciences, Lahore, Pakistan

Zainab Maqbool Department of Botany, Lahore College for Women University, Lahore, Pakistan

Shumaila Mehdi Department of Pharmacology, Faculty of Pharmaceutical Sciences, Government College University, Faisalabad, Pakistan

Malik Hassan Mehmood Department of Pharmaceutical Sciences, Government College University, Lahore, Pakistan

Mehmooda Munazir Department of Botany, Government College Women University Sialkot, Sialkot, Pakistan

Mamoona Munir Department of Botany, Rawalpindi Women University, Rawalpindi, Pakistan

Mubashrah Munir Department of Biological Sciences, University of Veterinary and Animal Sciences, Lahore, Pakistan

Ambreen Najaf Department of Botany, Rawalpindi Women University, Rawalpindi, Pakistan

Sidra Naseer Department of Botany, Rawalpindi Women University, Rawalpindi, Pakistan

Saba Nazir Dr. Muhammad Ajmal Khan Institute of Sustainable Halophyte Utilization, University of Karachi, Karachi, Pakistan

Madalina Neacsu The Rowett Institute, University of Aberdeen, Aberdeen, Scotland

Mirza Faisal Qaseem Department of Botany, Pir Mehr Ali Shah Arid Agriculture University, Rawalpindi, Pakistan

Rahmatullah Qureshi Department of Botany, Faculty of Sciences, Pir Mehr Ali Shah Arid Agriculture University, Rawalpindi, Pakistan

Muhammad Shafiq Department of Botany, University of Karachi, Karachi, Pakistan

Shabnum Shaheen Department of Botany, Lahore College for Women University, Lahore, Pakistan

Muddaser Shah Department of Botany, Bacha Khan University, Charsadda, Khyber Pakhtunkhwa, Pakistan

Naima Shahzadi Department of Botany, Government College Women University Sialkot, Sialkot, Pakistan

Bushra Shaukat Department of Pharmacology, Faculty of Pharmaceutical Sciences, Government College University, Faisalabad, Pakistan

Ayesha Siddiqa Department of Botany, Pir Mehr Ali Shah Arid Agriculture University, Rawalpindi, Pakistan

Sumayya Siddiqa Department of Botany, Rawalpindi Women University, Rawalpindi, Pakistan

Ruqaiyyah Siddiqui Institute of Biological Chemistry, Biophysics and Bioengineering, Heriot-Watt University, Edinburgh, Edinburgh, UK

Rida Siddique Department of Pharmacology, Faculty of Pharmaceutical Sciences, Government College University, Faisalabad, Pakistan

Shazia Sultana Department of Plant Sciences, Quaid-e-Azam University, Islamabad, Pakistan

Aftab Umar Department of Botany, University of Agriculture, Faisalabad, Punjab, Pakistan

You-Cai Xiong Agricultural Ecology and Sustainability Science, College of Ecology, Lanzhou, Gansu, China

Zubaida Yousaf Department of Botany, Lahore College for Women University, Lahore, Pakistan

Ammara Zafar Department of Botany, Government College Women University Sialkot, Sialkot, Pakistan

Noor-Ul-Ain Zafar Department of Botany, Pir Mehr Ali Shah Arid Agriculture University, Rawalpindi, Pakistan

Kulsoom Zahara Department of Botany, Pir Mehr Ali Shah Arid Agriculture University, Rawalpindi, Pakistan

Ali Zeeshan Department of Biological Sciences, University of Veterinary and Animal Sciences, Lahore, Pakistan

Abbreviations

ABR	Arganeraie Biosphere Reserve
ACE	Angiotensin-Converting Enzyme
ADRL	Autoregressive Distributed Lag
AES	Agri-Environmental Schemes
AI	Artificial Intelligence
ALA	Alpha-Linolenic Acid
AMPs	Aromatic and Medicinal Plants
APC	Agricultural Paddy Cultivars
BAP	Biodiversity Action Plan
BD	Biological Diversity
BEF	Biodiversity, Ecosystem Functioning
BGI	Blue-Green Infrastructure
Br	Brassinosteroids
BSCs	Biological Soil Crusts
CAP	Common Agricultural Policy
CBD	Cannabidiol
CBD	Convention on Biological Diversity
CC	Changing Climate
CETP	Cholesterol Ester Transfer Protein
CIDSE	Cooperation Internationale pour le Développement and la Solidarité
CK	Compound Fertilizer
COBD	Conservation On Biological Diversity
CR	Critically Endangered
CRBD	Completely Randomized Block Design
CS	Calf Serum
CSA	Community Supported Agriculture
CSR	Corporate Social Responsibility
CUPRAC	Cupric Reducing Antioxidant Capacity
CWR	Crop Wild Relatives
DHA	Docosahexaenoic Acid
DMR	Direct Matrix Ranking
DPPH	2,2-Diphenyl-1-Picrylhydrazyl

EAC	Ehrlich Ascites Carcinoma
EbA	Ecosystem-Based Adaptation
EFSA	European Food Safety Authority
EGCG	Epigallocatechin Gallate
EN	Endangered
eNOS	Endothelial Nitric Oxide Synthase
ENPs	Engineered Nanoparticles
EPA	Eicosapentaenoic Acid
ES-MS	Electrospray Ionization-Mass Spectrometry
F2F	Farm to Fork
FAO	Food and Agriculture Organization
FDA	Food and Drug Administration
FFA	Free Fatty Acids
FL	Fidelity Level
FOP	Flora of Pakistan
FSF	Future Smart Foods
FSN	Food Security and Nutrition
GC-MS	Gas Chromatography-Mass Spectrometry
GDP	Gross Domestic Product
GHG	Greenhouse Gases
GIS	Geographical Information Systems
GLA	Gamma-Linolenic Acid
GLP-1	Glucagon-Like Peptide-1
GLUT4	Glucose Transporter Type 4
GMEM	Glasgow Minimum Essential Medium
GWP	Global Warming Potential
HMG-CoA	Inhibition of 3-Hydroxy-3-Methyl-Glutaryl-Coenzyme A
HP	Highly palatable
HPLC-ICP-MS	High-Performance Liquid Chromatography–Inductively Coupled Plasma Mass Spectrometry
IAS	Invasive Alien Species
ICF	Informant Consensus Factor
ICP-MS	Inductively Coupled Plasma Mass Spectrometry
IFAD	International Fund for Agricultural Development
IKS	Indigenous Knowledge Systems
ILO	International Labor Office
INKOTA	Information, Koordination, Tagungen
IODs	Indian Ocean Dipoles
IPBES	Intergovernmental Science-Policy Platform on Biodiversity and Ecosystem Services
IPCC	Intergovernmental Panel on Climate Change
IPM	Integrated Pest Management
IR	Insulin Resistance
IUCN	International Union on Conservation of Nature
IVI	Importance Value Index

KMGBF	Kunming Montreal Global Biodiversity Framework
LC	Least Concern
LC/ESI-MS	Liquid Chromatography/Electrospray-Mass Spectrometry
LCA	Life Cycle Assessment
LC-DAD	Liquid Chromatography with Diode Array Detector
LDL	Low Density Lipoprotein
LMICs	Low- and Middle-Income Countries
LN	Liquid Nitrogen
LOC	Line of Control
LOS	Low-Oxygen Storage
LP	Less Palatable
LPL	Lipoprotein Lipase
LPS	Low-Pressure Storage
MEAs	Multilateral Environmental Agreements
MeJA	Methyl Jasmonates
MetS	Metabolic Syndrome
MOEFCC	Ministry of Environment, Forest and Climate Change
MP	Moderately Palatable
NASA	National Aeronautics and Space Administration
NBFs	Nano-Biofertilizer Formulations
NbS	Nature-Based Solutions
NBSAP	National Biodiversity Strategy and Action Plan
NCS	National Conservation Strategy
NFs	Nano Fertilizers
NGOs	Non-Governmental Organizations
NSDS	National Sustainable Development Strategy
NT	Near Threatened
NUCS	Neglected and Underutilized Crop Species
NUS	Neglected and Underutilized Species
NZNP	Net-Zero Nature-Positive
OECD	Organisation for Economic Co-operation and Development
PAs	Protected Areas
PCBs	Polychlorinated Biphenyls
PEPCK	Phosphoenolpyruvate Carboxykinase
PGR	Plant Growth Regulator
PGRFA	Plant Genetic Resource for Food and Agriculture
PGRs	Plant Genetic Resources
PGS	Prostaglandins
PR	Preference Ranking
PRA	Participatory Rural Appraisal
PUFAs	Polyunsaturated Fatty Acids
R&D	Research and Development
RAAS	Renin-Angiotensin-Aldosterone System
RCTs	Randomized Controlled Trials
RCU	Royal Commission for AlUla

REDD	Reduced Emissions from Deforestation and Degradation
RH	Relative Humidity
ROS	Reactive Oxygen Species
RP	Rarely Palatable
SAF	Sistema Agroflorestal
SDA	Stearidonic Acid
SDGs	Sustainable Development Goals
SDLDL	Small, Dense Low-Density Lipoprotein
SFP	Sahara Forest Project
SGLT2	Sodium-Glucose Cotransporter 2
SIBEs	Sites of Biological and Ecological Interests
SOC	Soil Organic Carbon
SOM	Soil Organic Matter
SPCS	Sarhad Provincial Conservation Strategy
SSE	Social and Solidarity Economy
TCM	Traditional Chinese Medicine
TCs	Tropical Cyclones
TDF	Tropical Dry Forests
TEK	Traditional Ecological Knowledge
TG	Triglycerides
THC	Tetrahydrocannabinol
TK	Traditional Knowledge
TKDL	Traditional Knowledge Digital Library
Tr	Triazoles
UA	Urban Agriculture
UN	United Nations
UNCCD	United Nations Convention to Combat Desertification
UNEP	United Nations Environment Programme
UNESCO	United Nations Educational, Scientific and Cultural Organization
VLDL	Very-Low-Density Lipoproteins
VU	Vulnerable
WEPs	Wild Edible Plants
WFPs	Wild Food Plants
WHO	World Health Organization
WP	Whole Plant
WRB	World Resource Base
WUE	Water Use Efficiency
WWF	World Wildlife Fund

Plant Biodiversity: A Key to Sustainable Agriculture and Climate Adaptation

1

Rahmatullah Qureshi, Noor-Ul-Ain Zafar, Saba Latif, and Ayesha Siddiqa

Abstract

Plant biodiversity contributes to the efficacy, stability, and resilience of agricultural land systems, especially in addressing climate change breakdown. This chapter conveys how diverse plant communities are obligatory for the equipping of ecosystems, such as nutrient cycling, soil carbon storage, pollination, natural pest and disease control. It shows how simplified, high-input farming systems have a greater vulnerability to environmental stresses. Combining above- and below-ground processes, this chapter demonstrates how plant–soil–microbe interactions build agroecosystem resilience and form the basis of agroecological intensification. The importance of favorable policies and participation by farmers is stressed in this report along with practical diversification strategies, i.e., agroforestry, intercropping, rotations, and cover cropping. In sum, it shows how to manage plant biodiversity in improving the climate resilience of agriculture and ensuring long-term food security in a practical and low-cost way.

Keywords

Plant biodiversity · Sustainable agriculture · Ecosystem functioning · Agricultural productivity · Ecological intensification

R. Qureshi (✉)
Department of Botany, Faculty of Sciences, Pir Mehr Ali Shah Arid Agriculture University, Rawalpindi, Pakistan
e-mail: rahmatullahq@uaar.edu.pk

N.-Ul-A. Zafar · S. Latif · A. Siddiqa
Department of Botany, Pir Mehr Ali Shah Arid Agriculture University, Rawalpindi, Pakistan

R. Qureshi (ed.), *Plant Biodiversity*, https://doi.org/10.1007/978-981-92-1471-6_1

1.1 Introduction

Plant diversity is essential for the functionality, viability, and adaptability of agricultural ecosystems, and it is increasingly recognized as a critical component of sustainable agriculture and climate adaptation. Biodiversity promotes and stabilizes the ecological services that promote food production, including primary productivity, nutrient cycling, soil carbon sequestration, pollination, biological pest control, and disease regulation (Cardinale et al. 2012; Isbell et al. 2017; Tilman et al. 2014). Biodiversity sustains ecological activities critical to the production of food, such as primary production, the cycling of nutrients, the retention of carbon in the soil, pollination, ecological pest management, and disease maintenance (Blesh et al. 2015; Cardinale et al. 2012; Hooper et al. 2005). Most modern agricultural systems, on the other hand, frequently employ crop combinations, intense use of external inputs, and homogenization of landscape aspects, all of which contribute to their severe sensitivity to biological stresses and adverse climates (Foley et al. 2011; Folke et al. 2004). Climate change worsens these risks by increasing the likelihood and strength of severe droughts, extreme heat, floods, and pest outbreaks that threaten worldwide food supplies (Change, 2014). Farming practices, such as changes in the usage of land, fertilizers, the use of pesticides, and cultivation, lead to loss of biodiversity and worldwide emissions of greenhouse gases (Bélanger et al. 2019; Rockström et al. 2017). Maintaining a balance among crop production and protecting biodiversity, in addition to global warming issues, is expected to be one of the most significant obstacles of this century.

1.2 Biodiversity, Ecosystem Functioning, and Agricultural Sustainability

Biodiversity plays an important role in ecosystem functioning and increasing ecological system functioning and quality among a range of ecological circumstances across varied plant species (Cardinale et al. 2012; Isbell et al. 2017). It improves the plant biomass production, nutrient retention, and plant adaptation to biotic stresses from diseases and pests (Isbell et al. 2011; Loreau et al. 2001). The selection effect arising from competitive species in association, with the complementary effect caused by differences in resource usage, seasonal fluctuations, and functional features of plants, boosts the utilization of resource and their efficiency (Matson et al. 1997). Along with these beneficial features, diversity in agriculture has declined. Monoculture, particularly when assisted by artificial fertilizers and pesticides, might be profitable, but the outputs are frequently barren, dirty, and less robust lands (Tilman et al. 2006; Tscharntke et al. 2012). Eutrophication and acidification of soil are caused by nitrogen runoff, declining plant diversity, and biological cycling stability (Willem et al. 2014). Chemical pesticides, which spread much beyond their target sites, can build up in soils and water bodies, impacting other living organisms and human health as well (Carvalho, 2006). Annual species are less profitable, but have been shown to be especially significant to the well-being of agricultural systems and biodiversity. Policy initiatives should favor living challenges, nitrogen-fixing legume species, and annual crops in farming systems.

1.3 Ecological Intensification and Climate Adaptation

Agriculture should shift from its current heavy use of input expansion toward ecological expansion. Production could be based on biologically driven services of the ecosystem rather than depending upon external inputs if this sector is to meet the growing demand of food at no cost to ecosystem sustainability, as shown in Fig. 1.1 (Bommarco et al. 2013; Tittonell, 2014). Diverse ecosystems include plantation like crops, trees, and a diverse landscape improve pollination, cycling of nutrients, regulation of pests, and soil health (Pretty et al. 2018). Biodiversity enhances resistance and adaptability to environmental disruption. According to the known insurance hypothesis states that diversified species assemblages increase the possibility that an ecosystem's function is retained in new conditions due to species-specific adaptations to climatic variations (Isbell et al. 2015; Yachi et al. 1999). Experimental research suggests that biodiversity mitigates such disturbance in agricultural productivity that is attributed to drought and other extreme climatic disruptions (Jactel et al. 2017; Tilman et al. 2006).

1.4 Above- and Below-Ground Pathways Linking Plant Diversity to Ecosystem Functioning

Even though initial studies on BEF were focused on above-ground processes, below-ground interactions, including soil microbes that have received increased attention as playing an important role in the functioning of agricultural ecosystems (Bardgett et al. 2014; Van Der Heijden et al. 2008). Plant diversity influences soil microbes' distribution and functionality through root features, rhizome deposition, and trash inputs (Bassi et al. 2023). Biologically diverse plant communities promote complex microbial networks that control breakdown, nitrogen cycling, the formation of soil aggregation, and the exchange of greenhouse gases (Berg et al. 2009). Increasing microbial activities stabilizes soil organic matter and captures carbon

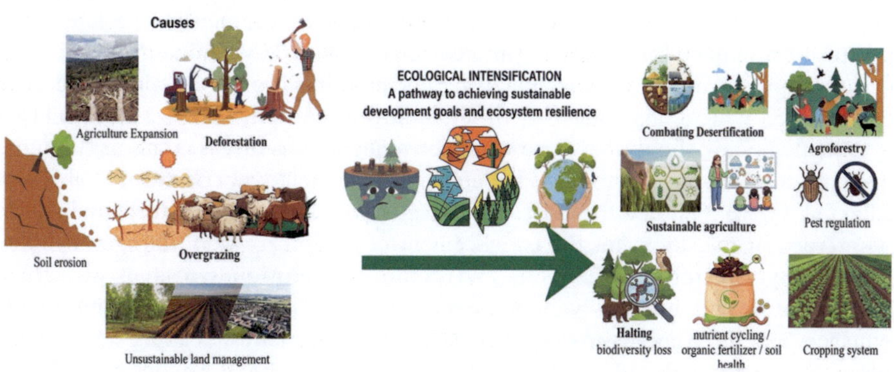

Fig. 1.1 Ecological intensification from degrading system to resilient system

(Delgado-Baquerizo et al. 2016; Kokhia et al. 2024). Plant-soil associations also refer to the causal relationships between plants and soil microbes and are fundamental to the adaptability and functionality of an ecosystem. Positive plant-soil feedback promotes functional microbial groups like arbuscular mycorrhizal fungi, whereas negative feedback comes from host-specific soil microbes. Furthermore, plant-soil connections will be dependent on, influenced by climate, soil qualities, and management techniques (Beattie et al. 2025; Lange et al. 2015).

1.5 Biodiversity-Based Diversification Strategies

Plant-soil feedbacks are the dual interactions that take place between plants and soil organisms, and which serve as the foundation for explaining the biodiversity functionality significance in agricultural ecosystems. Practical techniques to harness biodiversity include forestry, intercropping, crop rotation, cover cropping, and plant variety mixtures (Bever et al. 2010). Positive feedbacks occur when crops grow microbes that are beneficial and improve nutrient intake and inhibit diseases, whereas negative feedback occurs when host-specific pathogens grow (Kulmatiski et al. 2008). Multiple communities seem to reduce pathogens and enhance mutualists such as *Arbuscular mycorrhizal* fungi, which increase production and control diseases (Guo et al. 2025; Maron et al. 2011).

The amount and direction of feedback are context-dependent, influenced by environment, type of soil, and agricultural management (Li et al. n.d.). The impact of climate change may interrupt such processes; for example, drought or warmth may cause the microbial population to become less mutualisti (Castro et al. 2010). A complete understanding of these uncertainties is essential to developing sustainable agricultural systems.

1.6 Biodiversity-Based Diversification Strategies in Agriculture

When crop systems diversify in both temporal and spatial dimensions, a number of practical techniques for harnessing biodiversity arise. The techniques include agricultural forestry, intercropping, plant variety mixtures, cover cropping, and rotational crop, which take place at different levels (Altieri, 2019; Lin, 2011). A systematic review shows that there would be reductions in diseases and pest control, and enhancement of soil productivity, an increase in plant yield (Jeavons et al. 2023; Liu et al. 2025). On the other side, agroforestry practices combine trees and shrubs with crops or livestock, for the enhancement of carbon fixation, nitrogen cycling, biodiversity, microbes regulation, and control of erosion (Huss et al. 2022). Intercropping and in-plant variety mixtures: intercropping and cultivar mixture enhance the differences among species or plant varieties, leading to efficient resource use and reduced disease control (Leung et al. 2003; Snyder et al. 2020). Temporary diversification carried by cover crops and crop rotations disturbs pest and microbes' life cycles while enhancing soil texture and organic matter in the soil (Bennett 2017; Pan et al. 2009).

1.7 Trade-Offs, Multifunctionality, and Long-Term Perspectives

Agriculture based on biodiversity involves making trade-offs among environmental functions and the production of agriculture. Empirical evidence shows that richer ecosystems exhibit superior multifunction, producing a wide spectrum of ecosystem services at intermediate levels, whereas simplified, lower-diversity systems tend to normalize a single function at the expense of others, as found by (Manning et al. 2018). With time, the benefits of diversity increases biotic interactions, compared to the negative consequences related to monoculture, including pest proliferation (Bennett 2017; Tilman et al. 2006).

1.8 Social-Ecological Dimensions and Policy Implications

The transition to sustainable agriculture remains, basic social-ecological undertaking. The adoption of biodiversity-led farming is highly dependent on resources, knowledge, markets, and enabling environmental policies (Foley et al. 2011; Ostrom 2009). The small-scale farmers who conserve a large proportion of agricultural land, biodiversity resources have an important place in the maintenance of food security systems and climate stability, especially from a Global South perspective (Bélanger et al. 2019; Lin 2011). Diversification policies, as based on the basic principles of assurance of land, markets, and community involvement, serve as enabling incentives that align biodiversity with livelihood (Stratton et al. 2021). Resolving the issue serves for future agriculture policy frameworks that offer incentives toward diverse cropping systems, like polycultures, rotations in crops, and integrated crop-livestock systems. Supporting local markets, subsidy payments, and practices that advance biodiversity reinforce farmers' ability to sustain production (Change 2014; Pretty et al. 2018). Multidisciplinary research begins with the participation of stakeholders, is regarded as critical in social situations that enable the use of innovative agricultural ecology (Gaba et al. 2015; Wezel et al. 2009).

1.9 Future Directions: Integrating Biodiversity, Policy, and Agroecosystem Resilience

A shift to a sustainable agriculture ecosystem is a process that comprises environment and technology, as represented in Fig. 1.2. To have a good conservation of the biodiversity, there must be participation of farmers, policymakers, and consumers (Foley et al. 2011).

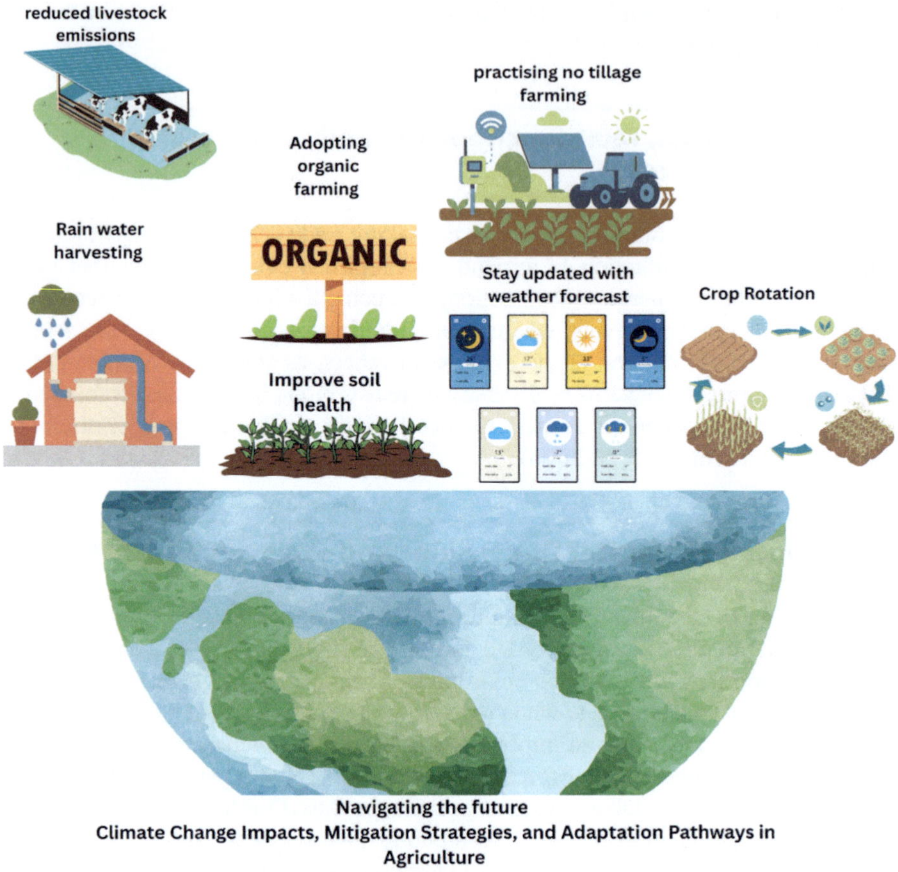

Fig. 1.2 Shows navigating the future: the climate change impact. Mitigating strategies and adaptation pathways in agriculture

On the other side, it has been shown that farm systems with help by using cooperative marketing or federal purchasing programs have richer on-farm biodiversity, self-cultivation of food, and access to soil nutrients (Bélanger et al. 2019; Hlpe, 2017). Another evidence is the support of the public policies on the conservation of production rights and a right to a healthy diet (Silva et al. 2025). Incentives on agricultural diversity, guaranteed domestic markets, and research at the farm level are some of the policies that could be used to promote environmentally friendly farming (Bommarco et al. 2013; Silva et al. 2025). The research priorities include: the restoration of positive plant-microbial interactions and plant-to-plant interactions, which have been damaged by the existing intense farming paradigm (Bardgett et al. 2014; Van Der Heijden et al. 2008). Balancing the farming practice, such that plant communities aim at ecological sustainability, will allow for maintaining multifunction and lessen greenhouse gas emission (Isbell et al. 2015).

The investigation of the potential of plant biodiversity is, by definition, free and a straightforward route to a sustainable agriculture and security in the face of climate change disruptions (Bardgett et al. 2014; Lal 2016). Plant species are useful in the sustainable agriculture of biodiversity. Despite the extensive research on the above-ground value of the plant species biodiversity, there is still a need to carry out more studies on the below-ground processes (Cardinale et al. 2012; Tilman et al. 2006). To be able to cope with the strategies that can be modified into the alternative settings, the uncertainties of the impact of the management practices under the various practices must be overcome. There is also growing evidence that management of plant biodiversity is a low-cost, sustainable environmental method for obtaining resilient agriculture.

1.10 Concluding Remarks

Plant diversity is a valuable tool for achieving sustainable, resilient, and climate-smart agriculture. The studies on ecosystem and biodiversity function show that diverse plant communities improve the supply of the important ecosystem services, including cycling of nutrients, soil carbon fixation, and control of pest and disease, while simplifier, input-intensive systems interfere with these processes and make the systems more prone to climate stresses. The resilience of agriculture does not depend only on the diversity above the ground. Moreover, the balancing of nutrients, physical traits of the soil, and long-term carbon storage are promoted by the complex interactions between plants, soils, and microorganisms that exist underground. Furthermore, such interactions facilitate ecological intensification by reducing resilience to external inputs. Agroforestry, intercropping, rotational crops, cover cropping, and cultivar combinations are all biodiversity-based strategies that provide practical and realistic choices for solving these situations. While there may be short-term trade-offs in yield, more diversified ones are more stable and poly-functional in the long-term compared to monocultures. The transition toward biodiversity-based agriculture is fundamentally a social-ecological transition, which will demand suitable policies and functionality of the market, as well as knowledge transfer and participating research, particularly so as to empower smallholder farmers whose contribution is essential in the process of conservation agrobiodiversity and sustainability to climate change.

References

Altieri MA (2019) Agroecology: principles and practices for diverse, resilient, and productive farming systems. In: Oxford Research Encyclopedia of Environmental Science

Bardgett RD, Van Der Putten WH (2014) Belowground biodiversity and ecosystem functioning. Nature 515(7528):505–511

Bassi L, Hennecke J, Albracht C, Broecher M, Solbach M, Schaller J, Ebeling A (2023) Leaf and root chemical and physical defence traits mediate monoculture yield decline in a grassland experiment. Authorea Preprints

Beattie GA, Edlund A, Esiobu N, Gilbert J, Nicolaisen MH, Jansson JK, Martiny J (2025) Soil microbiome interventions for carbon sequestration and climate mitigation. mSystems 10(1): e01129–e01124
Bélanger J, Pilling D (2019) The state of the world's biodiversity for food and agriculture. FAO
Bennett EM (2017) Research frontiers in ecosystem service science. Ecosystems 20(1):31–37
Berg G, Smalla K (2009) Plant species and soil type cooperatively shape the structure and function of microbial communities in the rhizosphere. FEMS Microbiol Ecol 68(1):1–13
Bever JD, Dickie IA, Facelli E, Facelli JM, Klironomos J, Moora M, Zobel M (2010) Rooting theories of plant community ecology in microbial interactions. Trends Ecol Evol 25(8):468–478
Blesh J, Wittman H (2015) "Brasilience:" assessing resilience in land reform settlements in the Brazilian Cerrado. Hum Ecol 43(4):531–546
Bommarco R, Kleijn D, Potts SG (2013) Ecological intensification: harnessing ecosystem services for food security. Trends Ecol Evol 28(4):230–238
Cardinale BJ, Duffy JE, Gonzalez A, Hooper DU, Perrings C, Venail P, Wardle DA (2012) Biodiversity loss and its impact on humanity. Nature 486(7401):59–67
Carvalho FP (2006) Agriculture, pesticides, food security and food safety. Environ Sci Pol 9(7–8): 685–692
Castro HF, Classen AT, Austin EE, Norby RJ, Schadt CW (2010) Soil microbial community responses to multiple experimental climate change drivers. Appl Environ Microbiol 76(4): 999–1007
Change IC (2014) Impacts, adaptation and vulnerability. Part A: global and sectoral aspects. Contribution of working group II to the fifth assessment report of the intergovernmental Panel on Climate Change, 1132
Delgado-Baquerizo M, Maestre FT, Reich PB, Jeffries TC, Gaitan JJ, Encinar D, Singh BK (2016) Microbial diversity drives multifunctionality in terrestrial ecosystems. Nat Commun 7(1):10541
Foley JA, Ramankutty N, Brauman KA, Cassidy ES, Gerber JS, Johnston M, West PC (2011) Solutions for a cultivated planet. Nature 478(7369):337–342
Folke C, Carpenter S, Walker B, Scheffer M, Elmqvist T, Gunderson L, Holling CS (2004) Regime shifts, resilience, and biodiversity in ecosystem management. Annu Rev Ecol Evol Syst 35(1): 557–581
Gaba S, Lescourret F, Boudsocq S, Enjalbert J, Hinsinger P, Journet E-P, Malézieux E (2015) Multiple cropping systems as drivers for providing multiple ecosystem services: from concepts to design: S. Gaba et al. Agron Sustain Dev 35(2):607–623
Guo H, Zhao Y, Zheng B, Huang Y, Chen X, Wang L (2025) Nitrogen deposition reshapes plant nutrient acquisition strategies: a meta-analysis within the root economics space. bioRxiv
Hlpe H (2017) Sustainable forestry for food security and nutrition. A Report by the High Level Panel of Experts on Food Security and Nutrition of the Committee on World Food Security
Hooper DU, Chapin FS III, Ewel JJ, Hector A, Inchausti P, Lavorel S et al (2005) Effects of biodiversity on ecosystem functioning: a consensus of current knowledge. Ecol Monogr 75(1): 3–35
Huss C, Holmes K, Blubaugh C (2022) Benefits and risks of intercropping for crop resilience and pest management. J Econ Entomol 115(5):1350–1362
Isbell F, Calcagno V, Hector A, Connolly J, Harpole WS, Reich PB, Van Ruijven J (2011) High plant diversity is needed to maintain ecosystem services. Nature 477(7363):199–202
Isbell F, Craven D, Connolly J, Loreau M, Schmid B, Beierkuhnlein C, De Luca E (2015) Biodiversity increases the resistance of ecosystem productivity to climate extremes. Nature 526(7574):574–577
Isbell F, Gonzalez A, Loreau M, Cowles J, Díaz S, Hector A, Duffy JE (2017) Linking the influence and dependence of people on biodiversity across scales. Nature 546(7656):65–72
Jactel H, Bauhus J, Boberg J, Bonal D, Castagneyrol B, Gardiner B, Brockerhoff EG (2017) Tree diversity drives forest stand resistance to natural disturbances. Curr For Rep 3(3):223–243

Jeavons E, Le Lann C, Van Baaren J (2023) Interactions between natural enemies and pollinators: combining ecological theory with agroecological management. Entomologia Generalis 43(2): 243–259

Kokhia M, Gorgadze O, Tsiklauri K, Lortkipanidze M, Nebieridze D, Kuchava M (2024) Earthworms (Oligochaeta: Lumbricidae) of Georgia, Caucasus: distribution and biodiversity. Zootaxa 5512(2):209–221

Kulmatiski A, Beard KH, Stevens JR, Cobbold SM (2008) Plant–soil feedbacks: a meta-analytical review. Ecol Lett 11(9):980–992

Lal R (2016) Soil health and carbon management. Food Energy Secur 5(4):212–222

Lange M, Eisenhauer N, Sierra CA, Bessler H, Engels C, Griffiths RI, Scheu S (2015) Plant diversity increases soil microbial activity and soil carbon storage. Nat Commun 6(1):6707

Leung H, Zhu Y, Revilla-Molina I, Fan JX, Chen H, Pangga I, Mew TW (2003) Using genetic diversity to achieve sustainable rice disease management. Plant Dis 87(10):1156–1169

Li X, Li M.-q (n.d.) Alterations in microhabitat can impact litter decomposition by reducing litter C/N and enhancing soil fungal activity. N and Enhancing Soil Fungal Activity

Lin BB (2011) Resilience in agriculture through crop diversification: adaptive management for environmental change. Bioscience 61(3):183–193

Liu K, Khakbazan M, Bandara M, Liang C, Machado PVF (2025) Diversifying wheat-based cropping systems with pulse crops enhances ecosystem services. Agron Sustain Dev 45(2):17

Loreau M, Hector A (2001) Partitioning selection and complementarity in biodiversity experiments. Nature 412(6842):72–76

Manning P, Van Der Plas F, Soliveres S, Allan E, Maestre FT, Mace G, Fischer M (2018) Redefining ecosystem multifunctionality. Nat Ecol Evol 2(3):427–436

Maron JL, Marler M, Klironomos JN, Cleveland CC (2011) Soil fungal pathogens and the relationship between plant diversity and productivity. Ecol Lett 14(1):36–41

Matson PA, Parton WJ, Power AG, Swift MJ (1997) Agricultural intensification and ecosystem properties. Science 277(5325):504–509

Ostrom E (2009) A general framework for analyzing sustainability of social-ecological systems. Science 325(5939):419–422

Pan Y, Birdsey R, Hom J, McCullough K (2009) Separating effects of changes in atmospheric composition, climate and land-use on carbon sequestration of US mid-Atlantic temperate forests. For Ecol Manag 259(2):151–164

Pretty J, Benton TG, Bharucha ZP, Dicks LV, Flora CB, Godfray HCJ, Morris C (2018) Global assessment of agricultural system redesign for sustainable intensification. Nat Sustainability 1(8):441–446

Rockström J, Williams J, Daily G, Noble A, Matthews N, Gordon L, Steduto P (2017) Sustainable intensification of agriculture for human prosperity and global sustainability. Ambio 46(1):4–17

Silva AR, de Sá MEG, Lelli Riga H, Norder LAC, Borsatto RS (2025) Agroecological transitions in traditional agricultural systems with agroforestry: outcomes in biodiversity hotspots. Agroecol Sustain Food Syst:1–23

Snyder LD, Gómez MI, Power AG (2020) Crop varietal mixtures as a strategy to support insect pest control, yield, economic, and nutritional services. Front Sustain Food Syst 4:60

Stratton AE, Wittman H, Blesh J (2021) Diversification supports farm income and improved working conditions during agroecological transitions in southern Brazil. Agron Sustain Dev 41(3):35

Tilman D, Reich PB, Knops JM (2006) Biodiversity and ecosystem stability in a decade-long grassland experiment. Nature 441(7093):629–632

Tilman D, Isbell F, Cowles JM (2014) Biodiversity and ecosystem functioning. Annu Rev Ecol Evol Syst 45(1):471–493

Tittonell P (2014) Ecological intensification of agriculture—sustainable by nature. Curr Opin Environ Sustain 8:53–61

Tscharntke T, Tylianakis JM, Rand TA, Didham RK, Fahrig L, Batáry P, Dormann CF (2012) Landscape moderation of biodiversity patterns and processes-eight hypotheses. Biol Rev 87(3): 661–685

Van Der Heijden MG, Bardgett RD, Van Straalen NM (2008) The unseen majority: soil microbes as drivers of plant diversity and productivity in terrestrial ecosystems. Ecol Lett 11(3):296–310

Wezel A, Bellon S, Doré T, Francis C, Vallod D, David C (2009) Agroecology as a science, a movement and a practice. A review. Agronomy Sustain Develop 29(4):503–515

Willem J, Leach A, Adams M, Agboola J, Ahmetaj L, Alard D, Ahmetaj L (2014) Nitrogen deposition effects on ecosystem services and interactions with other Pollutants and Climate Change (pp. 493–505)

Yachi S, Loreau M (1999) Biodiversity and ecosystem productivity in a fluctuating environment: the insurance hypothesis. Proc Natl Acad Sci 96(4):1463–1468

Plant Diversity and Food Security

2

Sidra Islam, Zainab Maqbool, Zubaida Yousaf, Arusa Aftab, and Rahmatullah Qureshi

Abstract

Plant biodiversity is closely connected with the idea of nutritional stability and global food security. The chapter discusses the importance of plant genetic resources, plant species and ecosystems in the development of resilient and sustainable food systems. It emphasizes the great importance of crop diversity in improving nutritional stability, covering the regional imbalances, and the danger of excessive dependence on a small pool of staple crops. The chapter highlights the significance of the wild food plants (WFPs), neglected and underutilized species (NUS), and traditional types of crops in increasing the dietary diversity, nutrition and health, especially among the vulnerable population. It also underlines the importance of the biodiversity in safeguarding the environment against stress and climate change, ensuring a steady crop, and ecosystem processes. The chapter concludes by highlighting the importance of biodiversity in terms of soil health and system sustainability by comparing the disadvantages of industrial monoculture with advantages of agroecological practices, such as crop diversification, agroforestry, and regenerative agriculture. In addition, destruction of habitats, climate change, and other irresponsible activities that cause the loss of biodiversity create serious dangers to the availability of food as well as the quality of food. The chapter ends with an important aspect on the urgent necessity of effective conservation plans that include seed banks, gene banks, and supportive policies that would combine traditional

S. Islam · Z. Maqbool · Z. Yousaf (✉) · A. Aftab
Department of Botany, Lahore College for Women University, Lahore, Pakistan
e-mail: Zubaida.Yousaf@lcwu.edu.pk

R. Qureshi
Department of Botany, Faculty of Sciences, Pir Mehr Ali Shah Arid Agriculture University, Rawalpindi, Pakistan
e-mail: rahmatullahq@uaar.edu.pk

R. Qureshi (ed.), *Plant Biodiversity*, https://doi.org/10.1007/978-981-92-1471-6_2

knowledge and modern science. It is necessary to move to food systems based on biodiversity in order to attain sustainable development concerns and the provision of sustainable food and nutrition security to the growing world population.

Keywords

Plant diversity · Food security · Plant genetic resources · Ecosystem functions · crop diversification

Abbreviations

APC	Agricultural paddy cultivars
FAO	Food and Agriculture Organization
FSF	Future Smart Foods
NUCS	Neglected and Underutilized Crop Species
NUCS	Neglected and Underutilized Crop Species
NUS	Neglected and Underutilized Species
NUS	Neglected and Underutilized Species
PGRs	Plant Genetic Resources
PGRs	Plant Genetic Resources
UN	United Nations
WEPs	Wild Food Plants
WFPs	Wild Food Plants
WHO	World Health Organization
WHO	World Health Organization

2.1 Plant Biodiversity a Larger Framework of Food Security

Crop and cattle genetic resources are crucial in sustainable agricultural activities particularly during climate change. Agricultural systems can be made more resilient through diversity in crops, livestock, and genetics and ecosystems, which will minimize the effects of climatic variability and assure food security (Loizou et al. 2025). When unique species and genetically diverse populations exist in large numbers, the value and reliability of biodiversity to aid food security is usually increased. Genetic, species, and environmental aspects of biodiversity are key to the livelihood improvement of the populations, enhancement of resilience, nutrition, food security, and the challenge of the various and changing agricultural systems. Some of the measures to enhance diversification are the use of multiple species, forestry, aquaculture, crops, and livestock, and habitat diversity at the landscape or seascape scale (Kumar et al. 2025).

2.2 Plant, Food, and Health

The world contains between half a million known plant species, of which around three hundred thousand are edible, and some of these edible plant species are wild food plants (WFPs), which grow naturally without any human intervention. WFPs refer to forest products that are not timber harvested in natural or managed settings as personal or commercial goods and which have cultural, social as well and religious implications. The plants are a main or secondary source of food in many cultures across the globe. The introduction of WFPs in our diets can improve food security and nutrition, which would benefit vulnerable groups like farmers and harvesters who will rely on these resources. Incorporation of wild plants into the human diet is an essential move in ensuring that some of the sustainable development goals of the United Nations (UN), such as no poverty, zero hunger, good health and well-being, and life on land are achieved (da Costa Santos et al. 2025).

Herbalism involves the employment of entire plants or certain parts of the plants to cure various diseases and is increasingly becoming popular in the developing nations. The extraction process and purification that applies to various herbal products are different, including dried materials, the raw parts of the plants and decoction like herbal tea. World Health Organization (WHO) statistics state that around 80% of the global population still has fundamental healthcare requirements met with the help of traditional medicine and herbs (Sharma 2025). In the province, a study revealed the presence of 47 plant species of 31 plant family with the presence of key antidiabetic plants such as *Momordica balsamina, Hypoxis hemerocallidea,* and *Aloe ferox*. The research reported botanical and regional names, parts of the plant used, methods of preparation, doses, frequency of references and the level of authenticity. The results highlight high levels of biodiversity of KwaZulu-Natal (KZN) and the important role that medicinal plants can play in the management of diabetes. The paper supports the need to conduct more pharmacological investigations to affirm the medicinal properties of these plants and determine their effectiveness. The cooperation between traditional healers, researchers, and policymakers will help to introduce these plants into modern healthcare systems and ensure the conservation of the medicinal plant species and the further use of safe and efficacious natural remedies in the modern healthcare (Yusuf et al. 2025).

2.3 Plant Biodiversity and Survival

2.3.1 Plant Genetic Diversity and Its Impact on Yield

The total genetic material found in a crop species and its wild relatives, including all of the alleles of different genes, is referred to as plant genetic resources (PGRs). They are a vital resource that people rely on to boost farming productivity and resilience. Therefore, as the global population grows, so will the need for genetic resources. To ensure long-term food security, the genetic variety of these precious resources must be preserved. All of the alleles of different genes found in a crop species, including horticultural and medicinal plants, as well as their wild cousins, are collectively referred to as plant genetic resources (PGRs). PGRs can also be described as any kind of plant species' vegetative or reproductive propagation material. Landraces, modern cultivars, obsolete cultivars, breeding stocks, wild forms, weedy forms, wild species of cultivated crops, newly developed varieties, cultivated crop varieties, and genetic stocks, such as current breeders' lines, elite lines, and mutants, are all considered PGRs. These serve as the foundation for genetically enhancing industrial and agricultural crops.

PGRs are also essential to the majority of the agro-processing and agro-industry sectors. These are the cornerstones of agricultural improvement initiatives, and the level of genetic variety found in PGRs determines global food security. Programs for crop improvement, especially those pertaining to varietal development, employ PGRs. Systematic study in fields like evolutionary biology, cytogenetics, biochemistry, physiology, phylogenetics, ecology, disease, molecular studies, etc., also makes use of these resources. PGRs include landraces, traditional cultivars, advanced breeding lines, and all cultivated and wild relatives of cultivated species. For crop types to be developed and genetically improved, PGR diversity must be maintained. The extinction rate of plant species is currently soaring, and anthropogenic activities and climate change are causing the sixth mass extinction event of life on Earth, which might result in ecological collapse (Teixeira and Huber 2021). Plant breeders can create or enhance crop varieties with desired traits thanks to diverse PGRs. Farmers' preferences, such as high-yielding varieties, quality, and resistance to diseases and insect pests, must be taken into account when creating new cultivars. In the past, people used the population's inherent genetic variability to choose desirable genotypes (Govindaraj et al. 2015). As the climate changes over time, so does the preference for the creation of new crop kinds. Plant breeders create climate-resilient cultivars that have all the desired characteristics, such as resistance to a range of biotic and abiotic stressors. Modern crop varieties are improved and developed through the utilization of genetic diversity, which can take the form of breeding stocks, wild species, mutant lines, etc. (Bhandari et al. 2017). Furthermore, it aids plant breeders in using genetically varied parents in a breeding program to increase the yield of horticultural and agricultural crop varieties (Raffard et al. 2019).

2.3.2 Biodiversity Is a Shield Against Environmental Stress and Climate Change

The level of variability in the flora growth, development, form, and reproduction are characterized by the particular climates. Climatic conditions also play relevant roles in adaptations. There is a variation in the habits of animals in other parts of the world and they have developed specific ways of living, dietary habits and reproductive behavior. Climate not only has a great influence in the human life, but also the local culture and the local traditions. Thus, it could be inferred that climate influences all spheres of human, animal, and plant life (Fig. 2.1). Six primary climates are discussed on Earth (Mashwani 2020). The processes of soil formation and soil

Fig. 2.1 Role of plant biodiversity on earth

conservation, water conservation and purification, regulation of hydrologic cycles, regulation of biochemical cycles, pollutants and waste products decomposed and decomposed, evaluation and regulation of natural climate all to take the role of maintaining ecosystem services and biodiversity (Rawat and Agarwal 2015). Climate change transforms the ecosystems, changes their operation, and biodiversity may reduce such impact. (Pires et al. 2018). Efforts by the government to tackle the problem of NCP degradation and biodiversity loss can also help to banish climate change. These co-benefits vary in accordance with the processes and natural elements that are the focus of management operations (Shin et al. 2022). It was a traditional belief that species-rich ecosystems offer insurance against severe droughts than the species-poor ecosystems based on the different rates at which different species would respond to the climatic disturbances. Diversity of species has been found to positively impact on resilience in grassland communities with moderate fluctuating climatic conditions. The species diversity has also been identified to increase the resilience of the global forest eco systems. The same has been observed in the case of tree seedlings in temperate forests in Europe and tropical areas, and this indicates that a high level of species diversity can alleviate the effect of drought (Yu'an 2025).

2.4 Importance of Traditional Crop Varieties

Neglected and underutilized species (NUS) like Amaranth, Bambara groundnut, Cassava, Fabirama, Okra, Moringa, Guinea sorrel, and sweet potato are required to solve the climate change and food insecurity issues and promote food sovereignty and agrobiodiversity (Rokka et al. 2025). It is essential that food security and sustainability can be improved through the development of climate-resilient food systems as Ginigaddara and Kodithuwakku (2024) note. Food systems are exposed to climate-related issues due to the overdependence on some major crops such as rice, maize, and wheat (Hossain et al. 2021). It is an opportunity that has been missed by ignoring nutrient-rich plants as part of a traditionally consumed food (Hunter et al. 2019). The concept of Neglected and Underutilized Crop Species (NUCS) is regarded as a potential opportunity, which will help satisfy the increased global food/nutritional needs. These crops have been described by the FAO as future smart foods (FSF) because of their resistance to climate, nutritional, and health-promoting bioactive compounds (Yesuph et al. 2025).

Agricultural paddy cultivars (APCs) have high nutritional and therapeutic values and they are resistant to severe stress. APCs are naturally tough and resistant to pests because they are soil and region-specific, organic and without pesticides. They are also the best in dealing with the problem of unpredictable rainfalls and insects thanks to their ability to withstand climate change. APCs are also less demanding to maintain and are more marketable than the high-yielding varieties. There are health conditions that have been treated with the traditional rice as well, such as blood sugar control, skin inflammation, digestive, and high blood pressure. Sustainable agriculture is essential in educating the farmers, seed breeders, and researchers on the significance of APCs and increasing their production fields (Patra et al. 2025). Traditional cereals have been ignored in food supply chains in the world even though they are nutritious and have diverse flavors. There is a growing need of the high quality bread, particularly in the urban markets, and this can give an opportunity to farmers who may be interested in the traditional kinds of bread as well as in sustainable methods of producing the bread. Nevertheless, there are issues like logistics, marketing, policies and seed supply. The proposed study will help to improve the value chain of traditional cereals in Portugal, recognizing the main opportunities, challenges, and developmental plans in the CERTRA project (Dinis et al. 2025). Certainly, plants have been used as medicine since ancient times, particularly in southwest Ethiopia, where people are dependent on herbs as a source of traditional medicine. Due to traditional medicine and low cost of the drugs, Ethiopian plants have proven effective in the treatment of different diseases in humans and animals. Medicinal plants are also popular and have healing qualities, which pose a great demand in Ethiopia (Awoke et al. 2025).

2.5 Role of Plant Biodiversity in Shaping Nutrition, Food Security, Health, and Trade

2.5.1 Integrating Biodiversity into Agro-Ecological Farming Practices

Natural agroecosystem processes are favored by the presence of different agriculture systems in the region, crop diversification, and associated practices. Crop diversification is being regarded as an important alternative to more simplistic production methods to increase the level of biodiversity in agricultural areas. Crop diversity has not been used extensively although its advantages toward agroecology are strong. Although scientific evidence of the benefits is overwhelming, the uptake of the approaches remains at a slower rate relative to what is required to address the needs of the society and the environment. The attitudes, intentions, perceived motives, control policy, incentives and market acceptability in relation to the farmers contribute largely toward its adoption. Agronomic interventions need to be redirected by redistributing subsidies to incentives that facilitate soil-based green revolutions, maintain high yields, and facilitate biodiversity to be successful (Gawdiya et al. 2025).

The loss of biodiversity in the era of the Green Revolution needs a fine-grained crop diversification and mosaic development of natural biotic populations at the landscape level to restore biodiversity. In introducing this review, we talk of the benefits of crop variety to aid in organic processes. These results are a result of interaction between individual or collective agency, including priorities, teamwork, and strategic choices, and structural influences, such as governance, cultural norms, and availability of resources, in intricate ways. This dynamic interaction determines crop diversification and how it is acquired in different agricultural situations and as well as how it promotes natural ecological processes. The dissemination of sustainable practices in farms and landscapes is more effective where there is a collaboration between people and communities held by robust networks and supported institutions. This collective action which is fueled by common knowledge, creativity and access to resources enhances the capacity to integrate ecological concepts that result in resilient and sustainable agroecosystems. Thus, the methods of crop diversification enhance the stability of yield, increase the adaptation to the environment, enhance the profitability in the long-term perspective, which is a win-win outcome. Diversification strategies need to be adopted at a bigger level to ensure that the biodiversity is preserved, that climate change effects are mitigated and that the food systems are transformed both in the local and the global context. The scientists need to identify this soon to enable better governance and evidence-based policymaking to bring a paradigm shift in the agriculture sector that relies on natural ecosystems (Gawdiya et al. 2025).

2.5.2 Plant Biodiversity Role in Regeneration of Soil Fertility

Regenerative farming practices have been promising in increasing the nutrient content of crops, increasing the health of the soil, and making the ecosystem more resilient. These methods comprise crop diversity, organic soil manipulation, less chemical application and agroecological practices (Fig. 2.2). The problem of food production and nutrition security can be solved sustainably by shifting to the regenerative agricultural model instead of the extractive, yield-maximizing one (Muhyidiyn 2025). Rubber agroforestry will be used to co-culture rubber trees with other appropriate crops or vegetation to maximize the utilization of the land, increase the soil fertility, and biodiversity (Tongkaemkaew et al. 2025). Sustainable

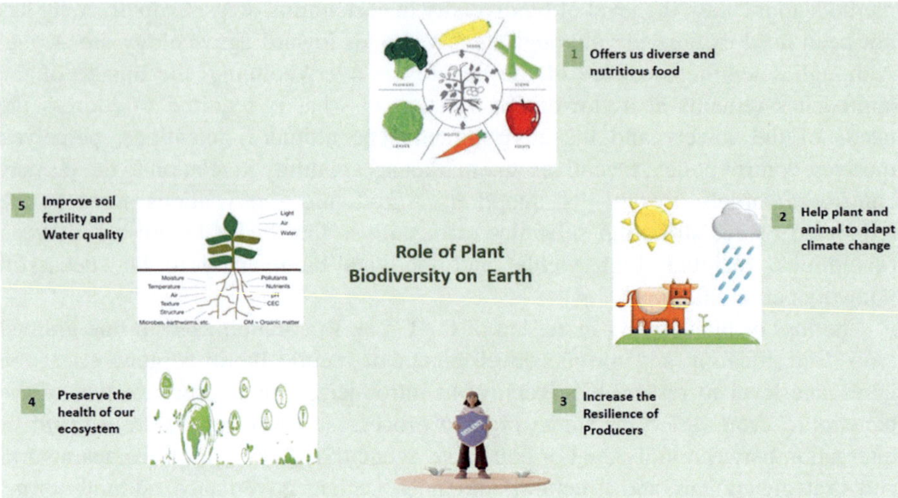

Fig. 2.2 Methods to improve soil fertility

agriculture involves soil health as an important aspect in order to guarantee sustainable production, environmental sustainability, and food security. Regenerative agriculture offers a holistic approach to restoring soil health by using methods in the form of cover crops, less-till farming, composting, agro forestry and rotating grazing (Kumar et al. 2025). Regenerative agriculture is an ever-growing field in which biocrust can take on new applications and be studied as an ecology because of the rising interest in using sustainable farming techniques. Biocrust, an intricate association of microbes crucial to the operation of an ecosystem, could be used in regenerative farming to augment the soil fertility and well-being. The study highlights the importance of knowledge gaps about the relationship between soil microbiomes and biocrusts and their effects on crop resilience and productivity due to the urgent necessity to enhance the fertility of soils and restore degraded lands (Gufwan et al. 2025).

2.6 Role of Agroforestry and Permaculture in Food Security

Agro forest incorporates trees and shrubs in fences and farmlands, which offer not only food but also wood products, crop, and cattle shade as well as environmental value such as pollinator habitat, water retention, and control of erosion (Fig. 2.3). The application of such practices as farmer-managed natural regeneration has

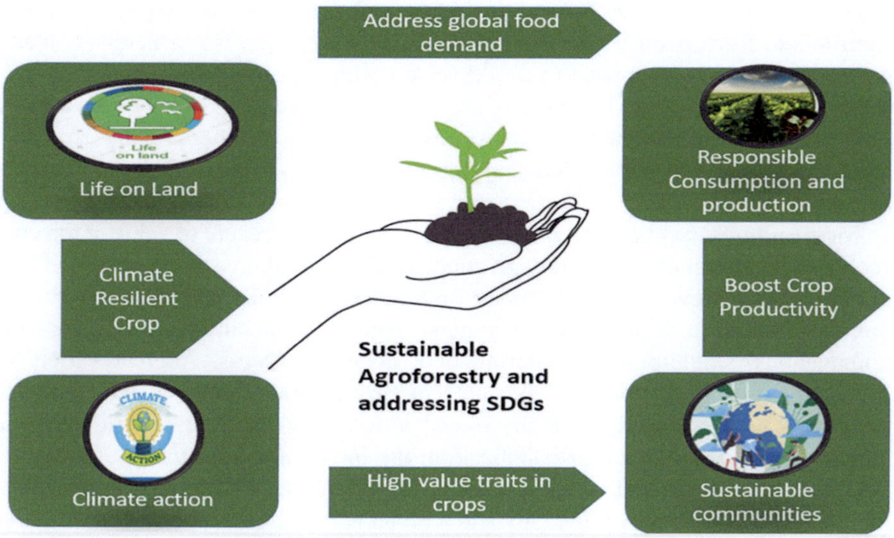

Fig. 2.3 Sustainable agroforestry and addressing SDGs

brought about more than five million hectares of degraded land in Niger resulting in drought alleviation and the generation of wood, food, and fodder. Permaculture is an ecological design method that incorporates crops, trees, livestock in close proximity with water collection and renewable energy to recreate natural ecosystems and make them self-sufficient and sustainable. Semi-arid Zimbabwe permaculture farms which have intercrop of leafy greens, beans and maize have produced up to four times above the conventional monocrops. Introduction of permaculture in more than 100 countries enables communities to achieve their needs in a sustainable manner. Permaculture describes a holistic approach to planning that tries to build resilient systems that recreate the connections and relationships that occur in the natural ecosystem (Saxena et al. 2025; Pazil et al. 2025).

2.7 Loss of Biodiversity and Its Impact on Food Systems

2.7.1 Consequence of Industrial Agriculture Monoculture

The industrial agriculture emerged due to a unique American culture of plenty land, successful industry, and a government that impacted the direction in scientific and agricultural research. The interdependency of the factors of industrial agriculture, the mechanization and chemical inputs, better seed, and scientific understanding of biological processes, and above all, the bureaucratic concepts of efficiency and rationality upon which they are used, is shown in both China and the Soviet Union. The reliance of industrial agriculture on fossil fuels is one of the primary reasons of climate change. The industrial livestock production is also another major contributor of greenhouse gases and cattle produce methane. Nevertheless, contemporary farming has been used to fit a diverse political agenda and social surroundings. In the 1980s, the Indonesian government subsidized 85% of the cost of pesticides, which has become politically acceptable in the country and has stimulated excessive application and subsequent loss of crops and resistance to bugs (McKittrick 2025).

Monocultures may be conceptual such as money as a measure of value or material such as crop monocultures. Although conceptual monocultures are exemplified by a universal worldview or mental frame of reference, which is offered as superior to, dominant over, or exclusive of all others, material monocultures encompass the homogenous, single-purpose ordering of a material process. A great variety of individuals and other forms of life are colonized with a range of monocultures associated with colonialism, or the forced introduction of a new worldview and way of life by the settlers, or colonization, the intentional adjustment of a natural system to a new, desired form, as is seen in agriculture. Material and conceptual monocultures, at a price of diversity and a broad and flexible repertoire of workable forms, functions and outcomes, make use of multiple control mechanisms to promote consistency, uniformity predictability and scalability. Monocultures are more vulnerable to systemic failure since their rigidity causes the control resources such as pesticides and antibiotics to become ineffective. By changing the inclination toward the prevalence of material and conceptual monocultures and toward ecological approaches that acknowledge a diverse array of life-supporting material and conceptual systems, it would enhance the long-term stability and integrity of socioecological systems (Garver and Derani 2025).

Food production in Denmark is dominated by industrial agriculture, which is increasingly being subject to criticism due to its harmful impact on the environment. Consequently, it is apparent that alternative food systems should be explored. An illustration of such would be permaculture which educates humans on how to live in harmony with the environment. This thesis explains the objectives of eight practitioners of Danish permaculture in their future gardens and their envisioned ways of disrupting industrial agriculture. The idea of Real Utopias proposed by Erik Olin Wright is implemented in order to make the visions not only desirable but also realistic. The research indicates that industrial agriculture has been prioritized in Denmark limiting the application of permaculture and other alternative food systems. Through information sharing, prioritizing good health over profits, building communities on the ground, growing food, and expanding their territory, the followers of permaculture intend to reverse this. This implies that permaculture may be a local alternative to industrial farming (Aarup 2025).

2.8 Risk Linked to Climate Change and Depletion of Genetic Resources

Climate change is altering the interactions of the plants and the pathogens and the effect is felt greatly on agricultural systems within the world. Increases in the levels of tropospheric ozoes brought about by climate change obstruct plant photosynthesis and predispose plants to biotic invasion (Kumar et al. 2025). Climate change is affecting global biodiversity through its effects on all taxa and habitats. Due to the fact that temperature is generally a determinant to their plants, mountain ecosystems are particularly susceptible to climate change. The study of the impact of future climate change on the plant diversity of the Irano-Anatolian global biodiversity hotspot, consisting of high altitudes and a rich endemic species, has excellent chances of inquiry.

Indeed, in recent decades, an increasing number of scientists worry about how the biodiversity of the planet is becoming affected by human activity on a global scale (Jaureguiberry et al. 2022). Although habitat loss and fragmentation are generally said to be the key causal factors, climate change has been cited as a significant factor in the current and future biodiversity crisis, threatening one in six species (Lajeunesse and Fourcade 2023). There are quite a number of but irregular studies done all over the world on the effects of climate change to the species composition of mountain areas and the high alpine ecosystem more so the European mountains (Auld et al. 2022). In particular, animal species in the European alpine systems are shifting their altitude ranges higher at an average rate of 2.7 m per year with the increase in temperature (Steinbauer et al. 2018).

Species are also experiencing significant range changes toward the poles and overall toward the north toward cooler and north-facing directions at an average of 6.1 km per decade. Climate sensitive species face the threat of going extinct due to the fact that these processes normally occur at a slower pace than the climate change. Although selection causes a decrease in genetic variety, it must be preserved by breeders in their tactics. Clustering of massive pools of genetic resources has in the past century represented great diversity. The use of the top kinds has not been resolved, however, because they do not work as well in creating variety. With the advancements of genotyping and statistical methods, they can also be utilized effectively with the help of diversity monitoring and bridging programs based on genomic prediction. The breeding of genetic variety inhibits sustainable genetic gain. Nevertheless, the depletion of diversity in closed systems is unavoidable in real numbers of the populations of the actual plant breeding, and there is a need to introduce suitable donors to the diversity.

The black plum (*Vitex doniana* Sweet), a wild plant belonging to the Lamiaceae family growing in sub-Saharan Africa, immensely improves the quality of life of the local people (women in particular). Nonetheless, its users exploit the use of this all round species. The two of the risky factors are bushfires and poor natural regeneration induced by a lack of seed germination due to seed dormancy. This over exploitation will lead to an ecological imbalance and degradation of the genetic material that will see the species extinct in their natural environment. Thus, to preserve this species, it is necessary to retain it in its natural environment, breed it, make it domestic, and create awareness of the population.

The increasing human population and the increased dependency on the medicinal plants puts constant pressure on the available resources and thus, some wild species continuously continue to decrease or even become extinct. The disappearance of natural wild vegetation is in the meantime taking place at an alarming pace. Thus, the conservation of medicinal plants species has been proposed in several ways worldwide (Mir et al. 2021).

2.9 Impact of Biodiversity Loss on Food Availability and Nutrition

The diversity of crop and livestock species and genetic diversity within a species is of paramount importance to sustainable agriculture, nutrition, and livelihood to millions of individuals. The direct connection involves food security, as it is closely associated with biodiversity of the plants and animals one consumes. Regrettably, the destruction of habitats (settlement, changing agricultural systems, deforestation, and industrialization) coupled with global warming and unrestricted propagation of invasive species has cost biodiversity, including genetic diversity, species diversity and ecosystem diversity a lot. This loss of biodiversity is further caused by pollution, nitrogen deposition, and shifts in precipitation (Cramer et al. 2017). The last half-century has seen the disappearance of indigenous and traditional food crops because of the focus on conventional cereal and horticulture-based crops as part of agriculture (Akinola et al. 2020).

Although more than 6000 plant species have been domesticated, nine of them yield 66% of all crops, which shows that there is monoculture farming. In the present state, 12–20 species constitute 80–90% of the human diet and rice, maize and wheat in isolation supply approximately 60% of the calories and plant-acquired proteins. The food security and human nutrition is under a real danger by heavy dependence on few species as a source of food which is compounded by loss of biodiversity through underutilization of the ecosystem and no conservation initiatives taken on them. Thus, farming must develop out of the period of 50 years of the Green Revolution technologies which were highly dependent on the increased inputs and genetic engineering. Though these practices have contributed to preventing massive famine, they have impacted some very negative effects like the extensive use of agrochemicals, ineffective irrigation, disappearance of additional beneficial biodiversity (e.g., pollinators, soil fauna), and less crop and variety diversity (Owino et al. 2022).

Climate change affects agricultural systems by changing the yield, biomass, food composition, and nutritional quality, influencing the global food production and human health (Panel 2020). Climate change has the potential to disorient food supply chains and logistics, resulting in the price variability of food and threatening human health, nutrition, and food security (FAO 2020). These disruptions affect more vulnerable populations, especially those with fewer resources (Niles et al. 2017). Attempts should also focus on transforming food systems into more nutrient-responsive and climate-resilient production-consumption (Bryan et al. 2019).

Sustainable food-based dietary guidelines can be used to support diets that are healthy to both human health and the environment (Wiseman et al. 2019). The effects of the climate change on human health may include the changes in the nutritional value of foods, and the rising level of CO^2 in the atmosphere (Dietterich et al. 2015). Increased CO^2 levels could cause the plants to grow faster but lower the amounts of plant protein and other minerals such as calcium, iron and zinc.

2.10 Preservation and Sustainable Use of Plant Resources

To give high quality information about conservation and sustainable use of the plant, ethnologists and especially ethnobotanists are still trying to catalogue the interface between the nature and culture. Plant conservation is the wise utilization of the available plant resources by the present generation in the name of the future generations. Since the surrounding communities consider themselves as the key stakeholders, the traditional conservation ethics protect the natural resources and plant diversity. Modern strategies of conservation of plants are supported globally, but not traditional ones. The use of taboos, totemism, rituals, domestication, reserves, secrecy, selective harvesting, sacred groves, etc., are some of the examples of traditional practices that are used to preserve plants. Sacred groves can be referred to as a piece of land that is believed to be sacred by the indigenous people and are maintained by sociocultural, economical, and religious means. These are such as traditional sacred grove, temple groves, burial and cremation sites, and the Kwedivikilo sacred forest at Tanzania. There are many ecological, social, and economic aspects that apply to these mostly informal conservation and use activities (Table 2.1). They have

Table 2.1 List of factors affecting our biodiversity presented by Muluneh (2021)

Factors	Effects on biodiversity
Habitat loss and fragmentation	Decrease in natural habitat, homogenization of species composition, fragmentation of landscapes, and soil degradation
Invasive alien species	Competition with and predation on native species Changes in ecosystem function Extinctions and homogenization
Overexploitation	Genetic contamination Extinctions and decreased populations Alien species introduced after resource depletion
Climate change	Homogenization and changes in ecosystem functioning Extinctions Expansion or contraction of species ranges
Pollution	Changes in species compositions and interactions Higher mortality rates Nutrient loading and acidification
Anthropogenic threats	Species extinction Habitat loss and conversion Degradation and fragmentation

also served to conserve plant species: *Ficus thonningii, Lippia javanica, Milicia excelsa, Adansonia digitata, Spathodea campanulata*, and *Ziziphus mucronata* (Ogwu and Osawaru 2022).

2.11 Policies and Global Efforts for Biodiversity Conservation: Approaches to Enhance Plant Biodiversity for Food Security

2.11.1 Role of Genetic Resources Centers and Seed Bank

Seed banks are important in the supply of wild and native species that cannot be found commercially because seed producers may be reluctant to supply it since the market demand may be unpredictable. The small number of species preferred in the market by independent seed collectors and commercial nurseries tends to be highly profitable, thereby making it difficult to fulfill the needs of a restoration program in terms of seed and intraspecific diversity, particularly when enhancement of species diversity is the goal.

Seed banks may work with development organizations to increase the supply of a variety of plant material to be used in restoration undertakings. Although seed banks may be able to provide seeds in a small amount to be used in the restoration process they can also be able to increase their own seed stock. Growing this would require technical capacity and infrastructure of the seed banks and availability of a favorable climate to help grow seeds of target species of restoration. This potential is yet to be exploited despite the growing awareness of the importance of seed banks in the supply of planting materials to restoration. Merritt and Dixon believe that seed banks should improve its techniques and enhance their capacity in the event that it can assist the restoration practices in terms of seed supply. Seed banks need to invest strategically in research and seed production capacities in order to satisfy the increasing demand of seeds (Wambugu et al. 2023).

The community genebanks and community seed banks may be confused and have limited scientific literature on the subject. One of the differences is in their main functions. A community seed bank is a common type of seed bank that is organized as a seed store which is used mainly as a storage site of crop production, though it is also known as an organized seed bank. Conversely, a community genebank is a genetic diversity preservation-service to the farming community, which functions as a backup mechanism of different local varieties and materials. As a result, the amount of seed that is distributed to a farmer by a community genebank is usually small as compared to a community seed bank. Genebank collections are typically comprehensive in terms of diversity of crops, with the emphasis on genetic diversity protection instead of economic growth or active crop cultivation. There are, however, other cases of these objectives overlapping, which would lead to a hybrid role merging both the characteristics of a community seedbank and community genebank.

2.12 Conclusion

Plant biodiversity plays a vital role in food security, nutrition, and agricultural resilience against climatic changes. Crop genetic diversity is crucial to creating high-resilience and nutritious varieties to climate change as well as farm and landscape-level diversity enhances ecosystem stability, soil fertility, and food choices. Reliance on select few important crops has predisposed food systems to vulnerability, as there is the necessity to include neglected species, traditional landraces, and wild edible plants into mainstream food systems and diets. The monocultures, habitat fragmentation, and over exploitation (which leads to loss of biodiversity) have threatened the food availability, quality, and long-term sustainability. In order to safeguard the plant genetic resources, conservation should focus on seed banks, community-based, and policy measures toward diversification. It is important to collaborate, make evidence-based policymaking as well as a balance between indigenous knowledge and scientific innovation. To develop an equitable, healthy, and climate-resilient food system, it is important to prioritize and invest in plant biodiversity to support the current and future generations.

References

Aarup TL (2025) An alternative path-permaculture visions as a response to Denmark's industrial agriculture. Master Thesis Series in Environmental Studies and Sustainability Science

Akinola R, Pereira LM, Mabhaudhi T, De Bruin FM, Rusch L (2020) A review of indigenous food crops in Africa and the implications for more sustainable and healthy food systems. Sustainability 12(8):3493

Auld J, Everingham SE, Hemmings FA, Moles AT (2022) Alpine plants are on the move: quantifying distribution shifts of Australian alpine plants through time. Divers Distrib 28(5): 943–955

Awoke A, Gudesho G, Chane K, Siyum Y, Tilahun W, Gebremedhin H, Tadesse A (2025) Traditionally used phytomedicines and their associated threats in Bita district, southwestern Ethiopia. J Ethnobiol Ethnomed 21(1):8

Bhandari HR, Bhanu AN, Srivastava K, Singh MN, Shreya HA (2017) Assessment of genetic diversity in crop plants—an overview. Adv Plants Agric Res 7:279–286

Bryan E, Chase C, Schulte M (2019) Nutrition-sensitive irrigation and water management. The World Bank

Cramer W, Egea E, Fischer J, Lux A, Salles JM, Settele J, Tichit M (2017) Biodiversity and food security: from trade-offs to synergies. Reg Environ Chang 17(5):1257–1259

da Costa Santos ÉM, Barbosa DM, Gomes DL, dos Santos GMC, de Almeida Caetano R, da Silva Queiroz F et al (2025) From forest to table: the role of product naming in consumer expectations of biodiversity-derived foods. Food Qual Prefer 122:105302

Dietterich LH, Zanobetti A, Kloog I, Huybers P, Leakey AD, Bloom AJ et al (2015) Impacts of elevated atmospheric CO_2 on nutrient content of important food crops. Sci Data 2(1):1–8

Dinis I, Santos D, Mendes-Moreira P (2025) Revitalising traditional cereals in Portugal: challenges, opportunities, and strategies for value chain development. Sustainability (2071-1050) 17(6)

FAO, U (2020) Toolkit for value chain analysis and market development integrating climate resilience and gender responsiveness-integrating agriculture in National Adaptation Plans (NAP-Ag) Programme

Garver G, Derani C (2025) Moving from monocultures to resilience. In: Ecological integrity and international law. Routledge, pp 165–179

Gawdiya S, Sharma RK, Singh H, Kumar D (2025) Crop diversification as a cornerstone for sustainable agroecosystems: tackling biodiversity loss and global food system challenges. Discov Appl Sci 7(5):1–16

Ginigaddara GAS, Kodithuwakku AN (2024) Building climate-resilient food systems in Sri Lanka through site-specific agricultural management. Eur J Agron 156:127148

Govindaraj M, Vetriventhan M, Srinivasan M (2015) Importance of genetic diversity assessment in crop plants and its recent advances: an overview of its analytical perspectives. Genet Res Int 2015:431487

Gufwan LA, Gufwan NM, Abdulraheem MI, Jaafar AAK (2025) Biocrust ecology and functionality for regenerative agriculture: a review of the opportunities and challenges facing microbial agroecosystem restoration. Arch Agron Soil Sci 71(1):1–28

Hossain A, Islam MT, Maitra S, Majumder D, Garai S, Mondal M, Islam T (2021) Neglected and underutilized crop species: are they future smart crops in fighting poverty, hunger and malnutrition under changing climate? In: Neglected and underutilized crops-towards nutritional security and sustainability. Springer Singapore, Singapore, pp 1–50

Hunter D, Borelli T, Beltrame DM, Oliveira CN, Coradin L, Wasike VW, Tartanac F (2019) The potential of neglected and underutilized species for improving diets and nutrition. Planta 250(3): 709–729

Jaureguiberry P, Titeux N, Wiemers M, Bowler DE, Coscieme L, Golden AS, Purvis A (2022) The direct drivers of recent global anthropogenic biodiversity loss. Sci Adv 8(45):eabm9982

Kumar N, Devi R, Kumar S, Kurbah I, Chandel A (2025) Empowering food security through indigenous biodiversity management. In: Ecologically mediated development: promoting biodiversity conservation and food security. Springer Nature Singapore, Singapore, pp 3–25

Lajeunesse A, Fourcade Y (2023) Temporal analysis of GBIF data reveals the restructuring of communities following climate change. J Anim Ecol 92(2):391–402

Loizou E, Spinthiropoulos K, Kalogiannidis S, Chatzitheodoridis F, Kalfas D, Tzilantonis G (2025) Enhancing climate resilience and food security in Greece through agricultural biodiversity. Land 14(4):838

Mashwani ZUR (2020) Environment, climate change and biodiversity. In: Environment, climate, plant and vegetation growth. Springer International Publishing, Cham, pp 473–501

McKittrick M (2025) Industrial agriculture. In: A companion to global environmental history. Wiley, pp 375–394

Mir TA, Jan M, Khare RK, Bhat MH (2021) Medicinal plant resources: threat to its biodiversity and conservation strategies. In: Medicinal and aromatic plants: healthcare and industrial applications. Springer International Publishing, Cham, pp 717–739

Muhyidiyn I (2025) Restoring agroecosystem biodiversity. International Congresses of Turkish Science and Technology Publishing, pp 854–865

Muluneh MG (2021) Impact of climate change on biodiversity and food security: a global perspective—a review article. Agric Food Secur 10:36. https://doi.org/10.1186/s40066-021-00318-5

Niles, M. T., Ahuja, R., Esquivel, J. M., Mango, N., Duncan, M., Heller, M., & Tirado, C. (2017). Climate change and food systems: assessing impacts and opportunities

Ogwu MC, Osawaru ME (2022) Traditional methods of plant conservation for sustainable utilization and development. In: Biodiversity in Africa: potentials, threats and conservation. Springer Nature Singapore, Singapore, pp 451–472

Owino V, Kumwenda C, Ekesa B, Parker ME, Ewoldt L, Roos N et al (2022) The impact of climate change on food systems, diet quality, nutrition, and health outcomes: a narrative review. Front Clim 4:941842

Panel G (2020) Future food systems: for people, our planet, and prosperity. Global Panel on Agriculture and Food Systems for Nutrition, London

Patra S, Saha A, Pal SC, Islam ARMT, Halder K, Srivastava AK et al (2025) Highlighting the role of traditional paddy for sustainable agriculture and livelihood: issues, policy intervention and the pathways. Discov Sustain 6(1):181

Pazil NSM, Muhamad SNN, Razak NI, Yais ZFYH (2025) Factors influencing the awareness to adopt permaculture practices among Harumanis farmers in Perlis. J Comput Res Innov 10(1): 146–156

Pires AP, Srivastava DS, Farjalla VF (2018) Is biodiversity able to buffer ecosystems from climate change? What we know and what we don't. Bioscience 68(4):273–280

Raffard A, Santoul F, Cucherousset J, Blanchet S (2019) The community and ecosystem consequences of intraspecific diversity: a meta-analysis. Biol Rev 94:648–661

Rawat US, Agarwal NK (2015) Biodiversity: concept, threats and conservation. Environ Conserv J 16(3):19–28

Rokka S, El Bilali H, Borelli T, Calabrese GJ, Lecci S, Otieno G, Acasto F (2025) Current policies for promoting neglected and underutilized crop species in Burkina Faso and Niger. Int J Agric Sustain 23(1):2459541

Saxena RB et al (2025) Role of sustainable agricultural practices in rural transformation

Sharma K (2025) Herbal pharmacopeias: bridging ancient traditions, nanotechnological innovation, and global regulatory cohesion for equitable healthcare. Pharmacol Res Nat Prod 8:100301

Shin YJ, Midgley GF, Archer ER, Arneth A, Barnes DK, Chan L, Smith P (2022) Actions to halt biodiversity loss generally benefit the climate. Glob Chang Biol 28(9):2846–2874

Steinbauer MJ, Grytnes JA, Jurasinski G, Kulonen A, Lenoir J, Pauli H, Wipf S (2018) Accelerated increase in plant species richness on mountain summits is linked to warming. Nature 556(7700): 231–234

Teixeira JC, Huber CD (2021) The inflated significance of neutral genetic diversity in conservation genetics. Biol Sci 118:e2015096118

Tongkaemkaew U, Commons M, Verhofste R, Phooi CL, Wonglom P, Bua-Khwan B (2025) Regeneration of soil fertility in relationship with the diversification of rubber agroforestry systems. ASEAN J Sci Technol Rep 28(2):e255018–e255018

Wambugu PW, Nyamongo DO, Kirwa EC (2023) Role of seed banks in supporting ecosystem and biodiversity conservation and restoration. Diversity 15(8):896. https://doi.org/10.3390/d15080896

Yesuph DS, Dalka DD, Baye MT (2025) Underutilized and neglected crop species and their role in enhancing household food security amid climate change, Wolaita zone, Ethiopia. Heliyon 11(3)

Yu'an W (2025) Plant biodiversity and drought resilience: insights from a resilience model study in North America. Ecol Front 45(2):507–513

Yusuf BO, Ajao AAN, Sabiu S (2025) Ethnobotanical and phytochemical investigation of antidiabetic plants in KwaZulu-Natal: bridging indigenous knowledge with modern healthcare. S Afr J Bot 184:769–779

Climate Change and Biodiversity in Agriculture

3

Bisma Javed, Mamoona Munir, Rahmatullah Qureshi, Sadia Abdullah, Mushtaq Ahmed, Sumayya Siddiqa, Maliha Asma, Shazia Sultana, Samreen Dawood, Maria Amin, and You-Cai Xiong

Abstract

Global food security, biodiversity, and agricultural sustainability are subject to a serious threat posed by climate change. The increase in greenhouse gas emission, deforestation, and land degradation has intensified the loss of biodiversity, deteriorated the ecosystem services including pollination, nutrient cycling, and pest regulation, and made crops and livestock more susceptible to environmental stresses and extreme events. As a result, there is a decline in agricultural systems in terms of productivity, increased instability, and increased risk of pests and diseases. This chapter focuses on the interaction and interdependence between the climate change and agricultural biodiversity and the need to consider genetic

B. Javed · M. Munir (✉) · S. Abdullah · S. Siddiqa
Department of Botany, Rawalpindi Women University, Rawalpindi, Pakistan
e-mail: mamoona.munir@f.rwu.edu.pk

R. Qureshi
Department of Botany, Faculty of Sciences, Pir Mehr Ali Shah Arid Agriculture University, Rawalpindi, Pakistan
e-mail: rahmatullahq@uaar.edu.pk

M. Ahmed · S. Sultana
Department of Plant Sciences, Quaid-e-Azam University, Islamabad, Pakistan

M. Asma · M. Amin
Center for Interdisciplinary Research in Basic Sciences, Faculty of Sciences, International Islamic University, Islamabad, Pakistan

S. Dawood
Department of Botany, University of Mianwali, Mianwali, Pakistan

Y.-C. Xiong
Agricultural Ecology and Sustainability Science, College of Ecology, Lanzhou, Gansu, China

MOE (Ministry of Education) Engineering Research Center for Dryland Agriculture and Ecological Conservation, Beijing, China

R. Qureshi (ed.), *Plant Biodiversity*, https://doi.org/10.1007/978-981-92-1471-6_3

diversity in crops and livestock as a major resource to create climate-resilient systems. It emphasizes the role of biodiversity in supporting the functioning of the ecosystem in addition to facilitating adaptation by the formation of varieties that resist drought, heat, salinity, and new biotic stresses. The chapter gives the strategic adaptation pathways such as climate-smart agriculture, agroecological practices, preservation of genetic resources, and incorporation of indigenous and traditional knowledge. It also highlights how underutilized crops, resource management that is sustainable, and innovation can help in enhancing resilience. Emphasis is placed on community-based strategies and involvement of smallholder farmers, which is backed by a system of policy and governance in line with global climate and biodiversity policies. The chapter highlights that conservation of agricultural biodiversity is necessary in order to attain resilient food systems, maintain rural livelihoods, and achieve a sustainable future in response to changing climatic conditions.

Keywords

Climate change · Agricultural biodiversity · Climate-resilient food systems · Agroecological approaches · Agricultural sustainability

Abbreviations

AI	Artificial intelligence
CAP	Common Agricultural Policy
CBD	Convention on Biological Diversity
CSA	Climate-Smart Agriculture
CSA	Community Supported Agriculture
CWR	Crop Wild Relatives
FAO	Food and Agriculture Organization
GHG	Greenhouse Gases
IAS	Invasive Alien Species
IODs	Indian Ocean Dipoles
IPCC	Intergovernmental Panel on Climate Change
IPM	Integrated Pest Management
LMICs	Low- and Middle-Income Countries
PAs	Protected Areas
SAF	Sistema Agroflorestal
TCs	Tropical Cyclones

3.1 Introduction

Climate change is a long-term change in the meteorological conditions, majorly as a result of human activities that change the composition of the atmosphere. Greenhouse gas levels have increased significantly since 1750; the amount of carbon dioxide emissions has increased to 36.14 billion metric tons in 2014 since the year 1990. Since 1975, the average global temperature has been rising at 0.15–0.20° C/ decade, and it is estimated that it will increase by 1.45–5.8 °C by the very end of this century. The level of CO_2 in the atmosphere has increased significantly with 65% of the emission being attributed to burning fossil fuels and industrial activities, 11% of emission being attributed to forestry, and the remaining 15% of the emission being attributed to methane, nitrous oxide and fluorinated gases (Mostafa et al. 2025).

The climate change causes extreme weather events that are disastrous to biodiversity such as floods, cyclones, and droughts. Natural catastrophes like earthquakes, extreme temperatures, volcanic eruptions, landslides, storms, wildfires, floods, and droughts are all factors that exacerbate the decrease in biodiversity affecting ecosystems and species diversity. The global climate-related disasters have grown by 7348 disasters between 2000 and 2019 and led to 1.23 million fatalities and economic damages of 2.97 trillion (Mostafa et al. 2025). In 2020–2024, droughts, wildfires, and extreme temperature became much more common. These conditions promote the bushfires, making them hotter and drier, the burning of biomass causes the emission of CO_2, and other greenhouse gases, which contribute to increasing global warming (Mostafa et al. 2025).

The mutual dependence of climate change and wildfires highlights the importance of successful management and mitigation measures that deal with the causes and effects. The intensity of storms has also been increased by climate change, which creates extreme rainfall and increases the amount of moisture in the atmosphere, creating the necessity to implement more climate resilience strategies and disaster preparedness, particularly in the areas prone to hazardous climate conditions. The IPCC stresses the necessity to decrease the emissions of the greenhouse gases, switch to the use of renewable sources of energy, and enhance energy efficiency (Mostafa et al. 2025). Other natural disasters caused by climate, like the melting polar ice caps, are irreversible tipping points which can only be mitigated to minimize the harm caused (Mostafa et al. 2025).

Ecosystem productivity, ecological balance, countering pollution, climate stabilization, water resources protection, and soil health are all possible with the help of biodiversity. It provides the key biological resources, such as medicines, food, decorative plants, wood products, and various genetic stock among species and ecosystems. In addition to ecological services, biodiversity also provides recreation, tourism, cultural values, education, and scientific research (Mostafa et al. 2025).

The United Nations has been trying to address climate change, but still there have been dire effects on biodiversity (Mostafa et al. 2025). Environmental degradation, habitat loss, over-exploitation of biological resources, and the introduction of alien species is a sixth mass extinction crisis which is caused by human-made environmental changes; some scientists have predicted that half of all species might become extinct over the next century (Mostafa et al. 2025). One of the main causes of this decrease is Africa (Mostafa et al. 2025). The rising global temperatures will surpass 1.5 °C compared to the pre-industrial temperatures in the next 20 years and species will have to adapt or face extinction (Mostafa et al. 2025). The climate change has already led to the migration of numerous plant and animal population toward the poles or toward the higher elevations (Mostafa et al. 2025). The populations which cannot migrate risk local extinction, which decreases their genetic diversity and makes them more susceptible to pests, diseases, and other stresses (Mostafa et al. 2025).

Extinction of species due to rising sea levels and extreme weather events is affecting some species such as the melomys, some of the Adélie penguin's colonies, and the western bumblebee (Mostafa et al. 2025). The factors lead to habitat destruction, development of new growth conditions, and increased vulnerability to pests and diseases (Mostafa et al. 2025). Agricultural productivity and food security have been threatened by almost 40% of the world plant species being on the brink of extinction (Mostafa et al. 2025). Invasive species also outcompete the native species, causing damages to the local biodiversity (Mostafa et al. 2025). Warming and acidification of oceans impact marine organisms, particularly corals, to a significant extent (Mostafa et al. 2025). Polar bears rely on sea ice and it is decreasing as the temperatures increase (Mostafa et al. 2025).

Climate change poses a high risk to agriculture and has a major economic impact. Climate effects are playing a role in domestic and international policies, trade patterns, the utilization of resources, regional planning, and human welfare. Temperature variation and variation in rainfall have drastic impacts on crop yields. The developing countries are also disproportionately impacted due to the fact that their economies depend largely on agriculture, there are fewer resources available to adapt, they are in warmer climate and experience more extreme events. In 2100, production of wheat, corn, and rice will decrease significantly. Since the 1900s, extreme weather events have become commonplace, with impacts on the yield of wheat. The frequency of drought is estimated to increase to 44.0% by 2100, as opposed to the 15.4% in 2025; this will lower the nutritional value of food grains like rice. Food production is also interrupted by flood and droughts and global warming influences the efficiency of pollination. The most susceptible sector is agriculture due to its reliance on weather conditions and the worsening of the hazardous conditions. The measures taken to mitigate the problem are sequestration of CO_2, reduction of emissions, growth of renewable energy, and encouragement of green energy production. Sustainable agriculture is a way to guarantee food security, safeguard natural resources, and establish human-nature harmony (Mostafa et al. 2025; Rădoi 2020).

3.2 Climate Change–Induced Extreme Events

The absolute warming of air on the global land surface has increased by 1.53 °C since the pre-industrial era (1850–1900) more than the sum of land-ocean warming (0.87 °C) (Ebi et al. 2021). It has resulted in more commonly, severe, and larger-scale extreme occurrences in most land locations, such as heavy precipitation, droughts, desertification, and dust storms. Climate change will likely perpetuate and intensify these tendencies; the severity, frequency, and length of extreme heat, heavy rains, and droughts are all expected to grow over the twenty-first century, as

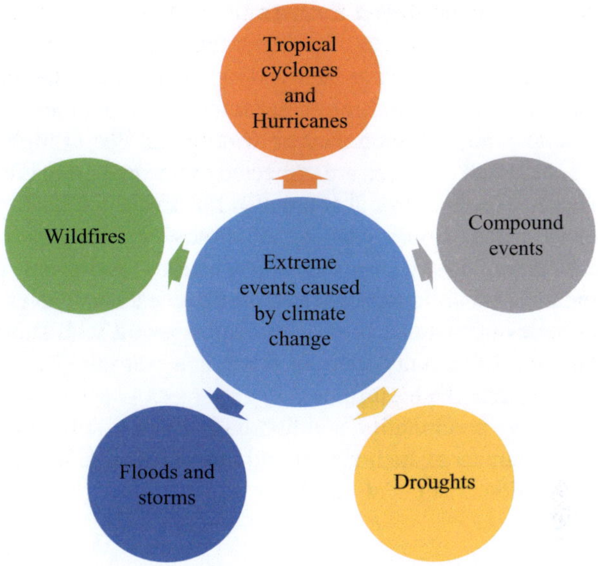

Fig. 3.1 Extreme events by climate change

shown in Fig. 3.1. Maximum temperatures on the hottest day of the year across regions differ between climate models; some areas might experience minimal change in the hottest day of the year as long as warming is in the lower quartile. Nevertheless, given that the Earth is likely to become warmer by 1.5 °C in 2030–2052, the next decades might be characterized by the massive and quick warming (Ebi et al. 2021).

3.2.1 Tropical Cyclones or Hurricanes

Tropical cyclones (TCs) are destructive and expensive processes that cause intense winds, flooding, and destruction of coastal infrastructure. TCs have cost the United States almost 1 trillion since 1980. Low- and middle-income countries (LMICs) incur fewer losses on an absolute basis, but the losses are a higher proportion of their domestic economies and have more welfare implications. TCs also harm human health: from 1963 to 2012, there were 2544 direct deaths in the U.S., mainly from TC-related forces. There are as many indirect deaths due to a storm worsening pre-existing conditions. The death toll of hurricane Katrina alone was 520 and 565 direct and indirect, respectively. New datasets on TC exposure enable long-term epidemiological studies of these diverse health impacts. Nonetheless, there is a low confidence in future TC projections; the most intense ones will probably become more common and with higher rates of precipitation, but it is unclear whether the overall number of TCs will rise or fall. There are also other trends, including the growing populations on the coastal regions and the rise in sea levels, which are exacerbating TC risks to communities (Ebi et al. 2021).

3.2.2 Compound Events

Other extremes such as heat waves may increase the ozone and other pollutants that pose harm to health through hazardous combinations of heat and air pollution. As an example, air pollution may be widespread due to high temperatures, wildfires, and smoke. Such episodes are projected to become more common due to climate change. Heat waves and droughts are also positively fed back, with heat and dryness being exacerbated. Some patterns of atmospheric waves have the potential to cause concurrent heat waves in many locations, increasing the chances of simultaneous breadbasket collapses. Most of the deadliest heat waves could actually be interpreted as series of extreme heat events interspersed with short periods of cooler climates. Historical incidents such as Hurricane Katrina (2005) and Hurricane Rita (2011) demonstrate the harmfulness of extremes of such compounds as tropical cyclones or excessive precipitation and flooding. There is still a lot more research to learn about the dynamics of such compound hydrological extremes, and to measure the risks to populations (Ebi et al. 2021).

3.2.3 Droughts

Global warming and altered precipitation patterns are increasing droughts frequency and severity. In the last 20 years, drought has afflicted over one billion people globally, with Africa being one of the most crucial examples due to the dire consequences on human health and society. Since 1980, 3865 deaths have been caused by droughts in the United States, mostly as a result of heat waves. The drought-health relationship pathways are diverse and difficult to follow; the most prevalent is less water to satisfy societal needs, as pollutants are increasingly concentrated. Stagnant, warm water that facilitates freshwater pathogens can also be caused by droughts. Due to the slow nature of droughts, disturbances may last a long time, and the health effects may manifest after a long delay including losses in agriculture and environmental degradation. Higher levels of particulate matter due to drought and high winds may lead to respiratory disease and fatalities, and transmit pathogens. Mental health issues and conflict have also been associated with drought (Ebi et al. 2021).

3.2.4 Floods and Storms

In the 1969–2018 period, 10,009 extreme weather events resulted in more than two million deaths and almost four million cases of diseases. Floods and storms were the most common extreme events worldwide, with storms, droughts, and floods being the most direct causes of health impacts. An analysis of health impacts from 1985 to 2014 found that these events affect health in various ways: injuries, infectious and parasitic diseases, exacerbations of non-communicable diseases, increased healthcare contact, and cardiopulmonary and skin complaints. Physical health issues and vulnerabilities vary in floods and storms; in Hurricane Harvey, the key problem was physical health issues that impacted those who did not evacuate. The risk of flood exposure is also not equal, and poorer people are at greater risk. Floods and hurricanes may augment post-traumatic stress disorder, depression, and anxiety and diminish the health-related quality of life. Floods may have a devastating effect on healthcare access and infrastructure, such as loss of records, effects on water supplies and laboratory operations, decreased access to care, and evacuations (Ebi et al. 2021).

3.2.5 Wildfires

The world has been experiencing more wildfires, and climate change is a contributing factor. Estimates of mortality burden of wildfire smoke range between 260,000 and 600,000, which could be lower with increase in population and severe wildfires in North America, Australia, Brazil, and other regions since 2012. Wildfire smoke emits chemicals that react with ground-level ozone; more research is needed on the health impacts of other smoke components. Exposure to wildfire smoke is associated with the most predictable negative respiratory health effects, including asthma attacks that result in hospitalization, emergency treatment, or doctor visit. It has been demonstrated that lung function in asthma patients does not necessarily decrease when exposed to smoke, perhaps due to the patients taking more medication. Older people are more affected by wildfire smoke. Wildfires may also translate to strains on health systems, growth in service demands, and destruction of infrastructure (Ebi et al. 2021).

3.2.6 Heat Waves

One of the most significant indicators of climate change is heat waves, or a time of excessively hot weather with high humidity. According to the definition of the World Meteorological Organization, a heat wave is 5 days or more during which the maximum temperature of the day surpasses the maximum temperature average by at least 5 °C (9 °F). Europe has had severe heat waves since the early twentieth century; the heat wave of 2003 resulted in 70,000 excess deaths in France and Italy, and the 2010 heat wave in Russia caused massive loss of crops, numerous wildfires, and approximately 55,000 excess deaths in Moscow. The heat waves normally appear when the high-pressure systems stand stationary and winds on the back side of the systems keep pumping the hot and humid air northeastward, creating extreme conditions. The rising severity and occurrence of heat waves cannot be attributed to natural climatic changes only; climate change caused by humans is a significant contributor. One of the factors that scientists speak about in relation to the mid-latitude weather in North America, Europe, and Asia is the weakening of the polar jet stream under the impact of global warming. Until the late twentieth century, heat waves were regarded primarily as a repeat of meteorological phenomena, with the main emphasis on the impact of drought on agrobiodiversity. Nevertheless, since 1950, unique alterations in severe climate and weather conditions have been noticed, and studies have evidently attributed these alterations to human intervention in the atmosphere with regard to the concentration of greenhouse gases. Climate extremities prove great susceptibility to global warming (Marx et al. 2021).

3.3 Impacts of Climate Change on Agriculture

The following model (Fig. 3.2) illustrates the overall effects of climate change on agriculture. In general, the figure underscores the fact that the issue of climate change affects agriculture in a very interrelated and systemic fashion and that there should be integrated measures that are adapted and mitigated toward climate change. The core

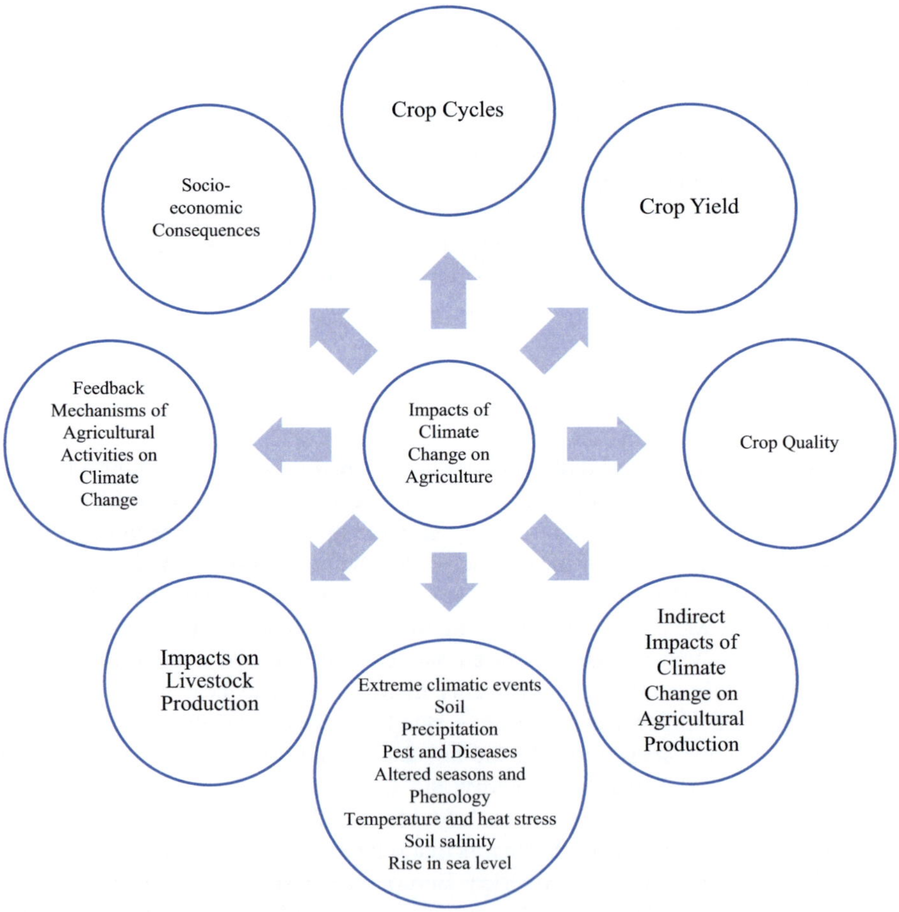

Fig. 3.2 Impacts of climate change on agriculture

climatic drivers that affect the overall agricultural systems are shown at the center and comprise a rise in temperature, changes in rainfall patterns, extreme weather, soil erosion, pest dynamics, and sea level rises. All these have direct impacts on crop production, which include changing crop cycles, decreasing crop yields, and affecting the quality of crops. Concurrently, indirect effects are manifested by the interruption of the agricultural systems, resource availability, and management practices. Climate change also has a huge impact on livestock production in the sense that it decreases the availability of food to feed them and increases the heat stress and disease risk. Notably, agriculture by itself is a source of climate change due to such feedbacks as greenhouse gases emissions and land use alterations, forming a vicious circle. The net effect of all these interactions is socio-economic effects whereby farmers earn less, are more vulnerable, and there are food security issues.

3.4 Cycling of Crops Affected by Climate Change

Climate change has reduced the time of crop growth, raised the change in temperatures and reduced the amount of sunlight, which may lead to insufficient nutrient content. It has impacted flowering and grain-filling processes as well, resulting in frost damage and drought stress. The study that integrated farmer decision models and biophysical crop models predicted crops cycles of maize, rice, sorghum, soybean, and wheat and included the impact of climate warming on crop yields. The research result was that planting dates and types could modify the beginning, ending, and middle of the growth, particularly high-temperature sensitive growth. Temperature is the main trigger of crop growth at a latitude of about 30° S and above. The other research study looked at the effects of global climate change on the water cycle and crop development, with the special consideration of winter wheat and summer maize, which were under intensive irrigation. The findings indicated earlier maturity and initiation of both crops, which resulted in increased end-of-season yields and overall biomass per day. This greatly reduces the length of the wheat and rice growing period, and changes the growth phases (Yuan et al. 2024).

3.4.1 Effects of Climate Change on Crop Yields

Climate change has a major effect on crop yields; high temperatures result in heightened evapotranspiration and water evaporation, resulting in soil drought and water shortage, which adversely impact yields. Global analysis has been done on four major cereals: maize, soybean, rice, and wheat. In other areas, smaller amounts of precipitation and increased temperatures have a compound heat-drought impact because they jointly result in a greater reduction in yields. Climate change is already affecting agricultural productivity sooner than was anticipated: high-latitude areas might experience yield gains whereas low-latitude tropical areas might suffer yield declines. An international survey of the yields of wheat, maize, millet, sorghum, and rice in extreme climatic conditions under no adaptations revealed that the simulated losses can be between 7% and 23%. Irrigation and nutrient management are the most effective adaptations, which demand a lot of investment but might not be possible in areas with water shortages. Consequently, the issue of food security in the regions and globally needs to be controlled through a multifactorial approach (Yuan et al. 2024).

3.4.2 Effects of Climate Change on Crop Quality

Crop quality, such as nutritional content and taste, is also a factor impacted by global climate change. Increasing CO_2 levels and temperatures have the potential to decrease protein levels and quality of staple crops, affecting consumer demand and agricultural production. Rice protein, iron, and zinc content decrease under constantly growing atmospheric CO_2, with an exception of vitamin E. Recent changes in vitamin content are attributed to the molecular level of nitrogen, which is a potential threat to the well-being of approximately 600 million populations across the world. Climate change has adverse effects on protein and yield of wheat. Climatic variation, including temperature variations, drought stress, and high temperatures, has a considerable impact on wheat protein, gluten percentage, and starch content. Climate change can have an impact on the quality of wheat by influencing the synthesis and accumulation of the wheat protein. These results emphasize the necessity of specific measures to tackle the issues climate change presents to agricultural production (Yuan et al. 2024).

3.5 Indirect Climate Change Effects on Agricultural Production

Climate change moderates' ecosystems, which has an indirect impact on agricultural productivity. Such indirect impacts are the alterations in the pest and disease occurrence, the frequency and severity of extreme weather events, deterioration of soil fertility, and the lack of water. These indirect effects may decrease agricultural production and even cause food security issues (Yuan et al. 2024).

3.5.1 Effect of Climate Change on Extreme Climatic Events

Extreme weather events are becoming more frequent and severe due to climate change and may lead to large-scale damage of crops or even crop failure. They can also destroy the agricultural infrastructure, increasing the cost of production. Droughts, hailstorms, and frosts are major factors that have impacted Brazil, which is highly dependent on agricultural production, leading to serious losses in national production, particularly in susceptible biological communities. With the adjustments to the farm subsidies and credit programs, the losses impact small-scale farmers, domestic and international commodity markets, and the national economy. Extreme weather patterns especially drought are serious threats to agricultural activities that lead to huge losses in yields and economic losses. Scholars have suggested novel measures to curb the danger of composite extreme events and climate change on crops and agriculture. Extreme weather events heavily affect the food supply chains due to their interconnectedness, with the fruits, vegetables, and livestock production being the most affected sectors (Yuan et al. 2024).

3.5.2 Effects of Climate Change on Soil

Climate change disturbs the soil organic carbon stocks, endangering the soil fertility and biodiversity and contributing to soil salinization worsening. Severe activities such as floods and droughts can make the soil less fertile and hasten the decomposition of soil organic matter, leading to the loss of nutrients. All these contribute to the decreased soil quality, which impacts crops. Climate change causes drought that has a complex set of ecological impacts, such as low plant species diversity, low photosynthesis, retarded root development, higher soil oxygen, and alterations in soil microbial community structure. The role of soil quality, fertility, and health in the context of biomass production is noteworthy as soil degradation and climate change impact its production. Modifications in the soil quality have a strong impact on the biomass production, implying that climate change can also have an impact on biomass production because of modifications in the quality of soil. Climate change, in particular, an increase in temperature and dryness, has adverse impacts on the multifunctionality of the soil, which may decrease the soil ability to sustain agricultural activities (Yuan et al. 2024).

3.5.3 Altered Growing Seasons and Phenology

Phenotypic variation among species helps preserve species coexistence in plant communities (Richardson et al. 2013; Chu et al. 2023). The length of the growing season is based on the senescence and growth start time (Richardson et al. 2013). Climate change tolerant crops may exhibit significant phenological events, such as flowering, fruit set, leaf drop, and winter hardening (Chu et al. 2023). Climate change influences the spring and the duration of the growing season. Plant phenology has changed in temperate and boreal areas, where the length of the growing season has increased. The phenological earlier spring and delayed leaf senescence results in extended leaf-on and could result in greater growth. However, responses vary among species: early-season species show more pronounced advances in spring phenology than late-season species. With the ongoing climate change, the differences might intensify, and the ecosystem structure and functioning could change (Richardson et al. 2013). Crop plants in boreal and temperate regions develop annually based on seasonal climatic conditions, but species-specific information about these changes is often lacking (Chu et al. 2023). The reducing day length and lower temperatures in fall induce bud set, leaf senescence, and dormancy, whereas temperatures are the major determinant of phenology in spring (Richardson et al. 2013). Low temperatures during early and cold springs, rather than their capacity to follow the temperature variations largely determine the phenological reactions of early-season species to global warming (Chu et al. 2023).

3.5.4 Temperature and Heat Stress

Another important abiotic factor that severely reduces crop yields and crop productivity is temperature and heat stress, which reduces farmer incomes. The plants have a preferable temperature range to experience normal physiological processes, growth, and development. Increasing temperatures around the planet are a critical risk to the development and yield of various plants, and it is predicted that the temperature will rise 2–3° C over the next half-century. Reduced life cycles, reduced reproductive periods, and reduced end yields can be the outcomes of faster development of growth stages in crops such as maize, wheat, and rice. Both extreme low and high temperatures may harm crop growth in the most vital growth phases, which will lower productivity. Research indicates that there is an inverse relationship between the yields of crops and average temperature rises above the optimum, and, on average, every 1° C higher temperature is estimated to decrease wheat production by 4–6%. The source-sink relationship can be negatively disrupted by high temperatures and reduce the final economic yields, and it is predicted that the loss in yield could reach 70% at the close of the twenty-first century (Alotaibi 2023).

3.5.5 Increasing Soil Salinity

Salinity is a significant issue in coastal agriculture, semi-arid, and arid regions of the global arena. Due to the last 25 years, the salinity level has risen by 33% in the coastal lowlands because of the rise of sea level. High temperatures and climate change have resulted in more drought and precipitation, causing more evapotranspiration in shallow water tables and intrusion of seawaters in coastal regions. The long-term effects of rising temperatures and low precipitation have led to the rise of salinity in arid soil, turning agricultural lands into salty ones at 10% annual rate. If soil salinization continues, over 50% of arable land will be salt-affected by 2050. The increase in the agricultural areas that are exposed to salt presents a major challenge to food security in the world; salinity due to climate change is likely to rise in arid lands like the northern areas of Algeria, Morocco, and Spain. However, soil salinity is expected to decrease or remain constant in dryland areas like the Horn of Africa, West Kazakhstan, the north-western United States, Turkmenistan, and Eastern Europe by 2100 (Alotaibi 2023).

3.5.6 Rise in Sea Level

It is forecasted that rising sea levels will cause the salinity of soils in the coastal regions to rise, resulting in the decrease of agricultural lands and low crop yields. Climate models assume that there will be more average precipitation, river runoff, water availability in the high-latitude and tropical areas, and a decrease in low mid-latitude and subtropical areas may contribute more to flooding and drought stress. The sea level is expected to rise by 50 cm by 2100, posing more threats of flooding in most parts of South and East Asia. Aquaculture, coastal fisheries, and fisheries that rely on spawning grounds in coastal wetlands and mangrove swamps are also threatened by sea level rise. There will be more and more intense tropical cyclones, and extreme storms and high-water events will take place and impact more inland. Sea flooding will impact more individuals by 2080 than sea level rise alone. Large, high-intensity cyclones from the Arabian Peninsula have occurred in recent decades, with the strongest and most intense ones occurring in the Indian Ocean during 2019. The rainy season in October to December 2019 was one of the wettest to hit East Africa and extreme positive Indian Ocean Dipoles (IODs) are likely to become more frequent with an increase in temperature (1.5 °C) and global warming (Alotaibi 2023).

3.6 Impacts on Livestock Production

Global warming, floods, and droughts contribute to climate change which has a significant impact on animal health and physiology, and consequently on livestock performance and welfare (Morgado et al. 2022; Cheng et al. 2022). Heat stress is a negative consequence of higher temperatures, which causes a decrease in heat production, inflammatory reactions, and a decrease in the immune system. This weakens the animal farming industry since nutrients are channeled to keep the body temperature but not to produce the products. Heat stress also influences the milk production, growth, reproduction, and carcass characteristics, which add to the healthcare expenses and result in extreme thermal stress particularly to lactating cows and animals without a shade. This stress causes the U.S. livestock industry an estimated annual loss of $1.7 to 2.4 billion as a result of respiration, pulse, heart rate, and body temperatures, reduced feed intake, milk production, reproduction efficiency, mortality, and immune system functioning (Cheng et al. 2022).

The environmental temperatures may cause serious impacts on some livestock such as ruminants, lactating dairy cows, goats, hogs, poultry, and dairy cattle. Such conditions may cause a decrease in feed consumption, a decrease in milk production, and reproductive problems. High heat stress decreases the appetite, rumination, and gut motility in ruminants. Heat stress related economic loss has a greater impact on the dairy industry, where milk production and feed dry matter intake decreases in dairy cows. Ruminants exposed to heat stress have a smaller body size, carcass weight, and fat thickness, and lower meat quality. Small ruminants, feedlot cattle, hogs, chicken, broilers, and eggshell strength are more heat-stress sensitive compared to other characteristics. It also affects reproduction: females are more likely to experience anestrus and embryonic mortality and males experience reduced estrous periods and fertility. In layers, environmental stress can slow ovulation, decrease yolk quality, and hatchability. Species, breed, geographical location, disease characteristics, and animal susceptibility are some of the factors through which climate change affects the health of livestock. The impact of heat stress on immune functions can adversely influence the susceptibility of livestock to diseases and their morbidity and mortality rates. The most influential factor is mortality, where more cows die among dairy cows (Cheng et al. 2022).

3.7 Indirect Effects

Climate and water supplies affect livestock feed, which is mostly made of the forages and grain/oilseed crop products. The impact of climate change on feed supply and water is indirect because of its impact on crop production. Grasses and legumes are examples of forage plants; climate change impacts on the amount and quality of forage, and its extent is determined by the livestock system, area, and species. Production variation is also affected by precipitation patterns and extreme weather events. Another factor that is influential is water supplies: C3 crops (soybeans, cotton, and wheat) grow better when the CO_2 level is higher, whereas C4 crops (corn and sorghum) might indirectly grow better when there is drought. The other climatic conditions that influence livestock feed are precipitation, temperature, and extreme weather conditions such as drought. More rain is favorable to corn, sorghum, rice, and soybeans, whereas C4 plants experience more advantages due to rise of temperature. Drought results in severe losses of crop production particularly in hot areas. The supply of grass forage is also impacted, and higher average temperature leads to drastic alterations in pasture structure, composition, and distribution of the biome (Cheng et al. 2022).

3.7.1 Forage Quality

Forage is an essential constituent in the diets of ruminants and plays an important part in weight gain, production, and reproduction. Nevertheless, the quality of forage is highly variable between different types of crops and animals, which requires finding the optimal way to provide adequate feed. Research on forage quality is concentrated on aspects such as digestibility, nutritive value, voluntary intake, and anti-quality compounds. High digestibility in plants means that forage will give additional energy from each unit of dry matter, and the nutritive values include neutral detergent fiber, acid detergent fiber, crude protein, and minerals. The climate influences the quality of forage by warming up the soil, thus reducing the content of water-soluble carbohydrates and nitrogen in forage. Increased CO_2 can boost the content of non-structural carbohydrates in forage but lower its digestibility (Cheng et al. 2022).

3.7.2 Water

Water scarcity is a global challenge where water used for agricultural purposes accounts for 69% of withdrawals of fresh water. With the increase in population, income level, and livestock product demands, water scarcity will emerge as a limiting factor in production agriculture. For the livestock industry, water will be utilized for drinking by animals, feed crop cultivation, and processing of products; the industry contributes 22% of all evapotranspiration and 41% of consumptive water use. Global warming and sea level rises will lead to changes in water availability and utilization in animal production through salinization. Competition in water usage between livestock, feed crops, and other sectors will necessitate efficiency in production.

3.7.3 Seasonal Variation

Climate change is changing the seasonal distribution of resources and yield of crops, which will influence livestock productivity. The rise of heat waves may lead to extra heat stress in animals. The difference in cattle muscle glycogen content and pasture conditions across different seasons will affect beef quality. The variation in forage availability makes grazing and livestock management difficult. Extreme droughts result less availability of forage coupled with prone to adaptability of forage species. Moreover, the alteration of snowmelt timing may influence water availability and feed availability. Generally, climate change is a threat to livestock production (Cheng et al. 2022).

3.8 Response Mechanisms of Agricultural Activities on Climate Change

The agricultural sector is an important aspect of the advancement of the world's economic and societal developments. The sector serves many purposes, including providing food products and materials for manufacturing, along with contributing to climate change. The agriculture sector is one of the major sources of greenhouse gases and can be a sector that can mitigate climate change. Studying the feedback mechanisms of agricultural activities on climate change can help us understand the role of agriculture in climate change and generate some important concepts for sustainable development (Yuan et al. 2024).

3.8.1 Greenhouse Gas Emissions

The agricultural sector plays a pivotal role in greenhouse gas emissions, mainly CO_2, methane (CH_4), and nitrous oxide (N_2O), which worsen global warming. The contribution from livestock to greenhouse gas emissions is 14.5%, and the emissions from enteric fermentation contribute 39% of that number. Decomposition of animal waste occurs under anaerobic conditions, emitting CH_4 and N_2O in intensive farming settings. Methane emissions from rice farming contribute another major source, as flooding leads to 11% of the global emission rate of CH_4. The duration of flooding impacts soil oxygen and soil microbial community composition, resulting in increased CH_4 production and emissions. Another factor that causes greenhouse gas emissions is the heavy reliance on nitrogen fertilizers in modern agriculture. Nitrogen fertilizers produce CO_2 as a result of fossil fuel consumption, and microorganisms in the soil undertake nitrification and denitrification reactions, leading to the formation of N_2O. Nitrous oxide has a global warming effect 283 times more potent than CO_2 (Yuan et al. 2024).

3.8.2 Deforestation and Land Use Change

Deforestation and land use change are key factors in climate change, which have affected about one-third of the world's total land area from 1960 to 2019. Not only do these factors decrease the ability of forests to act as a carbon sink, but they also lead to an increase in the concentration of CO_2 in the atmosphere. The trees in the forests absorb CO_2 through photosynthesis and store it in biomass and soil. After clearing, these forests release stored carbon back into the atmosphere, exacerbating global warming. In a study of the relationship between climate change and land use/cover change from 1966 to 2015, it was observed that the change in agriculture and pasture lands has an effect on CO_2 emissions. While deforestation causes a rise in atmospheric CO_2 levels, it is also responsible for the degradation of land, which in turn leads to soil degradation (Yuan et al. 2024).

3.8.3 Resilience and Disruptive Technologies in Agriculture

In agriculture, resilience is characterized by individuals and societies' capacity to prevent, anticipate, absorb, adapt to, and transform in relation to risks. Effective implementation is important in order to sustain vital functions without neglecting sustainable development, peace, security, realization of human rights, and well-being. To build resilience, the food system must withstand any shock, recover quickly from disruptions, and proactively realign itself in view of changing times. Technologies, innovations, and good governance, among other things, are needed in order to build resilience. Disruptive technologies have the potential of radically transforming agriculture for greater efficiency, transparency, data-driven decision-making, and timely response to problems. Disruptive technologies, especially CSA, can help achieve greater productivity through sustainable means and resilience against climatic hazards. By using precision farming and climate data, along with other technologies, it is possible to optimize resources and protect plants, making better decisions. Through traceability, resource management, and other means, disruptive technologies can help realize sustainability objectives like a circular economy in agribusiness (Harsanto et al. 2025).

3.8.4 Supply Chain Disruption

Climate change has a significant impact on the agricultural supply chain, which includes cold chain systems and food waste (Yuan et al. 2024; Harsanto et al. 2025). Cold chain systems preserve the quality and safety of perishable goods but involve significant energy consumption and greenhouse gas emissions. One-third of annual food production is wasted, resulting in greenhouse gas emissions such as methane and carbon dioxide. Contemporary food waste processing technologies, such as anaerobic digestion and incineration, can minimize emissions through nutrient and energy recovery. Still, avoiding food waste creation has more advantages for greenhouse gas emission reductions. In addition, climate change impacts crop production, technology, pest development, disease propagation, and storage operations. Elevated temperatures may negatively affect crops, whereas inconsistent precipitation results in droughts and floods. Any climatic alterations that affect microorganism dynamics could pose additional threats to food safety in the dairy sector. Higher humidity levels and fluctuations in temperature could result in food waste during storage processes, affecting transportation in agriculture. Climate change could also bring about food shortages, particularly among more vulnerable groups (Yuan et al. 2024). Knowledge on the agricultural supply chain is essential for enhancing agricultural productivity and food security (Yuan et al. 2024; Harsanto et al. 2025).

3.9 Socio-economic Impacts

Climate change is a worldwide environmental change that causes changes in weather, thus causing a variation in climatic parameters like temperature, precipitation, humidity, wind speed, solar radiation, and evaporation. Climate change has adverse effects on the agricultural sector, the major income generator for most people living in developing nations, because of increased temperatures and fluctuations in precipitation. The economy of a country is adversely affected by climate change and climate variations with expected economic costs at the global level reaching up to $23 billion per annum by the year 2050. High risk and uncertainties characterize developing countries owing to their underdeveloped status and limited adaptive capacity, thereby causing negative impacts on the economy of farmers and food security. Reduced agricultural output in developing countries is attributed to inefficient farming, over-exploited land, poor services, limited markets, and climatic factors including droughts and floods. About 60% of African smallholder farmers have limited access to climate change technology, which increases yield losses caused by climate change (Harsanto et al. 2025). Africa is one of the continent's most vulnerable to climate change, as its people rely on natural resources for their livelihoods, including agriculture, pastoralism, and fishing. Adaptation techniques, such as planting trees, soil conservation, and migration, indicate how the farmers have used their resources to combat the dangers posed by climate change. The perception of the farmers and their adaptations become vital in light of ongoing climate change (Harsanto et al. 2025).

3.10 Agricultural System Biodiversity

Biodiversity greatly benefits the agricultural system, since the stability of productivity and resilience as well as the capacity for the system to resist pest attacks, disease, and climate change can be increased through various means. Some of the examples are shown in Fig. 3.3. One-way biodiversity is known to increase the resilience of the agricultural system is by increasing the number of crops and animal species available within the system. For instance, crop rotation and cultivation of polycultures are ways in which the application of chemical inputs can be reduced by reducing the proliferation of diseases and pest infestations. Moreover, there are certain ecological processes that are aided by biodiversity, and these include pollination, nitrogen fixation, and soil fertility. For example, many beneficial insects and pollinators, without which crops may be unable to reproduce, are present in diverse ecosystems. In addition, diverse ecosystems provide a conducive environment for diverse microorganism populations, thereby increasing soil moisture retention and nutrient uptake (Diyaolu and Folarin 2024).

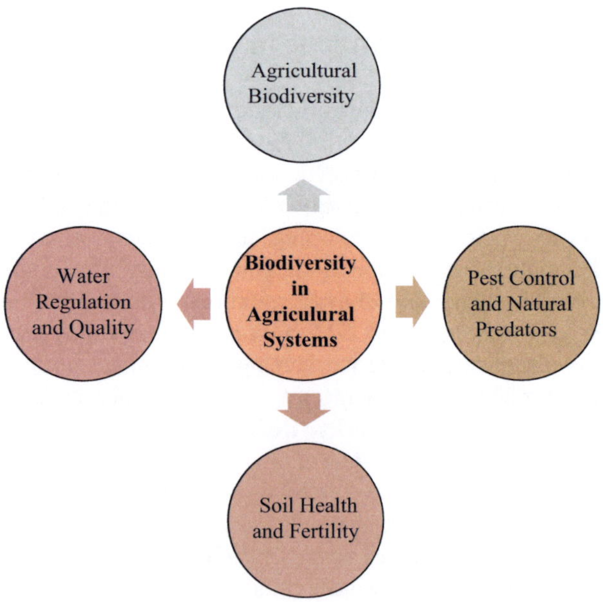

Fig. 3.3 Biodiversity in agricultural systems

3.10.1 Agrobiodiversity

The term agrobiodiversity is defined as the variety of crops and their genetic variation within different types of ecosystems. In other words, agrobiodiversity encompasses the biological resources found in agriculture, which include both wild-type and genetically engineered crops. Over the past millennia, humans have successfully adapted their societies to survive in various environments thanks to human intervention, such as selective breeding and farming practices. Unfortunately, agrobiodiversity loss has emerged as one of the most significant issues in modern times, affecting global food security because of industrial agricultural practices, monocultures, and low crop diversity. It is necessary to preserve a large gene pool and agrobiodiversity for adaptation to future problems, such as changes in climate conditions, crop diseases, and fluctuations in natural resources. Managers of agroecosystems can take advantage of biodiversity services for sustainable agriculture (Cadena-Zamudio et al. 2024).

The maintenance of biodiversity, which includes crop-associated biodiversity, is essential for ensuring that there will be sufficient food and other services provided by biodiversity in terms of evolution and crop breeding, production assistance, and general ecosystem services (Cadena-Zamudio et al. 2024). The fauna in soil and microorganisms, as well as the roots of the crops themselves, provide for nutrient cycling. Pests and diseases are controlled through predators and disease control microorganisms, while gene resistances in crop plants are maintained.

3.10.1.1 Genetic Diversity

The importance of genetic diversity in crops and livestock cannot be overstated, as it is necessary for the sustainability of agriculture. Genetic diversity ensures that organisms can adapt to changes in their environment, including alterations in climate, soil type, and pest infestation. The cultivation of crops with a broad range of genes offers farmers the opportunity to cultivate disease- and pest-resistant plants. For example, planting different genetic types of wheat leads to higher resistance against fungi and the subsequent reduction in chemical fungicide application (McFadden and Sasakuma 1987, as cited in Diyaolu and Folarin 2024). On the other hand, having a diverse gene pool among livestock also decreases the chances of epidemic outbreaks and helps animals survive under difficult climatic conditions, such as cold weather or famine. Livestock species with adaptive properties, such as drought-resistant cattle or sheep breeds, highlight the significance of genetic diversity in farming operations (Diyaolu and Folarin 2024).

The sustainability and resilience of agricultural systems require the presence of genetic diversity within crops and livestock. Adaptation to ever-changing climatic conditions, as well as variations in soils and pests, can be ensured due to the aforementioned phenomenon. The ability to choose crops that will have a certain resistance to pathogens or pests can be ensured through maintaining a wide genetic base. For instance, using genetically diverse varieties of wheat has shown higher resistance to fungi, which reduced the requirement for the use of fungicides. Genetic diversity among cattle is vital since it not only makes them more resistant to various natural factors, such as high temperatures or a shortage of food, but also helps prevent the emergence of an epidemic. Some breeds are suited for specific climatic conditions of different regions, such as sheep or cows capable of surviving droughts. Moreover, genetic diversity contributes to the sustainability of ecosystems because the process of food production is dependent upon ecological services like pollination, recycling of nutrients, and soil fertility, which function effectively in diversified agriculture. Additionally, genetic diversity ensures that if new challenges arise, there will be options available for the farmers to choose from, thereby helping them select breeds and varieties of crops based on their own requirements. Promoting diversity in livestock and crops benefits the overall health and productivity of the agriculture system, along with making it adaptive to change (Diyaolu and Folarin 2024).

3.10.1.2 Biodiversity of Species

Biodiversity is very important in increasing production in. The stability of agroecosystems and their ability to survive harsh conditions while providing sustainable food sources is due to the presence of different kinds of plants, animals, insects, and bacteria (Diyaolu and Folarin 2024).

3.10.1.2.1 Plant Diversity

Diversity of plants may be used in agriculture to improve ecosystem processes and make soil healthier. Polycultures, where different crops grow at the same time, increase nutrient recycling and prevent weeds from growing, and also decrease insect problems. According to studies, diverse species can use available resources better than monocultures, and therefore crop yield may become higher in ecosystems with diverse species. Besides, plants may form a habitat for useful organisms (Diyaolu and Folarin 2024).

3.10.1.2.2 Animal Biodiversity

The resilience of agroecosystems is enhanced by the integration of several livestock species. For example, having several types of animals, including sheep, chicken, and cattle, will promote diverse grazing behavior which, in turn, promotes efficient nutrient recycling and land use. Also, some livestock breeds are more resistant to certain diseases and parasites, which leads to healthier overall conditions (Diyaolu and Folarin 2024).

3.10.1.2.3 Insect Biodiversity

Insects play an important role in maintaining the productivity of agricultural systems, serving as natural pollinators and biological pest controls. The sufficient pollination of crops is crucial in producing fruits and seeds, while the abundance of insects guarantees that process. Furthermore, insect diversity may be important in controlling the pest population in a farm (Diyaolu and Folarin 2024).

3.10.1.2.4 Bacterial Diversity

Bacteria and fungi are soil microorganisms that are important for the breakdown of organic matter, the cycling of nutrients, and the prevention of disease. Different kinds of soil microorganisms contribute to the development and sustainability of plants through increasing soil fertility and soil structure. There are ways of increasing microbial diversity, such as using cover crops and reduced tillage, that can help achieve healthier soils and productive agroecosystems. Through supporting beneficial relations among plants, animals, insects, and soil microorganisms, increased species diversity will lead to more resilient and productive agroecosystems. Increasing different species is helpful for farmers to create resilient agroecosystems that are capable of overcoming problems due to environmental damage (Diyaolu and Folarin 2024).

3.10.2 Landscape Diversity and Its Impact on Agriculture

The variety of landscapes within a particular area, which includes forests, swamps, grasslands, and agricultural land, among others, is known as landscape diversity. The sustainable management of agricultural practices depends greatly on maintaining landscape diversity because it helps conserve biodiversity, build resilience, and enhance ecosystem services (Diyaolu and Folarin 2024).

3.10.2.1 Ecosystem Services

Diverse landscapes provide indispensable ecosystem services to agriculture. Swamps can purify water by filtering out surplus nutrients and chemicals before reaching the agricultural areas. Forests and hedgerows can be used as buffers to protect agricultural lands from soil erosion and also act as homes for beneficial insects such as pollinators and pest predators. Therefore, by ensuring that the soils and water remain healthy, these services contribute to sustainable farming practices (Diyaolu and Folarin 2024).

3.10.2.2 Biodiversity Support

The variation in the landscape creates habitats that foster biodiversity by creating spaces that support the existence of several species. The stability of the ecosystem increases due to increased biodiversity; therefore, making agriculture resilient to any pests, diseases, and climatic issues. For instance, landscapes that have high levels of biodiversity are able to host populations of beneficial insects that control pests, thereby reducing the use of synthetic pesticides (Diyaolu and Folarin 2024).

3.10.2.3 Adaptation to Change

By fostering diversity in landscapes, agricultural systems become more resilient. Diverse ecosystems are able to offer microclimate zones that protect crops from unfavorable climatic zones. Furthermore, by adapting the farming practices according to the unique qualities of various landscape zones, it is easier to implement land management methods that promote sustainability. Through this process, landscape diversity supports biodiversity, enhances adaptation and resilience to climatic and environmental changes, and ensures ecosystem services, thus making agricultural systems sustainable and productive. Sustainable and productive agriculture systems might be beneficial for farmers and the environment through diversity of the landscape (Diyaolu and Folarin 2024).

3.10.3 Biodiversity Assisted Key Ecosystem Services

3.10.3.1 Pollination

Agricultural production and food security largely depend on the process of pollination, which serves as an ecological service. In order to ensure the survival of pollinators including bees, butterflies, birds, bats, and other insects, biodiversity is needed. Many crops used by people for their consumption belong to many flowering plants depending on those organisms for their production (Diyaolu and Folarin 2024).

3.10.3.2 Importance of Pollinators

Approximately, 35% of food crops and 75% of flowering plants in the world require animal pollination. These crops are the sources of important nutritional components of human food, such as fruits, vegetables, and nuts. Improved productivity and better quality of fruits due to a higher likelihood of pollination are observed because of diverse communities of pollinators. Research demonstrates that diverse communities of pollinators contribute to improved productivity of crops, including almond and blueberry plants (Diyaolu and Folarin 2024).

3.10.3.3 Biodiversity and Pollinator Health

Biodiversity helps pollinators as there are different habitats and plant foods available for them. Some plants attract specific pollinators; a diverse habitat with various flower types provides nutrition for them during their growing periods. Moreover, diversified ecosystems could serve as a buffer against environmental factors that affect pollinators, such as habitat destruction and climatic shifts (Diyaolu and Folarin 2024).

3.11 Threats to Pollinators

Unfortunately, pollinators face threats to their numbers due to diseases, pesticides, habitat destruction, and climatic changes. These challenges may be intensified by declining biodiversity, especially on agricultural lands, since it leads to a shortage of pollen and places to nest. Pollinator abundance and agricultural productivity can be improved using habitat restoration and biodiversity-friendly practices such as planting wildflowers and cover crops. Biodiversity is crucial for pollination services, which are vital for sustainable food production. The productivity of farming, food security, and resilience of the whole ecosystem are all improved through a diverse ecosystem that promotes healthy pollination. The sustainability of food production in the future will be dependent on protecting biodiversity as agriculture faces increasing stress due to climate change and destruction of the environment (Diyaolu and Folarin 2024).

3.11.1 Pest Management and Natural Predators

Biodiversity plays a key role in pest control in agricultural ecosystems since biodiversity creates more predators. Through biodiversity, it is possible to create a conducive environment in which the use of chemicals is reduced as there will be organic elimination of pests through predators (Diyaolu and Folarin 2024).

3.11.2 Role of Natural Predators

In cases where there are multiple plant types living within a diverse environment, natural predators such as ladybugs, lacewings, and predatory wasps will thrive. As they adapt to changes in the environment and attack various pests, natural predators are often better at controlling the number of pests than artificial insecticides. As revealed in the findings, ladybugs possess the ability to consume thousands of aphids throughout their lifetime without any harmful side effects associated with chemicals (Diyaolu and Folarin 2024).

3.11.2.1 Biodiversity and Pest Control

Multiple natural enemies can be sustained with the existence of different habitats, including crop rotation, hedgerow, and cover crops. They provide the necessary micro-habitat, food, and breeding grounds for beneficial insects. Due to their ability to disrupt insect life cycle through complexity, biodiversity has been found to promote beneficial insects while suppressing insect pests on farms (Diyaolu and Folarin 2024).

3.11.2.2 Reducing Use of Chemicals

By encouraging biodiversity and associated natural pest management systems, farmers might reduce their dependency on chemical pesticides. Besides ensuring healthier soil and greater resistance in crops, this move is environmentally friendly as it decreases the amount of chemicals that end up in rivers and streams. Additionally, decreased use of pesticides allows farmers to sustain themselves economically as this decreases production costs. Thus, by sustaining natural predator populations, diversification greatly improves pest management in agricultural systems. Apart from protection of beneficial organisms and sustainable farming, the aforementioned eco-friendly pest management method decreases the harmful effects of chemical pesticides (Diyaolu and Folarin 2024).

3.11.3 Soil Health and Fertility

The maintenance of soil fertility and health is critical for long-term agricultural output that depends heavily on biodiversity. The soil contains numerous types of living organisms, ranging from bacteria and fungi to earthworms and other soil fauna; all these contribute to essential ecological functions of soil like organic matter decay, nutrient cycling, and soil structure formation (Diyaolu and Folarin 2024).

3.11.3.1 Nutrient Cycling

Nutrient cycling is facilitated by a diverse soil microbiome in terms of ensuring efficient supply of important minerals such as potassium, phosphorus, and nitrogen to the plants. Various microbial species are involved in nutrient cycling as well as in converting these elements in forms suitable for uptake by the plants. For instance, nitrogen fixing bacteria can be used to convert atmospheric nitrogen into usable plant fertilizer (Diyaolu and Folarin 2024).

3.11.3.2 Decomposition

A diverse soil microbiome significantly influences the process of decomposition. There are many microorganisms and detritivores including earthworms and fungi that decompose organic materials in order to add organic material back to the soil. Not only does this add valuable nutrients to the soil, but also facilitates proper root growth and drought resistance and improves water retention capabilities of the soil. Better root development and drought resistance are made possible by this organic matter, which also improves soil structure and water retention in addition to providing essential nutrients (Diyaolu and Folarin 2024).

3.11.3.3 Microbial Interactions

Soil health is enhanced by interactions between various microbial groups. As shown by the studies mycorrhizal fungi, for example, form symbiosis with the roots of the plants to increase the uptake of nutrients for the better growth of plants as well as get some carbohydrates provided by plants. Such interactions are crucial for the formation of a healthy soil community and maintaining agricultural sustainability (Diyaolu and Folarin 2024).

3.12 Water Regulation and Quality

Biodiversity helps maintain water regulation, quality, and management of floods in agricultural ecosystems. It is also very important to sustainable agricultural. Diversity can be achieved through the creation of riparian buffers and wetlands as well as growing a variety of plants (Diyaolu and Folarin 2024).

3.12.1 Water Management

Greater water absorption and storage are made possible by the increased root mass and structure in the soil caused by plant diversification. This is especially crucial when there is heavy rainfall, since different root systems can reduce floods by slowing down water flow and encouraging infiltration (Diyaolu and Folarin 2024).

3.12.2 Water Quality Protection

A wide variety of plants acts as natural filters, filtering out the nutrients and contaminants before they can get into any bodies of water. For instance, eutrophication can be stopped by blocking the flow of nutrients into lakes and rivers because of the absorption of excess nitrogen and phosphorus in buffer strips and cover crops (Diyaolu and Folarin 2024).

3.12.3 Flood Mitigation

Wetlands are ecosystems characterized by high biodiversity and function as natural sponges that take up extra water and mitigate the impact of floods. Such wetlands have the capacity to minimize peak flows during storms, and this helps in ensuring that flood hazards do not affect the lives of those around them or the farmland. For the sustainability of farming and the management of water resources, biodiversity within agricultural environments needs to increase (Diyaolu and Folarin 2024).

3.13 Agricultural Biodiversity Threats

3.13.1 Monoculture Farming

Intensive agriculture and mono-cropping are some practices which significantly jeopardize agrobiodiversity through reduction of ecosystem functioning and vulnerability to pests and diseases. Mono-cropping involves cultivation of a single plant species in vast tracts of land over many cropping seasons, and is widely used because of its efficacy in producing significant amounts of yield; however, there could be negative consequences in terms of biodiversity loss and agroecosystem resilience. Farmers who practice mono-cropping usually select few high-yielding varieties which suit their need for increased crop production and thus decrease biodiversity of agricultural ecosystems. The gene pool for combating diseases is severely reduced by such practice, as well as the flexibility of agrobiodiversity to adapt to changing conditions. Lack of diversity makes an agricultural system vulnerable to pest infestations, since specialist pests love monoculture systems, making farmers apply more pesticides against pests; moreover, it has detrimental effects on non-target organisms (Diyaolu and Folarin 2024).

3.13.2 Functioning of the Ecosystem

Intensive farming relies heavily on fertilizer and pesticide use, which harms soil health and disturbs the equilibrium between the soil microorganisms. While leading to soil erosion and depletion of nutrients, their use adversely affects species needed for pollination, pest control, and recycling of nutrients. Moreover, the homogeneity of the agricultural landscape limits the availability of habitats for insects, pollinators, and mammals, thus reducing ecosystem functions that contribute to food security. To conclude, mono-cropping and intensive farming result in reduced genetic diversity in crops, interfere with ecosystem functioning, and require more chemicals in agriculture, making farming systems vulnerable and threatening agricultural biodiversity (Diyaolu and Folarin 2024).

3.13.3 Chemical Inputs and Pesticides

Chemical inputs and pesticides are significant risks to biodiversity due to their negative impact on different living things in an ecosystem, including pollinators, soil organisms, and water bodies. Although these chemicals are intended for the eradication of specific pests and diseases, their widespread application often results in unexpected consequences for the ecosystem and other non-target organisms. Pesticides have been blamed for the reduction in population levels of pollinators, particularly bees and butterflies. These chemicals interfere with the nervous system of insects, which impairs their navigation and feeding abilities and limits their pollination activities crucial for agricultural production. Such a decrease in the number of pollinators increases the loss of biodiversity, not only by limiting the reproductive potential of different wild plants but also by reducing crop yields (Diyaolu and Folarin 2024).

3.13.4 Soil Microbes and Invertebrates

All the various organisms responsible for decomposition of organic matter, nitrogen cycle and maintenance of soil structure are dependent on soil health. The presence of these microorganisms in soil can be compromised by the use of artificial pesticides and fertilizers, causing reduction in soil diversity. Such disruptions lead to poor soil fertility, high soil erosion, and low water retention in soil, all which play important roles in sustainable farming methods (Diyaolu and Folarin 2024).

3.13.5 Water Ecosystems

Aquatic ecosystems may be negatively impacted through runoff water contamination into adjacent water sources where there is presence of chemical fertilizers and pesticides used on crops. Oxygen deficiency may cause the death of marine creatures because of algal blooms formed by such chemicals. These disturbances in aquatic ecosystem may adversely affect their balance, thus, impacting their biodiversity and ecological benefits. Due to extensive use of pesticides and chemicals in agriculture, there is a threat to pollinators, soil health, water systems and sustainability of agricultural practices (Diyaolu and Folarin 2024).

3.13.6 Climate Change

Climate change is another major factor that contributes significantly to the loss of biodiversity in agricultural systems, creating problems that threaten the sustainability of animals and plants. Warming in the temperature levels, precipitation changes, and increase in the occurrence of extreme climatic phenomena may upset the ecological balance, affecting the core of agricultural production. Crops grow under different conditions as the temperature rises, which can lead to low yields from traditional species, which do not match the climate changes. As a result, climate change can reduce genetic diversity due to monoculture or practices of farming involving limited crop varieties. The food security and stability of agricultural systems become endangered in the end because of decreased crop diversity since crops become vulnerable to diseases. In addition, animal health and productivity are affected by climate change. This is because high temperatures may reduce the reproductive rate and milk production ability of cattle, while weather may pose dangers to animals' habitats and their food supply. Additionally, climate change could facilitate disease transmission among cattle, exacerbating the problem and reducing the robustness of the animal husbandry system (Diyaolu and Folarin 2024).

3.13.7 Impact on Ecosystem Services

Two important ecosystem services such as pollination and pest control are adversely affected by climate change required by agriculture. The populations of the species responsible for providing these essential services to crops may suffer as a consequence of habitat loss as a result of climate change. This will adversely impact the overall productivity of agriculture. Thus, climate change not only leads to accelerated biodiversity loss but also lowers the robustness of agricultural systems by disrupting crop and livestock profitability and ecosystem services (Diyaolu and Folarin 2024).

3.14 Interaction Between Climate Change, Biodiversity Loss, and Agriculture

The domains of climate change, biodiversity loss, and agriculture are dependent on one another with implications stretching far beyond global food security, environmental balance, and human well-being (Wudu et al. 2023; Kremen and Merenlender

2018). It goes beyond just being linked but forms a connected web in which each element influences and requires the other elements (Wudu et al. 2023). While agriculture plays a crucial role in human existence, it remains a major driver of biodiversity loss due to changes in land use, habitat fragmentation, pollution from agricultural chemicals, and monoculture practices (Kremen and Merenlender 2018). On the other hand, loss of biodiversity in agroecosystems adversely impacts their ability to withstand escalating pressures due to climate change (Wudu et al. 2023). An understanding of these intricate interrelationships is necessary to create sustainable solutions for meeting the food requirements of an expanding population without compromising the biodiversity of our planet (Kremen and Merenlender 2018).

3.14.1 Impact of Climate Change on Biodiversity Loss in Agriculture

3.14.1.1 Alteration of Habitats and Loss of Wild Relatives

The phenomenon of climate change has an impact on the ecological environment of organisms, resulting in habitat alteration and destruction. These effects are especially evident in Crop Wild Relatives (CWRs), or wild plant relatives of agricultural crops, whose genes carry traits such as drought resistance, disease resistance, and pest resistance, which cannot be observed in standardized modern crop varieties. Rising temperature, changing rainfall patterns, and increasingly extreme weather conditions result in the destruction of the specific microclimatic and climatic environment to which CWRs have adapted. For example, the CWR that has been able to survive in cold and wet conditions in the mountains will experience competitive pressure from other species invading its niche as the optimal climate changes. This results in a significant reduction in genetic diversity, limiting the possibility of breeding new climate-resilient crops in the future. Species with poor dispersal capacity, specialized habitats, or those that exist in fragmented habitats (ubiquitous in agricultural landscapes) are most susceptible; the speed of climate change tends to outpace species' capacity to adapt and move into new habitats, resulting in population decline and local extinctions (Wudu et al. 2023).

Crops are dependent on many beneficial insects and microorganisms for increased yields. More than 75% of staple food crops require the service of pollinators, such as bees, butterflies, birds, and bats, to ensure successful fertilization. Natural predators like ladybugs, lacewings, parasitic wasps, and microbes keep insect pest populations in check, thus minimizing the need for pesticide use. However, the delicate synchronization between crops, the pollinators, the natural enemies, and the microbes is disrupted by climate change. The warming of springs triggers early plant flowering, but if pollinator emergence does not also change at the same rate, then there will be a phenological mismatch that limits successful pollination. Similarly, variations in temperature and humidity reduce the lifespan and effectiveness of natural predators of the pests, allowing the pests to proliferate. Loss of habitat via extreme climatic events, altered plant flowering periods, and plant physiological stress caused by heat and drought reduce the population and biodiversity of these species, thus leading to reduced agricultural productivity and dependency on unsound feedback loops (Kremen and Merenlender 2018).

3.14.1.2 Pressures on Water Resources and Habitats

The importance of climate change for the functioning of the hydrological cycle and its implications for freshwater resources and their ecosystems (Solheim et al. 2010; Ingrao et al. 2023; Mishra 2023) extend to biodiversity in the agricultural system (Solheim et al. 2010). The increased evapotranspiration that arises due to warming leads to decreased water availability. Changes in precipitation patterns, alongside more frequent and extreme droughts and floods, affect the state of water bodies (Ingrao et al. 2023). As far as agriculture lands' irrigation needs during dry seasons are concerned, there is an increased need to withdraw water resources from both surface and ground sources (Mishra 2023). This process directly impacts water-based ecosystems and their habitat diversity (Solheim et al. 2010). Drying out of rivers, lakes, and wetlands results in immediate death and disturbance of reproductive processes among species and loss of vital habitats for a variety of animals, starting from fish and amphibians and ending with moisture-requiring insects and birds (Solheim et al. 2010). While it is true that flooding seems to result in an increased supply of water, there will be a greater runoff of agricultural chemicals (pesticides, fertilizers) into the water supply, thus making water more toxic and impacting aquatic species diversity (Mishra 2023). Increased water temperature, owing to changes in the climate, also reduces oxygen levels in water bodies, thereby harming aquatic life and resulting in algal growth (Solheim et al. 2010). Contamination by chemicals and nutrient runoffs cause eutrophication of water bodies, which in turn harms aquatic organisms through direct toxicity and impacts their reproductive activities, altering food chains (Ingrao et al. 2023).

3.14.1.3 Climate Change and Spread of Invasive Species

Climate change is one of the key forces behind the introduction and spread of invasive alien species (IAS) which poses a higher risk for biodiversity loss in disturbed agroecosystems (Essl et al. 2020; Hulme 2017). An increase in temperatures would create favorable environmental conditions in previously inhospitable environments due to low temperature for some IAS, allowing them to expand into new areas (Essl et al. 2020). For example, some tropical plants could survive the winter season at temperate latitudes (Hulme 2017). Non-indigenous species will compete with native species for natural resources (sunlight, water, soil nutrients), change the structure of their habitat, unbalance the food chains, and bring with them various disease organisms (Essl et al. 2020). This would lead to a reduction in biodiversity due to decline or extinction of some native species (Hulme 2017). Natural disasters such as long-lasting drought, heavy rainfalls, and storm damages in indigenous ecosystems would create a disturbed ecosystem more prone to invasions (Essl et al. 2020). Native species that have already been affected by changing climates could become more vulnerable to the competition posed by invasive species (Hulme 2017). In farming areas, newly introduced weed species can reduce crop production and require increased management efforts (such as more herbicide applications), insect introductions can result in additional crop damage, and disease introductions can spread rapidly through monoculture crops, posing serious threats to food security (Essl et al. 2020). Climate change may also alter human behavior in ways that unintentionally facilitate the spread of IAS, such as by altering shipping lanes or agricultural practices (Hulme 2017).

3.14.2 Impact of Biodiversity Loss on Resilience of Agriculture to Climate Change

Natural resilience of agricultural systems to cope with shocks and stress associated with the effects of climate change becomes lower in case of loss of biodiversity (Jha and Dev 2024; Prakash 2021; Borrelli et al. 2020; Al-Tawaha et al. 2021).

3.14.2.1 Shrinkage of Genetic Pool for Developing Climate-Resilient Varieties

Biodiversity within crops in terms of genetic diversity can be understood as the stock of crop varieties with diverse genotypes that serves as a base for breeding of more advanced and productive varieties in agriculture (Finch et al. 2021; Jha and Dev 2024). The period of modernization of agriculture over the past decades was marked by the focus on development of several high-yielding varieties and breeds of domestic animals, leading to the reduction of the amount of genetic diversity in agriculture (Skendžić et al. 2021). For example, globally, there are only a few varieties of wheat, rice, and maize that dominate the world market, replacing the thousands of landraces that have been developed in each region. The same can be said for the small number of cattle breeds that are responsible for the majority of livestock production (Kisvarga et al. 2023).

In the event of climate shock, the inability to find genetic variation leads to similar responses from plants and animals, which are mostly detrimental (Finch et al. 2021). In such cases, the gene pool lacks the adaptive characteristics resistance to drought, heat shock, and diseases which breeders can utilize or develop through natural evolution (Jha and Dev 2024). This problem becomes more difficult to solve due to the destruction of habitats and changes in climate, which put the CWRs at risk. The CWRs serve as an important source of genes for breeding resilient crops to withstand climate shocks. The destruction of the population of CWRs would imply the destruction of a crucial source of genes (Kisvarga et al. 2023).

3.14.2.2 Deterioration in Ecosystem Services

Ecosystem services are the advantages that humans obtain from an ecosystem (Sutter and Albrecht 2016; Van der Sluijs and Vaage 2016). In agriculture, these services are critical for productivity and sustainability; loss of biodiversity directly disables these important functions (Uhl and Bruhl 2019). There are a variety of economically valuable crops that depend entirely on pollination by animals. The reduction of pollinator populations as a result of destruction of habitats, use of pesticides and climate change directly influences production (Yang et al. 2023). Reduction in biodiversity means that if one type of pollinators is negatively affected by climate conditions (such as sudden temperature change during the flowering period) there will be a lack of alternative pollinators. Hence, this leads to unpredictable yield and susceptibility to climate change (Sutter and Albrecht 2016). The farmer may have to spend additional money for renting the bees and manual pollination, which is inefficient (Van der Sluijs and Vaage 2016).

In healthy ecosystems, farmers have enough natural enemies like ladybugs preying on aphids, parasitoids fighting against caterpillars, beneficial organisms preventing crop diseases (Uhl and Bruhl 2019). These species provide "ecosystem-based pest control" and reduce dependence on artificial pesticides (Yang et al. 2023). The breaking of natural cycles of predator–prey interactions is possible when biodiversity is lost (Sutter and Albrecht 2016). For instance, when a monoculture exists, a huge supply of the same food source is provided for a certain pest, which allows its number to grow rapidly when there are no or few antagonists to that pest (Van der Sluijs and Vaage 2016). As a result, farmers tend to apply artificial pesticides that may have adverse effects and thus create a vicious circle (Uhl and Bruhl 2019). Changes in life cycles and distribution areas of pests in the context of climate change also require more natural control and make the problem more complicated (Yang et al. 2023).

3.14.2.3 Increased Vulnerability to Pests, Diseases, and Extreme Events

Diversity is critical in providing protection from various stressors, but a lack of biodiversity increases the vulnerability of agricultural ecosystems (Diyaolu and Folarin 2024; Kumar et al. 2024). With low levels of biodiversity, the presence of monoculture could lead to the formation of "green carpets," thereby exposing the whole crop to pest and disease attacks (Singh et al. 2025). Crop diversity provides "epidemiological breaks," which help slow the development of epidemics (Diyaolu and Folarin 2024). The effects of climate change are felt through the changes in the habitat of pests and diseases, which create opportunities for faster multiplication and dispersion (Kumar et al. 2024). The result is a lack of resilience to new threats because a reduced number of diversities implies more frequent outbreak, higher rates of crop losses, and dependence on costly and ecologically harmful control measures (Singh et al. 2025).

Ecological diversity is more perturbation resistant than crop diversity, thus making the diversified farm resilient to extreme climate events (Diyaolu and Folarin 2024). However, a mix of crops that have been diversified could yield in spite of such adverse conditions, and in the case of a simplified agricultural environment that does not support much biodiversity, just one such extreme event would be enough to wipe out the crop, having significant implications for food security and livelihood (Kumar et al. 2024).

3.14.2.4 Vicious Cycles and Feedback Loops

Climate change and biodiversity loss are connected issues that have created a vicious cycle of mutual support (Jha and Dev 2024; Prakash 2021). The process of climate change affects habitats and living beings, causing the degradation of ecosystems, altering the ranges of different species and making the environment favorable for invasive species (Borrelli et al. 2020). With decreasing biodiversity, agricultural lands are becoming less resilient to pests and diseases and more prone to extreme weather occurrences (Al-Tawaha et al. 2021). It results in agricultural activities becoming even harsher, involving the clearing of lands, use of synthetic fertilizers and pesticides, as well as irrigation with significant energy expenditures (Jha and Dev 2024). Such approaches, being driven by reduced resilience, cause greenhouse gas emissions and loss of biodiversity due to habitat destruction and pollution, which, in turn, negatively affects biodiversity through climate change (Prakash 2021). Breaking this vicious circle requires using holistic solutions, such as agriculture based on sustainable farming methods that promote biodiversity, genetic diversity, and emissions reduction as a solution to climate change (Borrelli et al. 2020).

3.15 Strategies for Building Climate-Resilient and Biodiversity-Rich Agricultural Systems

3.15.1 Agroecology and Permaculture

Agroecology and permaculture are two methods that can enhance agricultural resilience and sustainability. Agroecology involves cultivating a variety of crops within a region, which can reduce pests, diseases, and chemical pesticide use. Polyculture systems, which involve cultivating a range of crops within the same region, improve soil health by addressing different nutritional requirements and root systems. This also enhances resilience to harsh weather and climate change, allows farmers to sell a variety of products, and protects against market instability. Agroforestry is crucial for preserving water resources by improving water infiltration and decreasing runoff; tree roots create channels in the soil, allowing deeper penetration of precipitation and preserving soil moisture. Soil structure improvement is achieved through tree roots, which provide organic matter, increase nutrient availability, stimulate microbial activity, and retain nutrients and water (Diyaolu and Folarin 2024).

In combination, agroecology and permaculture can lead to sustainable food production, ecological balance, and increased resilience in agricultural landscapes; by combining these methods, farmers can create more robust and sustainable food production systems. Improved soil structure comes about as a result of tree root actions, which involve providing organic matter, enhancing nutrient availability, encouraging microbial activity, and keeping the nutrients and water. The use of agroecology alongside permaculture results in sustainable food production, environmental balance, and more resilient food production landscapes since farmers may integrate these practices to achieve this (Diyaolu and Folarin 2024).

3.15.2 Indigenous Species and Traditional Conservation Practices

Sustainable agricultural systems and biodiversity are reliant on the survival of indigenous plant species and traditional practices in agriculture. Indigenous plants are essential in the context of food security as they tend to possess unique traits that enhance resistance to local pests, diseases, and weather conditions. By preserving such plants, farmers can supply genetic diversity, which is a key element in adapting to changing environmental conditions. Moreover, the traditional knowledge possessed by local and indigenous cultures is also a major aspect of sustainable agriculture. Some of the examples of proven practices that enhance biodiversity and fall under this knowledge include agroecological practices, crop rotation, and intercropping. The techniques increase ecosystem service provision such as pollination and pest management as well as soil productivity and health. The traditional knowledge and indigenous species can be introduced into modern agricultural systems to offer a more sustainable and resilient agricultural practice; in the wake of problems that affect the entire world, such as climate change, we could save ecosystem services, enhance biodiversity, and ensure the continuity of agricultural practices by valuing and integrating these elements (Diyaolu and Folarin 2024).

3.15.3 Agrobiodiversity in Response to Climate Change

3.15.3.1 Crop Diversification and Varietal Selection

The climate change is influencing agricultural systems and creating the necessity to have more crop diversity and varieties (Ortiz et al. 2021; Hafeez et al. 2023). This has the potential to increase crop resilience to climate variability and extreme events, contain pest outbreaks, and delay disease transmission (Ortiz et al. 2021). Although the process of crop diversity adoption is still slow, the advantages of crop diversity are quite evident, but the dissemination of this approach has been complicated by biotechnology strategies, economic factors, and the idea that monocultures are more effective than diverse systems (Hafeez et al. 2023). Nevertheless, crop diversity can be implemented in many ways and on different scales, which means that farmers can select a technique that will increase their resilience and generate financial benefits (Ortiz et al. 2021). Climate-resilient crops refer to crops that can withstand biotic and abiotic stresses such as drought, floods, heat, chilling, freezing, and salinity; they are crops that are designed to remain unaffected or even increase crop yields in stressful situations, which secures food security of farmers and communities (Hafeez et al. 2023). Various types of climate-resilient crops include drought-tolerant, heat-tolerant, pest-resistant, salinity-tolerant, and flood-tolerant crops (Ortiz et al. 2021). Technological approaches like biotechnology have significantly contributed to developing these crops; they can also contribute to adaptation strategies for crop production and efficient resource management (Hafeez et al. 2023). Climate change poses unprecedented threats to global agriculture, and there is need to transition to growing climate-resistant crops (Ortiz et al. 2021).

3.15.3.2 Renewing Traditional and Underutilized Crops and Livestock Breeds

The reintroduction of old and unused crops and livestock breeds can help to increase food security, biodiversity, and resilience to climate change. Nutrient-based crops and livestock (NUFCs) like Hordeum vulgare), beans (Phaseolus vulgaris, etc.), black gram (*Cicer arietinum*), horse gram (*Macrotyloma uniflorum*), taro (*Colocasia esculenta*), yam (*Dioscorea* spp.), amala (Phyllanthus emblica), and jaman (*Syzygium cumini*) are often neglected and underutilized in the Himalaya region due to changes in local food systems, food habits, policy priorities, and other factors. NUFCs are crucial to food and nutrition security, in addition to income enhancement. Promoting NUFCs as a way of diversifying agricultural production systems creates potential to enhance the adaptation, mitigation, and the resilience of both natural and socio-economic systems to climate change. Compared to advanced cash crops and cereals, NUFCs are more robust to climate stresses, and genetic resources can be crucial in maintaining agriculture and coping with climate change. To rejuvenate such crops and livestock breeds, collaboration among researchers, policymakers, farmers, and consumers is required to create more sustainable, resilient, and fairer food systems (Adhikari et al. 2017).

3.15.3.3 Conserving Wild Relatives and Landraces

There are two key strategies used in conservation of landraces and wild relatives, in-situ conservation (retaining species in their natural environment) and ex-situ conservation (retaining species outside their natural environment) (Aldow et al. 2025; Ramirez-Villegas et al. 2022). The two are also critical to the agricultural diversity and the continued preservation of valuable genetic resources (Raggi et al. 2022). Crop Wild Relatives (CWRs) are taxa of wild plants closely related to crops that can improve agricultural systems' resilience and productivity. However, CWR diversity is often under-conserved, with an estimated 16–35% of species listed as endangered under the IUCN Red List Categories and Criteria. Climate change, population growth, and the increase in food consumption emphasize that there is a necessity to enhance food security, and one way to achieve it is to breed more crops with the help of CWRs (Aldow et al. 2025). Enhanced in-situ conservation of CWR may avail itself of more adaptive genetic diversity to the breeders. Hotspots of CWR diversity are not consistent with general biodiversity, or plant diversity hotspots, the highest density of naturally occurring CWRs occurs in the area of West Asia and North Africa, and in Northeast Africa (Ramirez-Villegas et al. 2022). Protected Areas (PAs) frequently receive emphasis on conservation of CWR, since they are also enjoying conservation practices and site management plans can be adapted to CWR conservation (Aldow et al. 2025).

Crop landraces are populations of cultivated plants whose cultivation, selection, and spread have been evolved by Indigenous or traditional agrarian societies. These genetically diverse populations are usually cultivated in a mosaic of various crop species and varieties, to maintain a local agricultural resilience, adaptive capacity, human nutrition, and cultural demands. The exchange and gene flow between populations of landraces based on farmed lands promote the emergence of new variation, whereas extended cultivation and selection result in the response to the local environmental and social conditions (Raggi et al. 2022). Modern crop breeding and agricultural domestication and breeding origination rely on crop landrace diversity (Ramirez-Villegas et al. 2022). The purpose of ex-situ repositories/gene banks is usually research access, but the landrace diversity is not usually thought of as being fully represented ex-situ (Raggi et al. 2022). The conservation of landraces is a priority of major international agreements, and European Commission has already taken steps toward improving in-situ conservation and utilization of landraces by promoting access to seeds (Ramirez-Villegas et al. 2022). The major steps to undertake to advance the conservation of European landrace diversity are to enhance systemic knowledge of landrace locations of cultivation, encourage the characterization of landraces on-farm, and help custodian farmers (Raggi et al. 2022).

3.15.3.4 Climate-Smart Agriculture (CSA) Practices

One of the solutions to climate change is to expand agricultural production and flexibility using water-efficient irrigation and smart technologies in farm management. Cutting-edge agricultural technology like unmanned aerial vehicle surveillance, remote sensing, and precision agriculture make it possible to better control crop growth and irrigation, which raises crop production and quality. Based on satellite data and deep neural networks, the deep drainage estimate model for irrigated farmland may help promote water-saving agricultural practices and evaluate the efficacy of different water-saving irrigation technologies. Drip irrigation and micro-sprinkler irrigation are examples of water-saving irrigation methods that help save water while ensuring crop development in hot and dry climates. One water-saving irrigation technique that addresses the unequal distribution and shortage of water resources is the plastic film drip irrigation system utilized in the study; it maximizes irrigation management and increases irrigation water use efficiency (Yuan et al. 2024).

3.15.3.5 Sustainable Closed-Loop Systems

Reducing waste production and contamination in the environment while maximizing resource recycling is the goal of developing sustainable closed-loop systems. Through the integration of waste management, energy generation, and agricultural production, closed-loop systems may decrease greenhouse gas emissions, lessen dependency on fossil fuels and chemical fertilizers, and transform agricultural waste into useful resources. Furthermore, irrigation water in closed-loop systems may be efficiently collected and recycled, giving crops sufficient protection from the elements. Chemicals have less of an influence on the environment and food safety is improved since plants are grown in controlled surroundings without the need for pesticides, fertilizers, or herbicides. Closed-loop ecosystem research throughout the world shows that this all-encompassing management strategy not only increases resource efficiency but also boosts agriculture output sustainability while offering crucial information and scientific assistance for upcoming space exploration. With major ecological and financial ramifications, the development of sustainable closed-loop systems provides creative ways to combat climate change and advance agricultural sustainability (Yuan et al. 2024).

3.15.3.6 Sustainable Closed-Loop Systems

The aim of creating sustainable closed-loop systems is the reduction of waste generation and the contamination of the environment, as well as the maximum use of resources in the form of recycling. By combining waste management, energy production, and agriculture, closed-loops can reduce greenhouse gas emissions, reduce reliance on fossil fuels and chemical fertilizers, and convert agricultural waste into valuable products. Furthermore, irrigation water in closed-loop systems may be efficiently collected and recycled, giving crops sufficient protection from the elements. Chemicals have less of an impact on the environment and food safety is improved since plants are grown in controlled environments without the need for pesticides, fertilizers, or herbicides. Closed-loop ecosystem research throughout the world shows that this all-encompassing management strategy not only increases resource efficiency but also boosts agricultural productivity and sustainability while offering crucial information and scientific assistance for upcoming space exploration. With significant ecological and financial implications, the development of sustainable closed-loop systems provides creative ways to combat climate change and promote agricultural sustainability (Yuan et al. 2024).

3.15.3.7 Sustainable Land Management

Conservation tillage and retention of residues are some of the crop management methods that can be used to reduce the negative impacts of climate change. Efficiency in water use and reduction of adverse impacts can be achieved through irrigation scheduling, drip irrigation use, and planting practices. Crop-livestock, livestock-forest, and crop-livestock-forest systems can enhance carbon accretion, fertility, and soil moisture conservation, but can affect crop resistance to drought and heat; nevertheless, these practices can decrease productivity. Crop rotation, cover crops, and intercropping are stress management strategies used by farmers (Benitez-Alfonso et al. 2023). Cover crops prevent soil erosion, whereas crop rotation helps to increase organic matter due to the decomposition of litter over time. The cultivation of crops and interplanting enhances the quality of soil, water storage, and heat stressing. Plant species diversity has such advantages as complementarity in resource utilization, overlap in a niche, and enhanced soil structure and chemistry. Intercropping system advantages drought-prone regions by enhancing soil fertility, carbon and water infiltration and moisture during dry seasons. Diversified crop regimes can enhance soil biomass microbial, enhance soil structure, organic matter, and water retention (Benitez-Alfonso et al. 2023).

3.15.3.7.1 Water Management Efficiency

To achieve agricultural sustainability in arid zones, efficient use of water is needed due to climate change adaptation. Drip irrigation and rainwater collection are popular water-saving methods in drylands. Different techniques can be employed to collect rainwater, such as engineering, agriculture, biology, chemistry, or integrated strategies. Rainwater collection techniques can be categorized as in-situ and ex-situ, with in-situ techniques that include furrows, plastic films, and infiltrations, and ex-situ techniques that use tanks to capture the surface runoff. Such techniques as micro-catchment and in-situ rainwater harvesting are more widespread than macro-catchments. Particularly common methods of drip irrigation such as partial root-zone drying, controlled deficit watering, and subsurface drip irrigation are frequently studied. Drip irrigation systems installed in the subsurface are more effective because they ensure the adequate content of soil water and ensure a rapid restoration of soil water. Crop yields can be boosted with the help of high-frequency irrigation. An integrated system that regulates plant growth, productivity, and development based on physiological responses and controls crop water usage is essential for adapting to climate change; comparing the benefits and drawbacks of these water-saving strategies is crucial, but research is still challenging (Zhang et al. 2021).

3.15.3.7.2 Organic Fertilizers and Nutrient Management

Traditional fertilizers are indispensable to the productivity of agriculture, yet are viewed as detrimental to the environment because of their long-term impacts on soil and microbial diversity. Organic manure, such as biochar, manure, and fermented grass, can enhance the health of the soil, as they enhance the physical, chemical, and biological characteristics of the soil. They promote the growth of helpful soil microorganisms, decrease heavy metals, enhance the uptake of nutrients, suppress the growth of pathogens in crops, and decompose xenobiotic organic compounds. Cover crops improve the soil health as well. Organic fertilizers and crop rotation can improve the health of soil and agricultural production that will not have negative effects on the environment; the combined approach will help decrease chemical fertilizers and their adverse effects on the environment. Climate change mitigation and sustainable agroecosystems can be achieved through organic fertilization and cover crops, which can also enhance and maintain microbial diversity, biomass, fixation of nitrogen and carbon in the soil. In this way, organic fertilization combined with cover crops may result in a sustainable food production and a more sustainable future (Khan et al. 2024).

3.15.3.7.3 Integrated Pest Management

The management of agricultural insect pests is becoming more difficult due to climate change, which is less effective using traditional methods. The changes necessitate the development of Integrated Pest Management (IPM) that is the combination of various methods to reduce the destruction of pests and minimize the harm to the environment. Nevertheless, as the climate patterns become more variable, the conventional IPM strategy will have to be modified to include the risks and uncertainties that are changing. Some of the strategies include changing the planting time of crops based on the climatic trends, studying local climatic trends, planting varieties that are more adapted to the changing climate, planting more biodiversity at field edges, and combining divergent crop rotations, intercropping, and augmenting floral diversity in agroecosystems. The use of pheromones and allelochemicals is also a key to IPM practices, though it is likely to reduce with the changing climate, and a synergist or adjuvant is needed to ensure their success (Subedi et al. 2023).

3.15.3.7.4 Livestock Management for Reduced Emissions and Increased Resilience

Climate change is significantly impacting cattle and livestock-based systems, necessitating strategies to adapt cattle to these changes. Breeding Native breeds and community-based participatory breeding can assist the cattle to be resilient to drought and global warming; this integrates only one breeding pool to enhance genetic improvement. Planting of grazing lands that tolerate hot and dry climates (sagebrush and legumes) can guarantee food production during drought periods when there is shortage. Methane emissions, a major contributor to agriculture, can be reduced by feeding ruminants higher-quality diets. Trees can be used in silvopastoral systems, a combination of productive African forage grasses, to heal degraded pasture lands and the trees retain carbon by their deep root systems. Increasing the amount of carbon that grasslands sequester can also help pastoralists adjust to climate change by increasing soil water retention and drought resistance (Ortiz 2011).

3.15.3.7.5 Biodiversity for Building Ecological Resilience

Genetic diversity of crops enhances the ability to withstand climate change because it aids crop varieties to perform optimally in diverse environmental stresses. This adaptation enhances the long-term adaptability of agricultural management and can be used for future breeding endeavors. An increase in species diversity in agricultural systems reduces the pressure of pests and diseases since species-rich agroecosystems are more ecologically complex. Applying multiple systems with close-row planting of several crop species can suppress pests and increase agricultural system stability; however, assessments of species diversity require precise analyses considering factors like farm management practices, economic conditions, and policies. The various ecosystems offer biophysical services such as nutrient recycling, soil formation, and water purification, which are crucial in the sustainability of agroecosystems. These ecosystems when incorporated into agriculture would be able to maintain production and ecological permanency in the future which is part of climate-smart farming. Functional diversity has an influence on nutrient cycling and the provision of ecosystem services in that crop rotation and cover crops may result in better nutrient cycling and reduced reliance on chemical fertilizers. However, functional diversity comes with an economic trade-off, with higher costs for farmers to realize these benefits (Loizou et al. 2025).

3.15.3.7.6 Farmer Led Innovation and Knowledge Sharing

The global food system resilience framework demonstrates the possibility of the agrobiodiversity to reduce climate change. It highlights the significance of more effective management and utilization of biodiversity in agricultural land management, but multiple challenges should be considered, such as the absence of awareness among farmers and policymakers, the necessity to rely on traditional knowledge and research, and the creation of proper policies and incentives. The CSA includes agroecological-specific practices and technologies, including climate-resilient crop varieties, conservation agricultural methods, agroforestry, precision farming, and water management methods, as well as better livestock management. By implementing these practices, triple win results can be achieved: increased productivity, enhanced resilience, and reduced emissions. Farmers, researchers, policymakers, and civil society organizations ought to help fill the gap of knowledge by offering empirical evidence on the conditions that lead to the susceptibility of agricultural systems to change in climate. An agroecology that is led by farmers and focused on ensuring biodiversity through continuity of farming would be more innovative and responsive to science and society; systemic collaboration and proactive action are the key to a thriving future of agriculture and the health of the communities in the world (Bohan et al. 2022).

3.16 Role of Farmers and Policymakers in Biodiversity Protection

3.16.1 Farmers as Stewards of Biodiversity

The management of agricultural landscapes is important to the resilience and well-being of ecosystems since the farmers are involved in their management. Sustainable farming will allow farmers to enhance the biodiversity, preserve ecosystem services, and make their business sustainable over the long-term. Diversification strategies in crops promote a variety of species on their farms, lessening the use of monocultures and fostering a healthy ecosystem. Furthermore, agroecological methods such as intercropping, cover crops, and hedges maintenance enhance biodiversity by providing a habitat to useful species such as pollinators and natural insect predators. The farmers are also involved in conservation through the maintenance of native plants and traditional methods of farming which aids in conservation of genetic diversity to adapt to the environment. Farmers, researchers and policymakers should collaborate to promote approaches to safeguard biodiversity; farmers can adopt innovative approaches balancing the sustainability of the ecological environment and agricultural production by sharing resources and knowledge (Diyaolu and Folarin 2024).

3.16.2 International Frameworks and National Policies

National and international policies play a role in biodiversity protection in agriculture because they offer a framework of sustainable practices and ecosystem well-being. Such policies have the potential to enhance food security and climate resilience and reduce adverse impacts of agricultural intensification. Governments of countries can implement legislation that promotes sustainable agricultural practices, including conservation agriculture, agroecology, and organic farming; they can be in the form of subsidies, tax credits, or cash benefits. The Common Agricultural Policy (CAP) of the European Union is designed to promote farmers to plant cover crops, hedges, and establish buffer strips to safeguard biodiversity. Countries can develop laws with the assistance of such international structures as the Convention on Biological Diversity (CBD); the countries that signed this convention are urged to develop national biodiversity plans, where agricultural biodiversity is treated as an essential component. In promoting the need to preserve genetic resources in agriculture, the Food and Agriculture Organization (FAO) supports the global strategy of plant conservation and global plan of action of animal genetic resources. These policies help in sustaining use and conservation of agricultural biodiversity to make sure that the current and future generations enjoy a diversity of genetic resources (Diyaolu and Folarin 2024).

3.16.3 Incentives and Support for Biodiversity-Friendly Practices

It is important to note that the governments (both public and private) have a significant role in encouraging farmers to engage in biodiversity-friendly agricultural activities through the provision of financial and technical support. These incentives minimize the cost of transition, and enhance the resiliency of ecosystems. Other governments such as the United States Department of Agriculture provide incentives to farmers who use organic farming systems, agroforestry and cover crops that enhance biodiversity. Biodiversity-friendly farming is also expanded through the efforts of private sector actors, including partnerships with businesses and NGOs. Many sustainable supply chain include businesses partners with farmers to implement best practices and supply resources and technical training. Significant programs that facilitate biodiversity and fair compensation to farmers who have adopted sustainable practices are rainforest alliance and fair-trade programs. These incentives and support systems can greatly help safeguard agricultural biodiversity and the development of sustainable food production systems in both the public and private sectors (Diyaolu and Folarin 2024).

3.16.4 Support for Research and Development for Varietal Resilience

One of the most important measures in the fight against climate change is the development of strong agricultural varieties. It was demonstrated that by changing the timing of plantings and utilizing the types of crops resistant to floods, the reduction of production loss could be reduced by 18%. Climate change impacts on agricultural productivity can be mitigated by growing resilient forms of crops, which may still grow and yield even in harsh climatic conditions. The development of new strong kinds of crops based on the technologies such as gene editing will therefore become a research priority. Through precision crop genome modification, gene editing technology has been shown to have potential benefits for enhancing food security in low- and middle-income countries. These technologies not only make crops more resistant to drought and disease, but also make the production of the seeds more affordable and available to small farmers (Yuan et al. 2024).

3.17 Successful Initiatives of Biodiversity and Resilience in Agriculture

3.17.1 Pheasant Farmers in Sub-Saharan Africa

In Sub-Saharan Africa, smallholder farmers are progressively implementing agro-ecological concepts to improve resilience and biodiversity in their farming operations. It has an estimated 500 million smallholder farms that are critical to the food production process although there are problems such as resource constraint, land degradation, and climate change (FAO 2019, as cited in Diyaolu and Folarin 2024). The strategy of diversified cropping has proved to be effective; farmers often plant a mixture of improved and traditional type of crops, which enhances genetic diversity and reduces the vulnerability to pests and diseases. An example of this is smallholder farmers in countries such as Ethiopia and Kenya have integrated cover crops and legumes in their crop rotations; in this case, the cover crops fix nitrogen, which enhances soil fertility and boosts food security by enabling farmers to produce multiple harvests in a year. Agroforestry strategies, such as adding trees to their farms, are also implemented by these farmers; various benefits of this practice exist, such as improved soil health, increased biodiversity, and climate change resistance. As an example, in production systems of coffee and cocoa, shade trees can be used to balance the microclimate and eliminate the effects of drought through the creation of habitat of diverse species. Also, smallholder farmers are adopting traditional knowledge and practices that have been transmitted over the years; farmers prefer local varieties of plants which are more adapted to their specific environmental conditions and they use this cultural heritage to enhance biodiversity. Besides enhancing biodiversity, native crops such as sorghum and millet should be encouraged to enhance nutritional security among the locals (Sustainable Development Solutions Network 2019, as cited in Diyaolu and Folarin 2024). By implementing these strategies, smallholder farmers in Sub-Saharan Africa are successfully protecting biodiversity and boosting agricultural resilience, proving that sustainable farming can flourish even under trying circumstances (Diyaolu and Folarin 2024).

3.17.2 Latin American Agroforestry

Agroforestry systems have also become a vital method of enhancing biodiversity and promoting sustainable agricultural practices in Latin America, particularly in countries such as Brazil and Costa Rica. Restoring degraded lands and enhancing ecosystem health in Brazil has been made possible by agroforestry practices including integrating native trees with crops and animals. An example of a system is the Sistema Agroflorestal (SAF), which involves planting assorted tree species along with cash crops; this system enhances soil fertility and resilience to environmental extremes and also increases sources of income. Costa Rica has also achieved significant progress in agroforestry through its Payment for Ecosystem Services (PES) program that incentivizes farmers to maintain and enhance biodiversity on their farms; in this program, farmers are encouraged to promote practices that enhance carbon capture and animal habitat by planting native tree species on their farms. Research shows that the agroforestry systems in Costa Rica support a diverse species of birds, insects, mammals, and are more biodiverse than monoculture systems. These agroforestry practices also contribute to sustainable food production by enhancing microclimate, decreasing soil erosion and enhancing water retention. An example is coffee plantations, which are favored by the shading effects of trees, enabling a diverse understory of biodiversity and enhancing quality of the beans, thereby resulting in healthier ecosystems and improved pests. Altogether, the example of agroforestry systems in Brazil and Costa Rica demonstrates how trees can be used to boost biodiversity besides ensuring sustainable production, and the way toward climate-resilient farming in the region (Diyaolu and Folarin 2024).

3.17.3 Organic Farming in Europe

In Europe, the use of organic farming methods has greatly improved agricultural resilience and helped conserve biodiversity. Organic farming promotes a holistic approach that reduces the adverse effects of conventional farming with a much-needed emphasis on biodiversity and natural processes. In countries such as Sweden, Germany, and France, large-scale organic farming initiatives have been implemented; the initiatives have led to healthier soils, greater species diversity, and greater stability of ecosystems. Organic farming is characterized by the prohibition of the use of artificial fertilizers and pesticides, as it contributes to the preservation of such useful animals as pollinators, natural insect predators, and soil microbes. Research has shown that organic farms have a higher sustainability rates of biodiversity compared to conventional farms and have more species and more of them, particularly wild plants and insects. This boosted biodiversity helps to improve the resilience of agricultural systems by promoting nutrient cycling, enhancing pollination services, and improving pest management. Moreover, organic farming techniques such as crop rotation, cover crops, and intercropping provide ecological niches to diverse species and increase complexity of habitats; the techniques enhance the fertility and structure of the soil, establishing farming systems that are more resilient to climate changes. Organic vineyards in France, such as the ones in the example, are proven to be more drought-resistant, as they retain water better and have healthier soil. The possibility of ecologically justified and sustainable food production is also evidenced by the popularity of organic farming in Europe, which can be used as an effective case study to increase biodiversity and agricultural sustainability (Diyaolu and Folarin 2024).

3.18 Obstacles in adopting the Biodiversity-Friendly Farming Practices

3.18.1 Economic Market Pressures

The trends in the global economy and consumer changes have tremendously influenced the way agriculture is practiced, with short-term gain taking precedence over long-term advantage of biodiversity. The economic incentives of high productivity can commonly cause intensive agricultural practices in the detriment of biodiversity. The high prices of commodities promote mono-cropping and the use of chemicals, which can be regarded as short-term productivity solutions; this potentially decreases the motivation of farmers to use biodiversity-friendly practices, which could lead to reduced money. Consumers tend to prefer homogenous, aesthetic products, which makes farmers grow a limited number of products and, as a result, genetic diversity in farming systems decreases. Smallholder farmers are also under pressure to prefer high-yield cultivars and chemical-intensive farming techniques over sustainable practices due to the emergence of large agribusinesses and corporate farming. The absence of economic incentives to protect biodiversity is also an additional barrier to the transition to sustainable farming practices; the adoption of farming practices that can support biodiversity is greatly impaired by market and financial limitations. The creation of market mechanisms and financial incentives that encourage the preservation of the biodiversity and encourage farmers to use a more resilient and environmentally friendly farming approach to promote sustainable agriculture is crucial (Diyaolu and Folarin 2024).

3.18.2 Lack of Knowledge and Access to Resources

Farmers do not have knowledge and resources that can assist them to adopt farming methods that will improve biodiversity. Most farmers, especially rural farmers are not aware of how biodiversity helps to make their farms more resilient to agricultural challenges; and may not be able to embrace effective biodiversity measures such as crop diversification, agroecological methods, and Integrated Pest Management without access to education and training on sustainable agricultural methods. The use of poor agricultural extension services, absence of high-quality seeds, and financial constraints are other resource constraints affecting farmers to adopt methods that promote biodiversity. Smallholder farmers can also be unable to invest in other crops or green technology; to apply these principles, proper tools, seeds, and support systems are critical. The lack of these resources can force farmers to use traditional practices that are more focused on short-term returns than on sustainability, continuing to destroy biodiversity. Addressing knowledge gaps and expanding resource access is crucial for promoting biodiversity-enhancing farming practices (Diyaolu and Folarin 2024).

3.18.3 Political and Institutional Barriers

There are institutional and political barriers that have slowed the adoption of sustainable farming practices that enhance biodiversity. The absence of a collective approach by the government and disproportional laws makes farmers unable to switch to sustainable practices. Farming policies usually favor short-term monetary returns at the cost of conserving biodiversity because farmers are not willing to make the requisite changes unless there are systems of long-term support, e.g., grants or subsidies aimed at biodiversity-friendly operations. The sustainable agricultural initiatives are either insignificant or temporary and farmers find it difficult to access the required knowledge and resources because of bureaucracies and lack of coordination among the government agencies. These barriers in the institutions and politics are critical toward facilitating sustainable agriculture and biodiversity conservation in agricultural systems (Diyaolu and Folarin 2024).

3.19 Future Directions

3.19.1 New Technologies in Biodiversity Monitoring and Management

There is an increasing recognition of the role of biodiversity in agricultural resiliency, with new technologies such as artificial intelligence (AI), machine learning, and remote sensing technologies transforming how scholars and agriculturalists assess biodiversity and its impact on agricultural production. Remote sensing technology is based on the principle that it involves the use of satellites or drones to capture information so that their real-time assessment of species diversity, vegetation wellness, and regional land use can be carried out; such data can help farmers prioritize conservation efforts and where biodiversity is concentrated. AI and machine learning enhance biodiversity surveillance by predicting tendencies and outcomes associated with agriculture practice and biodiversity; biodiversity technologies can replicate ecological interactions, assess how various agricultural methods affect biodiversity, and suggest optimal practices to maintain the ecological equilibrium and boost production. Citizen science platforms are also being designed to engage farmers and the general population in initiatives of monitoring biodiversity in regions, improving local biodiversity knowledge and promoting stewardship among agricultural stakeholders. This innovative technology will be integrated into agricultural activities, resulting in more sustainable and robust agricultural systems that will enhance the management of biodiversity (Diyaolu and Folarin 2024).

3.19.2 Integration of Biodiversity in National and International Food Policies

To implement food security and promote sustainable agricultural practices, the biodiversity should be part of the national and international food policy. Policymakers need to recognize the value of biodiversity in resilient agricultural systems and should focus on conserving it in their food security strategies (Diyaolu and Folarin 2024).

3.20 Policy Recommendations

3.20.1 Incorporation of Biodiversity in Food Security Strategies

The goals for biodiversity should be specifically included in national food security programs. To overcome the changing environmental requirements and climate change, there might be a need to build the frameworks enabling the preservation and sustainable utilization of the genetic resources, including native crops and livestock breeds (FAO 2019, as cited in Diyaolu and Folarin 2024).

3.20.2 Support of Agroecological Practices

Well-known examples of agroecological practices that enhance biodiversity in agriculture and can be supported by policy include crop diversification, intercropping, and organic farming. Governments can also associate agricultural production with the preservation of biodiversity by providing farmers who employ them subsidies, financial support, and technical support (Diyaolu and Folarin 2024).

3.20.3 Strengthening International Agreements

The Convention on Biological Diversity (CBD) and other international accords ought to give agricultural biodiversity priority. Countries have to commit to conserve biodiversity in agricultural landscapes and report on their efforts (Diyaolu and Folarin 2024). CBD (2020) suggests that this can be achieved by establishing specific objectives of protecting agricultural biodiversity that can be compared to objectives of climate action (as cited in Diyaolu and Folarin 2024).

3.20.4 Research and Knowledge Sharing

Research needs to be invested to understand the relationships between biodiversity and food production. Governments should also promote the exchange of information among academics, farmers, and policymakers to disseminate creative ideas and best practices to integrate biodiversity into agriculture. To enhance future generations resilience and food security, stakeholders can develop a more sustainable agricultural system by integrating biodiversity into national and international food policy (Diyaolu and Folarin 2024).

3.21 Conclusion

Biodiversity is significant to preserve the ecosystems which allow the production of food and it is critical to the long-term food security. It is associated with the ecological services of soil health, pollination, and pest control. Traditional farming practices, including climate change, chemical inputs, and intensive farming can contribute to soil degradation, biodiversity loss, and environmental change vulnerability. However, adopting biodiversity-friendly techniques like agroforestry, crop diversification, and agroecology can significantly increase crop resilience and production. The role of farmers and policymakers in conserving and promoting biodiversity in agricultural environments is very important; through sustainable production, which enhance ecosystem wellbeing, farmers become custodians of biodiversity. National and international policy must prioritize the preservation of biodiversity in agricultural plans; a framework to encourage sustainable behaviors and safeguard the natural resources must be provided offering incentives, research funds, and integrating biodiversity into food security strategies. Farmers need to be educated and inform them about the benefits of adopting biodiversity-friendly practices; this can be achieved by improving extension services and creating stakeholder engagement zones that can help speed up the process. The waning agricultural systems of the future will rely on novel technology to monitor biodiversity and adjust policies that take into account biodiversity. The studies of the interaction between biodiversity and food production will offer valuable insights to practice and policy. Biodiversity needs to be preserved and restored to provide food security, environmental wellbeing, and sustainable farming in the future.

References

Adhikari L, Hussain A, Rasul G (2017) Tapping the potential of neglected and underutilized food crops for sustainable nutrition security in the mountains of Pakistan and Nepal. Sustainability 9(2):291. https://www.mdpi.com/2071–1050/9/2/291

Aldow A, Brehm JM, Gaisberger H, Maxted N (2025) In situ and ex situ conservation gap analyses of crop wild relatives from Northeast Africa. In: Genetic resource and crop evolution. https://doi.org/10.1007/s10722-025-02337-8

Alotaibi M (2023) Climate change, its impact on crop production, challenges, and possible solutions. Notulae Botanicae Horti Agrobotanici Cluj-Napoca 51(1):13020

Al-Tawaha ARM, Gunal H, Krecek J, Zamfir RHC, Patel H, Vyas R, Zamfir CHC, Celik I, Khalid S, Aleksanyan A (2021) Soil fertility decline under climate change. In: Sustainable soil and land management and climate change. CRC Press, pp 127–145

Benitez-Alfonso Y, Soanes BK, Zimba S, Sinanaj B, German L, Sharma V, Bohra A, Kolesnikova A, Dunn JA, Martin AC (2023) Enhancing climate change resilience in agricultural crops. Curr Biol 33(23):R1246–R1261

Bohan DA, Richter A, Bane M, Therond O, Pocock MJ (2022) Farmer-led agroecology for biodiversity with climate change. Trends Ecol Evol 37(11):927–930

Borrelli P, Robinson DA, Panagos P, Lugato E, Yang JE, Alewell C, Wuepper D, Montanarella L, Ballabio C (2020) Land use and climate change impacts on global soil erosion by water (2015–2070). Proc Natl Acad Sci USA 117(36):21994–22001

Cadena-Zamudio JD, Ramirez-Mosqueda MA, Cadena-Zamudio DA (2024) Agrobiodiversity and food security: challenges and sustainable solutions. Agro Productividad 17(10):101–113

Cheng M, McCarl B, Fei C (2022) Climate change and livestock production: A literature review on. Atmosphere 13(1):140. https://www.mdpi.com/2073-4433/13/1/140

Chu X, Man R, Dang QL (2023) Spring phenology, phenological response, and growing season length. Front For Glob Change 6:1041369

Diyaolu CO, Folarin IO (2024) The role of biodiversity in agricultural resilience: protecting ecosystem services for sustainable food production. Int J Res Publ Rev 5(10):1560–1573

Ebi KL, Vanos J, Baldwin JW, Bell JE, Hondula DM, Errett NA, Hayes K, Reid CE, Saha S, Spector J, Berry P (2021) Extreme weather and climate change: population health and health system implications. Annu Rev Public Health 42:293–315. https://doi.org/10.1146/annurev-publhealth-012420-105026

Essl F, Latombe G, Lenzner B, Pagad S, Seebens H, Smith K, Wilson JR, Genovesi P (2020) The convention on biological diversity (CBD)'s post-2020 target on invasive alien species: what should it include and how should it be monitored? NeoBiota 62:99–121

Finch DM, Butler JL, Runyon JB, Fettig CJ, Kilkenny FF, Jose S, Frankel SJ, Cushman SA, Cobb RC, Dukes JS (2021) Effects of climate change on invasive species. In: Invasive species in forests and rangelands of the United States: a comprehensive science synthesis for the United States forest sector. Springer, Cham, pp 57–83

Hafeez U, Ali M, Hassan SM, Akram MA, Zafar A (2023) Advances in breeding and engineering climate-resilient crops: a comprehensive review. Int J Res Adv Agric Sci 2(2):85–99

Harsanto B, Kasumaningrum Y, Arviansyah MR, Siregar AY, Purnomo D, Freddy, Iskandar Y, Utama ID, Sari DI (2025) Leveraging disruptive technologies for food security: a systematic review on agricultural supply chain resilience to climate change. Curr Res Food Sci 10:101079. https://doi.org/10.1016/j.crfs.2025.101079

Hulme PE (2017) Climate change and biological invasions: evidence, expectations, and response options. Biol Rev 92(3):1297–1313

Ingrao C, Strippoli R, Lagioia G, Huisingh D (2023) Water scarcity in agriculture: an overview of causes, impacts and approaches for reducing the risks. Heliyon 9(8):e18507

Jha MK, Dev M (2024) Impacts of climate change. In: Smart internet of things for environment and healthcare. Springer, pp 139–159

Khan MT, Aleinikoviene J, Butkeviciene LM (2024) Innovative organic fertilizers and cover crops: perspectives for sustainable agriculture in the era of climate change and organic agriculture. Agronomy 14(12):2871. https://www.mdpi.com/2073-4395/14/12/2871

Kisvarga S, Horotan K, Wani MA, Orloci L (2023) Plant responses to global climate change and urbanization: implications for sustainable urban landscapes. Horticulturae 9(9):1051

Kremen C, Merenlender A (2018) Landscapes that work for biodiversity and people. Science 362: eaau6020

Kumar C, Kotra V, Kumar N, Singh K, Singh AK (2024) Biodiversity and bioresources: impact of biodiversity loss on agricultural sustainability. In: Biodiversity and bioeconomy. Elsevier, pp 165–198

Loizou E, Spinthiropoulos K, Kalogiannidis S, Chatzitheodoridis F, Kalfas D, Tzilantonis G (2025) Enhancing climate resilience and food security in Greece through agricultural biodiversity. Land 14(4):838. https://www.mdpi.com/2073-445X/14/4/838

Marx W, Haunschild R, Bornmann L (2021) Heat waves: a hot topic in climate change research. Theor Appl Climatol 146(1):781–800

Mishra RK (2023) Fresh water availability and its global challenge. Br J Multidiscip Adv Stud 4(3): 1–78

Morgado J, Santeramo F, Lamonaca E, Ciliberti MG, Caroprese M (2022) Meta-analysis and systematic literature review of climate change effects on livestock welfare. EFSA J 20:e200413

Mostafa AR, Owes AM, Ghoniem SA (2025) Interconnected impacts of climate change on biodiversity, agriculture and human health. Adv Basic Appl Sci 4(1):43–63

Ortiz R (2011) Agrobiodiversity management for climate change. In: Agrobiodiversity management for food security. CABI, p 189

Ortiz AMD, Outhwaite CL, Dalin C, Newbold T (2021) A review of the interactions between biodiversity, agriculture, climate change, and international trade: research and policy priorities. One Earth 4(1):88–101

Prakash S (2021) Impact of climate change on aquatic ecosystem and its biodiversity: an overview. Int J Biol Innov 3(2)

Rădoi MI (2020) Interconnections between sustainable development, climate change and agriculture. Revista de Ştiinţe Politice. Revue des Sciences Politiques 68:53–61

Raggi L, Pacicco LC, Caproni L, Alvarez-Muniz C, Annamaa K, Barata AM, Batir-Rusu D, Diez MJ, Heinonen M, Holubec V (2022) Analysis of landrace cultivation in Europe: a means to support in situ conservation of crop diversity. Biol Conserv 267:109460

Ramirez-Villegas J, Khoury CK, Achicanoy HA, Diaz MV, Mendez AC, Sosa CC, Kehel Z, Guarino L, Abberton M, Aunario J, Awar BA, Alarcon JC, Amri A, Anglin NL, Azevedo V, Aziz K, Capilit GL, Chavez O, Chebotarov D et al (2022) State of ex situ conservation of landrace groups of 25 major crops. Nat Plants 8(5):491–499. https://doi.org/10.1038/s41477-022-01144-8

Richardson AD, Keenan TF, Migliavacca M, Ryu Y, Sonnentag O, Toomey M (2013) Climate change, phenology, and phenological control of vegetation feedbacks to the climate system. Agric For Meteorol 169:156–173

Singh S, Kumar V, Kumar M, Adhikari BS (2025) Forest fire and climate change effects on shaping flora and fauna. In: Forest fire and climate change: insights into science. Springer, pp 101–118

Skendžić S, Zovko M, Pajač Živković I, Lešić V, Lemić D (2021) Effect of climate change on introduced and native agricultural invasive insect pests in Europe. Insects 12(11):985

Solheim AL, Austnes K, Eriksen TE, Seifert I, Holen S (2010) Climate change impacts on water quality and biodiversity, Background report for EEA European environment state and outlook report. Springer, pp 1–68

Subedi B, Poudel A, Aryal S (2023) The impact of climate change on insect pest biology and ecology: implications for pest management strategies, crop production, and food security. J Agric Food Res 14:100733

Sutter L, Albrecht M (2016) Synergistic interactions of ecosystem services: florivorous pest control boosts crop yield increase through insect pollination. Proc R Soc B Biol Sci 283(1824): 20152529

Uhl P, Bruhl CA (2019) The impact of pesticides on flower-visiting insects: a review with regard to European risk assessment. Environ Toxicol Chem 38(11):2355–2370

Van der Sluijs JP, Vaage NS (2016) Pollinators and global food security: the need for holistic global stewardship. Food Ethics 1(1):75–91

Wudu K, Abegaz A, Ayele L, Ybabe M (2023) The impacts of climate change on biodiversity loss and its remedial measures using nature-based conservation approach: a global perspective. Biodivers Conserv 32(12):3681–3701

Yang Y, Wu Y, Long H, Ma X, Shariati K, Webb J, Guo L, Pan Y, Ma M, Deng C (2023) Global honeybee health decline factors and potential conservation techniques. Food Secur 15(4): 855–875

Yuan X, Li S, Chen J, Yu H, Yang T, Wang C, Huang S, Chen H, Ao X (2024) Impacts of global climate change on agricultural production: a comprehensive review. Agronomy 14(7):1360. https://www.mdpi.com/2073-4395/14/7/1360

Zhang W, Sheng J, Li Z, Weindorf DC, Hu G, Xuan J, Zhao H (2021) Integrating rainwater harvesting and drip irrigation for water use efficiency improvements in apple orchards of Northwest China. Sci Hortic 275:109728

Traditional Knowledge and Ethnobotany

4

Sumayya Siddiqa, Mamoona Munir, Sidra Naseer, Sadia Abdullah, Mushtaq Ahmed, Bisma Javed, Shazia Sultana, Samreen Dawood, Maria Amin, and You-Cai Xiong

Abstract

This chapter states to the sensitive interface of ethnobotany with traditional knowledge, highlighting the central role they play in conserving biodiversity preserving culture and ensuring sustainable development. The chapter begins by defining ethnobotany and its domain, with emphasized importance being placed on traditional knowledge as a living reservoir of collective wisdom passed across generations. A brief history outlines the time-honored associations between indigenous communities and their environment. The chapter goes on to explore the foundations of traditional knowledge, including indigenous knowledge systems, oral tradition, and the pivotal roles of elders and local communities. An apex focus is on applied ethnobotany in traditional systems—particularly in medicine, nutrition, rituals, and culture practices. Sacred and symbolic uses of plants are also discussed, bringing out religious elements of plants human relationships. Due to the growing threats to both plant diversity and indigenous knowledge, urgent issues such as language loss, deforestation, and biopiracy were briefly discussed. It looks into current documentation strategies, challenges with keeping oral knowledge, and the role pf community-driven and digital archives. The conclusion recapitulates key points and call for integrative strategies for the preservation of traditional ecological wisdom in the modern world. Finally, the

S. Siddiqa · M. Munir (✉) · S. Naseer · S. Abdullah · B. Javed
Department of Botany, Rawalpindi Women University, Rawalpindi, Pakistan
e-mail: mamoona.munir@f.rwu.edu.pk

M. Ahmed · S. Sultana
Department of Plant Sciences, Quaid-e-Azam University, Islamabad, Pakistan

S. Dawood · M. Amin
Department of Botany, University of Mianwali, Mianwali, Pakistan

Y.-C. Xiong
Agricultural Ecology and Sustainability Science, College of Ecology, Lanzhou, Gansu, China

R. Qureshi (ed.), *Plant Biodiversity*, https://doi.org/10.1007/978-981-92-1471-6_4

future of ethnobotany and traditional knowledge encourage their revitalization through young people involvement, new technologies, and equitable partnerships.

Keywords

Traditional knowledge · Ethnobotany · Indigenous knowledge · Medicinal plants · Biodiversity conservation

Abbreviations

TEK	Traditional Ecological Knowledge
TK	Traditional Knowledge
TKDL	Traditional Knowledge Digital Library
TKS	Traditional Knowledge Systems
WHO	World Health Organization

4.1 Introduction

4.1.1 Definition and Scope of Ethnobotany

The study of the direct interactions between plants and people in their culture is known as ethnobotany. Man has relied on Mother Nature for all of his fundamental requirements from the beginning of time, and the variety of plants that surround him has always his interest. Man's first interest in plants sprang from his need for food, housing, and protection. Later, he turned his focus to using plants to treat wounds and illnesses (Rahman et al. 2019). The topic has a number of philosophical and transdisciplinary roots in the social and scientific sciences. The two words "ethno" and "botany," which denote the study of people and plants, respectively, can be used to broadly characterize ethnobotany. They cover a broad range of research topics, from history and archeology to the chemistry of natural products and the bioengineering of new crops, but they all center on the interactions between humans (and communities) and plants. According to Nolan and Turner 2011, ethnobotany plays a significant role in historical ecology as well as other natural and social science-based fields like political ecology, linguistics, cultural ecology, environmental ethics, conservation biology and ecological restoration, nutrition, and pharmacognosy.

Ethnobotany adds the understanding of the close relationships between humans and ecosystems, whereas many other historical ecological sciences comprehend broad perspectives of people-nature interactions, such as general land usage (Chauhan 2020). These are the attested facts based on studies and WHO-compatible sources of the significance of ethnobotany, specifically its contribution to global healthcare: Around 50,000 plant species of medicinal use are estimated to exist worldwide, out of millions of flowering plant species that have been reported. According to the World Health Organization (WHO), 80% of the world uses traditional medicine largely in the form of plant extracts for everyday health care, for new medications are still too expensive or out of reach for most of the world's people (Rahman et al. 2019). A vast majority of plant-based medicines currently in common use in medicine, today have their roots in early ethnobotanical experience; e.g., the majority of drugs regarded as essential by WHO are directly or indirectly derived from plants recorded in traditional practice. Additionally, over 400,000 tones medicinal and aromatic crops are commercially exchanged every year, including as many as 53,000 species in market highlighting the widespread application of plant resources on an extensive scale, much of which is facilitated by traditional demand (Martin 2010).

4.2 Scope of Ethnobotany

Ethnobotany is a highly multidisciplinary discipline that exists at the stage of natural and social sciences uniting botany, anthropology, pharmacology, ecology, economics, and linguistics to examine the way humans and plants interact and benefit societies (Pei et al. 2020) as shown in Fig. 4.1. The detailed elaborations are as follows:

- Knowledge and Cultural Practices: Accounts and records of indigenous and local people's use of plants for medicine, food, clothing, shelter, rituals, dyes, weaving, and tools.
- Medicinal Research and Drug Discovery: Studying plant medicines to discover, isolate, and screen bioactive compounds which can be developed as new drugs; it is projected that 25% of all existing prescription medications are derived from plants.

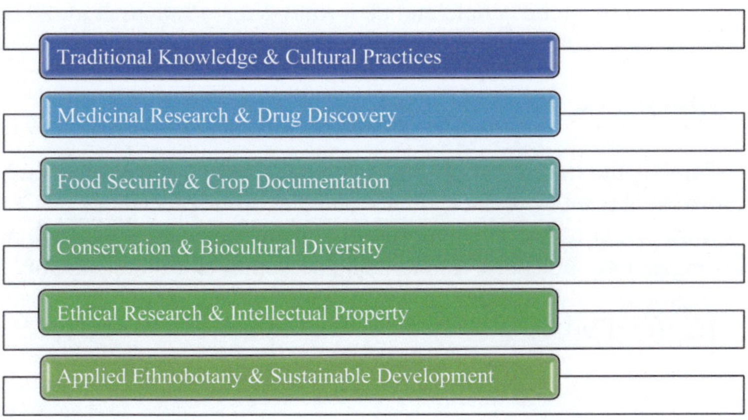

Fig. 4.1 Scope of ethnobotany

- Food Security and Crop Domestication: Uncovering neglected wild and edible plants for increasing food diversification, food sovereignty, and domesticating new crops for secure food production.
- Conservation and Biocultural Diversity: Facilitating conservation of biodiversity by recording plant species occurring in traditional cultures, supporting conservation of endangered species, and enhancing stewardship of cultural and sacred landscapes.
- Ethical Research and Intellectual Property: Interest in addressing ethical concerns like informed consent, benefit-sharing, and indigenous intellectual property rights protection in ethnobotanical research and bioprospecting.
- Applied Ethnobotany and Sustainable Development: Harness traditional plant knowledge to facilitate sustainable community-based development, environmental education, rural livelihoods (e.g., agroforestry, natural dyes, handicrafts), and ecological management strategies.

In essence, ethnobotany gives us the methodological tools and ethical guidelines to record, analyze, and utilize traditional plant-based knowledge in such a manner that promotes health, conservation, cultural preservation, and sustainable livelihoods (McClatchey et al. 2009).

4.3 Branches of Ethnobotany

Ethnobotany, which combines the fields of botany and anthropology, provides reliable knowledge about the uses of plant species by native people. It is an interdisciplinary field that combines ethnology with plant science. An ethnic group's botanical knowledge is referred to as a "multidisciplinary science" since it can be applied in a variety of ways. A number of fields, including ethnomedicine, ethnotaxonomy, ethnotoxicology, ethnoecology, ethnogynecology, ethnopharmacology, ethnopadiatrics, ethnoforestry, etc. (Fig. 4.2), have been applied in relation to ethnic people and their knowledge (Suthari et al. 2021).

4.3.1 Ethnomedicine

Ethnomedicine is the traditional healing systems of native and local societies. It is the composition of plant medicines, rituals, and diagnostic procedures that have been inherited from earlier times. The majority of contemporary medicines are made out of these fundamental practices. Ethnomedicine supports cultural identity maintenance and providing substitutes in areas with limited access to contemporary medicine (Gao et al. 2019).

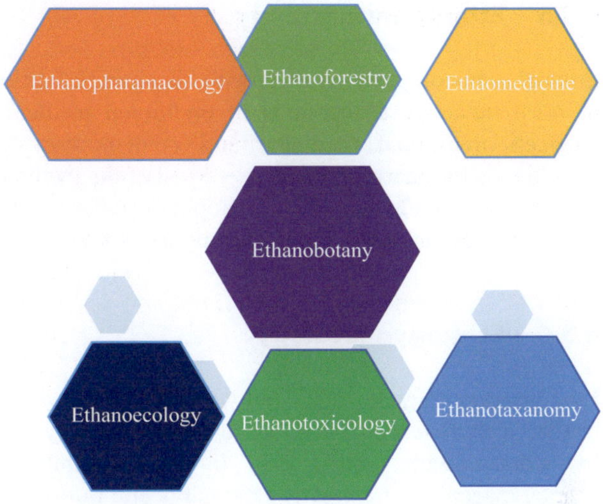

Fig. 4.2 Branches of ethnobotany

4.3.2 Ethnotaxonomy

Ethnotaxonomy refers to the investigation of how various societies categorize and name plants using local knowledge. It contrasts with scientific taxonomy since it is influenced by use, appearance, taste, or symbolic association. These classifications assist in recognizing edible, medicinal, or toxic species among communities. It is important in understanding human responses to biodiversity (Hidayati et al. 2018).

4.3.3 Ethnotoxicology

Ethnotoxicology investigates traditional knowledge regarding plant toxicity and how they can be applied without injuring individuals. It deals with how communities detoxify or prevent toxic plant constituents by means of detoxification strategies. Ethnotoxicology assists in plant toxicity evaluation and safe drug development. It is used in avoiding poisoning and in valorization of the culture of risk management (Eddouks et al. 2019).

4.3.4 Ethnoecology

Ethnoecology explores how local and indigenous people view, engage with, and organize their environment. It involves the knowledge of the habitats of plants, cyclical patterns, and harvesting that is sustainable. Ethnoecology maintains conservation by combining ecological principles and cultural practices. It is vital to ecosystem management and climate resilience measures (Sharma et al. 2024).

4.3.5 Ethnopharmacology

Ethnopharmacology is involved with the characterization and scientific examination of plant bioactive compounds of traditional medicinal use. Ethnopharmacology connects traditional medicine and contemporary pharmacology by isolating compounds for pharmaceutical use. Most of the medicines of today (e.g., morphine, quinine) have their origins in ethnopharmacology. Ethnopharmacology speeds up drug discovery and authenticates indigenous knowledge (Jauhari et al. 2024).

4.3.6 Ethnogynaecology

Ethnogynaecology describes old woman's knowledge and practice of reproductive health, such as fertility, delivery, and post-delivery care. Menstruation is regulated by women with plants, pain in labor is relieved, or infection is cured. Ethnogynaecology maintains maternal health information in remote regions. It also promotes acceptable interventions in females' health (Surendran et al. 2023).

4.3.7 Ethnopediatrics

Ethnopediatrics refers to the study of the indigenous plant medicines and methods employed in the management of childhood diseases. It encompasses treatment of the prevalent diseases such as colds, fever, gastrointestinal diseases, and skin conditions using herbal medications. They are safer and well-established among the people. Ethnopediatrics facilitated easier integration with the modern pediatric practice (Chandraiah et al. 2013).

4.3.8 Ethnoforestry

Ethnoforestry is a combination of scientific forestry and traditional knowledge for managing forests in a sustainable manner. It is native tree plantation, felling, and conservation of ecosystems. It offers people a way of enhancing biodiversity and health in the forest. It strengthens participatory forest management and deepens cultural respect in conservation policy (Bobsin et al. 2023).

4.4 Traditional Knowledge

As the term implies, traditional knowledge (TK) refers to the numerous customs and long-standing traditions of indigenous peoples. The culture and community knowledge that are passed down (usually orally) from one generation to the next are embodied is traditional knowledge. Traditional knowledge changes along with these cultures; it is never stagnant (Shivani and Mishra 2022). Traditional

Table 4.1 Comparison between traditional knowledge and ethnobotany

Aspects	Traditional knowledge	Ethnobotany
Definition	Knowledge passed through generation about environment, culture, and survival practices	Study how people of a particular culture use and perceive plants
Scope	Broad (agriculture, hunting, medicine, rituals, weather)	Narrower focuses specifically on human-plant relationships
Focus	Cultural and ecological knowledge (whole)	Botanical knowledge within cultural context
Origin	Indigenous and local communities	Academic and scientific field, but draws from traditional sources
Purpose	Preserving heritage, sustainability, survival and adaptation	Research, conservation, documentation of plant usage
Type of knowledge	Holistic, practical, spiritual, and oral	Scientific interpretation of traditional plant-related knowledge
Application	Agriculture, medicine, fishing, architecture, weather prediction	Medicine, conservation, pharmacology, agriculture
Documentation	Mostly oral and passed through practices	Mostly written in scientific literature and ethnographic records

Gerique (2006), Łuczaj (2023), and Lulesa et al. (2025)

knowledge (TK) is defined as indigenous people's knowledge, belief, and practice handed down through generations in relation to cultural memory, sharp observation of nature, and adaptive experience. Table 4.1 shows comparison between traditional knowledge and ethnobotany. TK encompasses not just ecological knowledge but also knowledge of food, medicine, shelter, tool, ritual, and land use. Examples are conventional practices such as burning land for preparation, seasonally determined use of resources, and sustainable resource harvest of priority species in an attempt to create well-being among humans (Gann et al. 2019).

Some examples of traditional knowledge are: (1) traditional Indian healing practices dating back over 3000 years, using plant-based remedies, dietary guidelines, yoga, and spiritual balance. This system integrates herbal treatments with knowledge of body constitution (doshas) and natural cycles. (2) Indigenous Australians have long used controlled burning (fire-stick farming) to manage vegetation, promote new growth, and prevent destructive wildfires. This method supports biodiversity and reduces fire hazards. (3) Indigenous communities in the Andes developed complex terracing systems to farm steep mountain slopes, preventing erosion and managing water efficiently knowledge still used today for sustainable agriculture. (4) The Inuit people of the Arctic have passed down detailed knowledge of ice patterns, animal migration, and weather forecasting through generations. This helps in safe hunting, traveling, and adapting to extreme conditions (Martin et al. 2010).

4.5 Historical Perspective of Traditional Knowledge and Ethnobotany

In 1895, American John William Hershberger officially coined the word "ethnobotany." Hershberger thought that ethnobotany could help to clarify the cultural position of the tribes that use plants for food, shelter, or clothing in an article titled "The Purposes of Ethnobotany" that was published in 1896. This clarification could then help to clarify the issue of plant distribution. Since the interaction with nature is only one aspect of a complex cultural system, many experts have disputed Hershberger's theory that it would be able to comprehend a complete society from how it used plants. We now know that understanding plants as a component of intricate social-ecological systems can improve our understanding of human relationships (Albuquerque et al. 2017).

Sumerian, Egyptian, Vedic, and Chinese civilizations all left voluminous amounts of data about plants, creating foundational but unscientific work. The nineteenth and early-twentieth centuries saw transition from anecdotal record to systematic observation by explorers and scholar, usually in a colonial environment, documenting indigenous plant uses. Richard Evans Schulte's was a leading figure in the mid-twentieth century, founding the discipline through his meticulous Amazonian work on ethnomedicinal and psychoactive species. Ethnobotany has come far from its beginnings as an interest in "primitive" cultures to an interest in all plant-human relationships within a large number of ecological and cultural situations. Modern ethnobotany is addressed by cross-disciplinary research founded on pharmacology, ecology, linguistics, and archeology. Ethnobotany provides the techniques for researching traditional ecological knowledge, bioprospecting for new drugs, and human-plant history reconstruction. This incremental progress reflects the role of ethnobotany in addressing global issues like conservation of biodiversity, cultural heritage, and quest for sustainable solutions.

The source of traditional knowledge (TK) is the millennia-old linkage between Indigenous peoples and nature, which was built upon thousands of years of observation, experimentation, trial and error, and adaptive learning. It was a holistic, place-based system of knowledge that arose from interconnections between nature (Berkes 2017) culture and religion for existence and sustainability. TK existed not through literature but through oral tradition, experiential knowledge, and group memory, typically held in rituals, folkloric narratives, and local language. This knowledge was employed by early societies to regulate agriculture, irrigation, animal domestication, and health care with remarkable accuracy. As an example, Andean farmers built terraced irrigation and soil husbandry systems appropriate for mountainous land. Simultaneously with this, Indigenous Australians had cultivated country by fire-stick agriculture to maximize biodiversity, where land use was not fixed but was dynamic over time, always adapted by ecological feedback. The development of TK also bears witness to collective worldview, in which nature, culture, and cosmology are integrated. Although they are diverse, but traditional knowledge system across the world have universal values of respect sustainability.

Its history attests to its worth as both intellectual heritage and as a living, dynamic system (Gammage 2012).

4.6 The Intertwined Paths of People and Plants

The expert gardener understands that there is no single method to interact with gardens. This understanding has results for scholars studying the advantages of people-plant interactions and practitioners looking to use them for personal and social development. Research suggests that people and plants connect in a variety of ways, including quiet abiding, tactile immersion, and physical exertion (Haller et al. 2019). Different types of interactions may not always be more successful. Interactions can be tailored to specific locations or groups, based on available space and resources (DelSesto 2020). Interactions between people and plants range from:

- Take part in sensory experiences with plant environments.
- Walk, move, or explore through plant environments.
- Engage in play or non-plant activities in plant environments.
- Count, sort, and examine plant materials.
- Practice table-top gardening with potted plants.
- Dig, till, plow, or work with soil.
- Create plant-based foods and products.
- Plant, weed, prune, or harvest in the garden.

4.7 Characteristics and Transmission of Traditional Knowledge

Traditional knowledge (TK) transmission has helped rural and indigenous people improve livelihoods during times of disruption and transition. Several researches have shown contradictory data on changes in TK, leading to diverse hypotheses to explain the findings. The decline in traditional knowledge (TK) can be attributed to several factors, including decreased plant diversity due to land use change, erosion of cultural practices and local languages, formalization of traditional education, and economic factors like migration, urbanization, and integration into market economies. The continual existence of traditional knowledge has been linked to ecosystem characteristics like low floristic diversity, the preservation of traditions and livelihood systems, and the commercialization of non-timber forest products. Different approaches and statistical techniques used in measuring TK lead to unpredictable results, as noted by other writers (Zent and Maffi 2010).

Investigations measuring cultural change frequently compare TK across age groups, inferring loss if older participants know more than younger ones. However, this strategy may result in an incorrect perception of cultural change among young versus older participants. Other research measures cultural change based on birth dates and links it to changes in wildlife abundance, education, livelihood shifts, and the development of transportation and communication infrastructure. Intensified globalization and economic development have an increasing impact on resilience, as they speed up information exchange and introduce foreign products, resulting in syncretic development and feedback cycles (McCarter et al. 2014). Understanding how these processes influence TK transmission and its ability to change and adapt is limited. Understanding the transmission of TK is crucial for understanding its conservation, loss, and dissemination. TK can be transmitted through three models: vertical, horizontal, and oblique (from older to younger). Not all TK domains are transmitted equally across generations, and identifying changes in distribution could be important in resource management. TK has contributed to biodiversity understanding and conservation strategies, making it essential to understand its distribution. Studies on the loss or alteration of Therapeutic Knowledge (TK) have focused on the intercultural context of these processes at larger scales. Camara-Leret et al.'s cross-cultural study examined medicinal palm use patterns across ecoregions, countries, human groups, communities, and individuals in north-western South America. Socioeconomic factors also influenced TK. These studies highlight the complexity of TK patterns and the need for standard protocols (Paniagua-Zambrana et al. 2016).

4.8 Diversity of Traditional Knowledge Systems

Traditional knowledge and traditional knowledge systems (TKS) are terms used to describe the understanding of resource and ecosystem dynamics and related management techniques that exist among members of communities that regularly and over extended periods of time interact with ecosystems for their livelihood and benefit. Information regarding the utilization of plants and animals by humans is not the only thing covered by the word. Instead, it comprises a system of resource usage and management, a set of actual facts about the local environment, and a system of classifications. It also covers beliefs in non-human entities (such as gods, ghosts, spirits, and ancestors) and their interactions with society (Reyes-García 2010).

The diverse TKS are not monolithic. Instead, they are profoundly heterogeneous within the frameworks of enacted worldviews, relations of power, dominion, and situated cultures. It is not the case that all communities are holistically or mechanistically inclined. It has been documented that different communities possess hybrid epistemic toolboxes which include mechanistic and holistic heuristics shaped by their metaphysical beliefs, social practices, and ethical values. The transmission pathways or oral storytelling, rituals, apprenticeships, and community proverbs pertaining to governance, healing, cosmology, and craftsmanship reflect vertical, horizontal, or oblique learning routes and vary from domain to domain which influences innovation and preservation in different ways. Decolonial scholarship has also expanded the boundaries of provenience and emphasizes the plurality of epistemologies alongside the call for epistemic justice or an ecology of knowledge consisting of multiple coexisting traditions that is equitable and free from Western epistemic dominance while fostering endogenous revitalization (Ludwig and Poliseli 2018).

4.9 Foundation of Traditional Knowledge

4.9.1 Indigenous Knowledge Systems

A diverse range of cultural knowledge, customs, and languages are possessed by indigenous tribes worldwide. There are over 7000 languages in the world, but many of them are indigenous and in danger of going extinct. As indigenous groups adjust to modern life, their languages and knowledge systems become less and less important. Many indigenous languages are today endangered, and their decline threatens the loss of important information. Nevertheless, these indigenous knowledge systems, which are frequently passed down through generations, are crucial for the preservation of distinctive identities and the sustainable management of ecosystems. A possible path for the revival and preservation of indigenous knowledge systems and languages in this digital age is to make use of digital media technology. This convergence causes major disruptions in conventional approaches and plans. In addition, digital media technologies are causing revolutionary shifts in a number of institutions and sectors, changing the way people and organizations communicate and work together. These technologies are revolutionizing our civilization by making it possible to preserve and disseminate indigenous knowledge through physical artifacts made from cutting-edge materials.

It is critical to recognize the impact of digital media technologies as they mark a significant age marked by significant technology breakthroughs that are changing global businesses and society (Oosthuizen et al. 2023).

Indigenous knowledge, which preserves oral histories and ancestral traditions, is an essential cultural, social, and historical resource for Indigenous people across the world. It is strongly ingrained in their customs and civilizations', frequently entwined with spiritual beliefs, language, and rituals. Indigenous knowledge must be preserved in order to support resistance to homogenizing pressures and ensure cultural continuity. Since they have coexisted peacefully with their natural surroundings for generations, indigenous groups offer valuable knowledge about how to mitigate climate change, preserve biodiversity, and maintain ecosystems. Traditional medicines and therapeutic methods are frequently included into their knowledge systems, which aids in the creation of novel medications and therapies (Vargas-Chaves 2024).

4.9.2 Oral Traditions and Cultural Transmission

The creation of social imaginaries is a kind of communication and an important part of traditional cultural expressions that originate with ethnic communities and indigenous peoples. In fact, these statements have a distinct meaning even though they are comparable to copyright-protected works since oral transmission is the primary mode of transmission and they are not incorporated or fixed in a media, as required by copyright.

The protection that copyright affords to things that are not embodied in a medium is a complicated matter that is up for debate in a number of discussion scenarios, even though the copyright regime gains relevance to promote the production and dissemination of culture by safeguarding the work put forth by authors and owners (Vargas-Chaves 2024).

4.9.3 Role of Communities and Elders

Social support-focused community-level treatments have demonstrated encouraging benefits in terms of enhancing mental health in older adult groups. According to Barreto, community therapy techniques, for example, establish forums for group sharing and support, promoting interpersonal relationships and resolving psychological issues. These therapies dramatically decreased the symptoms of anxiety and sadness in older adults living in rural Brazilian communities. Similar to this, peer support initiatives in rural Japan have shown promise in lowering loneliness and enhancing quality of life through mutual help networks and organized social events. These illustrations show how community-based; culturally appropriate methods can be used to treat mental health issues in older adult populations living in rural areas (Zhao et al. 2025).

4.10 Significance of Ethnobotany in Traditional Knowledge Systems

4.10.1 Plant Use in Medicine and Healing

The use of medicinal plants has always been a part of human evolution; they were among the earliest therapeutic resources made available to people and are now very important for maintaining human health. The World Health Organization (Qi 2013) estimates that over 80% of people in underdeveloped nations receive their main medical treatment through traditional medicine, which often involves the use of plant extracts or active chemicals (Carvalho and Silveira 2010). Medicinal plants, herbal medicines, or related products are traditionally utilized in primary care in a number of countries, according to WHO statistics. According to the WHO, a medicinal plant is any kind of plant that has a pharmacological effect when given to people. Ethnopharmacology's conclusions about the medicinal qualities of plants and common knowledge about their use have been offered as a basis for the advancement of technical scientific understanding (Cedillo-Cortezano et al. 2024).

In the daily lives of humans, plants have long been an integral part of existence. Numerous advantages have led to the traditional use of herbal remedies, including their affordability and accessibility as well as the lack of data about their efficacy or resistance to whole plant extracts (Barkaoui et al. 2017). As time goes on, this kind of traditional medical knowledge is passed down from generation to generation and is routinely used in homes. For many Moroccans, both rural and urban, using plants as a form of healing is still highly significant today. Through botanical surveys and laboratory biological tests on animal models, scientific research has shifted its focus in recent decades to plants used in traditional medicine to treat a variety of diseases. The goal is to identify species with medicinal qualities that could potentially replace certain chemical drugs with negative side effects (Kachmar et al. 2021).

Plants and plant extracts have been used for wound healing since the ancient period. Documents detail the application of plant extracts and extracts as poultices to halt bleeding and promote cicatrization. The intake of certain plants, which have systemic effects, has been linked to other usage. Therefore, the information gathered over time demonstrates that the only way modern medicine has evolved is through the transmission of traditional therapeutic techniques and the empirical knowledge related to them (Aquino et al. 2007). Around 100 million individuals worldwide have scars from burns, surgical procedures, or tissue ruptures brought on by various types of accidents each year; these scars need to be effectively and quickly treated. These figures suggest that wound healing is a contemporary therapeutic issue. The most effective organic and inorganic compounds in this respect are still a scientific mystery, despite several research that have attempted to enhance wound therapy by boosting the healing process. Different cellular structures are engaged in a number of intricate processes that make up healing. In order to prevent wound complications, the process starts with an enhanced immune response. This is accomplished through chemo-attraction, which supports the development of additional mediators required for later stages, including inflammation, cell proliferation, and re-epithelization, which ultimately result in wound closure (Ramalingam et al. 2022).

4.10.2 Plants Use as a Source of Food

Since the beginning of time, humans have relied on wild plants for sustenance as well as a variety of wild animals and their byproducts. In many places of the world, hunting and harvesting the wild plant has continued to be a source of income. Increased food production through plant domestication and cultivation resulted from the shift from collecting behavior to agriculture brought about by growing populations. Table 4.2 shows plants as a source of food. While truly domesticated plants comprise less than 200 species and only nine of them account for more than 66% of all crop production, thousands of wild food plant species are still used globally (Bélanger and Pilling 2019). The traditional eating of wild plants has been the subject of much research by anthropologists and ethnobotanists worldwide, especially in a number of African, American, European, and Asian nations. Wild food plants continue to be of interest because they are vital sources of fruits, vegetables, tubers, and nuts, all of which are necessary to balance the nutritional value of diets, and the food security of rural populations. It is true that wild food

Table 4.2 Plants as a source of food

Category	Plant name	Primary use	Region commonly grown	References
Cereal crops	Rice	Staple grains (Boiled flour)	Asia, Africa, Latin	Fukagawa and Ziska (2019)
	Maize	Bread, pasta, flour	Americas, Africa, Asia	Lulesa et al. (2025)
	Wheat	Staple food, oil	Europe, Asia, North America	Awulachew (2020)
Legumes	Lentils,	Dal, soups	Canada, South Asia	Fukagawa and Ziska (2019)
	Chickpeas	Curries, salads	India, Middle East	Wang et al. (2021)
Fruits	Banana	Fresh fruit	Tropical worldwide	Quan et al. (2024)
	Mango	Fresh fruit, juice, pickles	Africa, America, Latin	Minniti et al. (2023)
	Apple	Fresh, pie, cider	Temperate worldwide	Gayer et al. (2019)
Oil crops	Sunflower	Cooking oil, seeds	Eastern Europe	Schwingshackl et al. (2018)
	Olive	Olive's oil, pickled olives	Mediterranean countries	Fernandes et al. (2020)
	Ground nut	Oil, snacks, butters	India, China	Nunes et al. (2024)
Vegetables	Tomato	Cooking, sauces, salads	Global cultivation	
	Onion	Base flavors, salads	Global cultivation	Huang et al. (2021)

plant species from forests and farther-off places frequently have excellent nutritional contents (Punchay et al. 2020).

4.10.3 Fiber and Rituals Plants

Vegetables have been a significant source of natural fiber for centuries, which has been used to produce textiles, ropes, mats, and paper. Some of the more commonly utilized fiber-producing vegetables are cotton to produce soft fiber used in clothing; flax to produce linen; and jute to produce sacks and heavy weaves. Hemp is another crop used that contains good strong fiber to produce ropes and clothing, and coconut husk is also used to produce coir to produce brushes, mats, and mattresses. Table 4.3 shows plants used in manufacture of fiber. These fibers of plants are renewable and biodegradable, thereby making them crucial for the sustainable industries. Ramie is also a less popular plant fiber that can be blended with cotton for fabrics. Agave leaves give rise to sisal, which is highly prized in ropes and twines for its tensile strength. Fiber plants also provide a boost to rural economies and help encourage sustainable practices (Al-Fatimi 2024). Plants are part of traditions in the majority of the world's cultures. Table 4.4 shows plants used as a sign of rituals. The plants such

Table 4.3 Plants used in manufacture of fiber

Plant name	Fiber type	Primary uses
Cotton	Seed fiber	Clothing, textile, medical gauze
Flax	Bast fiber	Linen fabric, paper, rope
Ramie	Bast fiber	Fine textile, blending with cotton silk
Pina	Leaf fiber	Traditional textile
Sisal	Leaf fiber	Rope, mats, carpets, paper
Hemp	Bast fiber	Textile, paper, biodegradable plastics
Jute	Bast fiber	Bags, ropes, mats
Kenaf	Bast fiber	Ropes, textile paper, insulations

After Shahinur et al. (2022), Zimniewska (2022), and Akil et al. (2011)

Table 4.4 Plants used as a sign of rituals

Plant name	Region/culture	Ritual use	References
Cedar	Indigenous North America	Weddings and religious rituals, gift	Talin and Sanabria (2017)
Marigold	Mexico, India	Decoration of altars, symbol of spirit and divine connection	Raihandhany and Kurniawan (2023)
Betel leaf	India	Smudging and sweet lodge ceremonies	Ahuja and Ahuja (2011)
Cannabis	India, Jamaica	Spiritual elevation and meditation	Krol et al. (2022)
Sage	Native American	Smudging for purification and spiritual cleansing	Guedon (2000)
Ayahuasca	South America	Healing ceremonies	Anderson (2012)

as white sage are cleansed, tulsi is revered within Hindu tradition, and frankincense is incensed in Christian practices. Plants represent purity, protection, and spirituality.

4.10.4 Sacred and Symbolic Plants

Sacred and symbolic plants carry significant cultural, religious, and spiritual meanings in many traditions. For example, tulsi or holy basil is honored in Hinduism and is often planted in courtyards as a symbol of purity and protection. The lotus symbolizes purity and divine beauty in Buddhism and Hinduism and often it is linked with deities. In Christianity, the olive branch stands for peace and divine blessing. Table 4.5 shows sacred and symbolic plants. The bael tree is sacred to Lord Shiva and used in rituals and gift. Native Americans burn cedar and sage in ceremonies for purification. The bodhi tree, where Buddha reached enlightenment, represents wisdoms. These plants hold deep reverence and are essential to the spiritual lives of various communities (Jafari 2024).

Table 4.5 Sacred and symbolic plants

Plant name	Culture/religion	Symbolic meaning	References
Lotus	Hinduism, Buddhism, Egypt	Purity, spiritual, awakening, enlightenment	Iskandarovna (2024)
Tulsi	Hinduism	Protection, sacredness	Baliga et al. (2013)
Mistletoe	Celtic, norse	Immortality, fertility, protection	Thomas et al. (2023)
Frankincense	Ancient Middle East	Divinity, prayer, sacred presence	Hussain et al. (2022)
Bodha tree	Buddhism (India)	Enlightenment (tree under which Buddha attained nirvana)	Pokharel and Adhikari Pokharel (2021)
Ashoka tree	Hinduism, Buddhism (India)	Sorrow remover, love	Akhtar and Bashir (2021)
Sandalwood	Hinduism, Buddhism	Purity, calming, spiritual elevation	Reddy and Johar (2024)

4.11 Documentation and Preservation

Traditional knowledge and medicinal plants, which are frequently passed down orally, reflect a rich history of how people have interacted with nature. Its phytochemical contents, such as its secondary metabolites' antioxidant, antibacterial, and antifungal properties, make it a significant component of medical treatment. This expertise might be lost, though, as a result of globalization, changing lifestyles, and the aging of traditional practitioners. The answer provided by digitization is the systematic and easily available documentation and preservation of this priceless resource. In the domains of pharmacology, ethnobotany, and biodiversity protection, digitalization not only guarantees cultural preservation but also encourages creativity and cooperation by fusing ancient techniques with contemporary scientific methodologies (Abbasi et al. 2024).

Software documentation is essential to the creation of high-caliber software. This significance results from papers' capacity to help stakeholders use, comprehend, maintain, and improve a system more effectively (Aghajani et al. 2020). Information about the software system to which it belongs is communicated through a procedure known as software documentation. To put it another way, documentation helps all parties involved communicate and raises their level of project expertise with the goal of sharing and transferring information (Silva 2020). As a crucial component of the software development process, software documentation is used in many ways over the course of the software lifecycle, including as a guide for onboarding new developers, a communication tool between stakeholders, and a repository of information for keeping software current. Therefore, documentation is a collection of technical and general guides that are arranged as texts and comments using dictionaries, flowcharts, graphs, drawings, and other tools (Ding et al. 2014).

Document verification has several uses, ranging from confirming the legitimacy of financial records and contracts to confirming an individual's identification for online services and transactions. Strong document verification techniques are now more important than ever due to the widespread use of digital documents. Fraud, document forgery, and tampering present serious threats to businesses, people, and society at large. The need for effective and safe document verification methods is therefore growing (Brahma et al. 2021).

4.11.1 Signature Verification Techniques

In legal documents and financial transactions, signatures are frequently used as a form of personal verification. This part explores both automatic and manual approaches for verifying signatures. It examines the difficulties in verifying signatures, such as differences in writing style and the part played by dynamic elements like pressure and speed (Hsu 2020).

4.11.2 Verification Approaches

Extrinsic security characteristics that need to be verified include stamps, which are frequently seen on legal papers and certifications. Stamp verification techniques are the main topic of this part, with an emphasis on identifying authentic stamps and differentiating them from reproductions or counterfeits. It goes over how to improve stamp authenticity verification by using features and classifiers (Castelblanco et al. 2020).

4.11.3 Image Processing Methods

When managing scanned or digitized documents, image processing is very important for document verification. The image processing methods covered in this area include segmentation, feature extraction, and preprocessing approaches that improve the precision of systems for verifying documents (Roychoudhury et al. 2016).

4.11.4 Machine Learning Applications

Document verification has been transformed by machine learning, which makes it possible to automatically analyze and categorize documents. The machine learning applications in document verification are covered in detail in this area, including support vector machines, neural networks, and the processing of natural language (Garain and Halder 2008).

4.11.5 Comparative Analysis of Techniques

The comparative study paragraph helps readers to understand the advantages and disadvantages of various document verification methods. It assesses each method's performance, accuracy, and applicability, providing insight into which techniques are best suited for a certain document type and verification scenario.

In order to provide readers a thorough grasp of the many methods available for guaranteeing document authenticity and integrity, this section will examine these document verification procedures in great detail (Rusiñol et al. 2015). Even though document verification is essential in many fields, there are a number of issues that must be resolved for implementation to be dependable and successful. The drawbacks of document verification include scalability, accuracy constraints, security concerns, human variables, and real-world implementation.

4.12 Ethical Considerations in Ethnobotany: Respect, Reciprocity, and Rights

Ethnobotanical ethical principles are of top priority, particularly where indigenous peoples and local communities are involved. Some of the main principles are respect for traditional knowledge, fair return in research collaboration, and safeguarding intellectual property rights and rights to biodiversity (Vandebroek et al. 2025).

4.12.1 Respect for Indigenous Knowledge and Autonomy

Ethnobotanical research demands the respect of the cultural processes and intellectual sovereignty. What this entails is that informed prior consent be secured and that research practices align with the community values and expectations.

4.12.2 Reciprocity and Benefit-Sharing

Reciprocity is returning benefits to knowledge-sharing communities. It may come in forms of co-authorship, sharing of research data, or sponsorship of community initiatives. A study in the Mbe plain in Adamawa, Cameroon, depicts a variety of reciprocity forms, including material gifts and sharing information (Maxime et al. 2020).

4.13 Challenges and Opportunities for Traditional Knowledge in the Twenty-First Century

Traditional knowledge (TK) encompasses the knowledge, practices, and innovations developed by local and indigenous communities over generations. TK is under threat in the twenty-first century but provides huge opportunities for integration into modern systems (Ogar et al. 2020).

4.13.1 Globalization Erosion

Greater diffusion of Western technologies and ideologies has led to the marginalization of indigenous knowledge systems and the erosion of traditional practices and languages (Bihari 2023).

4.13.2 Overriding Cultures

Western culture, technology, and economics are likely to override indigenous knowledge and ways. This has the potential to de-emphasize and marginalize local languages, customs, and knowledge systems (Kehinde 2024).

4.13.3 Lack of Legal Protection

Traditional knowledge does not usually enjoy legal protection and recognition under international law, and hence it is prone to misappropriation and biopiracy (El Yazidi and Rijal 2024).

4.13.4 Integration Challenges

Integrating indigenous knowledge and contemporary scientific methods is challenging because of methodological and epistemological variations (Lubogo 2023).

4.13.5 Vulnerability to Climate Change

Indigenous peoples are the primary victims of the immediate effects of climate change, which threatened their ecosystems and, consequently, their indigenous knowledge (Jerez 2021).

4.13.6 Environmental Degradation and Climate Change

4.13.6.1 Disruption of the Traditional Practices

Global climatic change, ecological degradation, and habitat destruction have a direct effect on traditional land use, natural resource management, and ecological practices. Weather pattern changes, availability of traditional food, and loss of culturally significant species disrupt the delicate balance on which traditional knowledge depends (Ram and Shahzar 2024).

4.13.7 Socioeconomic and Political Pressures

4.13.7.1 Colonialism and Dispossession

Centuries of dispossession, discrimination, exploitation, and colonization have indeed resulted in the loss of traditional knowledge. Colonial policies tended to suppress indigenous practices, undermine traditional institutions, and deny access to traditional resources and lands (Nunn et al. 2024).

4.13.7.2 Lack of Respect and Recognition

Dominant scientific and political institutions devalue or discredit indigenous knowledge as "pre-logical" or "irrational." Lack of recognition discourages its incorporation into wider development and conservation frameworks (Penteado 2024).

4.13.7.3 Economic Pressures

Pressures for modernization and economic gain can lead to the discarding of customary livelihoods and practices for "more" modern alternatives, especially among generations that migrate to urban areas (Garrett et al. 2021).

4.13.7.4 Misappropriation and Biopiracy

Traditional knowledge, especially in medicine and agriculture, is frequently applied without the free, prior, and informed consent of the holders of the knowledge. Such "biopiracy" never brings about equitable benefits to such communities who have developed the knowledge, deterring its preservation and transmission (Sudheer and Bhadoria 2024).

4.13.7.5 Weak Legal Regimes

Lack of legal and policy regimes to protect traditional knowledge, as well as weak enforcement of already existing rights (e.g., the UN Declaration on the Rights of Indigenous Peoples), expose communities to loss of knowledge and exploitation (Lal et al. 2024).

4.14 Effects of Traditional Knowledge Loss

The loss and degradation of traditional knowledge have significant implications:

4.14.1 Loss of Cultural Identity

Traditional knowledge is at the core of indigenous and local peoples' identity, cultural heritage, and self-determination. Loss of traditional knowledge risks cultural disconnection, reduced cohesion of people, and empowerment of distinctive identities (El Yazidi and Rijal 2024).

4.14.2 Environmental Degradation

Traditional ecological knowledge (TEK) tends to contain the solutions for sustainable natural resource management, biodiversity conservation, and climate change adaptation. TEK loss results in unsustainable use, added vulnerability to the risks of the environment, and disturbance of ecosystems (Drissi 2022).

4.14.3 Loss of Original Solutions

The traditional knowledge provides valuable solutions to universal challenges like food deficiencies, global warming, and poor health. Loss means the loss of diversified innovations and techniques tried and optimized over centuries.

4.14.4 Health and Food Security Impacts

Production practices and indigenous agricultural knowledge are at the heart of the food security and health of numerous populations. Loss can result in increased dependence on outside, often less effective or culturally unsuitable, alternatives (Elendu 2024).

4.15 Opportunities of Traditional Knowledge

4.15.1 Sustainable Management of Resources

Indigenous practices provide valuable lessons on sustainable resource and land management, leading to climate resilience and conservation of biodiversity. Table 4.6 shows exploring traditional knowledge and ethnobotany (Suprayitno et al. 2024).

Table 4.6 Exploring traditional knowledge and ethnobotany

Component	Description	Ethnobotany connection	References
Traditional knowledge (TK)	An aggregative, locational body of know-how, practices, cultural values, and environmental knowledge accumulated across many generations in interaction with the environment	Provides the core knowledge base for ethnobotanical research initiating studies of plant uses in cultural, medicinal, ecological, and spiritual contexts	Reyes-García (2010)
Key knowledge domains	Domains encompass plant classification, medicinal/ cosmetic/culinary utilization, ecological management, spiritual beliefs, and management practices for the resource	Ethnobotanists record these sites in systematic fashion through interviews, participant observation, and fieldwork, maintaining existing structures of plant information	Christoph and Jensen (2017)
Documentation and survey methods	There are some traditional practices described through semi-structured interviews, focus groups, field walks, and direct interaction with healers and holders of traditional knowledge.	These methods transfer localized and oral knowledge to scientific data in an orderly form, record use-values, indices of consensus, and community histories	Berkowitz et al. (2022)
Scientific validation	Phytochemical screening, toxicity screening, nutritional/medicinal evaluation, and ethnomedicinal ratings are used to authenticate plant features and safety	Gives scientific validity to ethnobotanical uses, validates drug discovery, nutraceutical studies, and ecosystem management	Mondo (2024)
Challenges and ethical issues	Some of the issues are biopiracy, loss of TK to modernization, gender inequality in data collection, and absence of legal mechanisms to facilitate benefit-sharing	Ethnobotany is met with ethical approaches consent, fair collaborations, ISE code of ethics, and gender-sensitive research practices	Artal and Rubenfeld (2017)
Conservation and SDG applications	TK-informed ethnobotanical approaches promote biodiversity preservation, sustainable living, food security, and UN SDGs (e.g., health, climate action, partnerships) attainment	Leads community-conserved initiatives, promotes sustainable harvesting, drives policy and combines cultural heritage with contemporary agendas on sustainability	Xiao et al. (2018)

4.15.2 Integration with Modern Technology

Initiatives like India's AI-powered Traditional Knowledge Digital Library (TKDL) demonstrate how technology can aid in preserving and validating traditional medical systems.

4.15.3 Educational Inclusion

Incorporating traditional knowledge into education curricula can lead to respect for diversity and complement holistic learning strategies.

4.15.4 Policy Development

Identification and incorporation of traditional knowledge into policy-making can advance environmental management and community engagement (Jessen et al. 2022).

4.16 The Future of Ethnobotany: Innovation and Sustainability

Ethnobotany, the science of people and plants, is changing on a large scale. Being no anachronism, it is becoming a central discipline for dealing with problems facing the world today and specifically those of innovation and sustainability. Table 4.7 shows future directions of ethnobotany. The future of ethnobotany rests in its ability to reunite the old and the new, to promote concerted cooperation, and to step toward a sustainable future.

4.16.1 Bridging Traditional Knowledge and Modern Science

- Drug Discovery and Medicine: Ethnobotany has provided sources for new medicines (e.g., willow bark for aspirin, quinine) from Cinchona. Bioprospecting will increasingly be done in a more systematic, technological manner.
- Genomics and Metabolomics: Rapid identification of bioactive compounds and determination of their mode of action by novel techniques will speed up the drug discovery process.
- Personalized Medicine: Ethnographic understanding of variations in individuals' responses to plant medicines can guide personalized medicine strategies.
- Integration of Ethnomedicine: Integration of conventional practice into contemporary health care, where possible, can result in more comprehensive and culturally sensitive therapies (Yebouk 2025).
- Sustainable Agriculture and Food Security: Most traditional farming practices reflect sustainable values.
- Climate-Resilient Crops: Ethnobotanical information may be used for selecting resilient local landraces as well as traditional crops that have been adapted to the prevailing conditions, which are most important under climate change adaptation (Dean 2024).
- Agroecology and Permaculture: Ethnobotany guides designing sustainable agricultural systems that reflect the natural ecosystem, which supports biodiversity and minimizes the need for external inputs (Gillani et al. 2024).

Table 4.7 Future directions of ethnobotany

Theme	Innovations and trends	Sustainability and ethical practices	References
Digitization and LLMs	Application of machine learning and big language models (LLMs) for harvesting and analysis of traditional plant usage. LLMs enable the detection of latent uses and trends in taxonomy	Providing guarantees of data provenance, consent, privacy protection, and equitable benefit-sharing with source communities	Böhlen and Sujarwo (2020)
Drug discovery and biopharma	Screening plant natural products for antimicrobial and other therapeutic leads with the assistance of AI	Free intellectual property rights and accolades for indigenous knowledge providers	Chigozie et al. (2025)
Community-led conservation	Joint research with the locals to document and sustain local plant uses	Promoting community self-government, sustainable harvesting and renewable livelihoods	Hasan (2020)
SDGs and sustainability goals	Compliance with UN SDGs no poverty, zero hunger, health, sustainable consumption, climate action, biodiversity, and partnerships	Ethnobotanical information underpins sustainable ecosystems and food/health security for poor individuals	Kumar et al. (2021)
Education and curriculum integration	Integration of indigenous plant knowledge into institutions of learning and training throughout the world	Promotes cultural pride, facilitates knowledge transfer across generations	Ezeanya-Esiobu (2019)
Sustainable harvesting techniques	Implementation of old-fashioned selective and rotational methods of harvesting to reduce ecological footprint	Provides long-term stability to plant populations and local ecosystems	Girona et al. (2023)

- Nutritional Ethnobotany: Fieldwork among traditional food crops can uncover hidden sources of nutrition and introduce diversity in diets, fighting malnutrition, and promoting food sovereignty.

4.16.2 Technological Developments

- DNA Barcoding and Genetic Analysis: Such technologies allow for more precise and streamlined identification of plant species, adulteration detection in herbal supplements, and tracking of plant populations for conservation.
- Machine Learning and Artificial Intelligence: AI can parse enormous ethnobotanical data sets, recognize patterns, and forecast plant properties or potential medicinal applications, speeding up research.
- Remote Sensing and GIS: Using satellite and aerial images together with Geographical Information Systems, it becomes possible to monitor changes in the

distribution of plants, quantify the impact of climate change on plant populations and areas of high conservation value.

- Digital Databases and Platforms: The only way to access, share, and preserve ethnobotanical data is through online databases, such as Kew Royal Botanic Gardens' Plants of the World Online, Native American Ethnobotany Database, among others.
- Citizen Science: Local communities could be mobilized through mobile applications and other media to contribute data that flows into larger ethnobotanical research and monitoring. Table showing trends and sustainability of traditional knowledge in different areas.

4.17 Conclusion

This chapter concludes the deep link between traditional knowledge and ethnobotany as crucial to understanding the cultural use of plants. Traditional ethnobotanical knowledge (TEK), passed on through generations in local and Indigenous communities, provides the backbone of ethnobotany. It is such knowledge that supports sustainable use of medicinal and food plants and promotes conservation of biodiversity by maintaining the balance between human requirements and ecological balance. Indigenous knowledge, carried across generations through oral transmission and community guidance, holds immense ecological insight. Traditional uses of plants for healing, food, and rituals are central to many societies. However, this heritage faces threats from language loss, deforestation, and exploitation. Preserving it requires active documentation, knowledge sharing through various platform and respect for indigenous contributions from different areas. Community efforts and digital tool play a growing role in conservation. Engaging youth is vital to keeping these traditions alive. Modern science and traditional knowledge can work hand in hand od approached ethically. Policies must protect indigenous rights and ensure fair benefits-sharing. Ultimately, safeguarding traditional knowledge supports both cultural identity and global policies must protect indigenous rights and ensure fair benefit-sharing. Ultimately, safeguarding traditional knowledge supports both cultural identity and global biodiversity.

References

Abbasi TU, Ahmad M, Alsahli AA, Asma M, Mussagy CU, Abdellatief TM, Pastore C, Mustafa AJ (2024) Eco-friendly production of biodiesel from Carthamus tinctorius L. seeds using bismuth oxide nanocatalysts derived from Cannabis sativa L. leaf extract. Process Saf Environ Prot 191: 710–722

Aghajani E, Nagy C, Linares-Vásquez M, Moreno L, Bavota G, Lanza M, Shepherd DC (2020) Software documentation: the practitioners' perspective. In: Proceedings of ACM/IEEE 42nd international conference on software engineering. pp 590–601

Ahuja S, Ahuja U (2011) Betel leaf and betel nut in India: history and uses. Asian Agri-Hist 15:13–35

Akhtar J, Bashir F (2021) Sacred medicinal plants and their use by indigenous people for strengthening the immune system. In: Traditional herbal therapy for the human immune system. CRC Press, Boca Raton
Akil HM, Omar MF, Mazuki AM, Safiee S, Ishak ZM, Bakar AA (2011) Kenaf fiber reinforced composites: a review. Mater Des 32:4107–4121
Albuquerque UP, Ramos MA, Júnior WSF, De Medeiros PM (2017) Ethnobotany for beginners. Springer, Cham
Al-Fatimi MJ (2024) Traditional knowledge of wild plants on traditional tools, materials, products and economic practices in southern Yemen. J Ethnobiol Ethnomed 20:62
Anderson BT (2012) Ayahuasca as antidepressant? Psychedelics and styles of reasoning in psychiatry. Anthropol Med 23:44–59
Aquino DD, Silva RBL, Gomes VF, Araújo EC (2007) Nível de conhecimento sobre riscos e benefícios do uso de plantas medicinais e fitoterápicos de uma comunidade do Recife PE. Rev Enferm UFPE On-line:107–110
Artal R, Rubenfeld S (2017) Ethical issues in research. Best Pract Res Cl Ob 43:107–114
Awulachew MT (2020) The role of wheat in human nutrition and its medicinal value. GSC Adv Res Rev 2:50–54
Baliga MS, Jimmy R, Thilakchand KR, Sunitha V, Bhat NR, Saldanha E, Rao S, Rao P, Arora R, Palatty PL (2013) Ocimum sanctum L (Holy Basil or Tulsi) and its phytochemicals in the prevention and treatment of cancer. Nutr Cancer 65:26–35
Barkaoui M, Katiri A, Boubaker H, Msanda F (2017) Ethnobotanical survey of medicinal plants used in the traditional treatment of diabetes in Chtouka Ait Baha and Tiznit (Western Anti-Atlas), Morocco. J Ethnopharmacol 198:338–350
Bélanger J, Pilling D (2019) The state of the world's biodiversity for food and agriculture. FAO, Rome (IT)
Berkes F (2017) Sacred ecology. Routledge, London
Berkowitz CM, Ma J, Lowe J, Dolor RJ (2022) Assessing quality in advance care planning documentation: a survey of current methods. Am J Hosp Palliat Me 39:945–950
Bihari S (2023) Cultural heritage and indigenous knowledge: reviving traditions for future generations. Sustain Dev Goals SAARC Ctries 1:24–32
Bobsin CR, Bormann BT, Miller ML, Pelach BD (2023) Perspectives: ethnoforestry, ecosystem wellbeing, and collaborative learning in the Pacific Northwest. For Ecol Manag 529:120738
Böhlen M, Sujarwo W (2020) Machine learning in ethnobotany—a first experiment. arXiv preprint https://arxiv.org/abs/2008.01611
Brahma A, Goldberg DM, Zaman N, Aloiso MJ (2021) Automated mortgage origination delay detection from textual conversations. Decis Support Syst 140:113433
Carvalho AC, Silveira D (2010) Drogas vegetais: uma antiga nova forma de utilização de plantas medicinais. Brasilia Med 47(2):219–237
Castelblanco A, Solano J, Lopez C, Rivera E, Tengana L, Ochoa M (2020) Machine learning techniques for identity document verification in uncontrolled environments: a case study. In: Mex Conf Pattern Recognit, pp 271–281
Cedillo-Cortezano M, Martinez-Cuevas LR, López JAM, Barrera López IL, Escutia-Perez S, Petricevich VL (2024) Use of medicinal plants in the process of wound healing: a literature review. Plants 17:303
Chandraiah G, Sammaiah D, Devi U, Rajesham C, Ugandhar T (2013) Progress and prospects of medicinal plants of Ethnopediatric importance in Mahadevapur Reserve Forest of Karimnagar East Division of (AP) India. Biolife 1(3):140–148
Chauhan KJ (2020) Role of ethnobotany on Indian society: a review. J Adv Cult Philol 4:109–111
Chigozie VU, Ugochukwu CG, Igboji KO, Okoye FB (2025) Application of artificial intelligence in bioprospecting for natural products for biopharmaceutical purposes. BMC Artif Intell 1:1–21
Christoph D, Jensen RK (2017) Shea (Vitellaria paradoxa C.F. Gaertn.) at the crossroads: current knowledge and research gaps. For Ecol Manag 323:144–152

Dean M (2024) Exploring ethnobotanical knowledge: qualitative insights into the therapeutic potential of medicinal plants. Golden Ratio Data Sum 4:154–166

Delsesto MJ (2020) People–plant interactions and the ecological self. Plants People Planet 2:201–211

Ding W, Liang P, Tang A, Van Vliet H (2014) Knowledge-based approaches in software documentation: a systematic literature review. Inform Software Tech 56:545–567

Drissi S (2022) Indigenous peoples and the nature they protect. UN Environment Programme, Nairobi

Eddouks M, Ajebli M, Hebi M (2019) New indices for ethnotoxicological assessment of medicinal plants: example of Tafilalet region, Morocco. Cardiovasc Hematol Disord Drug Targets 14:127–139

El Yazidi R, Rijal K (2024) Science learning in the context of 'indigenous knowledge' for sustainable development. Int J Ethnosci Technol Ed 1:28–41

Elendu C (2024) The evolution of ancient healing practices: from shamanism to Hippocratic medicine: a review. Medicine 103:e39005

Ezeanya-Esiobu C (2019) Indigenous knowledge and education in Africa. Springer Nature, Singapore

Fernandes J, Fialho M, Santos R, Peixoto-Placido C, Madeira T, Sousa-Santos N, Virgolino A, Santos O, Carneiro AV (2020) Is olive oil good for you? A systematic review and meta-analysis on anti-inflammatory benefits from regular dietary intake. Nutrients 69:110559

Fukagawa NK, Ziska LH (2019) Rice: importance for global nutrition. J Nutr Sci Vitaminol 65:S2–S3

Gammage B (2012) The biggest estate on earth. In: The invisible thread: one hundred years of words. Halstead Press, Canberra

Gann GD, McDonald T, Walder B, Aronson J, Nelson CR, Jonson J, Hallett JG, Eisenberg C, Guariguata MR, Liu J (2019) International principles and standards for the practice of ecological restoration. Restor Ecol 27:S1–S46

Gao L, Wei N, Yang G, Zhang Z, Liu G, Cai C (2019) Ethnomedicine study on traditional medicinal plants in the Wuliang Mountains of Jingdong, Yunnan. China J Ethnobiol Ethnomed 15:41

Garain U, Halder B. 2008. On automatic authenticity verification of printed security documents. In: 6th Indian conference on computer vision, graphics, and image processing. pp 706–713

Garrett RD, Cammelli F, Ferreira J, Levy SA, Valentim J, Vieira I (2021) Forests and sustainable development in the Brazilian Amazon: history, trends, and future prospects. Annu Rev Environ Resour 46:625–652

Gayer BA, Avendano EE, Edelson E, Nirmala N, Johnson EJ, Raman G (2019) Effects of intake of apples, pears, or their products on cardiometabolic risk factors and clinical outcomes: a systematic review and meta-analysis. Curr Dev Nutr 3:nzz109

Gerique A (2006) An introduction to ethnoecology and ethnobotany. Theory and methods-integrative assessment and planning methods for sustainable agroforestry in humid and semiarid regions

Gillani SW, Ahmad M, Zafar M, Haq SM, Waheed M, Manzoor M, Shaheen H, Sultana S, Rehman FU, Makhkamov T (2024) An insight into indigenous ethnobotanical knowledge of medicinal and aromatic plants from Kashmir Himalayan region. Ethnobot Res Appl 28:1–21

Girona MM, Morin H, Gauthier S, Bergeron Y (2023) Boreal forests in the face of climate change: sustainable management. Springer Nature, Cham

Guedon M-F (2000) Sacred smokes in circumboreal countries: an ethnobotanical exploration. Northern Rev 21:11–26

Haller RL, Kennedy KL, Capra CL (2019) The profession and practice of horticultural therapy. CRC Press, Boca Raton

Hasan T (2020) Building on indigenous capacity: opportunities for self-determination through post-secondary education in Wasagamack first nation. Manitoba Law J 43(4):119–144

Hidayati S, Ghani BAA, Giridharan B, Hassan MZ, Franco FM (2018) Using ethnotaxonomy to assess traditional knowledge and language vitality. Ethnobot Lett 9:33–47
Hsu W-P (2020) Intelligent document recognition on financial process automation. In: International symposium on VLSI design, automation and test. p 1–1
Huang W, Tang G, Zhang L, Tao J, Wei Z (2021) Effect of onion on blood lipid profile: a meta-analysis of randomized controlled trials. Food Sci Nutr 9:3563–3572
Hussain H, Rashan L, Hassan U, Abbas M, Hakkim FL, Green IR (2022) Frankincense diterpenes as a bio-source for drug discovery. Expert Opin Drug Dis 17:513–529
Iskandarovna KS (2024) Pragmatic features of symbols in cross-cultural communication. World Econ J Lang Ed 2:50–54
Jafari M (2024) Sacred plants as signs of a bond between myths and divine stories. Turk J Res Arch 3
Jauhari S, Jauhari R, Rupanagunta GP, Nandave M, Upadhyay J, Joshi R (2024) Insights on the integration of ethnopharmacology and omics in medicinal plant research. In: Ethnopharmacology and OMICS advances in medicinal plants, vol 2. Springer, Cham
Jerez MM (2021) Challenges and opportunities for indigenous peoples' sustainability. DESA:1–10
Jessen TD, Ban NC, Claxton NX, Darimont CT (2022) Contributions of indigenous knowledge to ecological and evolutionary understanding. Front Ecol Environ 20:93–101
Kachmar MR, Naceiri Mrabti H, Bellahmar M, Ouahbi A, Haloui Z, El Badaoui K, Bouyahya A, Chakir S (2021) Traditional knowledge of medicinal plants used in the northeastern part of Morocco. Evid-Based Compl Alt 2021:6002949
Kehinde TB (2024) French colonial policies in West Africa: power dynamics, cultural impositions and economic legacies. Int J Adv Soc Sci Humanit 3:8–17
Krol A, Kokotkiewicz A, Luczkiewicz M (2022) White sage (Salvia apiana)—a ritual and medicinal plant of the chaparral: plant characteristics in comparison with other salvia species. Planta Med 88:604–627
Kumar A, Kumar S, Komal RN, Singh P (2021) Role of traditional ethnobotanical knowledge and indigenous communities in achieving sustainable development goals. Sustainability 13:3062
Lal R, Baeraiya B, Thakur R, Gautam RK, Narayan S, Yadav D, Pathak P, Singha S (2024) Traditional knowledge in drug development and the rights of indigenous peoples: a legal and ethical perspective. Law J:1–15
Lubogo I (2023) A right to protect indigenous languages. Social Sci Res Netw:1–12
Łuczaj Ł (2023) Descriptive ethnobotanical studies are needed for the rescue operation of documenting traditional knowledge. J Ethnobiol Ethnomed 19:37
Ludwig D, Poliseli L (2018) Relating traditional and academic ecological knowledge: mechanistic and holistic epistemologies across cultures. Biol Phil 33:43
Lulesa F, Alemu S, Kassa Z, Awoke A (2025) Ethnobotanical investigation of medicinal plants utilized by indigenous communities in the Fofa and Toaba sub-districts of the Yem zone, Central Ethiopian Region. J Ethnobiol Ethnomed 21:1–50
Martin GJ (2010) Ethnobotany: a methods manual. Routledge, London
Martin JF, Roy ED, Diemont SA, Ferguson BG (2010) Traditional ecological knowledge (TEK): ideas, inspiration, and designs for ecological engineering. Ecol Eng 36:839–849
Maxime LLG, Phalone KM, Dalatou T, Constantin AA, Laela ZZ, Rosette N, Guidawa F, Germo NJ, Néhémie DT, Gabriel AA (2020) Reciprocity in ethnobotanical research: case of a study carried out in the Mbe plain of Adamawa, Cameroon. Ethnobot Res Appl 20:1–12
McCarter J, Gavin MC, Baereleo S, Love M (2014) The challenges of maintaining indigenous ecological knowledge. Ecol Soc 19(3):39
McClatchey WC, Mahady GB, Bennett BC, Shiels L, Savo V (2009) Ethnobotany as a pharmacological research tool and recent developments in CNS-active natural products from ethnobotanical sources. Pharmacol Ther 123:239–254
Minniti G, Laurindo LF, Machado NM, Duarte LG, Guiguer EL, Araujo AC, Dias JA, Lamas CB, Nunes YC, Bechara MD (2023) Mangifera indica L., by-products, and mangiferin on cardio-metabolic and other health conditions: a systematic review. Lipids 13:2270

Mondo JM (2024) Southern Africa development community indigenous food development programme–scoping study on indigenous foods in DR Congo. WFP & FANRPAN, Johannesburg

Nolan JM, Turner NJ (2011) Ethnobotany: the study of people-plant relationships. Ethnobiology 9: 133–147

Nunes YC, De Oliveira Santos G, Machado NM, Otoboni AM, Laurindo LF, Bishayee A, Fimognari C, Barbalho SM (2024) Peanut (Arachis hypogaea L.) seeds and by-products in metabolic syndrome and cardiovascular disorders: a systematic review of clinical studies. Phytomedicine 123:155170

Nunn PD, Kumar R, Barrowman HM, Chambers L, Fifita L, Gegeo D, Gomese C, McGree S, Rarai A, Cheer K (2024) Traditional knowledge for climate resilience in the Pacific Islands. Wiley Interdiscip Rev Clim Chang 15:e882

Ogar E, Pecl G, Mustonen T (2020) Science must embrace traditional and indigenous knowledge to solve our biodiversity crisis. One Earth 3:162–165

Oosthuizen JH, Ungerer M, Volschenk JJ (2023) A fourth industrial revolution integrated intelligence taxonomy for top management. J Organ Chang Manag 20:404–443

Paniagua-Zambrana N, Cámara-Leret R, Bussmann RW, Macía MJ (2016) Understanding transmission of traditional knowledge across North-Western South America: a cross-cultural study in palms (Arecaceae). Bot J Linn Soc 182:480–504

Pei S, Alan H, Wang Y (2020) Vital roles for ethnobotany in conservation and sustainable development. Plant Divers 42:399

Penteado A (2024) Despite repeated warnings: a multidisciplinary approach of the shortcoming of numbers in indigenous knowledge and climate change. In: Traditional knowledge and climate change. Springer, Cham

Pokharel N, Adhikari Pokharel B (2021) A relationship between plants and their Hindu and Buddha cultures: an analysis Ficus religiosa (Pipal). In: Transcultural diplomacy and international law in heritage conservation. Springer, Cham

Punchay K, Inta A, Tiansawat P, Balslev H, Wangpakapattanawong P (2020) Traditional knowledge of wild food plants of Thai Karen and Lawa (Thailand). Genet Resour Crop Ev 67:1277–1299

Qi Z (2013) WHO traditional medicine strategy 2014–2023. WHO, Geneva

Quan Z, Li C, Zhao L, Cui D, Liu S, Yin Y, Tang Q, Zeng D, Song L, Fu X (2024) Effect of banana intake on serum potassium level in patients undergoing maintenance hemodialysis: a randomized controlled trial. Int J Nurs Stud 11:197–204

Rahman IU, Afzal A, Iqbal Z, Ijaz F, Ali N, Shah M, Ullah S, Bussmann RW (2019) Historical perspectives of ethnobotany. Clin Dermatol 37:382–388

Raihandhany R, Kurniawan B (2023) The utilization of marigold (Tagetes erecta L.) for traditional ceremonies in Indonesia: a systematic review. Global J Biol 2:130–137

Ram AL, Shahzar E (2024) Land, loss and liberation: indigenous struggles amid the climate crisis. World Economic Forum, Cologny

Ramalingam S, Chandrasekar MJ, Nanjan MJ (2022) Plant-based natural products for wound healing: a critical review. Curr Drug Res Rev 14:37–60

Reddy VVVV, Johar V (2024) Whispers of the sacred wood: a comprehensive review of Santalum album. Int J Pharm Res 4:01–07

Reyes-García V (2010) The relevance of traditional knowledge systems for ethnopharmacological research: theoretical and methodological contributions. J Ethnobiol Ethnomed 6:32

Roychoudhury S, Bellarykar N, Kulkarni V (2016) A NLP based framework to support document verification-as-a-service. In: IEEE 20th international enterprise distributed object computing conference. pp 1–10

Rusiñol M, Karatzas D, Lladós J (2015) Automatic verification of properly signed multi-page document images. International symposium on visual computing. pp 327–336

Schwingshackl L, Bogensberger B, Benčič A, Knüppel S, Boeing H, Hoffmann G (2018) Effects of oils and solid fats on blood lipids: a systematic review and network meta-analysis. J Lipid Res 59:1771–1782

Shahinur S, Sayeed MA, Hasan M, Sayem ASM, Haider J, Ura S (2022) Current development and future perspective on natural jute fibers and their biocomposites. Polymers 14:1445

Sharma A, Patel SK, Singh GS (2024) Revitalizing traditional knowledge of the Himalayan indigenous tribes: an ethnoecological and ethnobotanical approach. In: Traditional knowledge systems for environmental sustainability in the Himalayas. Springer, Cham

Shivani S, Mishra S (2022) Traditional knowledge: much more than what meets the eyes. Int J Health Sci 6:1570–1581

Silva ASDF (2020) Documentação de software: uma análise comparativa entre documentação tradicional e living documentation. Universidade Federal do Rio Grande do Norte, Natal

Sudheer N, Bhadoria G (2024) The intersection of intellectual property rights (IPR) and sustainable development goals (SDGs): examining how the former can either hinder or facilitate progress towards achieving the latter. Int J Law Manage Humanit 7:3126

Suprayitno D, Iskandar S, Dahurandi K, Hendarto T, Rumambi FJ (2024) Public policy in the era of climate change: adapting strategies for sustainable futures. Migr Lett 21:945–958

Surendran S, Prasannan P, Jeyaram Y, Palanivel V, Pandian A, Ramasubbu R (2023) Knowledge on ethnogynaecology of Indian tribes—a comprehensive review. J Ethnopharmacol 303:115880

Suthari S, Kota S, Kanneboyena O, Gul MZ, Abbagani S (2021) Ethnobotanical perspectives in the treatment of communicable and noncommunicable diseases. In: Phytomedicine. Elsevier, Amsterdam

Talin P, Sanabria E (2017) Ayahuasca's entwined efficacy: an ethnographic study of ritual healing from 'addiction'. Int J Drug Policy 44:23–30

Thomas PA, Dering M, Giertych MJ, Iszkuło G, Tomaszewski D, Briggs JJ (2023) Biological flora of Britain and Ireland: Viscum album: no. 303. J Ecol 111:701–739

Vandebroek I, Stepp J, Kunwar R, Hilgert N, Pulido Silva MT, Ladio AH, Clement CR, Ferreira WS Júnior, Towns AM, Borokini I (2025) Upholding ethical accountability in ethnobotany and ethnobiology research. Ethnobotany:1–12

Vargas-Chaves I (2024) Oral transmission in traditional cultural expressions, and the challenge of their protection from intellectual property rights. Rev Prop Intelect 9:101–112

Wang J, Li Y, Li A, Liu RH, Gao X, Li D, Kou X, Xue Z (2021) Nutritional constituent and health benefits of chickpea (Cicer arietinum L.): a review. Food Res Int 150:110790

Xiao W, Mills J, Guidi G, Rodríguez-Gonzálvez P, Barsanti SG, González-Aguilera D (2018) Geoinformatics for the conservation and promotion of cultural heritage in support of the UN sustainable development goals. ISPRS J Photogramm 142:389–406

Yebouk C (2025) The new face of ethnobotany: tradition meets scientific method. Ethnobot Res Appl 31:1–3

Zent S, Maffi L (2010) Methodology for developing a vitality index of traditional environmental knowledge (VITEK). Terralingua, British Columbia (CA)

Zhao S, Han L, Liu Y, Rui X (2025) Investigation and analysis of mental health status of the elderly in western rural areas. Front Public Health 13:1612600

Zimniewska M (2022) Hemp fibre properties and processing target textile: a review. Materials 15: 1901

Foods as Medicine

5

Shabnum Shaheen, Sumera Javad, and Rahmatullah Qureshi

Abstract

The idea of food as medicine dates back to the times and is still in the practice of disease prevention and health promotion. Fruits, vegetables, whole grains, and lean proteins contain nutrient-dense foods which provide essential bioactive compounds that aid immune processes, regulate inflammatory processes, and maintain metabolic homeostasis, such as antioxidants, vitamins, minerals, and phytochemicals. The components have been linked to lower risk of non-communicable diseases such as diabetes, cardiovascular diseases, and some types of cancers. Foods with high probiotics additionally increase the balance and general immunity of the gut microbiota. Nutritional science has clarified the action of major bioactives, e.g., curcumin and omega-3 fatty acids, which supports the use of functional foods and nutraceuticals in the contemporary diets. New technologies in the field of accurate nutrition allow customization of dietary therapy according to genetic and metabolic specifics of the individual. Nevertheless, issues exist such as strong clinical validation and increased access to nutritious foods. Evidence-based policies and awareness about nutrition would be important in integrating nutrition into healthcare systems so as to utilize the full therapeutic potential of diet in sustainable and preventative healthcare.

Keywords

Food · Medicine · Traditional healing · Therapeutic benefits · Antioxidants · Vitamins · Minerals · Nutrition

S. Shaheen (✉) · S. Javad
Department of Botany, Lahore College for Women University, Lahore, Pakistan

R. Qureshi
Department of Botany, Faculty of Sciences, Pir Mehr Ali Shah Arid Agriculture University, Rawalpindi, Pakistan
e-mail: rahmatullahq@uaar.edu.pk

R. Qureshi (ed.), *Plant Biodiversity*, https://doi.org/10.1007/978-981-92-1471-6_5

Abbreviations

ALA	Alpha Linolenic Acid
DHA	Docosahexaenoic Acid
EGCG	Epigallocatechin Gallate
EPA	Eicosapentaenoic Acid
FIM	Food is Medicine
LDL	Low-Density Lipoprotein
MF	Medicinal Foods
TCM	Traditional Food Medicine

5.1 Introduction to Medicinal Foods

Foods are dietary components that provide health benefits beyond basic nutrition. They include a wide range of natural foods, fruits, vegetables, whole grains, herbs, spices, and fermented products, as well as specialized formulations designed for clinical dietary management as shown in Fig. 5.1. Their incorporation into daily nutrition represents a holistic approach to maintaining physiological balance and improving health outcomes. These foods are enriched with bioactive constituents such as vitamins, minerals, antioxidants, and phytochemicals that contribute to disease prevention, health promotion, and therapeutic support. Daily consumption of medicinal foods improves both health and body equilibrium.

The concept of *Food is Medicine (FIM)*, although rooted in antiquity, has gained renewed scientific attention in contemporary healthcare discourse. Historically, the notion is often attributed to Hippocrates, who emphasized the therapeutic role of diet (Witkamp and van Norren 2018). Early religious texts, including Biblical accounts such as the "Daniel Diet," also allude to plant-based dietary patterns associated with improved physical and cognitive well-being (Bloomer et al. 2010).

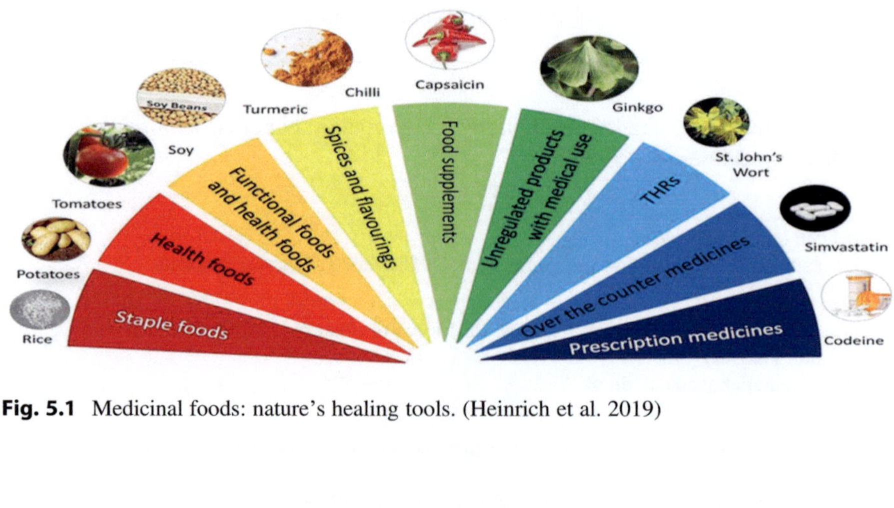

Fig. 5.1 Medicinal foods: nature's healing tools. (Heinrich et al. 2019)

5.2 From Nutrition to Medicine: The Emerging Power of Medicinal Foods

Modern nutrition and health science are experiencing a blurring effect between food classification and pharmaceutical substances. Hippocrates proclaimed through history that food should function as medicine, and medicine should serve as food. The significance of this principle grows vital in our present time because scientific evidence demonstrates how various foods maintain health and fight illness alongside their capability to control chronic diseases. Functional foods that people refer to as therapeutic foods or nutraceuticals or medicinal foods provide necessary nutrients while delivering natural compounds which enhance wellness and shield against sickness. The chapter reveals fundamental information about medicinal foods starting from their definition through their classifications and benefits and their current applications. Modern healthcare relies heavily on medicinal foods since these ingredients serve to protect health and integrate medical approaches. In recent decades, FIM has transitioned from a philosophical idea to an evidence-based approach increasingly integrated into mainstream medicine. Modern dietary guidelines from leading professional bodies advocate plant-forward dietary patterns characterized by high intake of whole grains, fruits, vegetables, and minimally processed foods, while limiting saturated fats and ultra-processed products (Arnett et al. 2019; Visseren et al. 2021). This paradigm shift is particularly relevant in the context of escalating global healthcare expenditures (World Health Organization 2023), where cost-effective and preventive strategies are urgently required.

The FIM framework offers a compelling solution by delivering significant health benefits at relatively low cost, while simultaneously contributing to environmental sustainability. In the domain of cardiovascular health, plant-based diets have consistently been associated with reduced blood pressure, weight loss, lower incidence of type 2 diabetes, and decreased cardiovascular morbidity and mortality (Freeman et al. 2018; Pikula et al. 2024).

5.3 Medicinal Foods: Nature's Healing Tools

Medicinal foods (MF) represent specific foods together with certain ingredients which extend medical advantages above their nutritional value. The list includes fruits, vegetables, herbs, seeds together with fermented foods and specially made dietary products which doctors may recommend for use (Fig. 5.1).

5.3.1 Key Characteristics

5.3.1.1 Contain Powerful Natural Compounds

Like flavonoids, alkaloids, phenolics, carotenoids, and glucosinolates that support health.

5.3.1.2 Help Specific Body Functions

Such as boosting immunity, improving digestion, supporting heart health, and controlling blood sugar.

5.3.1.3 Can be Part of Daily Meals

While it can be consumed regularly as part of a normal diet, it may also be taken in measured amounts under clinical guidance for specific therapeutic effects.

5.4 From Tradition to Today: The Legacy of Medicinal Foods

Ayurveda together with Traditional Chinese Medicine (TCM) as well as Unani depend heavily on medicinal foods for their healing practices. Human beings from various cultural backgrounds have used food as a natural healing method to support the body while achieving equilibrium for many centuries. Throughout history, food has been recognized not only as sustenance but also as an important contributor to health maintenance, disease prevention, and healing. Ancient societies treated food as medical treatment while the practice of therapeutic food usage has been transmitted across multiple generations. Ayurveda and Traditional Chinese Medicine (TCM) described food as the critical component in achieving body-nature balance. These systems established the critical link between food consumption, personal life choices, and health outcomes thus developing fundamental principles that modern people still use in their healthcare today. Scientific research currently validates the therapeutic value of practices which have been performed for generations.

5.5 Ayurveda: The Ancient Indian System of Healing

The medicinal system known as Ayurveda originated approximately 5.000 years ago within the Indian subcontinent as shown in Fig. 5.2. The medical system known as Ayurveda emerges from the Sanskrit combination of "Ayur" which means life and "Veda" which means knowledge thus translating to the knowledge of life. Food takes a central place in Ayurvedic medicine because it strengthens wellness and shields against sickness while healing illnesses (Tripathi and Chaturvedi 2007). Panchakarma detoxification in Ayurveda involves special dietary practices that help eliminate body toxins (Ama) called food for detoxification. The recommended foods for detoxification periods include mung beans along with rice and warm herbal teas that provide both lightness and easy digestion and detoxifying properties. The selected foods serve dual purposes of revitalizing digestive health and promoting wellness in the body.

Ayurvedic medicine groups food items according to their ability to impact the three doshas which the body contains. Food items possess distinct qualities consisting of heat and cold and dryness and moisture which affect the doshas in different ways. The natural bodily makeup known as Prakriti along with the current

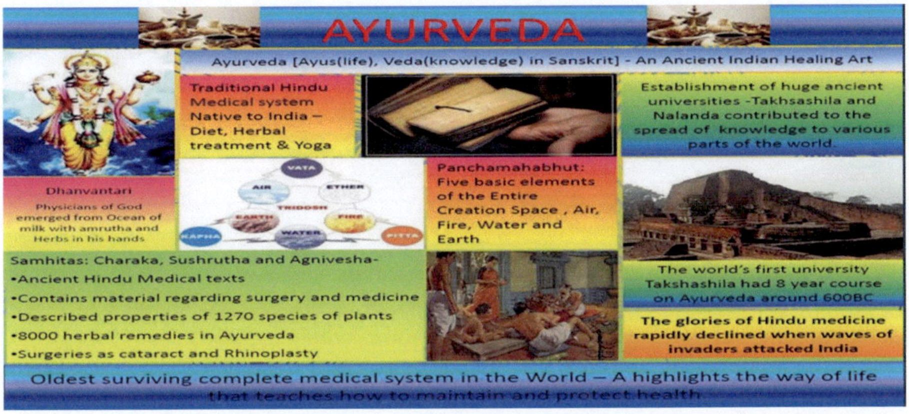

Fig. 5.2 Ayurveda: the ancient Indian system of healing. (Mukherjee and Wahile 2006)

state of imbalance known as Vikriti determines proper food choices for wellness (Patwardhan et al. 2005). Ayurvedic principles concentrate on maintaining equilibrium between three body energies named Vata, Pitta, and Kapha.

The Ayurvedic recommendation for *Vata* balance includes eating warm, oily, and ground foods which include soups together with root vegetables and dairy products. *Pitta* dosha balances fire and water principles but becomes unbalanced when it generates inflammatory reactions and increases acid production together with irritability symptoms. Pitta balance requires cool and hydrating foods which include dairy products, leafy greens, and cucumbers. The *Kapha* body type bearing characteristics of Earth and water elements may develop weight gain as well as sluggish behavior and respiratory congestion. An ideal diet for Kapha balance consists of spicy and drying foods, including ginger, beans, and bitter greens (Rastogi 2010).

Ayurvedic philosophy interprets food as more than mere sustenance since it acts as a pathway toward spiritual healing. A holistic therapeutic benefit from food depends on how it is prepared and presented as well as how it is consumed. According to Ayurveda, patients should practice mindful eating through eating their food in peaceful environments while enjoying each bite. People follow this ritual because it supposedly improves digestion as well as mental equilibrium (Gogte and Dole 2015).

5.6 Traditional Chinese Medicine (TCM): A Holistic Approach to Health

Traditional Chinese Medicine (TCM) has practiced food medicine for more than two thousand years as an essential element for health maintenance. Within TCM, patients focus on maintaining Qi energy balance between their body systems and external environmental forces featuring Yang and Yin elements. According to TCM, principles food items affect both the energies balance and the flow of Qi. Traditional Chinese Medicine organizes food substances into groups depending on how they affect body temperature so practitioners can use them to treat medical conditions through energy restoration.

5.6.1 Yin and Yang in Food

TCM organizes foods into Yin and Yang groups and creates dietary plans by analyzing individual medical situations as shown in Fig. 5.3. A person with Yang deficiency should follow dietary recommendations which include consuming ginger and garlic and lamb to help restore balance. The TCM treatment for someone with excessive Yang or internal heat would include giving them cooling foods such as watermelon, green tea, and cucumber to restore body balance (Qiu 2007).

5.6.2 Qi and Food Therapy

According to Traditional Chinese Medicine, the natural flow of Qi depends on the correct consumption of food because Qi functions as the life-sustaining force which exists within all living creatures. A person needs the continuous movement of Qi for their body to stay healthy. The health of the patient can be harmed by stagnated or deficient Qi. People who need their Qi strengthened should consume tonifying foods while those who require Qi stimulation should consume moving foods. Both ginseng and longan fruit work to strengthen the body but chili peppers and garlic aid in Qi movement through stimulation.

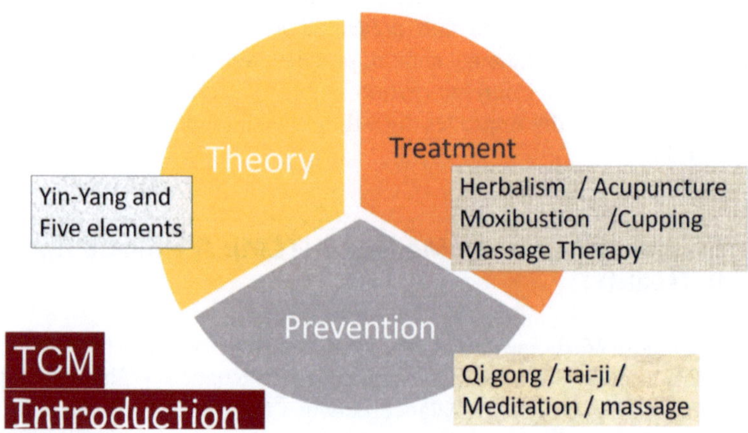

Fig. 5.3 Traditional Chinese Medicine: a holistic approach to health. (Rishabh 2024)

5.6.3 The Five Elements Theory

Food selection in TCM follows the fundamental principles of TCM which are the five elements: Wood, Fire, Earth, Metal, and Water as shown in Fig. 5.3. Dietary selection depends on how well different foods match the body elements according to the Chinese medical framework. Per TCM principles, the Earth element consists of sweet potatoes carrots and squash which provides grounding nourishment to the body. According to TCM beliefs, radishes and pears belonging to the Metal element help cleanse the body and specifically support the health of the lungs (Jiang et al. 2012).

The therapeutic impact of TCM therapies gets strengthened through the use of herb and food pairs. Congee functions as a traditional porridge made from rice that ensures different vegetable and herbal combinations support treatment of particular medical conditions like digestive problems and weariness. The combination of ginger and garlic serves dual purposes for digestion by warming the body while providing anti-infective properties.

TCM shares with Ayurveda the practice of using food as a method to prevent illnesses and preserve overall health. The traditional Chinese wisdom states that food should serve as both therapeutic treatment and therapeutic substance for promoting wellness. TCM practitioners suggest eating foods in accordance with environmental cycles because this dietary approach improves overall wellness. Traditional Chinese Medicine (TCM) along with Ayurveda has recognized the profound bond that exists between food and medicinal substances for a long period of time. The ancient approach to medicine sees food as more than mere sustenance because it holds the power to sustain health while preventing diseases while seeking restoration of equilibrium between body and mind. These traditional healthcare approaches address health issues at their core instead of treating symptoms through the use of herbs and carefully selected diets and spices. Modern science continues to confirm through research what ancient wisdom already recognized about medical applications in food. Ayurveda and TCM teachings guide modern healthcare practices since people choose natural preventative healthcare methods which demonstrate again and forever that food acts as medicine (Liu et al. 2015).

5.7 Categories of Medicinal Foods

The categorization of medicinal foods depends on their origin along with their functional purpose (Mine et al. 2009).

5.7.1 Natural Whole Foods: Fruits (Berries, Pomegranate)

Vegetables (e.g., cruciferous vegetables like broccoli).
Legumes and nuts (e.g., lentils, almonds).
Whole grains (e.g., oats, quinoa).

5.7.2 Herbs and Spices

The medicinal properties of cinnamon and ginger and black seed and fenugreek make them useful for blood sugar control and inflammation reduction and digestive health.

5.7.3 Functional and Fortified Foods

The health benefits from these foods include yogurts containing probiotics along with milk enriched with vitamins D and omega-3 and eggs fortified with omega-3.

5.7.3.1 Medical Foods

Medical foods exist for dietary management of particular diseases including PKU formula and gluten-free foods for celiac disease patients.

5.8 Health Benefits of Medicinal Foods

Scientific evidence demonstrates that medicinal foods play a crucial role in preventing disease and managing various medical conditions (Liu 2004).

5.8.1 Heart Health

The beta-glucan content in oats serves as a beneficial agent to reduce LDL cholesterol levels in the body. Green tea joins dark chocolate in offering polyphenols which help maintain blood vessels in good condition.

5.8.2 Digestive and Gut Health

Sauerkraut alongside kefir allows nutritious bacteria to grow within the digestive tract. Foods containing high fiber levels promote digestive health along with reducing the danger of developing colon cancer.

5.8.3 Anti-inflammatory and Antioxidant Effects

Berries and turmeric and ginger together reduce inflammation and oxidative stress which play major roles in developing chronic illnesses.

5.8.4 Metabolic Health

Blood sugar management is possible through the use of cinnamon and fenugreek along with bitter melon. Green tea complexes with apple cider vinegar to help people maintain their weight effectively.

5.8.5 Immune System Support

The immune system gets stronger through consumption of vitamin C in citrus fruits combined with Allicin from garlic and beta-glucans from mushrooms.

Clinicians are adopting food-based therapies at an increasing rate in present day medical facilities. Healthcare professionals now widely suggest dietary modifications together with medicinal foods through recommendations from doctors and dietitians and integrative health practitioners. The healthcare system can cut down its dependency on several medications through promotion of natural alternatives to synthetic drugs.

5.9 Personalized Nutrition

Research in genomics and microbiome science allows the development of personalized nutrition programs as a new industry trend. The method uses particular dietary suggestions based on each person's distinct genetic makeup together with their microbial composition for maximizing health benefits and disease prevention. Natural and preventive health approaches have created a rising demand that makes medicinal foods vital for future healthcare development (Martirosyan and Singh 2015).

5.10 Nature's Pharmacy: Exploring Bioactive Compounds in Food

Bioactive compounds are defined as chemical constituents of food which are natural and have certain biological effects on the health along with their nutritional benefits. These bioactive compounds are non-essential nutrients. It means they are not essentially required by the body, but they can enhance the health and wellness of the body. These can reduce the chances of chronic diseases like diabetes, cancer, cardiovascular diseases, and neurogenerative disorders. These are particularly found in grains, spices, fruits, legumes, teas, vegetables, etc. These include polyphenols, carotenoids, phytosterols, peptides, and glucosinolates as shown in Fig. 5.4. There are some bioactive compounds of animal origin as well like omega-3 fatty acids. There is a need of precise and a regular intake of plant-based and fermented foods that ensures the overall support of these compounds for good health and well-being. Threat of cardiovascular diseases is rising in developed countries due to increased

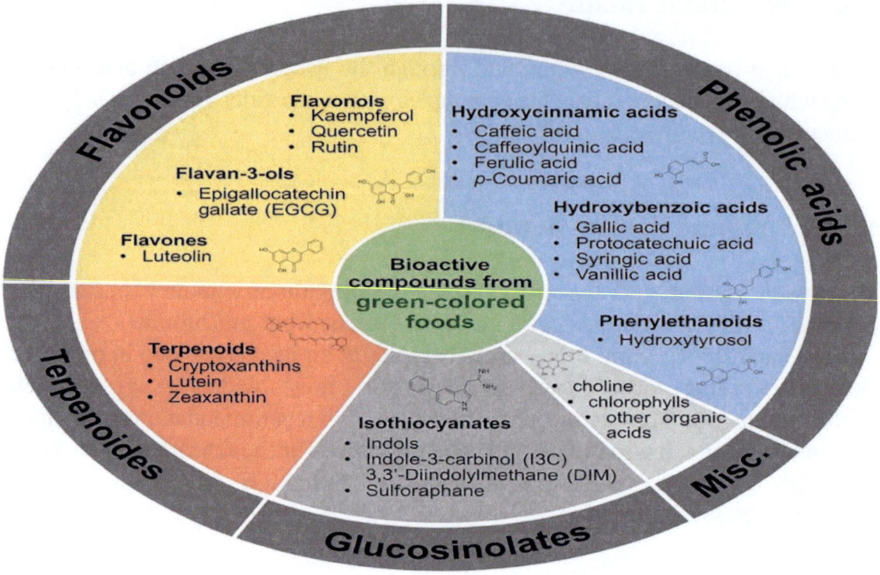

Fig. 5.4 Exploring bioactive compounds in foods (Bolte et al. 2021)

usage of processed foods. It has many risk factors including family history, age, ethnicity, smoking habits, blood pressure, obesity, and unhealthy diets. Bioactive components from food can be a good solution toward cardiovascular diseases.

5.10.1 Phenolic Compounds

Plant secondary metabolites are produced by plants for their own defense system having strong antioxidant and antimicrobial potential. More than 8000 phenolic compounds are known today with a variety of structure and function. The primary objectives of food processing are to improve nutritional profile, prolong shelf life, make products available during off-season, and inactivate viruses and enzymes. However, food processing may also be essential for the bio-accessibility of phenolic chemicals because it can have both positive and negative impacts. Both traditional processing methods like pasteurization, sterilization, grinding, cooking, baking, drying, and so forth, as well as nonthermal technologies like high-pressure processing, high-intensity pulsed electric field, and ultrasound, can affect the phenolic content and/or profile of the food materials. Data pertaining to the impact of fermentation and processing on phenolic compounds linked to fiber following in vitro gastrointestinal digestion are still scarce. Numerous fruits and vegetables have already been tested for the bio-accessibility of their phenolics after changes in structures brought on by processing.

Phenolic acids like gallic acid, ferulic acid, caffeic acid, ellagic acid, etc., are simplest form of phenolic compounds and their basic structure contains a phenolic ring, and they are of two main types of hydroxy benzoic acids and hydroxycinnamic acids. These food ingredients are biologically very active as for example gallic acid can be considered as a lead compound for treating neurodegenerative disorders (Bhuia et al. 2023), various cancers (Jiang et al. 2022), diabetes (Xu et al. 2021), and resistant bacterial infections. This phenolic acid occurs widely in nature in leaves, roots, fruits, and seeds of variety of plants and is part of our daily food (Dhiman and Mukherjee 2021). Similarly, caffeic acid is another phenolic acid that is part of various fruits, vegetables, and beans like coffee beans. It is well-known for its anti-cancerous (Alam et al. 2022), antibacterial (Khan et al. 2021), anti-viral (Taysi et al. 2023), anti-toxin (Ehtiati et al. 2023), and anti-inflammatory (Kim et al. 2023) activities. Phenolic lipids are another group of bioactive compounds obtained from poly-acetate pathways (Tomé-Carneiro et al. 2013a). These are hydrophobic compounds that have more than 11 carbon atoms in chains and also known as bioactive lipids (Tomé-Carneiro et al. 2013b). These have direct antimicrobial and antifungal characters (Kozubek et al. 2001).

5.10.2 Flavonoids

Flavonoids are predominantly found in the flowers, leaves, and seeds of plants, where they provide coloration to attract insects and animals, and protect the plant from UV radiation and herbivores. These compounds are highly ubiquitous in the plant kingdom. Most of the flavonoids are very common in plants (Shen et al. 2022). Plants rich in flavonoids include Glycine max, Medicago sativa, *Artemisia* spp., *Trifolium* spp., *Calendula* spp., *Mentha, Ocimum, Rosmarinus, Citrus* spp., *Polygonum, Fagopyrum, Vaccinium, Rosa, Malus, Prunus, Ficus, and Morus*.

5.10.3 Carotenoids

It is another group of phytochemicals that is an active ingredient of our plant-based food. All photosynthetic organisms have variable but definite quantity of carotenoids in their tissues. They are meant for photosynthesis, signaling, photoprotection, phytohormone synthesis in plants. They are very much important part of human nutrition as well. They are precursors of vitamin A (An important pillar of immune health and immune stability). α- carotenoids, β- carotenoids, lutei, and zeaxanthines are some examples of carotenoids and they are very effective against cancers, cardiovascular diseases, eye diseases related to age, etc. (Sauer et al. 2019; Eggersdorfer and Wyss 2018). Carotenoids, overall, have proven to be protectors of the liver, lung, and neurological systems of the body (Zia-Ul-Haq et al. 2021). Regular use of carotenoid rich foods also affects the aging patterns of the body. Carotenoids can lessen the risk of aging-related diseases of the human body. Aging negatively impacts the body by damaging cellular proteins, lipids, and DNA.

However, dietary carotenoids can mitigate the oxidative stress placed on these primary metabolites, thereby significantly reducing the rate of aging (Bakac et al. 2023).

5.10.4 Phytosterols

Phytosterols include β-sitosterol, campesterol, and stigmasterol. These are found in unrefined oils, vegetables, nuts, legumes, seeds, and whole grains. These sterols are of plant origin and are part of our food and resemble animal cholesterol in structure. But functionally they are different from animal sterols. These are mainly responsible for maintaining fluidity and integrity of membranes along with the defense system of plants (Salehi et al. 2021). These are known to have cholesterol-lowering effects in the human body. They reduce the absorption of cholesterol in the small intestine. Thus, the ratio of bad cholesterol in blood is decreased. They have also shown anti-cancerous and immunomodulatory actions. It is important to note that due to their health benefits, they are also part of various functional foods including milk, yogurt, and margarine (Poli et al. 2021; Nattagh-Eshtivani et al. 2022).

5.10.5 Glucosinolates (Thioglucosides)

These are secondary metabolites that contain sulfur and nitrogen in their structure. These are predominantly found in foods like cabbage, mustard, broccoli, Raphnus, kale, etc. (Fahey et al. 2001). They consist of a β-D-glucopyranose, a sulfure linking group, and a variable side chain derived from amino acids. These have shown anti-cancerous, antioxidant, and anti-inflammatory actions (Orouji et al. 2023). These also support liver detoxification (Latimer et al. 2021) and relieve the effects of chronic pains (Lucarini et al. 2022).

5.10.6 Saponins

These are another group of bioactive metabolites of plant origin. These are glycosidic compounds having foaming ability. They are considered as anti-nutritive but they have shown positive nutritive effects as well. There are more than 60 types of saponins discovered from seeds, flowers, leaves, and roots of various plants. Saponins possess antioxidant and anti-cancerous potential. They also lower the blood cholesterol level. They have also shown strong antimicrobial activities as well. They have been also used against rheumatism and other chronic diseases as well (Shen et al. 2017; Parama et al. 2020). Saponins are commonly found in fenugreek, tea, soyabeans, chickpeas, lentils, peas, oats, barley, Amaranthus, spinach, asparagus, licorice root, ginseng, and yam (Sharma et al. 2023).

5.10.7 Alkaloids

Food plants rich in alkaloids include potatoes, tomatoes, eggplants, chillies, black pepper, and celery. Overall human body is sensitive toward the dose of alkaloids. Some alkaloids are toxic and their smaller doses are alarming while some are tolerable by body at lower doses with additional health benefits. But they have prominent hepatoprotective anti-cancerous, hypo-glycemic, hypolipidemic, antidiabetic, antihypertensive, and antidepressant characters (Elizalde-Romero et al. 2021). They are also considered as potential drugs against epilepsy, schizophrenia, Parkinson disease, Huntington disease and Alzheimer disease (Jayakumar and Murugan 2016; Hussain et al. 2018; Contreras-Angulo et al. 2020).

5.10.8 Organosulfur Compounds

They are known for their strong antioxidant and anti-cancerous properties. Therefore, these are consumed excessively on long-term basis like garlic (Allicin), onion, and leeks (Fenwick and Hanley 1985). Along with the contribution to the taste and smell of food, these sulfur containing compounds are antibiotic, antiseptic, and antithrombotic in nature and have been in use in Indian, Chinese, and Egyptian local medicines (Block 1985). These bioactive compounds are redox cyclers, undergoing spontaneous oxidation to disulfides and returning to thiols. Therefore, these are very much effective against liver, kidney, heart, and lung damage. Allicin, diallyl disulfide, diallyl trisulfide, S-allyl cysteine, thiosulfonates, glucosinolates, isothiocyanates, indole-carbanols are some of the organosulfur compounds. Foods which are rich sources of organosulfur compounds include broccoli, brussels, cauliflower, cabbage, garlic, onion, shallots, chives, meat, eggs, asparagus, legumes, etc.

5.10.9 Bioactive Peptides

These are small pieces of peptides derived from proteins being taken as food (Dwivedi et al. 2014; Sionek and Szydłowska 2025). They have active biological activities, like immunomodulatory, antioxidant, hypocholesterolemic, memory improving activities (Fan et al. 2022). These can help in enhanced absorption of minerals in the body and cyto-immunomodulation (Maestri et al. 2016).

5.11 Healing Through Diet: The Role of Medicinal Foods in Chronic Disease Control

5.11.1 Chronic Diseases

A chronic disease is defined as any health condition that lasts for a year or more, limits a person's daily activities, and requires continuous medical care. It may include various heart issues, strokes, cancers, respiratory diseases, and neurodegenerative disorders. Inflammation and oxidative stress are the root cause of chronic diseases by damaging organs and tissues (Diet 2003). This oxidative stress may be higher in aging people, stressed people, people with unhealthy lifestyle, smokers, drinking alcohol, or lacking physical exercises. Many drugs can cause an increase in body ROS level. Higher meat consumption can also cause an increase in reactive oxygen species (ROS) in the intestine. Trans fatty acids are also a source (Valko et al. 2001; Jomova et al. 2023). These chronic diseases remain as one of the main causes of deaths annually. Some people may also suffer from multiple chronic conditions as well. Experts say that change of lifestyle and healthy food intake can decrease the chances of chronic diseases (Hacker 2024). An estimate of 2020 showed that 80% of global deaths were due to cardiovascular diseases (Roth et al. 2020). The European commission has declared obesity as a chronic disease as well (Burki 2021). Obesity is a metabolic disorder that is included in the list of chronic diseases now. It is also responsible for the onset of other non-communicable chronic diseases like diabetes. Research has shown over time that different chronic diseases are associated with multiple dietary patterns. Doctors have claimed that intake of healthy diet decreases the risk of chronic diseases. These results were observed across gender and ethnic groups (Wang et al. 2023).

5.11.2 Targeting Chronic Disease through Diet: Functional Roles of Medicinal Foods

Specific foods have strong antioxidants, anti-inflammatory, and antioxidant properties due to their natural bioactive compounds. These compounds are for plants' own protective system against stress, pathogen, and oxidative damage. When these foods are consumed by humans, their bodies get benefits out of it (Lobo et al. 2010; Aggarwal and Harikumar 2009). Anti-inflammatory compounds of plants provide them protection against harsh environments, deter herbivores, or heal tissue damage. These metabolites are eaten by human beings, and these modulate the immune response of the body and reduce pro-inflammatory cytokines. When consumed by humans, these metabolites modulate the body's immune response and reduce pro-inflammatory cytokines. Thus, they can be used to address the problems associated with chronic diseases like arthritis, diabetes, and cardiovascular diseases. Similarly, plants face exposures to oxidative damage of pollutants, and UV-rays, therefore, they produce antioxidants to neutralize them. These antioxidants, when added to human food, can protect cells from damage, aging, cancer, etc. Similarly, plants can't escape pathogen easily as they are sessile, so they produce chemicals as a defense system against fungi and bacteria, and these compounds can also help human beings for food preservation, wound healing, and gut health, etc. (Aggarwal and Harikumar 2009; Russell and Duthie 2011).

5.11.3 Food and Gut Flora

Food has a direct role in the diversity, structure, and function of gut flora that establishes the overall health of the gut. There are certain foods like fiber rich foods, vegetables, whole grains can provide prebiotics to the gut flora. These are non-digestible fibers that are fermented by the bacteria there and converted to short chain fatty acids. These fatty acids are essential for the health of gut lining. These are also playing roles in reducing gut inflammation and improving the immune function. Fermented foods like yogurt, cheese, kimchi can deliver probiotics that help in enhancing the diversity and resilience of the gut microbes (Shah et al. 2023). Foods like berries, olive oil, and leafy green vegetables can enhance the anti-inflammatory potential of the lower gut to avoid the chronic gut ailments (Bolte et al. 2021). Many food nutrients like zinc, glutamine, etc., maintain the gut barrier integrity. Therefore, we can say that overall microbial balance of guts may be established by a balanced and healthy diet that has anti-inflammatory, antioxidant, and antimicrobial action.

5.12 Food Processing and Its Impact on Medicinal Properties

Food processing includes processes, technologies, and methods that are opted to make food long lasting, palatable, and sometimes tastier. It helps to reduce food spoilage (Weaver et al. 2014). Even new products are formed by processing raw vegetables. There are many purposes involved in carrying out food processing. It includes (Knorr and Augustin 2021):

(a) To ensure food safety and to prevent food borne illnesses like pasteurization, canned, and irradiation.
(b) To extend the shelf life of the food like freezing, vacuum packaging, and drying.
(c) To enhance out of season supply.
(d) To enhance convenience in usage (to store, eat, and cook) like ready-to-eat meals, noodles, preserved fruit slices.
(e) To improve food appearance (roasting, smoking, fermenting, making jams and jellies).
(f) To increase functional properties of food (cooking, extrusion).
(g) To innovate new foods.
(h) To meet special dietary needs.

Nowadays, various ultra-food processing procedures are being used to make food flavored, tasty, palatable, shelf-stable, convenient to store and use but on the cost of nutrition and health. Ultra-processed foods are basically industrial food products that are heavily modified and original contents may be combined with many added ingredients like emulsifiers, colors, sweeteners, thickeners. Such processing badly affects the medicinal properties of the food. There are different methods of food processing (Clark et al. 2014) as explained in Table 5.1.

Table 5.1 Types of food processing common nowadays

Serial No.	Food processing category	Processing	Method of preparation
1	Thermal	Boiling	Cooking food with water at 100 °C
		Roasting/ baking	Cooking by dry heat
		Steaming	Cooking with steam
		Frying	Cooking in oil
		Pasteurization	Mild heating with pressure
		Sterilization	High temperature treatments to canned food
		Blanching	Quick heat treatment before freezing
2	Mechanical	Grinding	To reduce particle size like making flour out of seeds or grains
		Slicing, cutting, dicing	To prepare food for cooking and packaging
		Extrusion	To change the structure of food by high shear to make foods like cereals, etc.
		Mixing and blending	To combine ingredients uniformly like smoothies
3	Preservation	Freezing,	To slow down the microbial growth
		Drying	To remove water for slowing down the spoilage
		Canning	To seal food to increase shelf life
		Pickling	To use vinegar, brine and fermentation to preserve veggies
		Vacuum packing	To remove air to extend shelf life
4	Fermentation	Yogurt making Cheese making Sauerkraut Sourdough bread	To increase the digestibility of food, natural process of fermentation is used
5	Chemical	Salting	To add salt to food to preserve it
		Sugaring	To add sugar to food to preserve like jams
		Smoking	To add smoke to food to preserve it
6	Novel tech.	Microwave	Rapid heating by microwaves
		High pressure processing	High pressure to kill microbes
		Irradiation	Use radiations to kill microbes
		Ultrasonics	Using sound waves for preservation and making emulsification

5.13 Comparison of Raw and Processed Food

Raw foods have many advantages like retaining heat-sensitive nutrients and have retained natural enzymes, higher level of bioactive components, no processing related harmful compounds, lesser oxidation of nutrients, etc. But raw foods have usually lower digestibility and may have harmful microbes or some inactive bioactive components like lycopene, curcumin, etc., whereas processed food has enhanced digestibility, improved bioavailability of some compounds, microbial safety, flavor enhancement, longer shelf life, etc. But many important biochemicals are lost during food processing like Vitamin C and polyphenols. High temperatures treatments can add undesirable chemicals like acrylamide. Each method and each food have their own advantages and disadvantages. Therefore, we can't limit the food or processing method, but it is required to choose raw or cooked/processed food depending on the type of medicinal food and targeted health benefits.

5.14 Medicinal Properties of Common Foods

Health prevention and recovery functions as a powerful dual purpose alongside nourishment that food provides to the body. Multiple cultures throughout time have recognized the healing properties of normal dietary foods which traditional healthcare approaches including Ayurveda and Traditional Chinese Medicine (TCM) and native medical practices use for treatment (Slavin and Lloyd 2012). Scientific investigations during our current time period demonstrate that traditional food practices were correct because they contain bioactive compounds which promote health along with fighting off chronic diseases (Liu 2013). A list of therapeutic food properties accompanies the therapeutic roles for common foods which is shown in Fig. 5.5.

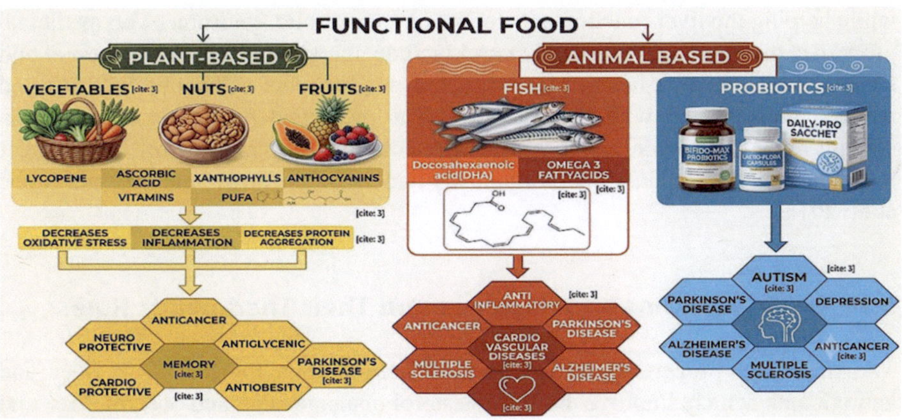

Fig. 5.5 Functional foods and their therapeutic effects (Liu 2013)

5.14.1 Fruits and Their Therapeutic Roles

The therapeutic properties of apples include pectin dietary fiber and vitamin C along with quercetin antioxidants which provide multiple health benefits. The berry family (Blueberries, Strawberries, Raspberries) contains anthocyanins, flavonoids, vitamin C, and multiple therapeutic compounds which provide strong antioxidant benefits and anti-inflammatory protection and help maintain brain and heart health (Basu and Rhone 2010). Oranges together with lemons and grapefruits provide many health benefits through their high content of vitamin C and flavonoids and limonoids.

5.14.2 Vegetables and Their Therapeutic Roles

The medicinal properties of garlic include Allicin and sulfur compounds together with flavonoids which provide antibacterial and antifungal protection and protect the heart system by reducing blood pressure and cholesterol levels. Onions possess two major therapeutic compounds including quercetin and sulfur compounds that help strengthen immunity while fighting cancer. The therapeutic components of leafy greens (Spinach, Kale, Moringa) include iron, calcium, folate, lutein, and fiber which deliver multiple health benefits such as improved blood health and digestive support as well as antioxidant protection (Böhm and Wahler 2012).

5.14.3 Spices Along with Herbs Offer Therapeutic Benefits for Human Health

Claudia Gomes from the National Cancer Institute stated that turmeric (Curcuma longa) has curcumin which functions as an anti-inflammatory agent while serving as an antioxidant. This plant matter demonstrates usefulness in decreasing joint pain while helping the liver function and controlling metabolic syndrome. The medicinal properties of ginger (*Zingiber officinale*) include abundant amounts of gingerol and shogaol which provide health benefits such as nausea relief and better digestion and reduced muscle pain with inflammation. The medicinal properties of cinnamon contain cinnamaldehyde along with polyphenols which help control blood sugar levels and defend against microorganisms and maintain heart health (Ranasinghe et al. 2013).

5.14.4 Whole Grains and Legumes and Their Therapeutic Roles

The medicinal properties of oats include beta-glucans together with fiber and antioxidants which improve both cholesterol management and weight loss and promote gut health. The medicinal value of legumes (lentils, chickpeas, beans) includes plant protein and fiber along with folate and iron. Health benefits include better heart health and stabilized blood sugar levels and reduced feelings of hunger (Didinger and Thompson 2021).

5.14.5 Various Fermented Foods Demonstrate Therapeutic Benefits for Human Health

There are various fermented foods which have healing effects. Such as:

Yogurt and Kefir Medicinal Properties: Contain probiotics like Lactobacillus and Bifidobacterium. These foods improve gut microbiota conditions and strengthen immune functions while breaking down lactose in the body.

Kimchi and Sauerkraut Medicinal Properties: Fermented vegetables rich in lactic acid bacteria. These are anti-inflammatory and help in digestive support and possible anticancer effects (Shahidi and Ambigaipalan 2015).

5.14.6 Nuts and Seeds and Their Therapeutic Roles

The medical properties of almonds and walnuts include omega-3 fatty acids and vitamin E along with fiber which provides health benefits for brain protection against inflammation and supports heart health. Flaxseeds along with Chia seeds contain alpha-linolenic acid (ALA) and lignans and soluble fiber in their medicinal properties. Health Benefits include better cholesterol management and improved digestive health and hormonal balance.

5.14.7 Beverages and Their Therapeutic Roles

Green Tea possesses catechins especially EGCG and antioxidants as medicinal components which support weight loss while improving brain function and reducing cancer risk. Apple cider vinegar contains two main medicinal components including acetic acid and polyphenols that benefit digestion and regulate blood sugar levels while fighting against microbial agents.

Food that people consume daily provides more than sustenance because these ingredients act as effective tools for healing patients and preserving good health. Various food types including fresh produce with herbs and spices and fermented products appear in nature to supply ingredients which enhance both physical health and mental well-being. Eating these foods regularly creates two benefits: It helps stop chronic diseases and it strengthens your current health condition. The scientific exploration of food-health relationships constantly confirms the ancient wisdom that nutrition acts as medical strength (Yang et al. 2018).

5.15 Animal-Based Foods and Their Therapeutic Roles

Multiple animal-derived products including fish, dairy items, eggs, and specific meat choices provide essential biomolecules that lead to better human health. Animal-derived foods serve an essential therapeutic function in balanced nutrition in addition to their well-known protective and preventive qualities of plant-based foods. Animal-based foods provide excellence proteins alongside healthy fats and necessary vitamins and minerals which assist with health condition management and disease prevention.

5.15.1 Fatty Fish (Salmon, Mackerel, Sardines, Trout)

Omega-3 fatty acids EPA and DHA in their most potent form exist naturally in fatty fish which serve as top food sources for these compounds. Scientific evidence proves that these essential fats both fight inflammation and protect the heart function (Calder 2017).

5.15.1.1 Therapeutic Benefits

5.15.1.1.1 Cardiovascular Health

People with cardiovascular issues benefit from omega-3 fats because these compounds reduce triglycerides and sustain normal blood pressure and enhance artery flexibility (Swanson et al. 2012).

5.15.1.1.2 Brain Function

The brain relies heavily on DHA to maintain cognitive function and develop proper memory capabilities along with neurological structures particularly during infancy and in older age groups.

5.15.1.1.3 Inflammation Reduction

Omega-3 supplements help reduce inflammation thus supporting patients who have rheumatoid arthritis asthma and inflammatory bowel diseases.

5.15.1.1.4 Mental Health Support

Research indicates that consuming omega-3s produces beneficial effects which reduce symptoms in patients with depression and anxiety and mood disorders.

5.15.2 Dairy Products (Milk, Yogurt, Cheese, Kefir)

The nutritional content of dairy products combines macronutrients and micronutrients through their calcium content as well as their vitamin D concentration and protein elements and probiotic benefits found in fermented dairy products (Rogers and Levine 2021).

5.15.2.1 Therapeutic Benefits

5.15.2.1.1 Bone Health

The combination of vitamin D with calcium serves as necessary elements for maintaining strong bones and fighting osteoporosis.

5.15.2.1.2 Digestive Wellness

The probiotic content in fermented dairy such as yogurt and kefir helps build digestive health through intestinal flora promotion and digestive aid functions.

5.15.2.1.3 Weight Management

Benefits from protein-rich dairy because it provides feelings of fullness which helps people maintain their body weight.

5.15.2.1.4 Blood Pressure Regulator

Different peptides in fermented dairy products provide natural blood pressure reduction benefits.

5.15.3 Eggs

The nutritional value of eggs includes every fundamental vitamin and mineral as well as premium protein and healthy fats together with choline and protective antioxidants such as lutein and zeaxanthin (Ruxton et al. 2010).

5.15.3.1 Therapeutic Benefits

5.15.3.1.1 Eye Protection

They come from both lutein and zeaxanthin because they defend against macular degeneration and cataracts.

5.15.3.1.2 Brain Development

The brain depends on choline to develop properly during fetal growth while also improving cognitive abilities overall.

5.15.3.1.3 Muscle Development

The complete protein content in eggs enables muscle growth while assisting muscle repair processes.

5.15.3.1.4 Appetite Control

The consumption of eggs helps people control their hunger while promoting either weight loss or maintaining a steady weight.

5.15.4 Organ Meats (Liver, Kidney, Heart)

The human body can access exceptional nutrition through organ meats which rank as some of the densest food sources. Today organ meats are less popular but contain noteworthy amounts of vitamins and minerals (Gunnars 2021).

5.15.4.1 Therapeutic Benefits

5.15.4.1.1 Vitamin A

Content in the liver is substantial because the organ provides preformed vitamin A which helps support immune functions and vision and promotes healthy skin growth.

5.15.4.1.2 Iron and Vitamin B_{12}

Crucial for red blood cell production and prevention of anemia.

5.15.4.1.3 Zinc and Selenium

Zinc together with selenium work as vital substances for building immune strength and managing thyroid health.

5.15.5 Bone Broth

The slow-boiled liquid made from animal bones and connective tissues contains collagen and gelatin alongside essential amino acids such as glycine and proline as well as trace minerals (Shah et al. 2019).

5.15.5.1 Therapeutic Benefits

5.15.5.1.1 Joint Support

The connective tissue structure of joints receives strength from collagen while the supplementation decreases both stiffness and pain experiences in joints.

5.15.5.1.2 Digestive Aid

Gelatin aids in repairing and protecting the digestive lining which serves people with leaky gut and digestive issues effectively.

5.15.5.1.3 Skin Rejuvenation

Amino acids together with collagen work as anti-aging agents which lead to skin becoming both hydrated and firmer.

5.15.5.1.4 Immune Enhancement

The trace minerals present in bone broth strengthen the immune system function.

5.16 Bridging Nutrition and Healing with Animal-Based Foods

Traditional healing systems together with modern nutritional science always recognize animal-based foods as essential dietary components. High-quality proteins with essential fatty acids and vitamins and minerals as well as bioactive compounds unite to provide therapeutic support for all vital body systems for cognitive health and cardiovascular health as well as bone strength and immune system defense. Consuming animal-based foods with moderation as part of dietary balance delivers healing abilities for disease prevention and total body health improvement. The health field benefits from these foods as therapeutic agents while they retain their role as food substances thus upholding the ancient understanding uniting nutrition and medical treatment (Mougou-Xanthopoulou et al. 2021).

5.17 Future Outlook and Challenges for Medicinal Food Adoption

Traditional medicine has incorporated medicinal foods for hundreds of years because these foods provide both healing effects and health benefits when consumed as food. Medicinal foods experience increased popularity because more people choose natural approaches for their wellness needs. Many obstacles continue to prevent widespread use of medicinal foods. The widespread adoption of medicinal foods faces obstacles from poor public understanding as well as restricted availability and expensive prices and various social barriers and scientific hurdles and legal barriers (Dalamagka 2024).

5.17.1 Issues in Popularizing Medicinal Foods

5.17.1.1 Awareness

5.17.1.1.1 Awareness Gaps and Insufficient Public Knowledge

The insufficient knowledge among the public about medicinal foods stands as a major impediment to wider adoption of these substances by the population. Numerous people fail to understand that certain foods offer benefits beyond nutrition as they hold healing properties which enhance wellness. People fail to understand proper methods of integrating medicinal foods into their meals for therapeutic purposes.

5.17.1.1.2 False and Incorrect Information

Social media together with false health claims has spread numerous incorrect facts about medicinal foods across platforms. Many consumers incorrectly believe that these foods possess curative properties for severe illnesses, despite scientific evidence to the contrary. This misconception fosters widespread confusion, ultimately undermining public trust in evidence-based medical advice. The result is increased confusion which decreases public trust in genuine healthcare directions.

5.17.1.1.3 Navigating Health Information Discrepancies

The healthcare sector presents numerous contradictory health messages which emerge from various treatment approaches such as supplements and medicines as well as conventional healthcare systems. People face difficulties determining whether they should select medicinal foods rather than alternative choices.

5.17.1.1.4 Culture Difference

The practice of using food as medicine differs between cultures because some traditions do not embrace this approach. People find medicinal foods strange and unappealing because of which they struggle to include them in their regular diet (Martirosyan and Singh 2015).

5.17.2 Accessibility

5.17.2.1 Geographical Limitations

The existence of medicinal foods occurs only within specific geographical areas or timeframes. Healing herbs and plants exist only within particular geographic areas as well as specific environmental conditions. People who live outside these specific areas encounter difficulties when acquiring these foods for dietary purposes.

5.17.2.2 Barriers in Supply Chain and Distribution

People encounter difficulties accessing medicinal foods even when such items exist within certain geographical areas. The combination of inadequate transportation networks with empty inventory in standard retail stores together with weak distribution systems creates obstacles when searching for medicinal foods. These special foods can currently be acquired only through specific retail locations that might be inaccessible or cost too much.

5.17.2.3 Cultural Barriers and Acceptance

People will not incorporate medicinal foods into their regular routines unless these foods are accepted by their cultural beliefs. The awareness and usage of turmeric along with garlic and ginger as medical ingredients differ between cultures since some cultures view them positively in medicine while others have limited knowledge about their medical properties.

The irregular and unstandardized monitoring of quality in medicinal food products presents a substantial obstacle to their accessibility. Problems with accessing medicinal foods exist because their quality levels and strength measurements along with purity standards differ between sources. Medical food safety and effectiveness depend on established rules with proper inspections which are currently unavailable. People find it difficult to depend on these foods as safe healthcare choices because of their unpredictable quality control practices.

5.17.3 Cost

5.17.3.1 Costly Medicinal Foods

A Barrier to Access: Certain organic medicinal foods along with those produced through unique cultivation methods tend to have high price tags. Only wealthy individuals can afford these medicinal products because their prices remain out of reach for low-income people.

5.17.3.2 Economic Influences on Market Demand

There are numerous medicinal foods classified as unique items rather than regular food items. Producing and distributing medicinal foods at the same rate as regular foods prove impossible because of their specialized nature which results in higher prices. Due to their lower market demand, their prices continue to rise until they reach a point where they cannot compete effectively with cheaper processed food products.

5.17.3.3 The Role of Insurance in Healthcare Access

Most health insurance programs exclude coverage for medicinal foods thus increasing their cost for patients. Insurance programs together with public health initiatives do not support the cost of these treatments thus restricting their use in treatment especially within communities who rely primarily on traditional medical systems.

5.17.4 Scientific and Regulatory Challenges

5.17.4.1 Insufficient Scientific Proof

The scientific research on medicinal foods shows growth yet remains at an initial stage. The majority of these foods remain untested through clinical trials because there is insufficient research into their safe and effective use. The insufficient evidence makes healthcare providers and general public question these products' validity.

5.17.4.2 Regulatory Framework and Standards

Most countries lack established and uniform regulations regarding medicinal foods. Since medicinal foods escape the strict regulatory framework which controls pharmaceutical medications, they can enter the market without any established standards for safety or dosage instructions. The absence of regulations creates quality disparities while confusing people about safe dietary implementation of these foods (Bigliardi and Galati 2013).

5.17.4.3 Sociocultural and Lifestyle Barriers

5.17.4.3.1 Dietary Preferences

Food preferences among people frequently diverge from incorporating medicinal foods into their dietary routine. Some cultural groups value taste and convenience over health advantages which makes it difficult to persuade members to integrate medicinal foods into their regular diet.

5.17.4.3.2 Food or Medicine

The Ongoing Debate: Present regulations face difficulties in classifying medicinal foods because the distinction between food and medicine remains vague. Different types of medicinal foods receive dual classifications either as food supplements or natural cures or regular meals. These unclear boundaries make it difficult for healthcare institutions to incorporate these foods into their system (Rao and Min 2017).

5.17.5 Market and Industry Challenges

5.17.5.1 Barriers to Commercial Growth

Due to their specialized nature, medicinal foods remain unpopular with manufacturers who avoid mass production of such products. The mainstream market rejects medicinal foods because big food companies choose to produce mass-scale processed foods that offer convenience and affordability.

5.17.5.2 Medicinal Food Companies Face Substantial Branding and Marketing Problems

Medicinal food companies face substantial branding and marketing challenges because their products lack sufficient market visibility and consumer awareness. The lack of branding together with insufficient advertising prevents many medicinal foods from gaining traction and attracting potential customers. Consequently, these foods remain confined to health food stores and niche markets due to a lack of targeted marketing efforts and broad awareness campaigns. The widespread acceptance of medicinal foods requires solutions for improving awareness levels and accessibility and cost reduction and breaking down cultural and regulatory barriers. The widespread acceptance of medicinal foods requires a comprehensive method which combines educational programs with policy adjustments research efforts and improved production systems and distribution networks. Efforts to introduce medicinal foods into mainstream society require extensive collaboration between social, economic, and scientific approaches (Granato et al. 2010).

5.17.6 Future Challenges and Strategic Priorities in Medicinal Foods Research

5.17.6.1 Scientific and Evidence-Based Testing and Validation

The proof of safety and effectiveness of medicinal foods for health requires thorough clinical research and testing. The process will generate solid proof to develop trust among medical professionals and consumer audiences. The development of standard procedures for health assessment of these foods should include protocols that produce uniform results across individuals and environments. i. The Science Behind Medicinal Food Benefits, ii. Absorption and Metabolism: Understanding the absorption pathways, metabolic processes, and physiological utilization of medicinal food ingredients is essential for optimizing their health benefits. These foods become more potent in disease prevention and health treatment by studying their biological impact on the human body. Identifying the foods that deliver maximum absorption benefits and health value requires us to discover efficient cooking techniques that enhance these properties.

5.17.6.2 Traditional Food Systems: Utilization and Healthcare Outcomes

Research examines medicinal food usage in combination with traditional medical treatments along with prescription drugs for healthcare applications. A research study should investigate whether medicinal foods have potential uses to either work alongside or instead of conventional chemical drugs particularly for disease prevention and long-term disease management.

5.17.6.3 Policy Development and Product Standardization

A standard rules system should be developed to label medicinal foods properly with quality and safety protocols which ensure their consistent safe nature during production. The inclusion of medicinal foods in health insurance and public health policies should become a priority to increase their availability.

5.17.6.4 Traditions, Beliefs, and Public and Social Acceptance

The analysis explores cultural perspectives on medicinal foods to determine obstacles to their acceptance together with strategies to increase acceptance. Such knowledge can develop strategies which promote the regular consumption of medicinal foods by people in their ordinary diets. The analysis examines educational and awareness programs that affect public perception of medicinal foods and promote their adoption.

5.17.6.5 Environmental Footprint of Medicinal Foods

The research investigates environmental effects of medicinal food cultivation by studying natural protective farm methods and techniques for minimizing environmental destruction. Research into efficient sustainable methods of producing medicinal foods should focus on areas with scarce resources.

5.17.6.6 Strategic Marketing and Cost Management

This section examines how medicinal foods compare to traditional medicines and supplements in terms of cost efficiency, and evaluates whether they offer viable economic value. The analysis of both costs and benefits could boost medicinal food usage by more people. The goal should be to establish strategies that lower the price of medicinal food products to make them accessible for low-income populations.

5.17.6.7 The Approach of Personalized Health Through Medicinal Food Dose Optimization

Further investigations need to occur regarding individualized nutrition practices that create customized medicinal foods for personal genetic makeup and lifestyle along with health requirements. The effectiveness of medicinal foods would increase while acceptance from people would improve if they were tailored specifically to individual needs. The study investigates how medicinal foods contribute to precision medicine systems for managing diabetes and high blood pressure along with heart diseases.

5.17.6.8 Emerging Novel Foods and Their Healing Potential

Scientists need to uncover medicinal foods together with active compounds that demonstrate potential healing properties. Research on unidentified herbs and vegetables together with fruits and fungi may uncover potential new medicinal therapies. The investigation of functional foods components that offer nutritional benefits with medical advantages will fulfill consumer demands for wellness solutions.

5.17.6.9 Trends in Consumer Food Preferences

Scientists conduct research to understand the reasons behind consumer selection between medicinal foods and conventional foods and medical treatments. The collected knowledge enables better product development as well as marketing approach optimization. The influence of branding as well as packaging design and availability decisions shape consumer choices regarding medicinal foods for their daily diet. The expansion of medicinal food research through exploration will help people learn about such foods thereby generating better global health outcomes (Sharma 2018).

5.18 Conclusion

Functional foods and nutraceuticals are an emerging and growing field in nutritional science, with a lot of potential regarding disease prevention and health promotion. Bioactive compounds that include antioxidants and omega-3 fatty acids in these products can be consumed to offer scientifically validated benefits in various physiological systems. Developments in mechanistic knowledge and clinical validation are required that can place research findings into some kind of practical health interventions. The understanding of the inter-individual variability is also essential as it creates the need to use individualized approaches to diet to maximize the results. An integrative and holistic strategy, i.e., the integration of scientific studies, clinical practice, and individualized nutrition will play a crucial role in realizing the full therapeutic value of functional foods and nutraceutical to sustainable health and well-being.

References

Aggarwal BB, Harikumar KB (2009) Potential therapeutic effects of curcumin, the anti-inflammatory agent, against neurodegenerative, cardiovascular, pulmonary, metabolic, autoimmune and neoplastic diseases. Int J Biochem Cell Biol 41(1):40–59

Alam M, Ahmed S, Elasbali AM, Adnan M, Alam S, Hassan MI, Pasupuleti VR (2022) Therapeutic implications of caffeic acid in cancer and neurological diseases. Front Oncol 12:860508

Arnett DK, Blumenthal RS, Albert MA, Buroker AB, Goldberger ZD, Hahn EJ, Himmelfarb CD, Khera A, Lloyd-Jones D, McEvoy JW, Michos ED, Miedema MD, Muñoz D, Smith SC Jr, Virani SS, Williams KA Sr, Yeboah J, Ziaeian B (2019) 2019 ACC/AHA guideline on the primary prevention of cardiovascular disease. Circulation 140(11):e596–e646

Bakac ER, Percin E, Gunes-Bayir A, Dadak A (2023) A narrative review: the effect and importance of carotenoids on aging and aging-related diseases. Int J Mol Sci 24(20):15199

Basu A, Rhone M (2010) Berries: emerging impact on cardiovascular health. Nutr Rev 68(3): 168–177

Bhuia MS, Rahaman MM, Islam T, Bappi MH, Sikder MI, Hossain KN, Akter F, Al Shamsh Prottay A, Rokonuzzman M, Gürer ES, Calina D (2023) Neurobiological effects of gallic acid: current perspectives. Chin Med 18(1):27

Bigliardi B, Galati F (2013) Innovation trends in the food industry: the case of functional foods. Trends Food Sci Technol 31(2):118–129

Block E (1985) The chemistry of garlic and onions. Sci Am 252:114–119

Bloomer RJ, Kabir MM, Canale RE, Trepanowski JF, Marshall KE, Farney TM, Hammond KG (2010) The Daniel fast: A plant-based dietary approach. J Int Soc Sports Nutr 7(1):1–9

Böhm V, Wahler D (2012) Antioxidant capacity of vegetables and their relation to phytochemicals. J Food Compos Anal 25(3):278–286

Bolte LA, Vila AV, Imhann F, Collij V, Gacesa R, Peters V, Wijmenga C, Kurilshikov A, Campmans-Kuijpers MJ, Fu J, Dijkstra G (2021) Long-term dietary patterns are associated with pro-inflammatory and anti-inflammatory features of the gut microbiome. Gut 70(7): 1287–1298

Burki T (2021) European Commission classifies obesity as a chronic disease. Lancet Diabetes Endocrinol 9(7):418

Calder PC (2017) Omega-3 fatty acids and inflammatory processes: from molecules to man. Biochem Soc Trans 45(5):1105–1115

Clark S, Jung S, Lamsal B (eds) (2014) Food processing: principles and applications. Wiley

Contreras-Angulo L, Emus-Medina A, Gutierrez-Grijalva E, Ambriz-Pérez D, Elizalde-Romero C, Heredia JB (2020) Pharmacological potential of Solanum genus. In: Souza AV (ed) Solanum: an overview, Plant science research and practices, vol 1. Nova Science, New York, p 199

Dalamagka MI (2024) Integrating traditional medicine into a modern health care system. Int J Sci Res Arch 12(01):2372–2375

Dhiman S, Mukherjee G (2021) Gallic acid (GA): a multifaceted biomolecule transmuting the biotechnology era. In: Prasad R, Kumar V, Singh J, Upadhyaya CP (eds) Recent developments in microbial technologies, Environmental and microbial biotechnology. Springer, Singapore. https://doi.org/10.1007/978-981-15-4439-2_8

Didinger C, Thompson HJ (2021) The role of pulses in improving human health: a review. Legume Science 3(1):e147

Diet, W.H.O (2003) Chronic diseases. World Health Organization, Geneva

Dwivedi S, Sahrawat K, Puppala N, Ortiz R (2014) Plant prebiotics and human health: biotechnology to breed prebiotic-rich nutritious food crops. Electron J Biotechnol 17(5):238–245

Eggersdorfer M, Wyss A (2018) Carotenoids in human nutrition and health. Arch Biochem Biophys 018(652):18–26

Ehtiati S, Alizadeh M, Farhadi F, Khalatbari K, Ajiboye BO, Rahimi VB, Askari VR (2023) Promising influences of caffeic acid and caffeic acid phenethyl ester against natural and chemical toxins: a comprehensive and mechanistic review. J Funct Foods 107:105637

Elizalde-Romero CA, Montoya-Inzunza LA, Contreras-Angulo LA, Heredia JB, Gutiérrez-Grijalva EP (2021) Solanum fruits: phytochemicals, bioaccessibility and bioavailability, and their relationship with their health-promoting effects. Front Nutr 8:790582

Fahey JW, Zalcmann AT, Talalay P (2001) The chemical diversity and distribution of glucosinolates and isothiocyanates among plants. Phytochemistry 56:5–51. https://doi.org/10.1016/S0031-9422(00)00316-2

Fan H, Liu H, Zhang Y, Zhang S, Liu T, Wang D (2022) Review on plant-derived bioactive peptides: biological activities, mechanism of action and utilizations in food development. J Fut Foods 2(2):143–159

Fenwick GR, Hanley AB (1985) The genus allium. Part 3. CRC Crit Rev Food Sci Nutr 23:1–73

Freeman AM, Morris PB, Adams C, Barnard ND, Esselstyn CB, Devries S, Ornish D, Ros E, Kris-Etherton P (2018) A plant-based diet for the prevention and treatment of cardiovascular disease. Am J Lifestyle Med 12(3):204–210

Gogte VM, Dole A (2015) Scientific aspects of Ayurvedic medicine. Anc Sci Life 34(3):156–161

Granato D, Branco GF, Nazzaro F, Cruz AG, Faria JAF (2010) Functional foods and nondairy probiotic food development: trends, concepts, and products. Compr Rev Food Sci Food Saf 9(3):292–302

Gunnars K (2021) Are organ meats healthy? Healthline

Hacker K (2024) The burden of chronic disease. Mayo Clinic Proceedings: Innovations, Quality & Outcomes 8(1):112–119

Heinrich M, Scotti F, Booker A, Fitzgerald M, Kum KY, Löbel K (2019) Unblocking high-value botanical value chains: is there a role for Blockchain systems? Front Pharmacol 10:396

Hussain G, Rasul A, Anwar H, Aziz N, Razzaq A, Wei WL (2018) Role of plant derived alkaloids and their mechanism in neurodegenerative disorders. Int J Biol Sci 14:341–357

Jayakumar K, Murugan K (2016) Solanum alkaloids and their pharmaceutical roles: a review. J Anal Pharm Res 3:00075

Jiang M, Lu C, Zhang C, Yang J, Tan Y, Lu A, Chan K (2012) Syndrome differentiation in traditional Chinese medicine: a review of case studies. J Ethnopharmacol 140(3):634–642

Jiang Y, Pei J, Zheng Y, Miao YJ, Duan BZ, Huang LF (2022) Gallic acid: a potential anti-cancer agent. Chin J Integr Med 28(7):661–671

Jomova K, Raptova R, Alomar SY, Alwasel SH, Nepovimova E, Kuca K, Valko M (2023) Reactive oxygen species, toxicity, oxidative stress, and antioxidants: chronic diseases and aging. Arch Toxicol 97(10):2499–2574

Khan F, Bamunuarachchi NI, Tabassum N, Kim YM (2021) Caffeic acid and its derivatives: antimicrobial drugs toward microbial pathogens. J Agric Food Chem 69(10):2979–3004

Kim CK, Yu J, Le D, Han S, Yu S, Lee M (2023) Anti-inflammatory activity of caffeic acid derivatives from Ilex rotunda. Int Immunopharmacol 115:109610

Knorr D, Augustin MA (2021) Food processing needs, advantages and misconceptions. Trends Food Sci Technol 108:103–110

Kozubek A, Arnowski RZ, Stasiuk M, Gubernator J (2001) Natural amphiphilic phenols as bioactive compounds. Cell Mol Biol Lett 6(2A):351–355

Latimer I, Chand R, Cridge B (2021) The in vitro toxicity of nitrile and epithionitrile derivatives of glucosinolates from swedes (Brassica napus) in human and bovine liver cells. N Z Vet J 69(3):165–173

Liu RH (2004) Potential synergy of phytochemicals in cancer prevention: mechanism of action. J Nutr 134(12 Suppl):3479S–3485S

Liu RH (2013) Health-promoting components of fruits and vegetables in the diet. Adv Nutr 4(3):384S–392S

Liu J, Wang S, Zhang Y, Fan HT, Lin HS (2015) Traditional Chinese medicine and cancer: history, present situation, and development. Thoracic Cancer 6(5):561–569

Lobo V, Patil A, Phatak A, Chandra N (2010) Free radicals, antioxidants and functional foods: impact on human health. Pharmacogn Rev 4(8):118

Lucarini E, Micheli L, Di Cesare Mannelli L, Ghelardini C (2022) Naturally occurring glucosinolates and isothiocyanates as a weapon against chronic pain: potentials and limits. Phytochem Rev 21(2):647–665
Maestri E, Marmiroli M, Marmiroli N (2016) Bioactive peptides in plant-derived foodstuffs. J Proteome 147:140–155
Martirosyan DM, Singh J (2015) A new definition of functional food by FFC: what makes a new definition unique? Funct Foods Health Dis 5(6):209–223
Mine Y, Young I, Yang M (2009) Recent advances in the understanding of the roles of food-derived bioactive peptides in preventing and managing chronic diseases. J AOAC Int 92(4):1023–1028
Mougou-Xanthopoulou C, Drouzas AD, Zampelas A (2021) The role of animal-based foods in human health and disease prevention: A focus on dairy, eggs, meat, and fish. Foods 10(7):1550
Mukherjee PK, Wahile A (2006) Integrated approaches towards drug development from Ayurveda and other Indian system of medicines. J Ethnopharmacol 103(1):25–35
Nattagh-Eshtivani E, Barghchi H, Pahlavani N, Barati M, Amiri Y, Fadel A, Khosravi M, Talebi S, Arzhang P, Ziaei R, Ghavami A (2022) Biological and pharmacological effects and nutritional impact of phytosterols: a comprehensive review. Phytother Res 36(1):299–322
Orouji N, Asl SK, Taghipour Z, Habtemariam S, Nabavi SM, Rahimi R (2023) Glucosinolates in cancer prevention and treatment: experimental and clinical evidence. Med Oncol 40(12):344
Parama D, Boruah M, Yachna K, Rana V, Banik K, Harsha C, Thakur KK, Dutta U, Arya A, Mao X, Ahn KS (2020) Diosgenin, a steroidal saponin, and its analogs: effective therapies against different chronic diseases. Life Sci 260:Article 118182
Patwardhan B, Warude D, Pushpangadan P, Bhatt N (2005) Ayurveda and traditional Chinese medicine: a comparative overview. Evid Based Complement Alternat Med 2(4):465–473
Pikula A, Sajdakowska M, Kowalczewski PL (2024) Plant-based diets and cardiovascular outcomes: a meta-analysis. Am J Clin Nutr 119(1):89–101
Poli A, Marangoni F, Corsini A, Manzato E, Marrocco W, Martini D, Medea G, Visioli F (2021) Phytosterols, cholesterol control, and cardiovascular disease. Nutrients 13(8):2810
Qiu J (2007) Traditional medicine: a culture in the balance. Nature 448(7150):126–128
Ranasinghe P, Pigera S, Premakumara GAS, Galappaththy P, Constantine GR, Katulanda P (2013) Medicinal properties of 'true' cinnamon (Cinnamomum zeylanicum): a systematic review. BMC Complement Altern Med 13:275
Rao AR, Min BC (2017) Barriers to functional food consumption: a review of consumer-related issues. Nutrients 9(10):1173
Rastogi S (2010) Building bridges between Ayurveda and modern science. Int J Ayurveda Res 1(1): 41–46
Rishabh G (2024) A brief history: traditional Chinese medicinal system. Pharmacol Res Mod Chin Med 10:1–10
Rogers MA, Levine D (2021) Dairy consumption and cardiovascular disease risk: a narrative review. Nutr Rev 79(2):177–194
Roth GA, Mensah GA, Johnson CO, Addolorato G, Ammirati E, Baddour LM, Barengo NC, Beaton AZ, Benjamin EJ, Benziger CP, Bonny A (2020) Global burden of cardiovascular diseases and risk factors, 1990–2019: update from the GBD 2019 study. J Am Coll Cardiol 76(25):2982–3021
Russell W, Duthie G (2011) Plant secondary metabolites and gut health: the case for phenolic acids. Proc Nutr Soc 70(3):389–396
Ruxton CH, Derbyshire E, Gibson S (2010) The nutritional properties and health benefits of eggs. Nutr Food Sci 40(3):263–279
Salehi B, Quispe C, Sharifi-Rad J, Cruz-Martins N, Nigam M, Mishra AP, Konovalov DA, Orobinskaya V, Abu-Reidah IM, Zam W, Sharopov F (2021) Phytosterols: from preclinical evidence to potential clinical applications. Front Pharmacol 11:599959
Sauer L, Li B, Bernstein PS (2019) Ocular carotenoid status in health and disease. Annu Rev Nutr 39:95–120

Shah BR, Li Y, Riedl KM (2019) Health-promoting bioactive compounds in bone broth: a review. Food Sci Nutr 7(7):2508–2516

Shah AM, Tarfeen N, Mohamed H, Song Y (2023) Fermented foods: their health-promoting components and potential effects on gut microbiota. Fermentation 9(2):118

Shahidi F, Ambigaipalan P (2015) Phenolics and polyphenolics in foods, beverages and spices: antioxidant activity and health effects – a review. J Funct Foods 18:820–897

Sharma R (2018) Shaping the future of medicinal foods: opportunities and challenges in research and development. J Funct Foods 46:314–324

Sharma K, Kaur R, Kumar S, Saini RK, Sharma S, Pawde SV, Kumar V (2023) Saponins: a concise review on food related aspects, applications and health implications. Food Chem Adv 2:100191

Shen X, Shi L, Pan H, Li B, Wu Y, Tu Y (2017) Identification of triterpenoid saponins in flowers of four Camellia Sinensis cultivars from Zhejiang province: Differences between cultivars, developmental stages, and tissues. Ind Crop Prod 95:140–147

Shen N, Wang T, Gan Q, Liu S, Wang L, Jin B (2022) Plant flavonoids: classification, distribution, biosynthesis, and antioxidant activity. Food Chem 383:132531

Sionek B, Szydłowska A (2025) Probiotics and prebiotics in the aspect of health benefits and the development of novel plant-based functional food. Appl Sci 15(6):3137

Slavin JL, Lloyd B (2012) Health benefits of fruits and vegetables. Adv Nutr 3(4):506–516

Swanson D, Block R, Mousa SA (2012) Omega-3 fatty acids EPA and DHA: health benefits throughout life. Adv Nutr 3(1):1–7

Taysi S, Algburi FS, Taysi ME, Caglayan C (2023) Caffeic acid phenethyl ester: a review on its pharmacological importance, and its association with free radicals, COVID-19, and radiotherapy. Phytother Res 37(3):1115–1135

Tomé-Carneiro J, Larrosa M, Yáñez-Gascón MJ, Dávalos A, Gil-Zamorano J, Gonzálvez M, García-Almagro FJ, Ruiz Ros JA, Tomás-Barberán FA, Espín JC et al (2013a) One-year supplementation with a grape extract containing resveratrol modulates inflammatory-related microRNAs and cytokines expression in peripheral blood mononuclear cells of type 2 diabetes and hypertensive patients with coronary artery disease. Pharmacol Res 72:69–82

Tomé-Carneiro J, Gonzálvez M, Larrosa M, Yáñez-Gascón MJ, García-Almagro FJ, Ruiz-Ros JA, Tomás-Barberán FA, García-Conesa MT, Espín JC (2013b) Grape resveratrol increases serum adiponectin and downregulates inflammatory genes in peripheral blood mononuclear cells: A triple-blind, placebo-controlled, one-year clinical trial in patients with stable coronary artery disease. Cardiovasc Drugs Ther 27:37–48

Tripathi YB, Chaturvedi N (2007) Ayurvedic concept of food and nutrition. Indian J Clin Biochem 22(2):201–208

Valko M, Morris H, Mazúr M, Rapta P, Bilton RF (2001) Oxygen free radical generating mechanisms in the colon: do the semiquinones of vitamin K play a role in the aetiology of colon cancer? Biochim Biophys Acta 1527(3):161–1666

Visseren FLJ, Mach F, Smulders YM, Carballo D, Koskinas KC, Bäck M, Benetos A, Biffi A, Boavida JM, Capodanno D, Cosyns B, Crawford C, Davos CH, Desormais I, Di Angelantonio E, Franco OH, Halvorsen S, Hobbs FDR, Hollander M, Jankowska EA, Michal M, Sacco S, Sattar N, Tokgozoglu L, Tonstad S, Tsioufis KP, van Dis I, van Gelder IC, Wanner C, Williams B (2021) 2021 ESC guidelines on cardiovascular disease prevention in clinical practice. Eur Heart J 42(34):3227–3337

Wang P, Song M, Eliassen AH, Wang M, Fung TT, Clinton SK, Rimm EB, Hu FB, Willett WC, Tabung FK, Giovannucci EL (2023) Optimal dietary patterns for prevention of chronic disease. Nat Med 29(3):719–728

Weaver CM, Dwyer J, Fulgoni VL III, King JC, Leveille GA, MacDonald RS, Ordovas J, Schnakenberg D (2014) Processed foods: contributions to nutrition. Am J Clin Nutr 99(6): 1525–1542

Witkamp RF, van Norren K (2018) Let thy food be thy medicine. . .when possible. Eur J Pharmacol 836:102–114

World Health Organization (2023) Global spending on health: coping with the pandemic. WHO Press
Xu Y, Tang G, Zhang C, Wang N, Feng Y (2021) Gallic acid and diabetes mellitus: its association with oxidative stress. Molecules 26(23):7115
Yang CS, Wang H, Sheridan ZP (2018) Studies on prevention of obesity, metabolic syndrome, diabetes, cardiovascular diseases and cancer by tea. J Food Drug Anal 26(1):1–13
Zia-Ul-Haq M, Dewanjee S, Riaz M (eds) (2021) Carotenoids: structure and function in the human body, vol 16. Springer Nature Switzerland AG: Springer International Publications, p 859

From Pond to Plate: Safety Perspectives on Algae in Food and Feed Systems

6

Mubashrah Munir, Mubeena Mansoor, Aqsa Ilyas, Ali Zeeshan, Adiba Akram, and Rahmatullah Qureshi

Abstract

Algae, ranging from microscopic microalgae to macroscopic seaweeds, represent one of the most promising bio-resources for the twenty-first century. Found in freshwater, brackish, and marine ecosystems, algae are rich in essential nutrients, including proteins, lipids, carbohydrates, vitamins, minerals, pigments, and bioactive compounds. These diverse biochemical constituents make them valuable for direct human consumption as functional foods and nutraceuticals, and as feed additives for fish, poultry, and livestock. Owing to their rapid growth rate, minimal land requirement, and ability to utilize wastewater or saline water, algae are a sustainable alternative to conventional crops. However, the safety and non-toxicity of algal species must be ensured before their large-scale incorporation into food and feed chains. This chapter reviews the nutritional profiles and safety aspects of selected seaweeds, viz., *Porphyra*, *Laminaria*, *Caulerpa*, *Hizikia fusiforme*, *Gracilaria vermiculophylla*, *G. edulis*, *G. verrucosa*, *Acanthophora specifera*, *Ulva*, *Undaria pinnatifida*, *Cladophora*, *Ascophyllum nodosum*, *Fucus*, *Palmaria palmata*, and *Enteromorpha prolifera* and microalgae such as *Dunaliella salina*, *Haematococcus pluvialis*, *Nannochloropsis*, *Tetraselmis*, *Spirulina*, and *Chlorella*. Studies indicate that eight seaweed species (*Porphyra*, *Laminaria*, *Caulerpa*, *Hizikia fusiforme*, *Gracilaria vermiculophylla*, *G. edulis*, *G. verrucosa*, and *Acanthophora specifera*) have shown potential toxic effects in humans when consumed

M. Munir · M. Mansoor · A. Ilyas · A. Zeeshan · A. Akram
Department of Biological Sciences, University of Veterinary and Animal Sciences, Lahore (Ravi Campus), Pakistan

R. Qureshi (✉)
Department of Botany, Faculty of Sciences, Pir Mehr Ali Shah Arid Agriculture University, Rawalpindi, Pakistan
e-mail: rahmatullahq@uaar.edu.pk

R. Qureshi (ed.), *Plant Biodiversity*, https://doi.org/10.1007/978-981-92-1471-6_6

excessively, while *Porphyra* has also been reported to cause toxicity in animal feed. Data on the safety of microalgae are limited but generally suggest low toxicity, with *Spirulina* and *Chlorella* Generally Recognized as Safe (GRAS) species. Despite this, contamination with heavy metals, persistent organic pollutants, or harmful algal metabolites can pose potential risks. Hence, systematic safety evaluation, species authentication, and adherence to food safety standards are essential before commercialization. The chapter emphasizes the need for toxicological validation, monitoring of cultivation environments, and harmonized regulatory frameworks to ensure consumer safety and promote algae-based food and feed industries.

Keywords

Microalgae · Seaweeds · Bioactive compounds · Pigments · Protein substitutes · Toxicity

Abbreviations

CS	Calf Serum
CUPRAC	Cupric Reducing Antioxidant Capacity
DPPH	2,2-diphenyl-1-picrylhydrazyl
EAC	Ehrlich ascites carcinoma
EFSA	European Food Safety Authority
ES-MS	Electrospray Ionization-Mass Spectrometry
FAO	Food and Agriculture Organization
GC-MS	Gas Chromatography-Mass Spectrometry
GMEM	Glasgow Minimum Essential Medium
HPLC-ICP-MS	High-Performance Liquid Chromatography–Inductively Coupled Plasma Mass Spectrometry
ICP-MS	Inductively Coupled Plasma Mass Spectrometry
LC/ESI-MS	Liquid Chromatography/electrospray- mass spectrometry
LC-DAD	Liquid Chromatography with Diode Array Detector
PCBs	Polychlorinated Biphenyls
PGS	Prostaglandins

6.1 Introduction

In the twenty-first century, the global food system confronts major challenges in delivering sustainable and nutritious food for a global population (7.6 billion), which is projected to approach 9.8 billion by 2050. This challenge is further exacerbated by climate change and urbanization, which threaten agricultural productivity and food security. As dietary preferences have shifted from starchy crops to protein-rich diets, concerns about the environmental sustainability and impact of animal-based protein have intensified. Global protein demand is expected to nearly double by 2050 (FAO 2022). However, the production of animal-based protein is linked to several issues, including environmental degradation, regulatory approval, agricultural runoff, animal welfare concerns, toxicity, and a rise in non-communicable diseases. To mitigate these challenges, the "Planetary Health Diet" encourages prioritizing plant-based foods over animal-based options, promoting both human health and environmental sustainability (Li et al. 2024). Among alternative protein sources, microalgae such as *Arthrospira* sp. (*Spirulina*) and *Chlorella* sp. have emerged as popular and commercially viable options. These species are widely consumed, marketed for human nutrition, and classified as (GRAS) (Ang et al. 2024; D'Amore et al. 2026).

Algae appeared as a viable solution because of their biomass production, minimal land and freshwater requirements and rich in organic component like proteins, lipids and bioactive compounds (Esmaeili et al. 2026). About 145 species of macroalgae are being used as human food throughout the world and in Europe approximately 500 tones algae is produced per year for direct consumption as food and indirectly as dietary supplement (Cheney 2016; Mendes et al. 2022). Seaweeds are major part of human food in Southeast Asian countries (Japan, China and Korea) and according to a report in many areas of China two billion kilograms seaweeds are consumed per year (Brown et al. 2014). Different species of red algae (*Asparagopsis taxi, Chondrus crispus, Gracilaria, Macrocystis, Undaria and Palmaria palmata*), green algae (*Codium, Caulerpa racemosa* and *Ulva*) (Borowitzka 1998) and brown algae (*Aschophyllum nodosus, Fucus vescicolosus, Himanthalia elongate* and *Laminaria digitata*) are commonly used as food (Desideri et al. 2016). Recently, authors have started to work on different algal species due to production of large number of compounds like carbohydrates, fatty acids, carotenoids, minerals, phycobiliproteins, vitamins and their use in different industries. Currently, the use of algae is increased due to overpopulation and as healthy human food (Molino et al. 2018). Algae is being eaten in fresh, dried or prickled form (Rao et al. 2018). Carrageenan and agar- agar obtained from macroalgae are being used in food industry as stabilizing, thickening and solidifying agents (Borowitzka 1998), phycobiliproteins, carotenoid and beta-carotenes are used as coloring agent in food, proteins are used in beverages and as food ingredient that prevent from cardiovascular diseases and obesity (Molino et al. 2018). Seaweeds are also used in different products such as dairy items and processed foodstuff (Tanna and Mishra 2019). The application of microalgae in animal feed and aquaculture is also increasing due to high protein content, carotenoids, long chain fatty acids especially omega 3 and vitamin contents. In salmon, shrimp and carp diet, *Hematococcus pluvialis* is used as coloring agent, two algal species (*Nannochloropsis* spp., *Dunaliella salina*) are used due to high carotenoid (Molino et al. 2018). *Chlorella vulgaris, Spirulina platensis, Ulva lactuca* and *Sargassum* spp. are used in fish and broiler chicken feed due to anti-inflammatory, antioxidant and anti-bacterial properties (Coudert et al. 2020). The use of algae as healthy food is becoming common due to low calories, enrichment of nutrients, high dietary fiber content and other health boosting effects (Cofrades et al. 2017). While on other hand in 2004, UK Food Standard Agency (FSA) declared use of an algal specie (*Hijikia fusiforme*) is unsafe for humans due to presence of high inorganic arsenic content (iAs) (Nakajima et al. 2006). Similarly, various poisonous metals such as arsenic, cadmium, mercury, lead, chromium in dried products of *Porphyra* sp. and *Laminaria* (Chen et al. 2021; Nakajima et al. 2006), high iodine content in humans due to intake of *Laminaria*, *Porphyra* and *Undaria* (Zava and Zava 2011), different harmful compounds like dioxins, polychlorinated biphenyls (García-Rodríguez et al. 2012) and micro-plastic accumulation had been reported (Gutow et al. 2016). Dioxins had been found in *Undaria* while sufficient data about its presence in literature is not available (Cheney 2016). The previous literature reveals that algae can uptake and accumulate polychlorinated biphenyls easily. In an experiment *Ulva rigida* absorbed polychlorinated biphenyls (3.9 mg per day) from contaminated water (Banach et al. 2020). Different components of algae like proteinaceous allergens cause allergies (James et al. 2022). The long lasting side effects of microalgae ingestion has been observed due to presence of contaminants and neurotoxic amino acids (Gantar and Svirčev 2008). The use of microalgae, macroalgae and their products is increasing to meet the demands of food and feed industry. While the accretion of heavy metals, toxins and allergens in the tissues of algae are a serious concern that should be addressed by further experiments.

Algae can grow in a wide range of habitat and climatic conditions especially along coastal areas without using agricultural land, freshwater, and fertilizers. During twentieth century, cultivation of algae was started in coastal areas and deserts where growth of crops was not possible due to harsh climatic conditions (Navarro et al. 2016). Overpopulation and climatic changes are global issues that exhausted resources and caused scarcity of food. There is a need to find out food crop substitutes to fulfill the food demand of increasing population (Bleakley and Hayes 2017; Klomjit et al. 2021; Mahadevan 2015).

During previous five decades, protein deficiencies in global health had been predicted. About 47–70% dry matter content of microalgae and macroalgae consists of protein (Becker 2007). Meat obtained from animals is major source of protein for humans in world that is not enough to fulfill the protein demands of increasing population (Mellor et al. 2022). Due to high quality of algal proteins and to lessen the overconsumption of meat, proteins obtained from algae can be used as animal-based protein substitute in human food, fish, poultry and livestock feed (Becker 2007; Mellor et al. 2022). The examples of species of algae being utilized in food and feed are given, and their safety for the human, fish, poultry, and livestock is discussed in literature. We found 12 reports (Chen et al. 2021; Cheney 2016; Devault et al. 2021; Hammann et al. 2016; Kumar and Sharma 2021; Lemée et al. 1993; Liu et al. 2020; Nakajima et al. 2006; Noguchi et al. 1994; Pérez et al. 2007; Vos and Koeman 1970; Yotsu-Yamashita et al. 2004) on toxicity of nine algae in food and one report (Vos and Koeman 1970) on toxicity of algae in feed. The main goal of this review article is to compile data available about the use of microalgae and seaweeds as food and feed. In addition, the purpose was to compile inventory of safe and unsafe species, their toxic compounds, and the number of studies dealing with these aspects.

6.2 Seaweeds as Human Food

6.2.1 *Porphyra*

Porphyra belongs to red algae and different names for it are Nori, zicai (Wei et al. 2003), purple laver, karengo, kim (Bito et al. 2017), hoshi-nori, and luche (Pérez et al. 2007). *Porphyra tenera* and *Porphyra yesoensis* are mainly cultivated as crop in Japan, China, and Korea while wild species of it are harvested in UK, Korea, Ireland, China, Canada, USA, and Japan. In Korea and China, *Porphyra dentata* and *Porphyra haitanensis* are grown, respectively. The shape of this alga is cylindrical with photosynthetic pigments such as chlorophyll and phycoerythrin. It is commonly used in form of dried flat cake as flavor enhancer in soups and with vinegar rice as sushi while available in markets as breads containing nori, toasted nori sheets and seasonal nori product (Fig. 6.1).

Dried nori products are good source of proteins, vitamins (K, C, Folate and B_{12}), dietary fibers, minerals (potassium, iron and iodine), lipids carbohydrates, and bioactive compounds (phenolic compounds, polysaccharides, peptides, phycobiliproteins and amino acids like mycosporine) (Table 6.1). It possessed hepato-protective, antiinflammatory, anti-tumor activities, immune-modulatory, antioxidative, anti-aging, anti-hyperlipidemia and other health boosting effects (Table 6.1). Due to high nutritional value, use of *Porphyra* sp. is beneficial for human (Bito et al. 2017).

During twentieth century toxic metal arsenic was detected in marine algae, and it caused cancer in kidney, liver, skin, lungs, and bladder. For determination of arsenic content in *Porphyra* information provided by HPLC-ICP-MS and ES-MS was used. Samples were collected from five different locations, and arsenic, predominantly in the form of arsenosugar-phosphate (As–PO_4), was identified in all extracted samples. To check stability, arseno-sugars were heated at 100 °C for short time and were stable at this temperature. Its stability was checked in humans after ingestion. For this purpose, urine samples of 6 volunteers were collected after ingestion of *Porphyra* sp., and results showed an increase in dimethyl-arsenic acid after metabolism of arseno-sugars. Other heavy metals such as lead, cadmium, mercury, and chromium had been detected in *Porphyra* sp. that caused skin diseases and damaged the central nervous system. No significant health risks associated with lead (Pb), mercury (Hg), or chromium (Cr) exposure were identified. However, given the high consumption of *Porphyra* spp., the potential health risks arising from cadmium (Cd) and arsenic (As) accumulation warrant careful consideration and should not be overlooked (Table 6.2). So, consumers should avoid the consumption of *Porphyra* sp. with high concentrations of heavy metals such as arsenic and cadmium (Chen et al. 2021; Wei et al. 2003). *Porphyra* is consumed in European countries than any other algae and is generally safe as human food due to its high protein, polysaccharides and bioactive compounds that have potential to prevent several disorders. In previous work, presence of arseno-sugars in *Porphyra* had been described; consumers should choose *Porphyra* with low concentration of arseno-sugars (Table 6.5). France has established national maximum levels (MLs) for

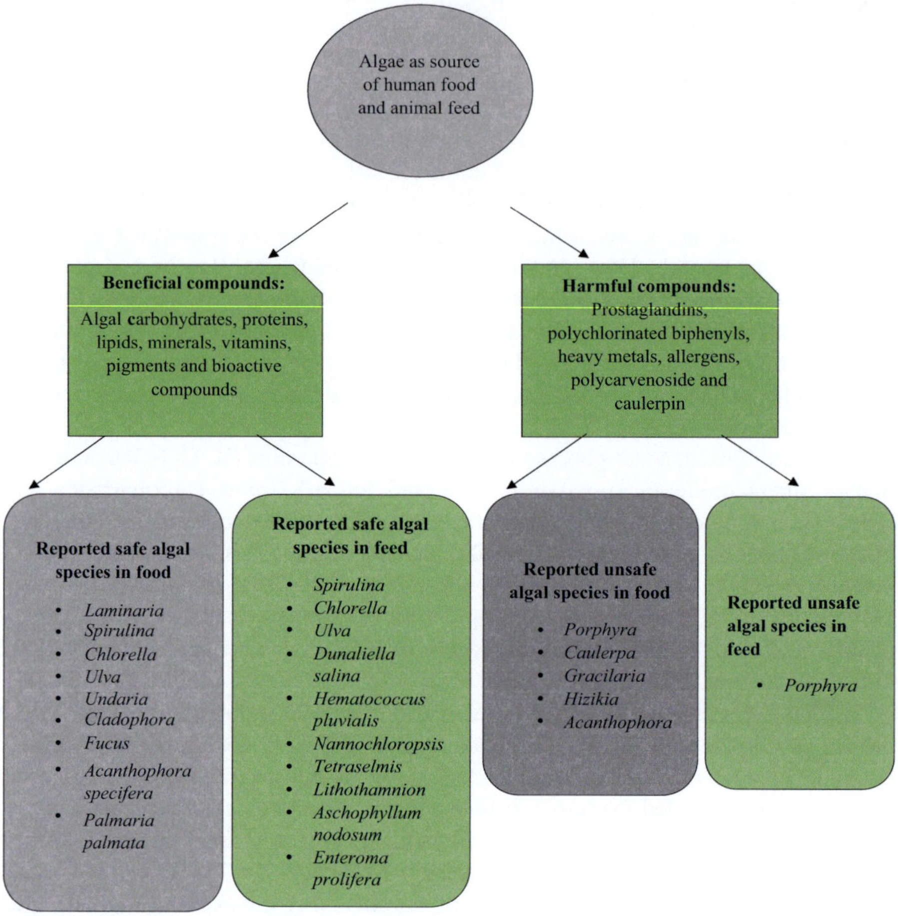

Fig. 6.1 Safe and unsafe algal species in food and feed reported in literature

inorganic arsenic, lead, cadmium, mercury, and iodine in edible seaweeds consumed as vegetables or condiments. With the growing popularity of seaweed as a food item, regulating arsenic content has become increasingly critical to address potential health risks associated with its consumption. Similarly, in China, the national food safety standard (GB2762-2022) specifies a maximum level of 0.3 mg/kg for inorganic arsenic in supplementary foods containing seaweed additives for infants, and 0.5 mg/kg for other applications (Guo et al. 2023).

6.2.2 *Laminaria*

Laminaria belongs to brown algae and different names for it are kelp, wakame, arame and kombu (Kumar and Sharma 2021). It is frequently used with soups and

Table 6.1 Useful compounds in seaweeds and their health benefits in humans

Sr. #	Botanical name/family/ common name	Use in food industry	Beneficial compounds	Health benefits	References
1	*Ulva*/sea lettuce /Ulvaceae	Used as human food in fresh or dried form, as vegetable, as food additive and in preparation of salads	Carbohydrates, pigments, fatty acids, macronutrients (C, N, P, Ca, Na, K, Mg), bioactive compounds, micronutrients (Fe, Zn, Cu, Mn, I) and amino acids	Anti-parasitic, anti-cancer, anti-microbial, anti-inflammatory, anti-hyperlepidemic, immunomodulatory, anti-coagulant and anti- oxidant properties	Bangmei and Abbott (1987), Li et al. (2018), and Yu-Qing et al. (2016)
2	*Undaria pinnatifida*/ Wakame/ Alariaceae	Used in salads, dried and processed food products, soups and as sweeting and coloring agent in food products	Proteins, vitamins, fibers, minerals (Ca, Na, K, Fe, and Mg), bioactive compounds (flavonoids, sterols, alkaloids) carbohydrate, lipids and amino acids	Antioxidant, anti-tumor, anti-diabetic, anti-coagulant, anti-bacterial, anti-viral, cardio-tonic, anti-obesity and anti-inflammatory activities	Nadeeshani et al. (2022) and Wang et al. (2018)
3	*Laminaria*/ kelp/ Phaeophyceae	Used with soups and stew like vegetable. Sodium alginate obtained from it is used as a stabilizer in dairy products such as chocolate milk and ice cream, in bakery products and canned food	Carbohydrate, proteins, dietary fibers (soluble as well as insoluble dietary fibers), vitamins, proteins, iron, iodine, magnesium, potassium, calcium, mannitol and polysaccharides such as alginates	Useful for bone tissues, enzyme immobilizers and possess anti-bacterial properties	Chapman et al. (1980), Cheng (1969), Kumar and Sharma (2021), and Tanna and Mishra (2019)
4	*Porphyra* sp./ Nori/ Bangiaceae	Used in the form of dried flat cake, in soups and with vinegar rice as sushi and also available as breads, toasted nori sheets and seasonal nori products	Proteins, vitamins (K, C, folate and B_{12}), dietary fibers, minerals (potassium, iron and iodine), lipids carbohydrates and bioactive compounds (phenolic compounds, polysaccharides, peptides, phycobiliproteins and amino acids like mycosporine)	Hepato-protective, antiinflammatory, anti-tumor activities, immune-modulatory, antioxidative, anti-aging, anti-hyperlipidemia and other health boosting effects	Bito et al. (2017) and Wei et al. (2003)

(continued)

Table 6.1 (continued)

Sr. #	Botanical name/family/ common name	Use in food industry	Beneficial compounds	Health benefits	References
5	*Caulerpa* spp./ sea grapes/ Caulerpaceae	Used in fresh form as vegetable salad, with sauce, onion, tomato and vinegar and also used in processed products such as prickles, soup and sweets	Proteins, carbohydrates (sulfate, galactose, glucose, arabinose, xylose, mannose, rhamnose, fucose), minerals (Ca, Mg, Na, P, K, Fe, Zn, Mn, Cu, Se, and Mo), vitamins (vitamin A, E, C, B complex), bioactive compounds (caulerprenylols A and B, sesquiterpenoid, di-terpenoid) and lipids	Hypo-cholesterolemic, antioxidant, anti-microbial, anti-hypertensive, immunomodulatory, prebiotic, mineral binder, anti-thrombotic, anti-diabetic, cardio-tonic, and anti-tumor activities	de Gaillande et al. (2017) and Tapotubun et al. (2020)
6	*Gracilaria edulis*/ Gracilariaceae	Used as gelling agent in pudding, cakes, jellies, jelly donuts and other desserts	Fatty acids (palmitic acid, margaric acid, stearic acid, oleic acid, linoleic acid, α-linolenic acid, stearidonic acid docosahexaenoic acid), bioactive compounds (steroids, phenols, terpenoid, bromo-phenols), proteins, carbohydrate, essential amino acids, minerals (Zn, Cu, Mg, Fe), vitamins (β-carotene, ascorbic acid, tocopherol), pigments (Chl a, b, carotenoids) and low lipid content	Antioxidant, anti-atherosclerotic, anti-cancer, anti-coagulant, immune-modulatory, cardio-tonic, anti-obesity, anti-microbial properties and improves the gastrointestinal functioning	Rosemary et al. (2019) and Sakthivel and Devi (2015)

7	*Palmaria palmata*/Dulse/	Used in soups, as condiment with bread, butter and milk, as thickening agent and as snacks, as substitute for fried bacon, as flavor enhancer	Carbohydrates, amino acids and proteins, fatty acids (high content of eicosapentaenoic acid), pigments (lutein, α- and β-carotene chlorophyll a/b) and other bioactive compounds	Anti-tumor, anti-microbial, anti-inflammatory, anti-viral, antioxidant, immune-modulatory, anti-diabetic and anti-hypertensive properties	Admassu et al. (2018), Lopes et al. (2019), and Mouritsen et al. (2013)
8	*Acanthophora specifera*/ Rhodomelaceae	Used raw, in salads, as additive and gelling or stabilizing agent in food	Fatty acids, phenolic compounds, sterols, enzymes, terpenoid, alkaloid, minerals, polysaccharides and vitamins	Anti-implantation, antioxidant, anti-cancer, anti-bacterial, anti-diabetic, anti-viral properties	Lavakumar et al. (2012) and Zakaria et al. (2011)
9	*Fucus* spp./ Fucaceae	Used in dairy products, pasta, bread, fish-derived food products and fish preservative	Carbohydrates, polysaccharides (alginic acid, fucoidan, laminarin), proteins and amino acids, lipids and fatty acids, minerals (Na, K, Ca, Mg, Fe, Cu, Zn, Mn), vitamins, fucoxanthin, phenolic compounds,	Antioxidant, anti-obesity, anti-aging, anti-diabetic, antioxidant, anti-inflammatory, anti-coagulant, anti-tumor, cardio-tonic, and anti-microbial properties	Catarino et al. (2018), Obluchinskaya et al. (2022), and Pozharitskaya et al. (2018)
10	*Cladophora*/ Mekong weed/ Thai/ Cladophoraceae	Used as food supplement	Carbohydrates, lipids, proteins, amino acids, fatty acids, pigments such as chlorophylls and β-carotene, lutein and zeaxanthin, minerals, vitamins, phenolic compounds	Anti-bacterial, anti-fungal, antioxidant, anti-coagulant, anti-diabetic, anti-inflammatory anti-cancer, anti-ulcer, immune-modulator, anti-hypertensive, anti-viral and cytotoxic properties	Michalak and Messyasz (2021) and Munir et al. (2019)

Table 6.2 Toxic compounds in seaweeds and their health effects in human

Sr. #	Botanical name/ family	Common name	Toxic compound	Effects	Reference
1	*Gracilaria vermiculophylla*/ Gracilariaceae	Ohmi	Prostaglandin E2, prostaglandin A2	Gastric disturbances such as nausea, vomiting and diarrhea in humans	Hammann et al. (2016) and Sousa et al. (2010)
4	*Gracilaria edulis*/ Gracilariaceae		Polycarvenoside A (PA)	Gastrointestinal, neurological disorders and death due to PA poisoning in humans	Yotsu-Yamashita et al. (2004)
2	*Caulerpa racemosa*/ Caulerpaceae	Sea grapes	Caulerpicin and caulerpin	Difficulty in breathing, numbness at tongue tips, depression, neuroblastoma and apoptosis in humans	Cheney (2016) and Tapotubun et al. (2020)
5	*Laminaria* sp./ Phaeophyceae	Kelp	Cadmium, potassium, copper and high iodine content	Liver damage, bone damage, nerve damage and interference in transport of ion in membrane, thyroid disorder in humans	Kumar and Sharma (2021) and Zava and Zava (2011)
6	*Hizikia fusiforme* (*Sargassum fusiforme*)/ Sargassaceae	Hijiki	Organic and inorganic arsenic	Arsenic poisoning, cancer in kidney, liver, skin, lungs and bladder in humans	Nakajima et al. (2006)
7	*Porphyra* sp./ Bangiaceae	Nori	Arsenic, cadmium, iodine rich and amphipod allergens	Cancer in kidney, liver, skin, lungs and bladder, goiter and inflammation in humans	Bito et al. (2017) and Wei et al. (2003)
8	*Gracilaria verrucosa*/ Gracilariaceae	Hudson	Arachidonic acid and prostaglandin E2	Nausea, diarrhea, vomiting, bleeding, hypertension, hypotensive shock and death in humans	Fleurence (2016) and Noguchi et al. (1994)

stew-like vegetables in China, Japan, and Korea. Sodium alginate obtained from it is used as a stabilizer in dairy products such as chocolate milk and ice cream, in bakery products and canned food (Cheng 1969). A gel is obtained from *Laminaria hyperborea* containing poly-uronic acid that is used in commercial products such as film formation, synthesis of low calorie fabricated cherries, potato chips, spaghetti, meatballs, and synthesis of flavored particles (Chapman et al. 1980). It is source of dietary fibers (soluble as well as insoluble dietary fibers), vitamins, proteins, iron, iodine, magnesium, potassium, calcium, mannitol and polysaccharides such as alginates. Alginates are useful for humans as material for bone tissue, enzyme immobilizers and possess anti-bacterial properties (Table 6.1). As food additive, *L. saccharina*, *L. cloustoni* and *L. japonica* are used that contains high concentration of iodine. Fucoidan, a sulfated polysaccharide present in *L. japonica*, was evaluated for its toxicity and mutagenic potential using bacterial reverse mutation assays, total arsenic content analysis, in vivo micronucleus assays, and chromosome aberration tests. The study demonstrated no evidence of genotoxicity or mutagenicity and concluded that fucoidan is safe for human consumption at a daily intake level of up to 2,000 mg (Kumar and Sharma 2021; Tanna and Mishra 2019).

To check the presence of various heavy metals such as lead, chromium, mercury, cadmium and arsenic in *Laminaria* sp., samples were collected from coastal areas of China. Plasma mass spectrometry (ICP-MS) was used to check total lead, cadmium, and arsenic content while atomic fluorescence spectrometry and liquid chromatography with ICP-MS were used to determine concentration of mercury and arsenic, respectively. The results showed that the concentrations of heavy metals in the samples were below the maximum permissible limits established by the European Union and Chinese regulatory authorities, indicating that the consumption of this alga is safe for consumers. While in another paper it deranged thyroid level due to excess iodine, kidney failure due to excess amount of potassium, liver damage, nerve damage due to high cadmium and chromium content (Table 6.2), and other side effects like allergic reactions, unusual bleeding, and cramping by using *Laminaria* sp. had been reported (Chen et al. 2021; Kumar and Sharma 2021). The use of *Laminaria* sp. is safe for humans while its excessive use can cause goiter, damage in liver and nervous system and allergy in humans due to presence of iodine, potassium, cadmium and chromium (Fig. 6.1).

6.2.3 *Caulerpa*

Caulerpa is a member of green algae and different names for it are sea grapes, *lat*, *lawi-lawi*, *umi budo*, *ar-arosep*, green caviar, latoh and *fiji* (Tapotubun et al. 2020). It belongs to green algae due to presence of chlorophyll. It grows in Maluku Kei Islands while its use as food is common in Indonesia, Taiwan, China, Japan, Vietnam, Malaysia, Thailand, Singapore, and Philippine. *Caulerpa lentillifera* and *Caulerpa racemosa* var. *turbinate* are commonly used as food in Indo-Pacific region due to enrichment of minerals. Due to attractive color and different taste, *Caulerpa* sp. are eaten fresh as vegetable salad, with sauce, onion, tomato and vinegar while also used in processed products such as prickles, soup, and sweets. The use of *Caulerpa* products is good for consumer health due to lack of food additives during processing and enrichment of proteins, carbohydrates (sulfate, galactose, glucose, arabinose, xylose, mannose, rhamnose, fucose), minerals (Ca, Mg, Na, P, K, Fe, Zn, Mn, Cu, Se, and Mo), vitamins (vitamin A, E, C, B complex), bioactive compounds (caulerprenylols A and B, sesquiterpenoid, di-terpenoid) and lipids. It possessed hypo-cholesterolemic, antioxidant, anti-microbial, anti-hypertensive, immunomodulatory, prebiotic, mineral binder, anti-thrombotic, anti-diabetic, cardio-tonic, and anti-tumor activities (Table 6.1) (de Gaillande et al. 2017; Tapotubun et al. 2020).

Caulerpin and caulerpicin have been reported in *Caulerpa racemosa, Caulerpa lentillifera, Caulerpa clavifera, Caulerpa taxifolia* that caused apoptosis, sensation of anesthesia, neuroblastoma and difficulty in breathing in consumer after ingestion (Table 6.2). The study was conducted on mice, mammalion cells and sea urchin to observe toxicity caused by caulerpin and caulerpicin. Secondary metabolites were obtained from *C. taxifolia*, suspended into 0.5% and 0.9% tween-20 and NaCl solution and 2 g/kg and extract was injected to mice. The results showed fresh *C. taxifolia* extract caused death in mouse during 24 h. The fibroblastic cell line BHK 21/C13 was kept in Glasgow Minimum Essential Medium (GMEM) with antibiotics and 10% Calf serum (CS) were planted into wells of micro-titer plates. The media was replaced with new medium (GMEM+ CS 5% + antibiotics) after 48 h and media components were diluted. After 24 h of incubation washing was performed with phosphate buffer saline. Then, wells were filled with new medium (CS 5% + supra-vital dye). After three hours media were removed and 0.5% formalin +1% $CaCl_2$ solution was added. After dye removal the plates were moved to micro-titer plate reader (IEMS, Lab system), at 450 nm absorbance was observed and 50% toxicity was observed. In sea urchin, delay in egg cleavage was observed. (Cheney 2016; Lemée et al. 1993). The use of *Caulerpa* sp. is increasing as a healthy food due to its nutritional value while there is a need to know about its toxic effects due to presence of caulerpin and caulerpicin.

6.2.4 *Hizikia fusiforme*

Hizikia fusiforme of family Phaeophyceae is commonly used in coastal areas of Japan, China, North America, United Kingdom and Korea. It is also known as *Sargassum fusiforme*, Hijiki, Hiziki,Yangqicai and is eaten fresh, with mussel, fish, bean curd, fried, as a filling in steam buns and filling for dumpling (Bangmei and Abbott 1987; Liu et al. 2020). *Hizikia* spp. are rich in minerals (iodine, potassium, sodium, chlorine) carbohydrates (alginates, laminarin, sulphated polysaccharides, mannitol, fucoidan, and cellulose), proteins, lipids (sterols, terpenoid, tocopherols and carotenoids) and polyphenols. It has anti-osteoarthritic, anti-viral, anti-cancer, antioxidant, immune-modulatory, neuro-protective, UV protective, anti-inflammatory and other health boosting properties. Due to consumption of Hijiki polychlorinated bio-phenyl contamination has been reported in women blood. *Hijikia* sp. contains high concentration of organic and inorganic arsenic, lead and iodine (Table 6.2). The limit of arsenic in Hijiki demonstrated by China, Korea and Japan is 40–80 ppm while inorganic arsenic content in it reaches up to 200 ppm. Arsenic content can be reduced (90%) by soaking and washing while in processed food an increase in arsenic contents had been observed. The recorded level of cadmium was in range of 0.5–2.0 ppm, however an increase in lead and iodine content had been observed (Liu et al. 2020). In 2004, UK Food Standard Agency (FSA) declared use of *H. fusiforme* is unsafe for humans due to presence of high inorganic arsenic content and warnings related to health risk such as increase in lung, kidney, bladder and skin cancer due to consumption of *H. fusiforme* had been reported (Devault et al. 2021; Nakajima et al. 2006). The use of *H. fusiforme* as human food is not safe due to high concentration of organic and inorganic arsenic contents.

6.2.5 *Ulva*

Ulva lactuca of Ulvaceae family is known as sea lettuce, green laver and string lettuce. In coastal areas of China, Guangdong, Indonesia and Southern Fujian *U. lactuca*, *U. conglobata*, *U. prolifera*, *U. fasciata* are mainly used as human food in fresh or dried form, as vegetable, food additive and in preparation of salads (Bangmei and Abbott 1987; Li et al. 2018). The use of *Ulva* sp. is beneficial for humans due to presence of a variety of nutrients (Table 6.1) such as carbohydrates, pigments, fatty acids, macronutrients (C, N, P, Ca, Na, K, Mg), bioactive compounds, micronutrients (Fe, Zn, Cu, Mn, I) and amino acids. It possessed anti-parasitic, anti-cancer, anti-microbial, anti-inflammatory, anti-hyperlepidemic, immunomodulatory, anti-coagulant and anti-oxidant properties (Yu-Qing et al. 2016). In Japanese population, major reason of acute encephalitis was flavivirus that caused inflammation and neurological disorders in humans. Proper treatment for this viral infection was not available. The experiment was conducted on Japanese encephalitis virus-infected C3H/HeN mice to check the potential of sulphated polysaccharide extract of *U. lactuca* against the flavivirus infection. The virus was unable to enter in cell because this algal extract blocked the attachment sites of virus. In mice, neurobehavioral defects were observed after five days of infection while on seventh day death of mice was recorded. An interruption in hind limb paralysis was observed in mice pretreated with algal extract that prevented death (Chiu et al. 2012). In another study, nutritional components and health risk of *U. prolifera* consumption due to presence of pollutants such as heavy metals, polycyclic hydrocarbons and pesticides had been analyzed. The level of heavy metals and polycyclic hydrocarbons that caused cancer was less than the USEPA permitted limit (Li et al. 2018). The use of *U. prolifera* as human food is safe due to enrichment of carbohydrates, proteins, essential and non-essential amino acid, unsaturated fatty acids and low number of heavy metals.

6.2.6 *Undaria pinnatifida*

Undaria pinnatifida of Alariaceae family is commonly known as Wakame. It is commonly used as food in France, Korea, Japan, New Zealand, Australia, China, Spain, USA and Argentina. It is used in salads, dried and processed food products, soups and as sweetening and coloring agent in food products (Nadeeshani et al. 2022). It contains a variety of nutrients such as proteins, vitamins, fibers, minerals (Ca, Na, K, Fe, and Mg), bioactive compounds (flavonoids, sterols, alkaloids) carbohydrate, lipids, and amino acids (Table 6.1). It exhibited antioxidant, anti-tumor, anti-diabetic, anti-coagulant, anti-bacterial, anti-viral, cardio-tonic, anti-obesity and anti-inflammatory activities (Nadeeshani et al. 2022; Wang et al. 2018). The *U. pinnatifida* samples were collected from two places of New Zealand to check the antioxidant properties of wakame Fucoxanthin. DPPH and cupric reducing antioxidant capacity (CUPRAC) assays were used to check antioxidant properties and results determined fucoxanthin of *U. pinnatifida* possessed strong antioxidant activity (L. Wang et al. 2018). In another study, nutritional value of *U. pinnatifida* was described and experiment was conducted on rats to determine the effects of Wakame feed. The rats were provided feed with 100 g/kg Wakame for 28 days and its effect on various organ weight and growth was recorded. The results showed in experimental group an increase in activities of γ-glutamyl trans-peptidase and di-saccharidase had been observed (Taboada et al. 2013). Recently, study was conducted to check the effects of Wakame ingestion on regulation of glucose level in human. The humans were provided rice with dried wakame and blood samples were collected before and after ingestion of dried Wakame with rice. The results determined a reduction in insulin and glucose level was observed after thirty minutes of ingestion that showed Wakame is beneficial for regulation of glucose level in human (Yoshinaga and Mitamura 2019). The consumption of *U. pinnatifida* is beneficial for humans due to presence of carbohydrates, proteins, lipids and bioactive compounds that possess various health-boosting effects.

6.2.7 *Gracilaria vermiculophylla*

Gracilaria vermiculophylla belongs to red algae and is commonly known as Ohmi. It is present in North America, Japan, East Asia, and Europe. It is used as raw, with fish, major source of agar that is used as gelling agent in pudding, cakes, jellies, and coating on fruits and vegetables (Sousa et al. 2010). Due to ingestion of *G. vermiculophylla* in humans poisoning had been reported (Table 6.2). The precursor of prostaglandin A2, prostaglandin E2, was arachidonic acid, and these prostaglandins (PGS) were responsible for poisoning. In this alga 45% arachidonic acid was present that became active in specific conditions and PGS formed. The presence of prostaglandins caused gastrointestinal disturbance (diarrhea and nausea) and nervous system disorders in human while hypotensive shock was reason of death (Hammann et al. 2016). *G. vermiculophylla* is mainly utilized as gelling agent in food industry and contains various nutrients such as proteins, carbohydrates, fibers, lipids, and minerals while activation of prostaglandins in specific conditions can cause poisoning in humans.

Since there is a need for guidelines on the limit and quality of seaweeds consumption, this work aims to provide a contribution to the collection of data requested by European Food Safety Authority (EFSA) on the levels of heavy metals in functional foods, such are seaweeds, in order to facilitate the evaluation of better risk assessments and potentially set official maximum levels for European legislation (Filippini et al. 2021).

6.2.8 *Gracilaria edulis*

Gracilaria edulis belongs to red algae and is mainly present in Indo-Pacific region. It is used in human food in Korea, China, Japan, and India. It is also known as *Gracilaria lichenoides* and *Polycavernosa tsudai* (Kumar and Sharma 2021; Yotsu-Yamashita et al. 2004). *G. edulis* is major source of agar that is used as gelling agent in pudding, cakes, jellies, jelly donuts, and other desserts (Table 6.1). It is most beneficial algal specie due to presence of valuable nutrients such as fatty acids (Palmitic acid, margaric acid, stearic acid, oleic acid, linoleic acid, α-linolenic acid, stearidonic acid docosahexaenoic acid),bioactive compounds (steroids, phenols, terpenoid, bromo-phenols), proteins, carbohydrate, essential amino acids, minerals (Zn, Cu, Mg, Fe), vitamins (β-carotene, ascorbic acid, tocopherol), pigments (Chl a, b, carotenoids), and low lipid content. It possess antioxidant, anti-atherosclerotic, anti-cancer, anti-coagulant, immune-modulatory, cardio-tonic, anti-obesity, anti-microbial properties and improves the gastrointestinal functioning (Rosemary et al. 2019; Sakthivel and Devi 2015). In 1991, three deaths were reported in Guam due to ingestion of *G. edulis*. The reason of death was presence of polycarvenosides (PA2, PA3, and PB2) in it. Similarly, eight deaths were reported in Philippine in 2002 due to ingestion of *G. edulis*. The toxic section in semi-purified form was obtained from this alga and was injected in mouse that caused death (Table 6.2). For detection of polycarvenosides LC-diode array detection (LC-DAD) and LC/electrospray-mass spectrometry (LC/ESI-MS) analysis were used and results determined presence of polycarvenoside A (PA). The total PA in *G. edulis* was 84 and 72 $nmolkg^{-1}$, while polycarvenoside A2 and A3 were below the limit to be described by using LC-diode array detection (LC-DAD) and LC/electrospray-mass spectrometry (LC/ESI-MS) analysis. The results determined the presence of polycarvenoside A is major reason of consumer's death due to ingestion of *G. edulis* (Yotsu-Yamashita et al. 2004). *G. edulis* possesses various health-boosting properties due to presence of beneficial nutrients while polycarvenoside limits its use. Further studies on polycarvenoside poisoning and its health effects on humans are suggested.

6.2.9 *Gracilaria verrucosa*

Gracilaria verrucosa belongs to red algae and is commonly known as Hudson. It is used in human food in Korea, China, Japan, South Africa, India and Namibia. It is eaten with potato, fish, onion rice, vinegar, sea bream, and source of agar that is used as gelling agent in pudding, cakes, jellies, jelly donuts, other desserts, and also used as raw material for bioethanol production. It is source of proteins, carbohydrates, fibers, lipids, and minerals (Fleurence 2016; Noguchi et al. 1994). (Noguchi et al. 1994) reported the food poisoning and death during 1993 in Japan, due to consumption of *G. verrucosa* (Table 6.2). Two samples of Hudson (remaining food of victims and fresh Hudson) were used to determine the reason of poisoning. In first sample, prostaglandin E2, H2, A2, E1, F2 were reported and due to addition of arachidonic acid PGE2 amount was increased because arachidonic acid was precursor of prostaglandins. Due to the inherent instability of PGH_2, its concentration could not be determined. However, PGA_2, PGF_2, and PGE_1 were detected at low levels in the victim's food sample, whereas the fresh Hudson sample contained these prostaglandins at a concentration of 22.4 ppm. The results showed presence of an enzyme prostaglandin endo-peroxide synthase that was responsible for production of PGE2 from arachidonic acid while 30 mg PGE2 and minor quantities of other PGEs had been reported in victim's body. The presence of prostaglandins caused gastrointestinal disturbance in victim and hypotensive shock was reason of death. *G. verrucosa* could be useful for humans due to its nutritional contents., while prostaglandins can cause food poisoning as well as human death in specific conditions. *G. verrucosa* is not more beneficial for human.

6.2.10 *Acanthophora specifera*

Acanthophora specifera of Rhodomelaceae family is present in Guam, India and Hawaii. It is commonly known as drift red algae, spiny seaweed and bulung tumbung bideng and is used raw, in salads, as additive and gelling or stabilizing agent in food. It contains a variety of nutrients such as fatty acids, phenolic compounds, sterols, enzymes, terpenoid, alkaloid, minerals, polysaccharides and vitamins. It possessed anti-implantation, antioxidant, anti-cancer, anti-bacterial, anti-diabetic, anti-viral and other health boosting effects (Lavakumar et al. 2012; Zakaria et al. 2011). The experiments were conducted on mice to check antioxidant and anti-cancer properties of alcoholic content of *A. specifera*. An oral dose of *A. specifera* ethanol extract 100 $mgkg^{-1}$ to 200 mg kg^{-1} was given to mice for treatment of Ehrlich ascites carcinoma (EAC) cells. The tumor volume, weight with mean survival days and different hematological parameters were measured after twenty-one days to check anti-cancer activity of *A. specifera* ethanol extract. An increase in body weight while reduction in tumor volume and weight in EAC mice treated with *A. specifera* extract (100–200 $mgkg^{-1}$) was reported. Similarly, methanol extract of this algae reduced oxidative stress in liver and increased antioxidant activities in mice (Lavakumar et al. 2012). In another study, gas chromatography-mass spectrometry (GC-MS) was performed to check composition of *A. specifera* hexane extract while disk diffusion method was used to evaluate antimicrobial properties of this extract. GC-MS analysis determined hexa-decanoic acid and oleic acid were main components of hexane extract and this extract showed antibacterial activity against *Methicillin-resistant Staphylococcus aureus* and *Pseudomonas aeruginosa* (Zakaria et al. 2011). In Philippine, during 2003 due to ingestion of *A. specifera* 36 poisoning cases and eight deaths were reported. The toxic section in semi-purified form was obtained from this alga and was injected in mice that caused death. For detection of polycarvenosides LC-diode array detection (LC-DAD) and LC/electrospray- mass spectrometry (LC/ESI-MS) analysis were used and results determined presence of polycarvenoside A (PA). The total PA in *A. specifera* was 84 and 72 $nmolkg^{-1}$, while polycarvenoside A2, A3, aplysiatoxin were below the limit to be described by using LC-diode array detection (LC-DAD) and LC/electrospray-mass spectrometry (LC/ESI-MS) analysis. The results showed that the presence of polycarvenoside A is major reason of consumer's death due to ingestion of *A. specifera*. The symptoms of poisoning included neurological disorder and gastrointestinal disturbance (Kumar and Sharma 2021; Yotsu-Yamashita et al. 2004). The use of this algal specie is beneficial for human due to its nutritional value and low level of heavy metals such as chromium, lead, zinc and copper.

6.3 Microalgae as Human Food

6.3.1 *Spirulina*

Spirulina belongs to blue-green algae and is used as food in Africa, Brazil, Spain, US, India, France, Canada, Egypt, Belgium, Ireland, Philippines and other countries. *Spirulina* is used in snacks as flavor enhancer, in dairy products such as yoghurt to make it homogeneous, in food and beverages to provide green color, in vegetable soups and in other food products such as dry cakes, pastries, chocolate bars, toasted bread and gluten-free pasta. In twenty-first century, Food and Agriculture Organization (FAO) declared it is best food due to presence of variety of nutrients such as proteins, vitamins (vitamin B_{12}, K, E, thiamin, niacin, biotin, folic acid), minerals (Ca, Cr, Cu, Fe, Mg, Mn, P, K), carbohydrates, natural pigments (phycocyanin, chlorophyll A, carotenoids) and fatty acids (Ali and Saleh 2012; Lafarga et al. 2020). *Spirulina platensis* and *Spirulina maxima* are mainly used as dietary supplement (Matufi and Choopani 2020). It possessed radiation protective, anti-cancer, antioxidant, anti-diabetic, anti-obesity, anti-inflammatory, immune-modulatory, hypolipidemic, hypoglycemic, hypotensive, anti-viral and anti-atherosclerotic properties (Belay et al. 1993b; Matufi and Choopani 2020; Reboleira et al. 2019; Saranraj and Sivasakthi 2014) (Table 6.3). To investigate the effect of *Spirulina* diet on hyperlipidemia, hypertension, diabetes, and renal toxicity, various clinical trials were conducted on rats, and results determined *Spirulina* diet was helpful to prevent renal failure while it reduced the cholesterol level, blood pressure and glucose level in rats. Similarly, clinical trials were performed on obese humans, and a decrease in body weight was observed due to ingestion of 2.8 g *Spirulina* three times in a day for 4 weeks. No serious health effects had been observed by using *Spirulina* while nausea, diarrhea, gastric problems, allergies, damage of skin and sleeping problems due to fat loss had been reported. (Belay et al. 1993b; Matufi and Choopani 2020). *Spirulina* is an excellent therapeutic food for humans due to presence of nutrients, bioactive compounds, and positive impacts on human health are reported by clinical trials.

6.3.2 *Chlorella*

Chlorella of family Chlorophyceae is commonly used as human food in Asia, China, Germany, USA, and Japan (Ahmad et al. 2020; Szabo et al. 2013). It is commonly

Table 6.3 Useful components in microalgae and their health benefits in humans

Sr. #	Botanical name/ family/common name	Use in food industry	Beneficial compounds	Health benefits	References
1	*Spirulina/* Oscillatoriaceae	Used in snacks, in dairy products such as yoghurt to make it homogeneous, in food and beverages, in vegetable soups and in other food products such as dry cakes, pastries, chocolate bars, toasted bread and gluten-free pasta.	Proteins, vitamins (vitamin B_{12}, K, E, thiamin, niacin, biotin, folic acid), minerals (Ca, Cr, Cu, Fe, Mg, Mn, P, K), carbohydrates, natural pigments (phycocyanin, chlorophyll A, carotenoids) and fatty acids	Anti-cancer, antioxidant, anti-diabetic, anti-obesity, anti-inflammatory, immune-modulatory, hypo-lipidemic, hypoglycemic, hypotensive and anti-atherosclerotic properties	Ali and Saleh (2012), Belay et al. (1993a), Lafarga et al. (2020), Matufi and Choopani (2020), and Saranraj and Sivasakthi (2014)
2	*Chlorella/* Chlorellaceae	Used in dairy and bakery products such as yoghurt, ice cream, cookies, bread, rolls, pastry, biscuits and in pasta, snacks, spaghetti, sweets and as food additive	Polyunsaturated fatty acids, sulfated extra-polysaccharides, carotenoids, proteins, minerals, vitamins, phenolic compounds and pigments	Anti-bacterial, anti-fungal, anti-cancer, anti-aging, anti-diabetic, anti-hypertensive, immune-modulatory, metal-binding and anti-inflammatory properties	Ahmad et al. (2020), Orantes Calleja (2022), Widyaningrum and Prianto (2021)
3	*Haematococcus pluvialis/* Haematococcaceae	Used as coloring agent and dietary supplement in functional food	Carbohydrates, lutein, vitamins, astaxanthin, carotenoids, minerals, canthaxanthin, and bioactive compounds	Antioxidant, anti-inflammatory, anti-cancer, immune-modulatory, cardio-tonic, radiation protective, anti-hypertensive, anti-diabetic and neuro-protective properties and also help in maintenance of eye health	Guerin et al. (2003) and Hussein et al. (2006)

4	*Nostoc* / Nostocaceae	Used as stabilizer and thickening agent in juices, jams, yoghurt and desserts	Protein, fatty acid, minerals (K, Mg, Na, Zn, Fe), vitamins, nucleic acids, carbohydrates, carotenoids, fatty acids, phycobiliproteins, lipo-peptides and polysaccharides	Antioxidative, anti-cancerous, anti-bacterial, immune-modulatory, anti-hypercholesteremic, anti-diabetic and anti-microbial	Choque-Quispe et al. (2022) and Zhu et al. (2022)
5	*Dunaliella salina*/ Chlorophyceae	Used in food industry as an additive	Polyunsaturated fatty acids, lipids, vitamins, proteins, minerals, carbohydrates, antioxidants, pigments (carotenoids, beta-carotenoids) and other bioactive compounds	Cardio-tonic, anti-cancer, immunomodulatory, anti-cancer and antioxidant properties, reduce chances of atherosclerosis, maintain gut and prevent from diarrhea	Hyrslova et al. (2022), Murthy et al. (2005), and Olivarría et al. (2010)

used in dairy and bakery products such as yoghurt, ice cream, cookies, bread, rolls, pastry, biscuits and in pasta, snacks, spaghetti, sweets and as food additive to improve flavor, color, quality and nutritional value. *Chlorella* is available in markets in powder form as a toxic binder that helps in removal of heavy metals (mercury, cadmium, lead) from body (Widyaningrum and Prianto 2021). *Chlorella* sp. is source of polyunsaturated fatty acids, sulfated extra-polysaccharides, carbohydrates, carotenoids, proteins, minerals, vitamins (vitamin K, B_1, E, B_2, B_3, B_5, and B_6), phenolic compounds and pigments (Ahmad et al. 2020; Silva et al. 2019). It exhibited anti-bacterial, anti-fungal, anti-cancer, anti-aging, anti-diabetic, anti-hypertensive, immune-modulatory, metal-binding and anti-inflammatory properties (Table 6.3). A reduction in anemia, edema and proteinuria was observed in pregnant women due to use of *Chlorella pyrenoidosa* in their diets (J. Silva et al. 2019; Widyaningrum and Prianto 2021). The use of *Chlorella* sp. is increasing due to its nutritional value, toxin binding capacity and positive impacts on human health while due to poor processing of these algae the gastrointestinal diseases had been observed.

6.4 Seaweeds as Animal Feed

6.4.1 *Porphyra*

Porphyra of red algae is used in fish, ruminants, and poultry feed due to its high protein contents. The study was conducted on Nile Tilapia to check the effects of *Porphyra dioica*, *Ulva*, and *Gracilaria* feed on the body composition, gut histology, and growth. 10% of whole diet was replaced by these seaweeds in three trials while diet without seaweed was used as control. These diets were provided to three groups of twenty-five Nile tilapia. After 84 days of trial, a reduction in fish growth and intestinal villi length was observed in fishes due to consumption of *Gracilaria* diet. The results showed use of 10% *P. dioica* and *Ulva* in fish diet for partial replacement of fish meal was useful with no negative effects on gut histology, body composition, and growth (D. Silva et al. 2015) (Table 6.4). Similarly, in another experiment, in sheep diet *Porphyra* was used as soybean meal substitute and its impact on growth, methane production and ruminal fermentation was observed. The results of in vivo trials on lambs revealed that there was no significant difference in production of volatile fatty acids, methane production and growth in both groups. So, protein contents of *Porphyra* and soybean meal was similar and can be used as soybean meal substitute in sheep diet (Lind et al. 2020).

The experiments were conducted on chicken to study the toxicity of polychlorinated biphenyls (Phenoclor DP6, Clophen and Aroclor 1260A) of *Porphyra*. In oral toxicity test, major differences in mean weight, mortality rate, mean survival rate, liver necrosis and edema formation were recorded after two months. The results showed Phenoclor DP6 and Clophen caused liver necrosis in chicken and were more toxic compounds than Aroclor 1260A. In all experimental groups spleen atrophy, increased amount of proto-porphyrin and coproporphyrin in

Table 6.4 Useful compounds in seaweeds and their health benefits in animals

Sr. #	Botanical name/Family	Feed substitute	Beneficial compounds	Health benefits	References
1	*Ulva*/ Ulvaceae	Soybean meal substitute and feed additive in fish meal	Essential amino acid, vitamins, chlorophyll, carotenoids and polysaccharides	Fishes: Positive impacts on growth, body composition, protein and lipid content Chicks: Improvement in dressing percentage and breast muscle	Abudabos et al. (2013), Costa et al. (2022), Elizondo-González et al. (2018), and Ergün et al. (2009)
2	*Cladophora glomerata*/ Cladophoraceae	Used as protein source in fish meal and livestock feed	Carbohydrates, lipids, proteins, amino acids, fatty acids, pigments such as chlorophylls and β-carotene, lutein and zeaxanthin, minerals (Mg, Ca, K, P and N) vitamins, phenolic compounds	Chicks: Zero mortality rate, improvement in meat quality and quantity Fish: With increase in amount of *C. glomerata* in feed growth rate, protein and lipid content was increased	Ali and Abid (2006), Appler and Jauncey (1983), and Michalak and Messyasz (2021)
3	*Laminaria* sp./ Phaeophyceae	Use as corn substitute in poultry feed, as fishmeal substitute in fish feed and as mineral supplement in ruminant feed	Carbohydrate, proteins, dietary fibers (soluble as well as insoluble dietary fibers), vitamins, proteins, iron, iodine, magnesium, potassium, calcium, mannitol and polysaccharides (laminarin, fucoidan, alginate and β-glucan)	Chicks: Mortality rate was reduced and meat quality was improved Fish: No negative effect on fish growth, body weight, survival rate, feed consumption, crude protein or lipid content and plasma chemistry Ruminants: Source of carbohydrates and minerals for ruminants	Costa et al. (2021), Hidayah et al. (2022), and Jeon et al. (2013)

(continued)

Table 6.4 (continued)

Sr. #	Botanical name/Family	Feed substitute	Beneficial compounds	Health benefits	References
4	*Porphyra* / Bangiaceae	Used as fishmeal and soybean meal substitute in feed	Proteins, vitamins (K, C, folate and B12), dietary fibers, minerals (potassium, iron and iodine), lipids carbohydrates and bioactive compounds (phenolic compounds, polysaccharides, peptides, phycobiliproteins and amino acids like mycosporine)	Fish: No negative effects on gut histology, body composition and growth Sheep: No significant difference in production of volatile fatty acids, methane production and growth by using *Porphyra* as protein meal substitute in sheep feed	Bito et al. (2017), Lind et al. (2020), and Silva et al. (2015)
5	*Gracilaria* sp. / Gracilariaceae	Used as feed additive	Fatty acids (palmitic acid, margaric acid, stearic acid, oleic acid, linoleic acid, α-linolenic acid, stearidonic acid docosahexaenoic acid), bioactive compounds (steroids, phenols, terpenoid, bromo-phenols), proteins, carbohydrate, essential amino acids, minerals (Zn, Cu, Mg, Fe), vitamins (β-carotene, ascorbic acid, tocopherol), pigments (Chl a, b, carotenoids) and low lipid content	Chicks: No adverse effects on chick growth, mortality, feed efficiency and feed conversion, immune response and body weight was improved Fish: No adverse effects on growth, performance, nutrient utilization and body composition of *Labeo rohita*, crude carbohydrates, proteins and fat content was increased	Al-Khalaifah and Uddin (2022) and Vadher et al. (2016)

(continued)

Table 6.4 (continued)

Sr. #	Botanical name/Family	Feed substitute	Beneficial compounds	Health benefits	References
6	Undaria pinnatifida/ Alariaceae	Used as feed additive	Carbohydrates, polysaccharides, amino acids, lipids, vitamins, organic acids, fatty acids and minerals, such as (Ca, Fe, Mg, Na, P, and Zn)	Ruminants: Dry matter digestibility, microbial growth, feed conversion rate and volatile fatty acid content was enhanced Shrimp: Improvement in lipid content, feed conversion rate, growth rate in Juvenile *Penaeus monodon*	Choi et al. (2020) and Niu et al. (2015)
7	Enteroma prolifera/ Ulvaceae	Used as feed additive	Proteins, lipids, fatty acids and cholesterol, high mineral contents (Na, K, Ca, Mg, P, Zn)	Broiler chicks: Improvement in feed intake, feed conversion rate and daily weight gain and meat quality while reduction in abdominal and subcutaneous fat Laying hens: Positive impacts on egg production, egg shell thickness, weight and yolk color, improvement in immune functioning Fish: Positive impacts on growth rate, feed efficiency, while protein retention was decreased	Aguilera-Morales et al. (2005), Asino et al. (2011), Wang et al. (2013a, 2013b)

(continued)

Table 6.4 (continued)

Sr. #	Botanical name/Family	Feed substitute	Beneficial compounds	Health benefits	References
8	*Ascophyllum nodosum*	Used as feed additive	Polysaccharides, proteins, pigments (xanthophyll, fucoidan, chlorophyll, beta-carotenes, rich in minerals (K, Na, I), and phenolic compounds (phlorotannins)	Livestock: Positive impacts on immunity and antioxidant properties in goats, sheep and cattle Lambs: Animal health and immunity was improved, animal was protected from oxidative stress induced due to transport	Kannan et al. (2007) and Makkar et al. (2016)
9	*Lithothamnion* sp./ calcified seaweeds/ Corallinaceae	Used as feed additive	Rich in minerals (Ca, Mg, P, S, B, Fe, Na, Mn, Co, Cu, Zn, Se, Mo)	Rabbits: No impact on growth, performance, diet digestibility but calcium amount was increased Dairy cows: Positive impacts on dry matter intake, energy balance, serum concentration of P and body condition score	Euler et al. (2008) and Neville et al. (2022)

fecal excretion was observed. Further experiments were conducted on rats and Japanese quail with Phenoclor DP6 compounds in order to confirm action of porphyrogens. Brain and liver of dead chickens were analyzed by Gas chromatography and results showed relationship of liver necrosis and chicken edema with different concentrations of PCBs (Vos and Koeman 1970).

In animal feed, *Porphyra* is used as partial substitute of soybean meal and had positive impacts on growth, body composition and methane production in fishes and ruminants. While in chickens' liver necrosis, spleen atrophy and edema formation caused by *Porphyra* polychlorinated biphenyls are one of major concern. PCBs can accumulate in *Porphyra* through environmental exposure, i.e., water bodies are contaminated by industrial discharge, agricultural runoff, or atmospheric deposition. So, there is a need to mitigate the PCB contamination by multi-faceted approach involving environmental management, feed processing improvements, and careful monitoring of the food supply chain.

6.4.2 *Ulva*

Ulva is used in livestock, poultry, and fish feed due to enrichment of essential amino acid, vitamins, chlorophyll, carotenoids, and polysaccharides (Costa et al. 2022; Janczyk et al. 2006). In an assessment of the nutritional value of *Ulva lactuca*, it was used as soybean meal substitute in chicken feed. Forty-five chicks were taken and divided into three groups. The chicken of the first group were given feed without *U. lactuca* while the second and third group chicken's feed was substituted with 1% and 3% *U. lactuca* respectively. At 12 and 33 days of chick's lives, the influence of substituting 1% and 3% *U. lactuca* with corn on performance, serum components and carcass characteristics of chicken was observed. There was no difference in feed ingestion, feed conversion rate, nutrient retention and body weight in control and experimental groups while reduction in cholesterol, lipid, uric acid content and alanine aminotransferase enzyme in serum was observed in experimental group. The use of 3% *U. lactuca* and corn diet improved dressing percentage and breast muscle in chicken (Abudabos et al. 2013). Similarly, in another study *Ulva* was used as feed ingredient and its impact on growth, nutrient consumption, and body composition of Nile tilapia was checked. Four isonitrogenous (CP 40%) feeds having 0% and 5% *Ulva*, 10% low lipid and 20% high lipid crude contents were provided to three groups of fishes for sixteen days. The results showed feed containing 5% *Ulva* with both high and low lipid contents improved growth, carcass quality, nutrient consumption, protein conversion rate, and body composition of Nile tilapia (Ergün et al. 2009). As feed additive *U. lactuca* 1%, 2%, and 3% was used in shrimp feed. The *U. lactuca* feed improved shrimp growth, carotenoid and lipids contents (Elizondo-González et al. 2018) (Table 6.4).

In various studies, the presence of arsenic, lead, cadmium, and mercury in European *Ulva* species had been reported, and in non-contaminated areas, the level of mercury and lead in *Ulva* sp. was higher than limit permitted by European Union that showed limitation in use of *Ulva* sp. in feed while its toxicity had not been investigated in animals (Bordean et al. 2016; Christiansen et al. 2017).

Ulva is used as soybean meal substitute and feed additive in fish meal and animal feed for its positive impacts on growth, body composition, protein, and lipid content. While high levels of mercury and lead had been reported in *Ulva* sp. that limit its use in feed being unsafe. There is a need to further investigate the concentration of heavy metals in *Ulva* sp (Table 6.4).

6.5 Microalgae as Animal Feed

6.5.1 *Spirulina*

Spirulina is used as dietary constituent in livestock, fish, pig, and poultry feed (Table 6.5). The use of *Spirulina platensis* is common in animal feed due to enrichment of vitamins, proteins, fatty acids and minerals and phycocyanin (Altmann and Rosenau 2022; Saranraj and Sivasakthi 2014). The use of *Spirulina* in poultry feed had positive impacts on growth of broiler chickens due to balanced amount of required amino acid while reduction in feed intake and weight loss in chickens had been observed due to high intake of *Spirulina* feed with unbalanced amino acid contents (Velten et al. 2018). Similarly, effects of *Spirulina* feed on meat quality of chicken had been investigated by using physio-chemical and sensory analysis. The results showed positive impacts on meat quality such as improvement in flavoring compounds, gamma linolenic acid and dark orange colored meat, while this feed reduced bioactive compounds and polyunsaturated fatty acid content in chicken (Martins et al. 2021). *Spirulina* can be used as soybean meal substitute in swine diet. In order to study the impact of *Spirulina* feed on swine and its meat quality the experimental group was provided feed with 9.5% ration and *Spirulina* while soybean meal was provided to control group. Physico-chemical and sensory analysis was used to check the fat and meat quality. The results showed an improvement in polyunsaturated fatty acid contents and reduction in mono-unsaturated fatty acids in experimental group than control group (Altmann and Rosenau 2022). In fish feed *Spirulina* is being used as fishmeal substitute. In another investigation carried out on brook trout, catfish, rainbow trout and brown trout to check the effects of *Spirulina* feed on growth and quality of fish product. Two types of balance diets were prepared. The animals of control groups were given fishmeal feed while the experimental groups animals were provided feed with *Spirulina*. The feed acceptance and conversion were good in all experimental groups except brown trout. Due to *Spirulina* feed, the growth rate, skin color of fishes, fish meat flavor was changed and reduction in long chain fatty acids in catfish had been observed (Rosenau et al. 2021, 2022). In another study *Spirulina* feed was provided to sheep, cattle, pig and rabbits. The *Spirulina* feed improved average daily gain, boar fertility in pigs while it increased feed intake, protein digestibility and reduced oxidative stress in rabbits. Similarly *Spirulina* feed had positive impacts on growth, milk production and meat quality of cattle and sheep (Holman and Malau-Aduli 2013).

Spirulina can be used as fishmeal, soybean meal, and dried skim milk substitute in animal feed. The high intake of *Spirulina* in livestock and fish feed with proper amino acid supplements is safe without adverse health effects while in chickens due to high intake of *Spirulina* feed weight loss and reduction in feed intake had been observed. Moreover, the use of *Spirulina* as fishmeal substitute is cost-effective and its implications are species dependent.

Table 6.5 Useful compounds in microalgae and their health benefits in animals

Sr. #	Botanical name/ Family	Feed substitute	Beneficial compounds	Health benefits	References
1	*Spirulina*/ Oscillatoriaceae	Fishmeal, soybean meal and dried skim milk substitute	Vitamins, proteins, fatty acids and minerals and phycocyanin	Rabbits: Increased feed intake, protein digestibility and reduced oxidative stress Pigs: Improvement in average daily gain and boar fertility Fish: Improvement in growth rate, change in skin color and meat flavor	Altmann and Rosenau (2022), Holman and Malau-Aduli (2013), Rosenau et al. (2022), Rosenau et al. (2021), and Saranraj and Sivasakthi (2014)
2	*Chlorella*/ Chlorellaceae	Used in mollusks, shrimp and fish feed	Carbohydrates, vitamins, proteins, polyunsaturated fatty acids, minerals, organic compounds and pigments	Anti-microbial, anti-tumor, immune-modulatory properties and positive impacts on animal growth and performance	Abdelnour et al. (2019), Akhmedkhanova et al. (2020), de Andrade and de Andrade (2017), and Janczyk et al. (2006)
3	*Dunaliella salina*/ Chlorophyceae	Feed supplement in chick and fish feed	Fatty acids, lipids, vitamins, proteins, minerals, carbohydrates, antioxidants, pigments and other bioactive compounds	Laying hens: Improvement in intestinal health, carotenoids production, liver metabolism and physico-chemical quality of eggs Chickens: Increase feed intake, body weight and feed conversion rate Fishes: Improvement in immunity, antioxidant enzyme activities, reduction in gene expression of endocrine hormone and protection from dioxin toxicity Shrimp: Increase in antioxidant properties, immunity and effective against white spot syndrome virus in aquaculture	Bassem et al. (2020), Fernandes et al. (2020), Madhumathi and Rengasamy (2011), Pratiwi (2020), and Sanderson (2018)

(continued)

Table 6.5 (continued)

Sr. #	Botanical name/ Family	Feed substitute	Beneficial compounds	Health benefits	References
4	*Hematococcus pluvialis/* Haematococcaceae	Dietary supplement in fish and chick feed	Carbohydrates, lutein, vitamins, astaxanthin, carotenoids, minerals, canthaxanthin, and bioactive compounds	Positive impacts on feed intake, growth, immune system and body weight in chick and fish	Awadh and Zangana (2021), Choubert and Heinrich (1993), and Ju et al. (2012)
5	*Nannochloropsis/* Monodopsidaceae	Feed supplement in chick and fish feed	Carbohydrates, proteins, lipids, amino acids, minerals and fatty acid contents	Laying hens: Positive impacts on egg pigmentation and omega 3 fatty acid contents in egg yolk Fish: Positive impacts on growth, weight gain, protein efficiency and muscular amino acid contents	Archibeque et al. (2009), Nitsan et al. (1999), and Qiao et al. (2019)
6	*Tetraselmis/* Chlorodendraceae	Feed supplement in fish and pig feed	Carbohydrates, proteins, lipids, fatty acids, carotenoids, phenolic compounds, vitamin E, terpenoid and flavonoids	Shrimps: High survival rate, growth rate, feed conversion ratio, lipid and protein content and reduction in antioxidant gene expression Fish: Improvement in body weight, nutrient digestibility and reduction in plasma cortisol response Pigs: Positive impacts on body weight, feed conversion rate and pigmentation	Chaves et al. (2021), Pereira et al. (2020), and Sharawy et al. (2020)

6.5.2 *Chlorella*

Chlorella is used in mollusks, shrimp, fish feed due to presence of a variety of nutrients such as carbohydrates, vitamins, proteins, polyunsaturated fatty acids, minerals, organic compounds and pigments (Abdelnour et al. 2019; de Andrade and de Andrade 2017). The positive impacts of *Chlorella* feed on animal growth and performance had been proved in literature (Akhmedkhanova et al. 2020; Janczyk et al. 2006; Kotrbáček et al. 2015). The use of *Chlorella* in piglets feed showed an increase in feed consumption and body weight. Similarly, due to ingestion of 0.5% and 0.3% *Chlorella* by chickens and ducks, an increase in phagocytic activity in chickens and reduced mortality rate in ducks had been observed. Anti-microbial, anti-tumor, anti-obesity, immune-modulatory and gamma irradiation resistance in rats had been reported due to consumption of C*hlorell*a extract. In egg laying hens and birds, 0.3–2% *Chlorella* feed improved carotenoid content of eggshell (Kotrbáček et al. 2015). A study was performed on female mice to study the effect of feed supplemented with 1.0% spray-dried *Chlorella vulgaris* powder on reproduction for three consecutive generations. They were randomly mated and gave birth to pups. At eighteenth day of pregnancy two female mice from control and experimental group were taken and slaughtered to observe prenatal fertility traits while remaining female mice gave birth to pups. The results showed there was no difference in fetuses and new born pups number in control and experimental group while pups of experimental group grew better and pre-weaning mortality rate was reduced (Janczyk et al. 2006). In lactating cow feed 1 or 1.5 liters/day of *C. vulgaris* was added that improved microelements, macro elements, protein, iodine and fat content of milk and milk processed products (Akhmedkhanova et al. 2020).

The use of *Chlorella* in animal feed is beneficial due to presence of a variety of nutrients such as carbohydrates, vitamins, proteins, polyunsaturated fatty acids, minerals, organic compounds and pigments that possessed anti-microbial, anti-tumor, anti-obesity, gamma irradiation resistance immunomodulatory and positively affected animal growth and performance.

6.5.3 *Dunaliella salina*

D. salina (Chlorophyceae) is added in fish, chick and animal feed for its fatty acids, lipids, vitamins, proteins, minerals, carbohydrates, antioxidants, pigments and other bioactive compounds (Fernandes et al. 2020; Pratiwi 2020). Six hundred laying hens were provided diets containing 0, 0.25, 0.50, 0.75 and 1% *D. salina* in completely randomized design. The diet containing *D. salina* improved intestinal health, carotenoids production and metabolism in liver and physico-chemical quality of eggs during 84 days (Fernandes et al. 2020). Similarly, another study concluded that diets containing 0.1% *D. salina* increased feed intake, body weight and feed conversion rate in chickens(Sanderson 2018).

The Nile tilapia feed was supplemented with *D. salina* extract for 30 days and was exposed to a pollutant known as dioxin. The *D. salina* diet enhanced immunity, antioxidant enzyme activities and reduced gene expression of endocrine hormone in Nile tilapia that protected fish against dioxin toxicity (Bassem et al. 2020). A study was conducted on *Penaeus monodon* shrimp to examine the impact of *D. salina* diet on white spot syndrome virus (WSSV). The feed pellets with different *D. salina* concentrations (0.5, 1 and 2%) were given to shrimp. The catalase, phenol-oxidase and superoxide dismutase assays showed that pellet feed supplemented with *D. salina* increased antioxidant properties, immunity in shrimp and was effective against WSSV (Madhumathi and Rengasamy 2011).

The use of *D. salina* is safe in animal feed due to presence of various nutrients such as fatty acids, lipids, vitamins, proteins, minerals, carbohydrates, antioxidants, pigments and other bioactive compounds that possessed antioxidant, immunomodulatory and anti-virus properties.

6.5.4 *Hematococcus pluvialis*

H. pluvialis contains carbohydrates, lutein, vitamins, astaxanthin, carotenoids, minerals, canthaxanthin, and bioactive compounds (Awadh and Zangana 2021; Choubert and Heinrich 1993). In shrimp diet defatted *H. pluvialis* meal was used as partial fishmeal substituent. This defatted meal was containing 40.3% and 0.9% of crude lipid and protein content. In diet 3%, 6%, 9%, and 12% defatted *H. pluvialis* meal were added and both (control and experimental) diets were containing similar protein and lipid contents. Diet was randomly provided to shrimps. The results revealed 12.5% defatted *H. pluvialis* meal can be used as fish meal substitute that increased growth and feed conversion rate (Ju et al. 2012). The astaxanthin powder of *H. pluvialis* improved immune system features of chick under both normal and high temperature conditions (Awadh and Zangana 2021).

The astaxanthin powder obtained from *H. pluvialis* was added in feed of pregnant mice to check its safety aspects. The mice of control group were provided olive oil with feed while the mice of experimental groups were given high oral dose of 20 g of astaxanthin. During 14 days of experiment body weight, morbidity and death rate was recorded in mice. Bone marrow micronucleus test was performed on mice and results revealed that 10 g of astaxanthin had no effect on mitotic apparatus and chromosomes in pregnant mice. In the repeat-dose toxicity test, various concentration of astaxanthin (100, 250, and 500 mg) were provided to mice, and results showed that use of astaxanthin in pregnant mice was safe because it had no negative effects on feed intake, hematology, body weight and behavior of mice (Niu et al. 2020). Astaxanthin obtained from *H. pluvialis* is used in fish and chicken feed as dietary supplement, and its positive impacts on feed intake, growth, immune system, and body weight had been observed. While, there is a need to conduct toxicological studies on astaxanthin.

6.5.5 *Nannochloropsis*

Nannochloropsis contains excellent dietary components in the form of carbohydrates, proteins, lipids, amino acids, minerals and fatty acid contents (Fig. 6.1) and added as supplement in the feed (Archibeque et al. 2009). Digestible proteins are components of commercial feed and test diets. *N. granulata* meal was for rainbow trout and white shrimp to study the digestible protein content. The average protein hydrolysis degree and protein apparent digestibility coefficients (2.09% and 74.6%) were almost same for all meals for white shrimp. Supercritical fluid extraction was used for processing of *N. granulata* meal at two different temperatures (70 and 90 °C). In processed *N. granulata* meal average protein hydrolysis degree and protein apparent digestibility coefficient were high than the unprocessed meal for rainbow trout. All *N. granulata* meal had moderate and high protein apparent digestibility coefficient for white shrimp and rainbow trout that indicated *Nannochloropsis* potential to be used in both shrimp and fish feed (Tibbetts et al. 2017). *Nannochloropsis* meal was used as fishmeal substitute in juvenile turbot feed. The fishes of control group were provided feed containing fishmeal while the experimental group fishes were provided feed with 2.5%, 5%, 7.5%, and 10% of *Nannochloropsis* sp. meal for 10 weeks. In experimental group, amino acid content in muscle, activities of pepsin, trypsin, lipase, superoxide dismutase and antioxidant properties were high while malondialdehyde content in liver was low. It represented 10% of *Nannochloropsis* sp. meal had positive impacts on growth, weight gain, protein efficiency, and muscular amino acid contents in fish (Qiao et al. 2019). Similarly, supplementation of *Nannochloropsis* in laying hens diet was useful due to its carotenoid and fatty acid contents that improved egg pigmentation and omega 3 fatty acid contents in egg yolk (Nitsan et al. 1999).

Nannochloropsis meal was used in rat feed in order to study its nutritive and toxic effects. The rats of control group were supplemented feed without *Nannochloropsis* meal while rats of experimental group were supplemented feed with 5% or 10% *Nannochloropsis* meal for 14, 21 and 28 days. The use of *Nannochloropsis* meal was safe because its ingestion had no negative effects on growth, hematology, and body weight while in experimental group rats the level of high density lipid and total lipid content was increased (Andrés et al. 1992). *Nannochloropsis* is excellent feed for white shrimp, rainbow trout, juvenile turbot, rat and chicks due to its dietary components. It is not toxic and improved animal growth, body weight, hematological parameters, enzyme activities, and amino acid content. Further investigations should be encouraged to study the effects of *Nannochloropsis* meal in other animals.

6.5.6 *Tetraselmis*

Tetraselmis suecica is green algal specie found in Mediterranean Sea, Egypt, Korea and Eastern harbor of Alexandria. *Tetraselmis* sp. are used in fish and poultry feed. Carbohydrates, proteins, lipids, fatty acids, carotenoids, phenolic compounds, vitamin E, terpenoid and flavonoids are nutrients in *T. suecica* thallus. Previously study was conducted on white shrimp (*Litopenaeus vannamei*) to study the influence of dietary supplementation of *T. suecica* on growth, chemical composition, feed consumption, superoxide dismutase, glutathione peroxidase and insulin-like growth factor-2 gene. Four tested meals were provided to shrimps three times in a day. The shrimp of control group were provided feed with basal diet while the experimental group shrimp were provided feed containing different levels (2.5, 5 and 7.5 g) of dried *T. suecica*. After 90 days of the trial, 7.5 g/kg of *T. suecica* feed exhibited high survival rate, growth rate, feed conversion ratio, lipid and protein content than the control group while lowered the antioxidant gene expression in shrimp (Sharawy et al. 2020).

In 61 days of trial, fish feed supplemented with 10% defatted *Tetraselmis* sp. increased body weight and nutrient digestibility in fishes compared to control diet with high protein content and 10% soybean meal. Furthermore acute confinement stress test showed fish feed supplemented with 10% defatted *Tetraselmis* sp. lowered plasma cortisol response in fish (Pereira et al. 2020). The addition of *Tetraselmis* sp. in piglets feed also increased body weight, feed conversion rate and pigmentation (Chaves et al. 2021).

Various studies revealed that *Tetraselmis* sp. addition in animal feed improved animal growth, body weight, nutrient digestibility, pigmentation and feed conversion rate. It also reduced the stress hormone in fish captivity. In shrimp it reduced the antioxidant gene expression. Further investigations on the toxicity, antioxidant activity, and other physiological parameters are suggested.

6.6 Conclusion

Algae have emerged as a sustainable and multifunctional resource capable of addressing global challenges in food and feed security. This chapter reviewed the published data regarding the use of algal species in human food, fish, poultry and livestock feed, along with their health benefits and safety aspects. Algal cells are recognized as a great source of nutrients such as carbohydrates, proteins, fatty acids, carotenoids, minerals, phycobiliproteins, vitamins, pigments and other bioactive compounds. The use of algal species in human food and animal feed is considered beneficial due to its nutritional value, rapid growth as crops and protein meal alternative. Algae are incorporated into food and feed to develop new healthy products, provide color, enhance flavor and improve the nutritional value and texture. Antioxidant, anti-atherosclerotic, anti-cancer, anti-coagulant, immune-modulatory, cardio-tonic, anti-obesity, anti-microbial, anti-hypertensive, anti-diabetic properties and improved gastrointestinal functioning have been attributed to algae in humans. Their rich nutritional composition, rapid biomass productivity, and adaptability to various environments make them a viable supplement or alternative to conventional resources. However, certain species contain bioactive or toxic compounds, such as heavy metal accumulators or arsenic-containing polysaccharides, which can pose health risks if not properly evaluated and processed. The absence of comprehensive safety data on many algal species,

Table 6.6 Toxic compounds in seaweeds and their health effects in animals

Sr. #	Botanical name/ family	Common name	Toxic compound	Effects	Reference
1	*Porphyra*/ Bangiaceae	Nori	Polychlorinated biphenyls	Liver necrosis, edema formation and spleen atrophy in chicks	Vos and Koeman (1970)

particularly microalgae, highlights the need for rigorous assessment protocols. Developing standardized methodologies for toxicity testing, quality assurance, and post-harvest handling is essential to safeguard public and animal health. Ensuring safety compliance will enhance the acceptance of algal products in global food and feed markets, thereby promoting sustainable bioeconomy initiatives. However, toxicity resulting from the use of microalgae in food had not been reported. Further studies on the toxicity of microalgae are recommended.

The incorporation of algae in feed has been found to improve growth rate, protein and lipid content, meat quality and quantity, egg production, egg shell thickness, weight and yolk color. Similarly, improvement in immune functioning, dry matter digestibility, microbial growth, feed conversion rate, volatile fatty acid content and antioxidant properties of ruminants have been observed due to addition of algae in feed. Positive impacts on body weight, nutrient digestibility and reduction in plasma cortisol response have been noted with the use of algae in fish feed. A single report on toxic effects of seaweed (*Porphyra*) in animal feed has been identified, while data on toxicity of microalgae in animal feed is currently unavailable (Table 6.6). Further research should be conducted to study the safety aspects of using microalgae and seaweeds in feed. Seaweeds like *Porphyra* and *Ulva* present a promising alternative to traditional feed ingredients, their current cost of production and regional availability pose challenges to replacing soybean meal on a large scale. However, if production costs decrease through technological advancements and government incentives, the environmental and potential health benefits make seaweed-based feed additives an appealing, sustainable option in the long term.Seaweed, or macroalgae, has been used as a staple food in Asian cuisine for decades and is increasingly being adopted as a food source in Europe and other Western countries. However, elements from the environment can be absorbed by seaweed, potentially resulting in exposure to toxic substances or excessive amounts of certain micronutrients. Given the rising demand for algal products, ensuring food safety in their use for human consumption and animal feed is considered crucial.

References

Abdelnour S, Abd El-Hack M, Arif M, Khafaga A, Taha A (2019) The application of the microalgae Chlorella spp. as a supplement in broiler feed. Worlds Poult Sci J 75(2):305–318

Abudabos AM, Okab AB, Aljumaah RS, Samara EM, Abdoun KA, Al-Haidary AA (2013) Nutritional value of green seaweed (Ulva lactuca) for broiler chickens. Ital J Anim Sci 12(2):e28

Admassu H, Gasmalla MAA, Yang R, Zhao W (2018) Bioactive peptides derived from seaweed protein and their health benefits: antihypertensive, antioxidant, and antidiabetic properties. J Food Sci 83(1):6–16

Aguilera-Morales M, Casas-Valdez M, Carrillo-Domınguez S, González-Acosta B, Pérez-Gil F (2005) Chemical composition and microbiological assays of marine algae Enteromorpha spp. as a potential food source. J Food Compos Anal 18(1):79–88

Ahmad MT, Shariff M, Md. Yusoff F, Goh YM, Banerjee S (2020) Applications of microalga Chlorella vulgaris in aquaculture. Rev Aquac 12(1):328–346

Akhmedkhanova R, Dzhambulatov Z, Gadzhaeva Z, Shabanov G, Alieva S (2020) *The influence of chlorella suspension on the quality of milk and its processing products*. Paper presented at the E3S web of conferences

Ali A, Abid R (2006) Cladophora glomerata (L.) Kutzing as feed supplement to broiler chicks. Int J Biol Biotechnol 3:425–428

Ali SK, Saleh AM (2012) Spirulina-an overview. Int J Pharm Pharm Sci 4(3):9–15

Al-Khalaifah H, Uddin S (2022) Assessment of sargassum sp., spirulina sp., and gracilaria sp. as poultry feed supplements: feasibility and environmental implications. Sustainability 14(14): 8968

Altmann BA, Rosenau S (2022) Spirulina as animal feed: opportunities and challenges. Foods 11(7):965

Andrés M, Raúl C, Luis L, Mariane L (1992) Evaluation of marine microalga Nannochloropsis sp. as a potential dietary supplement. Chemical, nutritional and short term toxicological evaluation in rats. Nutr Res 12(10):1273–1284

Ang WS, Kee PE, Lan JC-W, Chen WH, Chang J-S, Khoo KSJFB (2024) Unveiling the rise of microalgae-based foods in the global market: perspective views and way forward. Food Bioscience 66:105390

Appler H, Jauncey K (1983) The utilization of a filamentous green alga (Cladophora glomerata (L) Kutzin) as a protein source in pelleted feeds for Sarotherodon (Tilapia) niloticus fingerlings. Aquaculture 30(1–4):21–30

Archibeque S, Ettinger A, Willson B (2009) Nannochloropsis oculata as a source for animal feed. Acta Agronomica Hungarica 57(2):245–248

Asino H, Ai Q, Mai K (2011) Evaluation of Enteromorpha prolifera as a feed component in large yellow croaker (Pseudosciaena crocea, Richardson, 1846) diets. Aquac Res 42(4):525–533

Awadh IA, Zangana BS (2021) Effect of adding different levels of Astaxanthin extracted from an algae Haematococcus Pluvialis to the diet on some immunological characteristics of broilers reared under natural and elevated environmental conditions'. Paper presented at the IOP Conference Series: Earth and Environmental Science, 910

Banach J, Hoek-van den Hil E, van der Fels-Klerx H (2020) Food safety hazards in the European seaweed chain. Compr Rev Food Sci Food Saf 19(2):332–364

Bangmei X, Abbott IA (1987) Edible seaweeds of China and their place in the Chinese diet. Econ Bot 41(3):341–353

Bassem S, Abd El Tawab MI, Temraz T, Bassaly WK, El-Baz FK, Ali G, Abdel Gawad FK (2020) Dunaliella salina extract alleviates the toxic impact of dioxin induced endocrine disruption in Nile tilapia. Egypt J Chem 63(5):1787–1798

Becker EW (2007) Micro-algae as a source of protein. Biotechnol Adv 25(2):207–210

Belay A, Ota Y, Miyakawa K, Shimamatsu H (1993a) Current knowledge on potential health benefits of spirulina. J Appl Phycol 5:235–241

Belay A, Ota Y, Miyakawa K, Shimamatsu H (1993b) Current knowledge on potential health benefits of Spirulina. J Appl Phycol 5(2):235–241

Bito T, Teng F, Watanabe F (2017) Bioactive compounds of edible purple laver Porphyra sp. (Nori). J Agric Food Chem 65(49):10685–10692

Bleakley S, Hayes M (2017) Algal proteins: extraction, application, and challenges concerning production. Foods 6(5):33

Bordean D-M, Borozan AB, Cojocariu L, Horablaga A, Bostan C (2016) Ulva LACTUCA consumption: a "healthy risk"? Int Multidiscip Sci GeoConf 2:909–914

Borowitzka MA (1998) Algae as food. In: Microbiology of fermented foods. Springer, pp 585–602

Brown EM, Allsopp PJ, Magee PJ, Gill CI, Nitecki S, Strain CR, McSorley EM (2014) Seaweed and human health. Nutr Rev 72(3):205–216

Catarino MD, Silva AM, Cardoso SM (2018) Phycochemical constituents and biological activities of Fucus spp. Mar Drugs 16(8):249

Chapman V, Chapman D, Chapman V, Chapman D (1980) Algin and alginates. In: Seaweeds and their uses. Chapman and Hall, pp 194–225

Chaves AA, Martins CF, Carvalho DF, Ribeiro DM, Lordelo M, Freire JP, de Almeida AM (2021) A viewpoint on the use of microalgae as an alternative feedstuff in the context of pig and poultry feeding—a special emphasis on tropical regions. Trop Anim Health Prod 53:1–8

Chen Y, Liu Y-t, Wang F-h, Wen D, Yang H, Zhao X-l (2021) An investigation of toxic metal levels (Pb, Cd, Cr, As, Hg) in dried porphyra and laminaria collected from coastal cities, China. Biol Trace Elem Res 199(10):3987–3997

Cheney D (2016) Toxic and harmful seaweeds. In: Seaweed in health and disease prevention. Elsevier, pp 407–421

Cheng T-H (1969) Production of kelp-a major aspect of China's exploitation of the sea. Econ Bot 23(3):215–236

Chiu Y-H, Chan Y-L, Li T-L, Wu C-J (2012) Inhibition of Japanese encephalitis virus infection by the sulfated polysaccharide extracts from Ulva lactuca. Mar Biotechnol 14:468–478

Choi YY, Lee SJ, Lee YJ, Kim HS, Eom JS, Jo SU, Lee SS (2020) In vitro and in situ evaluation of Undaria pinnatifida as a feed ingredient for ruminants. J Appl Phycol 32(1):729–739

Choque-Quispe D, Mojo-Quisani A, Ligarda-Samanez CA, Calla-Florez M, Ramos-Pacheco BS, Zamalloa-Puma LM, Peralta-Guevara DE, Solano-Reynoso AM, Choque-Quispe Y, Zamalloa-Puma A (2022) Preliminary characterization of a spray-dried hydrocolloid from a high Andean algae (Nostoc sphaericum). Foods 11(11):1640

Choubert G, Heinrich O (1993) Carotenoid pigments of the green alga Haematococcus pluvialis: assay on rainbow trout, Oncorhynchus mykiss, pigmentation in comparison with synthetic astaxanthin and canthaxanthin. Aquaculture 112(2–3):217–226

Christiansen ER, Bruhn A, Sloth JJ (2017) Toxic metals in European Ulva spp.–evaluation of potential use in food and feed applications

Cofrades S, Benedí J, Garcimartin A, Sánchez-Muniz F, Jimenez-Colmenero F (2017) A comprehensive approach to formulation of seaweed-enriched meat products: from technological development to assessment of healthy properties. Food Res Int 99:1084–1094

Costa M, Cardoso C, Afonso C, Bandarra NM, Prates JA (2021) Current knowledge and future perspectives of the use of seaweeds for livestock production and meat quality: a systematic review. J Anim Physiol Anim Nutr 105(6):1075–1102

Costa MM, Pestana JM, Carvalho P, Alfaia CM, Martins CF, Carvalho D, Mourato M, Gueifão S, Delgado I, Coelho I (2022) Effect on broiler production performance and meat quality of feeding Ulva lactuca supplemented with carbohydrases. Animals 12(13):1720

Coudert E, Baéza E, Berri C (2020) Use of algae in poultry production: a review. Worlds Poult Sci J 76(4):767–786

D'Amore T, Akermi S, Amara FB, Smaoui S (2026) Toxicity and safety analysis of alternative and novel food sources. In: Health, nutrition and sustainability. Elsevier, pp 165–183

de Andrade CJ, de Andrade LM (2017) An overview on the application of genus Chlorella in biotechnological processes. J Adv Res Biotechnol 2:1–9

de Gaillande C, Payri C, Remoissenet G, Zubia M (2017) Caulerpa consumption, nutritional value and farming in the Indo-Pacific region. J Appl Phycol 29:2249–2266

Desideri D, Cantaluppi C, Ceccotto F, Meli MA, Roselli C, Feduzi L (2016) Essential and toxic elements in seaweeds for human consumption. J Toxic Environ Health A 79(3):112–122

Devault DA, Pierre R, Marfaing H, Dolique F, Lopez P-J (2021) Sargassum contamination and consequences for downstream uses: a review. J Appl Phycol 33:567–602

Elizondo-González R, Quiroz-Guzmán E, Escobedo-Fregoso C, Magallón-Servín P, Peña-Rodríguez A (2018) Use of seaweed Ulva lactuca for water bioremediation and as feed additive for white shrimp Litopenaeus vannamei. PeerJ 6:e4459

Ergün S, Soyutürk M, Güroy B, Güroy D, Merrifield D (2009) Influence of Ulva meal on growth, feed utilization, and body composition of juvenile Nile tilapia (Oreochromis niloticus) at two levels of dietary lipid. Aquac Int 17:355–361

Esmaeili N, Razavi SH, Amini Valashani A, Barzegar Marvasti F, Salehian R, Akbari NJSM, Biomanufacturing. (2026) How algae is shaping next generation food products. Syst Microbiol Biomanuf 6(1):15

Euler A, Ferreira W, Maurício R, Sousa L, Carvalho W, Teixeira E, Coelho C, Matos C (2008) In vitro gas production of diets with inclusion of seaweed (Lithothamnium sp.) flour for white New Zealand rabbits. Paper presented at the proceedings of the 9th world rabbit congress, Verona, Italy

FAO (2022) The state of food security and nutrition in the world. Food and Agriculture Organization of the United Nations. https://doi.org/10.4060/cc0639en

Fernandes RTV, Gonçalves AA, Arruda AMV d (2020) Production, egg quality, and intestinal morphometry of laying hens fed marine microalga. Rev Bras Zootec 49

Filippini M, Baldisserotto A, Menotta S, Fedrizzi G, Rubini S, Gigliotti D, Valpiani G, Buzzi R, Manfredini S, Vertuani SJC (2021) Heavy metals and potential risks in edible seaweed on the market in Italy. Chemosphere 263:127983

Fleurence J (2016) Seaweeds as food. In: Seaweed in health and disease prevention. Academic, pp 149–167

Gantar M, Svirčev Z (2008) Microalgae and cyanobacteria: food for thought 1. J Phycol 44(2): 260–268

García-Rodríguez D, Cela-Torrijos R, Lorenzo-Ferreira R, Carro-Díaz A (2012) Analysis of pesticide residues in seaweeds using matrix solid-phase dispersion and gas chromatography–mass spectrometry detection. Food Chem 135(1):259–267

Guerin M, Huntley ME, Olaizola M (2003) Haematococcus astaxanthin: applications for human health and nutrition. Trends Biotechnol 21(5):210–216

Guo Y, Lundebye A-K, Li N, Ergon Å, Pang S, Jiang Y, Zhu W, Zhao Y, Li X, Yao L, J. T. i. F. S., Technology (2023) Comparative assessment of food safety regulations and standards for arsenic, cadmium, lead, mercury and iodine in macroalgae used as food and feed in China and Europe. Trends Food Sci Technol 141:104204

Gutow L, Eckerlebe A, Giménez L, Saborowski R (2016) Experimental evaluation of seaweeds as a vector for microplastics into marine food webs. Environ Sci Technol 50(2):915–923

Hammann M, Rempt M, Pohnert G, Wang G, Boo SM, Weinberger F (2016) Increased potential for wound activated production of prostaglandin E2 and related toxic compounds in non-native populations of Gracilaria vermiculophylla. Harmful Algae 51:81–88

Hidayah N, Maulina N, Noviandi C, Astuti A, Dono N (2022) Chemical composition of brown and red algae from Kelapa Beach, Tuban, East Java and their potential as ruminant feed. Paper presented at the IOP Conference Series: Earth and Environmental Science

Holman B, Malau-Aduli A (2013) Spirulina as a livestock supplement and animal feed. J Anim Physiol Anim Nutr 97(4):615–623

Hussein G, Sankawa U, Goto H, Matsumoto K, Watanabe H (2006) Astaxanthin, a carotenoid with potential in human health and nutrition. J Nat Prod 69(3):443–449

Hyrslova I, Krausova G, Mrvikova I, Stankova B, Branyik T, Malinska H, Huttl M, Kana A, Doskocil I (2022) Functional properties of Dunaliella salina and its positive effect on probiotics. Mar Drugs 20(12):781

James CA, Welham S, Rose P (2022) Edible algae allergenicity–a short report. J Appl Phycol 35:1–14

Janczyk P, Langhammer M, Renne U, Guiard V, Souffrant W (2006) Effect of feed supplementation with Chlorella vulgaris powder on mice reproduction. Archiva Zootechnica 9:122–134

Jeon GH, Cho SH, Lee SM, Nam TJ, Kim DS (2013) Effects of the dietary inclusion of Porphyra and sea tangle Laminaria japonica on the growth, feed utilization, body composition, and plasma

chemistry of juvenile Korean rockfish Sebastes schlegeli. Korean J Fish Aquat Sci 46(5): 546–551
Ju ZY, Deng D-F, Dominy W (2012) A defatted microalgae (Haematococcus pluvialis) meal as a protein ingredient to partially replace fishmeal in diets of Pacific white shrimp (Litopenaeus vannamei, Boone, 1931). Aquaculture 354:50–55
Kannan G, Terrill T, Kouakou B, Galipalli S (2007) Blood metabolite changes and live weight loss following brown seaweed extract supplementation in goats subjected to stress. Small Rumin Res 73(1–3):228–234
Klomjit A, Sutthacheep M, Yeemin T (2021) Occurrence of microplastics in edible seaweeds from aquaculture. Ramkhamhaeng Int J Sci Technol 4(2):38–44
Kotrbáček V, Doubek J, Doucha J (2015) The chlorococcalean alga Chlorella in animal nutrition: a review. J Appl Phycol 27:2173–2180
Kumar MS, Sharma SA (2021) Toxicological effects of marine seaweeds: a cautious insight for human consumption. Crit Rev Food Sci Nutr 61(3):500–521
Lafarga T, Fernández-Sevilla JM, González-López C, Acién-Fernández FG (2020) Spirulina for the food and functional food industries. Food Res Int 137:109356
Lavakumar V, Ahamed K, Ravichandran V (2012) Anticancer and antioxidant effect of Acanthophora spicifera against EAC induced carcinoma in mice. J Pharm Res 5(3):1503–1507
Lemée R, Pesando D, Durand-Clément M, Dubreuil A, Meinesz A, Guerriero A, Pietra F (1993) Preliminary survey of toxicity of the green alga Caulerpa taxifolia introduced into the Mediterranean. J Appl Phycol 5:485–493
Li J-Y, Yang F, Jin L, Wang Q, Yin J, He P, Chen Y (2018) Safety and quality of the green tide algal species Ulva prolifera for option of human consumption: a nutrition and contamination study. Chemosphere 210:1021–1028
Li Y, He P, Shan Y et al (2024) Reducing climate change impacts from the global food system through diet shifts. Nat Clim Chang 14:943–953
Lind V, Weisbjerg MR, Jørgensen GM, Fernandez-Yepes JE, Arbesú L, Molina-Alcaide E (2020) Ruminal fermentation, growth rate and methane production in sheep fed diets including white clover, Soybean Meal or Porphyra sp. Animals 10(1):79
Liu J, Luthuli S, Yang Y, Cheng Y, Zhang Y, Wu M, Choi J i, Tong H (2020) Therapeutic and nutraceutical potentials of a brown seaweed Sargassum fusiforme. Food Sci Nutr 8(10): 5195–5205
Lopes D, Melo T, Meneses J, Abreu MH, Pereira R, Domingues P, Lillebø AI, Calado R, Domingues MR (2019) A new look for the red macroalga Palmaria palmata: a seafood with polar lipids rich in EPA and with antioxidant properties. Mar Drugs 17(9):533
Madhumathi M, Rengasamy R (2011) Antioxidant status of Penaeus monodon fed with Dunaliella salina supplemented diet and resistance against WSSV. Int J Eng Sci Technol 3(10):7249–7259
Mahadevan K (2015) Seaweeds: a sustainable food source. In: Seaweed sustainability. Elsevier, pp 347–364
Makkar HP, Tran G, Heuzé V, Giger-Reverdin S, Lessire M, Lebas F, Ankers P (2016) Seaweeds for livestock diets: a review. Anim Feed Sci Technol 212:1–17
Martins CF, Ribeiro DM, Costa M, Coelho D, Alfaia CM, Lordelo M, Almeida AM, Freire JP, Prates JA (2021) Using microalgae as a sustainable feed resource to enhance quality and nutritional value of pork and poultry meat. Foods 10(12):2933
Matufi F, Choopani A (2020) Spirulina, food of past, present and future. Health Biotechnol Biopharma 3(4):1–20
Mellor C, Embling R, Neilson L, Randall T, Wakeham C, Lee MD, Wilkinson LL (2022) Consumer knowledge and acceptance of "algae" as a protein alternative: a UK-based qualitative study. Foods 11(12):1703
Mendes MC, Navalho S, Ferreira A, Paulino C, Figueiredo D, Silva D, Gao F, Gama F, Bombo G, Jacinto R (2022) Algae as food in Europe: an overview of species diversity and their application. Foods 11(13):1871

Michalak I, Messyasz B (2021) Concise review of Cladophora spp.: macroalgae of commercial interest. J Appl Phycol 33(1):133–166

Molino A, Iovine A, Casella P, Mehariya S, Chianese S, Cerbone A, Rimauro J, Musmarra D (2018) Microalgae characterization for consolidated and new application in human food, animal feed and nutraceuticals. Int J Environ Res Public Health 15(11):2436

Mouritsen OG, Dawczynski C, Duelund L, Jahreis G, Vetter W, Schröder M (2013) On the human consumption of the red seaweed dulse (Palmaria palmata (L.) Weber & Mohr). J Appl Phycol 25:1777–1791

Munir M, Qureshi R, Bibi M, Khan AM (2019) Pharmaceutical aptitude of Cladophora: a comprehensive review. Algal Res 39:101476

Murthy KC, Vanitha A, Rajesha J, Swamy MM, Sowmya P, Ravishankar GA (2005) In vivo antioxidant activity of carotenoids from Dunaliella salina—a green microalga. Life Sci 76(12): 1381–1390

Nadeeshani H, Hassouna A, Lu J (2022) Proteins extracted from seaweed Undaria pinnatifida and their potential uses as foods and nutraceuticals. Crit Rev Food Sci Nutr 62(22):6187–6203

Nakajima Y, Endo Y, Inoue Y, Yamanaka K, Kato K, Wanibuchi H, Endo G (2006) Ingestion of Hijiki seaweed and risk of arsenic poisoning. Appl Organomet Chem 20(9):557–564

Navarro F, Forján E, Vázquez M, Montero Z, Bermejo E, Castaño MÁ, Toimil A, Chagüaceda E, García-Sevillano MÁ, Sánchez M (2016) Microalgae as a safe food source for animals: nutritional characteristics of the acidophilic microalga Coccomyxa onubensis. Food Nutr Res 60(1):30472

Neville E, Fahey A, Meade K, Mulligan F (2022) Effects of calcareous marine algae on milk production, feed intake, energy balance, mineral status, and inflammatory markers in transition dairy cows. J Dairy Sci 105(8):6616–6627

Nitsan Z, Mokady S, Sukenik A (1999) Enrichment of poultry products with ω3 fatty acids by dietary supplementation with the alga Nannochloropsis and mantur oil. J Agric Food Chem 47(12):5127–5132

Niu J, Chen X, Lu X, Jiang S-G, Lin H-Z, Liu Y-J, Huang Z, Wang J, Wang Y, Tian L-X (2015) Effects of different levels of dietary wakame (Undaria pinnatifida) on growth, immunity and intestinal structure of juvenile Penaeus monodon. Aquaculture 435:78–85

Niu T, Zhou J, Wang F, Xuan R, Chen J, Wu W, Chen H (2020) Safety assessment of astaxanthin from Haematococcus pluvialis: acute toxicity, genotoxicity, distribution and repeat-dose toxicity studies in gestation mice. Regul Toxicol Pharmacol 115:104695

Noguchi T, Matsui T, Miyazawa K, Asakawa M, Iijima N, Shida Y, Fuse M, Hosaka Y, Kirigaya C, Watabe K (1994) Poisoning by the red alga ‘ogonori’(Gracilaria verrucosa) on the Nojima Coast, Yokohama, Kanagawa Prefecture, Japan. Toxicon 32(12):1533–1538

Obluchinskaya ED, Pozharitskaya ON, Zakharov DV, Flisyuk EV, Terninko II, Generalova YE, Smekhova IE, Shikov AN (2022) The biochemical composition and antioxidant properties of Fucus vesiculosus from the Arctic region. Mar Drugs 20(3):193

Olivarría DF, Castillo LRM, López ÁM, Elías JAL (2010) Crecimiento y biomasa de Dunaliella sp. cultivada en medios limitantes en nitrógeno. Biotecnia 12(3):58–66

Orantes Calleja PD (2022) Optimización del cultivo de microalgas en agua residual sintética para la producción de biodiésel

Pereira H, Sardinha M, Santos T, Gouveia L, Barreira L, Dias J, Varela J (2020) Incorporation of defatted microalgal biomass (Tetraselmis sp. CTP4) at the expense of soybean meal as a feed ingredient for juvenile gilthead seabream (Sparus aurata). Algal Res 47:101869

Pérez AA, Farías SS, Strobl AM, Pérez LB, López CM, Piñeiro A, Roses O, Fajardo MA (2007) Levels of essential and toxic elements in Porphyra columbina and Ulva sp. from San Jorge Gulf, Patagonia Argentina. Sci Total Environ 376(1–3):51–59

Pozharitskaya ON, Shikov AN, Faustova NM, Obluchinskaya ED, Kosman VM, Vuorela H, Makarov VG (2018) Pharmacokinetic and tissue distribution of fucoidan from Fucus vesiculosus after oral administration to rats. Mar Drugs 16(4):132

Pratiwi DY (2020) A mini review-effect of Dunaliella salina on growth and health of shrimps. Int J Fish Aquat Stud 8:317–319

Qiao H, Hu D, Ma J, Wang X, Wu H, Wang J (2019) Feeding effects of the microalga Nannochloropsis sp. on juvenile turbot (Scophthalmus maximus L.). Algal Res 41:101540

Rao P, Periyasamy C, Kumar KS, Rao AS, Anantharaman P (2018) Seaweeds: distribution, production and uses. In: Bioprospecting of algae. Society for Plant Research, pp 59–78

Reboleira J, Freitas R, Pinteus S, Silva J, Alves C, Pedrosa R, Bernardino S (2019) Spirulina. In: Nonvitamin and nonmineral nutritional supplements. Elsevier, pp 409–413

Rosemary T, Arulkumar A, Paramasivam S, Mondragon-Portocarrero A, Miranda JM (2019) Biochemical, micronutrient and physicochemical properties of the dried red seaweeds Gracilaria edulis and Gracilaria corticata. Molecules 24(12):2225

Rosenau S, Oertel E, Dietz C, Wessels S, Tetens J, Mörlein D, Ciulu M (2021) Total replacement of fishmeal by spirulina (Arthrospira platensis) and its effect on growth performance and product quality of African catfish (Clarias gariepinus). Sustainability 13(16):8726

Rosenau S, Ciulu M, Reimer C, Mott AC, Tetens J, Mörlein D (2022) Feeding green: spirulina (Arthrospira platensis) induced changes in production performance and quality of salmonid species. Aquac Res 53(12):4276–4287

Sakthivel R, Devi KP (2015) Evaluation of physicochemical properties, proximate and nutritional composition of Gracilaria edulis collected from Palk Bay. Food Chem 174:68–74

Sanderson PJ (2018) Evaluation of biomass from the halotolerant microalga Dunaliella salina as an animal feed additive with immune-modulatory activity. University of Greenwich

Saranraj P, Sivasakthi S (2014) Spirulina platensis–food for future: a review. Asian J Pharm Sci Technol 4(1):26–33

Sharawy ZZ, Ashour M, Abbas E, Ashry O, Helal M, Nazmi H, Kelany M, Kamel A, Hassaan M, Rossi W Jr (2020) Effects of dietary marine microalgae, Tetraselmis suecica, on production, gene expression, protein markers and bacterial count of Pacific white shrimp Litopenaeus vannamei. Aquac Res 51(6):2216–2228

Silva D, Valente L, Sousa-Pinto I, Pereira R, Pires M, Seixas F, Rema P (2015) Evaluation of IMTA-produced seaweeds (Gracilaria, Porphyra, and Ulva) as dietary ingredients in Nile tilapia, Oreochromis niloticus L., juveniles. Effects on growth performance and gut histology. J Appl Phycol 27:1671–1680

Silva J, Alves C, Pinteus S, Reboleira J, Pedrosa R, Bernardino S (2019) Chlorella. In: Nonvitamin and nonmineral nutritional supplements. Elsevier, pp 187–193

Sousa AM, Sereno AM, Hilliou L, Gonçalves MP (2010) Biodegradable agar extracted from Gracilaria vermiculophylla: film properties and application to edible coating. Paper presented at the Materials science forum

Szabo NJ, Matulka RA, Chan T (2013) Safety evaluation of whole algalin protein (wap) from Chlorella protothecoides. Food Chem Toxicol 59:34–45

Taboada M, Millán R, Miguez M (2013) Nutritional value of the marine algae wakame (Undaria pinnatifida) and nori (Porphyra purpurea) as food supplements. J Appl Phycol 25:1271–1276

Tanna B, Mishra A (2019) Nutraceutical potential of seaweed polysaccharides: structure, bioactivity, safety, and toxicity. Compr Rev Food Sci Food Saf 18(3):817–831

Tapotubun AM, Matrutty TE, Riry J, Tapotubun EJ, Fransina EG, Mailoa MN, Riry WA, Setha B, Rieuwpassa F (2020) Seaweed Caulerpa sp position as functional food. Paper presented at the IOP Conference Series: Earth and Environmental Science

Tibbetts SM, Yasumaru F, Lemos D (2017) In vitro prediction of digestible protein content of marine microalgae (Nannochloropsis granulata) meals for Pacific white shrimp (Litopenaeus vannamei) and rainbow trout (Oncorhynchus mykiss). Algal Res 21:76–80

Vadher KK, Vyas A, Parmar P (2016) Effects of seaweed, Gracilaria sp. as feed additive in diets on growth and survival of Labeo rohita. J Exp Zool India 19:771–778

Velten S, Neumann C, Bleyer M, Gruber-Dujardin E, Hanuszewska M, Przybylska-Gornowicz B, Liebert F (2018) Effects of 50 percent substitution of soybean meal by alternative proteins from Hermetia illucens or Spirulina platensis in meat-type chicken diets with graded amino acid supply. Open J Anim Sci 8(02):119

Vos J, Koeman J (1970) Comparative toxicologic study with polychlorinated biphenyls in chickens with special reference to porphyria, edema formation, liver necrosis, and tissue residues. Toxicol Appl Pharmacol 17(3):656–668

Wang S, Jia Y, Wang L, Zhu F, Lin Y (2013a) Enteromorpha prolifera supplemental level: effects on laying performance, egg quality, immune function and microflora in feces of laying hens. Chin J Anim Nutr 25(6):1346–1352

Wang S, Shi X, Zhou C, Lin Y (2013b) Entermorpha prolifera: effects on performance, carcass quality and small intestinal digestive enzyme activities of broilers. Chin J Anim Nutr 25(6): 1332–1337

Wang L, Park Y-J, Jeon Y-J, Ryu B (2018) Bioactivities of the edible brown seaweed, Undaria pinnatifida: a review. Aquaculture 495:873–880

Wei C, Li W, Zhang C, Van Hulle M, Cornelis R, Zhang X (2003) Safety evaluation of organoarsenical species in edible Porphyra from the China Sea. J Agric Food Chem 51(17): 5176–5182

Widyaningrum D, Prianto AD (2021) Chlorella as a source of functional food ingredients: short review. Paper presented at the IOP Conference Series: Earth and Environmental Science

Yoshinaga K, Mitamura R (2019) Effects of Undaria pinnatifida (Wakame) on postprandial glycemia and insulin levels in humans: a randomized crossover trial. Plant Foods Hum Nutr 74(4):461–467

Yotsu-Yamashita M, Yasumoto T, Yamada S, Bajarias FFA, Formeloza MA, Romero ML, Fukuyo Y (2004) Identification of polycavernoside A as the causative agent of the fatal food poisoning resulting from ingestion of the red alga Gracilaria edulis in The Philippines. Chem Res Toxicol 17(9):1265–1271

Yu-Qing T, Mahmood K, Shehzadi R, Ashraf MF (2016) Ulva lactuca and its polysaccharides: food and biomedical aspects. J Biol Agric Healthc 6(1):140–151

Zakaria NA, Ibrahim D, Shaida SF, Supardy NA (2011) Phytochemical composition and antibacterial potential of hexane extract from Malaysian red algae, Acanthophora spicifera (Vahl) Borgesen. World Appl Sci J 15(4):496–501

Zava TT, Zava DT (2011) Assessment of Japanese iodine intake based on seaweed consumption in Japan: a literature-based analysis. Thyroid Res 4(1):1–7

Zhu, S., Xu, J., Adhikari, B., Lv, W., & Chen, H. (2022). Nostoc sphaeroides cyanobacteria: a review of its nutritional characteristics and processing technologies. Crit Rev Food Sci Nutr, 1–17

Harnessing Plant Biodiversity for Resilient Food Systems

7

Kulsoom Zahara, Yamin Bibi, and Ambreen Najaf

Abstract

The accelerating crisis of climate crisis, growth in population, environmental degradation, and food insecurity around the globe demand the need for sustainable and resilient food systems. Plants are able to be used as a transformative approach to enhance agricultural resilience because of their diversity and genetic variation. Because of the biodiversity and genetic variability of plants, they could be resistant against extreme weather conditions, pests, and diseases. Hence, this will provide stability in food production under fluctuating environmental conditions. The diverse cropping systems also improve soil health by reducing the reliance on chemical inputs. Nutritional security from plant biodiversity can be achieved by integrating these crops into the food chain, including underutilized crops and crops of neglected importance, which include all essential vitamins and minerals and show great potential to help staple crops in order to obtain a balanced diet. In addition, the incorporation of plant biodiversity into agricultural systems underpins several vital ecosystem services such as pollination, pest control, and regulation of the water resource that are essential for sustainable agriculture. Conservation of wild relatives of crops provides further opportunities for the creation of a genetic reservoir for breeding resilient crop varieties to ensure long-term agricultural sustainability. However, to tap the full potential of plant biodiversity, it is essential to develop supportive policy and innovative approaches, i.e., conservation of genetic resources, promotion of agrobiodiversity through research and extension services, and market development of diverse crops. While collaborative efforts are required from governments, institutions of research, farmers, and the private sector to develop an enabling

K. Zahara (✉)
Department of Botany, Pir Mehr Ali Shah Arid Agriculture University, Rawalpindi, Pakistan

Y. Bibi · A. Najaf
Department of Botany, Rawalpindi Women University, Rawalpindi, Pakistan

R. Qureshi (ed.), *Plant Biodiversity*, https://doi.org/10.1007/978-981-92-1471-6_7

environment for the sustainable use of plant biodiversity, harnessing plant biodiversity is not an option; it is a necessity for building up resilient food systems that can withstand challenges posed in the current times. By embracing the diversity of plant species, it would be possible to secure food production, improve nutritional quality, and encourage sustainable agricultural practices that contribute to the well-being of both the people and the planet.

Keywords

Biodiversity · Food · Ecosystem · Agrobiodiversity · Nutritional quality

Abbreviations

FAO	Food and Agriculture Organization
IFAD	International Fund for Agricultural Development
NGOs	Non-Governmental Organizations
NUS	Neglected and Underutilized Species

7.1 Introduction

Present era is marked with challenges such as the instability of climate, pressure from population growth, degradation of plants, and food insecurity. These challenges not only threaten agricultural production but also harm the stability of our food systems (Gawdiya et al. 2025). Conditions for growing food are becoming more volatile and uncertain: from an ongoing drought to unpredictable rainfall, the erosion of fertile soil, and a decline in pollinator populations (Tutwiler et al. 2017). As the global disruptions continue, the concept of resilient food systems is now at the forefront of efforts to address insecurity. A resilient food system is one that can survive changing circumstances, rebuild resources, and continue to offer sufficient, healthy, and appropriate food (DeClerck 2013).

Plant diversity has not been given sufficient coverage in the past, but it's very important for the establishment of resilient food systems. The broad diversity in plants, including crop species, wild relatives, landraces, and ecotypes, constitutes a valuable gene pool with the ability to improve the resilience of food systems through enhancing the nutritional content of foods derived from plants (Kennedy et al. 2022). It underpins such ecosystem services as nutrient cycling, pest control, and water purification. Importantly, a diversified farming system is tougher for pests and disease to upend and more resilient to climate-related stresses, given that its genetic diversity acts as a buffer against widespread failures (Ngongolo and Mmbando 2024).

Today's agricultural systems promote monocultures and homogeneous crop varieties that have high yields, ultimately resulting in a reduction of the gene pool.

This homogeneity makes our food systems vulnerable to disruptions, as was seen in the Irish Potato Famine and the known threats to global wheat from emerging strains of the rust disease (Kusuma et al. 2024).

Incorporating plant biodiversity into agriculture, breeding, and seed systems improves this. It assists farmers in selecting the right plant species or varieties suitable for the environment. It ensures the ability of farmers or communities to adapt to a predicted increase in climate stress. It assists in reviving knowledge about certain plant food products that the communities had previously abandoned.

7.1.1 Understanding Plant Biodiversity

Plant biodiversity encompasses the richness and diversity that exist in plants. It ranges from species diversity, genetic diversity, to ecosystem diversity. This diversity forms a fundamental element in enhancing the resilience and integrity of agricultural systems. Genetic diversity, in this case, allows plants to adapt, tolerate stress, resist disease, and thrive under changing environments; thus, genetic diversity forms a fundamental element in agricultural systems (Whitton and Rajakaruna 2001).

7.1.2 Historical and Ecological Perspectives

Conventional agriculture has always relied on the plant biodiversity. The native population and local farmers have been growing various crops suitable for their local climatic conditions. They have been able to conserve desirable genetic diversity by storing seeds or by following practices of crop rotation and growing local variants of crops known as landraces. From the biological perspective, various biological processes are sustained by diversity in plants. Plant diversity is responsible for increasing soil health by turning over nutrients, retaining soil moisture, sustaining healthy populations of pollinators, and securing the storage of carbon in forests (Whitton and Rajakaruna 2001).

7.2 The Role of Agrobiodiversity in Modern Agriculture

In comparison to outdated systems, Modern agriculture mostly relies on few genetically similar staple crops grown in monocultures (Altieri 2004). This can enhance efficiency in large-scale farming, while it also renders the system vulnerable to climate extremes, pests, and economic fluctuations. Agrobiodiversity provides the means toward stability and resilience. If crops within the system perform differently under climate stress, then total system is less likely to fail completely. Some wild crop varieties contain genes utilized for breeding varieties that are resistant to climate challenges. Consuming a diverse variety of plants enhances nutrition and health, while mixed cropping systems can prevent the complete loss of crops (Cook 2006).

Fig. 7.1 Dimensions of plant biodiversity in building resilient food systems

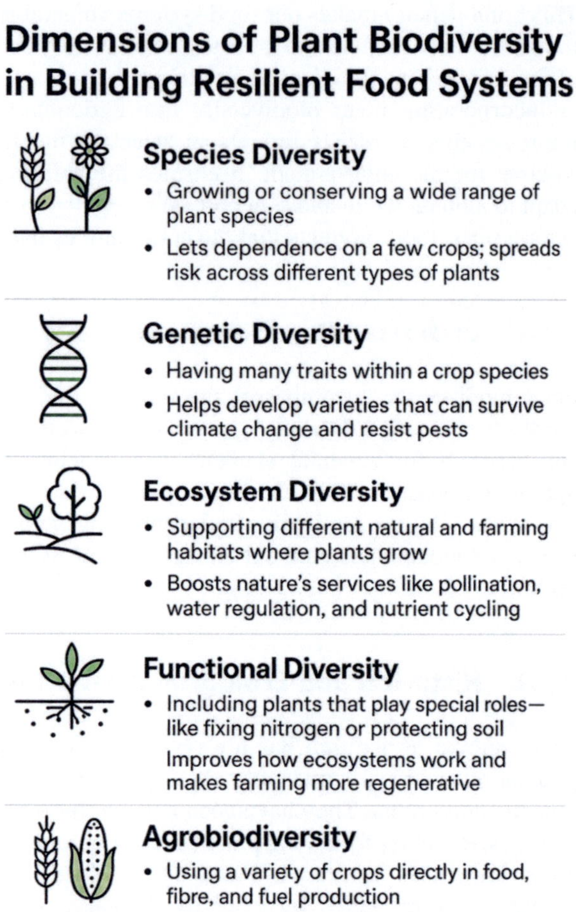

Traditional crops provide a safeguarding element to local culture and expertise. They also provide marketing avenues within niche markets, which will boost local enterprises. Further, they provide entry points within niche markets and support the resilience of the local economy (Crews et al. 2018).

However, the protection and enhancement of plant biodiversity are not merely ecological goals. These are also drivers for agricultural innovation (Fig. 7.1). Saving seeds by neighborhood farmers to the preservation efforts by large seed banks for rare species, the preservation of such genetic riches is a basic building block for a truly sustainable food system. It is well beyond nature preservation and the establishment of farms that are resilient enough to handle whatever the future holds (Crews et al. 2018).

7.3 Plant Biodiversity and Agricultural Resilience

Plant biodiversity entails both protection and restoration powers at the heart of that resilience, stabilizing food systems against environmental stress and biological threats (DeClerck 2013). This ensures that there are enhanced resilience capabilities for agricultural systems to withstand harsh weather conditions. However, with the rise in climate change events associated with weather conditions like droughts, flooding, heat waves, and storms, Modern agriculture will encounter more setbacks. Monoculture agricultural systems will be most affected, while agricultural systems that are more diverse will be less adversely affected because of the shielding effects of biodiversity (Fig. 7.2) (Whitton and Rajakaruna 2001).

Genetic diversity in plant species also induces factors such as tolerance to drought and floods, which help mitigate the risk of a complete crop failure. When plants grow with varied levels of root length and harvest maturities, the risk is diversified over space and time, with a naturally inherent risk management (Kusuma et al. 2024). In addition, perennial species are effective in enhancing soil structure and water-holding capacity, while diverse landscapes at the ecosystem scale steady temperatures, reduce erosion, and form microclimates to buffer against climate shocks (Gawdiya et al. 2025).

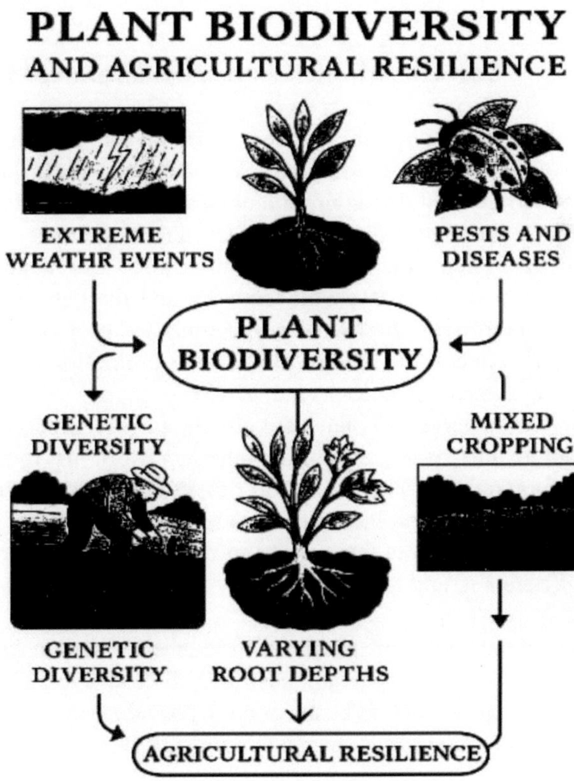

Fig. 7.2 Plant biodiversity bolsters agricultural resilience

Biodiversity also plays a critical role in minimizing vulnerability to pests and diseases (Girot 2012). In systems where everything is too similar, pests and diseases spread quickly because there's not enough genetic diversity to slow them down. When farmlands vary for all types of environments, genetic diversity buffering occurs: only the toughest species will survive and propagate resistance, and the overall ecosystem will be a healthy one. By increasing diversity, pests will have less chance to hit their targets, and any infestations will soon die out. By planting herbs along with vegetables and staggering them for insect-repelling properties, pests are easily deterred by smells and placement patterns. By providing a habitat for natural predators like birds and beneficial insects, or even frogs, we establish a natural and chemical-free defense system against pests (Altieri 1999).

Crop rotation helps ensure consistent production even when faced with challenging conditions because when one strain underperforms, another strain compensates for it. This functional diversity ensures ecological redundancy because when one species does not fulfill its role, others take over those functions because they are similar in functions and ensure the processes are not disrupted. Plant populations' diversity helps create microhabitats due to differences in temperature, humidity, and moisture content (Freer-Smith et al. 2017). Prominently, changing root systems and nutrient contributions increase soil health, supporting microbial activity and long-term fertility.

7.4 Soil Health and Reduced Chemical Dependency

A resilient food system not only relies on the diversity of crops above ground but also on the health of the soil (Freer-Smith et al. 2017). This impacts agricultural productivity and the environment as a whole. Current realities include the growing reliance on chemically intensive farming, which has caused the degradation of the soil ecosystem, disturbance of micro-life, and contamination of water sources (Usharani et al. 2019). Soil recovery and the enhancement of plant diversity, for example through the use of polyculture and crop rotation, are established methods for the reduction of chemicals and the promotion of more robust and sustainable natural systems (Cardoso et al. 2013).

When a variety of plants are grown, a lively underground ecosystem exists. Each plant exudes a variety of compounds from its roots, supporting a variety of microorganisms. The dense soil ecosystem supports nutrient cycling, controls soil-borne diseases, and increases disease resistance. At the same time, the varied shapes and depths of roots, ranging from tap roots to shallow roots, help decompress the soil, aerate it, and facilitate the flow of water through the soil (Fig. 7.3) (Awale et al. 2017).

Rotating different plant families in each season keeps nutrients in the soil in balance and helps disrupt pest and disease cycles without synthetic chemicals. Legume crops, such as beans or peas, provide symbiotic bacteria that fix atmospheric nitrogen to naturally enhance soil fertility (Freer-Smith et al. 2017).

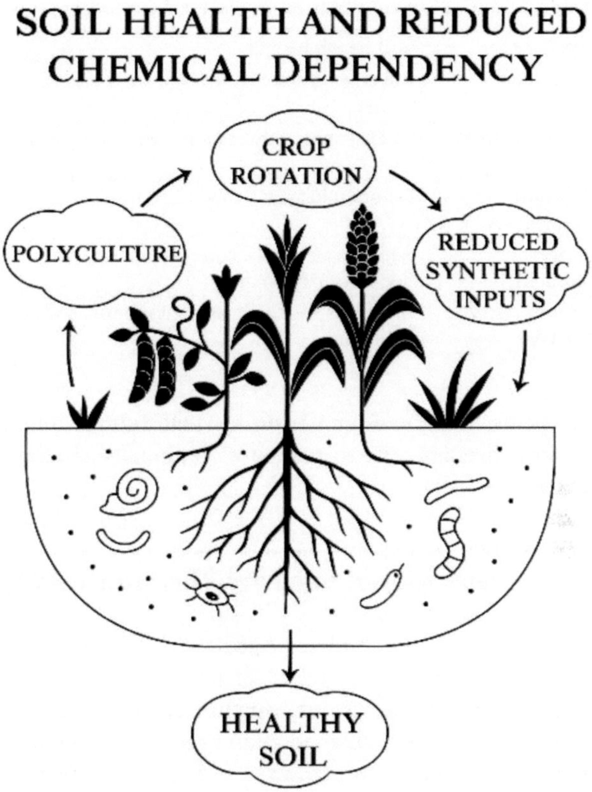

Fig. 7.3 Cycle of crop rotation, polyculture, and reduced synthetic inputs fostering healthy soil and sustainable agriculture

Such crop rotation prevents pest and disease buildup that is specific to one host, while the change in nutrient requirement avoids mineral depletion. For instance, maize-cowpea-sorghum rotation can restore soil nitrogen, break up pest cycles, improve soil structure, and as much as 40% a less need for fertilizer (Reddy 2017).

Working together, the effects of both polyculture and crop rotation create favorable conditions requiring minimal synthetic inputs. The role of legumes and cover crops here involves the biological fixation of nitrogen, thereby eliminating the need for fertilizer, and meanwhile, the increased plant biodiversity smothers weeds and enhances predator populations that control the pest insect population (Kibblewhite et al. 2008). Controlling the overflow of agrochemicals helps in keeping the water clean, the soil micro life healthy, and the nearby ecosystems and human health safe. Legumes and cover crops can easily add nitrogen instead of using agrochemicals, while the mixed crop can suppress the growth of weeds and entice natural pest-controlling agents to control insect pests (Table 7.1).

Rather, working with a range of plant species, not only those visible above the ground, but those that live below it as well, can assist farmers in improving soil health, reducing its chemical usage, and establishing efficient farming systems (Kibblewhite et al. 2008). Such biodiversity-based practices are the foundation of

Table 7.1 Biodiversity practices vs. chemical dependency

Practice	Soil health benefit	Reduction in chemical use
Polyculture	Enhances microbial diversity and organic matter	Reduces need for herbicides and pest control
Crop rotation	Replenishes nutrients and breaks pest cycles	Cuts synthetic fertilizer and pesticide requirements
Legume integration	Natural nitrogen fixation	Replaces nitrogen fertilizers
Cover cropping	Suppresses weeds, controls erosion, and boosts organic matter	Diminishes herbicide dependency
Agroforestry systems	Stabilizes microclimate and improves water retention	Minimizes pesticide drift and erosion-based runoff

regenerative agriculture, ensuring that future generations are left with healthy systems that are able to sustain a continually shifting environment (Perfecto et al. 2019).

7.5 Nutritional Security Through Underutilized Crops

Neglected and Underutilized Species (NUS) are gaining interest again as important allies in enhancing nutritional security. These species naturally thrive on difficult, marginal conditions. Reintroduction of NUS into farming systems and into local diets allows communities to address hidden hunger, regain some food sovereignty, and generate nutrition pathways resilient to climate change (Ali and Bhattacharjee 2023).

Such neglected crops are diversified, with different families, all contributing to offering various nutritional elements and endurance properties. Moringa, scientifically known as *Moringa oleifera*, and belonging to sub-Saharan Africa and South Asia, is known to be rich in vitamins A, C, and E, along with iron, calcium, and having a full protein spectrum (Ali and Bhattacharjee 2023). Amaranth (Amaranthus spp.), originally from Central and South America, outlooks for its high protein content mainly lysine alongside iron, magnesium, and antioxidant compounds. In West Africa, fonio (*Digitaria ex-ilis*) conveys easily digestible methionine and cysteine, while Ethiopia's tiny grain teff (*Eragrostis tef*) brings abundant fiber, iron, calcium, and a naturally gluten-free profile (Koréiss 2015). Taro (*Colocasia esculenta*), a staple in Asia and the Pacific Islands, enriches diets with potassium, dietary fiber, and vitamin C, and jackfruit (*Artocarpus heterophyllus*) from South and South-east Asia supplies potassium, vitamin C, B complex vitamins, and fiber (Chauhan et al. 2022). Extraordinarily, all of these species flourish with least inputs, exhibit natural pest resistance, and tolerate low-water conditions, making them ideal for smallholder farmers and regions facing climate stress.

Even though wheat, maize, and rice provide the majority of the world's calories, they are often inadequate in vital micronutrients. Incorporating NUS into regular daily meals can significantly enhance intake of iron, zinc, calcium, and vitamin A

Table 7.2 Regional implementation of Neglected and Underutilized Species (NUS)

Region	Crop(s)	Initiative	Key focus	Outcomes
Ethiopia	Teff	National nutrition strategy	Iron enrichment; drought resilience	Teff adopted in gluten-free markets and urban diets
Mali and Senegal	Fonio	NGO-led training on processing and marketing	Addressing iron/protein gaps; income streams	Increased fonio demand; higher farmer incomes
Philippines	Taro	Community-led revival projects	School nutrition; cultural heritage	Taro in school lunches; improved child health and cultural pride
India	Finger and pearl millets	"International year of millets" campaigns	Maternal/child nutrition; dryland farming	Millet-based programs; strengthened dryland agriculture
Andean region	Quinoa and Amaranth	FAO pilot programs and local cooperatives	Protein density; climate resilience	Yield stability; new export opportunities for smallholders
West Africa	Bambara groundnut	IFAD and local NGO support	Protein supplementation; drought tolerance	Expanded cultivation; reduced seasonal hunger
Pacific Islands	Breadfruit	Community agroforestry and reforestation	Staple food security; sustainable land use	Stabilized yields; improved household food reserves
South Asia	Moringa and drumstick greens	Government nutrition-garden schemes	Micronutrient supplementation; women's health	Reduced maternal anemia; increased adoption in home gardens

nutrients lacking in staples. Adding fiber-rich crops like taro and teff supports healthy digestion and helps communities reconnect with their traditional food ways (You and Henneberg 2016). This helps the food system recover better from disturbances and become less dependent on a few staple commodities. For example, in India, the distribution of finger millet and pearl millet in school lunches and local markets has been shown to increase the intake of iron and calcium in children (Perfecto et al. 2019).

Regional-specific programs demonstrate the actual impact that can be achieved when underutilized crops are integrated into focused programs. Teff, for example, has exceptionally high iron content and is drought-resistant; it therefore holds a prime position in national nutrition strategies of Ethiopia, while it feeds the expansion in gluten-free sectors and urban demand (Table 7.2). Mali and Senegal NGOs are working with farmers to commercialize fonio production, addressing iron and protein deficiencies, while providing new economic prospects. In the Philippines, national initiatives have promoted the revitalization of taro cultivation within native communities, integrating it into the school meal program, thus supporting the health of children. India's celebration of the 'International Year of Millets' has promoted nutritional programs for women and children centered on millets, thus enhancing the health of mothers and children, in addition to promoting dryland farming. All these examples illustrate the NUS importance for developing healthy, sustainable, and needy food systems.

7.6 Ecosystem Services Supported by Biodiversity

Plant diversity, besides its role in nutrient supply, is crucial in supporting ecosystem processes essential for increasing crop production, improving climate change resilience, and maintaining ecosystem integrity. Three essential ecosystem services include pollination, pest control, and water regulation.

7.6.1 Pollination

When agricultural land is interspersed with flowering plants, this results in nectar sources for bees and other flower-loving pollinators like butterflies, while some of these plants supply breeding sites for birds and other beneficial insects too! This results in increased levels of pollination too! aspects of staggering planting dates for various crops ensures that these beneficial creatures have a constant supply of these throughout the growing seasons, which goes a long way in maintaining their numbers. Reducing the overall use of broad-spectrum pesticides would ensure just that too! For example, when orchards are interspersed with hedgerow crops, there is increased fruit production too, courtesy of increased the activity (You and Henneberg 2016).

7.6.2 Natural Pest Suppression

In biodiversity-rich agricultural systems, the most effective mode of controlling pests takes place through natural predator-prey interactions. This can be achieved through a variety of plant growth that provides havens and source locations for beneficial organisms like ladybugs, spiders, parasitic wasps, and bird species that eat insects. This can happen through intercropping and crop rotation that pests are unable to survive. This happens with diverse vegetation that provides camouflage to host plants and baffles potential pests. Rice paddies planted with azolla, duckweed, and fish provide effective pest control with no need for agrochemicals (Ndakidemi et al. 2016).

7.6.3 Water Regulation

A planting scheme that involves species with varying root systems is extremely effective in the absorption capacity of the soil, thus reducing runoff and erosion.

Riparian buffers, agroforestry systems, among others, can be considered as natural filters that filter sediment, pollutants, and runoff, thus enhancing the recharge of groundwaters (Zhang et al. 2024). When there are hot and cold water flows, the landscape becomes more resilient due to the combination of species that utilize water in varying ways. Consider tropical agroforestry, for instance, where the stacked canopies, shrubs, and ground covers contribute to the stabilization of water tables and the prevention of soil erosion because of the diverse vegetation levels (McCluney et al. 2012).

7.7 Ecosystem Balance and Farm Sustainability

Whenever multiple species adopt the same ecological roles say, in the pollination of flowers, the breakdown of organic materials, or nitrogen fixation the system takes on redundancy. That means essential services happen even if one species struggles (Fig. 7.4). The feedback loop between plants, microbes, and animals supports the self-correcting process that absorbs the shock and triggers the regeneration process. Over time, the processes such as nutrient cycling, soil formation, and natural filtration systems come into a web to reinforce each other, resulting in agricultural landscapes that rebuild natural capital rather than depleting it (McCluney et al. 2012).

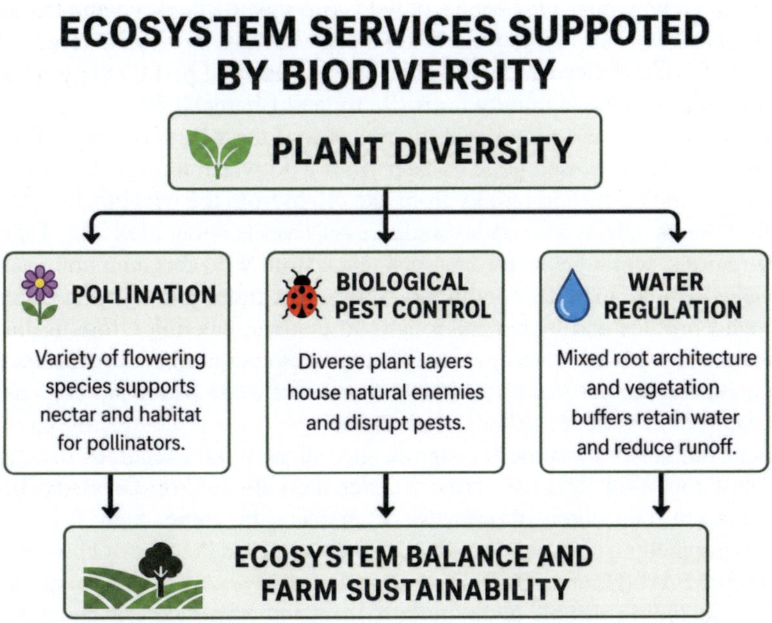

Fig. 7.4 Flowchart depicting how plant biodiversity underpins key ecosystem services pollination, biological pest control, and water regulation to sustain agricultural resilience

7.8 Conservation and Use of Wild Crop Relatives

Wild crop relatives are known as "cousins" in their wild state that are not domesticated but are related instead to our common food sources that we grow in farms today. These possess a vast genetic pool created over the course of many years from natural selection in the wild alone (Pretty and Bharucha 2014). These plants are designed to withstand drought, salt, insects, and extreme weather conditions. However, as a result of extensive breeding, most of these qualities are absent in current crop plants. Given the threat of climate change, erosion, and the potential for new diseases, the hardiness seen in natural plants is now needed to revamp the gene pool found in our crop plants (Handelsman 2021).

These wild crop relatives possess traits that can help improve domesticated crop species. These traits range from extreme resistance to drought and high temperatures, as well as resistance to high salt concentrations and crop pests. Since breeding has been done to create crop species that are similar and produce high yields but are susceptible to stress factors, many diverse traits are retained within wild crop species. Such traits can help domesticated crop species grow well under stressful conditions.

A two-fold approach is required to preserve the wild relatives of crop plants. Ex situ conservation involves protecting these species by preserving their genetic material in seed banks, botanical gardens, cryopreservation labs, or DNA banks to ensure the scientific community has access to these species in case they face threats in the wild. To maximize the benefits of wild crop species, conservation ex situ, or in situ conservation, is also employed. This strategy ensures the species remain healthy in the wild locations they can be found in, such as national parks, or agro-ecological land managed on the community basis (Pretty and Bharucha 2014).

Plant breeders have long utilized the richness of the wild relatives against grave challenges. They have used gene transfer from wild wheat to improve resistance to rust diseases and derived immunity from late blight from the wild potato (Table 7.3). Drought tolerance from wild barley and rice relatives is being bred into crops suited for dry regions, while genes for heat resilience from wild sorghum and millet help keep yields steadier in hotter conditions. At the same time, wild legumes bring richer amino acid profiles and higher micronutrient content, steering efforts to boost the nutritional value of cultivated pulses. Modern genomic tools and precision editing have quickened the pace and eased identification of those beneficial traits and their transfer into crop varieties (Zhang et al. 2024).

Global initiatives illustrate the significance of crop wild relatives for conservation. The Crop Wild Relatives Project under the Global Crop Diversity Trust has identified and conserved thousands of species in more than 20 countries, emphasizing their significance for adapting agriculture in the face of climate change, according to FAO (Hunter 2012). Despite such initiatives, there are several obstacles they face, including chronic underfunding, legal and ethical controversies related to the use of genetic resources, the effects of urbanization, and the decline of traditional knowledge. It is therefore imperative for farmers, researchers, government officials, and indigenous groups to forge stronger chains of collaboration in order to preserve the wild relatives of crops as a living source for innovation (Maxted et al. 2012).

Table 7.3 Conservation strategies for wild crop relatives

Conservation approach	Venue	Primary strengths	Key limitations	Illustrative example
In situ	Protected areas, community-managed farms	Preserves ongoing adaptation to local conditionsmaintains ecological interactions (pollinators, soil biota)	Susceptible to land-use change and habitat lossdependent on local governance and funding	Sierra Madre, Mexico: Farmers conserve wild teosinte alongside maize
Ex situ	Seed banks, botanical gardens, cryostores	Long-term genetic safeguardingready access for breeders and researchersrapid response to emerging threats	Detached from natural selection pressureshigh storage and maintenance costs	Svalbard global seed vault: Duplicate seed samples of wild relatives
Integrated	Combined in situ and ex situ networks	Combines evolutionary potential with practical accessibilityincreases resilience of conservation efforts	Requires cross-sector coordination and policy alignmentcomplex logistics and resource needs	Global crop diversity Trust's crop wild relatives project

7.9 Enabling Biodiversity-Driven Agriculture

Establishing a diverse plant species base for a resilient food system is a goal whose accomplishment through farm innovations alone is not sufficient but instead is achieved through a network of global agreements, policies, institutional arrangements, and market mechanisms (Diyaolu and Folarin 2024). Such international agreements as the Convention on Biological Diversity and the International Treaty on Plant Genetic Resources under the Food and Agriculture Organization are setting common obligations regarding the issue. It promotes countries to develop their national plans regarding biodiversity, take part in the global Multilateral Seed System, and incorporate farmers' rights to seeds and traditionally known (Idowu et al. 2018).

Within a nation, the governments ensure that these international agreements are upheld through the implementation of agrobiodiversity action plans, seed rights legislation, and financial support. The policies also include landraces and wild relatives topographic identification, hotspot designation for conservation, and adapting biodiversity goals to climate change and food security strategy frameworks. Financial support and tax incentives are used to encourage farmers to cultivate less-commercialized species, which has direct benefits to the environment (Idowu et al. 2018).

Local extension agencies, farmer field schools, community seed banks, and governance councils implement Policy by translating Policy into action. Extension personnel provide technical training and demonstration examples to farmers for growing crops through polyculture, intercropping, and the utilization of abandoned crops. Community seed banks preserve local crop varieties by entrusting the management to the care of local custodians, with benefit-sharing arrangements to ensure the resource remains with the community that uses the resource (Idowu et al. 2018).

For biodiversity-based agriculture to become more widely accepted, there has to be an economic incentive for farmers to do this kind of agriculture. Building value chains requires investing in post-harvest farms, packaging, certification for organically and fairly produced products, and storytelling efforts that re-establish a connection between products and their cultural heritage (Duru et al. 2015). Consumer interest increases when biodiversity is visible in shops and traceable through QR codes, and when schools and hospitals demand it through their purchasing programs. When such demand is combined with premium prices for green practices, subsidies for the diversification of crops, and for payments for ecosystem services, the diverse plant portfolio turns from a pure conservation objective into a viable commercial business model (Herzon et al. 2018).

7.10 Future Directions and Recommendations

The sudden a climate change progresses and the greater the movement of ecological systems, the higher the demand for food in the world, and agriculture needs to become more versatile and interlocked. Looking forward to the future, harnessing plant biodiversity will require the best and brightest of modern technology, novel educational and research paradigms, and widespread adoption of mixed crops designed to be stress-resistant. The recommendations below provide a coherent and data-driven agenda for the future of food systems.

New digital technologies are transforming the way we track and safeguard biodiversity. Detailed GIS mapping helps to pinpoint agroecological areas, biodiversity hotspots, and areas most vulnerable to climate change. AI-driven plant phenotyping enables rapid screening of thousands of plant species for desired traits such as drought and pest resistance. Space and mobile monitoring solutions provide real-time data on soil moisture, plant health, and changes in biodiversity patterns at any given time. Overall, these technological innovations close the gap between scientific discovery and informed decision-making in biodiversity conservation efforts.

Removing academic silos and reforming the way we teach is necessary in addressing the complex issues of biodiversity. Plant diversity relates not only to food, economic, and climate knowledge, but also to the study of society and indigenous knowledge systems. Moreover, there should be the formation of an interdisciplinary team in universities and research institutions, with the composition of geneticists, ecologists, agronomists, and experts in the social sciences, together with experts in indigenous knowledge systems. In the learning process from the

Table 7.4 Strategic pathways to biodiversity-driven agricultural resilience

Strategy	Tools and approaches	Impact on biodiversity and resilience
Technological innovation	GIS mapping, AI phenotyping, remote sensing	Enhances precision in crop selection and ecosystem monitoring
Collaborative research and education	Interdisciplinary programs, farmer-scientist partnerships	Fosters inclusive innovation and stakeholder engagement
Biodiverse cropping and climate adaptation	Polycultures, native species integration	Strengthens farm resilience and restores ecosystem functions
Data-driven decision support	Big data analytics, mobile advisory tools	Links real-time field data to actionable farming practices
Policy and institutional integration	Agroecology policies, biodiversity incentive schemes	Embeds plant diversity in mainstream agricultural policy

basic to the postgraduate levels, there should be an integrated topic involving biodiversity in all disciplines (Table 7.4).

By turning climate-resilient agriculture from concept to everyday practice, biodiverse farming systems must transition from pilot projects to evidence-informed norms and policies. Polish the notion of polyculture and visions of fields where crops of grain, legumes, vegetables, and native species can simultaneously thrive. These boost soil biodiversity and mitigate farm-level weather risks and the effects of weather extremes. Breeders can draw and cross their untapped and natural allies to develop adapted versions. Governments can invest in gene banks and establish economic incentives in the form of subsidies and/or payments for ecosystem services to incentivize farmers to prefer and practice agrodiverse and climate-resilient farming. Agroecology and regenerative practices will bring sustained results.

Such strategic visions transcend the bounds of more technical solutions, reflecting a paradigm shift, instead, that is attuned to ecological knowledge, more so, confers on the community, instead, enables a common resource stewardship, more so. The future for food security is expected, instead, to lie in diversity, and a consensual resource, instead, a state of sameness, more so.

7.11 Conclusion

Plant biodiversity is not only a conservation objective; it is, in fact, the foundation of resilient agro-ecosystems, providing sustenance for a growing population and healing degraded lands amidst a constantly shifting climate. It is, in essence, a means of cultivating a diversity of crops and plant life surrounding and within farm land in order to create a natural pest and disease barrier and heal the soil.

The path forward, as outlined in this book, is clear: increase the diversity of cropping systems, protect wild relatives, use digital and genetic technologies for precise breeding, and align markets and policies to favor biodiversity conservation. When these approaches are brought together by farmers, researchers, corporations, and policymakers, agriculture can transition from a trajectory of loss to one of renewal and resilience.

Acceptance of plant diversity means touching past one-size-fits-all clarifications and planting the seeds of ecological insight in every field. The result is a food system that can flex with uncertainty, support livelihoods, and protect our planet for future generations.

References

Ali A, Bhattacharjee B (2023) Nutrition security, constraints, and agro-diversification strategies of neglected and underutilized crops to fight global hidden hunger. Front Nutr 10:1144439

Altieri MA (1999) The ecological role of biodiversity in agroecosystems Invertebrate biodiversity as bioindicators of sustainable landscapes. Elsevier, pp 19–31

Altieri MA (2004) Linking ecologists and traditional farmers in the search for sustainable agriculture. Front Ecol Environ 2(1):35–42

Awale R, Machado S, Ghimire R, Bista P (2017) Soil health. Advances in Dryland Farming in the Inland. Pacific Northwest, pp 47–97

Cardoso EJBN, Vasconcellos RLF, Bini D, Miyauchi MYH, Santos CA d, Alves PRL et al (2013) Soil health: looking for suitable indicators. What should be considered to assess the effects of use and management on soil health? J Scientia Agricola 70:274–289

Chauhan VBS, Mallick SN, Pati K, Arutselvan R, Nedunchezhiyan M (2022) Status and importance of underexploited tuber crops in relation to nutritional security and economic prosperity. In: Compendium for winter school on "unexpected vegetables: unexplored treasure trove for food", pp 246–264

Cook RJ (2006) Toward cropping systems that enhance productivity and sustainability. Proc Natl Acad Sci USA 103(49):18389–18394

Crews TE, Carton W, Olsson L (2018) Is the future of agriculture perennial? Imperatives and opportunities to reinvent agriculture by shifting from annual monocultures to perennial polycultures. Global Sustain 1:e11

DeClerck, F. (2013). Harnessing biodiversity: from diets to landscapes Diversifying Food and Diets (pp. 17–34): Routledge

Diyaolu CO, Folarin IO (2024) The role of biodiversity in agricultural resilience: protecting ecosystem services for sustainable food production. Int J Res Public Rev 5(10):1560–1573

Duru M, Therond O, Martin G, Martin-Clouaire R, Magne M-A, Justes E et al (2015) How to implement biodiversity-based agriculture to enhance ecosystem services: a review. Agron Sustain Dev 35(4):1259–1281

Freer-Smith PH, Webber JF (2017) Tree pests and diseases: the threat to biodiversity and the delivery of ecosystem services. Biodivers Conserv 26(13):3167–3181

Gawdiya S, Sharma RK, Singh H, Kumar D (2025) Crop diversification as a cornerstone for sustainable agroecosystems: tackling biodiversity loss and global food system challenges. J Diverse Agric Syst 7(5):1–16

Girot, P. O. (2012). Plant disease, pests and erosion of biodiversity. Handbook of hazards and disaster risk reduction (pp. 384–395): Routledge

Handelsman J (2021) A world without soil: the past, present, and precarious future of the earth beneath our feet. Yale University Press

Herzon I, Birge T, Allen B, Povellato A, Vanni F, Hart K et al (2018) Time to look for evidence: results-based approach to biodiversity conservation on farmland in Europe. Land Use Policy 71: 347–354

Hunter D (2012) Crop wild relatives: a manual of in situ conservation. Routledge

Idowu AT, Ajirotutu RO, Erinjogunola FL, Onukogu OA, Uzondu NC, Olayiwola RK, Adio SA (2018) Biodiversity conservation and ecosystem services: a review of challenges and opportunities. Int J Adv Multidisc Res Stud

Kennedy G, Wang Z, Maundu P, Hunter D (2022) The role of traditional knowledge and food biodiversity to transform modern food systems. Trends Food Sci Technol 130:32–41

Kibblewhite MG, Ritz K, Swift M (2008) Soil health in agricultural systems hilos. Trans R Soc Lond B Biol Sci 363(1492):685–701

Koréiss Y (2015) Fonio (Digitaria exilis) in West Africa: towards improving nutrient quality. Wageningen University and Research

Kusuma I, Rahmadhia SN, Ma'arif AJ (2024) Exploring the role of biotechnology and biodiversity in achieving sustainable and nutritional food systems. J Sci Agrotechnol 2(1):21–29

Maxted N, Kell S, Ford-Lloyd B, Dulloo E, Toledo Á (2012) Toward the systematic conservation of global crop wild relative diversity. J Crop Sci 52(2):774–785

McCluney KE, Belnap J, Collins SL, González AL, Hagen EM, Nathaniel Holland J et al (2012) Shifting species interactions in terrestrial dryland ecosystems under altered water availability and climate change. Biol Rev Camb Philos Soc 87(3):563–582

Ndakidemi B, Mtei K, Ndakidemi P (2016) The potential of common beneficial insects and strategies for maintaining them in bean fields of sub Saharan Africa. Am J Plant Sci 7:3

Ngongolo K, Mmbando GS (2024) Harnessing biotechnology and breeding strategies for climate-resilient agriculture: pathways to sustainable global food security. Discov Sustain 5(1):431

Perfecto I, Vandermeer J, Wright A (2019) Nature's matrix: linking agriculture, biodiversity conservation and food sovereignty. Routledge

Pretty J, Bharucha ZP (2014) Sustainable intensification in agricultural systems. Ann Bot 114(8): 1571–1596

Reddy, P. P. (2017). Crop rotation Agro-ecological approaches to pest management for sustainable agriculture (pp. 229–242): Springer

Tutwiler, M., Bailey, A., Attwood, S., Remans, R., & Ramirez, M. (2017). Agricultural biodiversity and food system sustainability

Usharani K, Roopashree K, Naik DJ (2019) Role of soil physical, chemical and biological properties for soil health improvement and sustainable agriculture. J Pharmacogn Phytochem 8(5):1256–1267

Whitton J, Rajakaruna N (2001) Plant biodiversity. J E o B:56–64

You W, Henneberg M (2016) Cereal crops are not created equal: wheat consumption associated with obesity prevalence globally and regionally. AIMS Public Health 3(2):313

Zhang W-P, Surigaoge S, Yang H, Yu R-P, Wu J-P, Xing Y et al (2024) Diversified cropping systems with complementary root growth strategies improve crop adaptation to and remediation of hostile soils. Agric Food Scie Environ Sci 502(1):7–30

Hemp: A Potential Crop for Resilient Food Security

8

Rahmatullah Qureshi, Mehmooda Munazir, Chad A. Kinney, and Madalina Neacsu

Abstract

Hemp (*Cannabis sativa* L.) is an ancient multipurpose crop that has regained global attention due to its potential contributions to food security, environmental sustainability, and climate-resilient agriculture. This review highlights the nutritional, agronomic, industrial, and ecological significance of hemp. Hempseed is a rich source of high-quality protein, essential fatty acids, and bioactive compounds, making it valuable for functional foods and human nutrition. Beyond food applications, hemp serves as a renewable raw material for textiles, construction materials, bioplastics, cosmetics, and pharmaceutical products. The crop offers several environmental advantages, including high carbon sequestration capacity, phytoremediation potential, low pesticide requirements, and improvements in soil health. Recent regulatory reforms have stimulated renewed research interest and commercial cultivation worldwide. However, challenges related to seed retention, fiber quality, processing infrastructure, market development, and regulatory compliance continue to constrain its wider adoption. Overall, hemp represents a promising crop for supporting sustainable agriculture, circular bioeconomy initiatives, and diversified industrial and food systems, with significant potential to contribute to future environmental and economic sustainability.

R. Qureshi (✉)
Department of Botany, Faculty of Sciences, Pir Mehr Ali Shah Arid Agriculture University, Rawalpindi, Pakistan
e-mail: rahmatullahq@uaar.edu.pk

M. Munazir
Department of Botany, Government College Women University Sialkot, Sialkot, Pakistan

C. A. Kinney
Institute of Cannabis Research, Colorado State University Pueblo, Pueblo, CO, USA

M. Neacsu
The Rowett Institute, University of Aberdeen, Aberdeen, UK

R. Qureshi (ed.), *Plant Biodiversity*, https://doi.org/10.1007/978-981-92-1471-6_8

Keywords

Hemp · Food security · Circular bioeconomy · Sustainable agriculture · Environment friendly · Agroecological regions · Climate resilience

Abbreviations

CBD	Cannabidiol
FDA	Food and Drug Administration
GLA	Gamma-linolenic acid
GWP	Global warming potential
LCA	Life cycle assessment
PUFAs	Polyunsaturated fatty acids
SDA	Stearidonic acid
THC	Tetrahydrocannabinol

8.1 Introduction

Hemp (*Cannabis sativa* L.) is among the oldest domesticated crops and has a long history of cultivation of more than 6000 years with long periods of usage as food, fibers, oil, and medicine (Li 1974a). Due to its extraordinary versatility, the human societies in various agroecological locations have always used hemp specifically in its strong bast fibers and its rich seeds. Over the last few decades, hemp has resurfaced as a crop of critical strategic value due to increased concerns in the world about food security, global warming, ecological decline, and the desire to have sustainable agricultural systems. Its resistance to environmental stress factors, relatively low production input needs, and its ability to produce various high-value products make hemp a highly viable option in dealing with the issues of farming and food production in the future.

Although taxonomic definitions of Cannabis remain contentious, the genus Cannabis has historically been divided into *Cannabis sativa, C. indica,* and *C. ruderalis*. Traditionally, the Cannabis species were used as food, fiber, and medicine, although today they are loosely subdivided into two functional categories: marijuana and hemp, depending on the purpose of use and chemical structure. The psychoactive effects are the main reasons why marijuana is being cultivated, yet this drug has a growing therapeutic potential (Grotenhermen and Muller-Vahl 2016; Hill 2015). On the contrary, hemp is regarded as a source of non-intoxicating bioactive compounds, high-quality fiber, and nutritionally enriched seeds, and has been used to manufacture over 25,000 industrial and commercial products across the world (Johnson 2014).

Hemp (*Cannabis sativa* L.) is an annual that is short-day and fast-growing, botanically a member of the family Cannabaceae. It is also special in the sense that almost every single part of the plant can be effectively used as a product. The stem is a source of bast fibers and inner core of wood (hurds or shives), the seeds have good sources of oil and protein, and the flowers have a richness of phytochemicals, including non -psychoactive cannabidiol (CBD). Hemp is defined differently by regulations across the world, though the majority of European countries and North American countries legally define industrial hemp as Cannabis with a concentration of at most 0.2–0.3% D9-tetrahydrocannabinol (THC) on a dry-weight basis; a concentration that is not great enough to cause intoxication. Due to historical breeding differences, cultivation, and end uses, morphological and agronomic subtle, yet different, hemp and marijuana have developed, and this has enabled their separation in agricultural and scientific applications (Johnson 2014). Traditionally, hemp has been planted as a fiber crop or as a seed crop (grain) with each of the production systems being tailored to particular end uses. Recent policy changes, especially the Farm Bills of 2014 and 2018 of the USA, have seen hemp reintroduced as a legitimate research and commercial crop. Scientific interest in hemp agronomy, genetics, nutrition, and industrial applications has again become lively due to these regulatory changes and is now growing rapidly in both hemp-growing area and market openings.

Hempseed is of special interest in terms of its nutritional value. The seeds generally comprise about 30% protein, 25% carbohydrates (starch) and almost 30% oil. Hempseed oil is highly saturated with polyunsaturated fatty acids (more than 90% of total fatty acids) and has a good omega-6/omega-3 ratio, which makes it of great benefits to human and animal nutrition (Callaway 2004; Galasso et al. 2016). The oil is utilized in foods, cosmetics, and production of biofuels (Simopoulos 2002), and the leftover press cake is a livestock feed, a rich source of protein. Other than food and feed, hemp bast fibers were used in long to make good quality-paper and textile and the woody hurds were used as animal bedding (Karus and Vogt 2004). Recent developments in technology have further increased the use of hemp biomass since it can be used in bioplastics, carbon nanosheets, environmentally friendly construction composites, oil-absorbent substances, and in 3D-printing processes. Also, hemp is a source of over 100 naturally occurring cannabinoids, and cannabidiol (CBD) is the most thoroughly studied and commercially useful, because of its therapeutic potential (Mehmedic et al. 2010). Hemp has a rare set of nutritional, environmental and economic advantages in the context of growing population pressure, climate variability, and pressing demand of sustainable and climate-resistant crops. This chapter looks at the purpose of hemp as a multifunctional crop with a special focus on the possible contribution of this crop to the food security, the ecological sustainability, and the development of agro-industrial sector in the context of a transforming world.

8.2 Domestication, and Re-emergence of Hemp

Hemp is one of the oldest plants since ancient times, which has close relationship with the human culture. The great polymorphism of this species has resulted in an astonishing variability at the intra-species stage, which, consequently, resulted into a small number of species identification categories. Having a number of controversies regarding its complete classification, C. sativa (mostly dioecious species with a few monoecious lines, which are obtained through cross-breeding (Galasso et al. 2016)) has long been separated in hemp plants and marijuana plants because of the variability of chemical constituents and usage.

Low concentrations of D9-tetrahydrocannabinol (THC) (that is, less than 0.2–0.3%) make hemp none intoxicating and legal in most countries (Schluttenhofer and Yuan 2017). The main seed is made out of hemp and the nutrient-rich plants as well as the woody portion of the plant (hurds or shives). The hemp fruits are actually achene, single-seed, two cotyledons full of lipids and proteins, and surrounded by a thin indehiscent pericarp (Chandra et al. 2017). Although hemp can be grown in the form of fibers, seeds or both, there is further evidence that hemp was a major food crop during the ancient times. Historically, hemp seeds were discovered in the tombs of China many years ago more than 4500 years ago (Li 1974b). His history records that the hemp seeds were considered as one of the five staple grains in ancient Chinese food until their loss of popularity in the ninth century.

Hemp is native to a larger area that included Central, East, and South Asia. It was spread to temperate areas of all continents due to natural factors, including flowing water, and the contribution of animals to its spread, and its intentional distribution with human migration and trade (Small 2015; Clarke and Merlin 2016). Prehistoric records of hemp in East Asia have been made in the form of palynology and archaeobotany, whereby traces of hemp pollen have been found in Japan, circa 7000 BCE and in China circa 4500 BCE. Then, hemp found its way to Europe through the overland routes to the north and south and was firmly cultivated on the continent since around 500 BCE, in accordance to the researchers (Small and Marcus 2002; Clarke and Merlin 2016). The European explorations finally resulted in the introduction of hemp in the Americas where it started growing in South America in the mid-sixteenth century and in North America in the early seventeenth century. Hemp was mostly used as a fiber crop and secondary as an oilseed crop, but little attention was paid to the psychoactive properties of the plant (Small 2015).

In the last 100 years, the cultivation of hemp drastically became low across different regions of the world as a result of strict rules, particularly in North America. In the USA, hemp production was effectively halted by the Marijuana Tax Act of 1937, with a brief spurt in the production of hemp during World War II, when its fibers were in great demand. This was later revived in production but was limited until there was a change in legislation in the early 2000s. Regulations were relaxed with the Agricultural Act of 2014 allowing the pilot projects to research the industrial hemp and then the Agriculture Improvement Act of 2018 was passed that fully legalized the production of hemp by excluding this substance out of the Controlled Substances Act (Johnson 2014; Mark et al. 2020). Those modifications led to fast growth of hemp that started with four states with 755 acres in 2014 and reached 22 states with around 36,400 acres in 2018 (USDA Agricultural Statistics 2019). At the same time, land grant universities began developing academic programs in this new area, including the Global Hemp Innovation Center in Oregon State University, which deals with research and outreach to create sustainable value chains in the hemp industry.

The eventual expansion of the industry in Canada came as a result of a parliamentary act of 1998. The first one was fiber production but soon turned to the nutrition segment especially the hempseed and hemp oil. The industry was later extended with the Cannabis Act of 2018. In Canada, hemp yields average 1.1 tons per hectare of rain-fed fields to 3.3 tons per hectare on irrigated fields, and almost half of Canadian hemp is now grown in organic production (Small 2015; Health Canada 2020). Although the world had a prohibition in the cultivation of hemp, Europe was a major producer of hemp, particularly the eastern and southeastern regions, and other countries, such as France, Italy, and Poland.

The re-emergence of hemp cultivation in the European Union in the second half of the twentieth century and the early years of the twenty-first century was led by the legislation changes and the need to have more environmentally friendly raw materials. The number of hemp acres that are allocated in the European Union grew at least 75% between 2015 and 2019, increasing the total production significantly. France is currently the largest hemp producer in the EU zone (then Netherlands and Austria), and a number of other EU member countries (European Commission 2026). Currently, there are almost 100 hemp cultivars registered in the EU Plant Variety Database, which include fiber or grain or both purposes, which is a manifestation of high biodiversity of the crop (European Commission 2026). During the past decades, hemp has re-emerged as an important food security crop and a sustainable agricultural crop on an international scale. Hemp has the potential to become a crop of climate-resilient agriculture due to its flexibility to varying adverse factors, the comparatively low-input requirements, as well as its optimal nutritional content. Whether it is as food or as industrial, like in the cases of textiles, construction, medicine, and bioproducts, hemp is a general solution to the present demands in terms of agriculture yield, and viable conditions on the world level, particularly, as the requirements progress due to the climate change.

8.3 Hemp as a Nutritional Resource

8.3.1 Nutritional Composition of Hempseed

Hempseed is generally recognized as a food material that is rich in nutrients and is increasingly being called a nutrition powerhouse. Quantitative tests show that hempseed contains approximately 200–250 g of protein per kg, 250–350 g of fats per kg, 200–300 g of carbohydrates per kg, and 100–150 g of insoluble dietary fiber per kg, and contains considerable amounts of essential minerals (56–80 g per kg) (Hofmanova et al. 2014). This mineral composition also contains valuable amounts of iron, calcium, zinc, magnesium, and manganese, and they help in maintaining a balanced diet (Callaway 2004). Table 8.1 offers the detailed nutritional goods and products, hempseeds.

In addition to its exquisite macronutrient composition, hempseed is also abundant in numerous bioactive compounds which have been shown to have health benefit. The in vitro and in vivo studies have reported the presence of potent antioxidant activities and antihypertensive effects relating to hempseed components (Girgih et al. 2014). Interestingly, the concentration of anti-nutritional factors in hempseed is very low, which increases the bioavailability of proteins and makes it an alternative to the traditional sources of protein that are based on legumes (Russo and Reggiani 2015).

8.3.2 Protein Quality and Hempseed Functional Lipids

Hempseed is a protein food that consists mostly of albumin and edestin, both of which are non-allergenic storage proteins, highly digestible, and containing all the essential amino acids needed by human bodies (Mamone et al. 2019; Teh and Birch 2013). Hempseed protein is similar in terms of amino acid composition with both egg and soybean protein, which makes it one of the best-quality sources of plant-based protein (Vonapartis et al. 2015). Notably, the amino acid arginine that is especially high in hempseed is essential in the formation of nitric oxide and maintains the heart condition.

One of the nutritional features of hempseed oil is that this oil contains very high content of polyunsaturated fatty acids (PUFAs), which constitute approximately 80–90% of the total fatty acids. This oil contains linoleic acid (o-6) and a-linolenic acid (o-3), which are usually in a perfect proportion of 2:1 to 3:1 as required in promoting health (Teh and Birch 2013; Callaway 2004). Also, hempseed oil stands out as having biologically active long-chain fatty acid derivatives, including g-linolenic acid (GLA) and stearidonic acid (SDA), that are normally scarce in edible plant oils and add functional and therapeutic capacity to the oil.

8.3.3 Hempseed as Functional Food on Humans and Livestock

The history of hempseed goes back to a period of more than 3000 years where it has been used either as food in China and Nepal (Zuardi 2006). It is consumed either in its entire form as whole seeds and in several processed forms in the form of flour, protein powder, and oil. Despite the variation in the composition of hempseed depending on the varieties and the environmental conditions, the average hempseed mixture is characterized by 25–35% oil, 20–25% digestible protein, and 20–30% carbohydrates (Irakli et al. 2019).

Besides its general nutritional value, hempseed is full of numerous biologically active substances, such as antioxidant peptides, phenolic compounds, tocopherols, carotenoids, phytosterols. Phytosterols are especially important in the process of

Table 8.1 Applications and functional significance of hempseed concentrates in food products

Hempseed ingredient/ inclusion level	Product	Improved description of significance	References
5–10% defatted hempseed flour	Bread	The incorporation resulted in a slight decrease in the loaf volume and a considerable decrease in the formation of the heat-induced contaminants including hydroxymethylfurfural (HMF), aldehydes, and furfuryl alcohol, which means that baking was a safer process	Mikulec et al. (2019)
Hempseed oil press cake flour (0–40%)	Crackers	Gradual replacement generated increased nutritional density of the diet by increasing the levels of dietary fiber and at the same time decreasing the levels of digestible carbohydrates	Xu et al. (2022)
5–20% hempseed flour	Cookies	Addition of enrichment enhanced the protein, lipid, ash, total phenolic content and antioxidant ability, but gave the formation of soft texture and less hardness	Ertaş and Aslan (2020)
Hempseed flour	Biscuits	There was also an increase in the score of nutritional quality of the products, but the degree of sensory acceptance was relatively reduced, indicating a nutritional-palatability trade-off	Xu et al. (2022)
5–40% hempseed flour	Pasta	It enhanced the fat, ash, protein, and total dietary fiber levels, as well as facilitating desirable processing and extrusion properties	Teterycz et al. (2021)
Hempseed oil	Yogurt	The high antioxidant activity of fermented dairy matrix and the functional value was due to volatile bioactive compounds in the hempseed oil	Xu et al. (2022)
Hemp protein concentrate (20–60%)	Meat analogs	Had an increase in denaturation temperatures of proteins and decrease in water absorption leading to lower lightness (L) values and changed textural characteristics	Zahari et al. (2020)

(continued)

Table 8.1 (continued)

Hempseed ingredient/ inclusion level	Product	Improved description of significance	References
Hempseed, hemp flour, whole, and dehulled hempseed protein	Pork loaves	Inclusion enhanced hardness, dietary fiber, essential minerals (Mg, Fe, Cu, Mn) and polyunsaturated fatty acids, but there was a general reduced consumer acceptability; microbial stability was not affected	Zając et al. (2019)
Hulled *Fructus cannabis* (10–25%)	Bean curd (tofu)	The efficiency of processing was also significantly enhanced as the time spent in pressing milk was cut by half to three-quarters of what was necessary in traditional methods of production	Xu et al. (2022)
Hempseed oil	Yogurt	The volatiles of hempseed oil improved oxidative stability and antioxidant properties at storage and fermentation	Xu et al. (2022)
Defatted hempseed cake (≈60% fiber, 35% protein)	Extruded corn snacks	Enrichment made an increase in bulk density and mechanical strength as well as reduced tractability, cold-paste viscosity reduced, which showed better starch suspension stability, and enhanced solubility and water absorption indices	Leonard et al. (2020)
Hempseed flour (49% fiber, 30% protein, 8% fat) and protein concentrate (14% fiber, 49% protein, 12% fat)	Gluten-free bread	Gluten-free diets are also being enhanced through hempseed flour (significant increase in dietary fiber, 29.3 90.0 g/kg), and protein concentrates (15.2 to 61.0 g/kg) to enhance nutritional adequacy	Leonard et al. (2020)
Hempseed flour (52% fiber, 8% fat, 20% protein)	Gluten-free biscuits	The content of fiber, polyphenols, antioxidant activity, and protein content (41–143, 124, 40–122) improved significantly, and functional and nutritional characteristics were improved significantly	Mikulec et al. (2019)
Whole hempseed	Liver pâté	There was an increase in the quality of lipids especially in the form of more desirable fatty acid composition	Zając and Świątek (2018)

(continued)

Table 8.1 (continued)

Hempseed ingredient/ inclusion level	Product	Improved description of significance	References
Hempseed oil and hemp powder	Confectionery	Offer high-quality proteins and essential amino acids that add a unique nutty flavor profile to confectionery products	Xu et al. (2022)
Hempseed	Seasoning sauce	Sensory profile termed as slightly salty, slightly spicy, sour, aromatic, and slightly bitter; thermal processing has not caused any drastic changes in either color or taste	Xu et al. (2022)
Dehulled hempseed	Milk analog	Long fermentation improved the concentration of flavonoid and antioxidant activity considerably thus improving the quality of functional drinks.	Lai et al. (2021)

Modified after Thakur et al. (2025)

reducing the level of serum cholesterol by minimizing the absorption of lipids in the intestines and the formation of arterial plaque (Karche 2019). The properties help define hempseed as a functional food, which implies the possible benefits in the cardiovascular system and metabolism. Since players in the consumer market are getting more conscious of these benefits, numerous hemp-based food products have risen, including hemp butter, plant-based milk, protein supplements, and baking flour. Moreover, hempseed cake, a by-product of oil production is becoming widely accepted as a nutrient-rich feed supplement in animal farming. It has been found to improve the quality of the products by increasing the fatty acid composition of the milk and meat.

Hemp seed milk is a dairy substitution product that is prepared by using a combination of *Cannabis sativa* seedlings and has become a healthy and nutritious alternative dairy product due to its balanced macronutrients and functionality of the product (Thakur et al. 2025). Unlike traditional dairy, hemp milk is by nature lactose-free, soy-free, and gluten-free, which means it can be consumed by lactose intolerant and allergic individuals and is a complete protein with all the essential amino acids, which is highly valuable in vegan and vegetarian diets. This milk is interesting both in terms of its nutritional value and its organoleptic value due to the fact that the nutty taste and creaminess of the milk remain intact, and the proportionality of the fatty acids, especially the omega-3 and omega-6 polyunsaturated fats are in a perfect balance. Processing technologies such as wet milling, emulsification, and stabilization are significant in order to increase product stability and rendering it to be acceptable by the consumers, but there are still difficulties with the fragility of emulsions and the application of non-synthetic emulsifiers. In addition to alternative milk products, coagulated hemp seed milk-based tofu is a soy-free high-protein alternative to the conventional tofu, which is not yet fully developed in the review, but the principles are the same as the production of a hemp milk, i.e., high-protein and sustainable (Thakur et al. 2025). Overall, hemp milk (and its variations) can become the part of the plant-based nutrition sector in the future, and research needs to be conducted to enhance processing, shelf life, and consumer acceptability (Thakur et al. 2025).

8.3.4 Hempseed Oil: Nutritional and Functional Applications

Producing hempseed oil is typically carried out through cold pressing at a temperature of between 45 and 50 °C, as that of sunflower and linseed oils. The freshly pressed oil is green because of the naturally occurring chlorophyll in mature hemp seeds. A unique aspect of this oil is that it contains a high concentration of unsaturated fatty acids and approximately 80% of this is made up of tocopherols, which contribute to the oxidative stability and health benefits of this oil. Of the tocopherols, g-tocopherol is the most gaining most of the total tocopherol content about 89, and a-, b-, and d-tocopherols are also present in lesser proportions (Bagci et al. 2003). Phytosterols, GLA and SDA are found in hempseed oil and keep many of the known health benefits of this oil, including anti-inflammatory effects, cardiovascular protection, and skin health (Vogl et al. 2004). Consequently, hempseed oil is becoming more and more popular as a component of functional foods and nutraceutical products in the region of dealing with micronutrient deficiency and food insecurity.

8.3.5 Hemp Oil Personal Care, Cosmetics, and Industrial Applications

In addition to its nutritional uses, hemp oil has been incorporated into personal care and cosmetic products such as soaps, shampoos, balms, massage oils, and skin creams. Its elevated content of linoleic and linolenic acids helps in skin barrier, cell membranes stability, and immune regulation and thus is particularly useful in dermatology. Also, in the past, hemp oil has been used in other industrial sectors, such as inks, paints, varnishes, sealers, lubricants, and cleaning solutions. Its capabilities to absorb fast on the surface and have a high drying rate make it more useful in these applications. The possibilities of polymerized hemp oil have been explored as a sealant and coating material, and hemp bioplastics have potential as composite materials and plumbing materials (Vogl et al. 2004). Further research is required to increase the durability, recyclability, and biodegradability of industrial hemp products.

8.3.6 Hempseed and Hemp Oil, Feed Additives: Regulatory

The use of hempseed and products made using hemp in feeding animals is still under regulatory scrutiny because of the allegations of cannabinoid concentrations. Research evaluating the use of hempseed and hemp oil in the diet of cattle has produced mixed outcomes with some studies showing improvements in the nutritional value of the animal product and other studies showing detrimental effects such as reduction of feed consumption and behavioral changes when using high-cannabinoid plant products. Tetrahydrocannabinol (THC) predominantly occurs in the leaves and flowers, so there is a risk involved when giving animals whole hemp plants/silage. Consequently, the European food safety authority (EFSA) has introduced strict guidelines by limiting the levels of THC in the seed product feeds to a maximum of 10 mg kg-1. It also enables animal feeds to use hempseed, oil, expeller, flour, and fiber only when these components contain less than 0.2% of THC (Small 2015). Conversely, currently (as of March 2023) the use of hemp and hemp-derived products in animal feed and pet food remains illegal in the USA, because such use was not authorized in the 2018 Farm Bill. Regulatory issues in the USA are controlled by the Food and Drug Administration (FDA). In general, despite these pronounced nutritional benefits of hempseed and its products to human beings and livestock, the introduction of hempseed to livestock feeds has to be founded on rigorous scientific evaluation and in compliance with legal requirements (Viskovic et al. 2023).

8.4 Functional and Nutritional Significance of Hempseed-Based Food Products

The introduction of humped and its derivatives in most of the foodstuff and products depict the growing importance of the crop as a multifunctional, sustainable nutritional ingredient. Hempseed flour and concentrates are commonly utilized in cereal-based products, such as bread, crackers, cookies, pasta, and gluten-free formulations to enhance the quality of proteins and levels of dietary fiber, antioxidant ability, and mineral content. Using the example of the implementation of bread formulations with defatted hempseed flour, it has been proven that it reduces the formation of undesired heat-related contaminants (hydroxymethylfurfural and aldehydes) in addition to which its overall nutritional security is improved (Mikulec et al. 2019). This is also true of crackers, cookies, and biscuits because the inclusion of hempseed boosts fiber and phenolic and reduces carbohydrate, although it may not always make them S.A.-acceptable (Ertas and Aslan 2020; Xu et al. 2022). The analytical consideration of the hemp-based products is presented in Table 8.1.

Other effects that hempseed ingredients have on processing behavior and products structure, in terms of technology, include the product structure. An increased proportion of fat, ash, and fiber in pasta and extruded snacks causes improved extrusion, water uptake, and starch stability whereas elevated bulk density and hardness require optimization of the formulation (Teterycz et al. 2021; Leonard et al. 2020). Hempseed protein and flour are important sources of nutritional deficiencies in gluten-free diets, especially since gluten-free systems are generally less nutritionally dense than gluten systems due to their low fiber and protein levels.

Hempseed oils and proteins also mean that it can be used in dairy and meat alternatives. Yogurt and milk analogy using hempseed oil has proven the addition of volatile bioactive compounds to enhance antioxidant activity and oxidative stability and fermentation can further enhance flavonoid content and functional value

(Xu et al. 2022; Lai et al. 2021). Essential amino acids, minerals, and polyunsaturated fatty acids found in the meat analogue and processed meat products make hemp protein supplements have a healthier lipid profile, but a change in texture and color may lead to decreased acceptance by the consumer (Zajac et al. 2019; Zahari et al. 2020).

8.5 Industrial Hemp as a Resilient Crop

8.5.1 General Agronomic Characteristics

Hemp is regarded as an unconventional crop owing to its exceptional ecological adaptability, as it generally performs better than many established temperate crops across diverse temperature regimes and climatic conditions (Crini et al. 2020). Hemp is free of toxins and non-acidifying, environmentally friendly, and agronomically adapted (fast growth approximately 3–4 months), deep root system, and improves or alters the soil structure. Such characteristics have aided in increasing its relevance in both the commercial markets and scientific research (Adesina et al. 2020). The crop does not need much water and agrochemicals and has a high carbon uptake rate. Depending on the source, it is estimated that hemp can sequester up to five times the atmospheric CO2 per area than forest trees in its growth cycle, in part due to rapid biomass growth (Ranalli and Venturi 2004; Aytac 2018). Such attributes put hemp into the list of potentially effective alternatives in terms of stimulating environmental protection and climate-resilient agriculture (Bouloc 2013). Moreover, the re-emerging interest in the production of hemp in the world has been reinforced by the changing legalization policies and a wide range of potential industrial applications, such as food, textiles, cosmetics, bioplastics, pharmaceuticals, and building materials (Adesina et al. 2020).

8.5.2 Soil and Climate and Growth Requirements

Industrial hemp is incredibly adaptive to numerous environmental factors and grows well in the soils with a moderately acidic to neutral pH of between 6.0 and 7.5 that are well-drained and rich in nutrients (Bouloc 2013; Crini et al. 2020). The crop can also withstand moderate cold and heat and it is more adaptable than most crops that are grown in temperate areas (Adesina et al. 2020).

Planting is usually done in late spring when the soil temperature has already reached a high of say 10 °C, where germination of seeds and emergence of seedlings take a period of 3–7 days, unless the soil is very dry (Bouloc 2013). The optimum growth of the vegetation occurs between 13 to 25 °C. Hemp is a short-day plant and will begin flowering and maturing with the decrease in daylight hours in late summer and early autumn, so photoperiod sensitivity is a critical factor to consider when selecting varieties and timing of planting at different latitudes (Ranalli and Venturi 2004). The growth cycle in hemp is short and hemp is fully matured in

100–120 days. It has a fast biomass production rate, positive relations with the soil that do not result in acidic soil, and deep root development, which improves soil structure and aggregation that boosts its agricultural and ecological utility (Crini et al. 2020).

8.5.3 Sowing Practices and Density of Plants

The density of the plant is an important agronomic factor that influences the yield of hemp and depends on the purpose of the crop. Fiber and dual-purpose types are typically grown at more densities (between 100–120 plants m-1), and are planted with narrow row spacing (approximately 0.5 m). This dyestand planting facilitates vertical growth of the stem, increase of the biomass production, and effective management of the weeds due to the rapid canopy cover (Bouloc 2013; Ranalli and Venturi 2004).

Planted to yield grain or seed, on the other hand, are commonly spaced at a reduced plant density, typically 5–15 plants m −1, and sometimes 3–5 plants m −1, to encourage lateral branching, more inflorescence development and seed production per plant. At such densities intra-specific competition is also reduced and allows a greater spread of canopies and reproductive allocation, which is fundamentally important in grain productivity maximization. The given agronomic flexibility of hemp highlights the fact that farmers have been able to tune the density of their planting to any one end of a range of high-density, fiber-based production models and low-density, grain-based production models depending on the end-use objectives, the resources they have, and what their agroecology is able to accomplish (Adesina et al. 2020).

8.5.4 Nutrient Management

Nitrogen is the most important macronutrient that determines the growth, biomass increase, and seed production in hemp. The recommended rates of nitrogen impounding are normally 150,200 kg N per hectare of growing season (Bouloc 2013). The studies have shown that the increment of biomass and seed production can be significantly increased due to the nitrogen inputs of about 200 kg per hectare, but higher rates than 150 kg per hectare tend to result in the decreasing returns, and can negatively affect the quality of fibers (Ranalli and Venturi 2004). The phosphorus requirements of hemp are quite small compared to other agricultural crops, whereas the requirement level of potassium is greater, which is approximately 175 kg of K 2 O per hectare. Generally, hemp does not need as many nutrients as most common fiber crops such as cotton (Adesina et al. 2020). The nutrient uptake is greatest at the stage of vegetative growth and therefore it is recommended to apply split application of the fertilizer or slower formulations, which are related to the developmental stages of the crop. This strategy will be able to increase the efficiency of nutrient use and minimize environmental losses (Bouloc 2013).

8.5.5 Water Demand and Irrigation Control

Hemp has a moderate water demand and is characterized by a rather high water usage efficiency. Once it is established, the crop primarily relies on rainfall, but the adequate soil moisture is essential at the stage of germination and initial growth of seedlings (Ranalli and Venturi 2004). The industrial hemp water footprint is projected to be approximately 2700 liters of water per kilogram of dry biomass, which was much lower than the cotton projected water footprint that is approximately 10,000 liters per kilogram of dry biomass (Aytac 2018).

Hemp does well in temperate regions that are receiving 700–1000 mm of rainfall per year. Conversely, arid or semi-arid areas might need extra watering especially at the initial phases of development. Hemp is considered to be a viable crop to the agricultural systems that have limited water supply and are susceptible to climate change (Bouloc 2013; Crini et al. 2020).

8.6 Hemp and Environmental Sustainability

8.6.1 Climate Resilience and Ecological Adaptability

Hemp is also emerging as a friendly crop to the environment, which is a suitable crop in climate-resilient agricultural production. The increasing usage of the hempseed concentrates in the food systems is highly significant, especially in the climate change scenario. The fact that it can survive with relatively low external requirements including average fertilization, minimum irrigation, and minimum use of pesticides, makes it more ecologically advantageous than most of the conventional crops (Bouloc 2013; Adesina et al. 2020). Those qualities make hemp a viable option in low-input farming systems that consume few resources particularly due to climate change that creates unpredictability in the farming environment. Also, hemp has a natural resistance to numerous pests and diseases, reducing the use of chemicals to protect the plants by far, which stands in line with organic and integrated agriculture (Crini et al. 2020). Moreover, the hemp grows a deep and extensive root system that increases soil structure, aeration and water infiltration. This rooting trend enhances increased resistance to drought as plants are able to access deeper water reserves in the soil hence become more resistant to such water-deficient conditions.

Hemp is cultivated in diverse agro ecological systems in different parts of the world and has great degree of adaptability to climatic conditions which are comparable to those required in the farming of wheat. It grows best in temperate up to cooler weather and is well adaptable to well-drained and fertile medium and heavy soils such as silty loam, clay loam, and silty clay. Besides its capacity to adapt to farming, hemp is also quite capable of phytoremediation since studies have indicated that it can remove heavy metals in polluted soil even better than most common crops. Thus, hemp could be planted on polluted sites to clean up the soil gradually provided that the organization of the extraction and distribution of the contaminants extracted is done with caution. Hemp is best in mild temperate climates that have adequate moisture in manufacturing of fiber. A minimum of 0.65 m of rainfall in a year is vital in the period of seed germination and the early growth stages. Planting seeds spacing minimizes the chances of lodging and damage by the unfavorable climate conditions such as wind, rain, or hail. Moreover, the hemp does not need many pesticides and fertilizers, which is why it is a great candidate as the part of the crop rotation system that promotes the health of the soil and the sustainability of agricultural activities in the long term (Struik et al. 2000).

8.6.2 Circular Bioeconomy

The utilization of by-products of hempseeds such as press cakes and defatted meals complies with the concept of the circular bioeconomy since the waste is reduced and the utilization of resources becomes more efficient. Hemp-based proteins and fibers added to staple foods are an ecologically friendly approach to nutritional security by decreasing the use of production methods that require considerable resources and are based on animal protein (Mikulec et al. 2019; Ertas and Aslan 2020; Xu et al. 2022). Also, the diet diversification of which hemp-based functional foods input is one of the significant adaptation strategies that only grow in significance in the confrontation with climate disturbing interference in conventional crop systems. The nutrient density, antioxidant capacity, and functional stability of foods can be improved with hempseed ingredients and result in strong diets in the conditions of environmental stress and unstable food systems (Ertas and Aslan 2020; Xu et al. 2022).

Hemp has an extraordinary capacity to sequester carbon, and this is one of the greatest environmental advantages of hemp. Because of its fast growth and high biomass output, the hemp can capture atmospheric CO_2 at an estimated rate that is up to five times higher per unit area than forest trees during its growth (Ranalli and Venturi 2004; Aytac 2018). The carbon obtained with the help of photosynthesis is fixed in the plant tissues and when hemp biomass is utilized in long-lasting products such as construction materials, insulation, and biocomposites, the carbon can be stored over long periods. These attributes bring to light the potential role of hemp in a carbon neutral or even carbon-negative production system. With the rise of climate change effects on the world food, fiber, and material production, hemp can be considered a multifunctional crop that combines ecological stability with a limited impact on the environment and diverse industrial applications. Its integration into sustainable bio-based useful chains also makes it more significant as a strategic crop in terms of climate change reduction and the achievement of sustainable development objectives (Bouloc 2013).

8.6.3 International Growth and Industrial Importance

The global renaissance of hemp cultivation has been greatly ignited by the increasing legalization processes and the increase in demand of hemp products in the different industries including food, textiles, cosmetics, pharmaceuticals, bioplastics, and sustainable construction materials (Adesina et al. 2020). These trends enhance the significance of hemp as a source of agricultural products as well as the principal component of new bio-based and circulatory economies emphasizing sustainable development.

8.7 From Traditional Remedies to Modern Nutraceuticals

It was cultivated in most of the regions for more than 10,000 years and was transferred to India and the Far East during the Neolithic period, where it had been used in the food, fiber, medicine, oil, paper, and rope industries since then. *C. sativa* is adaptable and can be grown under different climatic conditions and soils though it grows well in warm and moist climatic conditions (Chaachouay et al. 2025). In recent decades, the therapeutic and nutritional usefulness of hempseed has received increased attention, though the medicinal use of hempseed is the most ancient of the traditional systems of the East, with its thousand years of history. Historically, hempseed was used in the treatment of skin disorders, pain, cough, jaundice, and digestive problems, although it is not clear whether the effects were caused by the action of the seeds, or a mixture of the seeds and the bracts which also contain cannabinoids and terpenes. Hempseed is also traditionally used to treat blood-related disorders and constipation. The Chinese development of a formula of Ma Zi Ren Pill, one of the traditional Chinese medicine formulas, where *Cannabis sativa* pollens are one of six ingredients works as an interesting example, and demonstrated valuable clinical results in the treatment of constipation by enhancing the full spontaneous movements of the bowel and reducing the number of straining motions without any serious adverse care (Cheng et al. 2011; Zhong et al. 2013). The hempseed-based medicine, Maltos-cannabis, gained popularity in Scandinavia in the early part of the twentieth century, and was used to treat an extensive list of disease, including lung diseases, anemia, gastritis, scrofula, neurasthenia and general weakness. Hempseeds started to be used in the poultry and feeding and in isolated instances in the human diet toward the end of the twentieth century. Today the hempseed products attract a new euphoria in other countries of the world, due to the recognized use of the seed in the production of medicines, nutraceuticals, and cosmeceuticals and suggest the shift in the historical perspective of using this seed in a form of fibers to the more extended use to nutritional and therapeutic advantages.

8.8 Hemp as an Organic and Sustainable Farm Crop

Hemp has broader plasticity in terms of above and belowground growth parameters including root characteristic amongst the plant populations. This developmental plasticity was further supported by the fact that the root characteristics remained identical year after year although changes in soil characteristics were apparent. When not fixed to compacted layers of soil, hemp roots may extend down to

200 cm reaching any groundwater present. Hemp has such characteristics and can be considered in the context of organic farming. The massive root biomass in this experiment and, particularly, its presence in the bottom layer of the soil provide further reason to the possible opportunities of hemp in regions of sustainable and resilient cultivation (Amaducci et al. 2008).

8.9 Environmental Impact Assessment of Fiber Hemp Production

Life Cycle Assessment (LCA) shows that Fiber hemp is a low-impact crop compared to other crops such as potato and sugar beet because Fiber hemp is less harmful to the environment in most impact measurements. The production of hemp has been estimated to hold a potential of approximately 20.5 kg PO43-equivalent of eutrophication potential, 2330 kg CO_2-equivalent of global warming potential (GWP), 9.8 kg SO_2-equivalent of acidification potential, 2.3 kg 1, 4-dichlorobenzene-equivalent of terrestrial ecotoxicity, 11.4 GJ of cumulative energy demand and 1.02 ha 6 year of land use per unit of hemp. These values are typically smaller or similar to the values of standard fiber and oilseed crops such as cotton, maize, and rapeseed which generally demand more energy, are more land-use intensive, and also have a far greater GWP and eutrophication effect due to the high fertilizer and pesticide inputs. The effects of marine eutrophication of hemp systems are nitrate ($NO3^-$) emissions, which can be regarded as up to 75–89% of the overall eutrophication potential of up to 75–89% of eutrophication of the fertilized fields, and this is also the case with nitrogen leaching and runoff. On the other hand, nitrous oxide (N_2O) (56–59% of total GWP) and carbon dioxide (CO_2) (40–43% of total GWP) are the single largest contributors to the potential with insignificant additional contributions by other gases. These percentages indicate how much the individual sources of emissions contribute to each category of impact rather than variability in aggregate levels of emissions and do not reflect the variability in impact of hemp farming in aggregate. The solution to the negative impact on the environment therefore needs to be to concentrate on mitigating eutrophication, followed by climate change, acidification and energy consumption. Although the eradication of tillage has clearly significant benefits because it reduces energy consumption, acidification, and climate change impacts, the replacement of mineral fertilizer with pig slurry is counterproductive as it encourages eutrophication, acidification, and terrestrial ecotoxicity. Mitigation to reduce the environmental impact of hemp farming includes leaching of agricultural soils with nitrates. Previous analyses indicated that by cutting the degree of nitrate draining by a half, impacts of eutrophication will be reduced by approximately 43% and the climate change impacts will be lowered by an estimated 10%. To reduce the nitrate leaching in the hemp-based systems, one can modify the rate of nitrogen fertilization to crop demand, transfer the nitrogen application in split applications at the moment of maximum uptake, and use slow-release or organic forms of nitrogen. Additional prevention can be encouraged through amelioration of the soil structure by incorporating organic matter, maintaining continuous soil cover by cover crops or intercropping and reducing tillage all of which enhance nitrogen and microbe retention and immobilization, respectively. The crop is particularly receptive to these nutrient management practices owing to early vigorous development of hemp, highly developed root system and nitrogen usage effectiveness which can then be utilized to attain improved nitrogen use and also the nitrate loss to groundwater and the surrounding water environment can be quantified (van der Werf 2004).

8.10 Performance of Hemp Varieties under Mediterranean Climatic Condition

Italian researchers compared the development of different hemp varieties in the North-East of the nation using trial techniques and simulations. Results had shown that the region has the potential to produce hemp biomass and stem can in the same way as the rest of the European regions, although seed yields on average were lower. It was noted that high temperatures, particularly above 30 °C during the grain filling stage was a major constraint and had a strong negative correlation with seed oil content and overall seed quality. In addition to this, the low biomass productivity was also noted to be caused by some of the soil properties, in addition to low rainfall. Among the cultivars, monoecious (Fedora and Ermes) performed better in multipurpose use where seed quantity was important, but a higher number were favored as dioecious (CS). However, they discovered that the quality of most of the cultivar seeds was poor (they contained low levels of oil and protein and high levels of fiber quality) thereby giving low processing yields as anticipated. The combination of field trial and simulation modeling was observed to be extremely helpful in terms of predicting crop performance, environmental impacts, and optimization of irrigation schemes, which also brought out further information of variability of yield, maturity period and water requirements across different climatic conditions (Baldini et al. 2020). Figure 8.1 and Table 8.2 illustrate the global production of hemp to fiber and seed between 1993 and 2013 and the nations producing hemp, respectively.

8.11 Role of Hemp in Sustainable Agriculture and Industry

Hemp has great promise in helping to support the sustainable crop system as well as other greater contribution to the sustainability within the society, yet to what extent is the contribution it will make actually evident. The analysis of the life cycle at late infancy showed that hemp production had both less energy demands as well as lesser climate impact than other important agricultural produce like wheat, corn, potato, and sugar beet and also less risk of eutrophication and acidification of water bodies (van der Werf 2004). Despite these advantages, hemp is an annual plant, and therefore must be planted on an annual basis and must be fertile enough to yield (van der Werf 2004). Production sustainability similar to other annuals depends on minimal disturbances of the soil and maximum nutrient input (Hobs et al. 2008). The establishment of hemp crops could be technically demanding and could require more

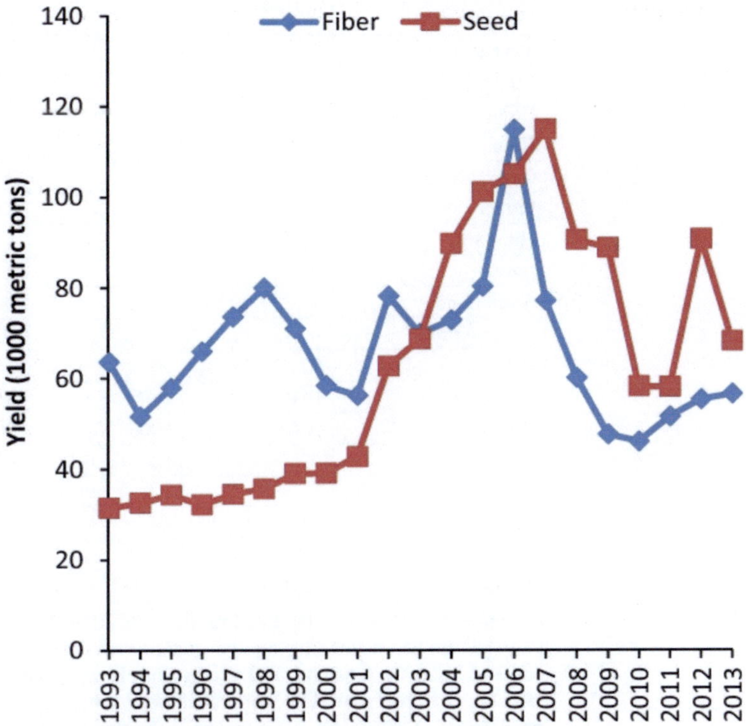

Fig. 8.1 Global production of hemp for fiber and seed from 1993 to 2013

research to optimize no-till and reduced-tillage methods of planting, which minimize soil erosion, improve soil health and might also elevate soil carbon sequestration (Fu et al. 2006; Langdale et al. 1979). More studies are also required to simplify the concept of an integrated pest management and to identify, in particular, how hemp can be used on different crops in the context of their various types of crop rotation (Britt et al. 2019). Although the final gauge of the on-farm merits of hemp remains to be taken into account, the crop is already demonstrating that it can be of enormous value beyond the farmland as the range of its industrial, nutritional and consumer applications is enormous. It was also noted earlier that hemp had the potential to be used in a total of more than 25,000 different products, a prospect that created tremendous optimism prior to its re-legalization (Anonymous 1938). Such expectations have not been fulfilled wholly, however, but in part this is to be attributed to the fact that during most of the twentieth century there has been no serious investment in the field of agronomic, industrial, and product-development research. Thus, much time, long-term investment and particular research will be required to ensure hemp becomes a powerful and useful constituent of modern agrarian systems and bio-based economies.

Table 8.2 Countries producing hemp

Countries	Seed (ha)	Fiber (ha)	Total (ha)
Austria	–	600	600
Canada	n/a	n/a	34,262
Chile	2200	4500	6700
China	5500	5000	10,500
Czech Republic	–	200	200
France	7706	600	8306
Hungary	1400	300	1700
Italy	–	425	425
Netherland	–	1284	1284
North Korea	–	20,000	20,000
Poland	60	70	130
Romania	1440	1600	3040
Russia	1644	4000	5644
South Korea	–	12	12
Spain	2	10	12
Turkey	1	20	21
Ukraine	2100	2000	4100
Total	22,053	40,621	90,906

Data was collected from the FAO for the year 2013. Total land area may be smaller depending on the quantity of dual-purpose (i.e., cultivars used for seed and fiber production) hemp planted. Annual statistics are from the FAO. Abbreviation: *ha* hectares. *Value is from Health Canada for the year 2015 (Schluttenhofer and Yuan 2017)

8.12 Laws and Growing Hemp Demand of Low-Tetrahydrocannabinol Varieties

The growing need of consumers to consume plant-based and nutrient-rich products has also boosted the rise in the consumption of hemp seeds in human diets and animal feeds (Bouloc 2013). Consequently, the development of non-pharmacological hemp plants with low 9-tetrahydrocannabinol (THC) is becoming stronger and stronger (Burton et al. 2022). According to the legal definition, industrial strains of Cannabis sativa that have low concentration of THC are considered hemp, and according to Commission Regulation (EU) 2022/1393, the highest concentration of 9-THC that can be used in hemp seeds, partially defatted seeds, and processed hempseed products is 3.0 mg/kg. The transparency in regulations helps in the addition of hemp-derived products in the food systems without any harm and in accordance with the international safety regulations.

8.13 Challenges for Hemp Cultivation and Utilization

8.13.1 Seed Shattering and the Post-harvest challenges in Hemp Grain Production

A large proportion of seed shattering causes yield loss in grain-grown hemp. The female and monoecious plants have their terminal inflorescences enwrapped in small bracts and the seeds are not evenly ripened- in the lower and earlier flowering sections to the top. This way a small number of the seeds can mature and dry well before the other ones and therefore before they even have a chance to mature and develop, they will prematurely shatter while the rest of the seeds are still in the dough stage or unfilled. Although the future breeding is estimated to increase uniformity in the ripening and in reduction of shattering, the existing methods of production force farmers to pick the crop earlier than the appropriate maturity date in order to cut on the losses and fibers wrapping around the harvesters (Soroka and Tarakan 1981). Premature harvesting, however, causes new post-harvest requirements: either the seeds have to be dried in a rush, either by a machine, or by letting them dry in the air; otherwise they will be spoilt or infected by bacteria, and even more cleaning is necessary to clean up the mess, which again increases the risk of hulls cracking and reducing the quality of the seed. All these activities bring about the expenses of labor, energy, and expense that ultimately resulted in increased carbon footprint of hemp grain production (Jeff and DW 2019; John et al. 2019).

8.13.2 Fiber Quality and Harvesting Standards of Textile Grade Hemp

It is proved that Hemp fibers are applicable in numerous industrial applications but the most economical way of applying these fibers is the field of textile production. Hemp must be harvested and processed to preserve the high-quality bast fibers in order to be useful as a complement, or even as a substitute to the cotton. Optimal production of fiber would be achieved when the plants would be harvested at, or immediately prior to flowering, the period when the average percentage of the female plants will have already begun their flowering. Structural support of the stem is activated by flowering to support the enlarged seed heads which causes lignification of bast fibers and secondary hurd fiber deposition (Ely et al. 2022). Both changes reduce the flexibility of fibers and complicate downstream processing and, therefore, are no longer as easy to employ in the textile industry (Westerhuis et al. 2019). The varieties of dual purpose, which are cultivated to be used as both grain and fiber, may still be necessary to furnish the other lower value products, but they will not be able to supply the finer textile that is required. Due to the extreme dependence of the quality and yield of fiber of hemp on the cultivar used and on the time of planting and harvesting, it is important to have detailed information about agronomic variables to optimize the production of textile grade hemp.

8.13.3 Sustainability Consideration, Processing Approaches, and Practices of Harvesting

When harvesting hemp as grain, fiber/floral compounds the methods required are different and they have varying energy requirements, and sustainability implications. The manufacture of grain is generally founded on the utilization of combine harvesters, notwithstanding the reality that it is commonly gathered at elevated moisture content to lessen the count of shattering losses, the recently picked grain has to be transported, dried, and cleansed within no time. Such after harvest programs which are often operated by natural gas or any other fossil fuel contribute significant carbon footprint to the crop. Flower production incorporates more techniques including the use of hand weeding up to the modernized systems. Drying of plants may be done in the air following harvest, or in controlled conditions of forced, heated air. They are growers where entire plants are clipped using forage harvesters which produce the material that may be bailed or extracted in silage. However, flower production is often more energy intensive than fiber ones though it occurs on a smaller scale.

Traditional fiber harvesting has largely relied on hay-type machinery that cuts hemp stalks and leaves them on the ground for drying. Modern harvesting equipment may additionally crimp or spread the stems to promote uniform drying and retting. In exceptionally tall or dense crops, specialized machinery may also be required to arrange the plants in a manner that facilitates efficient downstream handling and processing. More advanced harvesting technologies now include modified combine harvesters that are capable of harvesting multiple portions of the plants simultaneously, including harvesting leaves and flowers to obtain cannabinoids, and harvesting stalks to ret. and process fibers. Cannabinoids are also enhanced by the use of special equipment that is designed to separate floral tissues and stems. Such innovations are economical in situations where they reduce the number of laborers as well as assist in making processing cost-effective. A fiber production system requires controlled microbial degradation, called retting, that comes after the harvesting process and enables the separating process of bast fibers and the woody hurd. Field retting, in addition to increasing the separation of the fibers, results in nutrient recycling, in which the leaf litter, along with other plant residues, is returned to the soil, thereby reducing the external addition of fertilizers. However, it is highly weather-dependent process and unfavorable weather conditions, i.e., low temperatures, excessive rainfall, or the extended period of high moisture can disrupt the retting processes and negatively affect the quality of the fibers. More advanced processing technologies, in contrast, do not need the in-field retting process, use the entire plant biomass, with varying distributions of nutrients, and may have higher power input or result in wastewater in processing.

In the long term, production will tend to be more differentiated when the hemp market expands. Where the markets favor high-quality bast fiber over hurd, late harvesting may be a cost-effective option because the hurd will yield high quantity at the expense of quality bast fiber. Besides this, an improved understanding of how the timing of harvesting influences fiber properties can help the processors better align a fiber type with a specific end-use, and with the more valuable industrial use.

8.13.4 Challenges in Hemp Supply Chain and Market Volatility

There must be a reliable and stable supply chain of any agricultural commodity and hemp cannot be an exception. The biggest excitement when the legalization of hemp cultivation was announced in the USA was the high-value cannabinoid production. Very high prices of hemp flowers (at the beginning of 2019) encouraged many growers to enter the market hoping to generate a lot of money. However, by the close of the year this scenario changed immensely, as many manufacturers were unable to find buyers to their cultivated crop, and the value of hemp flowers dropped down to nearly 25,000 dollars per hectare. Some of the processing companies were forced out of business when the prices dropped because they were not able to conduct business profitably. The market is, literally, only beginning to level off but no chance of the market rejoining those high inflated prices but this time perhaps never again and even as the industry moves into a more mechanized production process. The absence of adequate processing capacity and extreme start-up risks are the major issues in addition to the fluctuation in price, which are impeding the evolution of hemp as a large-scale grain, or fiber crop. The emergence of a new industry can be viewed as a traditional chicken-or-egg scenario; farmers are hesitant to grow hemp, and processors are hesitant to invest in infrastructure before they have a minimum amount of the raw material. It is one of those uncertainties that will affect the entire supply chain. Though processors may strive to be innovative and develop new hemp-based materials, the processors may be unable to convince manufacturers to utilize new materials especially in instances where there are other competing materials, which can meet the required regulatory standards and they have already secured a market (Fike et al. 2008).

8.13.5 Legal Hurdles

The cultivation of industrial hemp is stipulated to possess a maximum amount of THC. The legal grow stipulates that the plant must not contain over 0.3% total THC and it is upon the farmers in India to demonstrate the same. Among the problems is that hemp and psychoactive marijuana nearly appear alike and there is no possibility that they can be distinguished by the naked eye unless they differ in their composition of THC. This is particularly a challenge to Indian hemp growers. The compulsory examination of THC can be rather costly and even traumatic as the crop that fails to meet the required standard will be burned by the police. Such compliance requirements may act as an incredible obstacle to the development of large-scale hemp production and research since it is likely to create stress, intimidation, and so on. However, with the favorable trends in the attitude of the population and the state toward hemp, a more liberal regulatory regime can emerge. This could encourage more agronomy research and development specifically the work to breed hemp varieties that would consistently produce less than 0.3% THC. A market already exists in the USA and some other countries on certified hemp varieties.

8.14 The Future of Hemp as a Commodity of the Future and Its Problems

Hemp has the potential to be one of the biggest commodity crops in the future, but it will require major concerted effort all along the supply chain. The ability to find appropriate varieties of hemp and to establish appropriate production timetables that fulfill the yield and quality standards in a certain eco-region will determine the successful introduction of hemp into the current farming modes, in the farm level. Also sustainable production methods like no-till methods, cover crops with good rotation of crops should be invented and refined. Moreover, the growers will be forced to adapt their processes to the interests of the processor, such as the harvest time, retting preferences, etc., which may impact the quality of products and systems sustainability. To ease this change introduction, there will be need to develop and then to select cultivars with high agronomic and processing traits. Despite technicalities involved in production of hemp, whether a farmer should grow hemp will be determined more on economic factors. Farmers will not use the alternative crop unless there are clear benefits of using the alternative crop as compared to the traditional methods of crop systems, unless that alternative can bring the obvious benefits in terms of increase in revenue or reduction in production costs or other benefits that can be felt. Without these benefits, hemp would not be massively adopted and popular as it is supposed to be.

8.15 Implications of Research, Industry, and Policy

Nutritional and functional merits of hempseed-based foods already have been proven, however, wider application of these foods needs to be optimized. The processing and phytochemical variability and sensory challenges suggest the need of having standard raw materials, selection of cultivar, and improved food processing methods. In terms of policy and sustainability, hemp planting as a climate-smart agriculture, value-added food chains as well as national nutrition plans also have the potential to boost food security, in addition to promoting environmental-friendly production systems.

8.16 Conclusion

Hemp truly is a multipurpose and resilient crop that could help solve severe problems in the development of agriculture, sustainability, and supply chain integrity. Its capacity to serve two basic needs i.e., food and fiber are incomparable to the existing crops. Hemp is unique to the agricultural sector of the world because of its nutritional value, fiber, industrial, as well as medicinal applications. Although hemp is a strong and ecologically flexible plant, it does not need many resources and is able to grow in a wide range of climatic conditions and improves the quality of soil which is not a minor aspect considering the fact that it provides a more environment friendly alternative to conventional crops. Nevertheless, to achieve its potential, considerable issues are associated with harvesting, processing, and regulatory issues, especially in terms of the level of THC, which should be properly addressed. Hemp may form an essential commodity crop in future, forming innovative production systems, agricultural economy diversification, reducing environmental damage, and food and industrial supply chains with targeted development efforts.

References

Adesina I, Bhowmik A, Sharma H, Shahbazi A (2020) A review on the current state of knowledge of growing conditions, agronomic soil health practices and utilities of hemp in the United States. Agriculture 10(4):129

Amaducci S, Zatta A, Pelatti F, Venturi G (2008) Characterisation of hemp (*Cannabis sativa* L.) roots under different growing conditions. Plant Soil 313(1–2):227–235. https://doi.org/10.1007/s11104-008-9695-0

Anonymous (1938) New billion-dollar crop. Pop Mech:238–239

Aytaç S (2018) An environmentally friendly plant in terms of oxygen supply: Hemp

Bağci E, Bruehl L, Aitzetmuller K, Altan Y (2003) A chemotaxonomic approach to the fatty acid and tocochromanol content of *Cannabis sativa* L. (Cannabaceae). Turk J Bot 27(2):141–147

Baldini M, Ferfuia C, Zuliani F, Danuso F (2020) Suitability assessment of different hemp (*Cannabis sativa* L.) varieties to the cultivation environment. Ind Crop Prod 143:111860. https://doi.org/10.1016/j.indcrop.2019.111860

Bouloc P (2013) Hemp: A plant with a worldwide distribution. CABI. https://doi.org/10.1079/9781845937935.0001

Britt KE, Pagani MK, Kuhar TP (2019) First report of brown marmorated stink bug (*Hemiptera: Pentatomidae*) associated with *Cannabis sativa* in the United States. J Integrated Pest Manag 10(1). https://doi.org/10.1093/jipm/pmz014

Burton RA, Andres M, Cole M, Cowley JM, Augustin MA (2022) Industrial hemp seed: From the field to value-added food ingredients. J Cannabis Res 4(1):45. https://doi.org/10.1186/s42238-022-00156-7

Callaway JC (2004) Hempseed as a nutritional resource: An overview. Euphytica 140(1):65–72. https://doi.org/10.1007/s10681-004-4811-6

Chaachouay N, Qureshi R, Benkhnigue O, Azeroual A, Zidane L (2025) Chapter 11: Marijuana (Cannabis sativa L. Cannabaceae). In: Chaachouay N, Azeroual A, Zidane L (eds) Comprehensive Guide to Hallucinogenic Plants. CRC Press, Florida. Taylor & Francis Group. Pages 20

Chandra S, Lata H, Elsohly M (2017). Cannabis sativa L.: Botany and biotechnology

Cheng CW, Bian ZX, Zhu LX, Wu JC, Sung JJ (2011) Efficacy of a Chinese herbal proprietary medicine (Hemp Seed Pill) for functional constipation. Am J Gastroenterol 106(1):120–129. https://doi.org/10.1038/ajg.2010.305

Clarke RC, Merlin MD (2016) Cannabis domestication, breeding history, present-day genetic diversity, and future prospects. Crit Rev Plant Sci 35(5–6):293–327

Crini G, Lichtfouse E, Chanet G, Morin-Crini N (2020) Traditional and new applications of hemp. In: Crini G, Lichtfouse E (eds) Sustainable agriculture reviews 42: Hemp production and applications. Springer, pp 37–87

Ely K, Podder S, Reiss M, Fike J (2022) Cannabis/hemp: Sustainable uses, opportunities, and current limitations. In: Cannabis/Hemp for Sustainable Agriculture and Materials, pp 59–87

Ertaş N, Aslan M (2020) Antioxidant and physicochemical properties of cookies containing raw and roasted hemp flour. Acta Sci Pol Technol Aliment 19(2):177–184

European Commission (2026). Hemp. Agriculture and rural development. Retrieved June 27, 2026, from https://agriculture.ec.europa.eu/farming/crop-productions-and-plant-based-products/hemp_en

Fike JH, Parrish DJ, Alwang J, Cundiff JS, Lamb M (2008) Challenges for deploying dedicated, large-scale bioenergy systems in the USA. CABI Reviews 28

Fu G, Chen S, McCool DK, Elliot WJ (2006) Modeling the impacts of no-till practice on soil erosion and sediment yield. Soil Tillage Res 85(1):38–49. https://doi.org/10.1016/j.still.2004.11.009

Galasso I, Russo R, Mapelli S, Ponzoni E, Brambilla IM, Battelli G, Reggiani R (2016) Variability in seed traits in a collection of *Cannabis sativa* L. genotypes. Front Plant Sci 7:688. https://doi.org/10.3389/fpls.2016.00688

Girgih AT, He R, Malomo S, Offengenden M, Wu J, Aluko RE (2014) Structural and functional characterization of hemp seed protein-derived peptides. J Funct Foods 6:384–394. https://doi.org/10.1016/j.jff.2013.11.005

Grotenhermen F, Müller-Vahl K (2016) Medicinal uses of marijuana and cannabinoids. Crit Rev Plant Sci 35(5–6):378–405

Health Canada (2020) Policy for the management of industrial hemp varieties on the list of approved cultivars. Government of Canada

Hill KP (2015) Medical marijuana for treatment of chronic pain. JAMA 313(24):2474–2483

Hobs J, Sayre K, Gupta R (2008) The role of conservation agriculture in sustainable agriculture. Philosophical Transactions of the Royal Society B

Hofmanová T, Švec I, Hrušková M (2014) Nutritional properties of non-traditional seeds. J Life Med 2:10–14

Irakli M, Tsaliki E, Kalivas A, Kleisiaris F, Sarrou E, Cook CM (2019) Effect of genotype and growing year on hemp seed properties. Antioxidants 8(10):491

Jeff K, Williams DW (2019) Hemp agronomy—grain and fiber production. In: Industrial hemp as a modern commodity crop, pp 58–72

John F, Williams D, Trey R, Jared N, Patrick F, Jeff K, Dong Z (2019) Industrial hemp as a modern commodity crop. ASA, CSSA, and SSSA

Johnson R (2014) Hemp as an agricultural commodity. Congressional Research Service 7-5700 www.crs.gov.RL32725

Karche T (2019) The application of hemp (*Cannabis sativa* L.) for a green economy: A review. Turk J Bot 43(6):710–723

Karus M, Vogt D (2004) European hemp industry: Cultivation, processing and product lines. Euphytica 140(1):7–12

Lai H, Wang S, Nguyen Z, Li B (2021) Process optimization and antioxidant activity changes of fermented hemp seed milk. Modern Food Technol 37(7):194–202

Langdale GW, Thomas AW, Leonard RA (1979) Reduction of soil erosion by the no-till system. Trans ASAE 22(1):82–86

Leonard W, Zhang P, Ying D, Fang Z (2020) Hemp seed in the food industry: nutritional value, health benefits, and industrial applications. Compr Rev Food Sci Food Saf 19(1):282–308

Li H-L (1974a) An archaeological and historical account of cannabis in China. Econ Bot 28(4):437–448

Li H-L (1974b) The origin and use of cannabis in eastern Asia. Econ Bot 28(3):293–301

Mamone G, Picariello G, Ramondo A, Nicolai MA, Ferranti P (2019) Production, digestibility and allergenicity of hemp protein isolates. Food Res Int 115:562–571

Mark T, Shepherd J, Olson D, Snell W, Proper S, Thornsbury S (2020) Economic viability of industrial hemp in the United States: a review of state pilot programs. Washington: United States Department of Agriculture Economic Research Service, p 83

Mehmedic Z, Chandra S, Slade D, Denham H, Foster S, Patel AS, ElSohly MA (2010) Potency trends of cannabinoids. J Forensic Sci 55(5):1209–1217

Mikulec A, Kowalski S, Sabat R, Skoczylas Ł, Tabaszewska M, Wywrocka-Gurgul A (2019) Hemp flour is a valuable component for enriching the physicochemical and antioxidant properties of wheat bread. LWT 102:164–172

Ranalli P, Venturi G (2004) Hemp as a raw material for industrial applications. Euphytica 140(1): 1–6

Russo R, Reggiani R (2015) Evaluation of protein concentration and antinutritional compounds in hempseed meal. Am J Plant Sci 6:14–22

Schluttenhofer C, Yuan L (2017) Challenges towards revitalizing hemp. Trends Plant Sci 22(11): 917–929

Simopoulos AP (2002) The importance of omega-6/omega-3 ratio. Biomed Pharmacother 56(8): 365–379

Small E (2015) Evolution and classification of *Cannabis sativa*. Bot Rev 81(3):189–294

Small E (2018) Dwarf germplasm: the key to giant Cannabis hempseed and cannabinoid crops. Genet Resour Crop Evol 65:1071–1107. https://doi.org/10.1007/s10722-017-0597-y.

Small E, Marcus D (2002) Hemp: a new crop with newuses for North America. In: Janick J, Whipkey A (eds) Trends in new crops and Newuses. ASHS Press, Alexandria, VA, pp 284–326

Soroka V, Tarakan N (1981) Anatomical structure of the bracts in hemp varieties

Struik PC, Amaducci S, Bullard M, Stutterheim N, Venturi G, Cromack HTH (2000) Agronomy of Fiber hemp in Europe. Ind Crop Prod 11(2–3):107–118

Teh S-S, Birch J (2013) Physicochemical characteristics of hemp seed oil. J Food Compos Anal 30(1):26–31

Teterycz D, Sobota A, Przygodzka D, Łysakowska P, Elfalleh W (2021) Hemp seed (cannabis sativa l.) enriched pasta: physicochemical properties and quality evaluation. PLoS One 16(3): e0248790

Thakur A, Morya S, Kasankala LM (2025) Hemp seed and its milk analog: A review on specialties, significance and emerging needs. J Nat Fibers 22(1). https://doi.org/10.1080/15440478.2025.2474136

USDA Agricultural Statistics (2019) A compilation of U.S. government publications is available from the Hemp Industries Association (HIA). http://www.hempology.org/ALLARTICLES.html

USDA (2019) U.S. Department of Agriculture, Risk Management Agency. Whole-Farm Revenue Protection Pilot Handbook, USDA Federal Crop Insurance Corporation FCIC-18160 (11-2019)

van der Werf HMG (2004) Life cycle analysis of Fiber hemp production. Euphytica 140(1):13–23

Visković J, Zheljazkov VD, Sikora V, Noller J, Latković D, Ocamb CM, Koren A (2023) Industrial hemp agronomy and utilization: A review. Agronomy 13(3):931

Vogl CR, Mölleken H, Lissek-Wolf G, Surböck A, Kobert J (2004) Hemp as a resource for green cosmetics. J Industrial Hemp 9(1):51–68

Vonapartis E, Aubin M-P, Seguin P, Mustafa AF, Charron J-B (2015) Seed composition of industrial hemp cultivars. J Food Compos Anal 39:8–12

Westerhuis W, van Delden SH, van Dam JEG, Pereira Marinho JP, Struik PC, Stomph TJ (2019) Secondary Fiber development in hemp. Ind Crop Prod 139:111493

Xu Y, Sismour E, Britland JW, Sellers A, Abraha-Eyob Z, Yousuf A, Rao Q, Kim J, Zhao W (2022) Physicochemical, structural, and functional properties of hemp protein vs several commercially available plant and animal proteins: A comparative study. ACS Food Sci Technol 2(10): 1672–1680

Zahari I, Ferawati F, Helstad A, Ahlström C, Östbring K, Rayner M, Purhagen JK (2020) Development of high-moisture meat analogs with hemp and soy protein using extrusion cooking. Foods 9(6):772

Zając M, Świątek R (2018) The effect of hemp seed and linseed addition on the quality of liver pâtés. Acta Sci Pol Technol Aliment 17(2):169–176

Zając M, Guzik P, Kulawik P, Tkaczewska J, Florkiewicz A, Migdał W (2019) The quality of pork loaves with the addition of hemp seeds, de-hulled hemp seeds, hemp protein, and hemp flour. Lwt 105:190–199

Zhong LL, Cheng CW, Chan Y, Chan KH, Lam TW, Chen XR, Bian ZX (2013) Chinese herbal medicine for functional constipation. Trials 14:366

Zuardi AW (2006) History of cannabis as a medicine: A review. Brazil J Psychiatr 28:153–157

Biodiversity and Sustainable Use of Egyptian Desert Plants: Economic Potential Under Climate Change

9

Monier M. Abd El-Ghani and Rahmatullah Qureshi

Abstract

Climate change encompasses significant, long-term alterations in the Earth's climatic system, including shifts in temperature regimes, precipitation patterns, and wind dynamics that define regional climates. These changes arise from both natural drivers, such as volcanic activity and variations in solar radiation, and anthropogenic influences. The increasing frequency and intensity of climate-related extremes are exerting profound impacts on ecosystems worldwide. Egypt has been particularly affected over the past five decades, experiencing pronounced manifestations of climate change, including prolonged droughts, declining precipitation, and a sustained rise in ambient temperatures. These conditions intensify environmental stresses by accelerating soil erosion, reducing the productivity of marginal lands, and promoting desertification. The impacts of climate change in Egypt are further compounded by land-use transformation, the proliferation of invasive alien species, and heightened extinction risks for numerous plant taxa already considered vulnerable. Located within the hyper-arid Sahara of North Africa, Egypt's three major desert regions such as the Eastern Desert, Western Desert, and the Sinai Peninsula rank among the most arid landscapes globally. Desert vegetation, dominated primarily by xerophytic shrubs and subshrubs, constitutes the prevailing plant life form and occupies more than 90% (one million km^2) of the country's land surface. Although each desert is characterized by distinct physiographic features and floristic assemblages, the combined effects of extreme aridity and escalating human pressures have resulted

M. M. Abd El-Ghani
Department of Botany and Microbiology, Faculty of Science, Cairo University, Giza, Egypt

R. Qureshi (✉)
Department of Botany, Faculty of Sciences, Pir Mehr Ali Shah Arid Agriculture University, Rawalpindi, Pakistan
e-mail: rahmatullahq@uaar.edu.pk

R. Qureshi (ed.), *Plant Biodiversity*, https://doi.org/10.1007/978-981-92-1471-6_9

in the decline and local disappearance of many economically valuable plant species from their natural habitats. Anthropogenic activities represent a major threat to Egypt's plant biodiversity, with overharvesting, urban expansion, overgrazing, waste disposal, road construction, and tourism-related development causing widespread degradation of natural ecosystems. This chapter assesses the current status of Egypt's natural plant resources through representative case studies from the three desert regions. It examines the traditional uses of desert plants for fodder, folk and herbal medicine, and household applications, and presents the biogeographical distribution of economically important plant species across Egypt's major ecological regions.

Keywords

Climate change · Hyper-arid desert · Egypt · Desert vegetation · Biogeography · Floristic diversity · Anthropogenic activities · Habitat loss · Folk medicine

Abbreviations

CC	Climate change
GHG	Greenhouse gases
IPCC	Intergovernmental Panel on Climate Change
RH	Relative humidity
UNCCD	United Nations Convention to Combat Desertification

9.1 Introduction

9.1.1 Climate Change: A Global Crisis

Climate change (CC) is one of the most critical challenges facing our era, with its profound impact on Earth's natural systems and human societies. Primarily in the form of fossil fuel combustion, deforestation, and intensive agriculture, it is driven by human activities since the Industrial Revolution, highlighting long-term shifts in temperature, precipitation patterns, and events of extreme weather (Chen et al. 2011). These changes are a direct result of escalating atmospheric concentrations of GHGs like CO_2, CH_4, and N_2O, trapping heat in the atmosphere. This increasing greenhouse effect has started a definite planetary warming, with clearer consequences—from receding glaciers to the increase in the aggression of storms—already very well visible, and global action focused on this issue is urgently needed (UNDP 2020).

The effects of this heating can already be seen in most parts of the world, though with different intensities. The world is seeing rising cases of climate-related disasters as the global average temperature reaches about 1.1 °C above the pre-industrial level (NDRR 2021). Recent tragic wildfires in Australia and California, unprecedented floods in Pakistan, and record heat-waves across Europe are proving moments of a changing climate easily (NOAA 2024). Growing sea levels have brought an existential threat to places along coastlines and put low-lying island nations like the Maldives in danger of loss.

This crisis demands a twin-track approach to both mitigate and adapt. Mitigation aims at lessening the level of GHG emissions at their source, through transition into renewable energy, enhanced efficiency in energy use, and the end of deforestation. Adaptation, on the other hand, involves adapting to current and unstoppable future impacts, for instance, deploying early warning systems for extreme weather and developing drought-resistant crop varieties to stabilize food supplies. The backbone of this effort is international cooperation, exemplified by the commitment within the Paris Agreement to limit heating to 1.5 °C. However, current country pledges under the Paris Agreement—nationally determined contributions—are insufficient. Emissions must fall by 45% by 2030 if the world is to stay on track, pointing to a significant ambition gap.

Substantial barriers impede progress, including persistent fossil fuel subsidies estimated at $5.9 trillion annually and political resistance that slows the energy transition. Global inequities further complicate these challenges as developing nations often lack the financial wherewithal to move toward green technologies and climate resilience (IEA 2021). Massive opportunities are revealed even amid such difficulties. The renewable energy sector is set to support 42 million jobs by 2050, making it well ahead of employment provided by the fossil fuel industry (IRENA 2021). This calls for financial mechanisms such as the Green Climate Fund to mobilize $100 billion each year to support the vulnerable countries (GCF 2023). Yet, there is a persisting shortfall in funding. Climate change is a current and existential crisis that requires urgent action from all parts of society. While the challenge is immense, a viable pathway forward exists. We can avoid the most catastrophic impacts if deep emissions cuts, significant investment in adaptation, and actual global equity are pursued. The IPCC stresses that every fraction of a degree of warming matters. Our collective future depends on sound policy, technological innovation, and relentless global solidarity (IPCC 2021–2023).

9.2 Climate Change Effects on Plant Biodiversity in Arid Ecosystems

Arid environments, occupying over 40% of the Earth's surface, are characterized by low and unpredictable precipitation, high temperatures, and high rates of solar radiation. These ecosystems support flora uniquely adapted to such conditions but are unusually sensitive to the myriad impacts of anthropogenic climate change. Given that Earth has already warmed by approximately 1.1 °C since the pre-industrial era and considering the last decade was the hottest on record from 2011–2020, arid biomes are in an acutely challenging situation for their special vegetation.

Direct impacts are from rising temperatures and altered precipitation patterns. Increasing global temperatures manifest as more frequent and intensive heat waves in arid zones that push desert plants which often exist near their physiological limits beyond their thermal tolerance. This stress can weaken photosynthesis, increase respiration, and lead to widespread mortality. Simultaneously, climate change is already disrupting hydrological cycles, with many arid regions projected to suffer more intense droughts coupled with erratic rainfall. Reduced water availability directly constrains plant survival and reproduction, as indicated by significant die-offs of iconic species, such as the saguaro cactus, during prolonged droughts in the Sonoran Desert.

These climatic drivers are having a fundamental impact on reshaping the composition and distribution of species. While some plants may shift their ranges to higher elevations or higher latitudes in order to track suitable climates, others—most notably endemic species with narrow niches and poor dispersal capability—are facing local extinction. For example, drought conditions are attributed to a decline in perennial shrubs in the Sonoran Desert, being replaced by less diverse, drought-tolerant annuals (Bowers 2005; Salguero-Gómez et al. 2012). Moreover, atmospheric CO_2 enrichment can further accelerate such changes (Le Quéré et al. 2020) due to disproportional benefits to invasive species, such as *Bromus madritensis* in the Mojave Desert, outcompeting native flora and decreasing biodiversity (Dukes and Mooney 1999).

Besides, climate change acts as a threat multiplier to the desertification process whereby productive land surfaces are degraded into deserts (Mirzabaev et al. 2019). This results in extensive habitat loss; the UNCCD estimates that 12 million hectares of productive land are lost annually, placing 500 million people at risk (UNCCD 2020). Desertification, caused by the synergy of climatic stressors such as prolonged droughts along with human activities like overgrazing and deforestation, causes soil erosion and salinization, further limiting the viability even of arid-adapted plants. For instance, in the Sahel region of Africa, declining rainfall has initiated a sharp decline in tree cover; this shows how devastating the process is to both flora and dependent fauna (Herrmann et al. 2005). Climate change disrupts crucial ecological interactions, and for this reason, plant phenology is of concern. The timing of life-cycle events such as flowering and fruiting, which are cued to temperature and moisture, may become desynchronized from the activity of specialist pollinators. Such mismatches can reduce seed set and hence reproductive success, as observed with Agave species in the Chihuahuan Desert and their bat pollinators. These disruptions, along with changes in plant community structure, also alter herbivore dynamics and have the potential to lead to the overgrazing of remaining palatable species, thus accelerating biodiversity loss (Forrest and Miller-Rushing 2010).

Climate change now poses a severe, multi-faceted threat to the distinctive plant diversity of arid lands. Such fragile ecosystems are being profoundly altered through direct physiological stress, changes in species composition, accelerated desertification, and the disruption of key ecological relationships. In the absence of decisive global mitigation to reduce emissions and targeted regional adaptation strategies—including habitat restoration, invasive species control, and assisted migration—

many endemic plants are likely to go extinct (UNEP 2023). The loss of this unique biodiversity would have cascading consequences for ecosystem function and the services these regions provide, highlighting the urgent need for concerted action to preserve the world's arid lands.

9.3 Role of Desert Plants in Arid Ecosystems

Arid environments are characterized by low and unpredictable precipitation, extreme temperature fluctuations, and intense solar radiation, which present formidable challenges for life (Noy-Meir 1973). Within these harsh landscapes, desert plants are far more than tenacious survivors; they are the foundational architects of ecological integrity. They function as primary producers, habitat engineers, soil conservators, and microclimate regulators, collectively sustaining the unique biodiversity and function of these fragile ecosystems.

9.3.1 Primary Production and the Food Web

As key producers, desert plants serve as the energetic base of desert food webs by photosynthetically harnessing solar energy to produce organic molecules (Körner 2018). These desert plants have developed a range of advanced adaptations to cope with water limitations, such as Crassulacean Acid Metabolism, extensive root depths, succulence, and drought-fixed leaves, ensuring increased food production at reduced water costs (Gibson and Nobel 1986). This plant-produced organic matter is in turn fuel for herbivores, including those as varied as insects and mammals, evolved in concert to efficiently tap already protected food sources. In turn, herbivores serve as food sources for carnivorous and omnivorous desert predators.

9.3.2 Habitat Provision and Modification

Apart from providing sustenance, arid plants are essential habitat modifiers. The physical attributes, for example, size, density, and topography, of arid plants are fundamental for providing habitat for diverse fauna. It must be noted that the shade offered by shrubs, tree, or even cacti helps to shield fauna from the effects of heat, as well as protect them from predators (Parker 1988; Polis 2023). Even the leaf litter of plants has a modifying effect as it helps to form micro-niches that are fundamental for the sustenance of detritivorous organisms, which help to improve the humidity of the soil. Therefore, the presence of plant structures has a fundamental positive correlation with the richness of fauna, as many faunal species are associated with particular plant structure forms for survival purposes (Fleming et al. 2001).

9.3.3 Soil Conservation and Nutrient Cycling

The service of desert plants in conserving and enriching the soils cannot be overemphasized. The root systems of desert plants play an important role in holding the soil, thus inhibiting erosion caused by wind and flash floods (Whitford and Duval 2019). This is essential in halting desertification (D'Odorico et al. 2013). Additionally, plants play an essential role in releasing nutrients into the soil through the turnover of litter-fall. Some legume plants in deserts can form associations with Nitrogen-fixers, an aspect that increases the nutritional value of poor soils. This action also leads to the formation of an "island of fertility" within these deserts, an essential site because the organic material increases nutrients, moisture, and microbial activity, creating an ideal sites for seedling growth (Schlesinger et al. 1990).

9.3.4 Microclimate and Hydrological Regulation

Desert plants play an equally important role in altering local microclimates and hydrological patterns. Shade cast by the canopy suppresses soil temperatures and rates of evaporation, thus helping to preserve valuable soil water (Nobel 2003). Vegetation also serves as an effective wind break that helps to protect the soil from loss due to desiccation. In addition, root systems of plants form routes that lead to increased water infiltration in the soil profile when it rains. Desert plants play the leading role in shaping and sustaining arid land biotic communities. Desert plants act as the backbone of an arid land biotic community. They act as the producers in initiating nutrient flow in the biotic community. They also protect water in the soil by conserving water through their roots. At the same time, they protect the soil from loss by increasing water evaporation due to reduced cover. They also protect arid lands by conserving the microclimate balance in arid biotic communities.

9.4 Hyper Arid Desert Environment of Egypt: Location, Physiography and Climatic Features

9.4.1 Location and Geography

Egypt covers the northeastern corner of the African continent, and its total area is just over one million square kilometers, which is about 3% of the African continent's total landmass (Abd El-Ghani et al. 2017). The country lies over 10° of latitude from 22°N to 32°N, and this means that Egypt is mainly located within the temperate region, except for its southern region, which is located north of the Tropic of Cancer. The country is arid, and this means that arid land covers 95% of its total landmass, and this is a big part of a large arid belt that stretches from the Atlantic Ocean, covering North Africa to the Arabian region. The country has low land, and this means that most of Egypt is below 500 meters, and this is an important contributor to its ecological diversity.

Geomorphological Structure

Egypt is fundamentally divided by the Nile River into two primary regions:

- The Western Desert: A vast, flat expanse representing the eastern extension of the Libyan Desert and one of the most extremely arid habitats on Earth.
- The Eastern Dissected Plateau: The land east of the Nile, which is further subdivided geographically by the Gulf of Suez into:
  - The Eastern Desert
  - The Sinai Peninsula, which acts as a land bridge connecting Africa to Asia.

This gives Egypt three distinct desert regions: the Western Desert, the Eastern Desert, and the Sinai Peninsula.

Phytogeographic and Biogeographic Significance

Egypt's location makes it a strategic crossroads for biotic exchange. Its Mediterranean coastline has historically connected it to Europe, while two primary corridors link it to tropical Africa: the Nile Valley and the Red Sea basin. These routes, along with the Sinai land bridge, have facilitated the movement of species over millennia and today serve as major flyways for migratory birds.

Based on its distinct environmental and climatic conditions, Egypt can be divided into five main phytogeographic regions:

- The Western Desert (including its oases and depressions)
- The Eastern Desert (encompassing Gebel Elba and the Red Sea coastal land)
- The Sinai Peninsula
- The Nile Land (the Nile Valley and Delta)
- The Western Mediterranean Coast

Egypt is a predominantly low-lying country located in the northeastern part of the African continent, covering a total area of slightly more than one million km^2, which represents approximately 3% of Africa's land surface (Abd El-Ghani et al. 2017). The country extends across nearly ten degrees of latitude, from 22°N to 32°N, placing most of its territory within the northern temperate zone, with less than one-quarter lying south of the Tropic of Cancer. Egypt forms an integral part of the extensive arid desert belt that stretches from the Atlantic coast across North Africa and into the Arabian Peninsula.

Geographically and biogeographically, Egypt occupies a strategic crossroads position. Its Mediterranean coastline links it with southern Europe, facilitating historical biotic exchanges during both glacial and interglacial periods. In addition, Egypt lies along major migratory routes, with several flyways for Palearctic–Afrotropical birds converging across its territory. Two principal biogeographical corridors connect Egypt with tropical Africa and regions further south: the Nile Valley and the Red Sea basin. The Sinai Peninsula serves as a natural land bridge between Africa and Asia, reinforcing Egypt's significance as a transitional zone between continents.

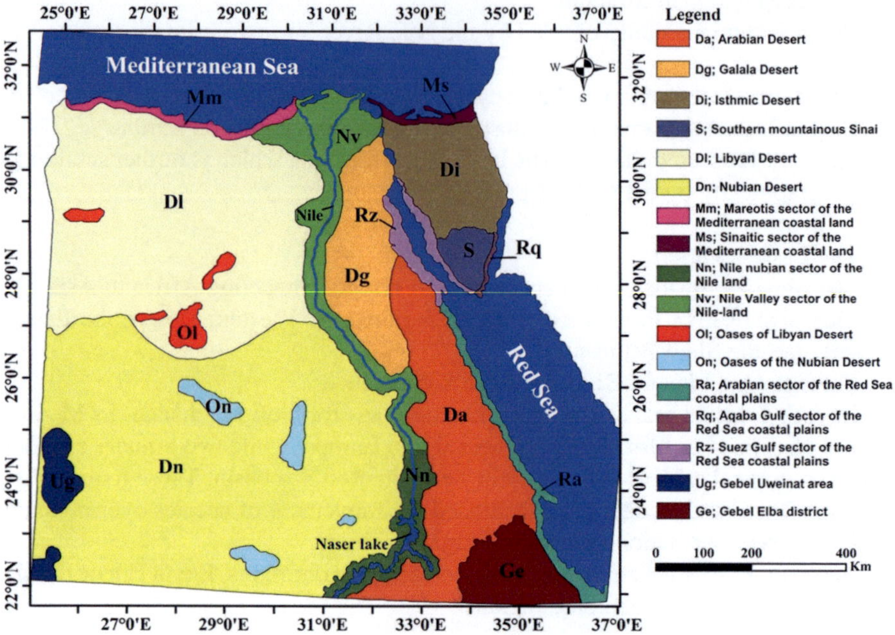

Fig. 9.1 Map of Egypt showing the different deserts, and biogeographic regions

Topographically, the majority of Egypt's land surface lies below 500 m above sea level, a feature that constrains overall habitat heterogeneity and potential biodiversity. Approximately 95% of the country is classified as desert, with the Western Desert representing one of the most extreme arid environments globally. The Nile Valley constitutes the dominant geomorphological feature of Egypt, effectively dividing the country into two major regions: the dissected eastern plateau and the expansive, relatively flat western plateau, which forms part of the greater Libyan Desert (Fig. 9.1).

Although the terrain east of the Nile Valley constitutes a single geomorphological unit, it is geographically subdivided into the Eastern Desert and the Sinai Peninsula, separated by the Gulf of Suez. Accordingly, Egypt's desert landscapes can be broadly categorized into three major regions: the Eastern Desert, the Western Desert, and the Sinai Peninsula. From a phytogeographical perspective, Egypt is traditionally divided into five principal regions: (i) the Western Desert, including its oases and depressions; (ii) the Eastern Desert, encompassing Gebel Elba and the Red Sea coastal plains; (iii) the Sinai Peninsula; (iv) the Nile Land, comprising the Nile Valley and the Nile Delta; and (v) the Western Mediterranean Coastal region.

9.5 Physiographic Features of the Deserts

9.5.1 The Western Desert

The Western Desert occupies nearly two-thirds of Egypt's total land area, covering approximately 681,000 km^2. It extends longitudinally from the Mediterranean coast in the north to the Sudanese border in the south over a distance of about 1073 km, and latitudinally from the Libyan border in the west to the Nile Valley in the east, spanning approximately 600–850 km. Climatically, the region is characterized by an extreme aridity gradient, with mean annual rainfall decreasing from about 150 mm along the Mediterranean coast to virtually zero in the southern and southwestern sectors, which are among the driest areas on Earth. In contrast to the Eastern Desert, the Western Desert lacks well-developed drainage systems or extensive wadi networks (Zahran and Willis 2009).

A defining geomorphological feature of the Western Desert, largely attributable to prolonged arid conditions, is the relative uniformity of its surface topography when compared with other regions of North Africa. Despite its hyper-arid nature, the Western Desert supports considerable floristic diversity in areas where water resources—derived from limited rainfall and, more importantly, groundwater are available. In this context, oases represent the most ecologically and physiographically distinctive features of the Western Desert, appearing as conspicuous green enclaves embedded within the surrounding arid landscape.

Five major inhabited oases occur in the Egyptian Western Desert: Siwa, Bahariya, Farafra, Dakhla, and Kharga (Fig. 9.1). These oases overlie the Nubian Sandstone Aquifer System, one of the largest fossil groundwater reservoirs in the world, which underpins agricultural activities and natural vegetation in the region (Abd El-Ghani et al. 2017). Irrigation in the oases is traditionally sustained by natural springs both thermal and non-thermal—or by groundwater extracted through wells. Water distribution is typically achieved via simple yet efficient systems of lateral channels designed to optimize limited water supplies. Consequently, both crop production and natural vegetation in the Western Desert oases are almost entirely dependent on groundwater resources (Bornkamm and Kehl 1990).

Inland salt marshes of the Western Desert occur predominantly as sabkhas surrounding lakes, springs, and wells within oases such as Siwa, Dakhla, Kurkur, and Dungul, as well as within major depressions including Qattara, Wadi El-Natrun, and Fayoum. These saline habitats are situated at lower elevations relative to the surrounding terrain and are therefore associated with shallow groundwater tables. In some locations, groundwater reaches the surface, forming lakes of brackish to highly saline water (Zahran and Girgis 1970). The development of these inland salines is primarily attributed to uncontrolled groundwater overflow, episodic flooding of flat plains, or the proximity of the water table to the soil surface. Under the prevailing arid conditions and in the absence of effective drainage systems, repeated flooding with slightly saline artesian water leads to rapid accumulation of salts in the soil profile. Unlike coastal salt marshes, these inland salines are considered secondary in origin.

The vegetation within these saline environments exhibits a highly patchy distribution, with each patch often dominated by a single species or, in some cases, by distinct ecotypes or growth forms of the same species (El Hadidi 1993). This spatial heterogeneity reflects fine-scale variations in salinity, soil texture, and groundwater depth and contributes to the unique ecological character of the Western Desert's inland wetlands.

9.5.2 The Eastern Desert

The Eastern Desert of Egypt encompasses approximately 223,000 km^2, representing roughly 21% of the country's total land area. This region comprises two principal ecological units, the Red Sea coastal plain and the inland desert, characterized by an extensive network of wadis. According to Zahran and Willis (2009), the inland desert can be subdivided into four main geomorphological and ecological regions: (1) the Cairo–Suez Desert, (2) the Limestone Desert, (3) the Sandstone Desert, and (4) the Nubian Desert.

The inland Eastern Desert stretches between the Red Sea coastal mountains in the east and the Nile Valley in the west (Fig. 9.1). It is dominated by a rocky plateau, dissected by numerous wadis, each comprising a primary channel with multiple tributaries. These wadis form discrete drainage basins, with most surface runoff directed westward toward the Nile Valley.

The Red Sea coastal mountain ranges constitute a distinct and ecologically significant habitat, hosting complex plant and animal communities that reflect floristic and faunal linkages between Egypt, Sudan, and Ethiopia. Among these, the Gebel Elba Mountains in southeastern Egypt are of exceptional ecological and biogeographical importance. This massif forms part of the granitic Red Sea highland complex extending between Egypt and Sudan and lies between 36°–37° E longitude and approximately 22° N latitude.

Gebel Elba supports an extraordinarily rich flora and fauna, including numerous endemic species and taxa that represent the northern limits of the Ethiopian highlands biota. Its floristic richness, combined with a biogeographical position at the intersection of the Saharo-Arabian and Sudanian regions, designates Gebel Elba as one of Egypt's principal phytogeographical zones (El Hadidi 2000). Vegetation here is more diverse than in other Red Sea coastal mountain groups due to the convergence of Palaearctic and Afro-tropical elements.

The flora of Gebel Elba also includes components of the Sahelian regional transition zone (White and Léonard 1991), representing the northernmost extension of this geo-element in Africa. Vegetation is particularly well developed on the northern and northeastern slopes, whereas southern and southwestern flanks exhibit lower diversity (Kassas and Zahran 1971). These unique ecological and geographical characteristics have fostered high levels of plant diversity, endemism, and the persistence of extensive woodland formations dominated by *Vachellia tortilis* subsp. *tortilis*, a vegetation type not observed elsewhere in the Eastern Desert (Zahran and Willis 2009).

Despite its biogeographical significance, Gebel Elba has been underrepresented in global biodiversity assessments (Heywood and Watson 1995). Of the 142 threatened woody perennial species listed in the Plant Red Data Book of Egypt (El Hadidi et al. 1992), 56 species (39.4%) are recorded from the Gebel Elba district. Consequently, the area was designated as Gebel Elba National Park in 1986 (Prime Ministerial Decrees 450/1986, 1185/1986, and 642/1995), encompassing approximately 35,600 km^2, with the primary objective of conserving biodiversity and promoting sustainable natural resource management.

Biodiversity conservation in Egypt is further supported by a national network of protected areas, currently comprising 21 sites that cover approximately 8% of the country's land surface, with an additional 19 areas proposed for protection. This network is organized based on natural regional classifications, aiming to preserve representative ecosystems across Egypt.

9.5.3 The Sinai Peninsula

The Sinai Peninsula occupies approximately 6% of Egypt's total land area and represents a Saharan-type desert that forms a biogeographical link between Africa and Asia. It constitutes a transitional zone between the Egyptian deserts and those of the Middle East and is of particular phytogeographical interest, lying at the intersection of the Mediterranean, Irano-Turanian, Saharo-Arabian, and Sudanian regions (Zohary 1973).

Climatic conditions in Sinai are highly variable, with mean annual precipitation decreasing from approximately 100 mm in the northern Mediterranean coastal region to 5–30 mm in the southern interior (Danin 1978). This climatic gradient, combined with diverse geological and edaphic features, supports a flora of roughly 900 species distributed across 200–300 vegetation associations (Danin 1986). Northern Sinai is largely dominated by sandy landscapes, whereas limestone hills and gravel plains prevail in the central region. Southern Sinai features a complex mosaic of landforms, including plains, wadis, oases, springs, salt marshes, and sand dunes. It also contains a rugged complex of igneous and metamorphic mountains, including Egypt's highest peaks: Gebel Saint Katherine (2641 m), Gebel Musa (2285 m), and Gebel Serbal (2070 m). The western coastal plain, known as El-Qaa, borders the Gulf of Suez.

These mountainous areas are rich in flora and fauna (Moustafa et al. 2001), supporting primarily Irano-Turanian steppe vegetation dominated by *Seriphidium herba-album* and associated with *Gymnocarpos decander*. Vegetation is generally sparse, comprising semishrubs confined to wadis, rocky slopes, and sandy plains, with trees being relatively uncommon (Danin 1986).

Southern Sinai represents a key center of plant endemism due to its relatively wetter microclimate and extensive exposures of smooth-faced rocks, which provide specialized habitats for rare species (Danin 1983). This region likely harbors the highest biodiversity in Egypt. A substantial portion of southern Sinai was declared a Protectorate in 1996, centered on the town of St. Catherine (1600 m a.s.l.), home to the historic sixth-century monastery situated at the traditional site of the biblical "burning bush."

From Mount St. Catherine, wadi systems drain eastwards toward the Gulf of Aqaba and westwards toward the Gulf of Suez. Although classified as hyper-arid (Zahran and Willis 2009), southern Sinai experiences occasional significant water flow in wadis, sometimes causing flash floods. Under typical conditions, most water remains underground, occasionally surfacing to sustain localized permanent or semi-permanent habitats that support dense vegetation and associated insect and animal communities.

9.6 Climatic Features

The deserts of Egypt are among the most arid regions of the world. Annual rainfall over most of the country does not exceed 10 mm, with the highest precipitation occurring along the Mediterranean coast, where mean annual rainfall reaches approximately 150 mm. Rainfall decreases rapidly southwards, declining to about 30 mm at Cairo, and further south to 3 mm per year or less (Table 9.1).

Egypt is thus characterized by a hot and predominantly rainless climate, with a national average annual rainfall of approximately 10 mm. Even in the narrow Mediterranean coastal belt, where precipitation is relatively higher, annual rainfall rarely exceeds 200 mm and declines sharply inland. This extreme aridity accounts for the predominance of barren and desert landscapes across the country.

Table 9.1 Annual mean (2000–2015) of climatic fractures in some meteorological stations in Egypt (Courtesy of Meteorological Authority in Egypt). Max = Maximum, Min = Minimum, RH = Relative humidity

Station	Biogeographic region	Temperature (°C)		Rainfall (mm)	RH (%)	Wind speed (km hr.$^{-1}$)
		Max	Min			
Aswan	Nile Valley	32.8	17.9	1	39	15
Hurghada	Red Sea coast	29.5	19	0.5	46.7	12.7
El Quseir		27.7	20.5	1	51.2	12.1
Safaga		31.9	20.2	0	46.33	17.58
Marsa Alam		28.4	21.2	0.5	45.9	14
Halaib and Shalateen		31.6	20.2	1.2	67.4	14.2
Sharm El-Sheikh	South Sinai	30.3	18.4	1	33.41	16.75
Dakhla Oasis	Western Desert	25	7.5	0	25.4	11
Kharga Oasis		24	8	0	25	13.9

9.7 Impacts of Climate Change on the Natural Plant Diversity

Climatic change, in a general sense, refers to the lack of balance in the typical conditions of temperature and rain/wind, among other elements, that distinguish each area of the Earth. This could be as a result of internal processes (volcanic eruption) or external (variation in the strength of the sun's rays). Essentially, the effects of the occurrence of climatic changes will be considerable on the diverse ecosystems on the surface of the earth (Loarie et al. 2009).

Egypt did not escape the climate change that took place around the world, particularly in the last fifty years, as evidenced in the periods of drought and lack of rainfall, as well as the increase in temperatures that would lead to an increase in soil erosion, thus making the planting of lands that could not previously grow crops more difficult. It could also experience changes in land use, introduction of foreign plants that would outcompete local plants for water and space, as well as extinction of many plant species that would have previously qualified to be endangered.

The three deserts in Egypt (Eastern Desert, Western Desert, and the Sinai Peninsular Desert) are some of the driest regions in the world because they are in the Sahara Desert in North Africa, which is a hyper-arid area with a very hot and almost rainfall-free climate. Thus, desert vegetation is clearly the dominant vegetation. This vegetation is also the most typical vegetation found in nature. It covers more than 90% of its total surface area (about one million km^2) and is composed largely of xerophytic shrubs and subshrubs. There is diversity in each of the three deserts in terms of distinct topography, including plant species.

The next parts will discuss case studies on the various effects of climate change on the current distribution and diversity of plants in Egypt.

9.7.1 Vegetation of the Western Desert Oases

The vegetation cover of Egypt's Western Desert is largely restricted to the five major inhabited oases like Siwa, Bahariya, Farafra, Dakhla, and Kharga which appear as prominent green islands within the hyper-arid desert landscape. These oases owe their existence to underlying groundwater sources, accessed through a network of old and recently dug wells or natural springs, which constitute the primary water supply sustaining both natural and cultivated vegetation.

The habitats within these oases can be classified into five major ecological zones (Abd El-Ghani et al. 2017; Fig. 9.2):

1. *Farmlands:* Comprising arable fields and orchards, these areas support both field crops and fruit trees. The combination of shade, cool microclimate, and localized moisture fosters the growth of several plant species.
2. *Canal Banks:* The vegetative composition along irrigation canals varies considerably depending on the size and intensity of water distribution in each location.
3. *Reclaimed Land:* These areas represent previously desertified tracts that have been converted to cultivation. Desert-adapted plants often colonize these lands

Fig. 9.2 Typical structure of an oasis (Bahariya Oasis, Western Desert)

along the boundaries of cultivated plots, leading to a mosaic of native and anthropogenic flora.

4. *Waste Lands:* These include previously fertile areas left fallow or rendered unproductive due to overuse or environmental constraints.
5. *Water Bodies:* Freshwater habitats such as irrigation channels, wells, reservoirs, and seasonal ponds support specialized riparian vegetation.

A total of 172 vascular plant species, representing 131 genera and 39 families, have been documented across the habitats of these five oases (Abd El-Ghani and Fawzy 2006). The highest species richness was recorded in Bahariya Oasis (143 species), followed by Siwa Oasis (79 species), while Dakhla Oasis, being the least impacted by human activity, hosted only 18 species.

The most diverse and dominant families included Poaceae, Fabaceae, Asteraceae, Cyperaceae, and Chenopodiaceae, collectively accounting for more than 50% of the total species richness. These families are characteristic of Mediterranean North African flora, reflecting both historical and ecological influences. Major genera included Euphorbia (5 species), Cyperus (4 species), and Avena, Chenopodium, Cuscuta, Medicago, Juncus, and Tamarix (3 species each).

9.7.1.1 Life-Form Spectrum

The life-form analysis indicates strong adaptation to aridity and intermittent water availability. Therophytes were dominant (57%) in the area, reflecting a strategy for survival in hot, dry environments with short periods of water availability. Hemicryptophytes constitute 12%, while Phanerophytes account for 9%. Chamaephytes and Geophytes each represent 6% of the flora. These distributions result from a combination of climatic extremes, topographic heterogeneity, human disturbances, and grazing pressure, as well as the ephemeral availability of water in oasis habitats.

9.7.1.2 Conservation Status and Threatened Species

Floristic surveys reveal that several taxa have disappeared from the oases flora, despite previously being widespread. For instance, *Nymphaea caerulea var.*

aschersoniana, endemic to Bahariya Oasis, and *Ranunculus rionii* are considered extinct within Egyptian oases. Similarly, *Ambrosia maritima*, once widespread in oasis farmlands, is now largely naturalized and considered invasive, often forming dense stands in shaded, moist orchard habitats. *Aster squamatus*, another recently introduced species, has similarly become dominant and invasive.

Several species are now rare and endangered due to their limited populations in characteristic habitats. These include *Rostraria rohlfsii* and *Stipagrostis vulnerans*. Conversely, taxa such as *Ammoides pusilla, Dianthus cyri, Diplotaxis harra, Ducrosia ismaelis, Melilotus serratifolius* and *Schmidtia pappophoroides* occur intermittently but are still present across the oases.

The primary drivers of species decline and habitat alteration include climatic fluctuations, expansion of human settlements, road construction, new well and irrigation development, and intensive grazing. These factors, coupled with the fragility of oasis ecosystems, underscore the urgent need for conservation measures to safeguard the remaining native flora of the Western Desert.

Two endangered pteridophytes, *Equisetum ramosissimum* and *Marsilea minuta*, are restricted to the Bahariya Oasis. These species occur in specialized microhabitats characterized by shaded and humid conditions, particularly along watercourses associated with old wells. Notably, Bahariya Oasis is regarded as the type locality of *Marsilea minuta*, which, until recently, had not been recorded elsewhere (Abd El-Ghani 1981). This unusual biogeographical phenomenon warrants further detailed ecological and taxonomic investigation. Several plant species commonly associated with rice-field ecosystems such as *Sphenoclea zeylanica*, *Cyperus fuscus*, *Pycreus polystachyos*, *Pycreus mundtii*, and *Eleocharis geniculata* were also recorded in the oasis. The occurrence of *Faidherbia albida* in Bahariya Oasis is considered doubtful (Abd El-Ghani 1981), although the species has been frequently reported from the Eastern Desert of Egypt (Kassas and Girgis 1971). *Sphenoclea zeylanica*, a species indigenous to the tropical regions of the Old World, represents a relatively recent introduction to Egypt. Its presence in Bahariya Oasis is likely of anthropogenic origin, probably introduced unintentionally with rice grains used for cultivation (Abd El-Ghani 1988). In the oasis, the species has been recorded from rice fields, where it appears to reach the northernmost limit of its continental distribution. Habitat preferences of the recorded species across different oases are summarized in Table 9.2.

As part of the Western Desert (Fig. 9.3), the physiographic feature of Wadi El-Natrun Depression (23 m below sea level) has been reflected in its flora and vegetation throughout millennia. Like many other parts in this country, climate change is pronounced in these areas, which can be reflected by the disappearance of certain plant species that were found and recorded in the last 7 decades. Variation patterns in the floristic composition of Wadi El-Natrun during the last 73 years (1938–2011) were thoroughly investigated, where the total number of recorded species varied from one inventory to another. The disappearance of 101 species from the flora of Wadi El-Natrun and the record of 61 species that considered new additions to its flora were confirmed (Abd El-Ghani et al. 2015). This may be attributed to dramatic changes in environment, particularly climatic attributes,

Table 9.2 Habitat preferences of the common species (>50% occurrences) recorded from different oases. Symbols of habitat types: H_1 = farmlands (include irrigated crop0fields and orchards), H_2 = canal banks, H_3 = reclaimed lands, H_4 = waste lands (moist land and abandoned salinized field plots) and H_5 = water bodies

H_1	H_2	H_3	H_4	H_5
Ambrosia maritima	*Centaurium pulchellum*	*Asphodelus tenuifolius*	*Aeluropus lagopoides*	*Lemna gibba*
Avena fatua	*Chenopodium ambrosioides*	*Bassia muricata*	*Alhagi graecorum*	*Ludwigia stolonifera*
Beta vulgaris	*Conyza bonariensis*	*Launaea capitata*	*Bassia indica*	*Ottelia alismoides*
Brachypodium distachyum	*Desmostachya bipinnata*	*Launaea nudicaulis*	*Carex divisa*	*Zannichellia palustris*
Brassica nigra	*Eclipta prostrata*	*Parapholis incurva*	*Cressa cretica*	
Calendula arvensis	*Pseudognaphalium luteoalbum*	*Polygonum equisetiforme*	*Cynanchum acutum*	
Centaurea calcitrapa	*Heliotropium lasiocarpum*	*Prosopis farcta*	*Cyperus laevigatus*	
Corchorus olitorius	*Imperata cylindrica*	*Pulicaria crispa*	*Eleocharis palustris*	
Digitaria sanguinalis	*Mentha longifolia*	*Thesium humile*	*Juncus hybridus*	

Fig. 9.3 Habitats of the oases of the Western Desert: (**a**) cultivated lands near the desert fringes, (**b**) date-palm orchards, (**c**) salinized lands, and (**d**) waste lands with dom-palms

human activities, urbanization, and land degradation (Akhtar-Schuster et al. 2011). Certain species showed more affinity to moist and canal bank habitats such as *Pluchea dioscoridis, Acacia nilotica, Mentha longifolia, Cyperus rotundus* var. *fenzelianus, C. difformis, Leptochola fusca, Limbarda crithmoides* and *Plantgo lagopus*. It also includes some salt-tolerant species such as *Atriplex halimus, A. lindleyi* subsp. *inflata, Tamarix tetragyna* and *Aeluropus littoralis*. Common weeds of croplands were also represented such as *Convolvulus arvensis, Oxalis corniculata, Euphorbia peplus, E. helioscopia, Capsella bursa-pastoris, Silene nocturna, Poa annua, Cichorium endivia* subsp. *divaricatum, Beta vulgaris, Rumex dentatus, Trifolium resupinatum* and *Portulaca oleracea*. In addition, the two most common parasitic weeds *Orobanche crenata* and *Cuscuta pedicellata* were recorded. Desert annuals and perennials such as *Eremobium aegyptiacum, Stipagrostis hirtigluma, Launaea mucronata* subsp. *cassiniana, Reichardia tingitana* and *Mesembryanthemum crystallinum* can be also mentioned.

In addition, Abd El-Ghani and Fawzy (2006) stated that floristic differences showed that 9 species were restricted to Farafra Oasis, namely: *Bromus diandrus, Polycarpon succulentum, Cutandia memphitica, Vicia lutea, Polycarpon tetraphyllum, Plantago amplexicaulis, Oligomeris linifolia, Nymphaea nouchali*, and *Avena sterilis*. Others 45 species were restricted to Bahariya Oasis, including: *Erucaria hispanica, Heliotropium lasiocarpum, Bromus diandrus, Allium ampeloprasum, Brassica juncea, Lotus halophilus, Chenopodium glaucum, Homognaphalium pulvinatum*, and *Geranium dissectum*. Additionally, 6 species were restricted to Dakhla Oasis, including: *Capparis aegyptia, Asphodelus fistulosus, Althaea ludwigii, Aerva javanica, Senna italica*, and *Abutilon pannosum*. Seven species were restricted to Kharga Oasis, including: *Erodium cicutarium, Sesbania sesban, Silene villosa, Conyza aegyptiaca, Rhynchosia minima, Stipagrostis scoparia* and *Kickxia elatina*.

9.8 The Wadis of the Eastern Desert: Case Studies

The wadis or valleys are the main feature in the interior part of the Eastern Desert when it receives runoff water that sustains different types of vegetation. The existence of many wadis or valleys in the Eastern Desert, indicating the different levels of dissection, shows that the region could not have always been in the same dry conditions as it is in the current period. Hence, wadis form the habitat for the plant resource in the Egyptian Desert. However, some of these wadis have been lately converted to major roads for the transport network from the Nile Valley to the Red Sea coasts (Fig. 9.4). In total, 497 vascular plant species were surveyed in 500 sites along 34 transects in the Eastern Desert areas in Egypt by Abd El-Ghnai et al. in 2014. Also, the most abundant families in terms of species numbers were Asteraceae, Poaceae, and Chenopodiaceae. As for the life forms, therophyta species dominated with about 40.12% from the total number of species, which were mainly represented by the Saharo-Arabian geo-element with 53%. However, the extreme effect caused by human activity could easily be recognized from the status reported

Fig. 9.4 Habitats of the Eastern Desert, (**a**) mangrove along the Red Sea coast, (**b**) inland desert wadi with *Acacia* trees, (**c**) coastal desert wadi north Hurghada, and (**d**) plant growth in Wadi Matuli

in the total number of species; about 18% or 92 species that were reported endangered, threatened, in addition to 25 species that were recorded single times with an extinct species named *Fagonia Täeckhomiana*. The most dominated species were *Zilla spinosa* or *Zygophyllum coccineum*, which could represent the more constant species in the area; hence, they could possibly act as a general term for all communities in the Desert areas in Egypt.

Pulicaria undulata, Ochradenus baccatus, Zygophyllum simplex, Acacia tortilis subsp. tortilis, Trichodesma africanum, Morettia philaeana, Farsetia aegyptia, and *Tamarix nilotica* demonstrated regional dominancy over some areas. Fifty6 highly common and common species included *Haloxylon salicornicum, Launaea nudicaulis, Echinops spinosus, Fagonia mollis, Atriplex halimus, Euphorbia retusa*, and *Calotropis procera*. Some species occurred occasionally (rare), and these included 91 species like *Zygophyllum decumbens, Heliotropium bacciferum, Capparis spinosa, Centaurea aegyptiaca, Launaea spinosa*, and *Kickzia aegyptiaca*. Some species rarely occurred (sporadic/very rare), and this comprised the largest number (247 species) of the recorded.

The result of the biogeographic study on the flora of the Eastern Desert over the last 100 years (1887 to 1979) was that 92 species (61 perennials, 31 annuals) were registered as historical records that were collected only once in the region (Abd El-Ghnai et al. 2014). There was nothing further known about them till this date. Contrary to this, 14 species, largely weeds, represented the current study that were not registered in any previous study nor in any literature. These can be identified as new additions to the flora of the Eastern Desert of Egypt. The weeds that were found to be registered included *Ammi majus, Convolvulus arvensis, Plantago major,*

Lolium rigidum, and *Rumex dentatus*. These five species are among the most popular weeds in the arable lands of Egypt. There would be no doubt that this significantly impacted species diversity was directly linked to the several aspects of climate change in the Eastern Desert over the years of dry conditions.

9.8.1 Phytogeographical Affinities and Vegetation of the Eastern Desert

Analyses of the phytogeographical composition of the Eastern Desert flora (Abd El-Ghani et al. 2017) indicate pronounced spatial variation in species affinities. Plant communities in the southern sector of the desert exhibit stronger influence from Sudanian and Irano–Turanian elements compared to the northern, eastern, and western sectors, reflecting historical biogeographic connectivity and climatic gradients. On the basis of these patterns, new phytogeographical divisions have been proposed, distinguishing the Eastern Desert proper from the Red Sea coastal belt, each of which may be further subdivided into finer floristic units that reflect local environmental and edaphic conditions.

The Gebel Elba mountain complex constitutes one of the three principal coastal mountain systems in southeastern Egypt, oriented along the Red Sea and extending between latitudes 24°50′ N and 22° N, near the Sudan–Egypt border (Fig. 9.5). The massif forms a biologically and ecologically unique region within the Eastern Desert, where complex interactions among topography, climate, and hydrology have fostered high levels of biodiversity. Over time, local communities have inhabited and interacted with these habitats, employing traditional ecological knowledge to utilize and modify plant and animal resources for fuelwood, construction materials, fodder, tools, handicrafts, and medicinal products. Small-scale agriculture along coastal plains, fishing in adjacent offshore waters, and cross-border trade with Sudan further illustrate the integration of human activity within this ecological landscape. Collectively, these practices have shaped ecosystem structure and, in some cases, contributed to the maintenance of genetic diversity within native plant populations.

The vascular flora of Gebel Elba National Park comprises 179 taxa distributed across 124 genera and 51 families. Over half of the taxa are concentrated within just ten families (Abd El-Ghani and Abdel-Khalik 2006). The families with the largest number of genera include Asteraceae, Poaceae, Fabaceae, and Caryophyllaceae (six genera each), followed by Asclepiadaceae, Brassicaceae, Scrophulariaceae, and Zygophyllaceae, all of which are characteristic of Mediterranean North African flora (Quézel 1978). Among the most species-rich genera are Euphorbia (six species), Launaea and Solanum (five species each), and Acacia and Convolvulus (four species each).

The composition and abundance of plant species in the park have been markedly shaped by anthropogenic pressures, including the intensive harvesting of trees and shrubs for domestic fuel and charcoal, as well as grazing by domestic and wild herbivores. These disturbances have limited the regeneration of *Acacia tortilis*

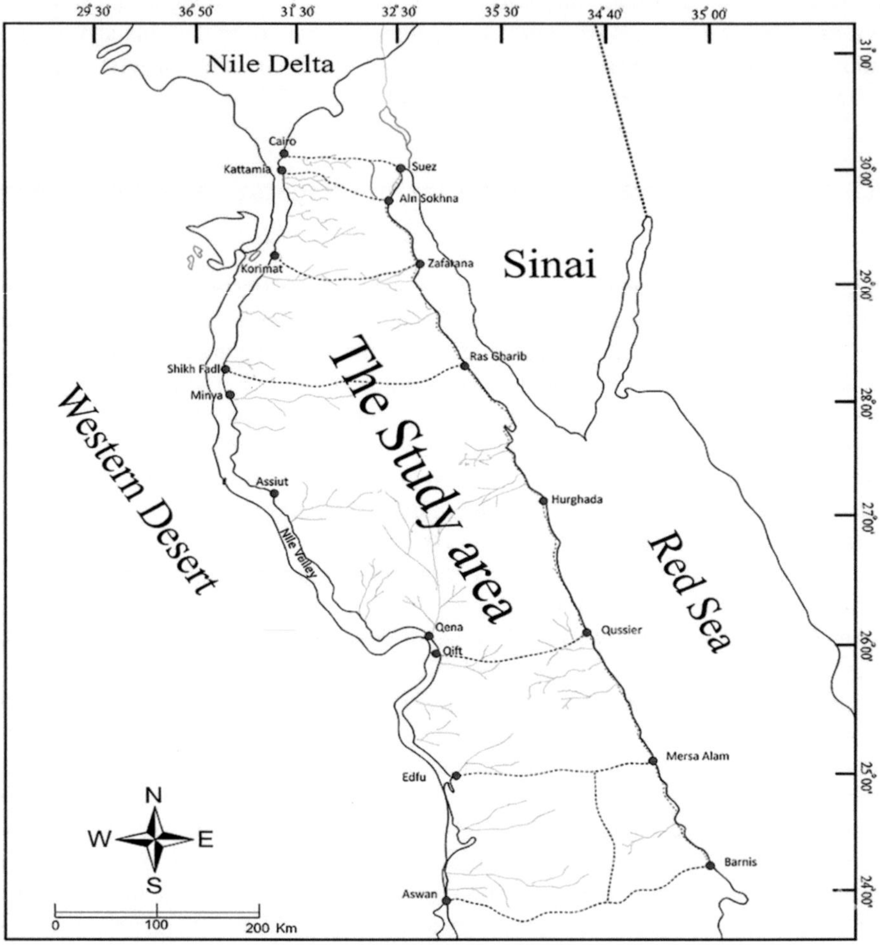

Fig. 9.5 Map of the Eastern Desert, showing the major desert roads

subsp. *tortilis*, *Balanites aegyptiaca*, and *Maerua crassifolia*, while promoting the proliferation of less palatable or un-browsed species, such as *Calotropis procera, Leptadenia pyrotechnica*, and *Senna italica*, which are now among the most frequently recorded taxa in the park.

9.8.2 Life-Form Spectrum and Adaptations

The life-form spectrum of Gebel Elba reflects the adaptations of plants to a hyper-arid environment. Therophytes dominate (48%), indicative of annual species that complete their life cycle rapidly in response to sporadic rainfall. Phanerophytes and chamaephytes each constitute 16.2% of the flora, while hemicryptophytes account for 13.5%. Most annuals are winter or cool-season species, whereas several taxa, including *Amaranthus graecizans* subsp. *graecizans*, *Corchorus depressus, Eragrostis ciliaris, Portulaca oleracea*, and *Setaria viridis*, are adapted to hot-season conditions. A few annuals display opportunistic, non-seasonal growth, emerging in response to rainfall events at any time of the year, such as *Chenopodium murale, Launaea capitata* and *Tribulus terrestris*.

The occurrence of parasitic, leafless, or functionally leafless species, such as *Cuscuta chinensis* and *C. pedicellata*, underscores the critical importance of water conservation strategies in this arid environment. Consistent with patterns observed in other desert ecosystems, vine species are relatively scarce and include *Citrullus colocynthis, Coccinia grandis, Cocculus pendulus, Cucumis prophetarum* subsp. *Prophetarum, Plicosepalus acaciae*, and *P. curviflorus*. Overall, the floristic and ecological characteristics of Gebel Elba reflect a unique convergence of Saharo-Arabian and Sudanian floristic elements, coupled with significant human-mediated modifications, producing a landscape of exceptional biodiversity, ecological resilience, and conservation value.

9.9 The Coastal and Inland Wadis of the Sinai Peninsula: Case Studies

Wadi Kid is the longest wadi located along the southern Sinai coastal belt, extending approximately 50 km in length (Fig. 9.6). The wadi generally provides a relatively accessible route for passage (Salama et al. 2013a), although rugged rock outcrops and steep sections in certain stretches can impede traversal. Its geomorphology, characterized by rocky substrates and occasional sandy deposits, creates a diverse range of microhabitats that support a distinctive desert flora.

A total of 70 vascular plant species were recorded in Wadi Kid, distributed across 56 genera and 33 families. Of these, 21 species were annuals, and 49 were perennials, reflecting a predominance of long-lived, drought-adapted taxa. The most species-rich genera included Fagonia and Zygophyllum, which are widely distributed across arid and hyper-arid environments.

The life-form spectrum indicated that therophytes were dominant, with 20 species representing 37% of the total flora, reflecting the importance of annual species adapted to ephemeral moisture availability. Other significant life forms included chamaephytes, phanerophytes, and hemicryptophytes, which together comprised the remainder of the flora. Among the recorded taxa, several species displayed broad ecological amplitude, notably *Ochradenus baccatus, Zygophyllum coccineum*, and *Zilla spinosa*, which were able to colonize a range of substrates and microhabitats. Conversely, the annual species *Diplotaxis acris*, *Matthiola livida,* and *Schouwia purpurea* exhibited higher relative abundance within the wadi's annual flora, highlighting their adaptive success under the prevailing climatic conditions.

Perennials accounted for 35 species (65% of the total flora), reflecting their relative stability and persistence within the wadi ecosystem. The presence of *Atriplex*

Fig. 9.6 Location of studied stands in Wadi Kid, south Sinai

leucoclada, *Salsola imbricata*, *Tamarix aphylla,* and *T. nilotica* indicated localized soil salinization, a characteristic feature of Wadi Kid's hydrological and edaphic conditions. These species are adapted to saline soils and contribute to the formation of salt-tolerant plant communities.

Geographically, Wadi Kid is distant from the mesic influence of the Mediterranean region, yet it does not fall within the core of the Sudano-Zambezian zone, placing it firmly within the Saharo-Arabian phytogeographical chorotype (Fig. 9.6). The floristic composition, dominated by drought- and salinity-tolerant taxa, reflects the wadi's hyper-arid environment, its geomorphological complexity, and the interplay of edaphic and microclimatic factors that govern species distribution.

9.9.1 Floristic and Ecological Characteristics of Feiran Oasis, Southern Sinai

Oasis-like geomorphological and ecological features are also present in the Sinai Peninsula, particularly along the eastern coastal margin of the Gulf of Suez (Abd El-Ghani et al. 2017). Among these, Feiran Oasis is the most prominent inhabited oasis in southern Sinai, functioning as a key ecological and hydrological hub. According to Zahran and Willis (2009), Feiran represents the largest oasis in the region, receiving substantial runoff from the extensive drainage system of Wadi Feiran. Wadi Feiran, the longest and broadest wadi in southern Sinai, extends approximately 59 km, descending from the high mountainous terrain surrounding the Monastery of St. Katherine in the south (around 2500 m a.s.l.) to approximately 750 m a.s.l. at Feiran Oasis in the west. The wadi system forms the principal hydrological conduit sustaining the oasis's unique vegetation and supporting its human settlements.

Floristic surveys of Feiran Oasis have recorded 70 vascular plant taxa, comprising 49 dicotyledons and 21 monocotyledons, distributed across 63 genera and 26 families (Abd El-Ghani and Fahmy 1998). The families with the greatest species richness included Poaceae, Chenopodiaceae, Asteraceae, and Euphorbiaceae, which collectively dominate the oasis flora. Notably, 33 species were recorded for the first time from Feiran Oasis, of which 13 were identified as invading xerophytes that colonize cultivated fields and orchards, behaving as agricultural weeds. The proliferation of these taxa is likely driven by the prolonged dry conditions observed over the preceding four years, reflecting the sensitivity of the oasis flora to climatic fluctuations.

Botanical exploration of Wadi Feiran also documented eleven of the most widespread weeds of arable land in Egypt (El Hadidi and Kosinová, 1971), highlighting the increasing anthropogenic influence on local plant communities. These findings suggest a shift in species composition, with native taxa being gradually displaced by opportunistic weeds in cultivated and disturbed habitats. In addition, three species *Parietaria alsinifolia* Delile, *Rostraria pumila* (Desf.) Tzvelev, and *Samolus valerandi* L., are now considered extinct in the oasis, emphasizing the vulnerability of the native flora under combined pressures of environmental stress, climatic variability, and human activity. Overall, Feiran Oasis exemplifies the ecological complexity and fragility of desert oases, where hydrology, geomorphology, and human land use interact to shape floristic composition. The interplay of these factors underscores the need for sustainable management and conservation strategies to protect the oasis's unique plant biodiversity in the face of ongoing environmental change.

9.9.2 Wadi Feiran Road, Basin, and Oasis Flora

Wadi Feiran Road constitutes the sole major transportation corridor connecting the Gulf of Suez coastal road in the west with St. Catherine City and the Gulf of Aqaba in the east. Functioning as a lifeline for local communities, this road traverses the Feiran Basin, the principal drainage catchment of the region, through which Wadi Feiran channels surface runoff toward the Gulf of Suez. Owing to the steep topography and episodic heavy rainfall, both the road and surrounding villages are frequently subjected to flash floods, typically occurring once or twice per year, particularly during spring and autumn. Over the past few decades, the frequency and intensity of these floods have increased, resulting in substantial damage to infrastructure, transport networks, and public facilities. Notable flood events in

1994 and 2005 caused widespread destruction, including significant economic losses, urban damage, and disruption of road systems.

The current floristic composition of Feiran Oasis shows only limited similarity to the historical weed flora of southern Sinai. Early surveys by El Hadidi et al. (1970) documented 156 plant species, whereas subsequent investigations by Abd El-Ghani and Fahmy (1998) identified only 70 species in common, highlighting a marked reduction and compositional shift over recent decades. These changes reflect environmental degradation driven by prolonged aridity, water scarcity, and anthropogenic pressures. The small spatial extent of the oasis, coupled with limited water availability, rapid development due to increased tourism and population pressures, and prevailing socio-economic constraints, poses a severe threat to the persistence of remaining natural habitats.

The life-form spectrum of Feiran Oasis mirrors that of surrounding desert environments, with therophytes dominating, followed by hemicryptophytes and chamaephytes. This pattern closely aligns with observations by Olsvig-Whittaker et al. (1983) in the Negev Desert at Sede Boqer, Israel, and comparable environmental gradients have been documented by Danin and Orshan (1990).

Cultivated orchards form the most conspicuous vegetation in the oasis, primarily composed of date palm (*Phoenix dactylifera*). Other cultivated fruit trees include olive (*Olea europaea*), grapevine (*Vitis vinifera*), and pomegranate (*Punica granatum*). Due to continuous depletion of water resources, date-palm orchards are classified into three main types: Old orchards, the most vulnerable, where shallow wells have dried, leading to abandoned or severely degraded plantations. Orchards established 8–10 years ago, supplied by wells with limited discharge (Fig. 9.7).

Newly established orchards, supported by deep wells dug within the past four years, which provide adequate water. In these newly irrigated areas, the ground between date palms is typically colonized by *Alhagi graecorum, Cynodon dactylon,*

Fig. 9.7 Date-palm orchards of Wadi Feiran challenging drought and severe lack of water

Oxalis corniculata, and *Zygophyllum simplex*, alongside numerous weedy and xerophytic associates (Abd El-Ghani 1998).

A comparative floristic analysis between the Eastern Desert and Sinai Desert revealed 141 species in common out of 1378 recorded taxa, yielding a similarity index of approximately 20.5%. This relatively low similarity reflects their distinct phytogeographical affiliations, with the Sinai Desert being part of the Irano-Turanian region and the Eastern Desert belonging to the Saharo-Arabian region. Among the recorded species, 335 were exclusive to Sinai, while 64 were characteristic of the Eastern Desert. Despite these differences at the species level, similarity increased at broader taxonomic ranks, reaching 56.5% at the family level and 64.6% at the generic level, indicating shared evolutionary lineages and biogeographical connections across the two desert regions.

9.9.3 Floristic Composition of Selected Inland Wadis in Southern Sinai

The inland wadis of southern Sinai, specifically Wadi Romana, Wadi Solaf, and Wadi Al-Akhdar, support a diverse assemblage of desert-adapted plants, reflecting the unique ecological gradients of this arid region (Fig. 9.8). Collectively, these wadis harbor 116 plant species, comprising 45 annuals and 71 perennials, distributed across 95 genera and 37 families. The most species-rich families, which together account for 57.7% of the total flora, include Asteraceae, Brassicaceae, Fabaceae, Zygophyllaceae, Boraginaceae, Chenopodiaceae, and Poaceae. At the generic level, Fagonia and Astragalus emerge as the most prominent, demonstrating both ecological versatility and wide distribution across the wadis.

Several species, such as *Fagonia mollis, Reseda pruinosa* and *Zilla spinosa,* exhibit broad ecological amplitudes, occurring across multiple habitats and elevations within the studied wadis. The vegetation composition reflects the full spectrum of desert life forms:

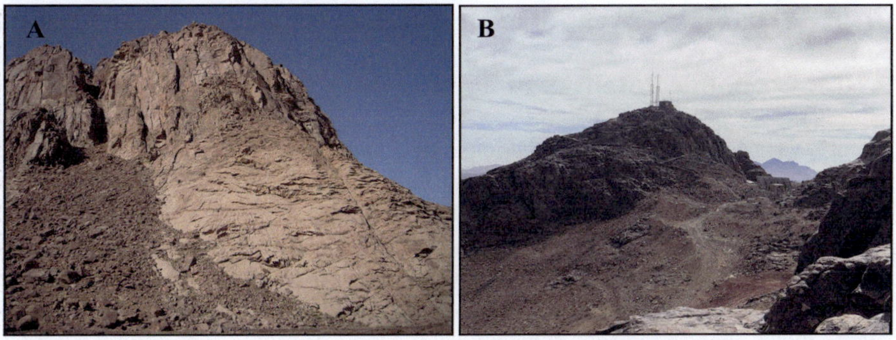

Fig. 9.8 Two highest mountain peaks in Sinai Peninsula: (**a**) Gebel Musa, (**b**) Gebel Saint Katherine

Trees: *Acacia tortilis* subsp. *tortilis*, *Retama raetam.*

Shrubs: *Aerva javanica, Anabasis articulata, Capparis spinosa, Ochradenus baccatus, Haloxylon salicornicum, Heliotropium arbainense, Zilla spinosa, Seriphidium herba-album.*

Perennial herbs: *Blepharis edulis, Citrullus colocynthis, Cucumis prophetarum, Erodium oxyrhynchum, Morettia philaeana.*

Annual herbs: *Asphodelus tenuifolius, Aristida adscensionis, Gypsophila capillaris, Launaea capitata, Diplotaxis harra, Reseda pruinosa.* The distribution and fidelity of species varied among the wadis, reflecting subtle differences in microhabitats, soil characteristics, and moisture availability. For instance:

Wadi Romana: Twelve species demonstrated strong fidelity, including *Cleome droserifolia, Crotalaria aegyptiaca, Caralluma europaea, Hyoscyamus pusillus, and Volutaria lippii.*

Wadi Al-Akhdar: Thirty species were strongly associated with this wadi, including *Astragalus eremophilus, Carduncellus eriocephalus, Crepis micrantha, Ephedra pachyclada, Fagonia schimperi, Pimpinella cretica*, and *Schimpera arabica.*

Wadi Solaf: Fifteen species were characteristic, including *Agatophora alopecuroides, Deverra tortuosa, Matthiola longipetala, Scabiosa palaestina, Schouwia purpurea*, and *Sisymbrium erysimoides*.

Overall, perennial species dominated the flora, accounting for approximately 60% of recorded taxa, which reflects the stability of long-lived desert-adapted plants in these relatively stable wadi habitats. Annuals, adapted to opportunistic germination following episodic rainfall, contributed to the remaining 40%, emphasizing the importance of climatic variability in shaping desert plant communities. The vegetation of these wadis not only represents Saharo-Arabian floristic elements but also highlights the ecological complexity of southern Sinai, where topography, soil heterogeneity, and ephemeral water availability combine to create a mosaic of microhabitats, each supporting distinctive assemblages of species (Salama et al. 2013b).

9.10 Economic Potential of Desert Plants under Climate Change

9.10.1 Economic Potential of Egypt's Desert Flora

Egypt's vast desert landscapes, encompassing over 90% of the country's land area (FAO 2020), host a remarkably diverse and resilient flora uniquely adapted to extreme arid conditions. These species have evolved a wide array of physiological and morphological adaptations to survive in harsh environments characterized by scarce and erratic rainfall, high temperatures, intense solar radiation, and saline or nutrient-poor soils (Ayyad and Ghabbour 1986). Beyond their ecological significance, these plants represent substantial untapped economic opportunities across multiple sectors, including pharmaceuticals, agriculture, bioenergy, cosmetics, and eco-tourism (Heneidy and Bidak 2004).

Given Egypt's pressing challenges—water scarcity, limited arable land, and dependence on imported fossil fuels (World Bank 2021)—the strategic utilization of desert plant resources offers a sustainable pathway toward economic diversification, enhanced food security, and environmental conservation. Desert flora thus holds immense potential as a renewable and low-input natural resource, particularly for arid and hyper-arid regions where conventional agriculture is often impractical (Orians and Solbrig 1977).

9.10.2 Adaptation and Economic Significance

Despite the severe climatic constraints, Egypt's deserts are home to a highly specialized and resilient flora, including both annual and perennial species. Many plants, shaped by millennia of adaptation, provide multiple ecosystem services, such as stabilizing soils, preventing desertification, and supporting local biodiversity. These species also possess traditional and modern economic utility—ranging from medicinal, fodder, and domestic uses to industrial applications—making them invaluable for sustainable development and livelihood improvement in arid regions.

Research has highlighted the importance of xerophytes (drought-tolerant plants) and halophytes (salt-tolerant plants) as viable resources in Egypt's deserts (Zahran et al. 1993; Shaltout and Ahmed 2012; Ahmad et al. 2020). In regions where freshwater is limited and groundwater is often brackish or saline, these plants can be cultivated without competing for scarce resources needed for conventional crops. Their tolerance to water and salinity stress makes them particularly valuable for the development of alternative cropping systems and sustainable agro-ecosystems.

9.10.3 Vegetation Structure and Ecological Roles

The native desert flora underpins Egypt's diverse desert ecosystems, which include mangroves, reed swamps, salt marshes, sand dunes, desert plains, wadis, and mountains. Perennial halophytes and xerophytes form the structural backbone of desert vegetation, while annual species, such as Zygophyllum simplex and Asphodelus tenuifolius, provide seasonal green cover during wet periods, supporting grazing for domestic and wild animals. These plant communities also serve critical ecological roles, such as nutrient cycling, soil stabilization, and microhabitat formation, which sustain the broader desert environment.

9.10.4 Economic Uses and Industrial Potential

Egyptian desert plants are classified into five primary economic categories (Zahran and Willis 2009):

- Fodder-producing plants: Species providing forage for livestock in marginal lands.
- Drug-producing plants: Medicinal flora used in traditional and modern pharmacopeias.
- Fiber-producing plants: Sources of textiles, ropes, and other materials.
- Oil and perfume-producing plants: Species yielding edible, industrial, or aromatic oils.
- Wood and fuel plants: Trees and shrubs supplying timber, fuelwood, and charcoal.

Many species exhibit multiple uses. For instance, halophytic mangroves (*Avicennia marina, Rhizophora mucronata*) and xerophytic trees (*Balanites aegyptiaca, Vachellia* spp.) provide food, fodder, medicinal compounds, and wood, while psammophytes stabilize mobile sand dunes, mitigating desertification and protecting human settlements.

9.10.5 Traditional and Contemporary Applications

Historically, local communities have relied on desert plants for medicinal remedies, nutritional supplements, and domestic materials, such as dyes, fibers, and fuelwood. These uses continue to hold relevance today, offering potential for income generation, ecological restoration, and sustainable natural resource management. Modern studies also indicate the potential for industrial applications, including bioenergy production, cosmetics, pharmaceuticals, and eco-tourism development, all of which could enhance economic resilience in arid regions.

Floristic surveys indicate a rich diversity of species with economic potential, distributed across Egypt's distinct biogeographic regions (Table 9.3). By leveraging this diversity, Egypt can achieve a sustainable balance between conservation, economic development, and resource utilization, transforming its deserts from marginal lands into productive and resilient landscapes.

9.11 Distribution of Economic Plants Among the Biogeographic Regions

The distribution of economic plants across Egypt's distinct biogeographic regions exhibits notable variation (Fig. 9.9). The Sinai Peninsula, the Mediterranean coastal region, the Nile Valley and Delta (Nile land), and the Oases collectively harbor the highest concentrations of these species, accounting for 77, 71, 68, and 44 recorded plant species, respectively. Together, these four regions contain approximately 70% of the total economic plant species documented.

Several species demonstrate a broad distribution range, being recorded in seven biogeographic regions, and are among the most commonly utilized desert plants. These included *Adiantum capillus-veneris*, *Citrullus colocynthis*, *Cyperus rotundus*,

Table 9.3 Distribution of economic plants in the biogeographical regions (see Fig. 9.1 for the abbreviations of regions). + = present, – = absent

Species	Biogeographic regions									
	N	O	M	D	De	Dw	R	GE	S	Uw
Species recorded in 7 regions										
Adiantum capillus-veneris L.	+	+	+	+	-	-	+	+	+	-
Citrullus colocynthis (L.) Schrad.	+	+	+	+	-	-	+	+	+	-
Cyperus rotundus L.	+	+	+	+	-	-	+	+	+	-
Hyoscyamus muticus L.	+	+	+	+	-	-	+	+	+	-
Phoenix dactylifera L.	+	+	+	+	-	-	+	+	+	-
Zilla spinosa (L.) Prantl	+	-	+	+	-	+	+	+	+	-
Ziziphus spina-christi (L.) Desf.	+	+	+	+	-	-	+	+	+	-
Species recorded in 6 regions										
Alhagi graecorum Boiss.	+	+	+	+	-	-	+	-	+	-
Capparis spinosa L.	-	+	+	+	+	+	-	-	+	-
Portulaca oleracea L.	+	+	+	+	-	-	-	+	+	-
Salvadora persica L.	+	+	-	+	-	-	+	+	+	-
Tribulus terrestris L.	+	-	+	+	-	-	+	+	+	-
Species recorded in 5 regions										
Balanites aegyptiaca (L.) Delile	+	+	-	+	-	-	-	+	+	-
Eruca vesicaria subsp. *sativa* (Mill.) Thell.	+	+	+	-	+	-	-	-	+	-
Ficus carica L.	+	+	+	+	-	-	-	-	+	-
Hyphaene thebaica (L.) Mart.	+	+	-	-	+	-	+	-	+	-
Salix subserrata Willd.	+	+	+	-	+	-	-	-	+	-
Trifolium alexandrinum L.	+	+	+	+	-	-	-	-	+	-
*Achillea fragrantiss*ima (Forssk.) Sch. Bip	-	+	+	+	-	-	+	-	+	-
Anastatica hierochuntica L.	-	-	-	+	-	-	+	+	+	+
Artemisia judaica L.	-	-	+	+	-	-	+	+	+	-
Avena sativa L.	+	+	+	+	-	-	-	-	+	-
Cleome droserifolia (Forssk.) Delile	-	-	-	+	-	-	+	+	+	+
Convolvulus arvensis L.	+	+	+	+	-	-	-	-	+	-
Hypericum perforatum L.	+	+	-	-	+	-	+	-	+	-
Juncus acutus L.	+	+	+	-	+	-	-	-	+	-
Olea europaea L.	+	+	+	+	-	-	-	-	+	-
Pluchea dioscoridis (L.) DC.	+	+	+	-	+	-	-	-	+	-
Salvia aegyptiaca L.	-	-	+	+	-	-	+	+	+	-
Senna alexandrina Mill.	+	-	-	-	+	-	+	+	+	-
Solanum nigrum L.	+	+	+	+	-	-	-	-	+	-
Urtica urens L.	+	+	+	-	+	-	-	-	+	-
Species recorded in 4 regions										
Coriandrum sativum L.	+	+	+	-	-	-	-	-	+	-
Ammi majus L.	+	+	+	-	-	-	-	-	+	-
Artemisia herba-alba Asso	-	-	+	+	-	-	+	-	+	-

(continued)

Table 9.3 (continued)

Blepharis edulis (Forssk.) Pers.	-	-	-	+	-	-	+	+	+	-
Brassica nigra (L.) W. D. J. Koch	+	+	+	-	-	-	-	-	+	-
Ephedra alata Decne.	-	+	+	+	-	-	-	-	+	-
Lepidium sativum L.	+	+	-	-	+	-	-	-	+	-
Orobanche crenata Forssk.	+	-	+	+	-	-	-	-	+	-
Papaver rhoeas L.	+	-	+	+	-	-	-	-	+	-
Plantago afra L.	-	-	-	-	+	-	+	+	+	-
P. major L.	+	+	+	-	-	-	-	-	+	-
P. ovata Forssk.	+	-	+	+	-	-	-	-	+	-
Rumex vesicarius L.	-	-	+	+	-	-	-	+	+	-
Silybum marianum (L.) Gaertn.	+	+	+	-	-	-	-	-	+	-
Withania somnifera (L.) Dunal	+	+	+	-	-	-	-	+	-	-
Zygophyllum coccineum L.	-	+	-	+	-	-	+	-	+	-
Species recorded in 3 regions										
Apium graveolens L.	-	-	+	-	+	-	-	-	+	-
Brassica rapa L.	+	+	+	-	-	-	-	-	-	-
Cyperus esculentus L.	+	-	+	-	-	-	-	+	-	-
Foeniculum vulgare Mill.	+	-	+	-	-	-	-	-	+	-
Malva sylvestris L.	+	-	+	-	-	-	-	-	+	-
Mentha spicata L.	+	+	-	-	-	-	-	-	+	-
Acacia nilotica (L.) Delile	+	+	-	-	-	+	-	-	-	-
Ajuga iva (L.) Schreber	-	-	+	-	-	-	-	-	+	+
Cichorium intybus L.	+	+	+	-	-	-	-	-	-	-
Colchicum autumnale L.	-	-	+	+	-	-	-	-	+	-
Convolvulus althaeoides L.	-	-	+	-	+	-	-	-	+	-
Marrubium alysson L.	-	-	+	+	-	-	-	-	+	-
Matricaria chamomilla L.	+	-	+	-	-	-	-	-	+	-
Peganum harmala L.	-	-	+	-	+	-	-	-	+	-
Petroselinum crispum (Mill.) Fuss	+	-	+	-	-	-	-	-	+	-
Solenostemma arghel (Del.) Hayne	-	-	-	-	+	-	-	+	+	-
Teucrium polium L.	-	-	+	+	-	-	-	-	+	-
Urtica pilulifera L.	+	-	+	-	-	-	-	-	+	-
Verbena officinalis L.	+	+	+	-	-	-	-	-	-	-
Species recorded in 2 regions										
Linum usitatissimum L.	+	+	-	-	-	-	-	-	-	-
Nigella sativa L.	+	-	+	-	-	-	-	-	-	-
Vicia faba L.	+	+	-	-	-	-	-	-	-	-
Ammi visnaga (L.) Lam.	+	-	+	-	-	-	-	-	-	-
Bistorta officinalis Delarbre	+	-	+	-	-	-	-	-	-	-
Cymbopogon schoenanthus (L.) Spreng. subsp. *proximus* (Hochst. ex A. Rich.) Mailler & Weiller	-	-	-	-	+	-	-	-	+	-
Drimia maritima (L.) Stearn	-	-	+	-	-	-	-	-	+	-
Hibiscus sabdariffa L.	+	+	-	-	-	-	-	-	-	-
Juniperus communis L.	-	-	-	-	+	-	-	-	+	-

(continued)

Table 9.3 (continued)

Leontice leontopetalum L.	-	-	+	-	-	-	-	-	+	-
Lupinus albus L.	+	+	-	-	-	-	-	-	-	-
Marrubium vulgare L.	-	-	+	-	-	-	-	-	+	-
Mentha pulegium L.	+	+	-	-	-	-	-	-	-	-
Moringa peregrina (Forssk.) Fiori	-	-	-	-	+	-	-	-	+	-
Narcissus tazetta L.	-	-	+	-	-	-	-	-	+	-
Nymphaea spp.	+	-	+	-	-	-	-	-	-	-
Polygonum aviculare L.	+	-	+	-	-	-	-	-	-	-
Ricinus communis L.	-	-	-	-	+	-	-	+	-	-
Sinapis alba L.	+	-	+	-	-	-	-	-	-	-
Sisymbrium officinale (L.) Scop.	+	-	+	-	-	-	-	-	-	-
Species recorded in ONE regions										
Capsella bursa-pastoris (L.) Medik.	+									
Cuscuta epilynum Weihe	+									
Cymbopogon citratus (DC.) Stapf	+									
C. nardus (L.) Rendle	+									
Datura stramonium L.	+									
Lepidium latifolium L.	+									
Triticum aestivum L.	+									
Zea mays L.	+									
Bryonia cretica L.			+							
Thymbra capitata (L.) Cav.			+							
Fumaria officinalis L.			+							
Pistacia lentiscus L.			+							
Posidonia oceanica (L.) Delile			+							
Stellaria media (L,) Vill.			+							
Ziziphus lotus (L.) Lam.			+							
Dodonaea viscosa Jacq.								+		
Ceratonia siliqua L.									+	
Convolvulus scammonia L.									+	
Crataegus monogyna Jacq.									+	
Cymbopogon schoenanthus (L.) Spreng.									+	
Ecballium elaterium (L.) A. Rich.									+	
Ferula communis L.									+	
Rhus coriaria L.									+	
Viola odorata L.									+	
V. tricolor L.									+	

Hyoscyamus muticus, *Phoenix dactylifera*, *Zilla spinosa*, and *Ziziphus spina-christi* (Table 9.3).

Conversely, 24 plant species exhibited a narrow distribution, often characterizing a specific biogeographic region. These are distributed as follows: eight species are

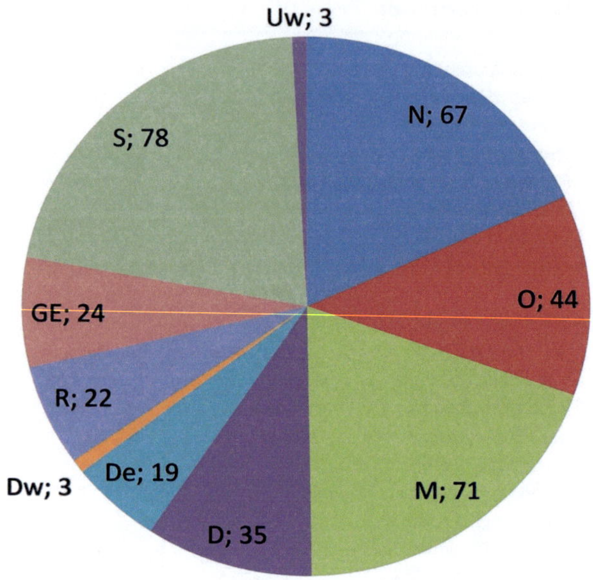

Fig. 9.9 Proportionate of the economic plant species in the biogeographical regions of Egypt (see Fig. 9.1 for more abbreviations)

primarily found in the Nile land, seven in the Mediterranean region, one in the Gebel Elba region, and nine in the Sinai Peninsula (Table 9.3).

Of particular concern were several species facing threats within their respective regions. For instance, *Bryonia cretica*, *Pistacia lentiscus*, and *Ziziphus lotus* are considered threatened in the Mediterranean region. It is also noteworthy that the nine medicinal plant species characteristic of the Sinai Peninsula are reportedly under threat. In response to these conservation concerns, various measures have been implemented to protect these valuable species (Abd El-Wahab et al. 2008).

9.12 Pharmaceutical and Medicinal Applications

Egypt's desert plants have long been utilized in traditional medicine, and many are now gaining recognition in modern pharmacological research (Aboul-Enein et al. 2012). Table 9.4 shows the uses of common desert plants used in folk medicine in Egypt. Plants were classified into trees, shrubs and herbs. *Aloe vera*, widely cultivated in Egypt's arid regions, possesses well-documented wound-healing, anti-inflammatory, and skin-moisturizing properties that have made it valuable in dermatological and cosmetic products (Hamman 2008). The global *Aloe vera* market, currently valued at over $1 billion annually, continues to expand, presenting significant export opportunities for Egyptian producers. Similarly, *Moringa oleifera*, a drought-resistant tree thriving in Upper Egypt, contains exceptionally high levels of vitamins, minerals, and antioxidants, with demonstrated potential in managing diabetes and hypertension (Abd El-Wahab et al. 2008). The plant's leaves and seeds also show promise in water purification applications (Ndabigengesere et al. 1995).

Table 9.4 Common medicinal desert plants and their uses, arranged alphabetically within their growth forms (trees, shrubs, and herbs)

Species	Family	Medicinal uses
Trees		
Balanites aegyptiaca (L.) Delile	Zygophyllaceae	Good for jaundice, intestinal worm infection, wounds, malaria, syphilis, epilepsy, dysentery, constipation, diarrhea, hemorrhoid, stomach aches, asthma, and fever
Calotropis procera (Aiton) W.T. Aiton	Apocynaceae	Respiratory system and antibacterial
Crataegus monogyna Jacq.	Rosaceae	Adjunct therapy for heart failure stage ll, relieve symptoms of temporary nervous cardiac complaints, relief of mild symptoms of mental stress and to aid sleep
Hyphaene thebaica (L.) Mart.	Arecaceae	Anti-inflammatory, antioxidant, antimicrobial, anticancer, and in the treatment of hypertension, bilharzias and as a hematinic agent. Fruits can reduce hyperlipidemia
Moringa peregrina (Forssk.) Fiori	Moringaceae	Treatment of a wide range of conditions, such as inflammation, gastrointestinal, hematological, cardiovascular, hepato and renal disorders. Also, it has been used for diabetes and hypertension as well as liver protection
Phoenix dactylifera L.	Arecaceae	The fruits were used traditionally as general tonic, for the treatment of liver diseases, memory disturbances, fever, inflammation, paralysis, loss of consciousness, nervous disorders and consumed by pregnant women before and after delivery. However, all parts of the plant were used for some purpose. Date fruits were considered a complete diet and a very important item of food, possessed many pharmacological effects including anticancer, antidiabetic, anti-inflammatory, antimicrobial, antiparasitic, antioxidant, anti-toxin, cardiovascular, hypolipidemic, gastrointestinal, immune-modullatory, neural, hepato and Reno-protective, reproductive and wound-healing effects
Rhus coriaria L.	Anacardiaceae	Antioxidant capacities that have ameliorative and therapeutic benefits for many common diseases including cardiovascular disease, diabetes, and cancer

(continued)

Table 9.4 (continued)

Species	Family	Medicinal uses
Tamarix nilotica (Ehrenb.) Bunge	Tamaricaceae	Antibacterial, carminative, astringent sudorif-ic, ulcer, expectorant, aphrodisiac, relieves headaches and reduces inflammation. Leaves and young branches are cooked for spleen edema, bark is used as a lotion
Vachellia nilotica (L.) P.J.H. Hurter & Mabb.	Fabaceae	Anti-cancer, antimutagenic, spasmogenic, vasoconstrictor, antipyretic, anti-asthamatic, cytotoxic, antidiabetic, anti-platelet aggregators, anti-plasmodial, molluscicidal, antifungal, inhibitory activity against Hepatitis C virus (HCV) and human immunodeficiency virus (HIV)-I, antioxidant, antibacterial, antihypertensive, antispasmodic activities
Vachellia seyal (Delile) P.J.H. Hurter	Fabaceae	Antibacterial, antimalarial, antimycobacterial, cyclooxygenase inhibition effect, molluscicidal activity, anticancer, hypoglycemic, antidiabetic, antioxidant, immunomodulatory, cytoprotective antiulcer, prebiotic properties
Zilla spinosa (L.) Prantl	Brassicaceae	Treatment of gall bladder and kidney stones and/or ailments, antifungal, anticancer, and anti-rheumatic
Ziziphus lotus (L.) Lam.	Rhamnaceae	Treating diabetes, obesity, dyslipidemia, ulcers, and spasms. Additionally, it was found to have an anti-urolithic effect, preventing the formation of kidney stones
Ziziphus spina-christi (L.) Desf.	Rhamnaceae	Antihyperglycemic, antibacterial, antifungal, antioxidant and antinociceptive activities
Shrubs		
Aerva javanica (Burm.f.) Juss. ex Schult.	Amaranthaceae	Toothache, snake, and insect's bites
Artemisia herba-alba Asso	Asteraceae	Antioxidants, anti-venom, antifungal, nematicidal, antibacterial, antispasmodic, anthelmintic, antileishmanial, neurological, hypoglycemic, and pesticidal activities
Artemisia judaica L.	Asteraceae	Anti-inflammatory, analgesic, antioxidant activity and high potential antiangiogenic activity
Ephedra alata Decne.	Ephedraceae	Anti-cancer, anti- asthmatic, also for arthritis, and rheumatoid

(continued)

Table 9.4 (continued)

Species	Family	Medicinal uses
Hyoscyamus muticus L.	Solanaceae	Antioxidant, and antimicrobial
Iphiona mucronata (Forssk.) Asch. & Schweinf	Asteraceae	Antimicrobial and antitumor
Ochradenus baccatus Delile	Resedaceae	Back pain and fistula
Peganum harmala L.	Nitrariaceae	Anti-inflammatory, analgesic, antibacterial, and possibly psychoactive effects
Salvadora persica L.	Salvadoraceae	Antibacterial, antiviral, antifungal, antioxidant, and even antiulcer effects, promising for dental and fields
Salvia aegyptiaca L.	Lamiaceae	Antiseptic, carminative, digestive, analgesic, antioxidant, anticancer, and antiproliferative effect
Senna alexandrina Mill.	Fabaceae	Remedies for waterborne diseases including typhoid, also joint, tendon ligaments and bone related diseases, pneumonia, reducing fever, bacterial, viral and fungal long-term infections, leukemia, jaundice, intestinal worms and use as healer for splenic enlargement
Solenostemma arghel (Delile) Hayne	Apocynaceae	Anti-inflammatory, antioxidant, hypoglycemic, hypolipidemic, and anti-obesity activities
Zilla spinosa (L.) Prantl	Brassicaceae	Treatment of gall bladder and kidney stones and/or ailments, antifungal, anticancer, and anti-rheumatic
Zygophyllum coccineum L.	Zygophyllaceae	The plant extract has efficient and safe anti-inflammatory potential, with promising antifungal and insecticidal actions. It was also reported to control elevated blood pressure and heart rate in induced rats
Herbs		
Achillea fragrantissima (Forssk.) Sch.Bip.	Asteraceae	Anti-inflammatory, antimicrobial, antidiabetic, sedative
Alhagi graecorum Boiss.	Fabaceae	Antioxidant, cardiovascular, antiulcer, hepatoprotective, antispasmodic, antidiarrheal, antinociceptive, antipyretic, anti-inflammatory, anti-rheumatic, antibacterial and antifungal activities
Anastatica hierochuntica L.	Brassicaceae	Hypolipidemic activity, antidiabetic, hepatoprotective, antioxidant, antiproliferative, gastroprotective, anti-inflammation, and cytotoxic activities
Bryonia cretica L.	Cucurbitaceae	Antimicrobial, antioxidant, antidiabetic, antinociceptive, and anti-inflammatory functions, along with an anticancer efficiency

(continued)

Table 9.4 (continued)

Species	Family	Medicinal uses
Capparis spinosa L.	Capparaceae	Used for rheumatism and gout, as diuretic, astringent, tonic, as appetizer, anti-diarrheic, treat hemorrhoids and spleen disease, and as expectorant, and for chest diseases, used in sciatica, dropsy, backache, and to relieve toothache, for ulcers, scrofula, and ganglions
Citrullus colocynthis (L.) Schrad.	Cucurbitaceae	Fruit is used to treat colds, diarrhea, parasitic worms, the expulsion of wind, tumors, ascites, leukoplakia, ulcers, asthma, bronchitis, diabetes insipidus, jaundice, splenomegaly, neck tuberculosis, constipation, anemia, throat diseases, lephantiasis, and joint pain; it is also used as an antipyretic
Cleome droserifolia (Forssk.) Delile	Cleomaceae	Treatment of hyperglycemia, accelerate wound-healing especially for diabetes mellitus patients, open sores, allergy, dermatitis, inflammation, scabies, as antimicrobial agent and antibiotic for wounds and burns
Convolvulus althaeoides L.	Convolvulaceae	Cytotoxic activity against various tumor cell lines
Cymbopogon schoenanthus subsp. *proximus* (Hochst. Ex A. Rich.) Maire & Weiller	Poaceae	Cardiovascular depressant, antidopaminergic and antiglutamic-aspartic acids in the antiemetic and anti-convulsant effects
Hypericum perforatum L.	Hypericaceae	Antidepressant, antioxidant, anti-convulsant, analgesic, anti-inflammatory, cytotoxic and antidiabetic activities
Lupinus albus L.	Fabaceae	Antidiabetic, anti-convulsant, antioxidant, antimicrobial and antihyperlipidemic activities. It shows efficiency in controlling serum glucose and dyslipidaemia
Marrubium alysson L.	Lamiaceae	Lipid lowering, anti-inflammation and anti-oxidation activities, managing Alzheimer's disease
Narcissus tazetta L.	Amaryllidaceae	Antibacterial and antifungal, antiviral, antimalerial, anticancer, antioxidant, dermatological, cardiovascular, immunomodulatory and acetylcholinesterase inhibitory effects
Papaver rhoeas L.	Papaveraceae	Treat nervousness, insomnia, digestive and respiratory disorders, baldness, eye infections, as well as measles treatment

(continued)

Table 9.4 (continued)

Species	Family	Medicinal uses
Plantago ovata Forssk.	Plantaginaceae	Lowering cholesterol, prevent constipation, improving cancer cells and reduce obesity
Rumex vesicarius L.	Polygonaceae	Antioxidant, antiproliferative and antimicrobial potential
Salvia aegyptiaca L.	Lamiaceae	Antiseptic, carminative, digestive, analgesic, antioxidant, anticancer, and antiproliferative effect
Silybum marianum (L.) Gaertn.	Asteraceae	Renal protection, hypolipidemic and anti-atherosclerosis activities, cardiovascular protection, prevention of insulin resistance, especially in cirrhotic patients, cancer, and Alzheimer prevention
Teucrium polium L.	Lamiaceae	Antioxidant, antibacterial, and anti-inflammatory activity
Tribulus terrestris L.	Zygophyllaceae	It has analgesic, diuretic and uricosuric effects, anti-hyperglycaemic, anti-inflammatory, antioxidant, and antibacterial.
Tricodesma africanum (L.)R. Br.	Boraginaceae	Abdominal pain, mouth ulcers, and chest congestion

Another notable species, *Vachellia nilotica* (Nile acacia), produces gum arabic, a versatile substance used globally in food processing, pharmaceuticals, and industrial applications. Neighboring Sudan generates approximately $100 million annually from gum arabic exports, suggesting substantial potential for Egyptian production. However, realizing this potential requires increased investment in biotechnology research, clinical validation of traditional medicinal uses (Gurib-Fakim 2006), and robust intellectual property protections to prevent biopiracy of Egypt's native plant species.

Balanites aegyptiaca (Desert Date) has various parts used for their hypoglycemic, anthelmintic, and antimicrobial properties (Chothani and Vaghasiya 2011), whereas *Artemisia judaica* is widely used as an antispasmodic and for treating gastrointestinal disorders, and *Citrullus colocynthis* (Colocynth) has been traditionally employed as a potent purgative, albeit with caution due to its toxicity.

9.13 Conclusions

To address the impacts of climate change on Egypt's plant biodiversity, a multi-faceted approach is essential. First, strengthening conservation efforts is critical. This includes expanding protected areas, particularly in biodiversity hotspots like Gebel Elba and the Sinai Peninsula, and enforcing stricter regulations to limit human activities such as overgrazing, urbanization, and tourism. Targeted conservation programs should be implemented for endangered species, focusing on habitat restoration, seed banking, and controlled cultivation. Additionally, sustainable water management practices must be developed, especially in oases and depressions, to ensure the survival of groundwater-dependent vegetation. Improving irrigation techniques and reducing water wastage will be key to mitigating the effects of prolonged droughts and rising temperatures.

Second, controlling invasive species is crucial to preserving native flora. Monitoring programs should be established to detect and manage the spread of invasive alien species, such as *Ambrosia maritima* and *Aster squamatus*, which threaten local ecosystems. Public awareness campaigns can educate communities about the risks of invasive species and the importance of using native plants in agriculture and landscaping. Furthermore, enhancing research and monitoring through long-term ecological studies will help track the impacts of climate change on plant biodiversity, particularly in vulnerable regions like the Western Desert and Sinai Peninsula. Regular floristic surveys and updates to the Egyptian Plant Red Data Book will ensure that the status of threatened species is accurately documented and addressed.

Finally, promoting sustainable land use and community involvement is vital. Encouraging sustainable agricultural practices in the Nile Valley and Delta, such as organic farming and reduced chemical inputs, will help minimize the impact on natural habitats. Urban planning strategies should also prioritize the preservation of natural vegetation in desert cities and along the Mediterranean coast. Engaging local communities in conservation efforts, particularly in areas like Gebel Elba and Sinai, will leverage traditional knowledge and provide economic incentives through eco-tourism and sustainable harvesting of medicinal plants. Strengthening environmental legislation and fostering international collaboration will further support these efforts, ensuring a coordinated approach to biodiversity conservation (Williams et al. 2020).

Glaucium corniculatum (L.) Rudolph; Saint Katherine Mountain, South Sinai. (Photos are property of the first author)

Citrullus colocynthis (L.) Schard.; Saint Katherine Mountain, South Sinai. (Photos are property of the first author)

Hyoscyamus muticus L. Gebel Musa, South Sinai. (Photos are property of the first author)

References

Abd El-Ghani MM (1981) Preliminary studies on the vegetation of Bahariya Oasis, Egypt. Unpublished M.Sc. Thesis, Cairo University

Abd El-Ghani MM (1988) Sphenocleaceae. In: El-Hadidi MN (ed) Flora of Egypt, vol 3. Taeckholmia Additional Series, pp 63–68

Abd El-Ghani MM (1998) Weed communities of date-palm orchards in the Feiran Oasis (South Sinai, Egypt). Fragmenta Floristica et Geobotanica 43(2):257–271

Abd El-Ghani MM, Abdel-Khalik KN (2006) Floristic diversity and phytogeography of Gebel Elba National Park, Southeast Egypt. Turk J Bot 30(2):121–136

Abd El-Ghani MM, Fahmy AG (1998) Composition of and changes in the spontaneous flora of Feiran Oasis, S. Sinai, Egypt, in the last 60 years. Willdenowia 28:123–134

Abd El-Ghani MM, Fawzy AM (2006) Plant diversity around springs and wells in five oases of the Western Desert, Egypt. International Journal of Agriculture and Biology 8(2):249–255
Abd El-Ghani MM, Salama FM, Salem B, El-Hadidy A, Abdel-Aleem M (2014) Biogeographical relations of a hyperarid desert flora in eastern Egypt. Afr J Ecol 52(2):173–191
Abd El-Ghani MM, Hamdy RS, Farahat AB (2015) Habitat diversity and floristic analysis of Wadi El-Natrun depression, Western Desert, Egypt. Phytologia Balacanica 21(3):351–366
Abd El-Ghani MM, Huerta-Martínez FM, Hongyan L, Qureshi R (2017) Plant responses to hyperarid desert environments. Springer. ISBN: 978-3-319-59134-6
Abd El-Wahab RH, Zaghloul MS, Kamel WM, Moustafa AA (2008) Diversity and distribution of medicinal plants in North Sinai, Egypt. Afr J Environ Sci Technol 2(7):157–171
Aboul-Enein AM, El-Ela FA, Shalaby EA, El-Shemy HA (2012) Traditional medicinal plants research in Egypt: studies of antioxidant and anticancer activities. J Med Plant Res 6(5): 689–703
Ahmad F, Hameed M, Ahmad MSA, Ashraf M (2020) Ensuring food security of arid regions through sustainable cultivation of halophytes. In: Handbook of Halophytes: From Molecules to Ecosystems towards Biosaline Agriculture, pp 1–21
Akhtar-Schuster M, Thomas RJ, Stringer LC, Chasek P, Seely M (2011) Improving the enabling environment to combat land degradation: institutional, financial, legal and science-policy challenges and solutions. Land Degrad Dev 22(2):299–312
Ayyad MA, Ghabbour SI (1986) Hot deserts of Egypt and Sudan. In: Evenari M et al (eds) Ecosystems of the world, 12B, hot deserts and arid Shrublands. Elsevier, Amsterdam, pp 149–202
Bornkamm R, Kehl H (1990) The plant communities of the western desert of Egypt. Phytocoenologia 19(2):149–231
Bowers JE (2005) Effects of drought on shrub and cactus communities in the Sonoran Desert, Arizona, 1991–2002. J Arid Environ 63(2):307–318
Chen IC, Hill JK, Ohlemüller R, Roy DB, Thomas CD (2011) Rapid range shifts of species associated with high levels of climate warming. Science 333(6045):1024–1026
Chothani DL, Vaghasiya HU (2011) A review on Balanites aegyptiaca Del (desert date): phytochemical constituents, traditional uses, and pharmacological activity. Pharmacogn Rev 5(9): 55–62
D'Odorico P, Bhattachan A, Davis KF, Ravi S, Runyan CW (2013) Global desertification: drivers and feedbacks. Adv Water Resour 51:326–344
Danin A (1978) Plant species diversity and ecological districts of the Sinai desert. Vegetatio 36:83–93
Danin A (1983) Desert vegetation of Israel and Sinai. Cana Publishing House, Jerusalem
Danin A (1986) Flora and vegetation of Sinai. Proceedings of the Royal Society of Edinburgh 89B:159–168
Danin A, Orshan G (1990) The distribution of Raunkiaer life forms in Israel in relation to environment. J Veg Sci 1:41–48
Dukes JS, Mooney HA (1999) Does global change increase the success of biological invaders? Trends Ecol Evol 14(4):135–139
El Hadidi MN (1993) Natural vegetation. In: Graig GM (ed) The agriculture of Egypt. Oxford Univ Press, Oxford, pp 39–62
El Hadidi MN (2000) The main features of the natural vegetation. In: El Hadidi MN, Hosni HA (eds) Flora Aegyptiaca, vol 1. Cairo University Herbarium and The Palm Press, pp 27–105
El Hadidi MN, Kosinova J (1971) Studies on the weed flora of the cultivated land in Egypt. 1. Preliminary survey. Mitteilungen der Botanischen Staatssammlung München 10:354–367
El Hadidi MN, Kosinová J, Chrtek J (1970) Weed flora of southern Sinai. Acta Universitatis Carolinae: Biologica 1969:367–381
El Hadidi MN, Abd El-Ghani MM, Fahmy AG (1992) The plant red data book of Egypt. I. Woody Perennials. Palm Press and Cairo University Herbarium, Cairo

FAO (2020) Climate change and food security: risks and responses. Food and Agriculture Organization of the United Nations

Fleming TH, Sahley CT, Holland JN, Nason JD, Hamrick JL (2001) Sonoran Desert columnar cacti and the evolution of generalized pollination systems. Ecol Monogr 71(4):511–530

Forrest J, Miller-Rushing AJ (2010) Toward a synthetic understanding of the role of phenology in ecology and evolution. Philosophical Transactions of the Royal Society B: Biological Sciences 365(1555):3101–3112

GCF (Green Climate Fund) (2023) Funded Activities. https://data.greenclimate.fund/public/data/projects

Gibson AC, Nobel PS (1986) The cactus primer. Harvard University Press

Gurib-Fakim A (2006) Medicinal plants: traditions of yesterday and drugs of tomorrow. Mol Asp Med 27(1):1–93

Hamman JH (2008) Composition and applications of *Aloe vera* leaf gel. Molecules 13(8):1599–1616

Heneidy SZ, Bidak LM (2004) Potential uses of plant species of the coastal Mediterranean region. Egypt Pak J Biol Sci 7(6):1010–1023

Herrmann SM, Anyamba A, Tucker CJ (2005) Recent trends in vegetation dynamics in the African Sahel and their relationship to climate and land use change. Glob Environ Chang 15(4):394–404

Heywood VH, Watson RT (1995) Global biodiversity assessment. UNEP, Cambridge

IEA (2021) World Energy Outlook 2021. International Energy Agency

IPCC (2021) Climate Change 2021: The Physical Science Basis. Contribution of Working Group I to the Sixth Assessment Report of the Intergovernmental Panel on Climate Change

IPCC (2022) Climate Change 2022: impacts, adaptation and vulnerability. Contribution of Working Group II to the Sixth Assessment Report of the Intergovernmental Panel on Climate Change. Cambridge University Press

IPCC (2023) Summary for Policymakers of Climate Change 2023: Synthesis Report of the IPCC Sixth Assessment Report

IRENA (2021) Renewable energy and jobs: annual review 2021. International Renewable Energy Agency

Kassas M, Zahran MA (1971) Plant life on the coastal mountains of the Red Sea, Egypt. Journal of the Indian Botanical Society Golden Jubilee Volume 50A:571–589

Körner C (2018) Concepts in empirical plant ecology. Plant Ecol Diversity 11(4):405–428

Le Quéré C, Jackson RB, Jones MW, Smith AJ, Abernethy S, Andrew RM et al (2020) Temporary reduction in daily global CO_2 emissions during the COVID-19 forced confinement. Nat Clim Chang 10(7):647–653

Loarie SR, Duffy PB, Hamilton H, Asner GP, Field CB, Ackerly DD (2009) The velocity of climate change. Nature 462(7276):1052–1055

Mirzabaev A, Wu J, Evans J, Garcia-Oliva F, Hussein IAG, Iqbal MH et al (2019) Desertification. In: Climate Change and Land: an IPCC special report on climate change, desertification, land degradation, sustainable land management, food security, and greenhouse gas fluxes in terrestrial ecosystems, pp 364–500

Moustafa AA, Zaghloul MS, Abd El-Wahab RH, Shaker M (2001) Evaluation of plant diversity and endemism in Saint Catherine protectorate, South Sinai, Egypt. Egypt J Botany 41(1):123–141

Ndabigengesere A, Narasiah KS, Talbot BG (1995) Active agents and mechanism of coagulation of turbid waters using *Moringa oleifera*. Water Res 29(2):703–710

NDRR (2021) Global assessment report on disaster risk reduction 2021: making development risk-informed. United Nations Office for Disaster Risk Reduction

NOAA (2024) Global Climate Report—Annual 2023. National Oceanic and Atmospheric Administration

Nobel PS (2003) Environmental biology of agaves and cacti. Cambridge University Press

Noy-Meir I (1973) Desert ecosystems: environment and producers. Annu Rev Ecol Syst 4:25–51

Olsvig-Whittaker L, Shachak M, Yair A (1983) Vegetation patterns related to environmental factors in a Negev Desert watershed. Vegetatio 54:153–165

Orians GH, Solbrig OT (1977) A cost-income model of leaves and roots with special reference to arid and semiarid areas. Am Nat 111(980):677–690

Parker KC (1988) Environmental relationships and vegetation associates of columnar cacti in the northern Sonoran Desert. Vegetatio 78:125–140

Polis GA (2023) The ecology of desert communities. University of Arizona Press

Quézel P (1978) Analysis of the flora of Mediterranean and Saharan Africa. Ann Missouri Botanic Garden 65:479–534

Salama F, Abd El-Ghani M, El-Naggar S, Aljarroushi M (2013a) Vegetation analysis and species diversity in the desert ecosystem of coastal wadis of South Sinai, Egypt. Journal of Biology and Earth Sciences 3(2):B214–B227

Salama F, Abd El-Ghani M, El-Naggar S, Aljarroushi M (2013b) Vegetation dynamics and species distribution patterns in the inland desert wadis of South Sinai, Egypt. Ecologia Mediterranea 39(2):93–110

Salguero-Gómez R, Siewert W, Casper BB, Tielbörger K (2012) A demographic approach to study effects of climate change in desert plants. Philosophic Trans Royal Society B: Biologic Sci 367(1606):3100–3114

Schlesinger WH, Reynolds JF, Cunningham GL, Huenneke LF, Jarrell WM, Virginia RA, Whitford WG (1990) Biological feedbacks in global desertification. Science 247(4946):1043–1048

Shaltout K, Ahmed DA (2012) Ecosystem Services of the Flora of southern Mediterranean Desert of Egypt. Ethnobotan Res Applic 10:403–422

UNCCD (2020) Update on the implementation of the voluntary land degradation neutrality targets and related implementation efforts. In Proceedings of the Committee for the Review of the Implementation of the Convention Nineteenth Session, Bonn, Germany, 15–19 March 2021

UNDP (2020) Human development report 2020: the next frontier—human development and the Anthropocene. United Nations Development Programme

UNEP (2023) Emissions gap report 2023. United Nations Environment Programme

White F, Lonard J (1991) Phytogeographical links between Africa and Southwest Asia. Flora et Vegetatio Mundi 9:229–246

Whitford WG, Duval BD (2019) Ecology of desert systems. Academic Press

Williams DR, Balmford A, Wilcove DS (2020) The past and future role of conservation science in saving biodiversity. Conserv Lett 13(4):e12720

World Bank (2021) Climate risk country profile: Egypt. World Bank Group. https://climateknowledgeportal.worldbank.org/sites/default/files/2021-04/15723-WB_Egypt%20Country%20Profile-WEB-2_0.pdf

Zahran MA, Girgis WA (1970) On the ecology of Wadi El-Natrun. Bull Inst Désert D'Egypte 18: 229–267

Zahran MA, Willis AJ (2009) The vegetation of Egypt, 2nd edn. Springer Science Business Media

Zahran MA, El-Demerdash MA, Mashaly IA (1993) On the ecology of *Juncus acutus* and *J. rigidus* as fiber producing halophytes in arid regions. Towards the rational use of high salinity tolerant plants: Vol. 2 Agriculture and forestry under marginal soil water conditions, pp 331–342

Zohary M (1973) Geobotanical Foundations of the Middle East, vol 1&2. Gustav Fischer Verlag, Stuttgart

From Forest to Table: Role of Wild Food Plants in Promoting Food Security and Biodiversity Conservation in Kabal Valley, Swat, Pakistan

10

Mohammad Ilyas and Rahmatullah Qureshi

Abstract

The use of wild food plants (WFPs) has become a topic of significant interest in recent years as the concept starts to be regarded as part of local food systems and important in terms of conserving plant biodiversity. With the increasing list of issues of the world today, including food insecurity, global warming, and the disappearance of biological and cultural diversity, WFPs now have become a means of ensuring sustainable rural development, food diversification, and ecosystem stability. The chapter is a document of the native knowledge, usage pattern and social-economic significance of wild food plants that were utilized by the locals in the Kabal Valley Swat District Khyber Pakhtunkhwa Pakistan. A 149 species of the vascular flora (64 families and 119 genera) were recorded to have been customarily consumed in an extremely wide repertory of dishes. These species were classed into six broad groups based on modes of consumption in that, vegetables (36.04%), cooked greens and tender shoots; fruits (30.96%), wild berries, nuts and drupes; salads (12.69%), raw and eaten as a food and source of flavor; herbal teas (10.66%), refreshing and medicinal; condiments (3.05%), natural spices and flavoring agents; and other (6.60%), e.g., gums, pickles and some of the WFPs also had great economic potential in addition to playing an important role in nutrition. The active trade of the species on local markets was twenty-one (14.1) that cost PKR 250-PKR 3000 per kilogram based upon the availability of the species and demand of consumers and the rarity of the species.

M. Ilyas
Department of Biology, Islamabad Model College for Boys, Islamabad, Pakistan

Department of Botany, Pir Mehr Ali Shah Arid Agriculture University, Rawalpindi, Pakistan

R. Qureshi (✉)
Department of Botany, Faculty of Sciences, Pir Mehr Ali Shah Arid Agriculture University, Rawalpindi, Pakistan
e-mail: rahmatullahq@uaar.edu.pk

R. Qureshi (ed.), *Plant Biodiversity*, https://doi.org/10.1007/978-981-92-1471-6_10

Commercially important taxa included, among others, *Bunium persicum* (black cumin), *Thymus linearis* (wild thyme), *Juglans regia* (walnut), *Zanthoxylum armatum* (winged prickly ash), and *Zizyphus jujuba* (wild jujube). Generally, WFPs are multidimensional when it comes to enhancing food, rural livelihood, and biodiversity security as highlighted in this chapter. It also shows that there is an urgent need to deploy indigenous knowledge systems and cultural heritage in ensuring sustainable exploitation of the wild plant resources, and also provides policy and research recommendations on how the WFPs can be incorporated into the national plans on food security, rural development, and biodiversity management in Pakistan.

Keywords

Wild food · Kabal valley · Ethnobotany · Food security · Biodiversity conservation · Use value index

Abbreviations

CR	Critically Endangered
EN	Endangered
FAO	Food and Agriculture Organization
IUCN	International Union for Conservation of Nature
LC	Least Concern
NT	Near Threatened
PRA	Participatory Rural Appraisal
VU	Vulnerable
WFPs	Wild Food Plants

10.1 Introduction

All human civilization and terrestrial habitats lie on the basis of plants. Not only are they the main producers of ecological food chains, but also important sources of food, medicine, shelter and income production. The importance of wild plant species within the group is tremendous, especially in rural and marginalized communities, where they can play a significant role in dietary diversity, nutritional sufficiency, and resilience with regards to subsistence (Shad et al. 2013).

A large human population in most of the developing countries experience chronic food insecurity, and in most cases, they are unable to sustain their daily caloric and micronutrient needs (Abbasi et al. 2013). This is aggravated by economic constraints, environmental erosion and climatic instability in the agricultural sector. The varied assemblage of wild flora, in these situations, has a central role in complementing household diets and in making them nutritionally adequate. Research has established that consumption of wild edible plants increases the resilience of the food system and empowers household coping mechanisms, particularly when agricultural supply is inadequate (such as times of agricultural failure), there is conflict (e.g., conflict), or when there is a natural disaster (FAO 2004; Asfaw et al. 2023).

10.2 Wild Food Plants and Their Global Significance

According to the Food and Agriculture Organization (FAO), wild food plants (WFPs) are species of plants that grow naturally or semi-naturally, and which can survive without the need to be cultivated by humans. These plants have one or many edible parts, i.e., leaves, stems, roots, fruits, flowers, or seeds, and when harvested at the right stage of plant development and correctly prepared they are fine to consume (Kallas 2010). The historical background of mankind is full of use of wild plants as the primary source of nutrition since pre-agricultural settings. Millennia of traditionally ecological knowledge (TEK) that has surrounded these species have created a cultural foundation of food sovereignty. Wild food plants are very important even in the modern world, as they are used in food security of rural people in several countries, especially those with low access to commercial food markets.

In other parts of the world, WFPs are becoming popular even in industrialized nations where they are being added to healthier and more natural diets. These plants have aesthetic value, phytonutrient content and culinary diversity (Redzic 2006). The world estimates that over one billion individuals habitually eat wild food with the massive majority being plant-based (Aberoumand and Deokule 2009). The ethnobotanical catalogs have shown that more than 7000 species of wild plants have been historically used by human consumption in different parts of the world (Grivetti and Ogle 2000). The practice of wild edible plants consumption is still present in Asian regions like China, India, Thailand, Bangladesh, and Pakistan in addition to traditional farming. These plants are also not only sources of food security but cultural and medicinal resources (Mazhar et al. 2007).

10.3 Pakistani Wild Food Plants in the Context

Due to topography and climate zones, which vary, including coastal mangroves and riverine forests, alpine meadows, and deserts, Pakistan is blessed with rich and diverse flora. The country harbors more than 6000 different species of vascular plants of which over half are endemic or of regional importance as it accommodates almost ten million hectares of ecologically diverse territory (Stewart 1972; Shinwari and Nasim 2008; GOP 2021). Nevertheless, in spite of this bountifulness of plants, Pakistan is experiencing chronic food insecurity. It is now only 109th on the Global Food Security Risk Index, which indicates the difference between the potential and the practice (Fahad et al. 2024).

According to the National Nutrition Survey (2018), food insecurity is approximately at 39.9% in Pakistan, and there is a significant shortage in the necessary micronutrients, i.e., iron, zinc, iodine, and vitamins A and D. In this case, wild food plants become an asset as a safety net, especially in isolated and underserved areas. The nutritional gap in the lean period when vegetables and fruits, which are grown locally, are unavailable or unaffordable can be addressed by the seasonal availability of WFPs (Ahmad and Pieroni 2016).

10.4 Case Study: Kabal Valley and Its Ethnobotanical Richness

Located in the hilly terrains of northwestern Pakistan, Kabal Valley is characterized by its rugged topography, ecological richness, and socio-economic marginalization. The valley's remote location and limited infrastructural development pose significant challenges for conventional agriculture and food distribution. Consequently, local communities have developed deep-rooted dependencies on the surrounding natural vegetation for food, fodder, fuelwood, and medicine.

Wild food plants in the Kabal Valley form a critical component of household subsistence strategies. These species are not only gathered for home consumption but are also occasionally traded in local markets, thereby contributing to household income. Traditional ecological knowledge surrounding the collection, preparation, and medicinal use of these plants is preserved and transmitted orally across generations, particularly among women, elders, and indigenous practitioners. Despite their importance, WFPs in the Kabal Valley remain poorly documented and under-researched. Scientific exploration into their taxonomy, seasonal availability, nutritional composition, and ethnobotanical significance is minimal. There is a pressing need for systematic documentation and validation of indigenous knowledge systems to support conservation strategies and promote sustainable use of these plant resources.

10.4.1 Objectives of the Present Study

By bridging the knowledge gap surrounding WFPs and highlighting their relevance in contemporary food systems, this chapter aims to contribute to broader discussions on climate-resilient agriculture, biodiversity conservation, and equitable food access. The work presents a comprehensive ethnobotanical study focused on wild food plants in the Kabal Valley. The primary objectives include:

(i) To document traditional knowledge regarding the identification, collection, and preparation of WFPs in local communities.
(ii) To analyze the seasonal availability and dietary significance of these plants in ensuring household food security.
(iii) To explore the cultural, medicinal, and economic dimensions of WFP usage.
(iv) To highlight the conservation challenges and propose strategies for the sustainable management of wild food biodiversity in the region.

10.4.2 Study Area

Kabal Valley is located in Tehsil Kabal in the northwestern region of District Swat, Khyber Pakhtunkhwa, Pakistan. It lies approximately 10 kilometers west of Saidu Sharif, the administrative capital of Swat. Geographically, the valley spans from 34° 43′ to 35°00′ N latitude and 72°07′ to 72°21′ E longitude, covering a diverse elevation range of 833 to 3012 meters above sea level. The valley features a characteristic U-shaped morphology that opens southward toward the Swat River, while the eastern, northern, and western boundaries are enclosed by an arc of hills that form part of the Hindu Kush mountain system (Ahmad and Sirajuddin 1996).

The lower reaches of Kabal Valley consist of fertile alluvial plains formed through riverine deposition, which support intensive agriculture, including staple crops, vegetables, and a variety of fruit orchards (Hussain and Ilahi 1991). The higher altitudes harbor rich montane flora and serve as grazing lands and sources of wild plant resources. Climatically, the valley falls within the temperate zone and experiences four well-defined seasons. Winters are prolonged and cold, often with snowfall in the upper hills, whereas summers are relatively short, mild, and conducive to agricultural productivity.

The total geographical area of Kabal Valley is approximately 40,026 hectares, of which 20,722 hectares are actively cultivated, while 19,304 hectares remain uncultivated due to steep topography, forest cover, or communal use. The human population of the valley is estimated at 480,800, comprising rural communities that are highly dependent on agriculture, livestock, and the sustainable use of local plant resources (GOP 2023).

10.5 Field Methods

To comprehensively document the traditional use of wild food plants (WFPs), an extensive ethnobotanical survey was conducted in Kabal Valley between 2020 and 2024. A combination of qualitative, quantitative, and conservation-based approaches was employed to ensure holistic data collection. The methods used adhered to standard ethnobotanical protocols (Martin 1995; Cotton 1996) and included the following:

10.5.1 Semi-structured and Structured Interviews

These were conducted with local inhabitants to gather detailed ethnobotanical knowledge about WFPs, including their vernacular names, plant parts used, methods of preparation, seasonal availability, and cultural significance.

10.5.2 Direct Observations and Participatory Rural Appraisal (PRA)

Field visits were conducted in collaboration with local communities, observing foraging, harvesting, processing, and consumption of WFPs in their natural or cultivated habitats.

10.5.3 Transect and Guided Field Walks

Knowledgeable informants accompanied researchers along forest trails, agricultural fields, and mountain slopes to locate, identify, and collect WFP specimens (Fig. 10.1). Emphasis was placed on habitats ranging from valley floors to mid-altitude hills.

10.5.4 Demographic Categorization of Informants

A total of 127 informants were randomly selected across different age groups, genders, and villages. The participants were categorized into three age brackets:

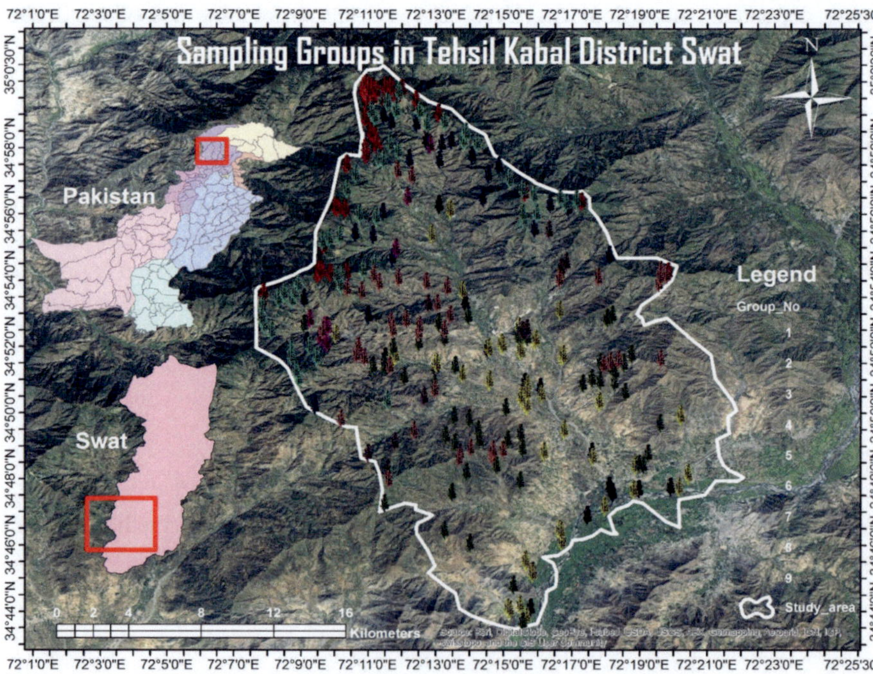

Fig. 10.1 Location map of the Kabal Valley, Swat, Pakistan

Table 10.1 Demographic features of informants

Characters	Male	Female	Total number	Percentage
Age				
15–25	21	2	23	18.11
26–50	40	5	45	35.43
> 50	36	23	59	46.46
Total	97	30	127	100.00
Marital status				
Married	86	27	113	88.98
Unmarried	11	3	14	11.02
Total	97	30	127	100.00
Literacy level				
Illiterate	11	4	15	11.81
Primary	59	18	77	60.63
Middle	14	5	19	14.96
High school	8	2	10	7.87
Undergraduate	4	1	5	3.94
Graduate	1	0	1	0.79
Total	97	30	127	100

15–25 years, 26–50 years, and above 50 years (Table 10.1). Preference was given to elders and traditional healers due to their retained knowledge of wild flora.

10.5.5 Local Market Surveys

Surveys of village bazaars and town markets were carried out to assess the economic role and commercialization potential of selected WFPs. Discussions were held with shopkeepers, herbal vendors, and foragers to record pricing, frequency of trade, and consumer preferences.

10.5.6 Plant Specimen Collection and Identification

Plant specimens were collected following standard botanical protocols and identified using the *Flora of Pakistan* (Nasir and Ali 1971–1995; Ali and Qaisar 1995–2012). Voucher specimens were assigned collection numbers and deposited in the Herbarium of the Department of Botany, PMAS-Arid Agriculture University, Rawalpindi, Pakistan, for future reference and verification.

10.5.7 Data Analysis

Free listing data were analyzed using descriptive statistical methods in Microsoft Excel. Wild food plants (WFPs) were categorized into six use-based groups following the classification by Nedelcheva (2013). For each species, the Use Value (UV) index was calculated as described by Phillips et al. (1994), and the frequency of citation (FC) was determined following Ahmad and Pieroni (2016).

10.6 Conservation Assessment Using IUCN Criteria

In addition to ethnobotanical documentation, the conservation status of each species was assessed based on a combination of field observations, ethnobotanical data, and literature review, and categorized according to the IUCN Red List Categories and Criteria (Version 3.1) (IUCN 2001). This involved:

10.6.1 Assigning IUCN Categories

Species were tentatively assigned conservation categories such as Least Concern (LC), Near Threatened (NT), Vulnerable (VU), Endangered (EN), Critically Endangered (CR), or Data Deficient (DD). These assessments are provisional and based on the best available local data.

10.6.2 Evaluation of Population Trends

Based on field observations, interviews with local foragers, and historical data, trends in population abundance were noted (increasing, stable, declining, or locally extinct).

10.6.3 Threat Identification

Key threats such as overharvesting, habitat fragmentation, grazing pressure, land-use change, and climate-related impacts were identified for each species.

10.7 Findings

10.7.1 Floristic Composition

Kabal Valley is a rural area and is associated with many challenges that translate to malnutrition such as population explosion, poverty, and natural and human-caused catastrophes. Agriculture and the natural resources around the people greatly support the livelihoods of its people. It is against this background that wild food plants (WFPs) have been an important, cheap and ready source of food. In the current research, 149 species of vascular plants were found to be employed by the native people to different nutritional purposes. These are species of 64 families and 119 genera (Table 10.2).

Table 10.2 Inventory of wild food plants used by local people of Kabal valley, Swat, Pakistan

Family	Plant species	Local name	Part used	Method of use	Use reports for food categories ($n = 127$)						Total use reports (N)	No. of Informants	UVI	FC	Conservation status
					Vegetable	Fruits	Salad	Herbal tea	Condiments	Other					
1. Alismataceae	1. *Alisma plantago-aquatica* L.	Jabai	Leaves	Young leaves are thoroughly cooked, often with *Oenanthe javanica* and served as a flavored potherb	13						13	15	0.87	0.10	LC
	2. *Sagittaria trifolia* L.	Taqay	Roots	The root tubers, once cooked and peeled, are subsequently roasted and utilized as a dietary component	15						15	15	1.00	0.12	LC
2. Alliaceae	3. *Allium ampeloprasum* var. *porrum* (L). Regel	Ogakai	Whole plant	The entire plant is consumed as a salad. The leaves are ground, mixed with corn dough, and baked on a hot griddle to prepare a traditional bread regarded as a delicacy. Additionally, the leaves serve as the primary ingredient in a local green rice dish known as "Chukanrh"	23		32			19	74	46	1.61	0.58	NT
3. Amaranthaceae	4. *Amaranthus graecizans* L.	Chalwaye	Shoot	The young shoots are utilized as a potherb in traditional culinary practices	60						60	67	0.90	0.47	LC

(continued)

Table 10.2 (continued)

Family	Plant species	Local name	Part used	Method of use	Use reports for food categories ($n = 127$)						Total use reports (N)	No. of Informants	UVI	FC	Conservation status
					Vegetable	Fruits	Salad	Herbal tea	Condiments	Other					
	5. *Amaranthus viridis* L.	Asil Chalwaye	Shoot	The leafy shoots are widely consumed as a staple green leafy vegetable in local cuisine	62						62	62	1.00	0.49	LC
	6. *Celosia argentea* L.	Soor Gwalay	Shoot	Although young shoots are used as a potherb, they are generally regarded as being of inferior quality compared to other edible greens	8						8	10	0.80	0.06	LC
	7. *Digera muricata* (L.) Mart.	Tandula	Shoot	Young shoots are commonly utilized as a potherb in traditional culinary applications	10						10	13	0.77	0.08	LC
4. Anacardiaceae	8. *Rhus punjabensis* J.L. Stewart ex Brandis	Titray	Fruits	The fruits having a slightly sore taste are commonly consumed by children, often as a seasonal snack		32					32	37	0.86	0.25	VU
5. Apiaceae	9. *Bunium persicum* (Boiss.) Fedtsch.	Toorey Zankai	Fruits	In local cuisine, the fruits serve as a condiment, enhancing the flavor of various dishes					25		25	28	0.89	0.20	CR

	10. *Oenanthe javanica* (Blume) DC.	Surkhai	Shoot	Young shoots are utilized as a potherb in traditional cooking	20						20	26	0.77	0.16	LC
	11. *Prangos pabularia* Lindl.	Zangali Kaga	Leaves	Approximately 10–15 fruits are traditionally boiled in cow's milk for bottle-fed infants, serving to improve both digestibility and flavor. Additionally, the raw leaves are chewed for their mouth-refreshing properties and to mitigate unpleasant oral odor			6	12	40	14	72	57	1.26	0.57	EN
	12. *Scandix pecten-veneris* L.	Gangahay	Shoot	The young shoots are eaten raw in salads and are also prepared as a potherb in traditional cuisine	6		7				13	20	0.65	0.10	LC
6. Araceae	13. *Acorus calamus* L.	Skha waja	Stem	The tender inner parts of young stems and developing inflorescences are traditionally eaten by children			5				5	51	0.10	0.04	CR
7. Asparagaceae	14. *Asparagus adscendens* Roxb.	Tindorhay	Shoot	Tender, freshly sprouted shoots are cut into small pieces and boiled in double their volume of cow's milk until the	65						65	67	0.97	0.51	VU

(continued)

Table 10.2 (continued)

Family	Plant species	Local name	Part used	Method of use	Use reports for food categories (n = 127)						Total use reports (N)	No. of Informants	UVI	FC	Conservation status
					Vegetable	Fruits	Salad	Herbal tea	Condiments	Other					
				liquid is reduced by half. The thickened mixture is subsequently fried in clarified butter (*desi ghee*), with sugar added to enhance flavor. The dish is regarded as a local delicacy											
8. Asteraceae	15. *Carthamus oxyacantha* Bieb.	Kareeza	Leaves	The young leaves are an essential component of a widely consumed local green rice delicacy referred to as "Chukanrh"	55						55	76	0.72	0.43	LC
	16. *Cichorium intybus* L.	Han	Leaves	The leaves are utilized as a green vegetable and are traditionally believed to possess hepatoprotective properties	51						51	67	0.76	0.40	LC
	17. *Helianthus tuberosus* L.	Alopech	Root tubers	The root tubers are consumed either raw or cooked as a food source	30						30	38	0.79	0.24	LC
	18. *Lactuca serriola* L.	Shawdapai	Leaves	The leaves are utilized in local cuisine, either eaten raw in salads or cooked and served as a potherb	8		7				15	10	1.50	0.12	LC

	19. *Onopordum acanthium* L.	Wrejakai	Seeds	Young, unripe seeds are traditionally consumed by children due to their rice-like taste		8					8	38	0.21	0.06	LC
	20. *Silybum marianum* (L.) Gaertn.	Ghata Wrejakai	Seeds	The seeds are traditionally consumed by children, often as a casual or seasonal snack		6					6	56	0.11	0.05	LC
	21. *Sonchus oleraceus* L.	Shawdapai	Leaves	The fresh leaves are commonly employed as a potherb in local diets	12						12	18	0.67	0.09	LC
	22. *Taraxacum officinale* F.H. Wigg	Toora Daal	Leaves	The leaves are used as a potherb in traditional cuisine, while the dried leaves are utilized for preparing herbal tea	8			7			15	47	0.32	0.12	LC
9. Berberidaceae	23. *Berberis lycium* Royle	Kwaray	Fruits	Immature, tender leaves are utilized in their raw state, while the ripe fruit is recognized as a wild edible with a characteristically sour flavor		67	22				89	70	1.27	0.70	VU
10. Brassicaceae	24. *Alliaria petiolata* (M. Bieb.) Cavara & Grande	Oga botay	Leaves	The young leaves are utilized as both a salad ingredient and a potherb, particularly during the winter and early spring seasons	12		8				20	19	1.05	0.16	LC

(continued)

Table 10.2 (continued)

Family	Plant species	Local name	Part used	Method of use	Use reports for food categories ($n = 127$)						Total use reports (N)	No. of Informants	UVI	FC	Conservation status
					Vegetable	Fruits	Salad	Herbal tea	Condiments	Other					
	25. *Brassica juncea* (L.) Czern. et Coss.	Awray	Leaves	The foliage is employed as a potherb in local culinary practices	45					8	53	45	1.18	0.42	LC
	26. *Brassica rapa* subsp. *campestris* (L.) Clapham	Sharhsham	Shoot	The young leaves are consumed as a potherb, while the immature flowering shoots are cooked as a traditional delicacy referred to as "*da ghandalo saag*" the oil extracted from the seeds is widely used in domestic cooking	76					45	121	123	0.98	0.95	LC
	27. *Descurainia sophia* (L.) Webb ex Prantl	Spin awray	Leaves	The leaves are occasionally cooked as a green vegetable and are noted for their slightly pungent flavor	8						8	13	0.62	0.06	LC
	28. *Eruca sativa* Mill.	Jawawa	Leaves	To reduce their pungency, young leaves are first boiled and the water is discarded. Subsequently, the leaves are fried in mustard oil along with garlic and tomatoes to produce a flavorful cooked green vegetable	26						26	28	0.93	0.20	LC

	29. *Lepidium apetalum* Willd.	Zangali Halam	Shoot	Traditionally, the young, tender shoots—known for their pungent, mustard-like taste—are consumed in salads and are reputed to support circulatory, respiratory, and digestive health			32				32	34	0.94	0.25	LC
	30. *Nasturtium officinale* R. Br.	Tarmeera	Shoot	The fresh leaves are utilized in salads, while the shoots are valued as a nutritious potherb, traditionally believed to exert stimulating and thermogenic effects on the body	56		47				103	80	1.29	0.81	LC
	31. *Sinapis alba* L.	Spin Sharhsham	Leaves/ Seeds	The leaves are consumed as a cooked green vegetable. The seeds yield a fixed oil that serves as a substitute for mustard oil	54					35	89	60	1.48	0.70	LC
11. Cactaceae	32. *Opuntia monacantha* Haw.	Zoqam	Fruits	Ripe fruits are traditionally consumed by children as a seasonal wild edible		28					28	34	0.82	0.22	LC
12. Caprifoliaceae	33. *Viburnum cotinifolium* D. Don	Shanglo	Fruits	The ripe fruit is considered edible and is occasionally consumed as a wild food source		56					56	62	0.90	0.44	NT

(continued)

Table 10.2 (continued)

Family	Plant species	Local name	Part used	Method of use	Use reports for food categories (n = 127)						Total use reports (N)	No. of Informants	UVI	FC	Conservation status
					Vegetable	Fruits	Salad	Herbal tea	Condiments	Other					
	34. *Viburnum grandiflorum* Wall. ex DC.	Khapyanga	Fruits	The ripe fruits are consumed as a wild edible		74					74	78	0.95	0.58	NT
13. Caryophyllaceae	35. *Silene conoidea* L.	Bashka mashorhgai	Fruits	The tender, fresh fruit possesses a mildly sweet taste and is traditionally eaten by children		46					46	54	0.85	0.36	LC
	36. *Stellaria media* (L.) Vill.	Olalai	Shoot	The fresh aerial parts of the plant are considered a valued potherb in traditional diets	46						46	75	0.61	0.36	LC
	37. *Vaccaria hispanica* (Miller) Rauschert	Ghwarha Bashka	Leaves	The leaves are utilized as a condiment in traditional culinary practices					12		12	18	0.67	0.09	LC
14. Chenopodiaceae	38. *Chenopodium album* L.	Sarmay	Shoot	Shoots are initially boiled and the decoction is discarded. The softened shoots are then roasted in oil along with *Allium sativum* (garlic) and table salt to produce a palatable cooked green vegetable	56						56	60	0.93	0.44	LC

	39. *Chenopodium murale* L.	Soor Sarmay	Shoot	The leaves are utilized as a potherb. They are typically boiled, with the water subsequently discarded, and then marinated in vinegar or yogurt. The marinated leaves are sautéed in oil along with garlic (*Allium sativum*) and ginger (*Zingiber officinale*) to prepare a traditional vegetable dish	12						12	44	0.27	0.09	LC
15. Convallariacere	40. *Polygonatum multiflorum* (L.) all.	Baramol	Shoot	Young, tender shoots that emerge during the spring season are harvested and boiled in milk with the addition of a small amount of sugar. This preparation is regarded as an esteemed delicacy in the local cuisine	48						48	50	0.96	0.38	VU
	41. *Polygonatum verticillatum* (L.) All.	Noor-e-Alam	Leaves, Rhizome	The young, tender leaves are either directly fried or first cooked in milk and subsequently fried in *desi ghee* (clarified butter), with a small amount of sugar added. This preparation is traditionally consumed	60		6				66	78	0.85	0.52	CR

(continued)

Table 10.2 (continued)

Family	Plant species	Local name	Part used	Method of use	Use reports for food categories ($n = 127$)						Total use reports (N)	No. of Informants	UVI	FC	Conservation status
					Vegetable	Fruits	Salad	Herbal tea	Condiments	Other					
				as a *salep*-like tonic and is regarded as a revitalizing food. The rhizome, which has a mildly sweet taste, is also eaten raw in salads. Additionally, it is boiled in cow's milk with the addition of sugar until the mixture attains a semi-solid consistency. This preparation is traditionally consumed as a nourishing and strength-enhancing food											
16. Cornaceae	42. *Cornus macrophylla* Wall. ex Roxb.	Khadang	Fruits, stem bark	The fruits are recognized as edible wild produce. A decoction prepared by boiling the stem bark is traditionally added to *halwa*—a confection made with clarified butter—and is utilized in local practices as a muscle-strengthening tonic		9					9	30	0.30	0.07	EN

17. Cucurbetaceae	43. *Cucumis sativus* L.	Badrang	Fruits	The fresh, tender fruit is commonly used as a salad ingredient and is also incorporated into *raita*, a traditional yogurt-based preparation				67			67	70	0.96	0.53	LC
	44. *Solena amplexicaulis* (Lam.) Gandhi	Kakorha	Fruits	The pulp of the ripe fruit is consumed as a food source		35					35	40	0.88	0.28	LC
18. Cupressaceae	45. *Juniperus communis* L. var. *saxatilis* Pallas	Gugarh	Berries	Fully ripe female cones are consumed fresh at the site of collection		23					23	30	0.77	0.18	EN
19. Dennstaedtiaceae	46. *Pteridium aquilinum* (L.) Kuhn	Baboozi	Young fronds	Young, tender shoots, harvested during the unfurling stage, are utilized as a green vegetable and potherb in traditional diets	27						27	29	0.93	0.21	LC
20. Dryopteridceae	47. *Dryopteris juxtaposita* H. Christ	Kwanjay	Young fronds	Young fronds, collected during the unfolding stage, are traditionally rubbed together to remove brown scales. The cleaned fronds are then chopped and cooked in whey, followed by frying in oil with garlic paste to prepare a flavorful green vegetable dish	57						57	60	0.95	0.45	VU

(continued)

Table 10.2 (continued)

Family	Plant species	Local name	Part used	Method of use	Use reports for food categories (n = 127)						Total use reports (N)	No. of Informants	UVI	FC	Conservation status
					Vegetable	Fruits	Salad	Herbal tea	Condiments	Other					
21. Ebenaceae	48. *Diospyros lotus* L.	Toor Amlok	Fruits	Following harvest, ripe fruits are kept under shade in a cool setting to allow natural ripening and browning. They are then subjected to air or sun drying to form a date-like dried fruit, which is traditionally consumed in the winter months		78					78	80	0.98	0.61	NT
22. Elaeagnaceae	49. *Elaeagnus umbellata* Thunb.	Ghanam ranga	Fruits	Ripe fruits are regarded as wild edibles and are commonly eaten immediately upon harvesting in their natural habitat		56					56	66	0.85	0.44	VU
23. Fabaceae	50. *Astragalus graveolens* Buch.-Ham.ex Benth.	Pichpach	Shoot	The tender shoots are employed as a potherb and incorporated into traditional diets	34						34	35	0.97	0.27	NT
	51. *Astragalus pyrrhotrichus* Boiss.	Mamol	Shoot	The young shoots are utilized as a potherb in traditional cuisine	12						12	15	0.80	0.09	LC
	52. *Lathyrus aphaca* L.	Kurkamanay	Shoot	Young shoots are utilized in traditional	24		15				39	28	1.39	0.31	LC

				cuisine, either consumed raw or prepared as a potherb											
	53. *Lathyrus sphaericus* Retz	Chilo	Pods	Young pods are traditionally fried and consumed as a vegetable, and the immature seeds are eaten raw		34		8			42	38	1.11	0.33	LC
	54. *Medicago lupulina* L.	Mardekakh	Shoot	The shoots are used as a potherb	8						8	13	0.62	0.06	LC
	55. *Medicago minima* (L.) L.	Kach Shpeshtaray	Shoot	The young shoots are traditionally utilized as a potherb in local culinary practices	12						12	15	0.80	0.09	LC
	56. *Medicago polymorpha* L.	Shpeshtaray	Shoot	Young shoots are widely employed as a potherb in traditional diets. They are either prepared fresh or preserved through drying for consumption during periods of limited availability	68						68	79	0.86	0.54	LC
	57. *Vicia bithynica* (L.) L.	Marghaikhpa	Shoot	The young shoots are consumed raw as a component of traditional salads			13				13	15	0.87	0.10	LC
24. Fagaceae	58. *Quercus baloot* Griff.	Breh/ Banjkarorhi	Seeds	The inner seed of the acorn is typically roasted and consumed as a dry fruit, valued for its		69					69	76	0.91	0.54	NT

(continued)

Table 10.2 (continued)

Family	Plant species	Local name	Part used	Method of use	Use reports for food categories (n = 127)						Total use reports (N)	No. of Informants	UVI	FC	Conservation status
					Vegetable	Fruits	Salad	Herbal tea	Condiments	Other					
				nutritional content and flavor. It is believed to have positive effect on urinary system function especially in children											
25. Hypericaceae	59. *Hypericum perforatum* L.	Shin Chai	Whole plant	The dried aerial parts of the plant are prepared as a herbal infusion and consumed as a green tea substitute to relieve mental depression and exhaustion				34			34	56	0.61	0.27	VU
26. Juglandaceae	60. *Juglans regia* L.	Ghuz	Fruits	The seed is a valued dry fruit, consumed either on its own or as an ingredient in confectioneries. The leaves are utilized for preparing herbal tea		120	7				127	127	1.00	1.00	CR
27. Lamiaceae	61. *Lamium amplexicaule* L.	Ghutyalay	Shoot	The tender shoots are prepared as a potherb and used in traditional dishes	24						24	66	0.36	0.19	LC
	62. *Mentha longifolia* (L.) L.	Enalay	Shoot	Fresh, tender shoots are consumed raw in salads and are also crushed to	35		47	15			97	63	1.54	0.76	LC

			prepare chutneys. Additionally, the shoots are used as a condiment in various leafy green vegetable dishes. In traditional practices, dried shoots are mixed with stored grains such as corn and wheat to deter rodents, including mice and rats											
63. *Mentha spicata* L.	Poodina	Shoot	Fresh twigs are used as natural mouth fresheners, incorporated into salads, and crushed to prepare chutneys. Dried and powdered shoots are used as a condiment and for making herbal tea			72	23	19		114	80	1.43	0.90	LC
64. *Nepeta erecta* (Boyle ex Benth.) Benth.	Pesho Botay	Shoot	The young shoots are used as a potherb	12						12	15	0.80	0.09	LC
65. *Origanum vulgare* L.	Shamakay	Shoot	The shoots, when dried, are used as a condiment. Fresh young shoots are traditionally employed as natural scrubbers for cleaning utensils, particularly the pots used for stirring whey and				12	29	30	71	65	1.09	0.56	LC

(continued)

Table 10.2 (continued)

Family	Plant species	Local name	Part used	Method of use	Use reports for food categories ($n = 127$)						Total use reports (N)	No. of Informants	UVI	FC	Conservation status
					Vegetable	Fruits	Salad	Herbal tea	Condiments	Other					
				yogurt during butter extraction; this method is believed to facilitate more efficient separation of butter from whey. Additionally, the shoots are rubbed inside earthen pots designated for beekeeping to attract honey bees and promote hive formation. Dried leaves are prepared as a refreshing herbal tea. Moreover, dried shoots are stored in kitchens to repel common household pests such as ants, cockroaches, and fruit flies											
	66. *Salvia lanata* Roxb.	Kianrh	Shoots	The young stems of the shoots are peeled and consumed raw as a salad. Additionally, the tender shoots are traditionally boiled in whey and then cooked with garlic, oil, and spices to prepare a	60		21				81	67	1.21	0.64	VU

				slightly sour, yet flavorful, dish											
	67. *Thymus linearis* Benth.	Da Ghra Sperkai	Shoot	Fresh shoots are eaten as salad. Green tea is prepared from the shade dried shoots for improving the function of respiratory, digestive and urinary systems			32	54			86	60	1.43	0.68	VU
28. Liliaceae	68. *Tulipa clusiana* DC	Ghantool	Bulbs	The bulbs are sometimes cooked as vegetable	6						6	9	0.67	0.05	LC
29. Malvaceae	69. *Hibiscus syriacus* L.	Gilasgolay	Flowers	The dried flower petals are used for making a pleasant tasting herbal tea				22			22	28	0.79	0.17	LC
	70. *Malva neglecta* Wallr.	Paneerak	Shoot	Young shoots are used as potherb with demulcent properties enhancing bowel movement	68						68	70	0.97	0.54	LC
	71. *Malva parviflora* L.	Sanchal	Leaves	The young leaves are used as an excellent potherb and are cooked or fried in various traditional preparations	77						77	82	0.94	0.61	LC
30. Moraceae	72. *Broussonetia papyrifera* (L.) L'Herit. ex Vent.	Kaghazi Toot	Fruits	The fruits when ripe are eaten by children		16					16	19	0.84	0.13	LC
	73. *Ficus palmata* Forssk.	Kach Inzar	Fruits	The fruit is consumed fresh, typically with a small amount of salt, and		112					112	119	0.94	0.88	LC

(continued)

Table 10.2 (continued)

Family	Plant species	Local name	Part used	Method of use	Use reports for food categories ($n = 127$)						Total use reports (N)	No. of Informants	UVI	FC	Conservation status
					Vegetable	Fruits	Salad	Herbal tea	Condiments	Other					
				is often followed by yogurt intake to prevent tongue irritation and diarrhea as the fruit is mild laxative											
	74. *Ficus sarmentosa* Bush: Ham. ex J.E. Smith	Da Zmakey Inzar	Fruits	The fruit are eaten by children		11					11	18	0.61	0.09	LC
	75. *Morus alba* L.	Spin toot	Fruits	Freshly sprouted twigs are used as a potherb. The ripe fruits are very sweet in taste and eaten fresh and dried	37	76					113	100	1.13	0.89	LC
	76. *Morus macroura* Miq.	Shatoot	Fruits, shoots	The tinder young shoots are cooked as potherb. The fruit is sweet and eaten fresh	39	78					117	89	1.31	0.92	LC
	77. *Morus nigra* L.	Toor toot	Fruits	The fruits when half ripe with red color are sore in taste and eaten raw. Fully ripe fruits are seldom eaten but cause diarrhea when eaten in excess		67					67	102	0.66	0.53	LC
31. Myrsineceae	78. *Myrsine africana* L.	Manrhgwaya	Fruits	Ripe fruits are slightly bitter but edible and utilized for removing intestinal worms		26					26	44	0.59	0.20	VU

32. Myrtaceae	79. *Myrtus communis* L.	Manrho	Fruits, Leaves	The leaves are added to tea for flavoring. The fruits are eaten fresh and dried		54		38			92	67	1.37	0.72	EN
33. Oleaceae	80. *Jasminum officinale* L.	Rambeel Chambeel	Flowers	The flowers when dried are used for making green tea				32			32	41	0.78	0.25	LC
	81. *Olea ferruginea* Royle	Khoona	Fruits	The fruit are eaten fresh and dried		78					78	118	0.66	0.61	LC
34. Oxalidaceae	82. *Oxalis corniculata* L.	Manzakay Tarookay	Shoot	The shoots are consumed raw as a sour-tasting salad that stimulates salivation. They are also chewed to neutralize the effects of sour foods, particularly by individuals with sensitive teeth			48			11	59	60	0.98	0.46	LC
35. Papaveraceae	83. *Papaver dubium* L.	Reday	Shoot, seeds	The young shoots are used as an ingredient in a traditional green rice dish known as "Chukanrh," which is believed to have a warming effect on the body. The seeds are commonly consumed by children	48	15		11			74	67	1.10	0.58	LC
36. Parnassiaceae	84. *Parnassia nubicola* ssp. *occidentalis* E.S-Temesy	Kamasla Mamera	Shoot	The plant is used as a potherb	14						14	21	0.67	0.11	VU

(continued)

Table 10.2 (continued)

Family	Plant species	Local name	Part used	Method of use	Use reports for food categories (n = 127)						Total use reports (N)	No. of Informants	UVI	FC	Conservation status
					Vegetable	Fruits	Salad	Herbal tea	Condiments	Other					
37. Phytolaccaceae	85. *Phytolacca latbenia* (Moq.) Walter	Gararha	Leaves	The young leaves are boiled and the water is removed after boiling. The boiled leaves are fried with garlic and condiments in mustard oil to prepare a delicious green dish	59						59	70	0.84	0.46	CR
38. Pinaceae	86. *Picea smithiana* (Wall.) Boiss	Kandal/ Mangazai	Shoot	The young shoots are used for making a green tea				9			9	56	0.16	0.07	VU
	87. *Pinus roxburghii* Sargent	Nakhtar	Seeds	The seeds are eaten like chilghoza		46					46	100	0.46	0.36	VU
39. Poaceae	88. *Alopecurus myosuroides* Huds.	Zozay	Anthers	Children commonly collect and consume the pendent anthers of the plant						12	12	27	0.44	0.09	LC
	89. *Avena fatua* L.	Jamdaray	Grains	Ground grains are boiled in milk to make a gruel which is believed to be a general body and nerve tonic						37	37	94	0.39	0.29	LC
40. Podophyllaceae	90. *Podophyllum emodi* Wall. ex Royle	Kakorha/ Gangorha	Fruits	The fruit is edible when fully ripe but cause purging when eaten in excess		16					16	67	0.24	0.13	EN

41. Polygonaceae	91. *Persicaria hydropiper* (L.) Delarbre	Palpolak	Shoot	To reduce their pungency, the young shoots are sometimes boiled in water and the decoction discarded. The shoots are then fried with garlic paste to prepare a green vegetable dish	8						8	55	0.15	0.06	LC
	92. *Polygonum aviculare* L.	Bandakay	Shoot	The shoots are used as a potherb. The leaves are used to make green tea	58			10			68	70	0.97	0.54	LC
	93. *Polygonum plebeium* R. Br.	Sur Bandakay	Shoot	The young shoots are used as a potherb	25						25	30	0.83	0.20	LC
	94. *Rumex dentatus* L.	Shalkhay	Leaves	The leaves are used as a potherb and salad	47		6				53	83	0.64	0.42	LC
	95. *Rumex hastatus* D. Don	Tarookay	Leaves	The leaves are used as an ingredient in various potherbs to give a sour and salty taste. The young leaves are used as a salad	56		34			22	112	67	1.67	0.88	LC
	96. *Rumex nepalensis* Spreng.	Da ghra Shalkhay	Leaves	The leaves are used as a potherb	26						26	36	0.72	0.20	LC
42. Pontederiaceae	97. *Monochoria vaginalis* (Burm. f.) Presl	Kacha Panrha	Leaves	The stalks of young leaves are eaten raw as salad and cooked as a potherb	11		6				17	15	1.13	0.13	LC
43. Portulacaceae	98. *Portulaca oleracea* L.	Warkharhay	Shoot	The shoots are used as a popular and preferred potherb cooked in a variety of ways often mixed with whey and lentils (*Lens caulinaris*)	79						79	89	0.89	0.62	LC

(continued)

Table 10.2 (continued)

Family	Plant species	Local name	Part used	Method of use	Use reports for food categories (n = 127)						Total use reports (N)	No. of Informants	UVI	FC	Conservation status
					Vegetable	Fruits	Salad	Herbal tea	Condiments	Other					
44. Primulaceae	99 *Anagallis arvensis* L.	Chichrha	Shoot	The young shoots are used as a potherb	34						34	50	0.68	0.27	LC
45. Punicaceae	100. *Punica protopunica* Balf. F.	Anangorhay	Fruits	The fruit mericarps have a slightly sour yet palatable taste. A decoction of the pericarp is traditionally mixed with milk to induce sourness and facilitate butter extraction		37				12	49	44	1.11	0.39	VU
46. Ranunculaceae	101. *Caltha alba* Camb.	Makanpat	Leaves	The young green leaves are used as potherb	50						50	54	0.93	0.39	VU
47. Rhamnaceae	102. *Sageretia thea* (Osbeck) M.C. Johnston	Mamanrha	Fruits	The tender young and fully ripe fruits are edible and eaten by children mostly		29					29	48	0.60	0.23	LC
	103. *Ziziphus jujuba* Mill.	Markhanrhai	Fruits	The ripe fruit are delicious and sweet. They are eaten fresh and dried		127					127	127	1.00	1.00	NT
	104. *Ziziphus nummularia* (Burm. f.) Wight & Arn.	Karkanda	Fruits	Fully ripe fruits are eaten having flour like starchy taste		58					58	60	0.97	0.46	NT
	105. *Ziziphus oxyphylla* Edgew.	Elanai	Fruits	Fruits are slightly sour and are eaten		49					49	60	0.82	0.39	VU

48. Rosaceae	106. *Agrimonia eupatoria* L.	Kamasla qarqara	Whole plant	The dried plant is used as a herbal tea with refreshing effects on the body				24			24	26	0.92	0.19	LC
	107. *Cotoneaster microphyllus* Wall. ex Lindl.	Khoonawrhay	Fruits	The fruits are eaten as wild fruits		22					22	26	0.85	0.17	LC
	108. *Cotoneaster racemiflorus* (Desf.) K. Koch	Kharawa	Fruits	The ripe fruits are delicious. The juice expressed from the fresh ripe fruits is taken as a body tonic and relieve stomach pain		47					47	56	0.84	0.37	NT
	109. *Duchesnea indica* (Andrews) Focke	Da Zmakey Toot	Fruits	The ripe fruits are eaten on the spot		49					49	50	0.98	0.39	LC
	110. *Fragaria nubicola* (Hook.f.) Lindl. ex Lacaita	Shadkaray	Fruits	The fruits when ripe are eaten on the spot. Excessive use or drinking water after it causes diarrhea		69					69	70	0.99	0.54	LC
	111. *Geum urbanum* L.	Qarqaranga	Roots/ Leaves	The leaves are used as flavoring agent in other green vegetables.The roots are used to make herbal tea	11			35		6	52	38	1.37	0.41	LC
	112. *Prunus cerasoides* D.Don	Annang	Fruits	The fruit is eaten on the spot		26					26	31	0.84	0.20	EN
	113. *Prunus cornuta* (Wall. ex Royle) Steud.	Changa/ Badara	Fruits	The fruit is edible and eaten on the spot. The		28				8	36	30	1.20	0.28	VU

(continued)

Table 10.2 (continued)

Family	Plant species	Local name	Part used	Method of use	Use reports for food categories (n = 127)						Total use reports (N)	No. of Informants	UVI	FC	Conservation status
					Vegetable	Fruits	Salad	Herbal tea	Condiments	Other					
				stem bark is used for milk curding											
	114. *Pyrus pashia* Buch.-Ham. ex D. Don	Gidarhtanga	Fruits	The fruit are edible only when fully ripe and start decaying and eaten by human beings only on the spot		57					57	60	0.95	0.45	LC
	115. *Rosa brunonii* Lindl.	Khorhach	Fruits	The ripe thalamus cup is eaten by children but causes stinging sensation on the tongue if the seeds inside are not removed carefully		13					13	24	0.54	0.10	LC
	116. *Rosa moschata* Herrm.	Palwarhey	Petals/ Shoots	The petals and young tender shoots are consumed raw as a salad. The petals are also slow-cooked with sugar to prepare a jam-like substance known as *gulqand*, traditionally used as a stomach tonic			48				48	54	0.89	0.38	LC
	117. *Rubus ellipticus* Sm.	Goraja	Fruits	The ripe fruits are eaten		56					56	60	0.93	0.44	VU
	118. *Rubus fruticosus* L.	Karwarha	Fruits/ Shoots	Young shoots when they emerge from soil in the		113	10				123	113	1.09	0.97	NT

			early spring are peeled of and eaten raw as a salad. The fruit are sore in unripe condition but become very delicious when fully ripe. The juice obtained by compressing the fruits is a coolant and general body tonic. The leaves are also used for mouth washing											
119. *Rubus niveus* Thunb.	Baganrha	Fruits/ Shoots	Young tinder shoots when they emerge from ground in early spring are eaten as a salad. The ripe fruits are very delicious and eaten raw		58	22				80	58	1.38	0.63	VU
120. *Rubus ulmifolius* subsp. *sanctus* (Schreb.) Sudre	Da Ghra Baganrhey	Fruits	The ripe fruits are juicy, delicious and eaten on the spot		46					46	48	0.96	0.36	NT
121. *Sanguisorba minor* Scop.	Zangali Salad	Leaves	Young leaves are used as a salad			12				12	13	0.92	0.09	LC
122. *Sorbus aria* Crantz	Doda	Fruits	The fruit becomes edible after undergoing bletting, for softening and partial decomposition that enhances palatability		12					12	14	0.86	0.09	CR

(continued)

Table 10.2 (continued)

Family	Plant species	Local name	Part used	Method of use	Use reports for food categories ($n = 127$)						Total use reports (N)	No. of Informants	UVI	FC	Conservation status
					Vegetable	Fruits	Salad	Herbal tea	Condiments	Other					
49. Rubiaceae	123. *Galium aparine* L.	Konay	Whole plant	The entire plant is used to prepare an herbal tea, which is traditionally believed to reduce body fats and have a slimming effect on the body				17			17	23	0.74	0.13	LC
	124. *Rubia cordifolia* L.	Karghan Mewa	Fruits	The ripe fruit are eaten by children on the spot		33					33	45	0.73	0.26	LC
50. Rutaceae	125. *Zanthoxylum armatum* DC.	Dambara	Fruits	The fruits are used as a condiment in chutney preparation and as an ingredient in mixed spice blends for frying fish			45		67		112	68	1.65	0.88	VU
51. Sapindaceae	126. *Cardiospermum halicacabum* L.		Leaves	Young leaves are used as vegetable	8						8	13	0.62	0.06	LC
52. Sapotaceae	127. *Monotheca buxifolia* (Falc.) A. DC.	Gwargura	Fruits	The ripe fruits are delicious and are regularly consumed		45					45	45	1.00	0.35	VU
53. Scrophulariaceae	128. *Mazus japonicus* (Thunb.) O. Kuntze	Spinmakhay	Leaves	Young leaves cooked as potherbs	23						23	40	0.58	0.18	LC
	129. *Veronica anagallis-aquatica* L.	Dada Tarkha	Shoot	The young shoots are used as potherb	36						36	43	0.84	0.28	LC

	130. *Veronica polita* Fries	Meekhakay	Shoot	The shoots are cooked as potherb	28						28	46	0.61	0.22	LC
54. Solanaceae	131. *Physalis divaricata* D. Don	Mangotey	Fruits	The fruits when ripe are bitter in taste but edible. Unripe fruit is cooked as vegetable	11	24					35	24	1.46	0.28	LC
	132. *Physalis peruviana* L.	Khog Batingarh	Fruits	The fruit is eaten raw and in dried form		36					36	36	1.00	0.28	LC
	133. *Solanum nigrum* L.	Kachmacho	Shoot/ Fruit	The shoots are used as potherb with a slightly better taste but considered a live tonic. The fruits when ripe and turned black are edible and eaten on the spot	57	25					82	60	1.37	0.65	LC
	134. *Solanum villosum* (L.) Moench	Sur Kachmacho	Shoot/ Fruit	The shoots are used as potherbs. The berries are eaten on the spot when fully ripe and turned orange red	35	27					62	56	1.11	0.49	LC
55. Taxaceae	135. *Taxus wallichiana* Zucc	Banrya	Berries	The female cones (Berries) when fully ripe are eaten		25					25	34	0.74	0.20	CR
56. Thymelaeaceae	136. *Daphne mucronata* Royle	Leoghonay	Fruits	The fruits when ripe are eaten but care should be taken to remove the small hair on fruit which cause irritation of mouth and tongue		27					27	43	0.63	0.21	LC

(continued)

Table 10.2 (continued)

Family	Plant species	Local name	Part used	Method of use	Use reports for food categories ($n = 127$)						Total use reports (N)	No. of Informants	UVI	FC	Conservation status
					Vegetable	Fruits	Salad	Herbal tea	Condiments	Other					
57. Tiliaceae	137. *Corchorus olitorius* L.	Malkheya	Leaves	The leaves are used as a potherb. A herbal tea is made from the leaves	36			21			57	40	1.43	0.45	LC
	138. *Grewia optiva* Drummond ex Burret	Pastawoonay	Fruits	The fruits are sweet and eaten on the spot. It is considered a cooling agent		22					22	34	0.65	0.17	LC
58. Trilliaceae	139. *Trillium govanianum* Wall. ex Royle	Matarzeela	Leaves	The leaves are used as potherb	8						8	41	0.20	0.06	CR
59. Typhaceae	140. *Typha latifolia* L.	Lokha	Roots/ shoot	The roots and the young inner part of shoots are used as vegetable	12						12	27	0.44	0.09	LC
60. Ulmaceae	141. *Celtis caucasica* Willd.	Tagha	Fruits	The fruits when ripe are eaten by children		23					23	37	0.62	0.18	LC
	142. *Celtis eriocarpa* Decne.	Niney Tagha	Fruits	The fruits are sweet and eaten on the spot. It is advised that the seeds should be spitted out because they can cause colic and intestinal irritation		45					45	50	0.90	0.35	LC
	143. *Celtis tetrandra* Roxb.	Ghat Tagha	Fruits	The fruits when ripe are eaten		11					11	24	0.46	0.09	LC

61. Urticeaeca	144. *Debregeasia saeneb* (Forssk.) Hepper & J.R.I. Wood	Ajlai	Fruits	The fruits are eaten on the spot		57					57	60	0.95	0.45	LC
	145. *Urtica dioica* L.	Seezonkai/ Jalbang	Leaves	The young leaves are used as a potherb. They are typically boiled, with the water discarded to reduce bitterness, and then fried in mustard oil along with garlic and tomatoes to prepare a flavorful green vegetable dish. Additionally, a green tea prepared from the leaves is traditionally believed to have a warming effect on the body	19			11		9	39	26	1.50	0.31	LC
62. Verbenaceae	146. *Lantana camara* L.	Gul Ghuncha	Fruits	The fully ripe fruits are edible		8					8	17	0.47	0.06	LC
	147. *Verbena officinalis* L.	Skha Botay	Shoot	The dried shoots are used to make herbal tea				34			34	42	0.81	0.27	LC
63. Vitaceae	148. *Vitis jacquemontii* Parker	Gidarh Kwar	Fruits	The ripe fruits are sore and edible		47					47	49	0.96	0.37	VU
64. Xanthorrhoeaceae	149. *Hemerocallis fulva* (L.) L.	Gwal ghotai	Roots/ Flowers	The tubers are used as a vegetable the flowers are used as a salad and garnish	11		10				21	15	1.40	0.17	LC

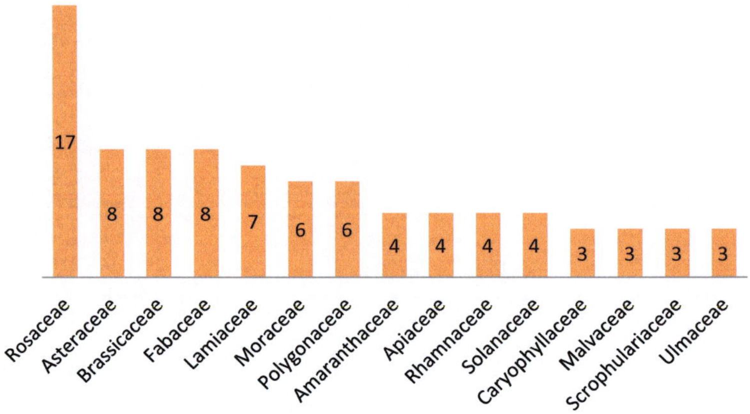

Fig. 10.2 Families distribution of wild food plants

The most represented family was the Rosaceae (17 species), followed by Asteraceae, Brassicaceae and Fabaceae (8 each), Lamiaceae (7 species), Moraceae and Polygonaceae (6 each) and Amaranthaceae, Apiaceae, Rhamnaceae and Solanaceae (4 each). A few other families like Caryophyllaceae, Malvaceae, Scrophulariaceae and Ulmaceae had 3 species each, and the rest of the families had 1 or 2 species (Fig. 10.2). Similar family dominance patterns have been seen in other ethnobotanical studies: Asteraceae, Rosaceae, and Lamiaceae were also predominant in Spain (Tardío et al. 2006), Bulgaria (Nedelcheva 2013), Kashmir (Ijaz et al. 2022), Malakand (Ibrahim et al. 2023), and Swat Valley (Khan et al. 2015), yet Fabaceae was the most significant one in the studies carried out in India (Lal et al. 2017) and Nepal (Acharya and Acharya 2010).

10.7.2 Edibility of Plant Parts

The parts of the plants that were consumed were analyzed and showed that fruits were the most used (60 species), shoots (49 species), leaves (30 species), and seeds and roots (5 species each) (Fig. 10.3). The usage of fruits and leaves is predominant, which is also in accordance with the results in Bulgaria (Nedelcheva 2013), Punjab (Waheed et al. 2023) and Swat (Khan et al. 2015). These sections are usually consumed crude or added to conventional meals to add flavor, garnish or as nutritional supplements.

Although they have a good nutritional profile, being good sources of vitamins, minerals, other bioactive compounds, most of these wild plants have been underutilized today. Much of the information about their collection, preparation and medicinal efficacy is preserved by the oral traditions of, mainly, the older generation. The ethnobotanical knowledge is becoming ever more Vulnerable to being lost as younger generations tend to change the approach to traditional diets and substance usage (Salvi and Katewa 2016).

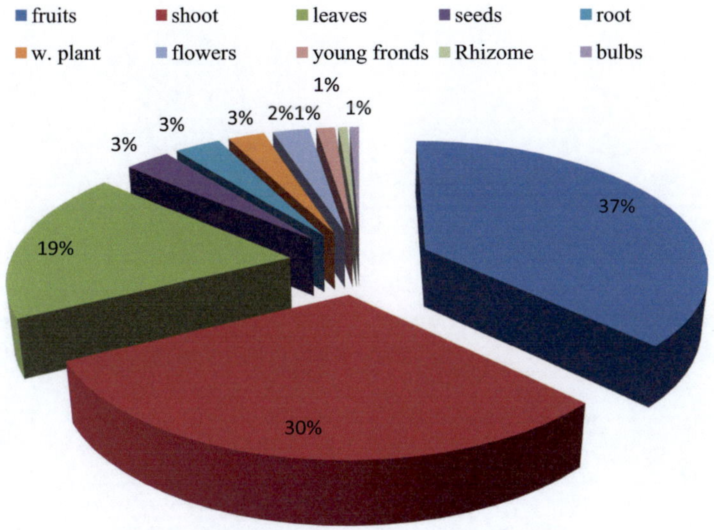

Fig. 10.3 Plants distribution based on edible parts

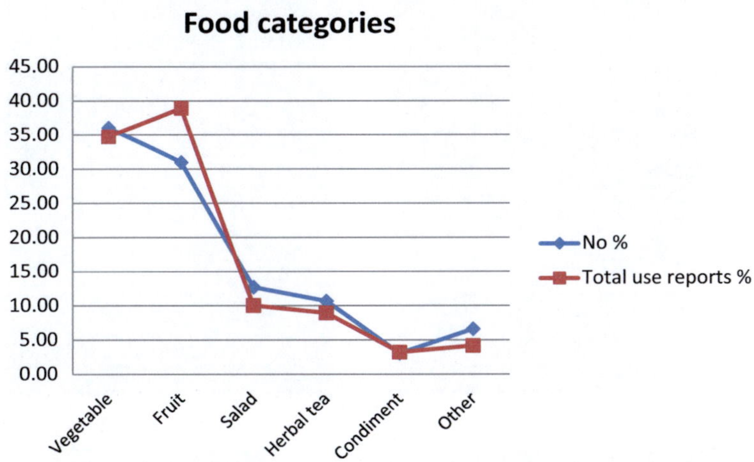

Fig. 10.4 Quantitative assessment of food categories and use reports

10.7.3 Cuisine and Food Covers

The wild food plants listed were categorized into six food groups according to their traditional food use (Table 10.1 and Fig. 10.4). The visual representation of some of the wild fruits and vegetables is in Plate 10.1. These were: vegetables (36.04%), fruits (30.96%), salads (12.69%), herbal teas (10.66%), condiments (3.05%) and other uses (6.60%). Most applications were in the form of vegetables and fruits,

Ficus arica

Asparagus adscendens

Duchesnea indica

Viburnum grandiflorum

Helianthus tuberosus

Nasturtium officinale

Taraxacum officinale

Digera muricata

Plate 10.1 Some of the wild fruits and vegetables of the study area

which is consistent with ethnobotanical studies in Bulgaria (Nedelcheva 2013), Parambikulam (Yesodharan and Sujana 2007), Nepal (Acharya and Acharya 2010), the Lesser Himalayas (Abbasi et al. 2013), and Punjab, Pakistan (Waheed et al. 2023). It is worth noticing that a number of species were used in more than one category so the number of uses reports was 197, which is more than the number of unique species (149). The multi-purpose use is reminiscent of findings by Tardío et al. (2006), who found similar patterns in Spain with species such as *Origanum vulgare*, *Nasturtium officinale*, and *Rubus ulmifolius* and by Shumsky et al. (2014) on *Amaranthus graecizans* in Kenya.

10.7.4 Quantitative Ethnobotanical Appraisal

Use value index (UVI) calculated according to the formula of Phillips et al. (1994) revealed that *Rumex hastatus* (1.67), *Zanthoxylum armatum* (1.65), *Allium ampeloprasum* var. *porrum* (1.61), *Mentha longifolia* (1.54), *Lactuca serriola* (1.50*), Urtica dioica* (1.50), *Sinapis alba* (1.48), *Physalis divaricate* (1.46), *Thymus linearis*, *Mentha spicata*, *Corchorus olitorius* (1.43 each) and *Hemerocallis fulva* (1.40) had highest use values among the WFPs. Lowest use values with respect to edibility were recorded for *Trillium govanianum* (0.20), *Picea smithiana* (0.16), *Persicaria hydropiper* (0.15), *Silybum marianum* (0.11) and *Acorus calamus* (0.10) as shown in Fig. 10.5.

Based on frequency citation (FC) values, *Juglans regia*, *Ziziphus jujuba* (1.00 each) *Rubus fruticosus* (0.97), *Brassica rapa* subsp. *campestris* (0.95), *Morus macroura* (0.92), *Mentha spicata* (0.90), *Morus alba* (0.89), *Ficus palmate*, *Rumex hastatus*, *Zanthoxylum armatum* (0.88 each) and *Nasturtium officinale* (0.81) were more well-known to the local people (Fig. 10.6). Slightly different values have been found for some of the common species from Lesser Himalayas,

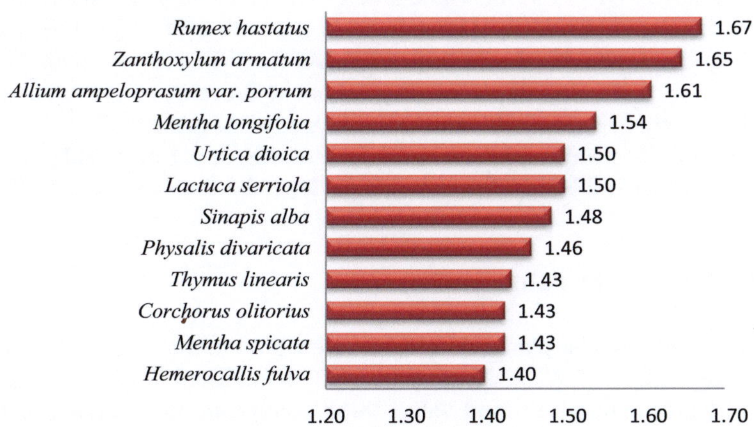

Fig. 10.5 Wild food plants with highest use values

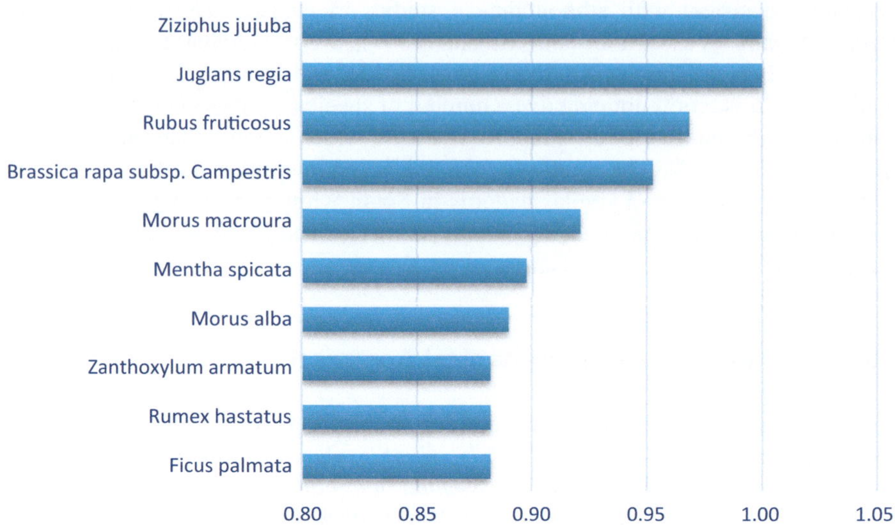

Fig. 10.6 Wild food plants with highest frequency citations

Pakistan (Abbasi et al. 2013), Takht-e-Suleiman Hills (Ahmad and Pieroni 2016) and Swat valley (Khan et al. 2015). Least known edible plants in the area were *Lantana camara* (0.06), *Silybum marianum* (0.05), *Tulipa clusiana* (0.05) and *Acorus calamus* (0.04).

10.7.5 Conservation Status

Among the 149 species of WFPs, 102 (68.46%) were in Least Concern (LC) category, 22 (14.77%) in Vulnerable (VU), 11 (7.38%) in Near Threatened (NT), 8 (5.37%) in Critically Endangered (CR), and 6 (4.03%) in Endangered (EN) categories in the study area (Fig. 10.7). As the WFPs are not endemic to the area and are wide ranging; therefore, they are not at the verge of extinction globally but still effective conservation measures are needed to prevent their loss locally. Somewhat similar results were obtained by Rostami and Khoshbakht (2012) for Bistoon area, Iran and North-Western Himalayas, India (Samant et al. 2007).

10.7.6 Marketable Value

Among the 149 vascular plants, 21 (14.1%) were sold in the local market having a price range between 250 ± 30 to 3000 ± 300 Pakistani Rupees per kilogram (Table 10.3). In addition to food value, these species are marketable and provide the opportunity to supplement household income. A study has shown that lower

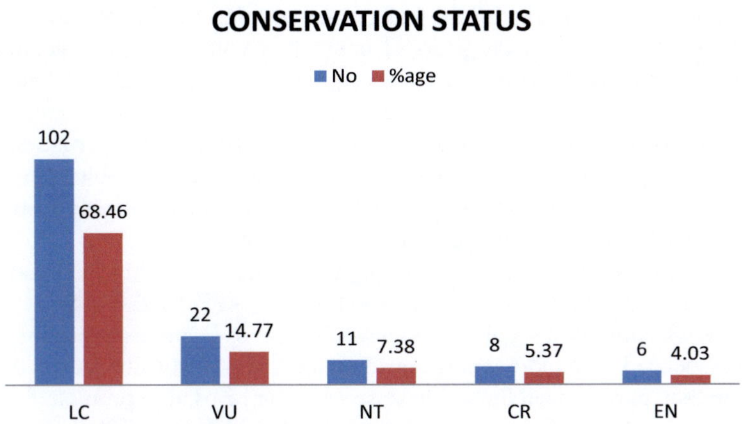

Fig. 10.7 Conservation status of wild food plants

Table 10.3 Marketable wild food plants of Kabal valley, Swat, Pakistan

S. #	Plant species	Price Kg^{-1}
1	*Acorus calamus*	500 ± 50
2	*Amaranthus viridis*	250 ± 30
3	*Asparagus adscendens*	600 ± 50
4	*Berberis lycium*	500 ± 40
5	*Bunium persicum*	3000 ± 300
6	*Cichorium intybus*	250 ± 30
7	*Diosperos lotus*	300 ± 30
8	*Dryopteris juxtaposita*	400 ± 30
9	*Ficus palmate*	500 ± 20
10	*Juglans regia*	1000 ± 100
11	*Mentha longifolia*	500 ± 50
12	*Mentha spicata*	500 ± 50
13	*Myrtus communis*	800 ± 70
14	*Pteridium aquilinum*	600 ± 50
15	*Punica protopunica*	700 ± 70
16	*Rubus fruticosus*	600 ± 50
17	*Rubus niveus*	700 ± 60
18	*Solanu nigrum*	600 ± 50
19	*Thymus linearis*	1500 ± 300
20	*Ziziphus jujuba*	900 ± 100
21	*Zanthoxylum armatum*	1000 ± 100

returns from farming necessitate the diversification of incomes from the sale of wild resources (Balemie and Kebebew 2006). Among these plants, *Bunium persicum* could generate maximum income followed by *Thymus linearis, Juglans regia*, *Zanthoxylum armatum,* and *Zizyphus jujuba*.

The traditional knowledge of the indigenous people of Kabal Valley is rich as it has been developed after many years of interaction with the local vegetation. This precious ethnobotanical wealth is on the verge of becoming extinct and needs to be properly recorded (Acharya and Acharya 2010). There is thus a great need to facilitate the documentation of local applications, practices, and expertise employing standard and quantitative ethnobotanical techniques. Sustainable development strategies should also be put in the forefront especially with the current unsustainable harvesting practices that are being witnessed in the area.

Future studies ought to incorporate phytochemical screening to determine the nutritional and medicinal values of wild food plants (WFPs) and determine their role in enhancing health, food security, and development of plant-based products. WFPs are also associated with biodiversity conservation and poverty alleviation in addition to the role they play in food diets. These species can be used to promote livelihoods through better diets and supplementing food supplies to the resource-poor households; as well as provide income-generation opportunities (Balemie and Kebebew 2006; Acharya and Acharya 2010).

Creating a comprehensive management system, aimed at conserving and sustainable usage of the WFPs, as well as discovering commercially viable species, may improve the standards of living of the local people to a higher degree. These societies are not just the custodians of the biological diversity of the region, but also a rich store of the traditional ecological knowledge (Samant et al. 2007). Conservation and development efforts should thus be based on effective involvement and collaboration of the local stakeholders (Ilyas et al. 2013). These inclusion methods will tend to provide a greater dedication to conservation and allow communities to gain more practical economic rewards with regard to the native plant materials.

10.8 Conclusion

The results of this study directly reflect on some of the United Nations Sustainable Development Goals (SDGs), including SDG 2 (Zero Hunger), SDG 12 (Responsible Consumption and Production), and SDG 15 (Life on Land). The ethnobotanical survey conducted in the Kabal Valley of Swat indicates evidently that the local people still depend on the wild food plants (WFPs) to be a source of nutritional sustenance, traditional medicine and other income. We identified 149 species of vascular plants (64 families, 119 genera) as components of local food systems, and very many of these species have multiple culinary applications (vegetable, fruit, herbal teas, condiments). Such a wealth of ethnobotanical knowledge that is largely retained by older members of the society shows tight rapport between individuals and local ecosystem and significance of ecological and cultural value of WFPs in dietary diversification and enhancement of food security. It is also notable that even twenty-one species are characterized by a clear market potential that is, livelihood can be better and rural economy diversified in the case of their rational use. The findings emphasize the role of the indigenous knowledge systems in the conservation of the plant biodiversity and sustainability of natural resources. In Pakistan, wild food plants when formalized, justified and integrated into national and regional policy systems, may serve as a food sovereignty, ecological and rural livelihoods. However, the study also reveals the emerging problems of the erosion of traditional knowledge primarily due to the changing generational patterns, modernization as well as changing socio-economic conditions. In addition, increasing human activities of over harvesting, land-use change, degradation of habitat and extraction of resources in an unsustainable manner are a threat to the long term sustainability of the wild plant resources. This will entail concerted efforts which will incorporate community-based conservation and creation of awareness and policy advocacy to make sure that the biological and the cultural heritage is conserved. It therefore requires constant research and participatory management techniques to ensure that the wild food plants are not discarded even to sustainable food systems and biodiversity conservation in the environment.

10.9 Recommendations

The following recommendations are feasible suggestions derived as a result of the research to promote sustainable use and conservation of wild food plants:

10.9.1 Documentation and Preservation of Indigenous Knowledge

To prevent cultural and ecological loss, there is urgent need to document ethnobotanical knowledge, particularly that of the senior knowledge holders. The information should be archived preserved in local languages to for safeguard in community-based education and intergenerational transfer.

10.9.2 Community-Based Conservation and Domestication

Local communities should be involved in conserving and planting of valuable and threatened species (e.g., *Bunium persicum*, *Thymus linearis*). There is need to promote agro-forestry and home gardening by involving WFPs.

10.9.3 Sustainable Harvest and Resource Management

There is need to develop and adopt sustainable harvesting methods and management practices by communities.

10.9.4 Integration into Nutrition and Health Programs

A value-chain system is need to be developed to preserve the ecology balance and secure the fair trade. Besides, the information may be incorporated into nutrition and health programs. The studied WFPs may be incorporated in school feeding schemes, community kitchens and maternal nutrition schemes. Besides, social health education programs can be launched to ensure knowledge on the health benefits of WFPs.

10.9.5 Research and Innovation

A comprehensive research studies must be initiated to evaluate the nutritional values, phytochemistry, pharmacology and safety of WFPs. Study on harvest and processing technologies and preservation to increase shelf life and marketability is also need to carried out.

10.9.6 Building Local Business

A small-scale businesses and co-operatives should be developed for the development and value addition of these herbs such as herbal teas, dried fruit, condiments, etc.

10.9.7 Policy and Institutional Support

The conservation measures are required for the WFPs and should be included in national biodiversity policies and food security policy. Universities, research and development agencies should be encouraged in the area to join in collaborative projects.

These recommendations will transform WFPs into food security, income, and biodiversity active bearers of underused traditional crops. The successful story of the sustainable development of wild vegetables in Kabal Valley can be used as an example to be pursued elsewhere in the same socio-ecological setting.

References

Abbasi AM, Khan MA, Zafar M (2013) Ethno-medicinal assessment of some selected wild edible fruits and vegetables of Lesser-Himalayas, Pakistan. Pak J Bot 45(SI):215–222

Aberoumand A, Deokule SS (2009) Studies on nutritional values of some wild edible plants from Iran and India. Pak J Nutr 8(1):26–31

Acharya KP, Acharya R (2010) Eating from the wild: indigenous knowledge on wild edible plants in Parroha VDC of Rupandehi district, Central Nepal. Int J Soc Forestry 3(1):28–48

Ahmad K, Pieroni A (2016) Folk knowledge of wild food plants among the tribal communities of Thakht-e-Sulaiman Hills, north-West Pakistan. J Ethnobiol Ethnomed 12(1):17

Ahmad H, Sirajuddin (1996) Ethnobotanical profile of Swat. In: Shinwari ZK, Khan BA, Khan AA (eds) Proceedings of the first training workshop on ethnobotany and its application to conservation. National Herbarium, NARC, Islamabad, pp 202–206

Ali SI, Qaisar M (1995–2012) Flora of Pakistan. Pakistan Agricultural Research Council, Islamabad

Asfaw A, Lulekal E, Bekele T, Debella A, Tessema S, Meresa A, Debebe E (2023) Ethnobotanical study of wild edible plants and implications for food security. Trees Forests People 14:100453

Balemie K, Kebebew F (2006) Ethnobotanical study of wild edible plants in Derashe and Kucha districts, South Ethiopia. J Ethnobiol Ethnomed 2(1):53

Cotton CM (1996) Ethnobotany: principles and applications. Published by Wiley, p 424

Fahad S, Basit A, Iqbal N, Akram N, Orakzai JK, Islam MU (2024) Evaluation of implementation strategies of food security in Pakistan. Khyber J Public Policy 3(1):178–218

FAO [Food and Agricultural Organisation of the United Nations] (2004) The state of food insecurity in the world: monitoring progress towards the world food summit and millennium development goals. Food and Agricultural Organisation of the United Nations, Rome

GOP (2021) Economic survey of Pakistan 2020–21. The Ministry of Finance

GOP. [Government of Pakistan] (2023) 7th population and housing census – detailed results. Pakistan Bureau of Statistics. Accessed at https://www.pbs.gov.pk/digital-census/detailed-results

Grivetti LE, Ogle BM (2000) Value of traditional foods in meeting macro-and micronutrient needs: the wild plant connection. Nutr Res Rev 13(1):31–46

Hussain F, Ilahi I (1991) Ecology and vegetation of lesser Himalayas, Pakistan. Department of Botany, University of Peshawar, p 187

Ibrahim M, Akhtar N, Wahab A, Alrefaei AF, Almutairi MH (2023) Exploring wild edible plants in Malakand, Pakistan: ethnobotanical and nutritional insights. Sustainability 15(17):12881

Ijaz S, Perveen A, Ashraf S, Abid R, Kousar S, Abbas Z, Arslan M (2022) Traditional knowledge of wild edible plants used by the people of Lawat, District Neelum, Azad Jammu & Kashmir, Pakistan. Ethnobot Res Appl 23:1–16

Ilyas M, Qureshi R, Shinwari ZK, Arshad M, Mirza SN, Haq ZU (2013) Some ethnoecological aspects of the plants of Qalagai hills, Kabal valley, swat, Pakistan. Int J Agric Biol 15(5): 801–810

IUCN [International Union for Conservation of Nature and Natural Resources] (2001) IUCN Red List Categories and Criteria. Version 3.1, IUCN species survival commission, vol 2. IUCN, Gland/Cambridge, p 30

Kallas J (2010) Edible wild plants: wild foods from dirt to plate. Gibbs Smith, Layton, p 416

Khan MPZ, Ahmad M, Zafar M, Sultana S, Ali MI, Sun H (2015) Ethnomedicinal uses of edible wild fruits (EWFs) in Swat Valley, northern Pakistan. J Ethnopharmacol 173:191–203

Lal S, Gupta DK, Dewangan B, Koreti D, Sunanda, Prakash K, Lal M, Yadav S (2017) Some edible plants of Bhoramdeo wild life sanctuary Kabirdham, Chattisgarh, India. Indian J Sci Res 13(2): 236–247

Martin GJ (1995) Ethnobotany: a methods manual, vol 1. Earthscan

Mazhar F, Buckles D, Satheesh PV, Akhter F (2007) Food sovereignty and uncultivated biodiversity in South Asia: essays on the poverty of food policy and the wealth of the social landscape. Academic Foundation, p 74

Nasir E, Ali SI (1971–1995) Flora of Pakistan. Department of Botany. University of Karachi, Karachi

Nedelcheva A (2013) An ethnobotanical study of wild edible plants in Bulgaria. Eur Asian J Biosci 7:77–94

NNS [National Nutrition Survey] (2018) Ministry of Health services Regulation & Coordination. NNS [National Nutrition Survey], Islamabad, p 50

Phillips O, Gentry AH, Reynel C, Wilki P, Gávez-Durand CB (1994) Quantitative ethnobotany and Amazonian conservation. Conserv Biol 8:225–248

Redzic JS (2006) Wild edible plants and their traditional use in the human nutrition in Bosnia-Herzegovina. Ecol Food Nutr 45(3):189–232

Rostami R, Khoshbakht K (2012) Conservation status and sustainable local use of crop and crop wild relative species in the Bistoon protected area/West of Iran. J Agrobiol 29(2):55–62

Salvi J, Katewa SS (2016) A review: underutilized wild edible plants as a potential source of alternative nutrition. Int J Bot Stud 1(4):32–36

Samant SS, Butola JS, Sharma A (2007) Assessment of diversity, distribution, conservation status and preparation of management plan for medicinal plants in the catchment area of Parbati hydroelectric project stage—III in Northwestern Himalaya. J Mt Sci 4(1):034–056
Shad AA, Shah HU, Bakht J (2013) Ethnobotanical assessment and nutritive potential of wild food plants. J Anim Plant Sci 23(1):92–97
Shinwari ZK, Nasim A (2008) Ethnobotany in Pakistan. In: Selin H (ed) Encyclopaedia of the history of science, technology, and medicine in non-Western cultures. Springer, pp 1–14
Shumsky S, Hickey G, Pelletier B, Johns T (2014) Understanding the contribution of wild edible plants to rural social-ecological resilience in semi-arid Kenya. Ecol Soc 19(4):34
Stewart RR (1972) An annotated catalogue of the vascular plants of West Pakistan and Kashmir. In: Nasir E, Ali SI (eds) Flora of West Pakistan. Fakhri Press, Karachi, p 1028
Tardío J, Pardo-de-Santayana M, Morales R (2006) Ethnobotanical review of wild edible plants in Spain. Bot J Linn Soc 152(1):27–71
Waheed M, Haq SM, Arshad F, Bussmann RW, Pieroni A, Mahmoud EA, Casini R, Yessoufou K, Elansary HO (2023) Traditional wild food plants gathered by ethnic groups living in semi-arid region of Punjab, Pakistan. Biology 12(2):269
Yesodharan K, Sujana KA (2007) Wild edible plants traditionally used by the tribes in the Parambikulam wildlife sanctuary, Kerala, India. Nat Prod Rad 6(1):74–80

Nano-fertilizers and Plant Biodiversity: A Pathway Toward Climate-Smart Agriculture

11

Muhammad Shafiq, Muhammad Zafar Iqbal, Mohammad Athar, Rahmatullah Qureshi, and Noor-Ul-Ain Zafar

Abstract

The growing global demand for food, coupled with declining soil fertility and environmental concerns associated with conventional fertilizers, necessitates innovative approaches to sustainable crop nutrition. Nano-fertilizers (NFs) have emerged as a promising technology for enhancing nutrient-use efficiency, improving crop productivity, and reducing the environmental footprint of agricultural systems. This chapter reviews the development, classification, mechanisms, and applications of nano-fertilizers in modern agriculture. Major categories discussed include nitrogen-, phosphorus-, potassium-, magnesium-, and micronutrient-based nano-fertilizers, as well as metal oxide nanoparticles and nano-biofertilizer formulations. Owing to their unique physicochemical properties, such as high surface area, controlled nutrient release, and targeted delivery, nano-fertilizers enhance nutrient uptake, stimulate plant growth, improve photosynthetic efficiency, and increase tolerance to biotic and abiotic stresses. The chapter further examines their role in promoting soil health, water-use efficiency, and climate-smart agriculture. Despite these advantages, concerns regarding nanoparticle toxicity, impacts on soil microbial communities, ecological safety, and regulatory challenges remain significant constraints to large-scale

M. Shafiq (✉) · M. Z. Iqbal
Department of Botany, University of Karachi, Karachi, Pakistan

M. Athar
California Department of Food and Agriculture, Sacramento, CA, USA

R. Qureshi
Department of Botany, Faculty of Sciences, Pir Mehr Ali Shah Arid Agriculture University, Rawalpindi, Pakistan
e-mail: rahmatullahq@uaar.edu.pk

N.-Ul-A. Zafar
Department of Botany, Pir Mehr Ali Shah Arid Agriculture University, Rawalpindi, Pakistan

R. Qureshi (ed.), *Plant Biodiversity*, https://doi.org/10.1007/978-981-92-1471-6_11

adoption. The chapter highlights recent advances in nano-fertilizer research and emphasizes the need for optimized application strategies, standardized safety assessments, and supportive regulatory frameworks. Overall, nano-fertilizers represent a transformative tool for sustainable agricultural intensification and food security, provided their agronomic benefits are balanced with environmental stewardship and long-term ecosystem health.

Keywords

Agriculture · Crop growth · Crop yield · Environmental benefits · Nano-fertilizers · Nutrients supply · Soil and smart agriculture

Abbreviations

AgNPs	Silver nanoparticles
Bio NFs	Nano based biofertilizers
CaP	Calcium phosphate
CeO_2	Cerium dioxide (ceria) nanoparticles
CK	Compound fertilizer (control)
CRBD	Completely randomized block design
CuO	Copper oxide nanoparticles
ENPs	Engineered nanoparticles
Fe_3O_4	Iron oxide (magnetite) nanoparticles
K NPs	Potassium nanoparticles
NBFs	Nano biofertilizer formulas
NF / NFs	Nano fertilizer(s)
NPs	Nanoparticles
NPK	Nitrogen, phosphorus, potassium
SiO_2	Silicon dioxide nanoparticles
SiNp	Silicon nanoparticles
TiO_2	Titanium dioxide nanoparticles
WUE	Water use efficiency
ZnO	Zinc oxide nanoparticles

11.1 Introduction

The food production in the world has continued to heavily rely on the application of fertilizers to support the increased demand of agricultural production and food productivity. With the constantly increasing population of the world, the synthetic fertilizer farming systems are daily confronted with the need to reduce the effectiveness of utilizing nutrients, soil degradation, and pollution of the environment (Singh et al. 2024). In this chapter, the inputs that have been developed through nanotechnology, primarily, nano-fertilizers (NFs), proved to be a good prospective innovation in the contemporary agriculture system (Yadav et al. 2020). Nano-fertilizers can

Fig. 11.1 Impact of nano-fertilizers on the agricultural sector

serve as an affordable solution to nutritional supplement of crop by increasing nutrient absorption, reducing loss, and alleviating the side effects on the environment associated with the higher utilization of chemicals (Dimkpa 2014; Liu and Lal 2015; Celebi 2020; Hameed and Ashfaq 2023; Rasool et al. 2023; Santás-Miguel et al. 2023; Tiwari et al. 2014).

Nano-fertilizers, also commonly known as smart fertilizers, are important in ensuring that food is produced to be consumed by human beings as well as in animal feeds. Their specific and controlled-release characteristics permit the nutrients to be discharged in line with the requirements of the crop and the nutrient condition of the soil leading to enhanced efficacy of using the fertilizers and more predictable crop development. Making a comparison with synthetic fertilizers, nano-fertilizers have demonstrated the ability to enhance both the physical and chemical properties of the soil, and at the time, reduce the loss of nutrients and environmental pollution, which contributes to the sustainability of agriculture (Kumari et al. 2024). The nano-fertilizers through innovative applications depending on the site-specific soil conditions and crop production is a significant development in precision agriculture (Fig. 11.1).

Recent studies have shown that the application of nano-fertilizers in the sustainable production of crops is mainly associated with distinctive physicochemical properties containing high surface activity, enhanced solubility, controlled release of nutrients, and application at low levels (Yadav et al. 2023). These properties enable nano-formulated fertilizers to deliver the nutrients more effectively than their conventional counterparts resulting in better nutrient availability, elevated quality of soil, improved plant growth, and higher crop yield in different agricultural systems. Consequently, there is a continuously growing scientific literature on the use of nanomaterials as a useful tool in enhancing the growth and productivity of plants, as well as their general performance in agriculture (Khota et al. 2012; Abbas et al. 2016).

Although nanoparticles have agronomic benefits, their universal application has generated significant issues over their toxicological impact especially on their overuse or the lack of appropriate regulations. Due to their very low size, high reactivity, and their persistence in the environment, some nanoparticles can sediment in soil and water, disrupt normal physiological functions of plants and eventually affect crop growth and production (Oloye 2005). ENPs may gain access to agroecosystems in a variety of ways, but wastewater treatment residues and their further use as sludge on farmlands are identified as the primary channels of environmental distribution (Bakshi et al. 2019).

Nanoparticles are very efficient toward binding and mobilizing metals, but the fact that they accumulate continuously in the soil and water basins of agriculture could disrupt plant metabolism and pose long-term hazards to soil microorganisms and food web trophic interactions (Emamverdian et al. 2023). As a response to these challenges, recent studies have centered on the preparation of nano- biofertilizer formulas (NBFs) which entail the incorporation of nanoparticles with beneficial biological agents. These hybrid formulas have demonstrated outstanding prospects to enhance growth of plants, enhance tolerance against metal induced stress, and sustainability of the ecosystem (Garg et al. 2023). On the whole, there is a lot of literature that has shown that nano-fertilizers can have a positive impact on crop productivity and health of agroecosystems (Table 11.1). Moreover, nano-fertilizers were also reported to enhance the plant resistance to biotic and abiotic stresses, and they can be viewed as the promising alternatives to the widely used synthetic fertilizers (Babu et al. 2022).

Nevertheless, the mass utilization of nano-fertilizers has not been widespread, especially in developing nations, because of such reasons as the lack of awareness among farmers, high inception costs, and the slow adoption of new modern technologies in agriculture (Baral et al. 2019). The objective of this chapter is therefore to give a detailed review of the application of nano-fertilizers in increasing crop productivity and sustenance.

11.2 Review Methodology

This chapter is based on a thorough literature review on recent literature published in the last 10 years (2004–2024) found in such major scientific databases as ScienceDirect, SciFinder, PubMed, MDPI, Google Scholar, the National Library of Medicine, and SCOPUS. The chapter is divided into parts on nanotechnology in agriculture, the principal categories of nano-fertilizers, their toxicity and tolerance,

Table 11.1 Phytotoxic effects of nanoparticles and environmental stressors on crop growth and soil health

Nanomaterial/stress factor	Plant species studied	Experimental conditions	Observed effects	Reference
Cerium oxide nanoparticles (CeO_2; nanoceria)	*Medicago sativa*, *Zea mays*, *Cucumis sativus*, *Solanum lycopersicum*	0–4000 mg L^{-1}	Significant reduction in seed germination; maize germination decreased by ~30% at 2000 mg L^{-1}; tomato and cucumber germination reduced by ~30% and 20%, respectively	López-Moreno et al. (2010)
Silver nanoparticles (PVP-AgNPs, GA-AgNPs) and $AgNO_3$	Wetland plant species (11 spp.)	1, 10, 40 mg Ag L^{-1}; direct vs soil exposure	Species-specific responses; PVP-AgNPs showed minimal effect, while GA-AgNPs at high concentration reduced germination in several species; root growth more sensitive than shoot growth	Yin et al. (2012)
Cobalt ferrite nanoparticles ($CoFe_2O_4$ NPs)	*Solanum lycopersicum*	0–1000 mg L^{-1} (hydroponic)	Altered nutrient uptake and induced toxicity at higher concentrations	López-Moreno et al. (2016)
Silver nanoparticles (AgNPs)	Wheat (*Triticum aestivum*)	1 and 5 mM	Reduced root and shoot growth; disrupted photosynthetic machinery; altered biochemical parameters including catalase, glutathione, flavonoids, and chlorophyll	Collins et al. (2012); Rastogi et al. (2019)
Zinc oxide nanoparticles (ZnO NPs) ± phenanthrene	Wheat	250–1000 mg L^{-1}; with/without 1 mg L^{-1} phenanthrene	Dose-dependent toxicity affecting biomass, flowering, shoot height, and root length; no significant difference between nanoparticle and bulk forms	Sabir et al. (2014); Zhu et al. (2019)
Carbon dots (C-dots)	*Cicer arietinum*, *Hordeum vulgare*, *Vigna radiata*, *Triticum aestivum*	Variable concentrations under ambient conditions	Influenced seedling growth, root and shoot elongation; altered chlorophyll content, particularly in mung bean	Chauhan et al. (2022)
Nanoparticle exposure (unspecified)	Tomato (*Solanum lycopersicum*)	450 mg L^{-1}	Germination rate reduced to below 80%	Acosta-Slane et al. (2024)
Nitrogen fertilizers (NP, NPK, NK, NPCa, NPKCa, NKCa)	Soil system	Continuous application	Decrease in soil pH (from 6.07 to ~4.25–4.89); increased exchangeable aluminum, indicating soil acidification	Tadesse et al. (2024)
Salinity stress (NaCl)	*Cucurbita pepo*	9.38 dS m^{-1}	Significant decline in plant growth, yield, and photosynthetic efficiency	Rady et al. (2023)

and prospects and is backed by three figures and one table. The synthesized information will be used to help researchers, farmers and environmental managers to implement sustainable nano-enabled solutions that enhance agricultural productivity and long-term food security. It is also worth noting that nano-based approaches including chitosan-tripolyphosphate nanoparticles loaded with paraquat herbicide have been reported as being environmentally safer as alternatives in managing weeds, another example of how nano-fertilizer technologies can be useful (Grillo et al. 2014).

11.3 Food Production

Food production across the world continues to rely heavily on fertilizers to produce food and achieve the set agricultural output and satisfy their growing demand (Singh et al. 2024). In this respect, the inputs of nanotechnology stability especially nano-fertilizers have become necessary elements of the contemporary agricultural systems thanks to their affordability and high productivity. Nano-fertilizers also lower the adverse environmental effects of traditional agrochemicals and enhance nutrient uptake, pest control efficiency and crop productivity (Dimkpa 2014; Liu and Lal 2015; Celebi 2020; Hameed and Ashfaq 2023; Rasool et al. 2023; Santás-Miguel et al. 2023; Tiwari et al. 2023). Nano-fertilizers are also commonly known as smart fertilizers and are essential in enhancing food production both to feeding the people and livestock feeds. They have demonstrated their use in the agricultural systems to overcome the productivity limit by improving the physical and chemical characteristics of the soil and mitigating environmental degradation relative to the conventional fertilizers (Kumari et al. 2024). Notably, precision and sustainable agricultural practices have a noteworthy developmental step of nano-fertilizers which are targeted using crop nutritional requirements and soils characteristics (Fig. 11.1).

11.4 Nano-fertilizers and Their Role in Sustainable Agriculture

Various types of nano-fertilizers are being developed and implemented in agricultural systems including nano-size preparations of all the essential macronutrients, including nitrogen, phosphorus, potassium, calcium, magnesium and sulfur, metallic nanoparticles such as silver (Ag), copper (Cu), zinc (Zn), silicon (Si), and platinum (Pt). All of these nano-enabled inputs or combined, have several agronomic benefits that include growing plants faster, enhancing crop yield, and minimizing environmental risks of traditional fertilizers (Fig. 11.2). Nano-fertilizers are produced by both chemical and biological methods and hence, they can be customized and personalized in terms of formulation and application according to the regional needs of the agriculture (Zulfiqar et al. 2019).

Various delivery options are available so as to maximize the efficacy of nano-fertilizers among them being seed nano-priming, seed coating, foliar application, and soil injection. Of these, seed nano-priming has received specific interest because it can be used to improve the initial development of seedlings and accumulation of

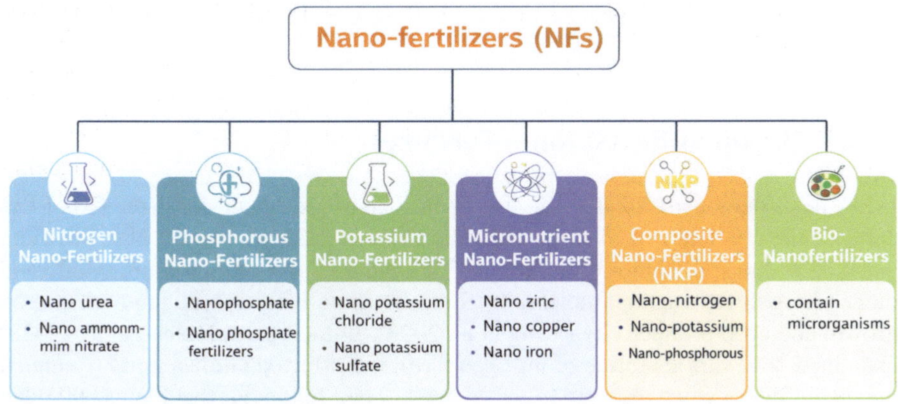

Fig. 11.2 Some different types of nano-fertilizers (NFs)

nutrients with little input (Igiku et al. 2024). Metallic nanoparticles, particularly silver, gold, copper, and zinc have also demonstrated high potential in enhancing the plant growth and wellness. Their use does not only promote physiological functions but also offers antimicrobial defense, which reminds better crop growth (Francis et al. 2024). When it comes to the normal condition, researchers demonstrated that the productivity of crops can be improved by using metallic nanoparticles by up to 20% and the disease rate can also be reduced by up to 50%.

The overall benefits of nano-fertilizers have a direct relationship with the capacity of the nano-fertilizers to deliver essential nutrients in a regulated and effective form to guarantee that plant nutritional demands are met at different growth and development phases. This application of the nutrient to its target produces sustainable growth of the plant and development, eventually leading to high and healthier yields of the plant. Moreover, slow-release micro-fertilizers and particularly polymers-laden ones help to control the availability of nutrients and keep down eliminations through absorption and volatilization. Despite the fact that most polymer coating are artificial in nature, the recent development has been aimed at using biopolymer-based nano-fertilizers that are better biodegradable and ecofriendly. Anitha et al. (2024) emphasized the opportunities of nano-fertilizers composed of biopolymers to nano-biotechnology as an efficient and environmentally friendly approach to the protection of plants and environmentally friendly agricultural production.

Chemical, biological, and nano-fertilizers application in the sphere of agricultural economy is attracting the attention of the regulatory bodies and research organizations all over the world (Arora et al. 2022; Tang et al. 2023). Using CK (compound fertilizer), T1 (compound fertilizer +0.3% nano carbon synergist), T2 (compound fertilizer +0.3% nano calcium carbonate synergist), T3 (compound fertilizer +0.3% composite nano synergist), T4 (70% compound fertilizer +0.3% nano carbon synergist), T5 (70% compound fertilizer +0.3% nano calcium carbonate synergist) and T6. The outcomes showed that in CK, the N accumulation of T1, T2, T3, T4, T5, and T6 increased by 40–50, 30–40, 55–65, 20–30, 15–20, and 30–40, respectively; and the N use efficiency by 12–19, 9–18, 16–22, 5–17, 4–16, and 10–20, respectively.

11.5 Nitrogen-Based Nano-Fertilizer

It has been demonstrated that the use of nitrogen-based nano-fertilizer (NFs) can enhance the efficiency of nitrogen utilization with a lesser likelihood of the nutrient leaching in the soil. Nano-formulations like nanourea, nano-ammonium nitrate and other nitrogen-rich nanoparticles proved to be quite effective in improving crop growth and crop productivity (Yang et al. 2023). In their case, Rashad et al. (2023) examined how various forms of nitrogen fertilizers affected Guinea grass (*Panicum maximum* Jacq.) when exposed to water stress. They found out that plants receiving nitrogen in nanoparticle form showed considerable increase in overall yield, plant morphology, leaf structure, and biochemical composition of modern macro and micronutrients, total carbohydrates, plant pigments, and crude protein, relative to plants receiving traditional NPK fertilizers at suggested doses. Likewise, it has been discovered that the application of engineered nanomaterials used individually or in concert with the NPK nutrients affect the development and yield of French bean crops with the seed priming and foliar application techniques showing varying effects (Abdel-Aziz et al. 2019). These findings point to the possible advantage of nitrogen nano-fertilizers to improve crop growth both in normal and stress conditions and to help to maintain more sustainable nutrient management strategies.

11.5.1 Phosphorous Nano-fertilizers

Application of phosphorus-based nano-fertilizers has been found to have positive effects of boosting crop production. Hydroxyapatite nanoparticles are one of these, as they are not very toxic and biocompatible, and the products of their degradation are a phosphate and calcium ion, which are naturally found in soils (Noruzi et al. 2023). Surveys of wheat (*Triticum aestivum* L.) growing in various media such as alkaline and acidic soil, peat moss, perlite mixtures and cocopeat revealed that phosphorus present in the hydroxyapatite nanoparticle was effectively distributed in the various regions of the plant. The results also showed that either in its suspension or powder form, hydroxyapatite nanoparticles can be used as an alternative to the conventional synthetic phosphorus fertilizers, primarily in acidic soils.

11.5.2 Nanopotassium Fertilizer

An important macronutrient required by plants is potassium (K). K is an important nutrient and it is very vital in the growth of plants. Drought tolerance and potassium uptake by the use of potassium NFs were enhanced. Growing of crop plantations on high potassium (K) demands soils by the region that are in the humid tropics and are made of laterite. The amount of K is usually low to medium in soils (Subramanian et al. 2024). In this regard, researchers propose Nano-fertilizers because they recommend that this is a good alternative to synthetic fertilizers since it contains nutrients that are released gradually over time to increase the use efficiency of the used nano-fertilizers. Split plots, which served as a completely randomized block design CRBD with four replicas, were conducted to know the effects of potassium nano-fertilizer on the soybean and water use efficiency (WUE) under the influence of various irrigation conditions in Moghan plain, Iran in 2014–2016 (Ajirloo and Amiri 2022). The findings proved that the application of potassium nano-fertilizer at all cutting irrigation stages reduced the impact of drought stress. Besides, the maximum yield (3.17 ton/ha) and yield components, physical properties, and WUE (3 kg/ha of potassium nano-fertilizer yield 0.112 kg/mm^{-1}) were recorded in the normal irrigation. The study involved the role of potassium nanoparticles (K-NPs) in different concentrations (0, 50, 100, and 200 mg/l) on onion development, production, pigments, and chemical content in two successive agriculture seasons in 2021 and 2022 in a private farm in Zagazig, Sharkia Governorate, Egypt (Salama et al. 2024). The findings revealed that spraying onion plants with K-NPs (200 mg/l) would be effective to increase vegetative characteristics of onion plant growth and production as well as the plant pigments and the levels of carbohydrate, oil, total indole, and phosphorus in onion plant bulbs. The onion plant yield and quality were proved to be stable with the use of K-NPs over the foliar surface, and it could improve the sustainability of agriculture.

11.5.3 Nano-fertilizers of Potassium and Magnesium

Magnesium is an essential nutrient during the formation of plants, as it is the central atom on the chlorophyll molecule and facilitates the action of different enzymes that are vital in photosynthesis (Guo et al. 2016; Arora et al. 2024). The use of nano-magnesium through foliar application that is usually mixed with chitosan has demonstrated positive results with regard to sesame (*Sesamum indicum* L.) plants in water-limited conditions. It has been also reported to increase physiological processes, overall yield, and affect biochemical parameters including the level of proline and total sugars, antioxidant enzyme activities in forms of peroxidase and catalase and ascorbate peroxidase, photosynthetic pigments, seed, and oil production (Khordadi et al. 2020). Moreover, magnesium oxide nanoparticles have been identified to enhance magnesium absorption and growth in tomato plants, implying its possible use to enhance nutrient utilization and crop growth (Cai et al. 2020).

11.5.4 PHARMAX Nano-fertilizer

The use of PHARMAX nano-fertilizer, its concentration (0, 1, 3, and 5 ml/L), as well as when the foliar sprays were applied (tillering stage and 2 weeks after the first application) was observed to affect the growth, yield, and essential oil content of coriander (*Coriandrum sativum* L.). The experiment revealed that the frequency of foliar treatments and their concentration greatly influenced the height of the plants, the umbels per plant, umbels per square meter, umbels per seed, grain yield, straw and single plant grain, biomass and the percentage but yield of essential oils. The authors came to the same conclusion as they decided that bi-application of the nano-fertilizer at 5 ml/L concentration yielded the finest outcomes, enhancing both the physical feature and the general yield of coriander (Nejatzadeh 2024).

11.5.5 Nano-fertilizer Derived Using Banana Peels

Nano-fertilizer based on banana peels was experimented on tomatoes and fenugreek and the outcome showed definite increase in seed germination with increase in dose of extract. The germination of tomatoes increased significantly when compared to the untreated control where germination rate was 14%, and on average it was 97% 7 days after planting. The same pattern was found to be true of fenugreek whereby germination was 25% of the control to 93.14% when using the nano-fertilizer made of banana peel (Hussein et al. 2019).

11.5.6 Nano-micronutrient Fertilizers

Zinc, iron, copper, boron, and manganese micronutrients can be added to nano-fertilizer to supply plants with essential elements required to grow in a healthy way and help prevent deficiencies of micronutrients. Some of these nutrients, e.g., silica, zinc, copper, and iron, have been synthesized at the nano level with success to boost the plant development and in general growth (Suriyaprabha et al. 2012; El-Saadony et al. 2021).

11.5.7 Metal Oxide NPs [Nanoiron, SiNp, ZnO, CuO, and Palladium Nanoparticle]

Phenolic extracts of different plants have been used to successfully prepare iron oxide nanoparticles (Fe_3O_4 NPs) and it was found that a variation in phenolic composition and structure can affect the distribution of particle size (Salgado et al. 2019; Ali et al. 2016). In addition to synthesis, iron oxide nanomaterials are also commonly studied with respect to their antibacterial activity (Behera et al. 2012; Arakha et al. 2015; Prabhu et al. 2015; Vihodceva et al. 2021). Magnetic Fe_3O_4 nanoparticles with eco-friendly properties are also produced by using extracts of seaweed (*Sargassum muticum*) and Mimosa pudica roots because of approaches called green synthesis (Mahdavi et al. 2013; Niraimathee et al. 2016).

The use of nano-iron has been proved to be useful in crop nutrient uptake and quality. An example of this is nano-iron which enhanced the nutrient accumulation and the quality of two onions (*Allium cepa* L.) cultivars (Joghatay et al. 2015). Nano-iron (III) oxide seed treatment also promoted seed germination, seedling growth, and salinity in sorghum (Maswada et al. 2018). On the same note, phenolic and flavonoid metabolism in butterhead lettuce and sweet pepper seedlings was regulated by cerium oxide nanoparticles (Kalisz et al. 2021). Fe_3O_4@HA nanoparticles coated with humic acid improved tomato plant height, dry biomass, and iron by 31, 97, and 247%, respectively, which is why they have the potential to enhance plant growth (Ma'mani et al. 2022).

Nanoparticles made of silicon and chitosan have also been used to enhance post-harvest and stress tolerance. A 1% chitosan film with 0.04% nano-silicon dioxide as a coating on the jujubes slowed down the rate of respiration, weight loss, and decay of the fruits in 32 days, and increased the content of flavonoids, which is a promising method of food preservation (Yu et al. 2012). Nano-silicon-treatments have facilitated increase in Changbai larch seedlings (Lin et al. 2004) and alleviate chromium (VI) toxicity in pea seedlings (Tripathi et al. 2015).

Another category of nano-fertilizers with much research is the zinc and copper nanoparticles. As nanoparticles, zinc oxide (ZnO) has shown antibacterial action against Escherichia coli O157:H7 (Liu et al. 2009), and in agricultural practice, ZnO nanoparticles have been found to enhance seedling vigor, chlorophyll content, root and shoot development, flowering and crop yield in crops like peanuts (Prasad et al. 2012). Nanoparticles made out of ZnO using plant components, e.g., quercetin of *Combretum ovalifolium* or *Calotropis procera* leaves, have also been used in environmental applications, e.g., in photodegradation and in preserving harvested produce (Jeyaleela et al. 2020; Gawade et al. 2017; Sogvar et al. 2016). ZnO, CuO, and SiO_2 nanocomposites based on chitosans have antifungal properties against such pathogens as *Fusarium solani* and *Alternaria solani* (Krumova et al. 2024).

Application of copper nano-fertilizers on cowpea increased physical characteristics, chlorophyll level and antioxidant property. The copper nanoparticles have been proven to enhance the growth of plants in field trials with soil and foliar application under different nano-fertilizer concentrations (Mustafa et al. 2024) and when the soil microbial communities were exposed to the long-term effects of Cu $(OH)_2$ nanoparticles (Peixoto et al. 2021). Lastly, palladium nanoparticles prepared through *Rosa damascena* flower have potential promising in vitro and in vivo uses, which further demonstrates the wide range of plant-mediated nano-fertilizers in improving crop growth, stress-resistance, and food security (Bi and Srivastava 2023).

11.5.8 The Cerium Oxide Nanoparticles (Nanoceria)

It has been demonstrated that the flower extract of *Calotropis procera* has antibacterial properties when applied to the synthesis of cerium oxide nanoparticles (Muthuvel et al. 2020). A study conducted by Rossi et al. (2017) showed that cerium oxide nanoparticles (CeO_2 NPs) treatment of *Brassica napus* L. affected the ability of this plant to endure salt stress through alteration of the development of root apoplastic barriers. In a similar manner, Liu et al. (2021) found out that due to the ability of the CeO_2 nanoparticles to help the cotton to maintain a desirable cytosolic K^+/Na^+ level under saline conditions, the nanoparticles were capable of enhancing salt tolerance in cotton. Cerium oxide nanoparticles were also reported to have a positive impact on *Calendula officinalis* L. growth and physiological functionality by improving photosynthetic pigments, antioxidant enzyme activity, compatible osmolytes, and reducing electrolyte leakage on foliar application (Jahani et al. 2019). In cilantro (*Coriandrum sativum* L.) CeO_2 nanoparticles affected the activity of key enzymes increasing catalase and ascorbate peroxidase activities in shoots and roots, respectively, where 125 mg kg^{-1} was found to be more effective than 500 mg kg^{-1} when applied to organic soil (Morales et al. 2013).

11.5.9 NiO and Silver Nanoparticles AgNPs

Nejatzadeh-Barandozi et al. (2014) illustrated the impact of the application of different amounts of silver nitrate (0, 100, 200, and 300 ppm) and silver nanoparticles (0, 20, 40, and 60 ppm) on basil crops at the seedling propagation stage. Their results indicated that shoot silver levels were familiar in comparing 100 ppm of silver nitrate to 60 ppm of nano silver. This was however not the case as increasing silver nitrate levels (100–300 ppm) resulted in a decrease in seed yield and on the other hand an increase in nano silver concentrations (20–60 ppm) resulted in an increase in seed yield, implying that nano silver may be an effective replacement of conventional silver compounds. Moreover, the antimicrobial properties of silver nanoparticles were stronger against *Escherichia coli* and *Candida albicans* and other species of *Streptococcal* species than *Staphylococcus aureus* (Youssef et al. 2014).

Suwan et al. (2018) have used water-soluble extracts of various parts of rice grains as reducing agents and Nigella sativa seed extracts as stabilizers to produce silver nanoparticles, and nanoparticles were demonstrated to be covered with bioactive compounds of the extracts (Boudiaf et al. 2021). There are also studies on possible cytotoxic properties of silver and titanium dioxide nanoparticles on plants (Cox et al. 2017). In addition to plant growth, biosynthesized metallic nanoparticles have also been found to have antibacterial, antifungal, and antioxidant properties with synergistic actions against gastrointestinal pathogens and candidal species (Patra and Baek 2017). It is important to note that the treatment with silver nanoparticles was identified to stimulate growth and promote anti-inflammatory effects in *Brassica juncea* (Sharma et al. 2012).

11.5.10 TiO_2 Nanoparticles

Nanoparticles of titanium dioxide (TiO_2) are used extensively across numerous industries and due to this, they are a common occurrence in wastewater and they are also deposited in sewage sludge (Venkatachalam et al. 2017). Such nanoparticles are also investigated to enhance the soil, and they can react with crops (Acosta-Slane et al. 2024). By way of example, the impact of sewage sludge with nano-TiO_2 on tomato plant were measured with the reference to physiological characteristics, biochemical reactions, composition of macromolecules and accumulation of titanium use within the lifecycle of tomato. It was found that amendments containing nano-TiO_2-enriched sludge had a profound positive impact on the growth of the plant, improving the biomass of the leaves by 142% and fruit yield by 102%, and leaving most of the biochemical parameters unaffected except the reduction of the contents of leaf tannins by 43% (Bakshi et al. 2019).

The effect of nano- TiO_2 (Yang et al. 2006) on the physiology of the plants was also identified in earlier studies. Its effect was studied within the chloroplast aging of spinach in the light conditions (Hong et al. 2005). The study by Gao et al. (2006) established the improvement of the photosynthetic carbon reactions in spinach using nano-anatase TiO_2 and Qi et al. (2013) established that nano-TiO_2 enhanced photosynthesis in tomato leaves under mild heat stress. Besides, Mahmoodzadeh et al. (2013) noted that nano-sized particles of titanium dioxide influenced germination and initial development in *Brassica napus*, as well. On the whole, these results indicate that TiO_2 nanoparticles, when used attentively, can increase plant growth and photosynthetic performance without significant biochemical imbalance and become an active implement in sustainable agriculture.

11.5.11 Composite Nano-fertilizers

Fertilizers produced by nano-compost improve the accessibility of nutrients required and when developed into composites such as NPK give a balanced source of nutrients to the crops as they grow and develop. They have also demonstrated the ability to promote favorable rhizosphere microorganisms using these nano-composite fertilizers that enhance the synthesis of secondary metabolites that enhance root colonization and general plant development (Panichikkal et al. 2019). The growing world food demand has resulted in the massive application of fertilizers (Nongbet et al. 2022). Not only do nanocomposites-based smart fertilizers increase mechanical strength, barrier properties but also raise their environmental sustainability, which increases crop productivity and sustainable agricultural practices (Chakraborty et al. 2023).

11.5.12 Nanobio-fertilizers

An alternative way of improving the growth of plants is the use of nano-based biofertilizers (Bio-NFs) in conjunction with complementary beneficial microbes, including the ones that can fix atmospheric nitrogen. These recipes usually contain a combination of bioactive and non-structural compounds sustaining crop growth. The green synthesis of nano-fertilizers using microbes also has a number of benefits over the chemical or physical process, such as it is safer, more reliable, and eco-friendlier (Kumar et al. 2023). Microbial nanotechnology also uses bacteria, fungi, algae, and even viruses to efficiently manufacture nanoparticles to offer sustainable solutions to a wide range of applications in agriculture (Shaikh et al. 2024).

One of the new categories of biofertilizers is calcium phosphate (CaP) nano-fertilizers, e.g., which not only activate the activity of microorganisms but can also increase their ability to fix nitrogen, solubilize phosphorus, synthesize plant growth-promoting compounds, and strengthen the immunity of plants (Majeed et al. 2023). These nano-biofertilizers help in the general growth of the plants, nutrient control and the preservation of crops (Dixit et al. 2023; Hameed and Ashfaq 2023). Besides this, nano-natural fertilizers will also offer an environmentally friendly and affordable option to the soluble fertilizers that are currently used. Their slow-release nature means that the nutrients are accessible all through the growth cycle of the plant, improving the quality and quantity of crops and reducing the environmental risks (Jha et al. 2023).

11.6 Potential Toxicological and Tolerance

Agriculture is a crucial aspect to the food security of a country, economic progress, and general well-being of the country. The use of agricultural methods in Pakistan has also changed drastically over the years, in large part because of the wide adoption of chemical fertilizers that have caused damage to the soil quality, water sources, and the ecosystem in general (Hadri et al. 2024). Although the use of engineered nanofertilizers has potential advantages, there are some disadvantages that have been reported with excessive use such as toxicity, lack of crop tolerance, growth inhibition, physiological system disruption, alteration of microbial communities, and deterioration of soil health and environment (Fig. 11.3). Furthermore, plants consistently experience various biotic and abiotic stressors that may constrain their growth and development, which ultimately decreases the yields of

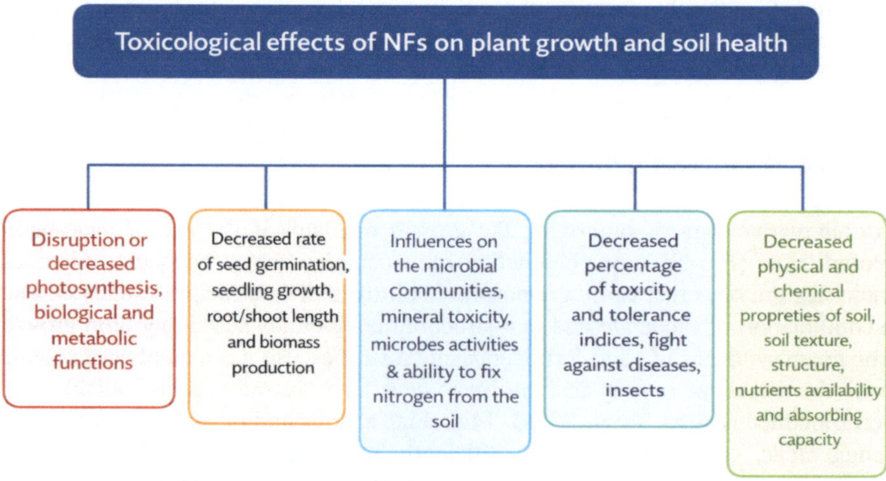

Fig. 11.3 Toxicological effects of NFs on plant growth and soil health at higher concentrations

crops, endangering food security, and resulting in losses related to economics (Maqsood et al. 2023).

There have been some research studies that have expressed concern on the possible negative impact of nano particles on agricultural products. This is mostly because of their capacity to resemble biological molecules, high mobility, and accumulation in soils, waters, and ecosystems (Oloye 2005; Awwad and Salem 2012; Lade et al. 2019; Yew et al. 2020; Grillo et al. 2021; Mubeen et al. 2023). As an example, the water-insoluble nanoparticles were tested using plant agar tests to examine the toxicity and bioavailability of copper nanoparticles in mung bean (Phaseolus radiatus) and wheat (Triticum aeustivum) (Lee et al. 2008). On the same note, Ghodake et al. (2010) have examined the negative impact of carbon nanotubes on *Brassica juncea* and *Phaseolus mungo*. Other studies reported studies on maize (Hu et al. 2021), tomato (Khodakovskaya et al. 2012), sesame (Khordadi et al. 2018), mung bean (Li et al. 2016), wheat (Wang et al. 2012), rice (Wang et al. 2021). Other significant factors that reduce crop yields the world over are stressors of the environment, including salinity. Nano-agrochemicals are major environmental pollutants in the recent years with an increasing number of studies trying to understand their fate and long-term effects in the environment (Zhang and Goss 2022).

It has been established that the application of nanoparticles affects the growth and development of different seedlings of crops. As an illustration, zinc oxide nanoparticles (ZnO NPs) in microscopic suspension influenced the early growth of mung bean (*Vigna radiata*) and chickpea (*Cicer arietinum*) seedlings in case of the plant agar method (Mahajan et al. 2011). On the same note, it has been discovered that bulk and nanoscale titanium dioxide (TiO_2) can affect wheat seed germination and seedling development (Feizi et al. 2012). Iron oxide nanoparticles were also found to exhibit significant physiological effects on the watermelon plants due to its magnetic properties (Li et al. 2013). Nanosilicon dioxide (SiO_2 NPs) has been also found to alleviate salt-induced stress in *Cucurbita pepo* L. and enhance tomato seedling establishment to a large extent (Siddiqui et al. 2014; Siddiqui 2014). SiO_2 nanoparticles increased growth and photosynthetic pigments content (Zhong et al. 2022) in maize (*Zea mays* L.) as reported by Sharifi-Rad et al. (2014). Zinc nanoparticles also had a similar effect of affecting developmental parameters in chickpea seedlings (Pavani et al. 2014), and the zinc oxide nanoparticles onto the soil-grown soybean (*Glycine max* cv. Kowsar) enhanced seed yield and antioxidant defenses (Yusefi-Tanha et al. 2020).

Recent studies have also depicted the impacts of TiO_2 nanoparticles on the growth of lucerne, tomato, and pepper plant at various levels of 450, 900, and 1800 mg L^{-1}. Findings indicated that propagation of tomatoes was reduced a little at 450 mg L^{-1} which indicated slight toxication and physical alterations were observed in all the three crops and primarily in roots. It turned out that tomatoes were the most affected by the decrease in the root length and width, and peppers were slightly affected by the change in the root shape (Acosta-Slane et al. 2024). Altogether, zinc oxide nanoparticles used in green house and field environments have been referred to as having profound implications on the physiology of plants and their growth and development (Hanif et al. 2024).

11.7 Difficulties and Future Prospects of NFs

Nanotechnology fertilizers can be used to enhance the health of soils and help in the production of crops better than the synthetic fertilizers. Nano-fertilizers (NFs) have several benefits such as regulated and gradual release of nutrients which reduces loss of nutrient in the form of leaching and plants are able to absorb beneficial elements in the long run. To achieve the most out of these fertilizers, this requires continuous research to create new formula, use, and procedures. Also, one should research on the safety and tolerance of nano-fertilizers to various crops. The education and awareness of farmers about their appropriate use and benefits will also be paramount because of the widespread awareness.

11.8 Conclusion

Nano-fertilizers represent an extremely effective and targeted delivery unit of nutrients that can have a significant effect on productivity of crops, their performance in relation to nutrient utilization, and soil welfare under agroclimatic-stressed conditions. Their controlled-release properties and improved bioavailability minimize waste of the nutrients, decrease the reliance on the traditional agrochemicals, and favor accuracy in agriculture, especially within resource-constrained areas. It has been experimentally proven that nano-fertilizers increase the growth, physiologic performance as well as stress tolerance of plants, and nano-bio-fertilizer forms enhance the microbial activity and sustainability of the ecosystem. Their effects are however concentration dependent where the over-use or uncontrolled use may lead to phytotoxicity, destabilize soil microbial balance, and inhibit plant metabolic activity. Hence, a promising future of climate-smart farming is through nano-fertilizers, which must be used safely and efficiently, with optimized dosage, long-term ecological risk, standard regulations, and enhanced awareness of farmers. The sparing agronomic and environmental safety balance will be used to achieve sustainable adoption to have resilient and productive agroecosystems.

References

Abbas SS, Haneef M, Lohani M, Tabassum H, Khan AF (2016) Nanomaterials used as a plants growth enhancer: an update. Int J Pharma Sci Rev Res 5:17–23

Abdel-Aziz HMM, Hasaneen MNA, Omer AM (2019) Impact of engineered nanomaterials either alone or loaded with NPK on growth and productivity of French bean plants: seed priming vs foliar application. S Afr J Bot 125:102–108. https://doi.org/10.1016/j.sajb.2019.07.005

Acosta-Slane D, Gonzalez-Franco AC, Hhernandez-Huerta J, Castillo-Michel H, Reyes-Herrera J, Sanchez-Chavez E, Valles-Aragon MC (2024) Titanium dioxide nanoparticles (TiO2-NPs) effect on germination and morphological parameters in alfalfa, tomato, and pepper. Notulae Botanicae Horti Agrobotanici Cluj-Napoca 52(2):13634. https://doi.org/10.15835/nbha52213634

Ajirloo RAA, Amiri E (2022) Effects of Nano-potassium fertilizer on yield and water use efficiency of soybean under water deficit conditions (Case study: Moghan plain, Iran). Commun Soil Sci Plant Anal 53(12):1542–1551. https://doi.org/10.1080/00103624.2022.2060247

Ali A, Zafar H, Zia M, ul Haq I, Phull AR, Ali JS, Hussain A (2016) Synthesis, characterization, applications, and challenges of iron oxide nanoparticles. Nanotechnol Sci Appl 9:49–67. https://doi.org/10.2147/NSA.S99986

Anitha S, Lakshmi BS, Manojkumar S, Gnanadesigan M, Chandrasekaran R, Kapildev G (2024) Chapter 7: Biopolymers based nanofertilizers. In: Abd-Elsalam KA (ed) Nano biotechnology for plant protection. Nanofertilizer synthesis. Elsevier, pp 109–128. https://doi.org/10.1016/B978-0-443-13535-4.00014-6

Arakha M, Pal S, Samantarrai D, Panigrahi TK, Mallick BC, Pramanik K, Jha S (2015) Antimicrobial activity of iron oxide nanoparticle upon modulation of nanoparticle bacteria interface. Sci Report 5(5):1–12

Arora S, Murmu G, Mukherjee K, Saha S, Maity D (2022) A comprehensive overview of nanotechnology in sustainable agriculture. J Biotechnol 20(355):21–41. https://doi.org/10.1016/j.jbiotec.2022.06.007. Epub 2022 Jun 22. PMID: 35752390.

Arora PK, Tripathi S, Omar RA et al (2024) Next-generation fertilizers: the impact of bionanofertilizers on sustainable agriculture. Microb Cell Factories 23:254. https://doi.org/10.1186/s12934-024-02528-5

Awwad AM, Salem NM (2012) A green and facile approach for synthesis of magnetite nanoparticles. Nanosci Nanotechnol 2:208–213

Babu S, Singh R, Yadav D, Rathore SS, Raj R, Avasthe R, Yadav S, Das A, Yadav V, Yadav B, Shekhawat K, Upadhyay P, Yadav DK, Singh VK (2022) Nanofertilizers for agricultural and environmental sustainability. Chemosphere 292:133451. https://doi.org/10.1016/j.chemosphere.2021.133451

Bakshi M, Clarisse L, Bedolla DE, Stein RJ, Kaegi R, Sarret G, Pradas del Real AE, Castillo-Michel H, Abhilash PC, Larue C (2019) Assessing the impacts of sewage sludge amendment containing nano-TiO2 on tomato plants: a life cycle study. J Hazard Mater 369:191–198. https://doi.org/10.1016/j.jhazmat.2019.02.036

Baral BR, Pande KR, Gaihr YK, Baral KR, Sah SK, Thapa YB (2019) Farmers' fertilizer application gap in rice based cropping system: A case study of Nepal. SAARC J Agric 17(2):267–277. https://doi.org/10.3329/sja.v17i2.45311

Behera SS, Patra JK, Pramanik K, Panda NN, Thatoi H (2012) Characterization and evaluation of antibacterial activities of chemically synthesized iron oxide nanoparticles. World J Nano Sci Eng 02:196–200

Bi S, Srivastava R (2023) *Rosa damascena* flower mediated phytofabrication of palladium nanoparticles, in-vitro and in-vivo applications. Mater Today Proc. https://doi.org/10.1016/j.matpr.2023.07.305

Boudiaf M, Messai Y, Bentouhami E, Schmutz M, Blanck C, Ruhlmann L, Bezzi H, Tairi L, Eddine MD (2021) Green synthesis of NiO nanoparticles using *Nigella sativa* extract and their enhanced electro catalytic activity for the 4-nitrophenol degradation. J Phys Chem Solids 53:110020. https://doi.org/10.1016/j.jpcs.2021.110020

Cai L, Feng H, Jia H, Yang H, Sun X (2020) Mechanism of magnesium oxide nanoparticles effectively stimulate the Mg absorption and growth of tomato plants. J Plant Nutr Fertil 26:1318–1327. https://doi.org/10.11674/zwyf.19403

Celebi O, Cinislia TK, Cinisli K, Celebi D (2020) Nano farming. Mater Today Proc 45:3805–3808. https://doi.org/10.1016/j.matpr.2020.12.1244

Chakraborty R, Mukhopadhyay A, Paul S, Sarkar S, Mukhopadhyay R (2023) Nanocomposite-based smart fertilizers: a boon to agricultural and environmental sustainability. Sci Total Environ 863:160859. https://doi.org/10.1016/j.scitotenv.2022.160859

Chauhan P, Kumari M, Chaudhary S, Chaudhary GR, Umar A, Baskoutas S (2022) Seed germination studies on Chickpeas, Barley, Mung beans and Wheat with natural biomass and plastic waste derived C-dots. Sci Total Environ 837:155593. https://doi.org/10.1016/j.scitotenv.2022.155593

Collins D, Luxton T, Kumar N, Shah S, Walker V, Shah V (2012) Assessing the impact of copper and zinc oxide nanoparticles on soil: a field study. PLoS One 7:e42663. https://doi.org/10.1371/journal.pone.0042663

Cox A, Venkatachalam P, Sahi S, Sharma N (2017) Reprint of: silver and titanium dioxide nanoparticle toxicity in plants: a review of current research. Plant Physiol Biochem 110:33–49. https://doi.org/10.1016/j.plaphy.2016.08.007

Dimkpa CO (2014) Can nanotechnology deliver the promised benefits without negatively impacting soil microbial life? J Basic Microbiol 54:889–904. https://doi.org/10.1002/jobm.201400298

Dixit M, Bhuyan M, Ghoshal D, Mandal A, Sahoo D (2023) Mode of action of nanofertilizers and their role in nutrient budgeting: quality and attribute perspectives. In: Nanofertilizer synthesis, pp 375–398. https://doi.org/10.1016/B978-0-443-13535-4.00020-1

El-Saadony MT, ALmoshadak AS, Shafi ME, Albaqami NM, Saad AM, El-Tahan AM, Desoky EM, Elnahal AS, Almakas A, Abd El-Mageed TA, Taha AE, Elrys AS, Helmy AM (2021) Vital roles of sustainable nano-fertilizers in improving plant quality and quantity-an updated review. Saudi J Biol Sci 28(12):7349–7359. https://doi.org/10.1016/j.sjbs.2021.08.032

Emamverdian A, Barker J, Pehlivan N, Ghorbani A (2023) Role of nanomaterials for alleviating heavy metal(oid) toxicity in plants. In: Nanotechnology for abiotic stress tolerance and management in crop plants, pp 289–306. https://doi.org/10.1016/B978-0-443-18500-7.00019-3

Feizi H, Moghaddam PR, Shahtahmassebi N, Fotovat A (2012) Impact of bulk and nanosized titanium dioxide (TiO_2) on wheat seed germination and seedling growth. Biol Trace Elem Res 146:101–106. https://doi.org/10.1007/s12011-011-9222-7

Francis DV, Abdalla AK, Mahakham W, Sarmah AK, Ahmed ZF (2024) Interaction of plants and metal nanoparticles: exploring its molecular mechanisms for sustainable agriculture and crop improvement. Environ Int 190:108859. https://doi.org/10.1016/j.envint.2024.108859

Gao F, Hong F, Liu C, Zheng L, Su M, Wu X, Yang F, Wu C, Yang P (2006) Mechanism of nano-anatase TiO_2 on promoting photosynthetic carbon reaction of spinach. Biol Trace Elem Res 111:239–253. https://doi.org/10.1385/BTER:111:1:239

Garg D, Sridhar K, Stephen IB, Chawla P, Tripathi M, Sharma M (2023) Nano-biofertilizer formulations for agriculture: A systematic review on recent advances and prospective applications. Bioengineering (Basel) 10(9):1010. https://doi.org/10.3390/bioengineering10091010. PMID: 37760112; PMCID: PMC10525541

Gawade VV, Gavade NL, Shinde HM, Babar SB, Kadam AN, Garadkar KM (2017) Green synthesis of ZnO nanoparticles by using Calotropis procera leaves for the photode gradation of methyl orange. J Mater Sci Mater Electron 28:14033–14039

Ghodake G, Seo Y, Park D, Lee DS (2010) Phytotoxicity of carbon nanotubes assessed by *Brassica juncea* and *Phaseolus mungo*. J Nanoelectron Optoelectron 5:157–160. https://doi.org/10.1166/jno.2010.1084

Grillo R, Pereira AE, Nishisaka CS, de Lima R, Oehlke K, Greiner R et al (2014) Chitosan/tripolyphosphate nanoparticles loaded with paraquat herbicide: an environ mentally safer alternative for weed control. J Hazard Mater 27:163–171. https://doi.org/10.1016/j.jhazmat.2014.05.079

Grillo R, Fraceto LF, Amorim MJ, Scott-Fordsmand JJ, Schoonjans R, Chaudhry Q (2021) Ecotoxicological and regulatory aspects of environmental sustainability of nanopesticides. J Hazard Mater 404:124148. https://doi.org/10.1016/j.jhazmat.2020.124148

Guo W, Nazim H, Liang Z, Yang D (2016) Magnesium deficiency in plants: an urgent problem. Crop J 4(2):83–91

Hadri SH, Afzaal A, Saeed L, Arshad A, Nazeer S, Akram M (2024) Recent advances in the development of nanoparticle based fertilizers for different kinds of crops: a review. Biocatal Agric Biotechnol 58:103194. https://doi.org/10.1016/j.bcab.2024.103194

Hameed A, Ashfaq M (2023) Nano-biofertilizers: plant growth promotions and protections. In: Nanofertilizer synthesis, pp 37–54. https://doi.org/10.1016/B978-0-443-13535-4.00007-9

Hanif S, Javed R, Cheema M, Kiani MZ, Farooq S, Zia M (2024) Harnessing the potential of zinc oxide nanoparticles and their derivatives as Nanofertilizers: trends and perspectives. Plant Nano Biol:100110. https://doi.org/10.1016/j.plana.2024.100110

Hong F, Yang F, Liu C, Gao Q, Wan Z, Gu F, Wu C, Ma Z, Zhou J, Yang P (2005) Influences of Nano-TiO_2 on the chloroplast aging of spinach under light. Biol Trace Elem Res 104:249–260. https://doi.org/10.1385/BTER:104:3:249

Hu Y, Zhang P, Zhang X, Liu Y, Feng S, Guo D, Nadezhda T, Song Z, Dang X (2021) Multi-Wall carbon nanotubes promote the rowth of Maize (*Zea mays*) by regulating carbon and nitrogen metabolism in leaves. J Agri Food Chem 69:4981–4991. https://doi.org/10.1021/acs.jafc.1c00733

Hussein HS, Shaarawy HH, Hussien NH, Hawash SI (2019) Preparation of nano-fertilizer blend from banana peels. Bull Natl Res Cent 43(1):1–9. https://doi.org/10.1186/s42269-019-0058-1

Igiku V, Adetunji CO, Oyewole OA, Akinhanmi FO, Ameh SS, Eniola T, Yerima MB (2024) Various techniques used in the application of nanobiofertilizers on crops after synthesis. In: Handbook of agricultural biotechnology, pp 233–252. https://doi.org/10.1002/9781394211548.ch9

Jahani S, Saadatmand S, Mahmoodzadeh H, Khavari-Nejad RA (2019) Effect of foliar application of cerium oxide nanoparticles on growth, photosynthetic pigments, electrolyte leakage, compatible osmolytes and antioxidant enzymes activities of *Calendula officinalis* L. Biologia 74:1063–1075. https://doi.org/10.2478/s11756-019-00239-6

Jeyaleela GD, Vimala JR, Sheela SM, Agila A, Bharathy MS, Divya M (2020) Biofabrication of zinc oxide nanoparticles using the isolated flavonoid from combretum ovalifolium and its anti-oxidative ability and catalytic degradation of methylene blue dye. Orient J Chem 36:655–664

Jha A, Pathania D, Damathia B, Raizada P, Rustagi S, Singh P, Rani GM, Chaudhary V (2023) Panorama of biogenic nano-fertilizers: a road to sustainable agriculture. Environ Res 235:116456. https://doi.org/10.1016/j.envres.2023.116456

Joghatay A, Afshari H, Masoud SJ (2015) Effect of nano iron and Solupotasse fertilizers on accumulation of nutrient elements and quality of two onion (*Allium cepa*) cultivars. J Crop Ecophysiol 9:491–502

Kalisz A, Huska D, Jurkow R, Dvork M, Klejdus B, Caruso G, Sekara A (2021) Nanoparticles of cerium oxide. Iron, and silicon oides change the metabolism of phenols and flavonoids in butterhead lettuce and sweet pepper seedling. Environ Sci Nano J 7:1945–1959. https://doi.org/10.1039/D1EN00262G

Khodakovskaya MV, Kim BS, Kim JN, Alimohammadi M, Dervishi E, Mustafa T, Cernigla CE (2012) Carbon nanotubes as plant growth regulators: effects on tomato growth, reproductive system, and soil microbial community. Small 9:115–123. https://doi.org/10.1002/smll.201201225

Khordadi VJ, Fanoodi F, Sinaki JM, Rezvan S, Damavandi A (2018) Physiological response of sesame (*Sesamum indicum* L.) to application of chitosan and magnesium-nano fertilizers under irrigation cut-off in a sustainable agriculture system. Iran J Plant Physiol 9:2629–2639

Khordadi VJ, Fanoodi F, Masoud Sinaki J, Rezvan S, Damavandi A (2020) Foliar application of chitosan and nano-magnesium fertilizers influence on seed yield, oil content, photosynthetic pigments, antioxidant enzyme activities of sesame (Sesamum indicum L.) under water-limited conditions. Not Bot Horti Agrobot Cluj-Napoca 48(4):2228–2243

Khota LR, Sankarana S, Majaa JM, Ehsania R, Schuster EW (2012) Applications of nanomaterials in agricultural production and crop protection: a review. Crop Prot 12(35):64–70. https://doi.org/10.1016/j.cropro.2012.01.007

Krumova E, Benkova D, Stoyancheva G, Dishliyska V, Miteva-Staleva J, Kostadinova A, Ivanov K, El-sayed K, Staneva G, Elshoky HA (2024) Exploring the mechanism underlying the antifungal activity of chitosan based ZnO, CuO, and SiO_2 nanocomposites as nanopesticides against *Fusarium solani* and *Alternaria solani*. Int J Biol Macromol 268:131702. https://doi.org/10.1016/j.ijbiomac.2024.131702

Kumar N, Samota SR, Venkatesh K, Tripathi S (2023) Global trends in use of nano-fertilizers for crop production: advantages and constraints – a review. Soil Tillage Res 228:105645. https://doi.org/10.1016/j.still.2023.105645

Kumari PS, Ramkumar S, Seethalaxmi M, Rekha T, Abiyoga M, Baskar V, Sureshkumar S (2024) Biofortification of crops with nutrients by the application of nanofertilizers for effective agriculture. Plant Physiol Biochem 212:108772. https://doi.org/10.1016/j.plaphy.2024.108772

Lade BD, Gogle DP, Lade DB, Moon GM, Nandeshwar SB, Kumbhare SD (2019) Chapter 7: Nanobiopesticide formulations: application strategies today and future perspectives. In: Koul O (ed) Nano-biopesticides today and future perspectives. Academic, pp 179–206. https://doi.org/10.1016/B978-0-12-815829-6.00007-3

Lee WM, An YJ, Yoon H, Kweon HS (2008) Toxicity and bioavailability of copper nanoparticles to the terrestrial plants mung bean (*Phaseolus radiatus*) and wheat (*Triticum aestivum*): plant agar test for water-insoluble nanoparticles. Environ Toxicol Chem 27:1915–1921. https://doi.org/10.1897/07-481.1

Li J, Chang P, Huang J, Wang Y, Yuan H, Ren H (2013) Physiological effects of magnetic iron oxide nanoparticles towards watermelon. J Nanosci Nanotechnol 13:5561–5567. https://doi.org/10.1166/jnn.2013.7533

Li W, Zheng Y, Zhang H, Liu Z, Shi S (2016) Phytotoxicity, uptake, and translocation of fluorescent carbon dots in mung bean plants. ACS Appl Mater Interfaces 8:19939–19945. https://doi.org/10.1021/acsami.6b07268

Lin B, Diao S, Li C, Fang L, Qiao S, Yu M (2004) Effect of TMS (nanostructured silicon dioxide) on growth of Changbai larch seedlings. J For Res 15:138–140. https://doi.org/10.1007/bf02856749

Liu R, Lal R (2015) Potentials of engineered nanoparticles as fertilizers for increasing agronomic productions. Sci Total Environ 514:131–139. https://doi.org/10.1016/j.scitotenv.2015.01.104

Liu Y, He L, Mustapha A, Li H, Hu ZQ, Lin M (2009) Antibacterial activities of zinc oxide nanoparticles against *Escherichia coli* O157:H7. J Appl Micro 107:1193–1201. https://doi.org/10.1111/j.1365-2672.2009.04303.x

Liu J, Li G, Chen L, Gu J, Wu H, Li Z (2021) Cerium oxide nanoparticles improve cotton salt tolerance by enabling better ability to maintain cytosolic K^+/Na^+ ratio. J Nanobiotechnol 19:153. https://doi.org/10.1186/s12951-021-00892-7

López-Moreno ML, De La Rosa G, Hernández-Viezcas JA, Peralta-Videa JR, Gardea-Torresdey JL (2010) X-ray absorption spectroscopy (XAS) corroboration of the uptake and storage of CeO_2 nanoparticles and assessment of their differential toxicity in four edible plant species. J Agric Food Chem 58:3689–3693. https://doi.org/10.1021/jf904472e

López-Moreno ML, Avilés LL, Pérez NG, Irizarry BÁ, Perales O, Cedeno-Mattei Y, Román F (2016) Effect of cobalt ferrite (CoFe2O4) nanoparticles on the growth and development of *Lycopersicon lycopersicum* (tomato plants). Sci Total Environ 550:45–52. https://doi.org/10.1016/j.scitotenv.2016.01.063

Mahajan P, Dhoke SK, Khanna AS (2011) Effect of Nano-ZnO particle suspension on growth of mung (*Vigna radiata*) and gram (*Cicer arietinum*) seedlings using plant agar method. J Nanotechnol 2011:696535.1–696535.7. https://doi.org/10.1155/2011/696535

Mahdavi M, Namvar F, Ahmad MB, Mohamad R (2013) Green biosynthesis and characterization of magnetic iron oxide (Fe_3O_4) nanoparticles using seaweed (*Sargassum muticum*) aqueous extract. Molecules (Basel, Switzerland) 18:5954–5964

Mahmoodzadeh H, Nabavi M, Kashefi H (2013) Effect of nanoscale titanium dioxide particles on the germination and growth of canola (*Brassica napus*). J Ornamental Hortic Plants 3:25–32

Majeed Z, Manzoor E, Javed MR (2023) Calcium phosphates nanofertilizers: synthesis and biosafety. In: Nanofertilizer synthesis, pp 129–143. https://doi.org/10.1016/B978-0-443-13535-4.00022-5

Ma'mani L, Chorom M, Moezzi A (2022) Improved iron use efficiency in tomato using organically coated iron oxide nanoparticles as efficient bioavailable Fe sources. Chem Biol Technol Agric 9(1):1–16. https://doi.org/10.1186/s40538-022-00318-y

Maqsood MF, Shahbaz M, Khalid F, Rasheed Y, Asif K, Naz N et al (2023) Biogenic nanoparticles application in agriculture for ROS mitigation and abiotic stress tolerance: a review. Plant Stress 10:100281. https://doi.org/10.1016/j.stress.2023.100281

Maswada HF, Djanaguiraman M, Prasad PVV (2018) Seed treatment with nano-iron (III) oxide enhances germination, seeding growth and salinity tolerance of sorghum. J Agron Crop Sci 204: 577–587. https://doi.org/10.1111/jac.12280

Morales MI, Rico CM, Hernandez-Viezcas JA, Nunez JE, Barrios AC, Tafoya AT, Flores-Marges JP, Peralta-Videa JR, Gardea-Torresdey JL (2013) Toxicity assessment of cerium oxide nanoparticles in cilantro (*Coriandrum sativum* L.) plants grown in organic soil. J Agric Food Chem 61:6224–6230. https://doi.org/10.1021/jf401628v

Mubeen I, Mfarrej FBM, Razaq Z, Iqbal S, Naqvi SAH, Hakim F, Mosa WF, Moustafa M, Fang Y, Li B (2023) Nanopesticides in comparison with agrochemicals: outlook and future prospects for sustainable agriculture. Plant Physiol Biochem 198(1076):70. https://doi.org/10.1016/j.plaphy.2023.107670

Mustafa M, Azam M, Haq NB, Khan A, Zafar L, Abdul MRA (2024) Green fabrication of copper nano-fertilizer for enhanced crop yield in cowpea cultivar: a sustainable approach. Biocatal Agric Biotechnol 56:102994. https://doi.org/10.1016/j.bcab.2023.102994

Muthuvel A, Jothibas M, Mohana VS, Manoharan C (2020) Green synthesis of cerium oxide nanoparticles using calotropis procera flower extract and their photocatalytic degradation and antibacterial activity. Inorg Chem Commun 119:108086

Nejatzadeh F (2024) Effect of foliar application frequency and different levels of nano fertilizer on growth and development of coriander (*Coriandrum sativum* L.). Heliyon 10(11):e31732. https://doi.org/10.1016/j.heliyon.2024.e31732

Nejatzadeh-Barandozi F, Darvishzadeh F, Aminkhani A (2014) Effect of nano silver and silver nitrate on seed yield of (*Ocimum basilicum* L.). Organic Med Chem Lett 4:1–6

Niraimathee VA, Subha V, Ravindran RE, Renganathan S (2016) Green synthesis of iron oxide nanoparticles from Mimosa pudica root extract. Int J Environ Sustainable Dev 15:227–240

Nongbet A, Mishra AK, Mohanta YK, Mahanta S, Ray MK, Khan M, Baek H, Chakrabartty I (2022) Nanofertilizers: a smart and sustainable attribute to modern agriculture. Plants 11(19): 2587. https://doi.org/10.3390/plants11192587

Noruzi M, Hadian P, Soleimanpour L, Shahbazi K (2023) Hydroxyapatite nanoparticles: an alternative to conventional phosphorus fertilizers in acidic culture media. Chem Biol Technol Agric 10(1):1–18. https://doi.org/10.1186/s40538-023-00437-0

Oloye OI (2005) Chemical profile of unripe pulp of carica papaya. Pak J Nutr 4:379–381

Panichikkal J, Thomas R, John JC, Radhakrishnan EK (2019) Biogenic gold nanoparticle supplementation to plant beneficial pseudomonas monteilii was found to enhance its plant probiotic effect. Curr Microbiol 76:503–509. https://doi.org/10.1007/s00284-019-01649-0

Patra JK, Baek KH (2017) Antibacterial activity and synergistic antibacterial potential of biosynthesized silver nanoparticles against foodborne pathogenic bacteria along with its anticandidal and antioxidant effects. Front Microbiol 8:167. https://doi.org/10.3389/fmicb.2017.00167

Pavani, K.V., V. Divya., I. Veena., M. Aditya. And G.V.S. Devakinandan. 2014. Influence of bioengineered zinc nanoparticles and zinc metal on *Cicer arietinum* seedlings growth. Asian J Agric Biol, 2: 216–223

Peixoto S, Henriques I, Loureiro S (2021) Long-term effects of $Cu(OH)_2$ nanopesticide exposure on soil microbial communities. Environ Pollut 269:116113. https://doi.org/10.1016/j.envpol.2020.116113

Prabhu YT, Rao KV, Kumari BS, Kumar VSS, Pavani T (2015) Synthesis of Fe_3O_4 nanoparticles and its antibacterial application. Int Nano Lett 5:85–92

Prasad T, Sudhakar P, Sreenivasulu Y, Munaswamy V, Reddy KR, Sreeprasad TS (2012) Effect of nanoscale zinc oxide particles on the germination, growth and yield of peanut. J Plant Nutr 35(6):905–927

Qi M, Liu Y, Li T (2013) Nano-TiO_2 improve the photosynthesis of tomato leaves under mild heat tress. Biol Trace Elem Res 156:323–328. https://doi.org/10.1007/s12011-013-9833-2

Rady MM, Mossa AH, Youssof AM, Osman S, Ahmed SM, Mohamed IA (2023) Exploring the reinforcing effect of nano-potassium on the antioxidant defense system reflecting the increased yield and quality of salt-stressed squash plants. Sci Hortic 308:111609. https://doi.org/10.1016/j.scienta.2022.111609

Rasool A, Krismastuti FSH, Zulfajri M, Meliana Y, Sudewi S (2023) A smart way to increase the growth and productivity of crops through nano-fertilizer. In: Molecular impacts of nanoparticles on plants and algae, pp 333–346. https://doi.org/10.1016/B978-0-323-95721-2.00013-0

Rashad HM, Wahab A, Mahmoud M, Alsamadany H, Alzahrani Y, Seleem EA, Ibrahim HMS (2023) Evaluation of nano-nitrogen fertilizers and other nitrogen sources on the performance of Guinea grass plants grown in newly reclaimed soil under water deficiency. Plant Stress 10: 100282

Rastogi A, Zivcak M, Tripathi DK, Yadav S, Kalaji HM (2019) Phytotoxic effect of silver nanoparticles in Triticum aestivum: improper regulation of photosystem I activity as the reason for oxidative damage in the chloroplast. Photosynthetica 57:209–216. https://doi.org/10.32615/ps.2019.019

Rossi L, Zhang WL, Ma XM (2017) Cerium oxide nanoparticles alter the salt stress tolerance of *Brassica napus* L. by modifying the formation of root apoplastic barriers. Environ Pollut 229: 132–138. https://doi.org/10.1016/j.envpol.2017.05.083

Sabir S, Arshad M, Chaudhari SK (2014) Zinc oxide nanoparticles for revolutionizing agriculture: synthesis and applications. Sci World J 8:925494. https://doi.org/10.1155/2014/925494

Salama DM, Khater MA, Abd El-Aziz ME (2024) The influence of potassium nanoparticles as a foliar fertilizer on onion growth, production, chemical content, and DNA fingerprint. Heliyon 10(11):e31635. https://doi.org/10.1016/j.heliyon.2024.e31635

Salgado P, Márquez K, Rubilar O, Contreras D, Vidal G (2019) The effect of phenolic compounds on the green synthesis of iron nanoparticles (FexOy-NPs) with photocatalytic activity. Appl Nanosci 9:371–385

Santás-Miguel V, Arias-Estévez M, Rodríguez-Seijo A, Arenas-Lago D (2023) Use of metal nanoparticles in agriculture a review on the effects on plant germination. Environ Pollut 334: 122222. https://doi.org/10.1016/j.envpol.2023.122222

Shaikh A, Jamla M, Joshi S, Patil S, Oak U, Kumar V (2024) Microbial nanotechnology for producing stress smart crops. Plant Nano Biol 7:100063. https://doi.org/10.1016/j.plana.2024.100063

Sharifi-Rad J, Karimi J, Mohsenzadeh S, Sharifi-Rad M, Moradgholi J (2014) Evaluating SiO_2 nanoparticles effects on developmental characteristic and photosynthetic pigment contents of *Zea mays* L. Bull Environ Pharmacol Life Sci 3:194–201

Sharma P, Bhatt D, Zaidi MGH, Saradhi PP, Khanna PK, Arora S (2012) Silver nanoparticle-mediated enhancement in growth and antioxidant status of *brassica juncea*. Appl Biochem Biotechnol 167:2225–2233. https://doi.org/10.1007/s12010-012-9759-8

Siddiqui M, Al-Whaibi MH (2014) Role of nano-SiO_2 in germination of tomato (*Lycopersicum esculentum* seeds Mill.). Saudi J Biol Sci 21:13–17. https://doi.org/10.1016/j.sjbs.2013.04.005

Siddiqui MH, Al-Whaibi MH, Faisal M, Al SAA (2014) Nano-silicon dioxide mitigates the adverse effects of salt stress on *Cucurbita pepo* L. Environ Toxicol Chem 33:2429–2437. https://doi.org/10.1002/etc.2697

Singh S, Singh R, Singh K, Katoch K, Zaeen AA, Birhan DA, Singh A, Sandhu HS, Singh H, Sahrma LK (2024) Smart fertilizer technologies: An environmental impact assessment for sustainable agriculture. Smart Agric Technol 8:100504. https://doi.org/10.1016/j.atech.2024.100504

Sogvar OB, Saba MK, Emamifar A, Hallaj R (2016) Influence of nano-ZnO on microbial growth, bioactive content and postharvest quality of strawberries during storage. Innov Food Sci Emerg Technol 35:168–176

Subramanian KS, Gopal M, Bhat R, Nair SS, Radhakrishnan M, Rajesh MK (2024) Nano-potassium intercalated composted coir pith: A slow-release fertilizer suitable for laterite soils of humid tropics of India. Biocatal Agric Biotechnol 57:103054. https://doi.org/10.1016/j.bcab.2024.103054

Suriyaprabha R, Karunakaran G, Yuvakkumar R, Rajendran V, Kannan N (2012) Silica nanoparticles for increased silica availability in maize (*Zea mays* L.) seeds under hydroponic conditions. Curr Nanosci 8:902–908. https://doi.org/10.2174/157341312803989033

Suwan T, Khongkhunthian S, Okonogi S (2018) Green synthesis and inhibitory effects against oral pathogens of silver nanoparticles mediated by rice extracts. Drug Discov Ther 12(4):189–196. https://doi.org/10.5582/ddt.2018.01034

Tadesse KA, Lu Z, Shen Z, Daba NA, Li J, Alam MA, ... Huimin Z (2024) Impacts of long-term chemical nitrogen fertilization on soil quality, crop yield, and greenhouse gas emissions: with insights into post-lime application responses. Sci Total Environ 944:173827

Tang Y, Zhao W, Zhu G, Tan Z, Huang L, Zhang P, Gao L, Rui Y (2023) Nano-pesticides and fertilizers: solutions for global food security. Nanomaterials (Basel) 14(1):90. https://doi.org/10.3390/nano14010090

Tiwari DK, Dasgupta-Schubert N, Villaseñor CLM, Villegas J, Carreto ML, Borjas GSE (2014) Interfacing carbon nanotubes (CNT) with plants: enhancement of growth, water and ionic nutrient uptake in maize (*Zea mays*) and implications for nanoagriculture. Appl Nanosci 4:577–591. https://doi.org/10.1007/s13204-013-0236-7

Tiwari A, Maity D, Saha S (2023) Hybrid nanofertilizers: synthesis and applications. In: Nanofertilizer synthesis, pp 75–91. https://doi.org/10.1016/B978-0-443-13535-4.00005-5

Tripathi DK, Singh VP, Prasad SM, Chauhan DK, Dubey NK (2015) Silicon nanoparticles (SiNp) alleviate chromium (VI) phytotoxicity in *Pisum sativum* (L.) seedlings. Plant Physiol Biochem 96:189–198. https://doi.org/10.1016/j.plaphy.2015.07.026

Venkatachalam P, Jayaraj M, Manikandan R, Geetha N, Rene ER, Sharma NC et al (2017) Zinc oxide nanoparticles (ZnONPs) alleviate heavy metal induced toxicity in *Leucaena leucocephala* seedlings: a physiochemical analysis. Plant Physio Biochem 110:59–69. https://doi.org/10.1016/j.plaphy.2016.08.022

Vihodceva S, Šutka A, Sihtmäe M, Rosenberg M, Otsus M, Kurvet I, Kasemets K (2021) Antibacterial activity of positively and negatively charged hematite (α-Fe_2O_3) nano particles to escherichia coli, staphylococcus aureus and vibrio fischeri. Nanomaterials 11:652

Wang X, Han H, Liu X, Gu X, Chen K, Lu D (2012) Multi-walled carbon nanotubes can enhance root elongation of wheat (*Triticum aestivum*) plants. J Nanopart Res 14:841. https://doi.org/10.1007/s11051-012-0841-5

Wang Y, Deng C, Rawat S, Cota-Ruiz K, Medina-Velo I, Gardea-Torresdey JL (2021) Evaluation of the effects of nanomaterials on rice (*Oryza sativa* L.) responses: underlining the benefits of nanotechnology for agricultural applications. ACS Agric Sci Technol 1(2):44–54

Yadav RK, Singh NB, Singh A, Yadav V, Bano C, Khare S, Niharika S (2020) Expanding the horizons of nanotechnology in agriculture: recent advances, challenges and future perspectives. Vegetos 33:203–221

Yadav A, Yadav K, Abd-Elsalam KA (2023) Exploring the potential of nanofertilizers for a sustainable agriculture. Plant Nano Biol 5:100044. https://doi.org/10.1016/j.plana.2023.100044

Yang F, Hong F, You W, Liu C, Gao F, Wu C, Yang P (2006) Influence of nano-anatase TiO_2 on the nitrogen metabolism of growing spinach. Biol Trace Elem Res 110:179–190. https://doi.org/10.1385/BTER:110:2:179

Yang M, Dong C, Shi Y (2023) Nano fertilizer synergist effects on nitrogen utilization and related gene expression in wheat. BMC Plant Biol 23(1):26. https://doi.org/10.1186/s12870-023-04046-9

Yew YP, Shameli K, Miyake M, Khairudin MBBA, Mohamad SEB, Naiki T, Lee KX (2020) Green biosynthesis of superparamagnetic magnetite Fe_3O_4 nanoparticles and biomedical applications in targeted anticancer drug delivery system: a review. Arab J Chem 13:2287–2308

Yin L, Colman BP, McGill BM, Wright JP, Bernhardt ES (2012) Effects of silver nanoparticle exposure on germination and early growth of eleven wetland plants. PLoS One 7:e47674. https://doi.org/10.1371/journal.pone.0047674

Youssef AM, Abdel-Aziz MS, El-Sayed SM (2014) Chitosan nanocomposite films based on Ag-NP and Au-NP biosynthesis by *Bacillus subtilis* as packaging materials. Int J Biol Macromol 69: 185–191

Yu Y, Zhang S, Ren Y, Li H, Zhang X, Di J (2012) Jujube preservation using chitosan film with nano-silicon dioxide. J Food Eng 113:408–414

Yusefi-Tanha E, Fallah S, Rostamnejadi A, Pokhrel LR (2020) Zinc oxide nanoparticles (ZnONPs) as a novel nanofertilizer: influence on seed yield and antioxidant defense system in soil grown soybean (*Glycine max* cv. Kowsar). Sci Total Environ 738:140240. https://doi.org/10.1016/j.scitotenv.2020.140240

Zhang Y, Goss GG (2022) Nanotechnology in agriculture: comparison of the toxicity between conventional and nano-based agrochemicals on non-target aquatic species. J. Hazard Mater 439: 129559. https://doi.org/10.1016/j.jhazmat.2022.129559

Zhong X, Wen H, Zeng R, Deng H, Su G, Zhou H, Zhou X (2022) Zein-functionalized mesoporous silica as nanocarriers for nanopesticides with pH/enzyme dual responsive properties. Ind Crop Prod 188:115716. https://doi.org/10.1016/j.indcrop.2022.115716

Zhu J, Zou Z, Shen Y, Li J, Shi S, Han S, Zhan X (2019) Increased ZnO nanoparticle toxicity to wheat upon co-exposure to phenanthrene. Environ Pollut 247:108–117. https://doi.org/10.1016/j.envpol.2019.01.046

Zulfiqar F, Navarro M, Ashraf M, Akram NA, Munné-Bosch S (2019) Nanofertilizer use for sustainable agriculture: advantages and limitations. Plant Sci 289:110270. https://doi.org/10.1016/j.plantsci.2019.110270

Climate Change, Plant Biodiversity, and Food Security: A Case Study from the AlUla Region of Saudi Arabia

12

Rahmatullah Qureshi and Monier M. Abd El-Ghani

Abstract

AlUla, a hyper-arid oasis in north-western Saudi Arabia, exemplifies the critical interplay between plant biodiversity, climate change, and food security in dryland ecosystems. This chapter synthesizes ecological, ethnobotanical, and agronomic evidence to demonstrate how native and cultivated plant diversity underpins ecosystem services, traditional livelihoods, and climate resilience. Despite receiving less than 90 mm of annual rainfall, AlUla sustains over 60 native medicinal, forage, and wild edible species (*Phoenix dactylifera, Moringa peregrina, Calligonum comosum*), alongside cultivated fruit systems comprising more than 3.1 million date palms and 260,000 citrus trees. These biodiversity-rich agroecosystems enhance soil stability, microclimate regulation, pollination, and nutrient cycling functions increasingly threatened by rising temperatures, erratic precipitation, groundwater depletion, and land-use change. The chapter documents how traditional water management (qanat systems), multi-layered oasis agroforestry, and indigenous knowledge have historically buffered climatic extremes. However, ongoing climate warming (+0.4–0.6 °C per decade in the region) is accelerating habitat degradation, reducing forage availability, and altering the distribution of keystone species. Conservation and restoration initiatives led by the Royal Commission for AlUla (RCU) including native seed banks, protected areas (Sharaan Nature Reserve), and sustainable agriculture programs demonstrate a pathway to reconcile biodiversity conservation with

R. Qureshi (✉)
Department of Botany, Faculty of Sciences, Pir Mehr Ali Shah Arid Agriculture University, Rawalpindi, Pakistan
e-mail: rahmatullahq@uaar.edu.pk

M. M. Abd El-Ghani
Department of Botany and Microbiology, Faculty of Science-Cairo University, Cairo University Herbarium, Giza, Egypt

R. Qureshi (ed.), *Plant Biodiversity*, https://doi.org/10.1007/978-981-92-1471-6_12

food production. The chapter concludes that AlUla serves as a living laboratory for climate-resilient food systems, offering transferable lessons for arid lands globally. Aligning national strategies (Saudi Vision 2030) with international frameworks (CBD, SDGs) and community-based adaptation is essential to safeguard biodiversity as a foundation for long-term food security.

Keywords

AlUla region · Arid ecosystems · Plant biodiversity · Cultural landscape · Climate change · Food security · Oasis agriculture

Abbreviations

FAO	Food and Agriculture Organization
IPBES	Intergovernmental Science-Policy Platform on Biodiversity and Ecosystem Services
IPCC	Intergovernmental Panel on Climate Change
IUCN	International Union for Conservation of Nature
OECD	Organisation for Economic Co-operation and Development
RCU	Royal Commission for AlUla
UNCCD	United Nations Convention to Combat Desertification
UNEP	United Nations Environment Programme
UNESCO	United Nations Educational, Scientific and Cultural Organization

12.1 Introduction

AlUla (also written AlUla or AlUla) is an ancient oasis area in the Madinah Province of the north-western part of Saudi Arabia, which lies between the latitude of 26° 37′ and the longitude of 37° 55′ (Fig. 12.1). The area is characterized by the landscape of a fertile valley measuring almost 20 km between parallel mountain ranges of sandstone, which is the ecological and cultural center of the region (Anonymous 2023a). It is an oasis located in the more comprehensive zone of Hejaz, a region of great historical significance, as it has been associated with the earliest Semitic civilizations, and it has played a significant role throughout the history of the formation of the Islamic culture (Anonymous 2023b; Qureshi et al. 2024).

AlUla region, in general, is almost 8490 km^2 in size and is characterized by arid deserts, impressive cliffs composed of sandstones, volcanic outcrops and mountainous terrain. Although the environment was harsh, the availability of dependable groundwater and seasonal runoff favored the rise of the fruitful oases and made them be able to remain the population of millennia. These natural resources which made AlUla an important station on the ancient Incense Road which linked South Arabia, India, and the Arabian Gulf to the Mediterranean world and Europe. Consequently, AlUla thrived as an economic hub, agricultural land, and a cultural hub, whereas the

Fig. 12.1 Location map of AlUla, Saudi Arabia

long-term factors, including floods, volcanic eruptions, and desertification, gradually defined its unique geomorphology and landscape beauty (Anonymous 2023c; Qureshi et al. 2024).

AlUla is a meeting point of civilizations. Since the seventh century BCE, the area experienced alternate periods of political and cultural might, which started with the Dedanite kingdom, then the Lihyanite state, the Nabataean civilization, and then the Roman rule. One of the inscriptions used in the AlUla oasis is the Harran texts, which document the campaigns of the Babylonian king Nabonidus in northern Arabia in the mid-sixth century BCE (Ghazanfar and Fisher 1998). AlUla became a prominent administrative and business hub and the territory was then governed by the Kingdom of Lihyan between around the fifth and second centuries BCE (Abd El-Ghani et al. 2025).

The Nabataean period saw AlUla being rediscovered and re-prominent, especially with the development of Hegra (modern Mada'in Salih), which was some 22 km north of the oasis. Hegra served as the capital of the southern part of the Nabataean Kingdom, and it was the key to controlling the trade on the Incense Route. The site is almost covered with 100 monumental sunken rock-cut tombs excavated out of sandstone outcrops bearing architectural designs that are more or less similar to those of Petra in Jordan today. Hegra is assumed to have been the southernmost position of Rome in the Arabian region after the Roman conquest of the Nabataean Kingdom in 106 CE (Sawicki 2016).

In the present day, Hegra is considered the first UNESCO World Heritage site in Saudi Arabia and one of the most important archeological landscapes of the area (Osman et al. 2014). Besides its cultural heritage, AlUla is also famous in terms of its ancient traditions of water management, irrigation and land use. Indigenous systems, such as wells, canals, and highly controlled agricultural plots, were managed by communities to cultivate crops for sustaining livelihoods in the arid environment for centuries. Such kind of practices not only supported food production but also contributed to the establishment of a stable and productive oasis environment, which was a result of the adaptive ingenuity of the AlUla inhabitants over successive generations (Burckhardt 2013). The combination of the geography of the region, its history, and the traditional land-use systems underscores the exceptional role of AlUla as a living archive of human-environment interactions in arid landscapes.

12.2 AlUla at the Crossroads of Biodiversity and Food Security

Plant biodiversity supports ecosystem processes that support agriculture and climate control. Plant diversity and wild vegetation enhance the fertility of the soil, pollination, and carbon retention, provide a buffer against pests and insect weather (Benton 2021; Cappelli et al. 2022). On the other hand, contemporary intensive food systems are significant causes of loss of biodiversity and greenhouse gases. As an illustration, about 40–55 percent of the habitable land is already under agricultural use (Boakes et al. 2024; Ellis et al. 2010) and produces approximately 21–37% of anthropogenic emissions of CO2-equivalent. This has caused unprecedented species losses, and the main reason is agriculture (Anonymous 2021; Benton 2021). These responses imply that food systems are responsible for causing and contributing to climate change: the increase in global warming and extreme weather poses challenges to yields and the diversity of plants, and the expansion of agriculture contributes to a total of about ~1 °C of global warming by 2100 (Rosenzweig et al. 2020).

12.3 Arid and Hyper-Arid Regions and Food Security

Drylands such as arid and hyper-arid areas are significant food suppliers which occupy more than 40% of the land on the planet and sustain approximately 2–2.5 billion people and almost half of the global livestock (FAO 2011; UNCCD 2018). Although drylands receive little rainfall, they produce a significant portion of the world's crops (produce almost 44% of the world cereal) and provide staple food items such as dates, grains and pulses. Their agrobiodiversity can consist of distinctive agrobiodiversity—e.g., drought-tolerant landraces and crop wild relatives—which are crucial genetic resources. Nevertheless, climate modeling always reveals that a warming (+ +°C) will have a drastic negative impact on productivity in dry regions. Pasture and livestock productivity is expected to decline in most semi-arid regions of the Middle East and Africa (IPCC 2021; Rojas-Downing et al. 2017). According to the IPCC, food security effects will be faced in the drylands of Africa and Asia. The lack of resources is already severe; therefore, water levels are decreasing and soils are being eroded in most desert oases. With climate change, such as rising heat and aridity, farming innovations, such as water-efficient irrigation, resistant varieties of crops, and agroforestry, are becoming important. It is therefore important to exploit the existing arid areas with more food production instead of changing the natural habitats to ensure security as well as biodiversity (UNCCD 2018; FAO 2011). The arid landscapes should be part of the core of climate-adaptive food systems, where their native endemic vegetation and local expertise can be used to respond to the warmer, drier climate.

12.4 AlUla—A Natural Climatic Laboratory for Resilient Agriculture

AlUla, the mountain oasis in north-western Saudi Arabia, is a natural landscape of an ancient desert-agriculture system that is currently being utilized as a living laboratory to conduct sustainable farming in harsh conditions. It has a rich cultural background of traditional qanat irrigation in which various types of fruits (dates, citrus, figs, etc.) are cultivated. To revive this heritage in a climate-sensitive mode, recently, the Royal Commission for AlUla (RCU 2023) has initiated the cultural revival of a cultural oasis project. RCU, 2023 documented more than three million date palms and dozens of local varieties (Qureshi et al. 2024).

It has been estimated that the water table has been decreased over the past decades and agronomic practices caused the abandonment of 85% of the old farmlands (Anonymous 2025a). To remedy this problem, there is a need to reintroduce local farming for enhancing water governance through integrating modern drip irrigation and not-so-new falaj channels, and restoring soil health (RCU 2023). The native plant seed bank currently contains more than 80 desert species of flora (Anonymous 2025a), providing the genetic diversity to revegetate the area and conserve the heirloom crops. These activities demonstrate the contribution of AlUla toward climate adaptation. Scientists and planners can experiment to understand the performance of desert-adapted crops and agroforestry systems of operation in warmer climates, and to compare conventional and high-technology practices. To illustrate, RCU (2023) is actively involved in the implementation of indigenous knowledge, such as annual date and citrus festivals, organic farming methods and modern science and policy (Sliman 2022). Such kind of experimentation is essential for sustainable oasis agriculture in line with Saudi priorities.

According to Saudi Vision 2030, emphasis has been given toward the use of renewable water in agriculture and the creation of strategic food reserves in connection with the use of soil restoration and environmental agritourism strategies being implemented in AlUla (Anonymous 2025b). Primarily, AlUla represents heritage-based, biodiversity-rich agricultural systems that can lead future food systems in drylands. The region serves as a living laboratory for climate-resilient practices such as polyculture, agroforestry, and indigenous seed conservation that balance agricultural productivity with ecosystem restoration and long-term sustainability (Anonymous 2023e).

In this chapter, the global and local perspectives of plant biodiversity, climate change, and food security are combined and AlUla is offering an example of climate-resilient agriculture. It seeks to: (a) evaluate connections between the diversity of plants and ecosystem well-being, climate change and food production; (b) highlight the strategic significance of arid and hyper-arid areas to future sustainable food systems and (c) evaluate how the past practices of AlUla and future restoration projects can inform climate-sensitive agriculture. The chapter affiliates global scientific frameworks (IPCC, FAO, and UNCCD) to the national and regional plans, especially Saudi Vision 2030 and RCU, 2023 plans, to show the convergence of priorities. It contextualizes AlUla in relation to the larger food security and biodiversity discussion based on the peer-reviewed literature and official reports. Structurally, it deals with biodiversity-food system linkages around the globe, challenges and opportunities of dryland agriculture, AlUla as a natural laboratory of resilient agriculture, and policy and practice implications (Saudi Vision 2030; RCU 2023).

12.5 Agroecological and Geographical Environment of AlUla

12.5.1 Geographic Location and Topographical Landscape

AlUla is located in the northwest of the Kingdom of Saudi Arabia, in Al Madinah Al Munawwarah Province, and marks one of the most ecologically and culturally unique landscapes of the Arabian Peninsula. The geographical boundaries of the area are large sandstone cliffs, separate rocky outcrops and wide desert lowlands. These land formations are cut through by a system of wadi valleys—seasonal watercourses which route the rare rainfall into fertile alluvial soils. The sandstone valleys of AlUla have been used since time immemorial as pathways of moisture, deposition of sediment and planting up, forming the niches where vegetation can survive in a hyper-arid environment. The volcanic harrats (lava fields) in between these valleys make the region more complex in terms of geology, and they provide a favorable environment to support the specific soil type that affects the distribution and variety of plants (Qureshi et al. 2024). This contrast between the soft sediments and wind-eroded rocks and basalt of volcanic origin adds to the edaphic (soil) mosaic of the area, creating a variety of microhabitats that sustain native shrub, herbs, and perennial plants not very common in the rest of the Arabian interior (Edgell 2006; FAO 2018). These dry landscapes contain oasis systems, which are patches of permanent soil moisture and nutrient supply. These oases have been historically and still being the agricultural and settlement centers since the upwelling of the groundwater and the intermittent wadi flow are concentrated to provide support to the orchards of date palms, citrus trees, figs, and other plants adapted to both seasonal extremes (Qureshi et al. 2024; RCU 2023). Vegetation layering, such as the deep-rooted trees in the orchard to understory species in the valley margins, improves microclimates, decreases evapotranspiration, and stabilizes soils (FAO 2018; Qureshi et al. 2024).

12.5.2 Climate Profile and Climate Change Trends

The climate of AlUla is that of a hyper-arid desert, which is characterized by extreme heat, very little precipitation, and extremely high evapotranspiration levels. Summer temperatures tend to be above 40 °C in the day with strong day night variations especially in high and open desert regions, where night temperatures may drop drastically. There is little rainfall, which is also very erratic and annual average rainfall is usually about 20–90 mm. In cases where precipitation does occur, the presence of short high events in the form of a storm is the main contributor to the generation of rapid surface runoff in the form of wadis as well as to the episodic filling of shallow alluvial aquifer (Qureshi et al. 2024; Saudi Meteorological Authority 2021).

The climate in the area of AlUla is also marked with long, hot and dry summers and comparatively short and cool winter, which is typical of the north-western deserts in Arabia. The summer season is characterized by clear skies and prolonged heat, with average maximum temperatures per day often exceeding 34 °C and reaching maximum levels in the late summer. August is the hottest period of the year; whereas, winter months remain dry but much colder, with nights being colder and moderate daytime temperatures. The coldest month is January, when temperatures can even go to single digits at night. By and large, AlUla holds a high average yearly temperature, which highlights the exposure to thermal stress and lack of water as persistent issues in the region (Abd El-Ghani et al. 2025; Burckhardt 2013; Osman et al. 2014).

12.6 Past Interactions of Humans and Plants

The settlement of people in AlUla dates back several thousand years and was mainly influenced by its location on old trade routes, especially, the Incense Road. The availability of stable supply of groundwater, periodic flow of Wadi Rivers and fertile alluvial soils allowed the communities to set up permanent settlements in a rather harsh environment of the desert. In order to support agriculture, residents invented advanced water harvesting and water distribution systems such as primitive aqueducts, diversion channels, and earthworks that could capture and store limited water resources. These ancient irrigation methods show a high level of awareness of environmental limitations and show how human beings adapted to the arid climatic extremes in the long term (Ringrose 1996).

The multi-layered cropping system based on oasis was indeed one of the distinguishing features of traditional agriculture in AlUla. The upper canopy was made of date palms to provide shade and to protect against wind, and under the date palms, cereals, and vegetables were planted and forage crops were planted. This vertical structure controlled extreme temperatures, loss of soil moisture and improved the cycling of nutrients by way of organic litter inputs. These types of systems are more reminiscent of early types of agroforestry, which demonstrates how biodiversity and agronomy were combined in the past to enhance resilience and sustainability during desert periods (FAO 2018).

The archaeobotanical evidence also attests to the extreme pre-eminence of the use and cultivation of plants in AlUla. Carbonized seeds, pollen and phytoliths have been discovered in excavations throughout the region that suggest domestication and control of cereals, legumes and fruit-bearing plant species. These data indicate that organized agrarian cultures existed already in the Bronze Age and developed during the Dadanite, Nabataean and the Islamic age. All of these records demonstrate a dynamic history of human-plants relationships which have been influenced by ecological constraints as well as cultural invention (Khalil 2017; Riehl 2010).

12.6.1 Existing Human Activities and Use of Land

Currently, heritage-based tourism, oasis farming, and conventional pastoralism are the elements that define land use in AlUla. The region has turned to become one of the leading tourist activities mainly due to the superb archaeological and cultural resources such as Hegra (Mada'in Salih) which is a UNESCO World Heritage Site. The Royal Commission for AlUla (RCU 2023) has made tourism one of the key pillars of regional development with significant investments in infrastructure development, with focus on environmental and cultural preservation (Qureshi et al. 2024).

Farming is an important part of the local livelihood due to the growing water shortage and the lack of arable land. The irrigation water is located in oases, and this is where cultivation is maximized. Date palms are also still a staple crop that serves household consumption, as well as commercial markets. Other crops of economic value are the mangoes, citrus fruits, and Arabic moringa, whose popularity can be attributed to increased demand and the efforts put into its cultivation. There were also historical records of the presence of cereals (barley and wheat) and legumes (lentils and peas) in the local systems of agriculture and this is also an indicator of a diversified agrarian foundation that was able to adhere to the seasonal circumstances (Abd El-Ghani et al. 2025; Burckhardt 2013; Ghazanfar and Fisher 1998; Osman et al. 2014; Zohary and Dārôm 1982). Crops have been supplemented with animal husbandry, which has remained the traditional production of livestock (sheep, goats and camels) that offers food, income, and movement to local people. Nevertheless, decreased water supply and a decrease in grazing land have resulted in a slight decrease in the number of herds. Land use is still industrial and has not expanded too much, leaving most of the land with the natural and cultural landscape of the region.

These traditional ways of human actions are being compounded by new types of human activity. Sustainable tourism projects tend to compromise economic development and the safety of natural ecosystems and heritage areas. There is support for renewable energy, especially solar and wind power, to sustain the energy needs as well as to decrease reliance on fossil fuels. Meanwhile, the RCU, 2023-backed research and development initiatives are aimed at conserving water, agricultural innovation, restoring the ecosystem, and reducing the environmental impact of tourism. These changing activities, combined, represent an integrated way of development in which long-term ecological and socio-economic sustainability is given high importance (Abd El-Ghani et al. 2025).

12.6.2 Botanical Exploration

There was systematic botanical work in AlUla starting in the early nineteenth century, with Jean-Louis Burckhardt being the most notable example of this activity, making records of plant species he saw in the area during his travels. The discovery was supplemented by European botanists, such as Eduard Glaser and Adolf Engler, who learned more about the desert plants of the AlUla area, including those species that would flourish in extreme arid conditions (Engler 1895; Khalil 2017). Another study carried out by Zohary and Orshan (1956) explored endemic species and traditional plant utilization and correlated floristic diversity with historical land-use activities. The ecology of AlUla has become a subject of interest in recent decades, and there is an increase in awareness of its importance in the ecological restoration. Modern research has been aimed at determining native species which can be used as landscapes, habitat rehabilitation, and the restoration of the ecosystem. One effort was a notable catalogue of 58 native plant species that were chosen to be included in a restoration and landscape design to provide representation of a wide range of growth forms and niches in deserts, wadis and oasis habitats (Qureshi et al. 2024). The AlUla Oasis being declared a UNESCO World Heritage site has also increased the international recognition of the natural heritage of the area. The current research in botanical studies is now focused on documenting all the diversity of the plants, to understand ecology and to create science-based conservation management in line with the sustainable development objectives.

The history of human habitation in AlUla can be traced back to thousands of years ago in terms of serving as one of the strategic crossroads of ancient trade routes like the Incense Road. Oasis farming became a source of settled settlement with pioneer farmers taking advantage of groundwater, wadi runoff as well as new irrigation methods to practice agriculture in an arid environment. Early types of aqueducts and earthworks that harvest and supply limited amounts of water; traditional water management systems are indicative of advanced adaptation to the extremes of climate in the area (Lightfoot 1996). The traditional oasis agriculture in AlUla was based on a plowing system of multilayers, date palms that were a canopy over cereals and vegetables in the understory. This vertical stratification moderated the temperatures, slowed down soil evaporation, and improved the cyclic circulation of nutrients by input of organic litter. These are demonstrative of the early agroforestry ideas that biodiversity and agriculture are mutually dependent and enhanced to increase resilience and productivity (FAO 2018).

Archaeobotanical records support the antiqueness of the plant-rearing and exploitation in AlUla and other parts of the area. Early agriculture. According to archaeological excavations, carbonized seeds, pollen grains and phytoliths testify to the early domestication of cereals, pulses, and fruit species—showing that the domestication and control of plants were already well organized by the Bronze Age and extended into Nabataean and Islamic eras (Khalil 2017; Tengberg 2012). The records highlight a history of long-term participation in human-plant interactions with local biodiversity of plants, both influenced by environmental barriers and cultural innovation.

The traditional ecological knowledge has been transmitted across generations, related to planting calendar, seed picking, and water gathering, which have created a cultural memory of the plant-environment interaction, which is inherent to the explanation of the agroecological development of AlUla. Modern studies, including (Qureshi et al. 2024) emphasized the role of such knowledge, combined with scientific restoration, in biodiversity restoration and sustainable land use in the conditions of a changing climate and the continuity of past activities and future strategies of adaptation.

12.7 Habitat-Specific Distribution and Ecosystem Types in AlUla

The distribution of plant species in AlUla is largely associated with the heterogeneity of the habitat regulated by the landform, soil texture, availability of moisture, and variation of the microclimatic conditions. Wadis, rocky escarpments (Fig. 12.2), agricultural terraces, and sandy desert plains have distinct plant assemblages that lead to unique ecological conditions that result in the composition of the species and functional diversity (Ghazanfar and Fisher 1998; Mandaville 2013).

Wadis and ephemeral drainage channels are the most moisture-preferable habitats in the area and they sustain rather thick vegetation in comparison to the rest of the landscape (Fakhry and Anazi 2017). These are habitats of deep-rooted trees and bushes like *Acacia tortilis* and *Tamarix nilotica* (Fig. 12.3), which are adjusted to episodic flooding and shallow groundwater supply. Wadis also serve as important

Fig. 12.2 Natural landscape of AlUla. (Photo by Prof. Dr. Rahmatullah Qureshi)

Fig. 12.3 *Tamarix nilotica* in AlUla. (Photo credit by Prof. Dr. Rahmatullah Qureshi)

ecological corridors in the hyper-arid environment since seasonally occurring runoff contributes to seed dispersal, soil nutrient replenishment and short-term alluvial aquifer recharge (Chaudhary 1999; Vincent 2008).

The rocky slopes and sandstone shelves carry a separate flora of mainly drought-resistant shrubs and perennial herbs that have evolved on shallow skeletal soils, intense sun radiation, and excessive changes in temperature. These habitats have frequently served as refugia of stress-tolerant and locally endangered species remaining in the crevices of the rocks, in shade slopes, and in the runoff accumulation habitats. These microhabitats play an unreasonable role in terms of plant species richness in the region as compared to their size (Mandaville 2013; Zohary and Dārôm 1982).

Agricultural terraces and native and cultivated date-palm groves are semi-natural environments that have the coexistence of both native and cultivated plants. These controlled natural ecosystems serve as bridges between the natural vegetation and human food production. The agroforestry system that has been established over a long period of time and is dominated by *Phoenix dactylifera* increases the complexity of the habitat (Fig. 12.4), understory crops and wild species and maintains livelihood through conservation of biodiversity at the farm (FAO 2018; Ghazanfar and Fisher 1998).

The geological composition of the area around the oasis consists of extensive plains of sandy deserts with dune systems, which are low on fertility of soils, ability to retain moisture, and susceptible to wind erosion. The vegetation cover of such places is low and is composed of xerophytic grasses and shrubs like *Haloxylon*

Fig. 12.4 Date palm grooves on the foothills of AlUla. (Photo by Prof. Dr. Rahmatullah Qureshi)

persicum, Retama raetam, and *Lycium shawii*. Even though the levels of the species richness are rather low, the utilization of the specified plant communities contributes to the stabilization of the soil, the feeding of pastoral systems, and the preservation of the desert ecosystem functioning (Chaudhary 1999; Mandaville 2013).

All these types of habitats together create a composite ecological mosaic, which is the foundation of the biodiversity value of AlUla. The habitat diversity supports the source of food, shelter, and breeding places of flora and fauna, further strengthening the ecological and cultural hotspot of the north-western region of Saudi Arabia (Charloux et al. 2022; Ghazanfar and Fisher 1998). Localized refugia of moisture-dependent woody species occur along wadis and ephemeral drainage channels, such as *Acacia tortilis* and *Tamarix nilotica*, on gains of episodic runoff and shallow alluvial aquifers. The rocky escarpments and mountainside slopes carry a unique collection of drought-tolerant shrubs and localized endemics that are able to survive on shallow soils as well as extreme thermal conditions. Sandy desert plains, on the other hand, are characterized by the predominance of xerophytic grasses and shrubs, which are very important in soil stabilization and seasonal forage to pastoral systems. Transitional areas are constituted by traditional agricultural terraces and date-palm groves, where indigenous vegetation lives in coexistence with the crops and other species that have been cultivated since ancient times, connecting agrobiodiversity and ancient food production systems and conservation values (Hausleiter et al. 2021).

12.7.1 Ecosystems of the AlUla Region

There are two major types of ecosystems in the AlUla region, viz., (1) the desert ecosystem and (2) the oasis ecosystem. These systems form a mosaic of habitats together that maintain high ecological heterogeneity and provide plant and animal communities with high diversity, making AlUla a significant ecological hot spot in north-western Arabia (Abd El-Ghani et al. 2025).

12.7.1.1 The Desert Ecosystem

AlUla is characterized by vast arid plains in the desert, mountainous sandstone formations, and mobile sand dune complexes that illustrate the ecosystem of the AlUla region. The vegetation cover is low density and highly specialized due to the low rainfall, high rates of evaporation, and low nutrient content substrates (Bornkamm and Kehl 1990; Ghazanfar and Fisher 1998). The area has several types of dunes, with the most common ones being barchans and transverse dunes and relatively rare ones that are the star and parabolic dunes. Wind regimes, the supply of sands, and topography regulate their distribution and morphology, and thus, the landscape is very dynamic. The floristic composition of this ecosystem is represented by some hard woody and shrub taxa, including *Acacia gerrardii* subsp. *negevenensis, A. raddiana, Capparis cartilaginea, Haloxylon persicum, Lycium shawii, Maerua crassifolia, Moringa peregrina, Retama raetam,* and *Ziziphus spina-christi*. This conservation effort, such as the Sharaan Nature Reserve, is meant to protect these desert environments and the wildlife found within them to ensure that a greater value is being placed on them as both ecologically and culturally significant environments (Boakes et al. 2024).

12.7.1.2 The Oasis Ecosystem

Located in the middle of the desert environment is the AlUla oasis, a very productive agroecological system supported by groundwater resources. The oasis is covered with massive plantations of date palm (*Phoenix dactylifera* L.), between which there are smaller areas with crops, and an escarpment surrounds the oasis. This system has a heavy reliance on fossil groundwater which is obtained through springs and shallow wells to support vegetation and agriculture (Abd El-Ghani 1998; Bornkamm and Kehl 1990).

Water management in AlUla was centuries old and was based on a complex qanat-based irrigation system that consisted of wells, canals and gravity-based channels. The role of springs like Ain Tidal to provide water to the old town and farmlands was a central one in the past. Usually, 10–20 m deep wells were created to hold large amounts of water, and operated through either animal-powered systems or simple pumps (Osman et al. 2014). The canal systems spread water evenly in palm groves and arable lands and averted inequalities and reduced losses (Engler 1895).

Over the past decades, some of the traditional qanat systems have been deserted or destroyed in the process of modernizing and restructuring of cities, where motorized pumping has taken the place of gravity-based irrigation. In spite of this change, the oasis irrigation heritage of AlUla has cultural values and has been taken into the UNESCO list as a heritage of the iconic date-palm cultivation cultures of the Arab world (Zohary and Orshan 1956). Other arid areas such as the Sinai Peninsula and the Western desert, Egypt, have also been reported to have used similar oasis irrigation systems, highlighting the applicability of the traditional water management methods used in AlUla (Abd El-Ghani and Fawzy 2006).

12.8 Plant Biodiversity at AlUla and Food Security

12.8.1 Floristic Diversity and Vegetation Types

The heterogeneous topographies of AlUla are the reason behind the presence of a great deal of plant species even under its arid to hyper-arid climate due to the diverse canyons of sandstone as well as volcanic harrats, alluvial plains, and oasis systems. These diverse land formations harbor different microhabitats which carry desert shrubs, annual ephemeral, perennial grasses, and planted fruit trees. Floristic surveys have reported a large percentage of drought-tolerant taxa of the families Poaceae, Fabaceae, Amaranthaceae, Zygophyllaceae, and Asteraceae which are significant in stabilizing the soils and supporting food and fodder regimes in drylands (Abd El-Ghani et al. 2025).

In AlUla, vegetation may be broadly categorized into such agroecosystems as oasis, wadi, rocky flora of the escarpment, and volcanic plateau. The oasis vegetation consists of date palm (*Phoenix dactylifera*), subordinate crops, and the related fruit trees, creating multi-layered agroforestry systems that increase the productivity and control of the microclimate. Wadis encourage relatively more species richness owing to the episodic water supply, whereas harrat ecosystems are typified by stress-tolerant shrubs and grasses, which are adapted to shallow soil and extreme temperatures (RCU 2023; Abid et al. 2025; Edgell 2006).

12.9 Wild Plant Genetic Resources

Wild edible plants form a significant, but underutilized, element of food security in AlUla, especially when there is climatic stress. Some indigenous species have edible fruit, seed, leaf, or tuber that have been used to supplement local products in the past (Table 12.1). Being highly adapted to water stress and adverse soils, these plants tend to be an excellent source of micronutrients and bioactive compounds and will be a valuable source of genetic material in the future food system resilient to climate change (FAO 2017).

12.9.1 Medicinal, Wild Food, and Forage Flora

The variety of medicinal, wild edible, and forage plant species in AlUla region is not only a source of traditional knowledge and cultural heritage but also a collection of biological resources that is important in climate adaptation and management of the

Table 12.1 List of plant species having medicinal and other economic benefits to be used in food security

S. No.	Species	Family	Arabic name	Habit	Traditional uses	References
1	*Abutilon fruticosum* Guill. & Perr.	Malvaceae	Shrubby Indian Mallow (Al-Hatmi)	Perennial Herb	The decoction of leaves and flowers is used to treat respiratory disease, especially cough and mild cold, due to their demulcent action. Stem fiber are used as cordage and the plant is used as supplementary forage when there is a shortage of fodder.	Gouda et al. (2022) and Hassan et al. (2021)
2	*Aerva javanica* (Burm.f.) Juss. ex Schult.	Amaranthaceae	Kapok Bush (Al-Tarfh)	Shrub	The inflorescences have silky floss with a traditional use as stuffing material. Aerial parts are topically used in minor wounds and dermatological conditions medicinally. In the ecological aspect, the species aids in stabilizing the soil and provides browse to livestock in arid rangelands.	Phondani et al. (2015) and Suleiman (2019)
3	*Aizoon canariense* L.	Aizoaceae	Canary Aizoon (Al-Daa'a)	Perennial Herb	The young leaves are eaten as a leafy vegetable and are eaten raw or slightly cooked. The plant is customarily considered to be cooling and soothing, and it serves as an understorey to minimize soil erosion, and it has low grazing value.	El-Amier and Al-Hadithy (2020) and Yonbawi et al. (2021)

(continued)

Table 12.1 (continued)

S. No.	Species	Family	Arabic name	Habit	Traditional uses	References
4	*Anthemis deserti* Boiss.	Asteraceae	Desert Chamomile (Oqhwan)	Annual Herb	Flower and leaf infusions have also been extensively used to treat gastrointestinal irritability and non-inflammatory fevers of low grade. The species is fragrant and attracts pollinators and its smell is similar to that of chamomile, as well as its medicinal use.	Ali (2019) and Chemsa et al. (2018)
5	*Arnebia hispidissima* (Sieber ex Lehm.) A.DC.	Boraginaceae	Prophet's Flower (Al-Mulleh)	Annual Herb	Its natural red dye found in roots is used in textiles and cosmetics. Besides, root extracts are used in classic dermatological preparations though contacts should be wary as it may cause skin irritation.	Dulta et al. (2023) and Mseddi et al. (2021)
6	*Artemisia monosperma* Delile	Asteraceae	White Wormwood (Al-Ather)	Subshrub	The aromatic aerial parts are traditionally used to treat digestive disorders, respiratory infections, and also to fumigate or cleanse. Livestock also at times browses on the plant in the event of limited forage.	Alshahri et al. (2022) and Elshibani et al. (n.d.)
7	*Artemisia scoparia* Waldst. & Kit.	Asteraceae	Red stem Wormwood (Ash-Sheeh Al-Ahmar)	Annual Herb	The plant is used as a stimulant to digestion, as an antiseptic and is also used in incense and aromatic preparations. It is highly bioactive and is therefore normally used in low doses.	Aati et al. (2020) and El-Seedi et al. (2022)

8	*Asphodelus fistulosus* L.	Xanthorrhoeaceae	Hollow Asphodel (Al-Brooq)	Perennial Herb	The paste of bulb is applied externally on skin ailments. Scarcely grazed by the livestock	Mohammed et al. (2025) and Mseddi et al. (2021)
9	*Asteriscus graveolens* (Forssk.) less.	Asteraceae	Arabian Starwort (Al-Niqidan)	Perennial Herb	The decoction of aerial parts are used to respiratory and digestive ailments. The essential oil of the plant is also used in perfumery and incense with low forage value.	Amtaghri et al. (2025) and Youssef (2013)
10	*Bassia eriophora* (Schrad.) Asch.	Amaranthaceae	Woolly Bassia (Al-Qattinah)	Annual Herb	It is mostly a forage plant, and it is appreciated due to saline and degraded soil tolerance. There are few medicinal uses, which are essentially extrinsic.	Al Ramahi and Al Bahadly (2020) and Grabowska et al. (2023)
11	*Cakile arabica* Velen. & Bornm.	Brassicaceae	Arabian Sea Rocket (Al-Salih)	Annual Herb	Leaves are also sometimes eaten by young ones and the traditional cure applies to minor digestive problems with the use of the plant. It is significant in the stabilization of sand in the coastal and desert environments.	Amin and Wafa'a (2017) and Wafa'a (2022)
12	*Calendula tripterocarpa* Rupr.	Asteraceae	Marigold (Al-Hanwah)	Annual Herb	Paste of flowers is applied as a topical agent on wounds, inflammation, and irritation of the skin. The species is also used as an ornament and an ecological activity, as it aids in pollination.	Al-Rifai (2018) and Mohammed-Ibtisam and Doka (2018)

(continued)

Table 12.1 (continued)

S. No.	Species	Family	Arabic name	Habit	Traditional uses	References
13	*Calligonum comosum* L'Hér	Polygonaceae	Arta (Al-Arta)	Shrub	Flowers are traditionally used as a potherb. Roots and stems are used in digestive problems. It plays a very important role ecologically in stabilizing dunes and as a source of fuelwood and shelter.	Al-Otaibi (2015)
14	*Capparis cartilaginea* Decne.	Capparaceae	Caper Bush (Al-Shaflah Al-Jabali)	Climber	The fruits and flower buds are eaten as condiments. Medically, different sections are applied in musculoskeletal pain, urinary and digestive issues.	Al-Goufi and Sonbol (2018) and Moharram et al. (2018)
15	*Capparis spinosa* L.	Capparaceae	Spiny Caper (Al-Shaflah)	Liana	It is commonly used in food and traditional medicine, especially against liver, digestive and anti-inflammatory effects. The species also helps in stabilizing the slope of the rocky slopes.	Aljawdah et al. (2025) and Kdimy et al. (2022)
16	*Cenchrus ciliaris* L.	Poaceae	Buffel Grass (Al-Sabat)	Perennial Herb	It is a very desirable perennial grass that is nutritious and can be used as forage, rangeland rehabilitation, erosion control, and soil conservation.	Qureshi and Bhatti (2005)
17	*Cenchrus divisus* (J.F. Gmel.) Verloove, Govaerts & Buttler	Poaceae	Desert Grass (Al-Halfa)	Perennial Herb	It is mainly used as fodder to cattle and to control sandy soils in dry areas.	Hussain et al. (2021) and Osman et al. (2024)

18	*Chrozophora tinctoria* (L.) A. Juss.	Euphorbiaceae	Dyer's Croton (Al-Tanoom)	Perennial Herb	It was used in the past as a source of blue dye. External use of parts in folk medicine is used in skin diseases.	Alnadif et al. (2022) and Dalia and AlQahtani 2022)
19	*Citrullus colocynthis* (L.) Schrad.	Cucurbitaceae		Subshrub	The fruit is used in constipation,bleedingpiles, diabetes and leucorrhea	El-Absy (2022) and Qureshi and Bhatti (2008)
20	*Citrus aurantiifolia* (Christm) Swingle	Rutaceae	Lemon	Tree	The fruit is used as an analgesic, immune modulator, anti-diarrheal, antiulcer, and anthelmintic.	Izah et al. (2023)
21	*Cleome droserifolia* (Forssk.) Delile	Cleomaceae	Samwa	Perennial herb	The decoction of whole plant is given to treat diabetes.	Moustafa et al. (2019)
22	*Convolvulus spicatus* Peter ex Hallier f.	Convolvulaceae	Spicate Bindweed (Al-Rakhama)	Perennial Herb	As a mild laxative, roots and leaves are taken. The plant is also used to protect the soil and as nectar to the pollinators.	Alsobeai (2024)
23	*Cucumis prophetarum L.*	Cucurbitaceae	Wild Cucumber (Al-Shari)	Herb	The fruit and roots are used as part of a strong purgative and dermatological treatment. The use is limited to experienced traditional practitioners due to its toxicity.	Alsayari (2014) and Shabana et al. (2021)
24	*Dodonaea viscosa* subsp. *angustifolia* (L.f.) J.G. West	Sapindaceae	Sand Olive (Al-Shath)	Tree	The use of leaves as antiseptics to the leaf is topical. The shrub is a source of fuelwood, material of tools and is a good windbreak.	Al-Namazi et al. (2020) and Hossain (2019)
25	*Eremobium aegyptiacum* (Spreng.) Asch. ex Boiss.	Brassicaceae	Egyptian Desert Cress (Al-Tarrbah)	Perennial Herb	Digestion is usually stimulated with the help of leaves and seeds. It is a well-adapted species to arid areas and can be used as minor forage.	Al-Sodany et al. (2025), Alshahri et al. (2022) and Qari et al. (2021)

(continued)

Table 12.1 (continued)

S. No.	Species	Family	Arabic name	Habit	Traditional uses	References
26	*Farsetia burtoniae* Oliv.	Brassicaceae	Burton's Farsetia (Al-Maqar)	Perennial Herb	The plant is used as forage and well adopted to desert environment	Ansari et al. (2022) and Moawed and Ansari (2015)
27	*Ferula sinaica* Boiss.	Apiaceae	Sinai Ferula (Al-Kalakh)	Perennial Herb	Gives a resin that is comparable to asafoetida, which is taken in small amounts to aid digestion and inflammatory conditions.	Alansari and Seleem (2024) and Alzahrani et al. (2026)
28	*Ficus palmata* Forssk. subsp. *palmata*	Moraceae	Fig (Al-Teen Al-Barri)	Tree	The fruits are consumed raw, whereas the latex and the leaves are utilized in a topical dressing. The tree can be used as a source of shade and environmental benefits.	Ahmed et al. (2022) and Alouffi et al. 2024)
29	*Ficus salicifolia* Vahl	Moraceae	Willow-leaf Fig (Al-Athab)	Tree	The fruit is edible. The plant is grown as shade tree	Alwahibi et al. (2017)
30	*Gomphocarpus sinaicus Boiss.*	Apocynaceae	Sinai Milkweed (Al-Ushr AlSinai)	Perennial Herb	Latex is applied externally over the skin conditions and seed floss is utilized as stuffing material. It is used internally because it is toxic.	Al-Sodany et al. (2025) and Al Masoudi (2024)
31	*Haloxylon persicum* Bunge	Amaranthaceae	White Saxaul (Al-Ghadha)	Shrub	One of the significant trees of the desert is employed to produce fuelwood and charcoal. Young shoots offer camel fodder in time of need and the species contributes significantly to soil stability.	Alghanem (2018)
32	*Haloxylon persicum* Bunge	Chenopodiaceae	White Saskaul (Ghada)	Shrub	Browsed by the livestock. The plant is sand dune stabilizer. Branches are used as fuelwood source	Elmefregy and El-Sheikh (2020) and Kürschner and Neef (2018)

33	*Haplophyllum tubeRCU, 2023latum* (Forssk.) A. Juss.	Rutaceae		Shrub	Juice obtained from leaves is used for strengthening weak muscles. It is also used in headache, arthritis, abdominal flatulence, and muscle spasms.	Elshibani et al. (2024) and Hamdi et al. (2018)
34	*Hyparrhenia hirta* (L.) Stapf	Poaceae	Common Thatching Grass (Al-Sanfi)	Perennial Herb	Fruits are eaten either raw or made into drinks. Mat, basket and rope weaving are done using leaves and fibers.	Al-Robai et al. (2019) and Osman et al. (2014)
35	*Hyphaene thebaica* (L.) Mart.	Arecaceae	Doum Palm (Doum)	Tree	The fruit is edible and used in beverages, seeds used in recreational item and prayer beads. Leaves are woven to make mats, basket and ropes. The wood is used as fuel source.	Aboshora et al. (2014)
36	*Kickxia pseudoscoparia* V. W.Sm. & D.A. Sutton	Plantaginaceae	False Scoparia (Safeeirah)	Perennial Herb	Traditional uses limited; may have medicinal properties but not well documented; grazed by livestock; found in sandy and rocky habitats	Al-Robai et al. (2017)
37	*Lavandula coronopifolia* Poir.	Lamiaceae	Lavender (Al-Wathahah)	Perennial Herb	Herbal teas, perfumery and headache remedies are made out of aromatic parts. Flowers are useful as a source of nectar	Ahmad et al. (2022) and Hassan et al. (2014)
38	*Lavandula pubescens* Decne.	Lamiaceae	Downy Lavender (Al-Khuzama)	Perennial Herb	It possesses the same use like previous species. Essesntial oil is extracted for medicinal use such as respiratory problems. Support honey bee browsing, ornamental.	Adgaba et al. (2015) and Ansari et al. (2014)
39	*Leptadenia pyrotechnica* (Forssk.) Decne.	Apocynaceae	Khipp	Shrub	Pods are cooked as vegetable. Stem is woven to make ropes and used in thatching of conical huts. Plant is used as fuelwood sources	Qureshi et al. (2024)

(continued)

Table 12.1 (continued)

S. No.	Species	Family	Arabic name	Habit	Traditional uses	References
40	*Lycium shawii* Roem. & Schult	Solanaceae	Desert Thorn (Al-Awsaj)	Shrub	Berries are also good to eat and they are also used in folk medicine as energy providers and metabolites. The shrub has a great deal of tolerance to arid conditions.	Alruways and Khan (2022) and Asiri et al. (2026)
41	*Maerua crassifolia* Forssk.	Capparaceae	Bushveld Shepherd's Tree (Al-Hamr)	Tree	Leaves are browsed by the livestock. The decoction of leaves and roots is used to treat digestive problem and as diuretic.	Saad and Asalam (2015) and Elaidarous et al. 2022)
42	*Moltkiopsis ciliata* (Forssk.) I.M. Johnst	Boraginaceae	Hairy Moltkiopsis (Al-Halamahl)	Perennial Herb	Fruit is used as purgative agent. Externally applied for parasites.	Al-Turki et al. (2000) and Alyemeni (2000)
43	*Moringa peregrina* (Forssk.) Fiori	Moringaceae	Moringa (Al-Baan Al-Arabi)	Tree	He edible and cosmetic oil extracted is of high quality and is obtained by the use of seeds. The leaves and the pods are also nutritious and have been used as food and medicines.	Al-Khalifah and Shanavaskhan (2015) and Alaklabi (2015)
44	*Ochradenus baccatus* Delile	Resedaceae	Sweet Mignonette (Al-Ghardaq)	Subshrub	The plant is grazed by the livestock. Fruits are consumed by birds and human also used as medicine as anti-inflammatory agent.	Al-Bakre (2025) and Petrie (2007)
45	*Pergularia tomentosa* L.	Apocynaceae	Hairy Periploca (Al-Ghalkah)	Liana	Latex used in traditional medicine for skin diseases and wound healing; leaves used as poultices; fibers from stems used for making ropes; grazed by livestock with caution due to potential toxicity.	Ads et al. (2021) and Rayyan et al. (2018)

46	*Phoenix dactylifera* L.	Arecaceae	Date Palm (Nakheel Al-Balah)	Tree	Dates are a popular food and weak medicinal supply. The palm is used to make almost everything: Palm tree is used in the construction, crafts, and sap for making traditional therapies.	Al-Khalifah and Askari (2003) and Assirey (2015)
47	*Pulicaria incisa* (Lam.) DC.	Asteraceae	Incised Fleabane (Al-Rabbal)	Perennial Herb	This aromatic plant is used in incense, and traditionally used as anti-inflammatory and respiratory ailments. Grazed by the livestock.	El-Said et al. (2021) and Mohammed et al. (2021)
48	*Retama raetam* (Forssk.) Webb & Berthel.	Fabaceae	White Broom (Ar-Ratam)	Shrub	This is an aromatic plant and flowers are used in perfumery. Externally applied for skin diseases and it is also providing sahe in drier regions.	Al-Sharari et al. (2020)
49	*Rumex vesicarius* L.	Polygonaceae	Bladder Dock (Al-Humayd)	Annual Herb	Leaves consumed as a leafy vegetable with sour taste; used in salads and soups; traditional medicine uses for detoxification and digestive aid; rich in vitamins and minerals; grazed by livestock.	Gomaa and Saleh (2014) and Sulieman et al. (2024)
50	*Salvadora persica* L.	Salvadoraceae	Toothbrush Tree (Al-Araak)	Tree	Fruits are edible. Due to antimicrobial properties of twigs, these are traditionally used to clean the mouth. Digestive and dental problems are treated with leaves and roots	Sher et al. (2010) and Sher et al. (2011)
51	*Searsia tripartita* (Ucria) Moffett	Anacardiaceae	Sumac (Al-Artan)	Shrub	The plant is culinary in nature and used as a spice (sumac). Medicinally it is used in digestive problems. Tannin	Ghareb and Gabr (2017) and Llewellyn et al. (2011)

(continued)

Table 12.1 (continued)

S. No.	Species	Family	Arabic name	Habit	Traditional uses	References
					extracted from the plant is used for leather.dye.	
52	*Senna italica* Mill.	Fabaceae	Italian Senna (Al-Eshreq)	Subshrub	Leaves are used as laxative agent	Alshallali (n.d.) and Alswat et al. (2023)
53	*Solenostemma arghel* (Delile) Hayne	Apocynaceae	Argel (Al-Hargal)	Perennial Herb	The dried parts are used in incense. The aerial parts are used as herbal tea for treating kidney and liver problems as well as respiratory issues.	Dhshan et al. (2025) and Soliman et al. (2022)
54	*Stipagrostis ciliata* (Desf.) De Winter	Poaceae	Feather Grass (Al-Saliyan)	Perennial Herb	Aerial parts are grazed by the livestock, used in sand dune stabilization.	Al-Homaid et al. (1990) and Sayed and Masrahi (2025)
55	*Tamarix nilotica* (Ehrenb.) Bunge	Tamaricaceae	Nile Tamarisk (Al-Athal)	Tree	The decoction of bark and leaves is used in digestive problems. Wood is used in construction material, fuelwood as well as windbreaker and soil stabilization	Al-Harthi (2025) and Arafah (2015)
56	*Tephrosia purpurea* subsp. *apollinea* (Delile) Hosni & El Karemy	Fabaceae	Purple Tephrosia (Al-Tiqtaq)	Perennial Herb	The aerial parts are used in respirtaory ailments and as an anti-inflammatory agent.	Al-Ghamdi (2013) and Alhashim et al. (n.d.)
57	*Tribulus terrestris* L.	Zygophyllaceae	Puncture Vine (Al-Qataf)	Annual Herb	Fruits are made into powder and traditionally used in the treatment of urogenital, cardiovascular, and metabolic diseases, especially as a tonic and diuretic. The plant is grazed by the livestock	Abdel-Kader et al. (2016) and Al-Hemaid and Thomas (1996)
58	*Trigonella stellata* Forssk.	Fabaceae	Star Fenugreek (Al-Naffl)	Annual Herb	The aerial parts are used as condiment and cooking for digestion purpose as fenugreek.	El-Ghanim et al. (2010) and Sheweita et al. (2020)

59	*Vachellia gerrardi* var. *najdensis* (Chaudhary) Ragup, Seigler, Ebinger & Maslin	Fabaceae	Najd Acacia (As-Sunt Al-Najdi)	Tree	Pods and leaves are browsed by the livestock. Wood is used as fuelwood source and tool making.	Ragupathy et al. (2014) and Alabdli et al. (2025)
60	*Vachellia tortilis* (Forssk.) Galasso & Banfi	Fabaceae	Umbrella Thorn Acacia (Al-Sammar)	Tree	Leaves and pods are browsed by the livestock. Wood is used as fuelwood and charcoal formation. gum is used in medicine. Bark and leaves are used for treating colds and skin ailments	Elfeel et al. (2014)
61	*Vachellia tortilis* subsp. *raddiana* (Savi) Kyal. & Boatwr.	Fabaceae	Raddiana Acacia (Al-Sayyal)	Tree	Same used as above species	Suleiman et al. (2018) and Vincent et al. (2024)
62	*Withania somnifera* (L.) Dunal	Solanaceae	Ashwagandha (Al-Abaab)	Shrub	Roots are taken as an adaptogenic tonic to increase stress resistance, vitality and general physiological balance	Alfaifi et al. (2016) and Alwadie (2002)
63	*Ziziphus nummularia* (Burm. f.) Wight & Arn.	Rhamnaceae	Jujube (Al-Sider Al-Barri)	Shrub	The fruits are edible and the leaves are fodder. The fruits are used in traditional medicine as digestive and skin related illnesses	Alqarni et al. (2021) and Zerrouk et al. (2021)
64	*Ziziphus spina-christi* (L.) Desf.	Rhamnaceae	Christ's Thorn Jujube (Al-Sidr)	Tree	Fruits are eaten fresh or dried. Stem, leaves and hair care leaves serve in skin and hair care, and flowers form an important source of honey, of high quality.	Alansi et al. (2016) and Almalki and Alzahrani 2018)
65	*Zygophyllum indicum* (Burm. f.) Christenh & Byng	Zygophyllaceae		Subshrub	The decoction of plant is given orally to teat skin itching/ irritation. A bathe of plant is taken for the same purpose acting as blood purifier	Qureshi et al. (2010) and Alzahrani and Albokhari (2018)

Table 12.2 Pictorial view of plant species of the AlUla Region

	*Abutilon fruticosum* Guill. & Perr
	*Aerva javanica* (Burm.f.) Juss. ex Schult
	*Aizoon canariense* L
	*Anthemis deserti* Boiss

(continued)

Table 12.2 (continued)

	*Arnebia hispidissima* (Sieber ex Lehm.) A.DC
	*Artemisia monosperma* Delile
	*Artemisia scoparia* Waldst. & Kit
	*Asphodelus fistulosus* L

(continued)

Table 12.2 (continued)

	*Asteriscus graveolens* (Forssk.) Less
	*Bassia eriophora* (Schrad.) Asch
	*Cakile arabica* Velen. & Bornm
	*Calendula tripterocarpa* Rupr

(continued)

Table 12.2 (continued)

	*Calligonum comosum* L'Hér
	*Capparis cartilaginea* Decne
	*Capparis spinosa* L
	*Cenchrus ciliaris* L

(continued)

Table 12.2 (continued)

	*Cenchrus divisus* (J.F. Gmel.) Verloove, Govaerts & Buttler
	*Chrozophora tinctoria* (L.) A. Juss
	*Citrullus colocynthis* (L.) Schrad
	*Citrus aurantiifolia* (Christm) Swingle

(continued)

Table 12.2 (continued)

	*Cleome droserifolia* (Forssk.) Delile
	*Convolvulus spicatus* Peter ex Hallier f
	*Cucumis prophetarum* L
	*Dodonaea viscosa* subsp. *angustifolia* (L.f.) J.G. West

(continued)

Table 12.2 (continued)

	*Eremobium aegyptiacum* (Spreng.) Asch. ex Boiss
	*Farsetia burtoniae* Oliv
	*Ferula sinaica* Boiss
	*Ficus palmata* Forssk. subsp. *palmata*

(continued)

Table 12.2 (continued)

	*Ficus salicifolia* Vahl
	*Gomphocarpus sinaicus Boiss*
	*Haloxylon persicum* Bunge
	*Haplophyllum tubeRCU, 2023latum* (Forssk.) A. Juss

(continued)

Table 12.2 (continued)

	*Hyparrhenia hirta* (L.) Stapf
	*Hyphaene thebaica* (L.) Mart
	*Kickxia pseudoscoparia* V.W.Sm. & D.A. Sutton
	*Lavandula coronopifolia* Poir

(continued)

Table 12.2 (continued)

	*Lavandula pubescens* Decne
	*Leptadenia pyrotechnica* (Forssk.) Decne
	*Lycium shawii* Roem. & Schult
	*Maerua crassifolia* Forssk

(continued)

Table 12.2 (continued)

	*Moltkiopsis ciliata* (Forssk.) I.M. Johnst
	*Moringa peregrina* (Forssk.) Fiori
	*Ochradenus baccatus* Delile
	*Pergularia tomentosa* L

(continued)

Table 12.2 (continued)

	*Phoenix dactylifera* L
	*Pulicaria incisa* (Lam.) DC
	*Retama raetam* (Forssk.) Webb & Berthel
	*Rumex vesicarius* L

(continued)

Table 12.2 (continued)

	*Salvadora persica* L
	*Searsia tripartita* (Ucria) Moffett
	*Senna italica* Mill
	*Solenostemma arghel* (Delile) Hayne
	*Stipagrostis ciliata* (Desf.) De Winter

(continued)

Table 12.2 (continued)

	*Tamarix nilotica* (Ehrenb.) Bunge
	*Tephrosia purpurea* subsp. *apollinea* (Delile) Hosni & El Karemy
	*Tribulus terrestris* L
	*Trigonella stellata* Forssk

(continued)

Table 12.2 (continued)

	*Vachellia gerrardi* var. *najdensis* (Chaudhary) Ragup, Seigler, Ebinger & Maslin
	*Vachellia tortilis* (Forssk.) Galasso & Banfi
	*Withania somnifera* (L.) Dunal
	*Ziziphus nummularia* (Burm.f.) Wight & Arn

(continued)

Table 12.2 (continued)

	*Ziziphus spina-christi* (L.) Desf
	*Zygophyllum indicum* (Burm.f.) Wight & Arn

landscape on sustainable grounds. New information about the spatial distribution, relationships in habitats and functional roles of the plant diversity within the AlUla Oasis has been offered in recent literature, such as a thorough study by (Abd El-Ghani et al. 2025; Qureshi et al. 2024). Their work is useful to explain how particular assemblage of habitats promotes resilience during climatic stress as it consolidates the functional importance of desert biodiversity that was earlier reported in the ethnobotanical and ecological literature (Tables 12.1 and 12.2).

12.9.2 Medicinal Flora and Climatic Stress Health Resilience

AlUla records medicinal species, including *Artemisia* spp., *Lavandula coronopifolia, Salvadora persica*, and *Withania somnifera,* which are physiologically adjusted to thermal and water stress extremes (Table 12.3). (Abd El-Ghani et al. 2025) also describe how these species are closely connected to certain microhabitats (oasis gradients, rocky escarpments, and wadis) that moderate the climatic extremes and ensure the availability of moisture within a local environment. These modifications are consistent with more general observations of arid environments where drought-resistant medicinal plants tend to increase the concentration of secondary metabolites when subjected to climatic stress and may increase their pharmacological performance (Selmar and Kleinwächter 2013).

Table 12.3 Medicinal, wild edible, and genetic resource plants of AlUla Region with conservation and policy relevance

Botanical name	Family	Local/ common name	Plant part used	Traditional medicinal uses	Wild edible/ genetic resource value	IUCN status[a]	Conservation priority	Key references
Ephedra alata Decne.	Ephedraceae	Desert joint-fir	Stems	Respiratory ailments, asthma, and ciRCU, 2023latory disorders	Medicinal genetic resource adapted to extreme aridity	LC	Medium	Alshaqhaa et al. (2025)
Commiphora kua (Royle) Vollesen	Burseraceae	Arabian myrrh	Resin, bark	Wound healing, skin infections, and anti-inflammatory	Cultural and medicinal resin; drought-tolerant	NE	High	Qureshi et al. (2024)
Kleinia odora (Forssk.) DC.	Asteraceae	Medicinal succulent	Leaves, stems	Antimicrobial, antioxidant, antidiabetic	Succulent genetic resource for drylands	NE	Medium	Qureshi et al. (2024)
Calligonum comosum L'Hér.	Polygonaceae	Fire bush	Young shoots, flowers	Anti-inflammatory decoctions	Emergency wild food; sand-dune stabilizer	LC	High	Abid et al. (2025) and Alshaqhaa et al. (2025)
Phoenix dactylifera L.	Arecaceae	Date palm	Fruits, seeds	Digestive aid, energy source	Keystone oasis crop and genetic resource	LC	Very high	Qureshi et al. (2024)
Moringa peregrina (Forssk.) Fiori	Moringaceae	Desert moringa	Leaves, seeds	Anti-inflammatory, antioxidant	Nutrient-rich edible leaves; oil crop	VU	Very high	Qureshi et al. (2024)
Anastatica hierochuntica L.	Brassicaceae	Rose of Jericho	Whole plant	Gastrointestinal and reproductive health	Herbal infusion; symbolic medicinal plant	LC	Medium	Alshaqhaa et al. (2025)
Matricaria aurea (Loefl.) Sch.Bip.	Asteraceae	Golden chamomile	Flowers	Digestive aid, sedative tea	Popular wild herbal tea	LC	Medium	Alshaqhaa et al. (2025)

Teucrium polium L.	Lamiaceae	Felty germander	Aerial parts	Antidiabetic, digestive stimulant	Medicinal herb with high ethnovalue	LC	High	Alshaqhaa et al. (2025)
Withania somnifera (L.) Dunal	Solanaceae	Winter cherry	Roots, leaves	Adaptogenic, stress relief	Medicinal genetic resource	LC	High	Qureshi et al. (2024)
Lawsonia inermis L.	Lythraceae	Henna	Leaves	Skin disorders, antimicrobial	Cultural plant	LC	Medium	Alshaqhaa et al. (2025)
Mentha spp.	Lamiaceae	Wild mint	Leaves	Digestive, carminative	Widely consumed wild edible herb	LC	Medium	Anonymous (2025a)

[a]Legend: *LC* Least Concern, *VU* Vulnerable, *NE* Not Evaluated (IUCN categories)

Medicinal plants are used by the local communities as indigenous health resources in the context of climate change, particularly in situations where the traditional health services are affected by extreme weather, heat waves, or disrupted supply chains (Abd El-Ghani et al. 2025; Organization 2024; Qureshi et al. 2024) Note that a common feature of oasis species can act as refugia of microclimatic conditions in which cooler temperatures and gradients of groundwater can support the growth of the species, a trend that not only promotes oasis species survival but also allows the extraction of medicinal biomass even in conditions of warming.

Alongside taxa that stimulate the consumption of edible products, AlUla harbors a wide range of medicinal plants that have been used traditionally in the treatment of gastrointestinal diseases, infections, and inflammatory disorders. These species are ethnobotanically important as a measure of human-plant relationships dating back over time and indigenous knowledge systems. As noted by (Qureshi et al. 2024), most of these species also have an indirect impact on food security, as they improve the health of the ecosystem, pollination, and soil fertility in degraded landscapes.

As a meeting point between biodiversity conservation, traditional knowledge, and food security, especially under hyper-arid systems, the wild edible and medicinal flora of AlUla is a combination of these three areas. Some of the species such as *Phoenix dactylifera* and *Moringa peregrina* serve as cornerstones to oasis agriculture, oasis nutrition, and climate-robust livelihoods (Qureshi et al. 2024). Phytopharmaceutical and nature-based health solutions in the area have not been properly utilized on a scale yet, as medicinal plants such as *Ephedra alata, Teucrium polium*, and *Withania somnifera* demonstrate (Alshaqhaa et al. 2025).

Conservation-wise, vulnerable and culturally significant species can be effectively conserved through combined in situ and ex situ conservation policies, which are in line with the priorities of Saudi Vision 2030 regarding the restoration of ecosystems, sustainable tourism, and bio-economy development (Qureshi et al. 2024; RCU 2023). Incorporating these plants into the restoration agendas and community-based businesses as well as climate-intelligent agricultural activities can enhance the resilience of the ecological system and the rural economy.

The oasis-related species, especially the *Phoenix dactylifera* and *Moringa peregrina* are the foundation of traditional food systems, and they are an ideal example of how biodiversity helps to maintain food security in desert environments (Qureshi et al. 2024). Wild herbs like *Matricaria aurea* and *Mentha* spp. have a positive effect on the diversity of the diet and health of humans (Alshaqhaa et al. 2025). These plants are together a vital bio-cultural legacy to AlUla and have potentials to provide conservation-based livelihoods, climate-resistant agriculture and even pharmacological research in the future (RCU 2023).

12.9.3 Food Insecurity: Wild Food Plants and Nourishment During Landscape Warming

Table 12.1 is showing list of plants containing edible plants native to hyper-arid environments of the AlUla region that include *Phoenix dactylifera* (date palm),

Moringa peregrina, Ficus palmata, and *Lycium shawii*. As stated in (Abd El-Ghani et al. 2025), oasis cores and related wadi systems also have the greatest species richness and edible biomass potential because of deeper soils and permanent water supply, which are also climate buffers. This has been observed to be an identical pattern in other desert locations where perennial food plants are essential in providing food diversity, intake of micronutrients and seasonal food security (Bharucha and Pretty 2010).

Climate simulations forecast rising temperature and erratic rain on the Arabian Peninsula; with such a scenario, the perennial and low-input food crops will be more stable and resilient in yield compared to annual staple crops (Calvin et al. 2023). The statistical evidence provided by (Abd El-Ghani et al. 2025) helps to highlight that wild food plants are more prone to exist in the microhabitats where water is more predictable as a phenomenon - a distribution pattern that supports the aesthetic usefulness of oasis corridors and habitat connectivity in providing food resources during the changing climatic conditions.

12.9.4 Forage Plants, Pastoral Livelihoods and Rangelands

Traditional pastoral systems in AlUla have been based on the use of hardy forage grasses and shrubs. Plant species such as *Cenchrus ciliaris, Cenchrus divisus, Haloxylon persicum* and *Calligonum comosum* prefer sandy plains, dune regimes, and transitional rocky conditions, where morphological and physiological attributes allow them to live when experiencing a considerable drought (Fig. 12.5). This is in line with previous rangeland research where native forage species are found to be more resilient to grazing stress and climatic stress compared to exotic pastures (El-Ghani et al. 2017).

Fig. 12.5 Author (Prof. Dr. Rahmatullah Qureshi) studying vegetation of AlUla Region (*Haloxylon persicum*)

Moreover, the new work emphasizes the significance of heterogeneity of the habitat, i.e., mosaic of wadis, escarpments, and gravelly slopes that promotes niche partitioning and dietary variety among livestock. This diversity is also necessary because climate change increases the occurrence of droughts and decreases the supply of forage in the homogeneous landscape. These forage plants also help to stabilize the soil, capture carbon, and decrease wind erosion, which strengthens the ecosystem services that are crucial to pastoral resilience on an ecosystem perspective (Baumer 2022; Iseman and Miralles-Wilhelm 2021).

12.9.5 Ecosystem Services, Biodiversity and Climate Adaptation Pathways

Coupled with the above, the patterns of distribution and consumption of medicinal, wild, and forage plants highlight the functional significance of biodiversity in the ecosystems of AlUla. (Abd El-Ghani et al. 2025) provide evidence that not only does the species richness depend on the habitat condition and plant assemblages of the place, but also the supply of significant ecosystem services, such as microclimate regulation, nutrient cycling, and landscape stability. These services are pivotal to climate reduction as well as adaptive capacity in the hyper-arid zones.

The role of biodiversity as a natural buffer against climatic extremes is increasingly being identified as a component of climate adaptation (IPBES 2019; FAO 2018). The AlUla case experience shows that the interplay of traditional plant usage, strong ecological experience, and landscape heterogeneity is one of the possible ways to sustain food security and human well-being in the conditions of climate hazards. Incorporating these components of biodiversity into climate-smart policy, conservation planning, and sustainable tourism might contribute to the increased resilience value of these resources as well as protect cultural heritage and local livelihoods (Abid et al. 2025; Khayat et al. 2023).

12.10 Agricultural Sustainability, Ecosystem Services, and Plant Biodiversity

The plant biodiversity of AlUla supports a large ecosystem service use base in sustainable agriculture such as soil stabilization, nutrient cycling, pollination and microclimate control. Ecological diversity increases the resilience of the ecosystems through less susceptibility to pests, diseases and climatic extremes. Native perennial plants are especially valuable in arid environments to decrease the erosion rate, preserve moisture, and preserve organic matter in the soil (Canton 2021; IPBES 2019).

The importance of native plant species in restoring damaged lands and promoting the production of sustainable food products is gradually being appreciated in restoration efforts in AlUla. Qureshi et al. (2024) suggested that reintroduction of native plants would have positive results regarding the enhancement of biodiversity, ecosystem restoration, and the establishment of preconditions that would support climate-smart agriculture. These nature-based solutions are in tandem with the long-term food security objectives in dryland areas.

12.11 Plant Biodiversity Threats and Implications of Food Security

There is increasing pressure on plant biodiversity in AlUla because of climate change, intensification of land use, overgrazing, invasion by invasive species, and unsustainable water extraction despite its ecological significance. The increase in temperatures and decreased rain are changing species composition and lowering the availability of palatable forage and wild edible plants. The changes have direct impacts on the local livelihoods and agricultural productivity (Legg 2021; Vincent 2008).

The destruction of habitats and loss of biodiversity may undermine the services of the ecosystem and the adaptability of the agricultural systems. Qureshi et al. (2024) emphasize that the native vegetation loss poses a threat to biodiversity, but it also compromises restoration and the resilience of the food system. To curb the threats, conservation and agricultural plans that take into consideration the protection of biodiversity, apart from producing food, must be integrated.

12.12 Associating Biodiversity Conservation with Food Security Strategies

Plant biodiversity conservation in AlUla provides enormous opportunities to enhance food security in the climate change. The presence of natives in agroecosystems, the encouragement of native crops and the preservation of wild relatives of cultivated crops can increase the genetic diversity and adaptation potential. These measures are in line with international demands of bio-based food systems that can sustain climatic uncertainty (Soomro et al. 2023; Stenseke 2024).

The AlUla region offers an interesting example of how biodiversity conservation can be associated with sustainable agriculture with the support of policy, research, and community. Plant-based restoration and conservation projects will positively and concurrently restore the integrity of the ecosystem, agricultural output, and food security in the long term, and thus biodiversity is a pillar to climate-resilient development in the region, as highlighted by (Qureshi et al. 2024).

12.13 Effects of Climate Change on Food Systems and Plant Biodiversity in AlUla

12.13.1 Effect on Arid Agroecosystems

The ongoing climate change is increasing the level of environmental pressure in arid and hyper-arid areas with significant consequences on food systems and plant biodiversity. At AlUla, the process of ecological changes and agricultural productivity is being transformed by increasing mean temperatures, increasing incidence of heat waves, shifting rainfall patterns, and extended droughts. These climatic stressors have a direct impact on the physiological performance, phenology, and survival of plants, especially moisture-sensitive species and traditional crops grown in oasis systems (Allan et al. 2023).

Especially vulnerable are arid ecosystems since their biological communities are already at tolerance limits. A minor shift in temperature or precipitation may cause an unproportional ecological reaction, such as loss of vegetation cover, soil erosion, and less biomass synthesis. Further worsening water scarcity in AlUla, decreasing the recharge of groundwater and increasing the rate of evapotranspiration puts even greater stress on natural vegetation and cultivated plants (Abid et al. 2025; Mekouar 2018).

12.13.2 Effects on Native Flora and Ecosystem Stability

The main resilience in the vegetation cover of the AlUla ecosystem consists of native vegetation, but the distribution, abundance, and regeneration potential of these species are threatened by climate change. Higher levels of aridity and thermal stress can cause changes in the species composition, which will be dominated by taxa very resilient to stress conditions, pushing the less tolerant ones to the periphery. These alterations may decrease the total biodiversity and undermine the ecosystem processes like soil stabilization, nutrient circulation, and provision of habitats (IPBES 2019; Vincent 2008). The destruction of native vegetation due to climate stress compromises the success of the restoration process and promotes land degradation in AlUla. Extinction of keystone plant species may trigger response whereby further losses are caused in the ecosystem, making it more vulnerable to erosion, invasive species, and desertification. Conserving native flora is thus not only important and a measure of conservation of biodiversity, but also to maintain ecosystem services that form the foundation of food security (Qureshi et al. 2024).

Agricultural systems in AlUla are directly impacted by climate variability because the systems are influenced by crop yields, growing seasons, and water availability. The traditional oasis agriculture, which was traditionally resistant, is now facing the challenge of increasing temperatures and reducing water supplies. Heat stress is capable of decreasing photosynthetic efficiency, flowering and fruit set, and augmenting crop water demands and, thus, decimate productivity of major fruit crops, date palm, citrus, and mango (Banadka et al. 2025; Food 2017).

The variability of rainfall also influences the soil moisture and nutrient supply and makes people more dependent on irrigation and foreign inputs. Devoid of adaptive management, such pressures can cause the local food system to decline in output and be more susceptible. According to (Qureshi et al. 2024), the agricultural stress due to climate changes highlights the necessity of incorporating the concept of biodiversity-based technologies, which include diversified planting and the use of native species, to preserve the productivity levels within the changing environments.

12.13.3 Impact on Forage Resources and Wild Food Plants

In arid areas, wild edible resources and natural forage resources offer a significant safety net to food and livestock systems. Climate change poses a threat to these resources due to changes in the growth cycles, decrease in biomass supply, and a decline of palatability. Extremes of heat waves and lengthy dry seasons in AlUla have the potential to significantly decrease the presence of wild food crops and grazing animals to influence pastoral livelihood and nutritional diversity (IPBES 2019; Nicolétis et al. 2019).

The scarcity in forage supply may result in excessive grazing of the leftover vegetation, which may further hasten the deterioration of the ecosystem. A list of forage/wild food resources of AlUla is provided in Table 12.4. Qureshi et al. (2024) emphasized that the vulnerability of the rangeland vegetation to climate not only affects the livestock productivity but also affects ecosystem restoration programs. The buffering effect of effects of climate on food systems therefore, requires the sustainable management of wild plant resources.

The importance of wild forage and edible plant resources to the ecology of arid landscapes and the livelihood of livestock and human populations in the AlUla region, northwest Saudi Arabia, is fundamentally important. These resources fulfill various functions: they assist grazing of domestic animals, supplement local food with seasonal food, and provide genetic diversity, which makes them resistant to climatic extremes (Table 12.4). Even though the recorded floristic lists of AlUla are being synthesized, the data collected in the region and the Saudi Arabian sources of ethnobotanics give a solid basis to identify the major forage and wild edible plants (Abid et al. 2025; Anonymous 2025c).

The desert ecosystems are based on arid-adapted shrubs and semi-woody perennials that are often preferred by pastoral animals like camels, goats and sheep. Examples of species that are typical of Arabian dune and gravel include *Calligonum comosum* (fire bush) and *Rhanterium epapposum*, which thrives on palatable green growth after infrequent rainfall that supports the nourishment of herds (Anonymous 2025c). *Calligonum comosum* has been especially appreciated as a forage crop, but also due to its desert-stabilizing effects, as the deep roots and flowers hold the moving sands together and also as food to be eaten by grazers (Anonymous 2025d). Similarly, *Rhanterium epapposum* has been known to persistently regenerate fast following rainfall, an occurrence in arid zones of the Saudi

Table 12.4 Forage resources and wild food plants of AlUla region

Botanical name	Family	Forage/ edible part	Use (forage or wild food)	Notes/traditional uses
Calligonum comosum	Polygonaceae	Leaves, stems, flowers	Forage for livestock; edible flowers	Food for grazing animals such as camels and sheep; flowers reportedly edible; also stabilizes sand dunes
Rhanterium epapposum	Asteraceae	Leaves	Major forage shrub	Produces leaves quickly after rainfall and is eaten by camels and sheep in desert systems
Ochradenus baccatus	Resedaceae	Berries, foliage	Forage and wild food	Berries eaten by birds; foliage grazed by goats; traditional medicinal uses reported
Hyphaene thebaica	Arecaceae	Fruit	Wild food	Edible fruit; leaves used for crafts; typical of arid oases and water-associated habitats
Various edible herbs (regional list)	Multiple	Leaves, stems, fruits	Wild food greens	Includes *Amaranthus viridis*, *Rumex vesicarius*, *Portulaca oleracea*, *Lactuca* spp., *Eruca sativa*, etc., with leaves eaten cooked or raw in similar Saudi environments
Moringa peregrina	Moringaceae	Leaves, seeds	Wild food and forage	Edible leaves and seeds; seed oil prized; leaves likely browsed by livestock
Crop species in oasis systems	Arecaceae, Rutaceae etc.	Fruits (dates, citrus)	Wild food crops	Date palms and citrus trees produce fruit consumed by people and animals; traditional fodder from fruit waste

Arabia and the Arabian Peninsula resulting in it becoming a primary forage crop in nomadic and pastoral forage systems (Anonymous 2025d).

Other wild plants besides the shrubs offer edible resources that are of direct benefit to the human diets. An example of this is *Ochradenus baccatus*, which has small berries that are opportunistically eaten by both the local people and birds in the area, and the leaves of which are eaten by livestock (Anonymous 2025d). Despite the lack of ethnobotanical records on the AlUla wild edible herbs, surveys of Saudi Arabian arid flora suggest that plants like *Amaranthus viridis, Rumex vesicarius* and *Portulaca oleracea* have been traditionally eaten, either cooked or raw, so that its green leaves, and that these species are likely to achieve appropriate microhabitats in the wadis and disturbed soils of AlUla (Anonymous 2025c).

The forage and wild food also include palms and oasis crops in the region. In the oases of AlUla, the *phoenix dactylifera* (date palm) is the most common cultivated plant that yields fruit that is a primary ingredient in the local diet and a notable source of calories and micronutrients (Anonymous 2025c). Fallen fruit and date by-products can also be used as livestock forage, and agricultural production has to coexist with pastoral use. On the same note, orchard crops like citrus trees are produced in the oasis environment, the fruits of which enrich the food supply of people and animals and stress the interrelation of the cultivated food system and wild foraging (Anonymous 2025c).

Forage and wild food sources are not only important to nutrition and grazing, but also important genetic and ecosystem diversity. Wild vegetation is a genetic reservoir of elements like drought resistance, salt resistance and phenological flexibility—elements that are becoming more significant, based on projected climate changes on the Arabian Peninsula. Some of the desert shrubs and semi-woody perennials such as *Calligonum comosum* and *Rhanterium epapposum* are ecologically significant because of their use as forage to the large herbivores in the Arabian Desert habitat. *Calligonum comosum* develops fleshy branches and flowers, which are consumed by the camels and the other grazing animals, whereas *R. epapposum* is a consistent source of nutritious leaves after the occasional incidences of rainfall, which makes it an important desert grazer crop. These species are therefore vital in sustaining grazing lifestyles and sustain reintroduced wildlife to the safeguarded reserves in AlUla, including Arabian oryx and gazelles (Anonymous 2025c).

12.13.4 Wild Food Plants and Edible Resources

Besides livestock fodder, there are several species of shrubs and fruiting trees that add to the wild food. *Ochradenus baccatus* is also a berry species that bears sweet berries that are wild animals love to feed on and has long been used as opportunistic food; *Hyphaene thebaica* (doum palm) is also a fruit-ectofauna that has historically been used as famine food and supplement due to its edible fruits. Common edible greens in Saudi Arabia and other arid regions like *Amaranthus viridis* and *Rumex vesicarius* are widespread in the region and could potentially be applied in AlUla (Anonymous 2025c), but no specific ethnobotanical information was recorded in AlUla.

12.13.5 Clustering of Oasis Crops and Agro-Forage

The oasis agriculture is a component of AlUla, which is a cultivated food crop that is supported by ancient aquifers like date palms (*Phoenix dactylifera*) and citrus species. The crops have direct culinary applications, as well as forage by-products and habitat structure that are used by domesticated and wild herbivores (Anonymous 2025c).

12.14 Climate Change, Loss of Biodiversity, and Food Security Relations

Climate change, loss of biodiversity, and loss of food security have a close relationship that is especially noticeable in arid areas such as AlUla. The loss of plant diversity decreases the adaptability of agroecosystems to climatic extremes and biotic strains. On the contrary, biodiversity-rich systems are more likely to be stable and more productive in changing environmental conditions (Legg 2021; Nicolétis et al. 2019).

Plant biodiversity conservation in AlUla is an emerging concept that has been identified as a strategic way of responding to climate change. By focusing on native and climate-resilient species in restoration and conservation efforts, ecosystems could be made to work better to enable sustainable food production. (Qureshi et al. 2024) stated that climate adaptation, together with biodiversity conservation, is one of the ways to enhance food security, ecological resilience, and long-term sustainability in the area.

12.15 Plant Biodiversity and Food Security in AlUla

12.15.1 Native and Endemic Plant Diversity Conservation

Preservation of endemic and native flora is the key to maintaining the ecological services as well as food security in the arid regions like AlUla. AlUla has native plants that have been specifically adapted to deal with severe arid conditions, inappropriate soil, and elevated temperature levels and, therefore, these species are essential genetic resources to enable climate-resistant agriculture and ecological recovery. Nevertheless, growing anthropogenic stress, overgrazing, and climate change present enviable risks to the plant diversity in the area, and special conservation actions are required (IPBES 2019; Qureshi et al. 2024). The conservation of these species is not only essential in biodiversity conservation but also in the stability of the ecosystems, e.g., in the stabilization of the soil, cycling of nutrients, and control of microclimate.

The in situ conservation measures, such as protecting areas, habitat recovery and controlled grazing schemes, have been found to be effective in conserving the AlUla plant diversity. To promote ecological resilience, the Royal Commission for AlUla (Abid et al. 2025) has launched landscape-scale conservation initiatives in order to restore the degraded habitats and also reintroduce native plant species. These efforts fit the international best practices in conserving drylands and focusing on climate change adaptation based on ecosystems (Abid et al. 2025; Restoration 2021).

12.15.2 Sustainable Oasis Agriculture

Traditional oasis agriculture is an example of sustainable land-use system that combines food production with biodiversity conservation. Historically, oasis systems in AlUla reinforced diverse crop schemes (date palms, cereals, legumes and vegetables) which formed multi-layered agroecosystems to maximize the utilization of water and make them more resilient to climatic extremes. Agroecological modes of revitalizing these systems present huge opportunities toward sustainable food security in the context of a changing climate (Altieri and Nicholls 2017; Li 2021).

Agroecological interventions like diversification of crops, organic soil additions, irrigation technologies with low water use, and planting of native species can be used to increase the productivity and biodiversity can be preserved. The work of AlUla proves that indigenous plants create better soil and lower their susceptibility to drought stresses, which supports the contribution of biodiversity to sustaining food systems (Nicolétis et al. 2019; Qureshi et al. 2024). These activities are in tandem with the objectives of the Saudi Arabia Vision 2030 of sustainable agriculture and environmental care.

12.15.3 Agrobiodiversity Potential

AlUla's flora includes a variety of drought-tolerant wild relatives of domesticated crops, which are crucial for breeding programs aimed at improving climate resilience (Qureshi et al. 2024). Species such as *Vigna* spp., *Cicer* spp., and *Allium* spp. exhibit traits including drought tolerance, salt tolerance, and pest resistance, which can be introgressed into cultivated varieties (Baena et al. 2023). These wild relatives also serve as repositories of secondary metabolites with medicinal and nutritional value.

In addition to wild crop relatives, native perennial forage species such as *Panicum turgidum, Stipagrostis plumosa,* and *Rhanterium epapposum* contribute to sustainable livestock production. By reducing reliance on imported feed, these species help maintain food security under increasingly variable climatic conditions (Morabito et al. 2022).

12.15.4 Plant Biodiversity and Food Security: Wild Plants and Fruit Cultivation in AlUla

Nestled in north-western Saudi Arabia, AlUla is far more than a dramatic desert landscape—it is a living testament to how people and nature have learned to thrive together under extreme arid conditions. Framed by sandstone valleys, volcanic plateaus, and fertile oases, the region reflects thousands of years of human settlement shaped by careful stewardship of land and water. Through a blend of traditional farming knowledge and modern sustainability initiatives, AlUla has developed a diverse and productive fruit-based agricultural system that supports local livelihoods, nutrition, and cultural identity. In the face of climate change and growing food security concerns, this agricultural heritage has gained renewed relevance and aligns closely with the sustainability goals of Saudi Vision 2030 (Anonymous 2026a, b).

Despite receiving very little and highly irregular rainfall, AlUla sustains remarkable plant diversity and resilient agroecosystems. Historically, farmers engineered ingenious water-harvesting and irrigation systems such as underground channels and oasis-based networks—to make the most of limited groundwater. These systems laid the foundation for date-palm-centered agroforestry, where tall palms created cooler, shaded microclimates for other crops growing beneath them. This natural layering reduced water loss, protected soils, and enhanced overall productivity, representing

Fig. 12.6 Lemon groove in AlUla. (Photo by Prof. Dr. Rahmatullah Qureshi)

early forms of climate-smart agriculture that remain highly relevant today (Anonymous 2026c).

At the heart of AlUla's agricultural landscape stands the date palm (*Phoenix dactylifera* L.), the cornerstone of local food security and rural livelihoods. The region is among Saudi Arabia's leading date-producing areas, with more than 3.1 million trees spread across approximately 16,579 hectares and annual production exceeding 116,000 tons. Locally adapted varieties such as Barni are particularly valued for their tolerance to heat, drought, and salinity. Beyond their economic importance, date palms stabilize oasis ecosystems, enrich soil structure, support intercropping, and create habitats that enhance biodiversity, making them essential allies in building resilience to climatic variability (Anonymous 2026d).

Citrus cultivation adds further diversity to AlUla's farming systems and plays an important role in improving dietary quality. Spanning over 650 hectares with more than 260,000 trees, citrus orchards yield oranges, citrons, and sweet lemons that are harvested mainly between December and March (Fig. 12.6). These fruits provide vital micronutrients, especially vitamin C, during the cooler months and help bridge seasonal nutritional gaps. Their cultivation also supports local markets and agritourism initiatives, contributing to broader economic diversification in the region (Anonymous 2026e).

Mango cultivation highlights AlUla's ability to innovate while respecting its ecological limits. Introduced more recently than dates and citrus, mangoes now cover about 125 hectares, with nearly 50,000 trees producing over 1100 tons annually. Varieties such as Zabda, Sanara, and Kate have adapted well to the region's microclimates. Harvested between July and September, mangoes enhance food availability during the hottest months and provide farmers with an additional income stream at a time when climate risks are highest (Anonymous 2026f).

Alongside these major crops, AlUla continues to nurture a range of indigenous and traditional fruit species, including *Moringa peregrina* (Arabian Ban or Yasar), figs, grapes, and pomegranates (Fig. 12.7). Although grown on smaller scales, these

Fig. 12.7 *Punica granatum.* (Photo by Prof. Qureshi)

species are vital for conserving agrobiodiversity and cultural heritage. *Moringa peregrina*, native to the region, is particularly valued for its nutritional, medicinal, and oil-producing properties (Fig. 12.8). Figs, grapes, and pomegranates, harvested from mid-summer to early autumn, further extend the fruiting calendar and strengthen year-round food availability while supporting ecological balance within oasis systems (Anonymous 2026g). *Moringa peregrina* has historically been valued in Arabian agriculture for its edible seeds and oil, which have culinary and traditional medicinal uses. Its adaptability to low water environments makes it suitable for cultivation on degraded or marginal lands where more water-demanding crops fail, positioning it as a sustainable perennial tree crop for food-security portfolios.

Fig. 12.8 *Moringa peregrina*. (Photo by Prof. Qureshi)

Moringa oleifera, although not native, has been identified as a promising species for Saudi Arabia's arid agricultural landscapes due to its rapid growth, nutrient density (leaves rich in proteins, vitamins, and antioxidants), and multipurpose uses in food, feed, and medicine. Government initiatives in the Kingdom are increasingly promoting moringa cultivation, and pilot plantations have been initiated in different provinces (Al-Khalifah and Shanavaskhan 2015).

Beyond direct human nutrition, moringa also contributes to animal nutrition and ecosystem services. Recent studies have shown that moringa biomass can serve as a sustainable forage alternative in livestock diets, potentially enhancing feed quality under arid conditions and reducing reliance on imported feeds—an important consideration for integrated farm systems in pastoral-agricultural ecotones (Abdoun et al. 2025). Moringa's high biomass production and leaf nutrient content further enhance soil cover and organic inputs, supporting soil biological activity and carbon sequestration.

Moringa represents an emerging opportunity for enhancing food security and agricultural resilience in AlUla. Two species are particularly relevant: *Moringa peregrina*—a native wild species with deep desert adaptations—and *Moringa oleifera*, a globally recognized "superfood" species introduced for cultivation in arid regions (Banadka et al. 2025). Both demonstrate characteristics favorable for hyper-arid agriculture: fast growth, drought tolerance, and high nutritional value.

Taken together, AlUla's fruit cultivation system demonstrates how biodiversity-based agriculture can sustain food security, livelihoods, and ecosystem health in some of the world's harshest environments. The careful integration of traditional knowledge, crop diversity, and modern conservation efforts has enabled the region to endure climatic stress while maintaining productive agroecosystems. Looking ahead, continued investment in biodiversity conservation, efficient water management, and farmer-centered development will be essential to protect and sustain this living agricultural heritage under future climate scenarios (Anonymous 2026h).

12.15.5 Traditional Food Plants and Ethnobotanical Knowledge

Local communities have relied on native plants to supplement their diet. For instance, fruits of *Ziziphus spina-christi* (Sidr) are rich in vitamins and minerals, providing a staple source of nutrition during periods of food scarcity (Qureshi et al. 2024). Flowers, buds and leaves of *Capparis spinosa* (Caper) are edible, providing micronutrients and culinary diversity (Anonymous 2023d). *Cleome gynandra* is traditionally consumed as a leafy vegetable, that provides essential proteins and minerals, particularly in the dry season when cultivated vegetables are limited (Swiderska et al. 2022).

In addition to food, many species serve medicinal purposes, reinforcing the interconnection between biodiversity and human health. For example, *Pulicaria undulata, Achillea fragrantissima*, and *Teucrium polium* are used to treat digestive and respiratory ailments (Morabito et al. 2022). Maintaining these plants within agricultural and natural landscapes ensures the continuation of traditional knowledge systems critical for community well-being.

12.15.6 Policy Frameworks and Supporting Institutions

Policy frameworks and institutional coordination are important in ensuring effective conservation and food security. The Vision 2030 of Saudi Arabia attaches a lot of importance to environmental sustainability, conservation of biodiversity, and sustainable food systems, which is a strategic basis of integrated land-use planning in places like AlUla. The national policies aimed at restoring the ecosystem, adapting to climate change, and practicing sustainable agriculture are more and more in line with the international environmental treaties, such as the Convention on Biological Diversity and the UN Sustainable Development Goals (Essl et al. 2020; Khayat et al. 2023).

Due to the regional level, the Royal Commission for AlUla is playing a central role in converting national policies into locally crafted concepts. RCU, 2023 has facilitated a comprehensive strategy in conservation of biodiversity and resiliency of the food system through science-based planning, stakeholder involvement and investment in research and capacity building. These institutional arrangements are necessary to achieve the trade-off between conservation and socio-economic development in weak arid ecosystems (Abid et al. 2025; Buckle et al. 2020).

12.15.7 Community Participation and Indigenous Knowledge Systems

The programs of community involvement and indigenous knowledge are the most important factors of sustainable biodiversity management and food security. The AlUla communities are rich in traditional knowledge on the use of plants, grazing, water management, and selection of crops that were formed over the centuries of adaptation to the arid conditions. Combining such an understanding with the current scientific methods would improve the efficacy and social acceptability of conservation interventions and agricultural interventions (Berkes 2017; Qureshi et al. 2024).

Participatory conservation models where the local stakeholders are involved in decision-making process lead to stewardship and long-term sustainability. Arid areas study also shows that natural resource management within communities enhances better biodiversity performance as well as livelihood and food security. In AlUla, these strategies can help address the connection between cultural heritage, biodiversity protection, and climate-resilient food systems (Baena et al. 2023; Restoration 2021).

12.16 Integrating Climate Adaptation, Food Security, and Plant Biodiversity

The facts provided throughout this chapter highlight that plant biodiversity is the basis of climate adaptation and food security of arid and hyper-arid areas including the AlUla. Both native and conventionally managed plant systems have been known to provide ecological stability as well as diversified food production, nutritional security and resilience to livelihood amidst climate stress. Biodiversity conservation and food systems are being integrated as a necessity and not a choice due to the rapidly increasing levels of climate change and land degradation (IPBES 2019; Li 2021). AlUla is an illustration of how agroecosystems can be resilient to extreme conditions using the concept of biodiversity.

The combination of ecological, agroecological, and socio-cultural aspects brings about the fact that the problem of food security in AlUla cannot be resolved only with technological intensification. Rather, it needs an approach that is systems-based and one that appreciates ecosystem services, genetic diversity as well as long-term human-plant interactions. AlUla studies show that the restoration of native vegetation and the rejuvenation of traditional oasis systems can both increase biodiversity and soil health as well as ensure food production resistant to climatic variability (Altieri and Nicholls 2017; Qureshi et al. 2024).

12.16.1 Implications to Arid Land Management and Policy

The AlUla results have strong implications on national and global policy on arid land management. Due to the growing intensity of climate change, which results in temperature extremes and water shortages, arid zones are taking center stage as a topic in food security discussions. AlUla shows that arid landscapes that are biodiverse can be used fruitfully as food-growing ecosystems but not as marginal lands. It disputes traditional accounts of deserts as agriculturally unproductive and emphasizes the necessity of adopting policy systems that place ecosystem-based adjustments first (Change 2001; Restoration 2021).

Policy wise, the only way that biodiversity conservation could be aligned to the agricultural development is through cross-sectoral coordination, long-term planning, and long-term investment in research and capacity building. The environment of Saudi Arabia supporting such integration is indicated by its Vision 2030 that highlights the focus on environmental sustainability, the restoration of the degraded lands, and climate resilience. The policy vision can be implemented at the landscape level through institutional leadership efforts directed by such entities as the Royal Commission for AlUla (Abid et al. 2025; Alzamil 2024).

12.16.2 AlUla as a Model of Climate-Ready Food Systems

AlUla can potentially succeed in becoming a regional and global model of applied climate-resilient food systems in arid environments. The ecological diversity of the region, its long history of agricultural activities, and the ongoing conservation efforts provide a great learning experience to other areas of drylands that experience comparable difficulties. AlUla shows that food production can be attained without compromising the ecological integrity by incorporating the diversity of native vegetation into the agricultural planning (Citaristi 2022; Qureshi et al. 2024).

Additionally, the fact that AlUla is defined as a living cultural and ecological landscape makes it possible to use experimental and adaptive management methods. Transferable knowledge can be created on the sustainability of drylands across the globe through long-term studies of the dynamics of vegetation, effects of climate and agricultural productivity. This body of evidence-based models is critical in shaping the strategies to transform the food systems around the world in the context of climate change (IPBES 2019; Li 2021).

12.17 Future Direction and Research Priorities

To overcome the interdisciplinary nature of the problems of climate change, biodiversity loss, and food insecurity in AlUla, further research and innovations are needed. There are several research gaps in realizing the full potential of plant biodiversity in AlUla as far as food security is concerned. Nutritional, functional, and genetic characterization of native and underutilized plant species needs to be done in detail to evaluate the suitability of these species in climate-resilient diets and sustainable agriculture. The areas of priority involve long-term observing of vegetation diversity, determining the effects of climate change on agroecosystems, and the advent of climatically stable crop varieties with native genes. Ecological and agronomic research has been especially necessary to help comprehend the intricate interactions between humans and the environment in the area (Change 2001; Qureshi et al. 2024). Moreover, the capacity building and the exchange of knowledge should be strengthened by reinforcing the regional and international research partnerships. Research-based policies can be useful in creating more local food security, as well as globally by making food systems adaptable to climate change in dryland areas, by establishing AlUla as a living laboratory of arid-land sustainability (IPBES 2019; Li 2021).

The interdisciplinary methods by incorporating archaeology, ethnobotany, agronomy, and climate science should also be a priority in future research. The archaeobotanical data in AlUla indicates that the domestication and management of plants have a long history and may provide insights into the sustainable practices that can guide the adaptation strategies in the modern context. Enhancing global partnerships and data-sharing platforms will help even more to improve the scientific basis of food systems based on biodiversity in arid areas (Berkes 2017; Swiderska et al. 2022).

12.18 Conclusion

The AlUla region of north-western Saudi Arabia provides a compelling case study of how plant biodiversity can sustain food security and human well-being under extreme aridity, even as climate change intensifies environmental pressures. The evidence presented in this chapter leads to the following conclusions:

12.18.1 Biodiversity as a Foundation of Dryland Resilience

AlUla's native and cultivated flora—ranging from drought-adapted shrubs (*Haloxylon persicum, Calligonum comosum*) to traditional oasis crops (*Phoenix dactylifera, Citrus* spp., *Moringa peregrina*) supports essential ecosystem services: soil stabilization, microclimate buffering, nutrient cycling, pollination, and provision of forage, food, and medicines. This functional diversity directly underpins local livelihoods and nutritional security, especially during climatic shocks.

12.18.2 Traditional Knowledge and Heritage Systems Are Adaptive Assets

Centuries-old practices of water harvesting (qanat/falaj), multi-layered agroforestry, and community-based land management have enabled AlUla's inhabitants to thrive in a hyper-arid environment. These systems reduce evapotranspiration, improve water use efficiency, and maintain genetic diversity. Revitalizing such indigenous knowledge alongside modern science is not a nostalgic exercise but a strategic climate adaptation tool.

12.18.3 Climate Change Is Already Eroding Biodiversity and Food Security

Observed warming, increased heat wave frequency, declining and erratic rainfall, and groundwater over-extraction are driving habitat degradation, shifts in species composition, reduced forage quality, and lower productivity of date palms and fruit trees. Wild food and medicinal plants, key safety nets during scarcity, are becoming less available, while pastoral livelihoods face mounting pressure from rangeland deterioration.

12.18.4 Conservation and Restoration Offer a Viable Response Pathway

Initiatives led by the Royal Commission for AlUla (RCU) including the native plant seed bank (>80 species), Sharaan Nature Reserve, and sustainable oasis agriculture programs demonstrate that biodiversity conservation can be integrated with food production. Reintroducing native perennials, rehabilitating traditional irrigation, and promoting agroecological practices (polyculture, organic soil amendments, low-water crops) enhance both ecological integrity and agricultural resilience.

12.18.5 Policy Alignment and Community Engagement Are Critical Success Factors

Saudi Vision 2030 provides a strategic umbrella for environmental sustainability, ecosystem restoration, and climate smart agriculture. However, effective implementation requires cross-sectoral coordination, sustained research funding, capacity building among local farmers, and participatory governance that respects traditional land-use rights and indigenous knowledge. Without such integration, top down interventions risk undermining the very social ecological systems they aim to protect.

12.18.6 AlUla as a Global Model for Arid Land Food Systems

The region's combination of rich biodiversity, long agricultural history, ongoing restoration science, and institutional commitment positions it as a living laboratory for climate-resilient food systems. Lessons from AlUla on native species selection, water wise cultivation, and community-based adaptation are directly transferable to other drylands across the Middle East, North Africa, Central Asia, and beyond. Future priorities include i) long-term monitoring of vegetation dynamics and climate impacts; ii) genetic and nutritional characterization of underutilized native food plants; iii) scaling up participatory conservation models; iv) strengthening regional and international research partnerships; and v) developing climate smart policy bundles that explicitly link biodiversity targets, food security indicators, and land-use planning.

In summary, AlUla demonstrates that plant biodiversity is not an obstacle to agricultural development but its essential ally. In an era of accelerating climate change, conserving and wisely using this biological heritage is a prerequisite for sustainable food security in arid lands a lesson that resonates far beyond the sandstone valleys of north-western Saudi Arabia.

References

Aati HY, Perveen S, Orfali R, Al-Taweel AM, Aati S, Wanner J et al (2020) Chemical composition and antimicrobial activity of the essential oils of Artemisia absinthium, Artemisia scoparia, and Artemisia sieberi grown in Saudi Arabia. Arab J Chem 13(11):8209–8217

Abd El-Ghani MM (1998) Environmental correlates of species distribution in arid desert ecosystems of eastern Egypt. J Arid Environ 38(2):297–313

Abd El-Ghani MM, Fawzy AM (2006) Plant diversity around springs and wells in five oases of the Western Desert, Egypt. Int J Agric Biol 8(2):249–255

Abd El-Ghani MM, Filimban FZ, Khan ZD, Baig MB, Al Shaikh AA (2025) Plant diversity and habitat composition of AlUla oasis: insights from Northwest Saudi Arabia. In: Food systems and biodiversity in the context of environmental and climate risks: dynamics and evolving solutions. Springer, pp 457–489

Abdel-Kader MS, Al-Qutaym A, Saeedan ASB, Hamad AM, Alkharfy KM (2016) Nephroprotective and hepatoprotective effects of Tribulus terrestris L. growing in Saudi Arabia. J Pharm Pharm Res 4(4):144–152

Abdoun KA, Al-Badwi MA, Alshamiry FA, Alsagan AA, Alsaiady MY, Samara EM, Al-Haidary AA (2025) Potential of Moringa oleifera and Moringa peregrina whole plants as sustainable forage alternatives in lamb (Ovis aries) nutrition. Scientific Reports

Abid MB, Elawady A, Shamim R (2025) Exploring AlUla, Saudi Arabia: a journey through history, heritage, resources, and tourism. Geo J 90(4):172

Aboshora W, Lianfu Z, Dahir M, Gasmalla MA, Musa A, Omer E, Thapa M (2014) Physicochemical, nutritional and functional properties of the epicarp, flesh and pitted sample of doum fruit (Hyphaene Thebaica)

Adgaba N, Al-Ghamdi A, Tadese Y, Shenkut A, Anwer AM, Mohammed J (2015) Floral phenology, Nectar secretion dynamics and honey production potentials of lavendula species (L. dentata, and L. pubescence) in Southwestern part of Saudi Arabia. J Apicultur Sci 59(2): 135–144

Ads EN, Abouzied AS, Alshammari MK (2021) Evaluation of cytotoxic effects of methanolic extract of Pergularia tomentosa L growing wild in KSA. Asian Pac J Cancer Prev 22(S1):67–72

Ahmad F, Kamaraj B, Mohamed JMM, Ebrahim D, Al Humoud SY, El-Sagheer AM, Emam ASA (2022) Ethnopharmacological and phytochemical study of major species from labiatae family in Saudi Arabia: a systematic review. Int J App Pharm 1:44–57

Ahmed SM, Al-Sodany YM, Seada ML, El-Kholy AS (2022) Biochemical and molecular markers differentiate Ficus species from four locations in Saudi Arabia. Egypt J Bot 62(3):917–934

Al Masoudi LM (2024) Floristic diversity and vegetation analysis of some Wadies Northeast of Makkah, Saudi Arabia. Sohag J Sci 9(4):429–438

Al Ramahi FKM, Al Bahadly ZKI (2020) The spatial analysis for Bassia eriophora (Schrad.) Asch. Plant distributed in all Iraq by using RS & GIS techniques. Baghdad Sci J 17(1):17

Alabdli AH, Elsafori AK, Alayafi AH (2025) Phyto-taxonomic insights into the Flora of Wadi Ganona: a case study from Southwestern of Saudi Arabia. Am J Plant Sci 16(4):453–469

Alaklabi A (2015) Genetic diversity of Moringa peregrina species in Saudi Arabia with ITS sequences. Saudi J Biol Sci 22(2):186–190

Alansari RM, Seleem AA (2024) Bioecological and molecular studies of Heliothis nubigera Herrich-Schäffer, 1851 and Trichoplusia ni (Hübner, 1803) associated with Ferula communis L. as a new host in AlUla, Saudi Arabia (Lepidoptera: Noctuidae). SHILAP Revista de lepidopterología 52(206):209–225

Alansi S, Tarroum M, Al-Qurainy F, Khan S, Nadeem M (2016) Use of ISSR markers to assess the genetic diversity in wild medicinal Ziziphus spina-christi (L.) Willd. Collected from different regions of Saudi Arabia. Biotechnol Biotechnologic Equip 30(5):942–947

Al-Bakre DA (2025) Ecological niche modeling of two co-occurring desert species in Saudi Arabia: implications for biodiversity and climate resilience. BMC Plant Biol 25(1):1706

Alfaifi MY, Saleh KA, El-Boushnak MA, Elbehairi SEI, Alshehri MA, Shati AA (2016) Antiproliferative activity of the methanolic extract of Withania somnifera leaves from Faifa mountains, Southwest Saudi Arabia, against several human cancer cell lines. Asian Pac J Cancer Prev 17(5):2723–2726

Al-Ghamdi FA (2013) Morphological diversity of some Tephrosia species (Fabaceae) in Saudi Arabia. Am J Plant Sci 4(03):543

Alghanem S (2018) Ecological and botanical diversity in Haloxylon persicum community at Al-Qassim region in Kingdom of Saudi Arabia. Am J Environ Prot 6:43–49

Al-Goufi ND, Sonbol HS (2018) Effect of Saudi Arabian Capparis cartilaginea fruit extracts on serum parathyroid hormone and 1α, 25-dihydroxyvitamin D3 levels in rats

Al-Harthi HF (2025) Antifungal and enzymatic activities of endophytic fungi associated with Tamarix nilotica. Plant Signal Behav 20(1):2439250

Alhashim N, Ferreira M, Siegert S, Walsh J, Williams C, Silveiro H, ... Lyons B (n.d.) First-year vegetation effects of protection and watering of hyper-arid lowland desert in Neom, Tabuk Province, Kingdom of Saudi Arabia Tabuk Province Kingdom of Saudi Arabia

Al-Hemaid F, Thomas J (1996) Review of the genus Tribulus L. in Saudi Arabia. Arab Gulf J Sci Res 14:415–444

Al-Homaid N, Khan MH, Sadiq M (1990) Ecology and some desert plant communities of the eastern province of Saudi Arabia. Arid Land Res Manag 4(4):253–260

Ali MA (2019) Molecular authentication of Anthemis deserti Boiss.(Asteraceae) based on ITS2 region of nrDNA gene sequence. Saudi J Biologic Sci 26(1):155–159

Aljawdah HM, Murshed M, Ammari AA, Maodaa SN, Al-Quraishy S (2025) Antioxidant and Anticoccidial activity of aqueous methanol extract of Capparis spinosa L. plant collected in Riyadh, Saudi Arabia. Ind J Anim Res 1:8

Al-Khalifah N, Askari E (2003) Molecular phylogeny of date palm (Phoenix dactylifera L.) cultivars from Saudi Arabia by DNA fingerprinting. Theor Appl Genet 107(7):1266–1270

Al-Khalifah N, Shanavaskhan A (2015) Moringa oleifera Lam., a promising crop species for arid conditions of Saudi Arabia and Moringa peregrina (Forssk.) Fiori, a native wild species for crop improvement. Paper presented at the I International Symposium on Moringa 1158

Allan RP, Arias PA, Berger S, Canadell JG, Cassou C, Chen D et al (2023) Intergovernmental panel on climate change (IPCC). Summary for policymakers. In: Climate change 2021: the physical science basis. Contribution of working group I to the sixth assessment report of the intergovernmental panel on climate change. Cambridge University Press, pp 3–32

Almalki RA, Alzahrani DA (2018) Morphological investigation of genus Ziziphus mill. (Rhamnaceae) in Saudi Arabia. Am J Plant Sci 9:2644

Alnadif AAM (2022) Biochemical composition and medicinal uses of chrozophora genus

Al-Namazi A, Al-Ammari B, Davy AJ, Al-Turki TA (2020) Seed dormancy and germination in Dodonaea viscosa (Sapindaceae) from South-Western Saudi Arabia. Saudi J Biologic Sci 27(9): 2420–2424

Al-Otaibi M (2015) Aspects of the autecology of Arta (Calligonum comosum L. Her) a medical plant from arid region of Saudi Arabia. J Biodivers Environ Sci 6(3):248–255

Alouffi S, Maarfi F, Khan MY, Khan S, Khan MS, Ahmad S (2024) Antiglycation and chemopreventive effects of leaf extracts of Ficus palmata Forssk. Found in the hail region of Saudi Arabia. Biotechnol Genet Eng Rev 40(3):1709–1725

Alqarni AS, Iqbal J, Raweh HS, Hassan AM, Owayss AA (2021) Beekeeping in the desert: foraging activities of honey bee during major honeyflow in a hot-arid ecosystem. Appl Sci 11(20):9756

Al-Rifai A (2018) Identification and evaluation of in-vitro antioxidant phenolic compounds from the Calendula tripterocarpa Rupr. S Afr J Bot 116:238–244

Al-Robai SA, Mohamed HA, Howladar SM, Ahmed AA (2017) Vegetation structure and species diversity of Wadi Turbah Zahran, Albaha area, southwestern Saudi Arabia. Ann Agric Sci 62(1):61–69

Al-Robai SA, Mohamed HA, Ahmed AA, Al-Khulaidi AWA (2019) Effects of elevation gradients and soil components on the vegetation density and species diversity of Alabna escarpment, southwestern Saudi Arabia. Acta Ecol Sin 39(3):202–211

Alruways MW, Khan N (2022) Antioxidant and antibacterial potential of Lycium shawii Roem. And Schult (Awsaj) and Tetraena qatarensis (Hadidi) Beier and Thulin leaves: A herbal approach to combat drug resistance and oxidative stress. Res J Pharm Technol 15(4):1543–1548

Alsayari AS (2014) Anticancer and antiviral activities of Cucurbitacins isolated from Cucumis Prophetarum var. Prophetarum Growing in the Southwestern Region of Saudi Arabia

Alshahri MS, Assaeed AM, Al-Rowaily SL, Abd-ElGawad AM (2022) Floristic composition of the plants associated with the Artemisia monosperma communities within three regions of Riyadh region, Saudi Arabia. Mansoura J Biol 57(4):50–56

Alshallali S (n.d.) Seasonal and Spatial Variability in Secondary Metabolites of Senna Italica Across Ecological Zones in Saudi Arabia. Available at SSRN 5358286

Alshaqhaa MA, Souid I, Korchef A, Alshehri MD (2025) Ethnobotanical study on medicinal plants used in the Aseer province, southwestern Saudi Arabia. J Ethnobiol Ethnomed 21(1):39

Al-Sharari N, Bakhashwain AS, Elfeel A (2020) Profiling and importance of underutilized neglected species of hyper arid climate of Saudi Arabia (Retama raetam—Retem): a review. Life Sci J 17(7):35–42

Alsobeai S (2024) Floristic diversity, life form, and plant life chorology at Al-Fawzan reserve, Kingdom of Saudi Arabia. Appl Ecol Environ Rese 22(4)

Al-Sodany Y, Abo Hatab A, El-Khalafy M, Shaltout S (2025) Ethnobotanical study of Saharo-Arabian endemic plants in the Egyptian flora. Egypt J Bot 65(3):401–423

Alswat AM, Jalal RS, Shami AY, Majeed MA, Alsaedi ZS, Baz L et al (2023) Investigating the metagenomics of the bacterial communities in the rhizosphere of the desert plant Senna italica and their role as plant growth promoting factors. Notulae Botanicae Horti Agrobotanici Cluj-Napoca 51(1):13053–13053

Altieri MA, Nicholls CI (2017) The adaptation and mitigation potential of traditional agriculture in a changing climate. Clim Chang 140(1):33–45

Al-Turki TA, Filfilan SA, Mehmood SF (2000) A cytological study of flowering plants from Saudi Arabia. Willdenowia:339–358

Alwadie H (2002) Ultrastructure of the pollen grains of Withania somnifera (L.) Dunal (Solanaceae), A study from Saudi Arabia. Taeckholmia 22(1):115–119

Alwahibi MS, Bukhari N, Al A (2017) Comparative morphological study on some Ficus species (Moraceae) in Saudi Arabia

Alyemeni MN (2000) Ecological studies on sand dunes vegetation in Al-Kharj region. Saudi J Biologic Sci 7(1):64–87

Alzahrani DA, Albokhari EJ (2018) Taxonomic revision of Saudi Arabian Tetraena maxim. And Zygophyllum L.(Zygophyllaceae) with one new variety and four new combinations. Bangl J Plant Taxonom 25(1):19–43

Alzahrani AM, Farid Ghazal AM, Al-Khulaidi AWA, Ahmed AI, Al-Gilani AA, Alghamdi FM et al (2026) New plant species records from Neom region of Saudi Arabia. J Taibah Univ Sci 20(1):2613986

Alzamil WS (2024) Sustainable urban development and climate Change challenges within Kingdom of Saudi Arabia's vision 2030. Paper presented at the International Conference on Geographic Perspectives on Climate Change Mitigation in Urban and Rural Environments

Amin SA, Wafa'a A (2017) Quantifying of plant species diversity. Composition and density at Dammam region, Estern province, Saudi Arabia. Int J Biodivers Conserv 9(12):389–398

Amtaghri S, Mokhetho KC, Okumu MO, Olofinsan KA, Iqbal S, Saayman JL et al (2025) Indigenous medicine applications, phytochemical and pharmacological properties of Asteriscus graveolens: A comprehensive overview. Fitoterapia 106863

Anonymous (2021) Food system impacts on biodiversity loss. https://www.unep.org/resources/publication/food-system-impacts-biodiversity-loss

Anonymous (2023a) Al-'Ula. https://en.wikipedia.org/wiki/Al-%27Ula#cite_note-Madina-4. Retrieved June 30, 2023

Anonymous (2023b) Heritage Sites in Al-'Ula, Saudi Arabia | ExperienceAlUla.com. Retrieved June 30, 2023

Anonymous (2023c) Paris exhibition unravels mysteries of Saudi Arabia's AlUla. https://www.arabnews.pk/node/1566616/saudi-arabia. Retrieved on July 2, 2023

Anonymous (2023d) Oasis of AlUla. https://www.experiencealula.com/en/discover-alula/nature-outdoors/alula-oasis. Retrieved on July 2, 2023

Anonymous (2023e) Royal Commission for AlUla hosts 'launching the vision for AlUla'. Date of access 8(7):2023. https://www.rcu.gov.sa/en/media-gallery/articles/royal-commission-for-alula-hosts-launching-the-vision-for-alula/

Anonymous (2025a) Plant Diversity of Saudi Arabia—Flora of Saudi Arabia: Renewable Natural Resources. Retrieved from https://plantdiversityofsaudiarabia.info/ethnobotany/

Anonymous (2025b) Comprehensive Regeneration: The AlUla Approach to Sustainable Development. rcu.gov.sa

Anonymous (2025c) (2025). What grows in AlUla? Retrieved from https://www.experiencealula.com/en/about/features/what-grows-in-alula?

Anonymous (2025d) Rhanterium epapposum. Retrieved from https://en.wikipedia.org/wiki/Al-%27Ula#cite_note-Madina-4. Retrieved Dec 30, 2025

Anonymous (2026a) AlUla Agriculture and Food Systems. Royal Commission for AlUla. Agriculture | Fruits of AlUla | The Royal Commission for AlUla

Anonymous (2026b) Agriculture and sustainable development initiatives in AlUla. Royal Commission for AlUla (RCU, 2022), Saudi Arabia. Agriculture | Fruits of AlUla | The Royal Commission for AlUla

Anonymous (2026c) Climate, aridity, and traditional water management systems in AlUla. Climate data and regional assessments. AlUla's Rock Cavities: Natural Reservoirs Preserving Rainwater

Anonymous (2026d) Date palm cultivation and production statistics in AlUla. Royal Commission for AlUla Agricultural Strategy Reports. Date Palm Area & Production

Anonymous (2026e) Citrus farming, seasonal production, and agritourism in AlUla. Regional Agricultural Development Reports Citrus Harvest in AlUla: A Significant Agricultural and Economic Event

Anonymous (2026f) Mango cultivation and diversification of fruit crops in AlUla. Agricultural Features and Market Development Reports. Over 50,000 Mango Trees Flourish in AlUla's Fertile Soil, Yielding over 1,000 Tons Annually

Anonymous. (2026g). Indigenous fruit species and agrobiodiversity conservation in AlUla. Biodiversity and Ethnobotanical Assessments. Agriculture | Fruits of AlUla | The Royal Commission for AlUla

Anonymous (2026h) AlUla Celebrates Key Milestone in Restoration: Closing Ceremony Highlights Visionary Ecosystem Strategy Draft. AlUla Celebrates Key Milestone in Restoration: Closing Ceremony Highlights Visionary Ecosystem Strategy Draft.—News | IUCN

Ansari AA, Abbas ZK, Saggu S, Rehman H, Moawed MM (2014) Growth responses of Lavandula pubescens to temperature regimes of Tabuk, Saudi Arabia. J Med Plant Stud 2(6):38–41

Ansari A, Siddiqui Z, Alatawi F, Alharbi B, Alotaibi A (2022) An assessment of biodiversity in Tabuk region of Saudi Arabia: A comprehensive review. Sustainability 2022, 14, 10564. In: s Note: MDPI stays neutral with regard to jurisdictional claims in published . . .

Arafah AM (2015) Investigation of antimicrobial and antioxidant properties of Tamarix nilotica (Ehrenb.) Bunge leaf extract against catheter-associated urinary tract infections (CAUTIs) in Al Kharj governorate. J Pharm Bioallied Sci

Asiri MMA, Ali MA, Alwahibi MS, Ahmed SS, Rahman MO, Mahato R et al (2026) The first complete chloroplast genome of Lycium shawii: genomic architecture, molecular Phylogenetics and evolutionary insights. Ecol Evol 16(1):e72972

Assirey EAR (2015) Nutritional composition of fruit of 10 date palm (Phoenix dactylifera L.) cultivars grown in Saudi Arabia. J Taibah Univ Sci 9(1):75–79

Baena PA, Brunel A, Fernández-de-Larrinoa Y, Martinez-Cruz TE, Milbank C, Way M (2023) In brief: the white/wiphala paper on indigenous peoples' food systems. Science and innovations for food systems transformation:229–259

Banadka A, Srinivas K, Bhavsar R, Sanjay S, Shaikh A, Murali N et al (2025) Genetic diversity of date palm (Phoenix dactylifera L.) and sustainable utilization. In: Genetic diversity of fruits and nuts. CRC Press, pp 187–226

Baumer C (2022) Rock art in Saudi Arabia–a general introduction and new findings. Adoranten:5–29

Benton MJ (2021) The origin of endothermy in synapsids and archosaurs and arms races in the Triassic. Gondwana Res 100:261–289

Berkes F (2017) Sacred ecology: Routledge

Bharucha Z, Pretty J (2010) The roles and values of wild foods in agricultural systems. Philos Trans Royal Soc B Biol Sci 365(1554):2913–2926

Boakes EH, Dalin C, Etard A, Newbold T (2024) Impacts of the global food system on terrestrial biodiversity from land use and climate change. Nat Commun 15(1):5750

Bornkamm R, Kehl H (1990) The plant communities of the Western Desert of Egypt

Buckle S, Karousakis K, Perry E, Ang G (2020) How to integrate biodiversity in economic and business decisions. Paper presented at the Annales des Mines-Responsabilité & environnement

Burckhardt JL (2013) Travels in Arabia: comprehending an account of those territories in Hedjaz which the Mohammedans regard as sacred. Routledge

Calvin K, Dasgupta D, Krinner G, Mukherji A, Thorne PW, Trisos C et al (2023) IPCC, 2023: climate change 2023: synthesis report. Contribution of working groups I, II and III to the sixth assessment report of the intergovernmental panel on climate change:35–115

Canton H (2021) Food and agriculture organization of the United Nations—FAO. In: The Europa directory of international organizations 2021. Routledge, pp 297–305

Cappelli SL, Domeignoz-Horta LA, Loaiza V, Laine A-L (2022) Plant biodiversity promotes sustainable agriculture directly and via belowground effects. Trends Plant Sci 27(7):674–687

Change IPOC (2001) Climate change 2007: impacts, adaptation and vulnerability. Genebra, Suíça

Charloux G, Crassard R, AlMushawh M, Albukaai D, Chung-To G, Depreux B et al (2022) Khaybar through time. First results of the Khaybar Longue Durée Archaeological Project

(2020–2021) in the light of historical sources. Paper presented at the Proceedings of the Seminar for Arabian Studies

Chaudhary S (1999) Flora of the Kingdom of Saudi Arabia. (vol. 1) Ministry of Agriculture and Water. National Agriculture and Water Research Center Kingdom of Saudi Arabia, Riyadh

Chemsa AE, Zellagui A, Öztürk M, Erol E, Ceylan O, Duru ME, Lahouel M (2018) Chemical composition, antioxidant, anticholinesterase, antimicrobial and antibiofilm activities of essential oil and methanolic extract of Anthemis stiparum subsp. sabulicola (Pomel) Oberpr. Microb Pathog 119:233–240

Citaristi I (2022) Specialized agencies and related organizations within the un system: Food and agriculture organization of the united nations—fao. In: The Europa directory of international organizations 2022. Routledge, pp 307–315

Dalia GG, AlQahtani R (2022) A review on: morphological, phytochemical and medical important of some wild Euphorbiaceae poisonous plants. Haya Saudi J Life Sci 7(11):299–313

Dhshan AI, Goda M, Ibrahim AK (2025) Solenostemma argel: A comprehensive review of its phytochemistry and pharmacological activities. Record Pharm Biomed Sci 9(2):121–133

Dulta K, Kaur G, Kumar K, Mazhar M, Bhardwaj G, Sharma A (2023) Species of arnebia genus found in the western himalayas: Arnebia euchroma (Royle ex Benth.), Arnebia benthamii (Wall. Ex G Don) Johnston, Arnebia guttata Bunge. Immunity Boosting Medicinal Plants of the Western Himalayas, pp 77–105

Edgell HS (2006) Arabian deserts: nature, origin, and evolution. Springer

El-Absy KM (2022) Effect of different habitats conditions on Citrullus colocynthis (L.) schrad. Growing naturally in Egypt and Kingdom of Saudi Arabia. J Adv Biol Biotechnol 25(2):8–29

Elaidarous AA, Osman HE, Galal TM, El-Morsy MH (2022) Vegetation–environment relationship and floristic diversity of wadi Al-sharaea, Makkah Province, Saudi Arabia. Rendiconti Lincei Scienze Fisiche e Naturali 33(1):169–184

El-Amier YA, Al-Hadithy ON (2020) Phytochemical constituents, antioxidant and allelopathic activities of Aizoon canariense L. on Zea mays (L.) and associated weeds. Plant Archiv (09725210) 20(2)

Elfeel AA, Asiry KK, Abohassan RA, Alanazi NA, Elsafori A (2014) Review of the genus acacia in kingdom of saudi arabia and its conservation status

El-Ghani MMA, Huerta-Martínez FM, Hongyan L, Qureshi R (2017) Plant responses to desert environment. In: Plant responses to hyperarid desert environments. Springer, pp 415–470

El-Ghanim WM, Hassan LM, Galal TM, Badr A (2010) Floristic composition and vegetation analysis in Hail region north of central Saudi Arabia. Saudi J Biol Sci 17(2):119–128

Ellis EC, Klein Goldewijk K, Siebert S, Lightman D, Ramankutty N (2010) Anthropogenic transformation of the biomes, 1700 to 2000. Glob Ecol Biogeogr 19(5):589–606

Elmefregy M, El-Sheikh M (2020) Ecological status of sand binder plant white saxaul (Haloxylon persicum) at the managed area of Al-Qassim, Saudi Arabia: plant associations and population structure. Appl Ecol Environ Res 18(2)

El-Said H, Ashgar SS, Bader A, AlQathama A, Halwani M, Ascrizzi R, Flamini G (2021) Essential oil analysis and antimicrobial evaluation of three aromatic plant species growing in Saudi Arabia. Molecules 26(4):959

El-Seedi HR, Kotb SM, Musharraf SG, Shehata AA, Guo Z, Alsharif SM et al (2022) Saudi Arabian plants: A powerful weapon against a plethora of diseases. Plants 11(24):3436

Elshibani F, Alamami A, Khan R, Sulaiman GM, Mohammed HA (2024) Haplophyllum tuberculatum (Forssk.) A. Juss essential oils: seasonal contents variation, bioactivity of the traditionally-favored, high-yield and frequent-use summer season oil. J Oleo Science 73(2): 263–273

Elshibani F, Eman Z, Elmiladi YN, Algilale S, Alhassi S, Husayn NM (n.d.) Pharmacological and biological activities of artemisia species in libya: a comprehensive review article

Engler A (1895) Die. In: Pflanzenwelt Ost-Afrikas und der Nachbargebiete, vol 5. Geographische Verlagshandlung Dietrich Reimer

Essl F, Latombe G, Lenzner B, Pagad S, Seebens H, Smith K et al (2020) The convention on biological diversity (CBD)'s post-2020 target on invasive alien species–what should it include and how should it be monitored? NeoBiota 62:99–121

Fakhry AM, Anazi HK (2017) Floristic diversity and vegetation structure of Al-Dissah area northwestern Saudi Arabia. Egypt J Exper Biol (Zool) 13(2)

FAO (2011) The state of Food and agriculture. Women in agriculture: closing the gender gap for development. Food and Agriculture Organization of the United Nations, Rome, p 160

FAO (2017) Food and Agriculture Organization of the United Nations (FAO). Climate-Smart Agriculture Sourcebook. FAO, Rome

FAO (2018) The state of the world's biodiversity for food and agriculture. FAO, Rome

Food F (2017) Agriculture organization of the united nations (FAO). Via delle Terme di Caracalla 100

Ghareb HE, Gabr M (2017) In vitro propagation of Searsia Tripartita (Ucria) Moffett As an endangered medicinal plant. Egypt J Desert Res 67(2):361–377

Ghazanfar SA, Fisher M (1998) Vegetation of the Arabian peninsula. Springer Science & Business Media

Gomaa SB, Saleh NM (2014) Medicinal attributes of Rumex vesicarius (Polygonaceae) growing in Sakaka, AlJouf, Saudi Arabia. Med J Cairo Univ 82(1):917–926

Gouda HM, Morsy AA, Youssef AK, Tolba IAE-M, Hassan GOO (2022) Phytochemical profile and antimicrobial assessment of Abutilon fruticosum Guill. & Perr. Growing in Gebel Elba, Egypt. Egypt J Chem 65(132):1299–1305

Grabowska K, Buzdygan W, Galanty A, Wróbel-Biedrawa D, Sobolewska D, Podolak I (2023) Current knowledge on genus Bassia All.: a comprehensive review on traditional use, phytochemistry, pharmacological activity, and nonmedical applications

Hamdi A, Viane J, Mahjoub MA, Majouli K, Gad MHH, Kharbach M et al (2018) Polyphenolic contents, antioxidant activities and UPLC–ESI–MS analysis of Haplophyllum tuberculatum A. Juss leaves extracts. Int J Biol Macromol 106:1071–1079

Hassan W, El Gamal A, El-Sheddy E, Al-Oquil M, Farshori N (2014) The chemical composition and antimicrobial activity of the essential oil of Lavandula coronopifolia growing in Saudi Arabia. J Chem Pharm Res 6:604–615

Hassan M, Haq SM, Rasool A, Fatima S, Ashraf A, Zulfajri M, Hanafiah MM (2021) Ethnobotanical properties and traditional uses of medicinal plant Abutilon theophrasti Medik. In: Medicinal and aromatic plants: healthcare and industrial applications. Springer, pp 271–285

Hausleiter A, Dinies M, Bouchaud C, Intilia A, Schimmel L, Zur A (2021) Al-Ula, Saudi Arabia. Archaeology and environment from the Early Bronze Age (3rd mill. BCE) onwards. Season 2019

Hossain MA (2019) Biological and phytochemicals review of Omani medicinal plant Dodonaea viscosa. J King Saud Univ Sci 31(4):1089–1094

Hussain MI, Akhtar N, Qureshi AS, Gallacher D (2021) Perennial forage grass production on the marginal Arabian Peninsula land. In Sustainable Agriculture Reviews 52 (pp. 279–308): Springer

IPBES W (2019) Intergovernmental science-policy platform on biodiversity and ecosystem services. Summary for policy makers of the global assessment report on biodiversity and ecosystem services of the intergovernmental science-policy platform on biodiversity and ecosystem services. IPBES Secretariat, Bonn

IPCC (2021) Intergovernmental Panel on Climate Change. Climate Change 2021: The physical science basis. Contribution of Working Group I to the Sixth Assessment Report. Cambridge University Press, Cambridge. https://doi.org/10.1017/9781009157896

Iseman T, Miralles-Wilhelm F (2021) Nature-based solutions in agriculture: the case and pathway for adoption. Food & Agriculture Org

Izah SC, Richard G, Odubo TC (2023) Citrus aurantifolia: phytochemical constituents, food preservative potentials, and pharmacological values. In: Herbal medicine phytochemistry: applications and trends. Springer, pp 1–26

Kdimy A, El Yadini M, Guaadaoui A, Bourais I, El Hajjaji S, Le HV (2022) Phytochemistry, biological activities, therapeutic potential, and socio-economic value of the caper bush (Capparis spinosa L.). Chem Biodivers 19(10):e202200300

Khalil E (2017) The potentials of maritime archaeology in the Kingdom of Saudi Arabia. In: Human interaction with the environment in the Red Sea. Brill, pp 346–361

Khayat RO, Felemban DN, Altowairqi TK, Aljahdali MO (2023) Measurement of Saudi society environmental awareness in relation to 2030 Saudi vision and the Saudi green initiative. (AJSP) ISSN 2663:5798

Kürschner H, Neef R (2018) Taymā' oasis (Saudi Arabia) and its surroundings–a first synthesis of the Flora, vegetation, natural resources, and floral history. In: In. JSTOR, pp 86–127

Legg S (2021) IPCC, 2021: climate change 2021—the physical science basis. Interaction 49(4): 44–45

Li L (2021) The state of the world's land and water resources for food and agriculture (SOLAW): systems at breaking point

Lightfoot DR (1996) Syrian qanat Romani: history, ecology, abandonment. J Arid Environ 33(3): 321–336

Llewellyn O, Hall M, Miller A, Al-Abbasi T, Al-Wetaid A, Al-Harbi R, Al-Shammari K (2011) Important plant areas in the arabian peninsula: 4. JABAL AJA'. Edinburgh J Botany 68(2): 199–224

Mandaville JP (2013) Flora of eastern Saudi Arabia: Routledge

Mekouar MA (2018) 15. Food and agriculture organization of the united nations (FAO). Yearbook of International Environmental Law 29:448–468

Moawed MM, Ansari A (2015) Wild plants diversity of Red Sea coastal region, Tabuk, Saudi Arabia. J Chem Pharm Res 7:220–227

Mohammed HA, Al-Omar MS, Khan RA, Mohammed SA, Qureshi KA, Abbas MM et al (2021) Chemical profile, antioxidant, antimicrobial, and anticancer activities of the water-ethanol extract of Pulicaria undulata growing in the oasis of central Saudi Arabian desert. Plants 10(9):1811

Mohammed HA, Al Ali A, Abd-El-Wahab MF, Almahmoud SA, Khan RA, Sulaiman GM et al (2025) Anti-microbial and anti-hepatitis-A activities of Asphodelus refractus boiss. Extract and pure isolated metabolites. J Mol Struct 1322:140597

Mohammed-Ibtisam M, Doka G (2018) Check list of flora and vegetation of Hafer Albatin region, northeastern Saudi Arabia. MOJ Ecol Environ Sci 3:138–143

Moharram BA, Al-Mahbashi HM, Saif-Ali R, AliAqlan F (2018) Phytochemical, anti-inflammatory, antioxidant, cytotoxic and antibacterial study of Capparis cartilaginea decne fromyemen. Int J Pharm Pharm Sci 10(6):38–44

Morabito L, Quartermaine J, Roberts K, Tuttle CA, Abu-Azizeh W (2022) Discovering al-ʿUlā (AlUla). Paper presented at the Proceedings of the Seminar for Arabian Studies

Moustafa ARAM, Mansour R, Alotaibi MO (2019) *Cleome droserifolia*: an Egyptian natural heritage facing extinction. Asian J Plant Sci Res 9(1):14–21

Mseddi K, Alghamdi A, Abdelgadir M, Sharawy S, Chaieb M, Miller T (2021) Phytodiversity distribution in relation to altitudinal gradient in Salma Mountains–Saudi Arabia. Global Ecol Conserv 27:e01525

Nicolétis É, Caron P, El Solh M, Cole M, Fresco LO, Godoy-Faúndez A et al (2019) Agroecological and other innovative approaches for sustainable agriculture and food systems that enhance food security and nutrition. A report by the High Level Panel of Experts on Food Security and Nutrition of the Committee on World Food Security

Organization, W. H. (2024) Report of a scoping meeting for the selection of indicators to monitor the impact of extreme heat on maternal, newborn and child health, Geneva, Switzerland, 24–25 April 2023: World Health Organization

Osman AK, Al-Ghamdi F, Bawadekji A (2014) Floristic diversity and vegetation analysis of Wadi Arar: A typical desert Wadi of the Northern Border region of Saudi Arabia. Saudi J Biol Sci 21(6):554–565

Osman H, Soliman M, El-Morsy M, Elaidarous A, Galal T (2024) Plant diversity and community composition along the prevailing environmental conditions in Wadi Murr Al-Zahran, southwestern Saudi Arabia. Appl Ecol Environ Res 22(4)

Petrie, J. M. (2007). Arabian desert primer: ornamental potential of hyper-arid adapted plants from Saudi Arabia

Phondani PC, Bhatt A, Elsarrag E, Alhorr YM (2015) Seed germination and growth performance of Aerva javanica (Burm. f.) Juss ex Schult. J Appl Res Med Aromat Plants 2(4):195–199

Qari SH, Alrefaei AF, Filfilan W, Qumsani A (2021) Exploration of the medicinal flora of the Aljumum region in Saudi Arabia. Appl Sci 11(16):7620

Qureshi R, Bhatti GR (2005) Nara Desert, Pakistan: Part I: Soils, Climate and Vegetation. Rangel, 27(5):27–31

Qureshi R, Bhatti GR (2008) Ethnobotany of plants used by the Thari people of Nara Desert, Pakistan. Fitoterapia 79(6):468–473

Qureshi R, Bhatti GR, Memon RA (2010) Ethnomedicinal uses of herbs from Nara Desert, Pakistan. Pak J Botany 42(2):839–885

Qureshi R, Arif IA, Asiri TKA-F, Alsheikh AA, Baig MB (2024) Role of plants in restoring degraded ecosystems to improve biodiversity in alula region—Saudi Arabia. In: Environment, climate, plant and vegetation growth. Springer, pp 205–227

Ragupathy S, Seigler D, Ebinger J, Maslin B (2014) New combinations in *Vachellia* and *Senegalia* (Leguminosae: Mimosoideae) for south and West Asia. Phytotaxa 162(3):174–180

Rayyan WA, Alshammari SA, ALSammary AM, AL-Shammari MS, Seder N, Abu-Qatouseh LF (2018) The phytochemical analysis and antimicrobial activity of Pergularia tomentosa in North East Kingdom of Saudi Arabia KSA. Biomed Pharmacol J 11(4):1763–1771

RCU (Royal Commission for AlUla) (2023) Shaping a sustainable future for AlUla. AlUla, Saudi Arabia

Restoration G (2021) Ecosystem restoration for people, nature and climate. UNEP & FAO, Nairobi

Riehl S (2010) Plant production in a changing environment: the archaeobotanical remains from Tell Mozan. Development of the environment, subsistence and settlement of the city of Urkeš and its region. Studien zur Urbanisierung Nordmesopotamiens A 3:13–158

Ringrose S (1996) The geomorphological context of calcrete deposition in the Dalmore Downs area, Northern Territory, Australia. J Arid Environ

Rojas-Downing MM, Nejadhashemi AP, Harrigan T, Woznicki SA (2017) Climate change and livestock: impacts, adaptation, and mitigation. Clim Risk Manag 16:145–163

Rosenzweig C, Mbow C, Barioni LG, Benton TG, Herrero M, Krishnapillai M et al (2020) Climate change responses benefit from a global food system approach. Nat Food 1(2):94–97

Saad AAM, Asalam AA (2015) Culture of leaves and immature fruits of *Maerua crassifolia in vitro*. Int J Inf Res Rev 2:496–498

Saudi Meteorological Authority (2021) Climate normals of Saudi Arabia (1991–2020). Ministry of Environment, Water and Agriculture, Riyadh

Sawicki P (2016) Les tombeaux nabatéens de Hégra

Sayed OH, Masrahi YS (2025) Grasses of Saudi Arabia: A review. Grass Forage Sci 80(1):e12705

Selmar D, Kleinwächter M (2013) Stress enhances the synthesis of secondary plant products: the impact of stress-related over-reduction on the accumulation of natural products. Plant Cell Physiol 54(6):817–826

Shabana, H. A., Mahmoud, T., Gairola, S., Al Ketbi, A., Aljasmi, M., & Al Sallani, M. (2021). Effect of storage conditions and sodium hypochlorite treatment on germination of *Cucumis prophetarum* (Cucurbitaceae) seeds from arid Arabian deserts

Sher H, Al-Yemeni M, Masrahi YS, Shah AH (2010) Ethnomedicinal and ethnoecological evaluation of Salvadora persica L.: a threatened medicinal plant in Arabian Peninsula. J Med Plants Res 4(12):1209–1215

Sher H, Al-yemeni MN, Wijaya L (2011) Ethnobotanical and antibacterial potential of Salvadora persica l: a well known medicinal plant in Arab and Unani system of medicine. J Med Plant Res 5(7):1224–1229

Sheweita SA, ElHady SA, Hammoda HM (2020) Trigonella stellata reduced the deleterious effects of diabetes mellitus through alleviation of oxidative stress, antioxidant-and drug-metabolizing enzymes activities. J Ethnopharmacol 256:112821

Soliman, M. S., Abdella, A., Khidr, Y. A., Osman, H. G., Aladadh, M. A., & Elsanhoty, R. M. (2022). Pharmacological activities and characterization of phenolic and flavonoid compounds in Solenostemma argel extract

Soomro AH, Marri A, Shaikh N (2023) Date palm (Phoenix dactylifera): A review of economic potential, industrial valorization, nutritional and health significance. Neglected plant foods of South Asia: Exploring and valorizing nature to feed hunger, pp 319–350

Stenseke M (2024) On IPBES' framing of the values of nature. Mistra BIOPATH

Suleiman MHA (2019) Ethnobotanical, phytochemical, and biological study of Tamarix aphylla and Aerva javanica medicinal plants growing in the Asir region, Saudi Arabia. Tropic Conserv Sci 12:1940082919869480

Suleiman M, Quoreshi A, Bhat N, Manuvel A (2018) Species identification of Vachellia pachyceras from Kuwait and its relatives Vachellia gerrardii and Vachellia tortilis, based on multilocus plastid gene sequences. Edinb J Bot 75(1):73–90

Sulieman AME, Abdallah EM, Alanazi NA, Idriss H, Adnan M, Jamal A et al (2024) Bioactive profiling of Rumex vesicarius L. from the Hail region of Saudi Arabia: A study on its phytochemical and elemental analysis, antibiofilm, antibacterial, antioxidant properties, and molecular docking analysis. Front Microbiol 15:1421143

Swiderska K, Argumedo A, Wekesa C, Ndalilo L, Song Y, Rastogi A, Ryan P (2022) Indigenous peoples' food systems and biocultural heritage: addressing indigenous priorities using decolonial and interdisciplinary research approaches. Sustainability 14(18):11311

Tengberg M (2012) Beginnings and early history of date palm garden cultivation in the Middle East. J Arid Environ 86:139–147

UNCCD (2018) Global land outlook. United Nations Convention to Combat Desertification (UNCCD), Bonn, Germany, pp 340

Vincent P (2008) Saudi Arabia: an environmental overview. CRC Press

Vincent B, Bourillon J, Gotty K, Boukcim H, Selosse M-A, Cambou A et al (2024) Ecological aspects and relationships of the emblematic Vachellia spp. exposed to anthropic pressures and parasitism in natural hyper-arid ecosystems: ethnobotanical elements, morphology, and biological nitrogen fixation. Planta 259(6):132

Wafa'a A (2022) Floristic diversity and vegetation of the az Zakhnuniyah Island, Arabian Gulf, Saudi Arabia. Heliyon 8(7)

Yonbawi AR, Abdallah HM, Alkhilaiwi FA, Koshak AE, Heard CM (2021) Anti-proliferative, cytotoxic and antioxidant properties of the methanolic extracts of five Saudi Arabian flora with folkloric medicinal use: *Aizoon canariense, Citrullus colocynthis, Maerua crassifolia*. Rhazya stricta and Tribulus macropterus Plants 10(10):2073

Youssef RS (2013) Medicinal and non-medicinal uses of some plants found in the middle region of Saudi Arabia. J Med Plants Res 7(34):2501–2513

Zerrouk S, Escuredo O, Rodríguez-Flores MS, Seijo MC (2021) Palynological characterisation of sedra honeys (Ziziphus lotus) produced in Algeria. Grana 60(1):69–80

Zohary M, Dārôm D (1982) Plants of the Bible: a complete handbook to all the plants with 200 full-color plates taken in the natural habitat. Cambridge Univ Press

Zohary M, Orshan G (1956) Ecological studies in the vegetation of the Near East Deserts II Wadi Araba: with 12 figures, 2 photographs and 9 tables. Vegetatio 7(1):15–37

Medicinal Plant Wealth of Pakistan with Potential to Combat Cardiometabolic Disorders

13

Rida Siddique, Rahmatullah Qureshi, Qadeer Ahmad, Shumaila Mehdi, Bushra Shaukat, Abdul Malik, Ruqaiyyah Siddiqui, Naveed Ahmed Khan, and Malik Hassan Mehmood

Abstract

Metabolic syndrome ((MetS) is a rapidly growing global health challenge characterized by obesity, hypertension, dyslipidemia, and insulin resistance, significantly increasing the risk of cardiovascular diseases and type 2 diabetes. The high prevalence of this syndrome in South Asian populations, including Pakistan, underscores the need for effective, affordable, and sustainable therapeutic strategies. This chapter explores the potential of medicinal plants indigenous to Pakistan as complementary and alternative interventions for the prevention and management of metabolic syndrome. Drawing upon Pakistan's

R. Siddique · Q. Ahmad · S. Mehdi · B. Shaukat
Department of Pharmacology, Faculty of Pharmaceutical Sciences, Government College University, Faisalabad, Pakistan

R. Qureshi
Department of Botany, Faculty of Sciences, Pir Mehr Ali Shah Arid Agriculture University, Rawalpindi, Pakistan
e-mail: rahmatullahq@uaar.edu.pk

A. Malik
Department of Pharmacology, College of Pharmacy, University of Sargodha, Sargodha, Pakistan

R. Siddiqui
Institute of Biological Chemistry, Biophysics and Bioengineering, Heriot-Watt University, Edinburgh, UK

N. A. Khan
School of Science, College of Science and Engineering, University of Derby, Derby, UK

M. H. Mehmood (✉)
Department of Pharmacology, Faculty of Pharmaceutical Sciences, Government College University, Faisalabad, Pakistan

Department of Pharmaceutical Sciences, Government College University, Lahore, Pakistan
e-mail: hassan.mehmood@gcu.edu.pk

R. Qureshi (ed.), *Plant Biodiversity*, https://doi.org/10.1007/978-981-92-1471-6_13

rich ethnobotanical heritage and biodiversity, the chapter reviews key medicinal species, including *Berberis brandisiana, Nigella sativa, Ajuga bracteosa, Viola odorata*, and *Ziziphus oxyphylla*, highlighting their bioactive constituents and mechanisms of action in regulating glucose metabolism, lipid homeostasis, oxidative stress, inflammation, and vascular function. The therapeutic potential of these botanicals is discussed in the context of their antioxidant, anti-inflammatory, antihyperglycemic, antihyperlipidemic, and cardioprotective properties. The chapter also addresses critical challenges related to phytochemical standardization, clinical validation, safety assessment, regulatory frameworks, and biodiversity conservation. It concludes that integrating scientifically validated medicinal plants into healthcare systems may provide cost-effective and culturally acceptable approaches for managing metabolic syndrome while promoting sustainable utilization of Pakistan's valuable plant resources.

Keywords

Metabolic syndrome · Cardiometabolic diseases · Medicinal plants · Ethnopharmacology · Insulin resistance · Bioactive compounds · Phytochemical standardization

Abbreviations

ACE	Angiotensin-Converting Enzyme
CETP	Cholesterol Ester Transfer protein
eNOS	Endothelial Nitric Oxide Synthase
FFA	Free Fatty Acids
GLP-1	Glucagon-Like Peptide-1
GLUT4	Glucose Transporter Type 4
HMG-CoA	Inhibition of 3-hydroxy-3-methyl-glutaryl-coenzyme A
IR	Insulin Resistance
LPL	Lipoprotein Lipase
MetS	Metabolic Syndrome
PEPCK	Phosphoenolpyruvate Carboxykinase
RAAS	Renin-Angiotensin-Aldosterone System
RCTs	Randomized Controlled Trials
ROS	Reactive Oxygen Species
SDLDL	Small, Dense Low-Density Lipoprotein
SGLT2	Sodium-Glucose Cotransporter 2
TG	Triglycerides
VLDL	Very-Low-Density Lipoproteins
WHO	World Health Organization

13.1 Introduction

The phenomenon of the metabolic syndrome can be considered a severe and growing national health threat characterized by a set of interrelated conditions clinical manifestations of central obesity, hypertension, dyslipidemia, and insulin resistance result in hyperglycemia. This comorbid convergence contributes highly toward the risk of acquiring cardiovascular diseases, type 2 diabetes, and other chronic conditions (Fig. 13.1). It has become pandemic in many places globally due to radical changes in lifestyles toward sedentary ones, high levels of processed foods and sugars in the diet, and growing urbanization (Shah et al. 2019). According to the latest International Diabetes Federation (IDF) Diabetes Atlas, approximately 11.1% of the global adult population (aged 20–79) is living with diabetes. The burden of the region, especially South Asia, is particularly worrying. The epidemiological transition in countries such as Pakistan is fast, and there is a high rate of non-communicable disease increases to match the rate of infectious diseases. In Pakistan, the metabolic syndrome has been proposed to have a prevalence of between 20% and 40% of adults due to genetic aspect, feeding habits that are high

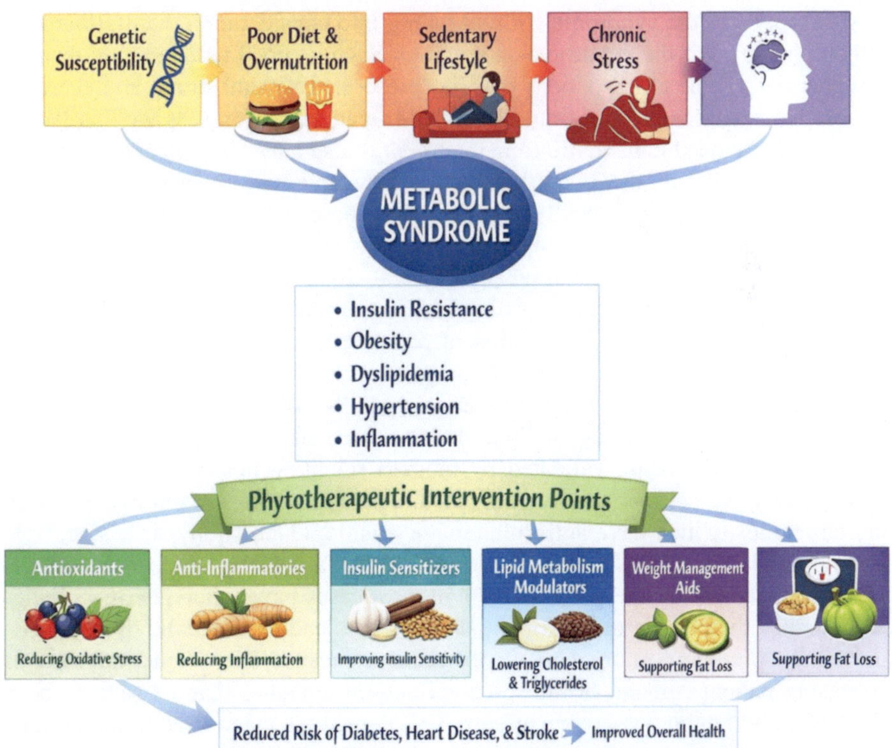

Fig. 13.1 Multifactorial pathogenesis of metabolic syndrome and phytotherapeutic intervention points

in refined carbohydrates and fats, and inadequate physical exercises. This influx puts a tremendous burden on existing strained healthcare systems struggling with resource shortages, resulting into more morbidity, death, and economical loss due to loss of productivity and high cost of treatment. The increasing trend of metabolic syndrome is not only a health issue but a socio-economic problem and requires effective and immediate intervention measures (Ansari et al. 2025).

The traditional pharmaceutical arsenal of metabolic syndrome management and treatment of its components such as statins to treat dyslipidemia, antihypertensives, metformin to treat insulin resistance, and more recent glucose-lowering drugs has certainly saved lives and beneficial clinical outcome. Nevertheless, chronic use of these therapy options is associated with severe constraints and side effects, completely ruining patient adherence and quality of life. Although useful, statins may develop myopathy, liver enzyme abnormality, and new-onset diabetes among some patients (Liaqat et al. 2023). Antihypertensives, e.g., beta-blockers and diuretics, can cause electrolyte imbalances, fatigue, and sexual dysfunction. Metformin chronic use is often associated to gastrointestinal discomfort, and current advances of SGLT2 inhibitors and GLP-1 receptor agonists are not only expensive but have numerous side effects, such as genitourinary infections, and uncommon but serious adverse events like pancreatitis. Moreover, these medications normally focus on single parts of the syndrome without taking into consideration the multifactorial pathophysiology of the syndrome as a whole. Such a siloed practice may result in polypharmacy, wherein a patient may be prescribed several drugs and there are higher chances of both drug-drug effects and financial toxicity, as well as lack of adherence. The need to have life-long treatment is also a psychological and financial burden, especially in the countries with a low and middle income where not everyone can afford such costly medications. These negative effects highlight a serious shortcoming in chronic disease treatment: More holistic, safe, and sustainable ways of treatment are required (Liaqat et al. 2023).

Considering these shortcomings of synthetic pharmaceuticals, the world has once again shown interest in the application of plants, both as therapeutic agents and as nutraceuticals in the management of complex chronic diseases such as metabolic syndrome. This renewed interest is based on a number of factors which are compelling. The ethnopharmacological knowledge which has been passed on through generations is now being validated by modern scientific investigation, which has spotted bioactive compounds in plants such as polyphenols, flavonoids, alkaloids, and terpenoids, which have anti-inflammatory, antioxidant, antihypertensive, lipid-lowering and insulin-sensitizing properties (Palla et al. 2021). Compared to single-target synthetic drugs, the plant extracts in most cases have a polypharmacological action, since they all act on multiple pathways that lead to metabolic dysregulation and hence provide a more holistic therapeutic means. This is in line with the increasing patient demand of the so-called natural remedies that are seen as being both safer and having lesser side effects. This demand is manifested in the boom of the global nutraceutical market (Palla et al. 2021). This paradigm shift is not a request to leave standard medicine but to seek integrative models in which evidence-based phytotherapy can be used as a supplementary or even preventive measure which may lead to a lower dosage and burden of synthetic drugs and counteract its side effects.

The role that medicinal plants play in the Pakistani healthcare environment is extensive and complex, and it exists in this global context. Pakistan has been blessed with biodiversity with more than 6000 species of wild flora, with many of them finding their way to traditional Unani, Ayurvedic, and folk medicine. Even in rural and peri-urban settings where the Pakistani population is composed of an overwhelming majority of populace with little or no access to allopathic medicine, herbal medicine still forms the initial mode of treatment of a plethora of disorders, including ones pertaining to metabolic syndrome (Sharif et al. 2020). *Momordica charantia* (Karela), *Allium sativum* (Garlic), *Trigonella foenum-graecum* (Methi), and *Cinnamomum verum* (Dalchini) are widely used plants in managing blood sugar, hypertension, lipids, diabetes, and insulin sensitivity. This is a unique opportunity. In the first place, the scientific validation and standardization of these locally available and culturally accepted remedies might result in the formation of the cheap, accessible, and effective adjunct medicine to metabolic syndrome that directly responds to the disease burden in the nation. Secondly, it holds socioeconomic advantages, as it forms value chains on the production, processing, and marketing of medicinal plants, which contribute to the rural livelihoods (Islam et al. 2021). Nonetheless, there are also considerable obstacles faced, such as the absence of strict clinical studies, quality control problems, extract standardization, sustainable harvesting to eliminate species extinction, and incorporation of such knowledge into the official healthcare system and its regulation. Carefully designed research, policy support, and parity between the pharmacopeia and the population could make it a cornerstone of community health resilience and a part of the global search for sustainable solutions to the pandemic of metabolic syndrome. Essentially, the utilization of the potential of medicinal plants is a culturally appealing, cost-effective, scientifically prospective, complementary approach to traditional health care in fighting one of the most urgent health problems in the country (Gutiérrez-Cuevas et al. 2024).

13.2 Metabolic Syndrome and Cardiometabolic Disorders: Pharmacological Targets

The metabolic syndrome (MetS) and the cardiometabolic disorders' pathophysiology are a complex, interlaced network of dysregulated biological pathways. It is essential to comprehend these discrete yet interrelated targets to develop effective therapeutic strategies. The development of modern pharmacology is characterized by a shift from treating isolated symptoms to targeting the underlying mechanisms, which is where plant-derived therapeutics are particularly promising, as they naturally possess the ability to act as polypharmacological agents. The essential targets may be systematized into insulin signaling, lipid metabolism, vascular functioning, and systemic inflammatory events (Younis et al. 2018).

13.2.1 Insulin Resistance and Glucose Dysregulation

In the center of MetS is insulin resistance (IR), whereby there is an abnormal insulin sensitivity of the major metabolic tissues that are mainly skeletal muscle, liver, and adipose tissue. This malfunction causes a series of metabolic imbalances. In muscle, insulin-mediated glucose uptake is impaired, thus resulting in poor postprandial glucose disposal. IR does not inhibit gluconeogenesis in the liver leading to overproduction of hepatic glucose. At the same time, IR worsens lipolysis in adipose tissue, overflowing the circulation with free fatty acids (FFAs), which further worsens the insulin signaling in a vicious circle (Alam et al. 2022).

IR mechanobiological basis comprises the breakage of the insulin receptor substrate (IRS)/phosphoinositide 3-kinase (PI3K)/protein kinase B (Akt) cascade, which is a fundamental signal transduction transducer of insulin metabolic activities. The possible pharmacological interventions in this case are insulin receptor sensitivity, downstream effector (Akt) stimulation, and glucose transporter type 4 (GLUT4) translocation to the cell membrane (Ullah et al. 2025). In addition, controlling of hepatic enzymes in the pathway of glucose production, including phosphoenolpyruvate carboxykinase (PEPCK) and glucose-6-phosphatase (G6Pase), is a proven approach. In addition to these direct actions, the lipid overload in non-adipose tissues (lipotoxicity) and the adipokine release of adipose tissue (e.g., the increase of adiponectin, the reduction of leptin resistance) are also the therapeutic objectives that are required to restore the insulin sensitivity of the body and normalize blood glucose (Anwar et al. 2022).

13.2.2 Dyslipidemia and Altered Lipid Homeostasis

The triad of elevated triglycerides (TG), low HDL cholesterol (HDL-C), and a high concentration of small, dense LDL (sdLDL) is clinically known as atherogenic dyslipidemia or the lipid triad and it serves as a powerful, direct driver of atherosclerosis and cardiovascular disease (CVD) in patients with metabolic syndrome. The basis of this lipid triad is mainly the overproduction of very-low-density lipoproteins (VLDL) by the liver caused by the intake of FFA into the liver by the insulin-resistant adipose tissue. The lipoprotein lipase (LPL) enzyme that clears the TG in the blood is usually suppressed (Alotaibi et al. 2021).

Pharmacological treatment has several interventions in lipid metabolism. The mechanism of action of statins is the inhibition of 3-hydroxy-3-methyl-glutaryl-coenzyme A (HMG-CoA) reductase, that is, the rate-limiting step in the production of cholesterol. Peroxisome proliferator-activated receptors (PPARs) are other important targets of the peroxisome proliferator-activated receptor, which are nuclear receptors that mediate lipid and glucose metabolism (Shamim et al. 2022). ATP-binding cassette transporters (ABCs) and Cholesterol ester transfer protein (CETP) are critical regulators of cardiovascular health and lipid metabolism. They dictate how cholesterol is cleared from peripheral tissues and processed by the liver. It is not only to decrease LDL-C but to restore the whole atherogenic lipid profile by decreasing lipoprotein containing TG, increasing functional HDL, and decreasing the production of sdLDL (Shamim et al. 2022).

13.2.3 Hypertension and Endothelial Dysfunction

In MetS, hypertension is, in most instances, salt-sensitive and highly interdependent on insulin resistance, overactivity of the chronic sympathetic nervous system, and, by nature, endothelial dysfunction. The lining of the blood vessels, the single-celled endothelium, becomes incapable of its normal regulation of the vascular tone when its usual production of the vasodilators such as nitric oxide (NO) is inhibited and vasoconstrictors such as endothelin-1 are stimulated (Khalid et al. 2020).

The main pharmacological targets are thus the renin-angiotensin-aldosterone system (RAAS), which is one of the major blood pressure and fluid regulation systems. Angiotensin-converting enzyme (ACE) inhibitors and angiotensin II receptor blockers (ARBs) represent examples of the gold standard of treatment, which do not only decrease blood pressure but also have direct protective properties on the endothelium and decrease vascular inflammation. The sympathetic nervous system is approached through the beta-blockers and alpha-agonists. More importantly, the direct target is the endothelial nitric oxide synthase (eNOS) pathway. Agents that can increase NO bioavailability via suppression of its inactivation through reactive oxygen species (ROS) or stimulation of its synthesis (e.g., L-arginine) are the main focus of the correction of endothelial dysfunction. Enhancement of endothelial health is therefore a dual-purpose outcome: It reduces blood pressure and meets one of the initiating events in the pathogenesis of atherosclerotic cardiovascular disease (Firdous et al. 2025).

13.3 Oxidative Stress and Chronic Low-Grade Inflammation

Oxidative stress, sterile, chronic low-grade inflammation has become the unifying pathological mechanisms which cause and result in the rest of the components of MetS, in a self-perpetuating cyclec. The surplus of nutrient flux (glucose and FFA) in cases of insulin resistance results in overloading of the mitochondria and the overproduction of ROS in cells. Reactive oxygen species (ROS) trigger innate immune responses by simultaneously activating the NF-κB pathway and the NLRP3 inflammasome. This leads to sustained release of pro-inflammatory cytokines including tumor necrosis factor-alpha (TNF- 6), interleukin-1 beta (IL-16), and interleukin-6 (IL-6) (Singhai et al. 2024).

Subsequently, these cytokines are able to cause insulin resistance of metabolic tissues and stimulate oxidative stress even further. In this area, important pharmacological endpoints are thus the triggers of ROS (e.g., NADPH oxidase enzymes), the pro-inflammatory transcription factor (NF-KB, NLRP3), and the pro-inflammatory cytokines. This pathogenic cycle is supposed to be interrupted with the help of strengthening endogenous antioxidant defenses (e.g., through the activation of the nuclear factor erythroid 2-related factor 2 Nrf2 pathway) and inhibition of the triggering of pro-inflammatory transcription factors. The strategy intends to heal the soil of the metabolic dysfunction as opposed to the seeds alone, which could have a far-reaching impact on all aspects of MetS (Hussain et al. 2020).

13.4 Molecular Targets for Plant-Derived Bioactives

The complex pathophysiology of MetS and its multiple interconnected targets is an ideal case when using plant-derived bioactives could be applied. In contrast to the majority of synthetic drugs, which are targeted on one of the above-mentioned pathways, phytochemicals tend to show pleiotropic properties, simultaneously influencing a number of these pathways. This polypharmacology is in perfect harmony with the complicated condition of MetS (Abid et al. 2024).

A huge number of plant compounds have been discovered to exhibit particular affinity with the major cardiometabolic disease molecular targets:

- To Enhance Insulin Resistance: Bioactive agents such as berberine (Berberis species) have been found to mobilize AMP-activated protein kinase (AMPK) which is a cellular energy sensor that enhances insulin sensitivity and suppresses hepatic gluconeogenesis. Curcumin (derived of turmeric) has the potential to block protein tyrosine phosphatase 1B (PTP1B), an insulin receptor signaling negative regulator, thus enhancing insulin receptors signaling.
- In Case of Dyslipidemia: Sirtuin 1 (SIRT1) and PPAR-γ coactivator-1alpha (PGC-1alpha) can be activated by polyphenols like resveratrol (in grape and berries) to induce fatty acid oxidation and mitochondrial biogenesis. Fenugreek saponins and other plants have the ability to impede the absorption of cholesterol in the intestine and regulate hepatic cholesterol production.
- In the Case of Hypertension and Endothelial Dysfunction: Flavonoids such as quercetin and catechins (abundant in onions, tea, and numerous fruits) are strong antioxidants which increase the eNOS activity and NO bioavailability and deactivate ROS that would otherwise inactivate NO. Garlic allicin has been documented to possess weak ACE-inhibitory protein and vasodilatory.
- In Oxidative Stress and Inflammation: Curcumin, resveratrol, and a great deal of other polyphenols are direct inhibitors of NF-KB and NLRP3 pathways. They are also very strong activators of the Nrf2 pathway, which increases the expression of endogenous antioxidants such as superoxide dismutase (SOD) and glutathione peroxidase. Gingerols of ginger and thymoquinone of black seed (Nigella sativa) are just a few examples of compounds with strong anti-inflammatory and antioxidant activity in experimental models of MetS.

These botanicals have a therapeutic potential because they provide a multi-targeted attack on the syndromic network of MetS. An example is that a single extract of cinnamon can contain compounds that can enhance the insulin sensitivity, prevent major gluconeogenic enzymes, exhibit antioxidant activities, and show slight anti-inflammatory activity. Such holistic pharmacology profile is especially applicable in areas such as Pakistan where well-developed tradition of medicinal plants provides culturally attractive and potentially inexpensive source of complementary medical therapy development (Ali et al. 2023). The future of metabolic syndrome treatment can perhaps rely on the clever utilization of these natural multi-target agents as either sole nutraceuticals or as complements to more traditional therapy to offer a more holistic and sustainable approach to this issue of worldwide health concern. The most important follow-up is to reconcile conventional knowledge with rigorous contemporary science—through standardized extracts, solid clinical trials, and definitive safety profiling—to make these plant-based solutions to be a complete armament in the evidence-based medical itinerary of cardiometabolic disease (Minniti et al. 2023).

13.5 Ethnomedicinal Background of Pakistani Medicinal Plants

The ethnomedicinal heritage of Pakistan stands as a vast and intricate repository of knowledge, deeply interwoven with the nation's ecology, culture, and healthcare practices. This foundation is built upon an exceptional diversity of flora, formalized and informal traditional healing systems, and the widespread, everyday use of plants for maintaining health and treating disease. Today, this rich tradition is undergoing a critical and necessary evolution, moving from anecdotal, experience-based use toward rigorous scientific investigation and evidence-based validation. This journey aims to preserve cultural wisdom while ensuring safety, efficacy, and integration into contemporary global health paradigms.

13.6 Botanical Diversity of Pakistan

Ethnomedicinal tradition of Pakistan is a massive and complex repository of knowledge, and it is completely connected to ecology, culture, and health care of the country. It is anchored on the outstanding range of flora, traditional systems of healing that are both formal and informal, and the everyday and daily use of plants to control health and treat illness (Rao 2018). The current critical and needed evolution of this rich tradition is that of leaving anecdotal, experience-based usage and making a shift to strict scientific research and evidence-based confirmation. The objective of this journey is to maintain cultural wisdom and safeguard, efficacy and fit in the current global public health paradigms.

The high geographical gradients of the country composed of the coastal areas of Arabian Sea, the fertile basin of the Indus River, and the highest mountain ranges on the earth, such as the Himalayas, the Karakoram, the Hindu Kush, generate a mosaic of ecosystems. These include desert-dry places and scrublands of small size to the alpine meadows and temperate forests, which have a unique set of plants. Such complexity of ecology supports a well-developed flora, and there are estimates of more than 6000 species of wild plants (Oluwamodupe and Adeleye 2023). Among this great phytodiversity, a smaller but still large percentage, commonly 400–600 species, is actively being used in traditional medicine. In this pharmacopeia, there are world-famous herbs and native riches. In the high-altitude North, the valuable species include *Podophyllum hexandrum* (Bankakri), a source of chemotherapeutic compounds that are very powerful and the dry plains and foothills are very rich in staples, *Withania somnifera* (Asgand), *Mentha arvensis* to digest and *Aloe vera* (Kanwar gandal) to treat skin diseases. This biodiversity is not just a passive resource, but an active, living library of possible therapeutics. Nevertheless, this natural wealth is facing a great danger through the rampant urbanization, unsustainable harvesting, agricultural growth, and widely spread impacts of climate change, and its protection is one of the issues of immediate priority, along with its exploitation (Butnariu et al. 2023).

13.7 Traditional Systems of Medicine and Herbal Practices

The exploitation of the plant biodiversity in Pakistan is informed by complex, traditional structures of medicine, mostly Unani Tibb, which are more influenced by Ayurveda and a strong underlay of folk knowledge. The system is known as Unani Tibb, a Greco-Arabic-based system that was extensively developed and established in the Indian subcontinent and is a practiced science. It is administered by trained practitioners called Hakims and is given on a holistic basis based on the four humors (akhlat: blood, phlegm, yellow bile, black bile) and the four qualities of heat, cold, moisture, and dryness. Medical prescriptions are elaborate polyherbal preparations, e.g., Ma'ajoon (electuaries), Jawarish (digestive tonics), and Sufoof (powders) depending on constitutions and are used in diabeties (Yaseen et al. 2024). Local practices are also influenced by Ayurvedic ideas of the balance of three bio-energies or doshas (Vata, Pitta, Kapha) particularly in areas with a historical cultural connection. These codified systems are based on a huge, far-reaching, and highly pragmatic folk medicine. It is composed of such simple and readily available preparations like a decoction of *Tinospora cordifolia* (Gilo) in the case of fever and immunity, a paste of *Azadirachta indica* (Neem) leaf in the case of skin infections, or even the use of *Justicia adhatoda* (Bhekkar) to cough and asthma. These are the product of decades of clinical experience and a vital and culturally ingrained kind of health literacy and self-sufficiency (Aslam et al. 2025).

13.8 Medicinal Plants as Primary Healthcare Resources

The medicinal plants in the socio-economic and infrastructural setup of Pakistan do not just limit to the domain of the complementary therapy but serve as essential primary healthcare provisions. Geographically inaccessible or simply beyond financial reach by a large majority of the population especially in rural and remote regions, local flora forms the first and most common line of defense against illness. The dependency on plants is promoted by a combination of availability, low cost, cultural knowledge, and the sense of security (Ahmad et al. 2025). Simple conditions are regularly treated using homemade remedies: bowel problems with fennel (*Foeniculum vulgare*) and ginger (*Zingiber officinale*); high blood pressure with garlic (*Allium sativum*) and olive leaves; diabetes with bitter gourd (Karela) and fenugreek (Methi); and injuries with turmeric (Haldi) paste. The practice offers a lifeline of serious safety, where there is a minimum of health care where institutional care is not in place. It is an extremely rational economic decision; plants may be grown in kitchen gardens, harvested in the wild cheaply, or bought in a local market at a fraction of pharmaceutical drug prices. Traditional medicine systems sustain the livelihoods of millions, acting as the foundation of an informal economy where wildcrafters and farmers gather raw materials, while local traders and practitioners process and distribute them. This unrecorded supply chain entrenches medicinal plants directly into local healthcare and rural economies (Abiri et al. 2023).

13.9 Transition from Folklore to Evidence-Based Validation

The modern need is to steer a responsible and fruitful process of traditional folklore turning into scientific and evidence-based validation. Such a change is necessary in order to reduce risks, including toxicity, adulteration, or ineffective use, but also to realize the full potential of these botanical resources to global and local health. The validation procedure is a multi-stage process. It starts with systematic ethnobotanical records, where data on the plant species, part that is used, preparation, dosage, and ailment of interest is recorded by traditional practitioners, thus generating knowledge and creating testable hypotheses to preserve it. Laboratory studies are done during subsequent preclinical research (Seidu et al. 2023). In vitro tests on cell cultures are used to measure a particular biological activity, e.g., the glucose-reducing effect of *Momordica charantia* extract on muscle cells or the anti-inflammatory effect of Nigella sativa oil on immune cells. This is then succeeded by in vivo experiments in diseased animal models that determine the holistic effectiveness, pharmacokinetics, and preliminary safety of the entire extract or the individual compounds. Clinical trial on human subjects is the most conclusive. Controlled trials (randomized controlled trials (RCTs)) offer maximum evidence of therapeutic efficacy and safety and advance more of a claim than a single, well-reported experience to a proven fact. Simultaneously, phytochemical studies are aimed at isolating, identifying, and characterizing the bioactive molecules that mediate the phenomena witnessed—e.g., the identification of berberine in *Berberis lycium* (Kashmal) as a compound that acts on AMPK to enhance insulin sensitivity. This rigorous scientific examination has several purposes: They protect the health of the population, give an opportunity to monitor quality and standardization of herbal medications, identify new leads in drugs, and, finally, promote respected traditional medicines to the level of evidence-based medicine on the international scale. To Pakistan, this will not mean a discarding of heritage but a continuation and empowerment of the same so that its ethnomedicinal wisdom can be part of modern healthcare solutions in a safe, effective, and credible manner (Saman et al. 2025).

13.10 Selected Medicinal Plants with Cardiometabolic Protective Potential

The extensive botanical legacy of Pakistan offers a plentiful foundation of identifying and discovering natural interventions that can be utilized to treat metabolic syndrome and cardiometabolic diseases. The chapter explores a carefully chosen collection of native medicinal plants, each having their own ethnobotanical lineage and accumulating scientific data of therapeutic use (Chiş et al. 2023). They are concerned with plants that are multi-target acting on the interdependent pathologies of the insulin resistance, dyslipidemia, hypertension, and systemic inflammation. It is possible to consider and value them as promising candidates of integrative and complementary cardiometabolic health care by exploring their botanical context, traditional uses, important phytochemicals, experimental evidence, and mechanistic insights (Derosa et al. 2024). Table 13.1 shows the key Pakistani medicinal plants for cardiometabolic health, which play a very important role in the human health.

Table 13.1 Some important medicinal plants of Pakistan for cardiometabolic health

Plant (common)	Key bioactives	Primary mechanisms	Evidence level
Berberis brandisiana (Kashmal)	Berberine (alkaloid)	Activates AMPK; inhibits NF-κB. Lowers glucose and lipids	Strong preclinical/ clinical
Nigella sativa (Kalonji)	Thymoquinone	Potent antioxidant; inhibits NF-κB. Improves glucose, lipids, inflammation	Robust clinical
Ajuga bracteosa (Kori Booti)	Diterpenoids (Ajugarin)	Enhances nitric oxide; anti-inflammatory. Lowers blood pressure	Established preclinical
Viola odorata (Banafsha)	Flavonoids, cyclotides	Antioxidant; protects endothelium. Reduces oxidative stress	Preclinical/ traditional
Ziziphus oxyphylla (Elanai)	Flavonoids, saponins	Vasorelaxant; antioxidant. Lowers oxidative stress and inflammation	Promising preclinical
Coriander seed (Dhania)	Linalool, flavonoids	Binds bile acids; antioxidant. Lowers cholesterol	Preclinical/pilot clinical
Fagonia indica (Dhamasa)	Flavonoids, saponins	Cardioprotective antioxidant. Reduces toxin-induced damage	Emerging preclinical
Cichorium intybus (chicory root)	Inulin, sesquiterpenes	Prebiotic; hepatoprotective. Improves glucose and lipid metabolism	Preclinical (extract)/clinical (inulin)

13.10.1 *Berberis brandisiana*

Berberis brandisiana is a spiny shrub that is native to the Himalayan and Karakoram area of Pakistan and is an important plant of the family. It is described to have obovate leaves, yellow flowers, and long red berries. Although a number of Berberis species are interchangeably used in the traditional medicine, B. brandisiana has a certain status in the local pharmacopoeias, especially in the Northern regions (Rafique et al. 2020).

The decoction and extracts of *Berberis brandisiana* (roots and stems) are used as a variety of metabolic and inflammatory conditions in the traditional practice. It is prescribed by Hakims and the local healers to treat diabetes (locally referred to as *ziabetus*) and liver ailments, fevers, and inflammatory joint pains. Its astringent flavor is linked to the traditional blood-cleaning and digestive-stimulating effects, which is in line with the Unani idea of normalizing the humoral imbalances, especially excess heat and inflammation (Mahwish et al. 2025).

The isoquinoline alkaloids are the main therapeutic agents in *B. brandisiana*, and the most visible and widely studied is berberine. There are also other related alkaloids including berbamine, palmatine, and jatrorrhizine that are also present and add to the total pharmacological profile. Berberine is a yellow crystalline substance that has attracted high research attention in the world due to its metabolic advantage (Mohanta et al. 2023). *Berberis brandisiana* extract and berbamine demonstrated significant anti-diabetic effects in diabetic rats by improving blood glucose, lipids, and insulin levels. This activity was mediated through enhancing antioxidant defenses, modulating inflammatory markers, and regulating key genes involved in glucose metabolism and insulin signaling (Mehdi et al. 2023).

The traditional use of Berberis species is confirmed by a large amount of experimental evidence based mainly on the use of berberine. In both animal models of metabolic syndrome and type 2 diabetes, berberine has shown strong antihyperglycemic activity in enhancing fasting glucose, postprandial glucose, and glycosylated hemoglobin (HbA1c) levels. It has also a hypolipidemic effect, which greatly decreases serum triglycerides, total cholesterol, and low denser lipoprotein (LDL) and increases high-density lipoprotein (HDL). Also, berberine has a solid anti-inflammatory potency, which inhibits the synthesis of pro-inflammatory cytokines (TNF- 6, IL-1) and pro-inflammatory colonies (IL-6) (Basheer et al. 2023).

13.10.2 *Viola odorata*

Viola odorata is ethnomedicinally applicable in Pakistan in such conditions as inflammatory and cardiovascular diseases. It has traditionally been used as a demulcent and expectorant in respiratory diseases such as bronchitis and asthma, which also contain inflammatory constituents. Related to the cardiometabolic more intuitively in its uses, it has found application as a mild cardiotonic, a diuretic to treat mild hypertension, and a blood cleanser of the skin to conditions related to internal inflammation, implying an intuitive historical perception of its systemic anti-inflammatory effects (Batiha et al. 2023).

Viola odorata is a phyto-prolific plant with a high amount of antioxidant and vasoactive compounds. It is rich in a number of flavonoids, including the glycosides of rutin and quercetin which are strong free radical scavengers. Its anti-inflammatory action is helped by the contribution of phenolic acids which include salicylic acid derivatives. Sweet violet is also a unique source of cyclotides, tiny, cyclic, peptides, with remarkable stability and a wide range of bioactivities, such as possible cardio-vascular protective properties (Ullah and Chen 2023).

These conventional uses are backed by experimental evidence. In vitro research of extracts of Viola odorata reveals a strong antioxidant effect that is capable of scavenging DPPH and other radicals. The extract has been demonstrated to have vasoprotective effects in animal models of vascular dysfunction, to maintain endothelial integrity and enhance vascular reactivity. Its anti-inflammatory effects are demonstrated by the decrease of edema in normal models and the release of inflammatory mediators (Samakar et al. 2022). Viola odorata seems to exert its protective effect on the cardiometabolism through the prevention of oxidative stress and endothelial dysfunction. The phenolic compounds and flavonoids directly counteract the reactive oxygen species (ROS) to limit the oxidative load that leads to inflammation and nitric oxide (NO) bioavailability. Oxidative damage to the endothelium is inhibited by these compounds and thus contributes to the normal production of NO and vasodilation which is extremely important in blood pressure regulation and atherosclerosis prevention. The combination of this mechanism and mild diuretic and anti-inflammatory effects is the basis of its future applications as a supportive treatment to retain vascular health in metabolic syndrome (Singh 2023).

13.10.3 *Ajuga bracteosa*

Ajuga bracteosa is found growing on the Himalayan foothills and the north plains of Pakistan It is a low-growing and hairy perennial herb. It can be well distinguished by its square stem (typical of the Lamiaceae family), serrated foliage, and thick spikes of bluish-purple flowers. It is a popular folk medicine in its scope (Shaukat et al. 2022a). The use Ajuga bracteosa is directly related to cardiometabolic disorders. It is most widely used in the treatment of hypertension, usually drinking it as a fresh leaf juice or a tea of herbs. It is also used to treat fevers, diabetes, and liver disorders—things that are usually related to metabolic derangements. The astringency and bitterness of it have been believed to cleanse the blood and cool down the heat in the traditional diagnostics (Sakhare et al. 2023). Phytochemical screening indicates that Ajuga bracteosa has been a rich source of bioactive diterpenoids, in particular, neo-clerodane diterpenes such as ajugains, which are considered to be some of the active compounds. In addition, the plant has major phenolic compounds, such as flavonoids and phenolic acids, iridoid glycosides, and essential oils, and this makes it a broad spectrum pharmacologically (Riaz et al. 2022).

There is a lot of preclinical evidence in favor of its traditional applications. Normotensive and hypertensive rat studies have demonstrated that aqueous and methanolic extracts of A. bracteosa cause significant dose-dependent decrease in arterial blood pressure. This hypotensive effect is also usually followed by a slight bradycardia (slowing heart rate). Also, the extracts have hypolipidemic properties, lowering serum cholesterol and triglycerides in dyslipidemic animal models, and have significant anti-inflammatory and analgesic effects in conventional tests (Wasti et al. 2024). The mechanistic understanding of the effects of *Ajuga bracteosa* is directed at the effects of changing the nitric oxide pathway and inflammatory mediators. Its hypotensive effect is partially explained by vasodilation by the endothelial NO pathway, perhaps, by the stimulation of endothelial nitric oxide synthase (eNOS). There is also some research indicating calcium channel blocking effect that relaxes the vascular smooth muscle. The diterpenoids and phenolics are strong inhibitors of the pro-inflammatory enzymes such as cyclooxygenase-2 (COX-2) and lipoxygenase (LOX) and inhibitors of cytokine synthesis such as TNF-a and IL-6. This twofold effect on vascular conduct and systemic inflammatory contend that A. bracteosa is an intriguing choice to organize the hypertensive and inflammatory elements of metabolic syndrome (Mahajan et al. 2025).

13.10.4 *Ziziphus oxyphylla*

Ziziphus oxyphylla (also known locally as Elanai or Ber Maloki) is a thorny shrub or small tree in the family Rhamnaceae, growing in dry and rocky areas of Balochistan, Khyber Pakhtunkhwa, and Punjab. It is identified by its zigzag branches, oval-shaped leaves that are small and unique drupe fruit. The traditional healing uses different sections of the plant such as fruit, leaves, and bark (Popstoyanova et al. 2024). *Ziziphus oxyphylla* has been used in traditional medicine in the promotion of cardiovascular and metabolic health. The fruit is taken as a general tonic and in cases of digestive problems, but decoctions of the bark and leaves are employed on the alleged hypertensive, anti-diabetic, and anti-inflammatory action. It also finds use as a wound healing agent as well as a skin infection agent suggesting its potential to act as an antimicrobial and tissue-repair agent (Edo et al. 2025).

The plant has a rich profile of phytochemicals that are antioxidants. It is full of phenolics, flavonoid (quercetin, rutin), saponins (jujubogenin glycosides), and ascorbic acid. This blend makes its extracts have an impressive free radical scavenging power as repeatedly shown in tests such as FRAP and DPPH which even exceeds common antioxidants (Naheed et al. 2025). The cardiometabolic potential of Z. oxyphylla extracts is supported by the experimental studies. Studies show that the extracts have the capacity to cause relaxation of pre-contracted aortic rings implying that they have a direct influence on vascular smooth muscle and endothelial activity which may reduce blood pressure. The strong antioxidant activity is converted to a strong decrease of oxidative stress markers in vivo, and tissues are not exposed to lipid peroxidation. Also, the extracts exhibit anti-inflammatory properties by reducing the concentration of C-reactive protein (CRP) and other inflammatory cytokines in chronic inflammation animal models (Jeon et al. 2024).

Ziziphus oxyphylla extract and kaempferol effectively lowered blood pressure in hypertensive rats. This action was achieved by improving nitric oxide and antioxidant levels while modulating key genes and inflammatory markers (Shaukat et al. 2022b). The potent effects of *Ziziphus oxyphylla* on vascular tone, oxidative stress, and inflammatory markers are the key mediators of the cardiometabolic benefits of *Ziziphus oxyphylla*. The mechanism of action of flavonoids on the ion channels and release of NO is likely to be the cause of the vasorelaxant effect. It enhances the natural defense mechanism by supplying the endogenous defense system with a large dose of dietary antioxidants, which are the core of the insulin resistance and endothelial dysfunction. At the same time, its anti-inflammatory effect can suppress the chronic low-grade inflammation of metabolic syndrome, and it provides a comprehensive protective measure of the cardiovascular system (Aafi et al. 2022).

13.10.5 *Nigella sativa*

Nigella sativa also referred to as black seed or kalonji in Pakistan is a flowering plant that is planted prolifically in the country every year. It yields thin-skinned blue or white flowers, capsules covered with tiny, hard, pungent seeds. It is the most salient medicinal plant in the Islamic world and is thought to be a staple of folk and Unani medicine, as it is worshiped as the panacea to all diseases except death in Prophetic tradition (Almasri et al. 2025). Application use of *Nigella sativa* is also quite extensive, yet its usage in metabolic and cardiovascular health is especially prominent. Conventionally, the seeds, or the oil contained within is used as a treatment of diabetes, high blood pressure, cholesterol, obesity and overall heart fortification. It is also a conventional cure to breathing illnesses, digestive diseases, and as an immune suppressant (Mashayekhi-Sardoo et al. 2024).

Thymoquinone (TQ), a benzoquinone in the essential oil of the seeds, is a main bioactive compound in a large part of its active action. Other quinones (thymohydroquinone, dithymoquinone), phenolic compounds, fatty acids (such as linoleic acid), and alkaloids, which work in synergy, are other important constituents (Abbas et al. 2024). Nigella sativa has the best experimental and clinical support of the discussed plants. The efficacy of it is supported through a large number of studies in animals and increasing amounts of human randomized controlled trials (RCTs). It has clinical meta-analytic support in terms of enhanced glycemic control (lowering fasting glucose and HbA1c), preferential lipid effects (lowering LDL and triglyceride, increasing HDL), and decreasing body weight and waist circumference. It also shows direct cardioprotective effects including attenuation of arterial stiffness and increased heart rate variability (Shafodino et al. 2022).

The anti-inflammatory and antioxidant actions of Nigella sativa, which are mainly facilitated by thymoquinone, are well explained. TQ is an effective free radical scavenger. It has a strong anti-NF-KB pathway effect, and, thus, the expression of inflammatory cytokines (TNF- alpha, IL-1 beta, IL-6) is suppressed. It also augments the action of the endogenous antioxidant enzymes such as glutathione peroxidase, superoxide dismutase. When applied to metabolism, it enhances the insulin sensitivity by increasing peripheral glucose uptake and maintains the pancreatic beta-cells functioning. This complex of attack on inflammation, oxidation, and metabolic disregulation makes it one of the best evidence-based botanicals in cardiometabolic syndrome (Wahab and Alsayari 2023).

13.10.6 Other Promising Medicinal Plants

Moreover, the species mentioned, the flora of Pakistan harbors many other species with strong cardiometabolic protective properties and is therefore a subject of future research. *Coriandrum sativum* (Cilantro/Coriander, Dhania) boasts a robust history in traditional medicine. Both its seeds and leaves are actively utilized, with its phytoconstituents acting as a holistic remedy for metabolic regulation, gastrointestinal relief, and fluid management. It has been scientifically demonstrated that coriander seed extract has important hypolipidemic properties, lowering the total cholesterol and triglycerides, and raising the HDL, probably in part because of its combination of linalool, polyphenols, and flavonoids. It is also shown to have hypoglycemic and antioxidant properties, which indicates that it may be a useful dietary supplement (Dar et al. 2023).

Caryopteris odorata and its compound coumarin effectively reduced obesity, hypertension, and insulin resistance in a rat model of cardiometabolic syndrome. The benefits were mediated by improving metabolic, inflammatory, and oxidative stress markers, as confirmed by tissue analysis (Ahmed et al. 2023).

Fagonia indica (Dhamasa) refers to a small spiny shrub of the arid regions, which is well known in folk medicine as a blood-cooling remedy, used in fevers and cancer-like disorders. There are new studies that point to strong antioxidant and cardioprotective effects. Extracts have been protective against cardiotoxicity caused by doxorubicin and have anti-angiogenic and pro-apoptotic effects against cancerous cells, indicating a multifaceted bioactivity that is applicable in the treatment of chronic disease (Sun and Shahrajabian 2023).

Cichorium intybus (Chicory, Kasni) is grown as a root crop (to produce coffee as a substitute) and as a vegetable in the form of leaves. It is traditionally a well-known hepatic tonic and diuretic. In the current studies, it has been emphasized that it has a high content of inulin (a prebiotic fiber that helps in the metabolism of glucose and lipids) and sesquiterpene lactones. It shows hepatoprotective, hypoglycemic, and anti-inflammatory properties and is therefore applicable in the hepatic aspect of the metabolic syndrome (Regassa et al. 2022).

The research requirements in the future are clear, strict phytochemical standardization of extracts, extensive dose-response and toxicity studies, effective preclinical trials in relevant metabolic syndrome models, and finally, effective clinical studies to turn the traditional wisdom into validated administering regimes. This scientific methodology will convert these potential ethnobotanical prospects into safe and dependable and useful ingredients of integrative cardiometabolic medicine (Antonio-Pérez et al. 2023).

13.11 Bioactive Constituents and Mechanisms of Action

The medicinal efficacy of medicinal plants on cardiometabolic diseases is not purely a lore issue but rather has some foundation on the advanced chemical structure of the medicinal plants themselves. The development of plant defenses and plant signaling has produced these botanical compounds that can react with human physiology in an amazingly deep fashion, providing polypharmacological interventions that are ideally appropriate to the complexity and multifactorial response of metabolic syndrome (Mihal et al. 2023). A symphony of protection processes is played by the major groups of bioactive constituent's alkaloids, flavonoids, terpenoids, and phenolic compounds. They all boil down to pathological centers: The oxidative stress is suppressed, the chronic inflammation is suppressed, the insulin pathways are sensitized, lipid metabolism is normalized, and the endothelial life is restored. This molecular communication between plant biochemistry and human biochemistry is important in understanding how to justify and maximize the use of plant-based therapy (Khan et al. 2022).

Chemical diversity of such bioactive classes offers a wide range of therapeutic arsenal. An example of alkaloids which are often nitrogen compounds with strong biological properties is berberine found in Berberis species. Flavonoids are ubiquitous polyphenols, which may be divided into such subclasses as flavonols (quercetin in Viola odorata) and flavones, known to possess antioxidant and signaling properties (Chiș et al. 2023). All terpenoids, which are constructed on isoprene units, include volatile essential oils and non-volatile diterpenes such as those present in *Ajuga bracteosa*, which frequently regulate enzyme and receptor activity. A large group of compounds with a phenol ring, which include a wide range of simple phenolic acids, to more complex tannins, are the focus of phenolic acid activity of plants such as *Ziziphus oxyphylla*. This is because of the phytochemical richness and a single plant extract can provide a multi-impact, synergistic approach, where multiple therapeutic targets are interacted with a reduced risk of the side effects that occur with high-dose, single-target synthetic medications (Fakudze et al. 2022).

More to the point, the nuclear factor erythroid 2-related factor 2 (Nrf2) pathway is activated by compounds like sulforaphane (in cruciferous vegetables) and curcumin and most importantly thymoquinone of Nigella sativa. Nrf2 is a master of cellular defense regulator; when activated, Nrf2 translocates to the nucleus and activates the expression of more than 200 genes which encode antioxidant and detoxifying enzymes. These plant compounds substantially enhance the body to protect itself, which can be sustained as an endogenous defense against oxidative damage against a primary cause of metabolic dysregulation (Martirosyan 2025).

The close association of oxidative stress is pathology of low-grade chronic inflammation, and in this case, phytochemicals have a potent ability to signal anti-inflammatory effects. They mainly act on the nuclear factor-kappa B (NF-κB) pathway, which is a central signaling node which triggers the expression of the pro-inflammatory genes. NF-KB activation can also be counteracted by bioactives such as berberine, thymoquinone, and resveratrol with high potency, blocking its translocation to the nucleus (Zargar et al. 2023). The downstream effect of this suppression is reduced production of inflammatory cytokines including tumor necrosis factor-alpha (TNF- 6), interleukin- 6 (IL-6), and interleukin- 1 beta (IL-1) 6. These cytokines are linked directly with the causes of insulin resistance in the adipose, muscle as well as liver tissues. In addition, numerous flavonoids and phenolic acids suppress actions of pro-inflammatory enzymes such as cyclooxygenase-2 (COX-2) and lipoxygenase (LOX) which produce inflammatory-mediating prostaglandins and leukotrienes. Plant compounds interrupt this vicious cycle where inflammation causes metabolic dysfunction and vice versa by damping down these inflammatory cascades at several different stages (Wink 2022).

Insulin resistance is the central issue in the metabolic syndrome, and one of the primary therapeutic effects of most botanicals is regulation of insulin sensitivity and glucose uptake. The alkaloids such as berberine have also become quite effective players in this field. They have a multifaceted mechanism. One of the well-established actions is an event that triggers adenosine monophosphate-activated protein kinase (AMPK), which is an enzyme that acts as a cellular sensor of energy. Activation of AMPK can increase glucose uptake in skeletal muscle by increasing the translocation of glucose transporter type 4 (GLUT4) to the cell membrane and recapitulates the action of insulin. AMPK functions as an antagonist of gluconeogenesis in the liver, which is the inappropriate generation of glucose which causes hyperglycemia during fasting (Mohamed Abdoul-Latif et al. 2023). In addition to AMPK, compounds such as curcumin and flavonoids enhance insulin signaling by enhancing phosphorylation of insulin receptor substrate (IRS) proteins and activation of downstream phosphoinositides 3-kinase (PI3K)/Akt pathway. Certain phytochemicals also block alpha-glucosidases in the intestines, preventing the rapid digestion of complicated carbohydrates and the post-meal testosterone spikes of glucose. This multi-targeted glucose homeostasis approach has a great benefit compared to unit-pathway pharmaceutical therapies.

The desirable impact of lipid metabolism and endothelial functioning fulfills the cardiometabolic protective profile. Critical causes of atherosclerosis are dyslipidemia and dysfunctional endothelium. Plant bioactives enhance lipid profiles in a number of ways. The saponins (fenugreek, Ziziphus) and soluble fibers have the potential to bind the dietary cholesterol and bile acid in the intestine, and they excrete it. Drugs, such as berberine, increase the amount of hepatic LDL receptors in the liver, enhancing the process of clearance of bad cholesterol in the blood. Others also cause PPAR-2 a nuclear receptor to stimulate fatty acid oxidation and lessen triglyceride synthesis (Sapkota et al. 2022). At the same time, protection and renewal of the endothelium, which is the important lining of blood vessels, are helped. Plant-derived flavonoids found in Viola odorata and Cichorium intybus support vascular health by activating endothelial nitric oxide synthase (eNOS). This enzyme enhances the production of nitric oxide a vital vasodilator which helps maintain vascular flexibility and regulate blood pressure. They also inhibit the destruction of existing NO by ROS. Besides, phytochemicals suppress the adhesion of inflammatory monocytes to the vessel wall (e.g., VCAM-1, ICAM-1) to prevent the formation of early critical steps in plaque formation. This bi-action on lipids and the vascular wall is a treatment of the risk factors in the bloodstream as well as the vulnerability of the arteries to pathology (Orimaye et al. 2024).

13.12 Evidence from Preclinical and Clinical Studies

The path taken between uses of medicinal plants in traditional medicine and their legitimate use as a therapy is supported by a strict gradation of scientific evidence. This is a compilation of evidence based on preclinical studies and clinical trials that assess the safety, efficacy, and mechanism of botanical interventions in cardiometabolic disorders systematically. The research pathway is progressive through regulated cellular and molecular experiments, complex whole-animal models that simulate human disease, and eventually to human clinical research. This multi-level strategy is necessary to shift potential ethnobotanical findings into evidence-based, reliable, and useful medical information that has direct implications on the health of the populace (Singla et al. 2023).

The first line of scientific validation is represented by in vitro models and molecular assays, which offer high-throughput controlled environments to determine bioactivity and clarify preliminary mechanisms. In research, certain plant extracts or individual compounds containing berberine, thymoquinone, or a fraction containing flavonoids are used on cultured cell lines. Important models are insulin-resistant hepatocytes (liver cells) to determine the rate of glucose production, adipocytes (fat cells) to examine lipid storage and adipokine release, and endothelial cell cultures to quantify the levels of nitric oxide and expression of inflammatory markers (Nagori et al. 2023). The effects are accurately determined by molecular assays. As an example, the insulin-sensitizing effect can be evidenced using a glucose uptake test in muscle cells, and the TNF-A or IL-6 release of immune cells provoked by an inflammatory stimulus can be measured using an ELISA test. The activation or inhibition of individual proteins, including the phosphorylated AMPK or Akt as part of insulin signaling, or the nuclear translocation of NF-0B is visualized by Western blot analysis. Such in vitro results are essential to affirm a direct pharmacological action and the localization of the molecular targets and to direct the choice of the most encouraging extracts in a more complex and expensive in vivo researches (Belete and Beyna 2025).

Figure 13.2 presents a proposed translational pipeline that outlines the essential stages and decision points for developing evidence-based therapies from ethnomedicinal leads. It emphasizes the need for multidisciplinary integration, robust validation, and regulatory science to overcome current bottlenecks.

Cellular knowledge is extended to in vivo models of metabolic syndrome and cardiometabolic disease, which give a comprehensive evaluation of treatment efficacy in the integrated physiological functions of a living organism. High-fat, high-

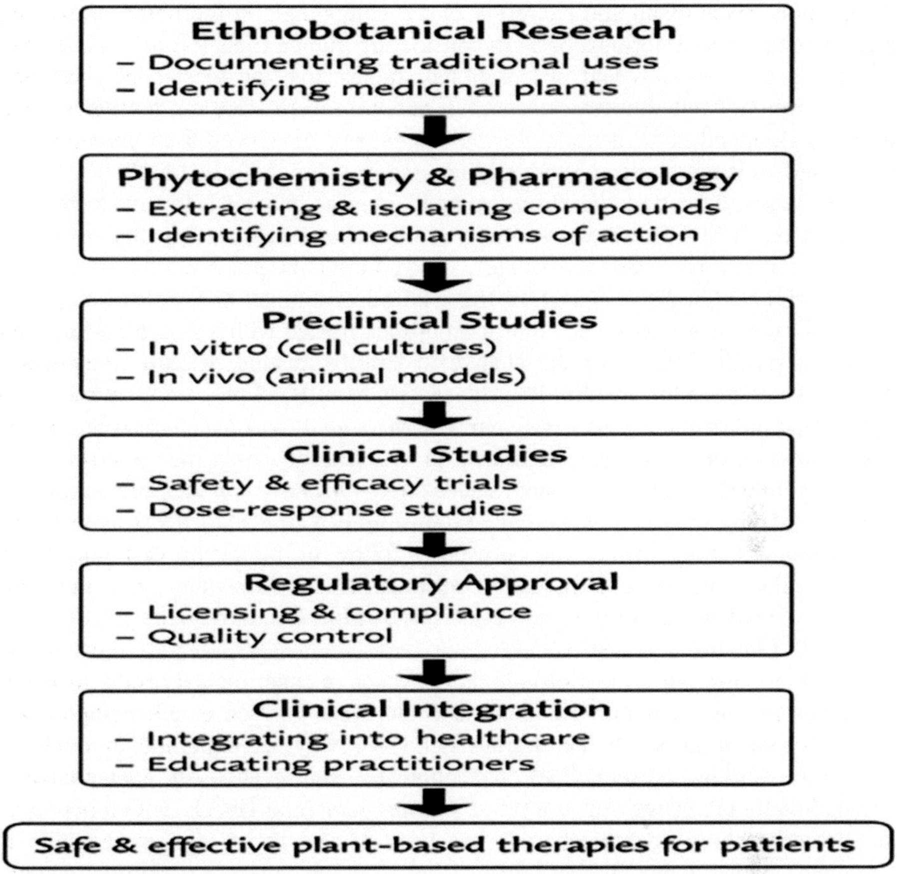

Fig. 13.2 Translational pathway for medicinal plant development: from ethnobotany to clinical integration

fructose, or high-sucrose diet-fed rodent is the gold-standard preclinical model. With weeks, these animals grow common characteristics of human metabolic syndrome which are central obesity, insulin resistance, dyslipidemia (high triglycerides and LDL, low HDL), hypertension, and systemic inflammation. It can also be based on genetic models like the Zucker diabetic fatty (ZDF) rats or the db/db mice (Salazar-Gómez and Alonso-Castro 2022). In these models, the researchers use standardized plant extracts and administer them during a specified duration and evaluate the endpoints which are of clinical value. They are fasting blood glucose and oral glucose tolerance tests (OGTT), serum lipid screen, blood pressure through tail-cuff or telemetry, and serum inflammatory cytokines. Tissues can be examined histopathologically to show a decrease in liver steatosis (fatty liver) or aortic plaque. They also give essential initial information on the pharmacokinetics (absorption, distribution, metabolism and excretion of the compounds in the body) and safety pharmacology which forms a basis in considering human trials (Nisar et al. 2023).

Human clinical trials and observational studies give the most conclusive evidence. Although ethnobotanical surveys and case series are supportive stories, properly designed clinical trials are the standard of evidence-based practice. The gold standard is randomized, double-blind, and placebo-controlled trials (RCTs). In the case of plants such as *Nigella sativa* and berberine, many RCTs have been done (Biagi et al. 2023). These are generally studies that entail participants who have metabolic syndrome, type 2 diabetes, or dyslipidemia. Intervention groups are treated with standardized dosage of the seed oil or extract, or isolated compound over 8–12 weeks or longer and control groups are treated with a placebo. The major outcomes usually consist of the change in HbA1c, fasting insulin, Homeostatic Model Assessment of Insulin Resistance (HOMA-IR), lipid, body mass index (BMI), and inflammatory (hs-CRP) markers. Meta-analyses of such trials on *Nigella sativa* are always statistically significant in bettering glycemic measures and lipid measures. In other promising plants, such as *Ajuga bracteosa* or *Ziziphus oxyphylla*, there is a gaping void as there is limited robust human clinical data available. These observational studies, which can be used to follow up the health of a population using regularly including herbs, may contribute to complementary real-world evidence of their long-term safety and effectiveness, but cannot be definitively causal like RCTs (Alharbi et al. 2023).

The final objective is the translational relevance of experimental results, which is the extent to which laboratory and clinic findings can be used in enhancing patient care and population health. To be translated effectively, the experimental evidence should be strong, reproducible, and clinically significant. An example of a translationally relevant result is a 0.5–1% improvement in HbA1c with a berberine supplement, which in clinical trials has been proven to be high as it has been associated with a lower likelihood of diabetic complications. In addition, translational relevance takes into account context. Although the supporting evidence of plants such as *Fagonia indica* and *Coriandrum sativum* have been strong in the preclinical models, they have to go through clinical trials and then we can come up with specific therapeutic recommendations. The results should also be incorporated within usable guidelines (Bajalia et al. 2022). This includes deciding on optimal dosing schedules, standardization of extracts to guarantee consistency in bioactive constituents, possible herb–drug interactions (with blood thinners or diabetes drugs), and health education of both those administering the product and the patients. In the case of Pakistan, which is a culturally tolerant country where these plants are already in use in many instances, the translational potential is enormous (Fig. 13.2). But to actualize this potential, a concerted national research agenda is needed to advance the key botanical candidates through the full validation continuum, i.e., the laboratory bench to the pharmacy shelf to the community clinic and thus turn traditional heritage into a modern, evidence-based health resource (Oswal et al. 2023).

13.13 Safety, Toxicology, and Standardization

The passage of a medicinal plant through traditional folklore to a reliable ingredient of contemporary medicine depends on its ability to pass the three most crucial steps of safety, toxicology, and standardization. Although the therapeutic potential of botanicals is very strong, their incorporation into evidence-based medicine requires a stricter examination than the efficacy. The only progress that can be made is an honest evaluation of the possible toxicities, determination of safe dosage levels, the elimination of serious variation in the plant materials, and a comprehensive understanding of the interactions of these botanical agents with traditional drug cutting. Overcoming these issues is not a slowing down point but rather the key to developing credible, reliable, and effective cardiometabolic health plant-based therapies (Kaggwa et al. 2022).

The basis of patient safety is toxicological considerations and safe dose ranges. All the pharmacologically active substances have a therapeutic window, i.e., a concentration between an ineffective and a toxic dose. This is difficult to establish in the case of plant extracts, which are complexes of numerous chemicals. Preclinical toxicology is necessary (Heinrich et al. 2022). The acute and sub-chronic toxicity tests on animal models determine the median lethal dose (LD50) and the No Observed Adverse Effect Level (NOAEL) and the target organs of toxicity (in many metabolic herbs, they are likely to be the liver and kidneys). As an example, although Nigella sativa oil is not usually harmful, in animal models very large doses of the product have been linked to hepatorenal toxicity. Pyrrolizidine alkaloids, present in plants belonging to the family Boraginaceae (sometimes as components of herbal preparations), are hepatotoxic and carcinogenic with a strong risk of not being avoided. Clinical trials do then assist in the determination of human equivalent dose. One of the most important problems is that the traditional prescriptions tend to suggest imprecise ones, such as a handful of leaves or a cup of decoction, which cannot be applied to present-day therapy (Alum et al. 2025). Determining the safe and useful range of dosage would entail systematic dose-escalation mono-human trials. As an illustration, clinical experience of berberine indicates that a therapeutic dose is between 900 and 1500 mg/kg/day, administered in doses with gastrointestinal discomfort being the most frequent but reversible side effect of high doses. In the absence of this scientific basis, the possibility of under-dosing (ineffectiveness) or overdosing (toxicity) is too great to compromise the results of patients, as well as the legitimacy of herbal medicine (Latif and Nawaz 2025).

The issue of inconsistency of phytochemical composition is a deep and far-reaching problem that directly influences efficacy and safety. The bioactive concentration of plant is not a predetermined number, but it is a dynamic value, which depends on a great amount of factors. Genetics of the plant, the exact chemotype that is being grown, the soil type and climate in which it is grown, time of harvest (diurnal and seasonal differences), which portion of the plant is harvested (root, leaf, seed), and post-harvest processing and storage conditions all radically change the end result of the chemical profile (Jan et al. 2025). A batch of roots of Berberis aristata collected in spring in one of the valleys in Kashmir might be abundant in a much higher concentration of berberine compared to a batch collected in autumn in a different region. Such variability implies that a patient taking a supposedly standard teaspoonful of powder herb may get a therapeutically effective dose, a non-therapeutic dose, or, in worst-case situations, an unexpectedly large dose of a drug of narrow therapeutically index. This internal inconsistency is one of the main causes of equivocal outcome of herbal remedies in clinical practice and the reason why the anecdotal evidence of their effectiveness is extremely hard to control in a systematic way. It poses a root cause to providing quality and repeatable healthcare (Oloya et al. 2022).

These differences are fed into the monumental task of challenges of standardization. Standardization involves ensuring that all individual preparations of a botanical preparation are of a known concentration of one or more marker compounds, the presence of which explains the therapeutic effect. It is the transition between the variability of the raw herb and the consistency of the batch-to-batch medicinal product. It can be summarized into a few steps: confirmation of the right plant species through morphological and genetic markers, controlled agricultural practices, and the utilization of state-of-the-art analytical methods such as the High-Performance Liquid Chromatography (HPLC) used to measure the major compounds. This extract is then concentrated to a certain level of drug to extract ratio to ensure that a particular amount of active is contained in a given dose (Hossain et al. 2022). A typical example is a standardized Nigella sativa extract which could be ensured to have 2–5% thymoquinone. However, challenges abound. The selection of the correct marker compound does not necessarily need to be simple—is thymoquinone the active in the black seed only or do we need to standardize dithymoquinone and other volatiles? With multifaceted polyherbal preparations, which are typical in Unani medicine, it is exponentially more challenging to standardize them. Moreover, quality control is expensive, regulatory frameworks are not very strict in most countries, and adulteration (intentional or unintended substitution by inexpensive, inactive, or toxic material) that is widespread in the informal sector is very destructive to changes (Alqahtani et al. 2022). In the absence of a high-quality standardization of clinical trials, the data will be hard to interpret, and it would be practically impossible to prescribe an accurate and reliable dose.

The most important safety issue in a polypharmacy era is perhaps the risk of concern of herb–drug interactions. Medicinal plants are not innocent they are multifaceted pharmacological agents capable of significantly changing the pharmacokinetics (what the body does to the drug) and pharmacodynamics (what the drug does to the body) of the synthetic drugs that are co-administered. These reactions may cause sub-therapy or fatal toxicity. Pharmacokinetic interactions usually take place through regulation of liver enzyme cytochrome P450 (CYP450) or drug transporters. As an example, one of the active inducers of CYP3A4 is compounds in Ginkgo biloba and *Hypericum perforatum* (St. John's Wort), which can increase the rate of metabolism and also decrease the plasma concentration of such medications as statins (simvastatin, atorvastatin) or antihypertensive drugs, making them ineffective (Anywar et al. 2022). On the other hand, the substances of grapefruit juice contain very strong inhibitory factors of the same enzyme, which can result in toxic levels of these drugs. Drug interactions are also of concern. The strongly hypoglycemic effects of plants, including *Momordica charantia* (bitter melon) or fenugreek, may interact with the action of insulin and sulfonylureas, which can result in serious hypoglycemia. Equally, anticoagulant herbs (e.g., *Ginkgo biloba* and high dose of garlic) may enhance risk of bleeding in association with warfarin or aspirin. In patients with cardiometabolic syndrome who are routinely taking a cocktail combination of antihypertensives, statins, and hypoglycemics, these interactions have ceased to be hypothetical and have become tangible and real (Maurya et al. 2026). This would require intensive education of health service providers and patients, disclosure of all the herbal supplements use in clinical consultations, and special studies that would be required to mapping these interaction patterns in a systematic manner of the popular medicinal plants.

13.14 Challenges and Future Perspectives

The road to the incorporation of Pakistan medicinal plant heritage into a contemporary and evidence-based system of fighting metabolic syndrome is fraught with major challenges closely linked together. Nevertheless, every challenge comes along with its own chance of opportunity which forms a vivid outline of future research, policy, and practice (Gupta et al. 2022). The ability to traverse this terrain successfully will see this form of traditional knowledge being a marginalized and informal asset or a validated and sustainable pillar of healthcare across the country and globe. The various challenges are discussed in Table 13.2.

The first is the disaggregation of the research and absence of translation pipelines. As sporadic university centers and single researches produce very useful phytochemical and pharmacological information, a strong gap exists between the academic

Table 13.2 Translational challenges and strategic solutions for medicinal plants of Pakistani

Key translational challenge	Proposed strategic solution
Fragmented research and lack of pipeline	Establish a National Multidisciplinary Consortium to prioritize species and coordinate research from field to clinical trial
Phytochemical variability and adulteration	Enforce mandatory standardization and authentication of commercial products using markers (e.g., % thymoquinone, berberine)
Scarcity of robust clinical trials	Direct targeted funding for phase II/III RCTs on priority plants (e.g., *A. bracteosa* for hypertension, *Z. oxyphylla*)
Threats to wild biodiversity	Promote sustainable cultivation programs with farmer incentives to ensure supply, conserve wild populations, and improve quality
Weak regulatory framework	Develop a dedicated herbal medicine policy classifying products as "herbal medicinal products" with required proof of safety and quality
Lack of clinical integration	Incorporate positive trial results into National Treatment Guidelines and medical education on integrative management of metabolic syndrome

information and real-world products of health or clinical guidelines. A lack of funding, regulatory roadmaps, and collaborations with industry to produce at expensive Good Manufacturing Practice (GMP) and Phase II/III clinical trials often causes research to stall at the preclinical animal study stage. This leaves a massive library of good in vitro and rodent data that do not amount to human good (Kenda et al. 2022). At the same time, there are threats to biodiversity and unsustainable activities looming. Overharvesting by wild harvesting is a result of the growing demand of common herbs as well as loss of habitat to deforestation and urbanization. This not only endangers the downfall of endangered species such as *Ajuga bracteosa* or *Berberis brandisiana* but also reduces the genetic diversity of such species which forms the basis of their medicinal power. This problem is further augmented by the informal and uncontrolled supply chain between the forager and the market vendor through adulteration and lack of quality control (Rehman et al. 2022).

Moreover, there is a vacuum of regulatory and standardization. Pakistan does not have a detailed, scientifically based regulatory system with respect to herbal medicines, which specifies strict standardization, safety, and efficacy to be demonstrated as require with synthetic drugs. The current policy landscape tends to confuse medicinal plants as with dietary supplements or traditional folklore, and the products of extremely diverse and unproven quality are flooded into the market (Amin and Park 2025). Such lack of standards leads to loss of consumer confidence, the risks of safety, and the inability to build an effective exporting business. Last, there exists a critical requirement of clinical validation and integration of healthcare. In the case of most native plants, there is a dire need of adequate human clinical trial evidence, particularly large-scale and randomized control trials (Parganiha et al. 2022). In the absence of this gold-standard evidence, physicians will have a legitimate reason to be reluctant to prescribe these remedies, and they cannot be officially incorporated into national treatment guidelines of diabetes, hypertension, or dyslipidemia.

The future vision to handle these challenges is stipulated and practical. The achievement of a nationwide multidisciplinary research consortium should be made the main priority. This would be a strategic way of liaising the work of botanists, pharmacologists, clinicians, chemists, and agronomists to focus on priority species, facilitate field to clinical trial research, and develop a database of centralized ethnobotanical and scientific information (Sadgrove 2022). Simultaneously with that, there should be a strong regulatory framework that should be created and implemented. This model must stipulate verification, conformity to active markers, and an affirmation of safety of any marketed herbal product to the rank of Herbal Medicinal Products (Dawan and Ahn 2022). This will safeguard the consumer and create both national and international market confidence.

Conservation and economic development requires the investment of sustainable cultivation and bioprospecting. By encouraging the growth of already desired medicinal species in controlled and agricultural settings with incentives given to farmers, this will not only take the strain off the wild population but will also offer a continuous supply of raw material and a source of livelihood to the rural population. Developed bioprospecting methods can also be used to discover new compounds and maximize production. Finally, the evidence-based integration into healthcare is the aim. This involves the allocation of substantial funds to multicentric clinical trials which are rigorous on the most promising plant candidates like standardized extracts of Nigella sativa or Berberis. The successful outcomes of such trials are then to be proactively published to the healthcare community and integrated into the medical education and population wellness efforts regarding the bulk disease management.

13.15 Conclusion

Medicinal flora of Pakistan represents an enormous and precious national treasure trove to face the growing epidemic of metabolic syndrome and the attendant cardiometabolic disease. The selected species such as *Berberis brandisiana, Nigella sativa*, and *Ajuga bracteosa* represent a pharmacopeia that is far more than the mere folklore and provides abundant sources of bioactive compounds that have been shown to be multi-target in their efficacy. These botanicals, through coordinated modulation of insulin sensitivity (through AMPK pathways), inhibition of oxidative stress and persistent inflammation (NF-kB inhibition), improved lipid metabolism, and improved endothelial performance are well aligned to the complex and network-based pathology of metabolic syndrome and offer a holistic therapeutic approach. In spite of this pledge, there has been limited translation of ethnomedicinal knowledge into mainstream clinical knowledge. Major obstacles are the relative lack of strong, large-scale clinical trials, inadequate standardization of phytochemical composition and therapeutic activity, and the lack of a broad regulatory framework in place to guarantee safety, quality control, and regulation of herb–drug interactions. The clinical integration of medicinal plants will not be comprehensive and uniform without billing such gaps. Going forward there is a need to have a concerted approach at national level. The clinical integration of medicinal plants requires a unified national approach prioritizing multidisciplinary research, sustainable production, phytochemical standardization, gradual clinical verification, and science-based regulatory policies. The end point is the evidence-based incorporation of proven botanical medicine in prevention and treatment cardiometabolic care. Conclusively, through a systematic combination of traditional knowledge with the scientific rigor, Pakistan can have a special chance to transform the botanical legacy into safe, effective, and available healthcare solutions. This type of strategy has the potential to enhance population health outcomes, minimize pressure on healthcare systems, enhance the maintenance of biodiversity, and contribute to sustainable socio-economic development.

References

Aafi E, Shams Ardakani MR, Ahmad Nasrollahi S, Mirabzadeh Ardakani M, Samadi A, Hajimahmoodi M et al (2022) Brightening effect of Ziziphus jujuba (jujube) fruit extract on facial skin: a randomized, double-blind, clinical study. Dermatol Ther 35(7):e15535

Abbas M, Gururani MA, Ali A, Bajwa S, Hassan R, Batool SW et al (2024) Antimicrobial properties and therapeutic potential of bioactive compounds in Nigella sativa: A review. Molecules 29(20):4914

Abid J, Alam I, Ahmad A, Saeed RF, Farooq U (2024) Dietary inflammatory potential of mediterranean diet versus a typical traditional Pakistani diet. Kurdish Stud 12(5):789–794

Abiri B, Amini S, Hejazi M, Hosseinpanah F, Zarghi A, Abbaspour F, Valizadeh M (2023) Tea's anti-obesity properties, cardiometabolic health-promoting potentials, bioactive compounds, and adverse effects: A review focusing on white and green teas. Food Sci Nutr 11(10):5818–5836

Ahmad RS, Siddique F, Elawady A, Pramanik A, Islam F, Imran A et al (2025) Food applications and ethnomedicinal properties of psyllium: a concise review. Cogent Food Agric 11(1):2487304

Ahmed MG, Mehmood MH, Mehdi S, Farrukh M (2023) Caryopteris odorata and its metabolite coumarin attenuate characteristic features of cardiometabolic syndrome in high-refined carbohydrate-high fat-cholesterol-loaded feed-fed diet rats. Front Pharmacol 14:1097407

Alam S, Sarker MMR, Sultana TN, Chowdhury MNR, Rashid MA, Chaity NI et al (2022) Antidiabetic phytochemicals from medicinal plants: prospective candidates for new drug discovery and development. Front Endocrinol 13:800714

Alharbi KS, Javed Shaikh MA, Imam SS, Alshehri S, Ghoneim MM, Almalki WH et al (2023) Role of flavonoids in management of various biological targets in Alzheimer's disease: evidence from preclinical to clinical studies. Curr Med Chem 30(18):2061–2074

Ali A, Kouvari M, Riaz S, Naumovski N, Liao L, Khan A et al (2023) Potential of Allium sativum in blood pressure control involves signaling pathways: a narrative review. Food Front 4(4): 1666–1680

Almasri RS, Bedir AS, Al Raish SM (2025) Comprehensive ethnopharmacological analysis of medicinal plants in the UAE: Lawsonia inermis, Nigella sativa, Ziziphus spina-christi, Allium cepa, Allium sativum, Cymbopogon schoenanthus, Matricaria aurea, Phoenix dactylifera, Portulaca oleracea, Reichardia tingitana, Salvadora persica, Solanum lycopersicum, Trigonella foenum-graecum, Withania somnifera, and Ziziphus lotus. Nutrients 17(3):411

Alotaibi BS, Ijaz M, Buabeid M, Kharaba ZJ, Yaseen HS, Murtaza G (2021) Therapeutic effects and safe uses of plant-derived polyphenolic compounds in cardiovascular diseases: a review. Drug Des Devel Ther Volume 15:4713–4732

Alqahtani AS, Ullah R, Shahat AA (2022) Bioactive constituents and toxicological evaluation of selected antidiabetic medicinal plants of Saudi Arabia. Evid Based Complement Alternat Med 2022(1):7123521

Alum EU, Manjula VS, Uti DE, Echegu DA, Ugwu OP-C, Egba SI, Agu PC (2025) Metabolomics-driven standardization of herbal medicine: advances, applications, and sustainability considerations. Nat Prod Commun 20(8):1934578X251367650

Amin A, Park S (2025) Chemotaxonomy, an efficient tool for medicinal plant identification: current trends and limitations. Plants 14(14):2234

Ansari P, Reberio AD, Ansari NJ, Kumar S, Khan JT, Chowdhury S et al (2025) Therapeutic potential of medicinal plants and their phytoconstituents in diabetes, cancer, infections, cardiovascular diseases, inflammation and gastrointestinal disorders. Biomedicine 13(2):454

Antonio-Pérez A, Durán-Armenta LF, Pérez-Loredo MG, Torres-Huerta AL (2023) Biosynthesis of copper nanoparticles with medicinal plants extracts: from extraction methods to applications. Micromachines 14(10):1882

Anwar R, Rabail R, Rakha A, Bryla M, Roszko M, Aadil RM, Kieliszek M (2022) Delving the role of Caralluma fimbriata: an edible wild plant to mitigate the biomarkers of metabolic syndrome. Oxidative Med Cell Longev 2022(1):5720372

Anywar GU, Kakudidi E, Oryem-Origa H, Schubert A, Jassoy C (2022) Cytotoxicity of medicinal plant species used by traditional healers in treating people suffering from HIV/AIDS in Uganda. Front Toxicol 4:832780

Aslam M, Tufail T, Almehmadi Y, Abualamah WA, Alzahrani AR, Shahid I (2025) Synergistic effect of carbohydrate-controlled diet and Cuminum Cyminum herbal infusion on metabolic syndrome. Front Nutr 12:1623478

Bajalia EM, Azzouz FB, Chism DA, Giansiracusa DM, Wong CG, Plaskett KN, Bishayee A (2022) Phytochemicals for the prevention and treatment of renal cell carcinoma: preclinical and clinical evidence and molecular mechanisms. Cancer 14(13):3278

Basheer S, Malik I, Awan F, Sughra K, Roshan S, Khalil A et al (2023) Histological and microscopic analysis of fats in heart, liver tissue, and blood parameters in experimental mice. Genes 14:515

Batiha GE-S, Lukman HY, Shaheen HM, Wasef L, Hafiz AA, Conte-Junior CA et al (2023) A systematic review of phytochemistry, nutritional composition, and pharmacologic application of species of the genus Viola in noncommunicable diseases (NCDs). Evid Based Complement Alternat Med 2023(1):5406039

Belete TM, Beyna AT (2025) Review on the ethnopharmacological use of medicinal plants and their anticancer activity from preclinical to clinical trial. Nat Prod Commun 20(4): 1934578X251333027

Biagi MM, Rigillo G, Sarill M, Collotta D, Di Giacomo S, Di Sotto A et al (2023) From preclinical to clinical evidence. Exploring the multiple perspectives and healing power of Boswellia serrata Roxb. ex Colebr. Pharmadvances 5(1):29–53

Butnariu M, Fratantonio D, Herrera-Bravo J, Sukreet S, Martorell M, Ekaterina Robertovna G et al (2023) Plant-food-derived bioactives in managing hypertension: from current findings to upcoming effective pharmacotherapies. Curr Top Med Chem 23(8):589–617

Chiş A, Noubissi PA, Pop O-L, Mureşan CI, Fokam Tagne MA, Kamgang R et al (2023) Bioactive compounds in Moringa oleifera: mechanisms of action, focus on their anti-inflammatory properties. Plants 13(1):20

Dar RA, Shahnawaz M, Ahanger MA, Majid I (2023) Exploring the diverse bioactive compounds from medicinal plants: a review. J Phytopharm 12(3):189–195

Dawan J, Ahn J (2022) Application of DNA barcoding for ensuring food safety and quality. Food Sci Biotechnol 31(11):1355–1364

Derosa G, D'Angelo A, Maffioli P, Cucinella L, Nappi RE (2024) The use of Nigella sativa in cardiometabolic diseases. Biomedicine 12(2):405

Edo GI, Nwachukwu SC, Makia RS, Jikah AN, Yousif E, Gaaz TS et al (2025) Unveiling the Chinese or red date (Ziziphus jujuba); its phytochemical, botanical, industrial and pharmacological properties: a review. Phytochem Rev 24(5):4237–4270

Fakudze NT, Aniogo EC, George BP, Abrahamse H (2022) The therapeutic efficacy of Punica granatum and its bioactive constituents with special reference to photodynamic therapy. Plants 11(21):2820

Firdous A, Ahmad U, Naeem S, Mallhi IY, Rasool S (2025) Estimating the therapeutic potential of Withania coagulans chapatti to treat hyperglycemia and hyperlipidemia. J Health Wellness Commun Res: e687–e687

Gupta SK, Jadhav S, Gohil D, Panigrahi GC, Kaushal RK, Gandhi K et al (2022) Safety, toxicity and pharmacokinetic assessment of oral Withaferin-A in mice. Toxicol Rep 9:1204–1212

Gutiérrez-Cuevas J, López-Cifuentes D, Sandoval-Rodriguez A, García-Bañuelos J, Armendariz-Borunda J (2024) Medicinal plant extracts against cardiometabolic risk factors associated with obesity: molecular mechanisms and therapeutic targets. Pharmaceuticals 17(7):967

Heinrich M, Jalil B, Abdel-Tawab M, Echeverria J, Kulić Ž, McGaw LJ et al (2022) Best practice in the chemical characterisation of extracts used in pharmacological and toxicological research—the ConPhyMP—guidelines. Front Pharmacol 13:953205

Hossain CM, Gera M, Ali KA (2022) Current status and challenges of herbal drug development and regulatory aspect: a global perspective. Asian J Pharm Clin Res 15(2):31–41

Hussain B, Sajid MA, Nabeel M, Saeed A, Shah AA (2020) Study of assessment of knowledge of patients and health care professionals about the use of nutraceuticals. World J Pharm Res 9(6): 49–119
Islam SU, Ahmed MB, Ahsan H, Lee Y-S (2021) Recent molecular mechanisms and beneficial effects of phytochemicals and plant-based whole foods in reducing LDL-C and preventing cardiovascular disease. Antioxidants 10(5):784
Jan A, Adil S, Ali T, Ahmed B, Ahmed Z, Hussain Z (2025) Desert and medicinal plants as novel sources of antimicrobial agents for crop protection. Planta Animalia 4(3):197–218
Jeon Y-J, Park SJ, Lee H, Kim H-Y, Jung D-H (2024) Deep learning-based model for effective classification of Ziziphus jujuba using RGB images. Agri Eng 6(4):4604–4619
Kaggwa B, Kyeyune H, Munanura EI, Anywar G, Lutoti S, Aber J et al (2022) Safety and efficacy of medicinal plants used to manufacture herbal products with regulatory approval in Uganda: a cross-sectional study. Evid Based Complement Alternat Med 2022(1):1304839
Kenda M, Kočevar Glavač N, Nagy M, Sollner Dolenc M (2022) Medicinal plants used for anxiety, depression, or stress treatment: an update. Molecules 27(18):6021
Khalid S, Arshad M, Saad B, Imran M, Ul HB, Ain TT et al (2020) Cardio protective effects of dates. Int J Biosci 16:110–123
Khan AW, Farooq M, Haseeb M, Choi S (2022) Role of plant-derived active constituents in cancer treatment and their mechanisms of action. Cells 11(8):1326
Latif R, Nawaz T (2025) Medicinal plants and human health: a comprehensive review of bioactive compounds, therapeutic effects, and applications. Phytochem Rev 25:1–44
Liaqat N, Jahan N, Rahman KU, Tahseen I, Anwar T, Qureshi H (2023) Investigation of phytotherapeutic potential of herbal mixtures and their effects on salbutamol induced cardiotoxicity and hyperlipidemia in rabbits. Bot Stud 64(1):23
Mahajan N, Gupta P, Manhas R (2025) Ethnobotany, phytochemistry, pharmacology, phytotoxicity, and industrial applications of Ajuga bracteosa Wall. ex Benth.: a review. Genetic Res Crop Evol 72:1–25
Mahwish, Imran M, Naeem H, Hussain M, Alsagaby SA, Al Abdulmonem W et al (2025) Antioxidative and anticancer potential of Luteolin: a comprehensive approach against wide range of human malignancies. Food Sci Nutr 13(1):e4682
Martirosyan D (2025) Functional food science and bioactive compounds. Bioact Compd Health Dis Online 8(6):218–229. ISSN: 2574-0334; Print ISSN: 2769-2426
Mashayekhi-Sardoo H, Sepahi S, Rahimi VB, Askari VR (2024) Application of Nigella sativa as a functional food in diabetes and related complications: insights on molecular, cellular, and metabolic effects. J Funct Foods 122:106518
Maurya NK, Yadav L, Kumari N, Maurya P (2026) Administrative regulations for herbal supplements. In: Herbal medicines and nutritional supplements for health benefits. Apple Academic Press, pp 283–297
Mehdi S, Mehmood MH, Ahmed MG, Ashfaq UA (2023) Antidiabetic activity of Berberis brandisiana is possibly mediated through modulation of insulin signaling pathway, inflammatory cytokines and adipocytokines in high fat diet and streptozotocin-administered rats. Front Pharmacol 14:1085013
Mihal M, Roychoudhury S, Sirotkin AV, Kolesarova A (2023) Sea buckthorn, its bioactive constituents, and mechanism of action: potential application in female reproduction. Front Endocrinol 14:1244300
Minniti G, Laurindo LF, Machado NM, Duarte LG, Guiguer EL, Araujo AC et al (2023) Mangifera indica L., by-products, and mangiferin on cardio-metabolic and other health conditions: a systematic review. Life 13(12):2270
Mohamed Abdoul-Latif F, Ainane A, Houmed Aboubaker I, Mohamed J, Ainane T (2023) Exploring the potent anticancer activity of essential oils and their bioactive compounds: mechanisms and prospects for future cancer therapy. Pharmaceuticals 16(8):1086

Mohanta YK, Mishra AK, Nongbet A, Chakrabartty I, Mahanta S, Sarma B et al (2023) Potential use of the Asteraceae family as a cure for diabetes: A review of ethnopharmacology to modern day drug and nutraceuticals developments. Front Pharmacol 14:1153600

Nagori K, Nakhate KT, Yadav K, Ajazuddin, Pradhan M (2023) Unlocking the therapeutic potential of medicinal plants for Alzheimer's disease: preclinical to clinical trial insights. Fut Pharmacol 3(4):877–907

Naheed S, Hussain K, Wajid M, Zubair H, Shehzadi N, Zulfiqar U et al (2025) Activity-guided isolation, characterization, and antibacterial evaluation of Ziziphus Oxyphylla leaves extract: molecular docking, MD simulations, and ADMET studies. Chem Select 10(43):e01871

Nisar A, Jagtap S, Vyavahare S, Deshpande M, Harsulkar A, Ranjekar P, Prakash O (2023) Phytochemicals in the treatment of inflammation-associated diseases: the journey from preclinical trials to clinical practice. Front Pharmacol 14:1177050

Oloya B, Namukobe J, Ssengooba W, Afayoa M, Byamukama R (2022) Phytochemical screening, antimycobacterial activity and acute toxicity of crude extracts of selected medicinal plant species used locally in the treatment of tuberculosis in Uganda. Trop Med Health 50(1):16

Oluwamodupe C, Adeleye AO (2023) Targeting retinol-binding protein 4 (RBP4) in the management of Cardiometabolic diseases. Cardiovasc Toxicol 23(9):285–294

Orimaye OE, Ekunseitan DA, Omaliko PC, Fasina YO (2024) Mitigation potential of herbal extracts and constituent bioactive compounds on salmonella in meat-type poultry. Animals 14(7):1087

Oswal M, Varghese R, Zagade T, Dhatrak C, Sharma R, Kumar D (2023) Dietary supplements and medicinal plants in urolithiasis: diet, prevention, and cure. J Pharm Pharmacol 75(6):719–745

Palla AH, Amin F, Fatima B, Shafiq A, Rehman NU, Haq I u, Gilani A-u-H (2021) Systematic review of polyherbal combinations used in metabolic syndrome. Front Pharmacol 12:752926

Parganiha R, Tripathi A, Prathyusha S, Baghel P, Lanjhiyana S, Lanjhiyana S, Sarkar D (2022) A review of plants for hepatic disorders. J Comp Med Res 13(46):10–5455

Popstoyanova D, Gerasimova A, Gentscheva G, Nikolova S, Gavrilova A, Nikolova K (2024) Ziziphus jujuba: applications in the pharmacy and food industry. Plants 13(19):2724

Rafique S, Hassan SM, Mughal SS, Hassan SK, Shabbir N, Perveiz S et al (2020) Biological attributes of lemon: A review. J Addict Med Ther Sci 6(1):030–034

Rao GH (2018) Diabetes and cardiovascular disease in south Asians: A global perspective. J Clin Prevent Cardiol 7(4):161–167

Regassa H, Sourirajan A, Kumar V, Pandey S, Kumar D, Dev K (2022) A review of medicinal plants of the himalayas with anti-proliferative activity for the treatment of various cancers. Cancer 14(16):3898

Rehman MHU, Saleem U, Ahmad B, Rashid M (2022) Phytochemical and toxicological evaluation of Zephyranthes citrina. Front Pharmacol 13:1007310

Riaz S, Hussain I, Ibrahim M, Ishtiaq M, Ali Q, Muazzam Ali M, Sajid Hamid Akash M (2022) Extraction and optimization of active metabolites from cluster bean: an in vitro biological and phytochemical investigation. Dose-Response 20(2):15593258221098992

Sadgrove NJ (2022) Honest nutraceuticals, cosmetics, therapies, and foods (NCTFs): standardization and safety of natural products. Crit Rev Food Sci Nutr 62(16):4326–4341

Sakhare VKM, Pathak VK, Richa M, Bhaskar N, Kola M (2023) Ajuga bracteosa: a review on endangered indian medicinal plant. Eur Chem Bull 12(12):1380–1398

Salazar-Gómez A, Alonso-Castro AJ (2022) Medicinal plants from Latin America with wound healing activity: Ethnomedicine, phytochemistry, preclinical and clinical studies—a review. Pharmaceuticals 15(9):1095

Samakar B, Mehri S, Hosseinzadeh H (2022) A review of the effects of Urtica dioica (nettle) in metabolic syndrome. Iran J Basic Med Sci 25(5):543

Saman F, Zahid MTB, Noor V, Raza U, Shahid A, Maheen U, Talha M (2025) Unlocking the healing power of beetroot: nutritional profile & evidence-based health benefits. J Health Wellness Commun Res:e999–e999

Sapkota B, Devkota HP, Poudel P (2022) Citrus maxima (Brum.) Merr. (Rutaceae): bioactive chemical constituents and pharmacological activities. Evid Based Complement Alternat Med 2022(1):8741669

Seidu BS, Osman H, Seidu S (2023) Lifestyle or pharmacotherapy in cardio-metabolic disease prevention. Ther Adv Cardiovasc Dis 17:17539447231177175

Shafodino FS, Lusilao JM, Mwapagha LM (2022) Phytochemical characterization and antimicrobial activity of Nigella sativa seeds. PLoS One 17(8):e0272457

Shah SMA, Akram M, Riaz M, Munir N, Rasool G (2019) Cardioprotective potential of plant-derived molecules: a scientific and medicinal approach. Dose-Response 17(2): 1559325819852243

Shamim T, Asif HM, Shaheen G, Sumreen L, Ayaz S, Qureshi T et al (2022) Anti-diabetic potential of indigenous medicinal plants of Cholistan Desert, Pakistan: a review. Rev Diabetic Stud 18(2): 93

Sharif H, Akash MSH, Rehman K, Irshad K, Imran I (2020) Pathophysiology of atherosclerosis: association of risk factors and treatment strategies using plant-based bioactive compounds. J Food Biochem 44(11):e13449

Shaukat B, Mehmood MH, Murtaza B, Javaid F, Khan MT, Farrukh M et al (2022a) Ajuga bracteosa exerts antihypertensive activity in l-NAME-induced hypertension possibly through modulation of oxidative stress, proinflammatory cytokines, and the nitric oxide/cyclic guanosine monophosphate pathway. ACS Omega 7(37):33307–33319

Shaukat B, Mehmood MH, Shah S, Anwar H (2022b) Ziziphus Oxyphylla hydro-methanolic extract ameliorates hypertension in L-NAME induced hypertensive rats through NO/cGMP pathway and suppression of oxidative stress related inflammatory biomarkers. J Ethnopharmacol 285:114825

Singh B (2023) Ethnopharmacological properties, biological activity and phytochemical attributes of medicinal plants, vol 1. CRC Press

Singhai H, Rathee S, Jain SK, Patil UK (2024) The potential of natural products in the management of cardiovascular disease. Curr Pharm Des 30(8):624–638

Singla RK, Wang X, Gundamaraju R, Joon S, Tsagkaris C, Behzad S et al (2023) Natural products derived from medicinal plants and microbes might act as a game-changer in breast cancer: a comprehensive review of preclinical and clinical studies. Crit Rev Food Sci Nutr 63(33): 11880–11924

Sun W, Shahrajabian MH (2023) Therapeutic potential of phenolic compounds in medicinal plants—natural health products for human health. Molecules 28(4):1845

Ullah I, Chen F (2023) Exploring medicinal plants and their nutraceuticals attributes from norther areas of Pakistan. Food Sci Appl Microbiol Rep 2(2):78–92

Ullah F, Rehman AU, Khan RD, Mahmood I, Khan A, Khan K (2025) Impact of lifestyle interventions on cardiovascular risk factors in patients with metabolic syndrome. Indus J Biosci Res 3(5):210–215

Wahab S, Alsayari A (2023) Potential pharmacological applications of nigella seeds with a focus on nigella sativa and its constituents against chronic inflammatory diseases: progress and future opportunities. Plants 12(22):3829

Wasti Y, Muntaqua D, Majid M, Naz I, Zafar A, Khan SU et al (2024) Characterization and comparative evaluation of wound healing potential of Ajugarin I and Ajuga bracteosa Wall. ex Benth. Front Chem 11:1325578

Wink M (2022) Current understanding of modes of action of multicomponent bioactive phytochemicals: potential for nutraceuticals and antimicrobials. Annu Rev Food Sci Technol 13(1):337–359

Yaseen HS, Zubair HM, Jamal A, Farrukh M, Mikrani R, Shaukat B et al (2024) Naringin: Cardioprotective properties and safety profile in diabetes treatment. Fitoterapia 176:106011

Younis W, Schini-Kerth V, Junior AG, Majid M (2018) Cardioprotective effect of Asphodelus tenuifolius Cav. on blood pressure and metabolic alterations in glucose-induced metabolic syndrome rats–an ethnopharmacological approach. J Ethnopharmacol 214:168–178

Zargar S, Wani TA, Rizwan Ahamad S (2023) An insight into wheat germ oil nutrition, identification of its bioactive constituents and computer-aided multidimensional data analysis of its potential anti-inflammatory effect via molecular connections. Life 13(2):526

14 Forage Plants and Palatability Preferences in the Nara Desert: Ethnobotanical, Ecological, and Conservation Perspectives

Rahmatullah Qureshi

Abstract

This chapter was aimed at assessing the year-round availability and palatability of local forage plant species in Nara Desert, Sindh Pakistan. It was observed that the palatability and availability of forage species did change significantly over the course of the growing season. One hundred and sixty species of plants were recorded with 151 species, 128 genera and 43 families used as livestock feed. These were grouped into the very palatable (49.01%), moderately palatable (26.49%), less palatable (18.54%), and rarely palatable (5.96%) groups. These were analyzed to show that 64.14% of species were eaten as entire plants, 21.38% of their leaves and 11.03% of their shoots. The most diverse diet was recorded among goats which fed on 124 species, then cows (100 species), camels (85 species), and sheep (82 species). It is also important to note that all types of livestock browsed 41 species. There was a seasonal deficiency of palatable forage between November and February thereby forcing livestock to consume dried or non-palatable species. The results offer a basis on which additional research regarding nutritional and mineral content of palatable species can be conducted to aid in rangeland management as well as livestock productivity.

Keywords

Forage plants · Palatability preference · Nara desert · Native flora · Dryland afforestation

R. Qureshi (✉)
Department of Botany, Faculty of Sciences, Pir Mehr Ali Shah Arid Agriculture University, Rawalpindi, Pakistan
e-mail: rahmatullahq@uaar.edu.pk

R. Qureshi (ed.), *Plant Biodiversity*, https://doi.org/10.1007/978-981-92-1471-6_14

Abbreviations

HP	Highly Palatable
LP	Less Palatable
MP	Moderately Palatable
RP	Rarely Palatable
WP	Whole Plant

14.1 Introduction

Rangeland covers about half of the global land and it is also a vital part of the world agricultural system; especially in such countries as Pakistan. The lands do not only provide grazing to the 142.9 million livestock in the country, as well as a source of fuelwood, timber, and wildlife habitats. The conditions in these regions are mostly not conducive to traditional farming as a result of the dry and semi-dry weather and lack of irrigation facilities. Nonetheless, this large natural resource, which covers more than 60% of the land area in Pakistan, has a great potential in livestock grazing and dry land afforestation (Qureshi and Bhatti 2007a).

The desert regions of Pakistan particularly Nara Desert, are locations of grazing to feed the rural people. A vegetation pattern, which is characterized by drought-tolerant trees, shrubs, grasses, and herbaceous plants, is determined by scarcity of water and climatic extremes.

The Nara Desert, which is low in primary productivity, is the place where the human and livestock population is high and estimated to be 1.05 million and 1.25 million, respectively (Anon 1992). The pastoral societies, which occupy this area, largely depend on livestock as a means of earning their-living and as a result, it has prompted a tendency of raising the number of livestock with time (Qureshi and Bhatti 2007a). The Nara Desert is a rangeland with numerous and nutritious malleable, palatable, and nutritious species of herbs, shrubs, and trees. Flora, vegetation, ecology, and ethnobotany of the project area were studied well previously (Bhatti et al. 1998, 2001a, b, 2002; Qureshi 2008b, 2011, 2012, 2018, 2020; Qureshi and Bhatti 2005a, b, 2014).

Pakistan on the other hand has limited studies that have examined the native fodder and forage species and their palatability. According to Wahid (1990), sheep and goats grazing in different rangelands of Balochistan were fed on 53–81% shrubs. Noor (1978) compared grazed and un-grazed vegetation in sub- alpine area of Sari, Kaghan. On summer pastures of the Nasirabad Valley, Hunza, Pakistan, Hussain and Mustafa (1995) documented 131 species of pastures grouping to 42 families with 12 species not being palatable. In total, they described 27 species as very palatable, 68 as moderately palatable, 20 as less palatable and 4 as never palatable.

Omera et al. (2006) stated that spring season in the dry temperate rangelands of north Pakistan recorded high forage production. Hussain and Durrani (2009) examined seasonal occurrence, palatability and preference of livestock in the Harboi arid rangelands in the Kalat region where 129 species were found palatable. Fifty-four point 4% (65 species) of these were highly palatable and 41.1% (53 species) were mostly palatable, 4.65% (6 species) were less palatable and 3.87% (5 species) were rarely palatable. The other contributions, which are also related, are those by Hussain and Chughtai (1984), Khan (1996) and additional studies by Hussain and Durrani (2007, 2008).

Since rearing livestock is the only means of livelihood of the nomadic people in this desert area, the current research was considered critical to determine palatability of traditional flora. This study records seasonal availability of fodder and forage species, palatability of different components of plants among animals that graze and species preferences among grazing animals. The results will have a significant contribution to the work of range ecologists and land managers in terms of the development of the strategies to use and enhance the condition of this rangeland sustainably.

14.2 Study Area: Nara Desert

Nara Desert, in the south eastern province of Pakistan, Sindh, is between the latitude of 26o-28o N and longitude of 68o-70o E with an elevation of 50 and 115 meters (Fig. 14.1). The eastern side of this desert is now bordered by Rajasthan, India i.e., Jodhpur and Jaisalmer. The region has been described as hot arid to semi-arid sandy desert with an annual rainfall of between 88 and 135 mm, mostly during the months of July and September. Mean minimum and maximum temperatures are about 20 °C and 45 °C respectively and the highest temperature is 47–51 °C during the month of May to July. The lowest temperatures are usually during January and they are 20–28 °C. This desert is mostly characterized by aridity and the period of wet and dry years (Qureshi and Bhatti 2005a).

14.2.1 Topographical and Ecological Features

The Nara Desert has three physical elements of the topography, which include sandy hills, steep slopes, and large flat areas at low altitude (Fig. 14.2). The sands accumulating to form large masses form sand dunes or sand ridges that have three structural components which are always consistent; the crest (the top part), the swale or slope (the middle part) and the foot (the bottom part). The space between two consecutive dunes is known as interdunal valley (Qureshi and Bhatti 2005a). The terrain is mostly composed of sandy plains of gently rolling terrain with sand dunes, the number and size of which are larger to the west than to the east.

The majority of the territory is covered by hyper-arid to arid regions that are characterized by long dry summers, severe drought, strong wind speed, low humidity, high rates of evapotranspiration, and low precipitation. The southern part of the area (including places like Khipro and Sanghar) is slightly semi-arid and bears a

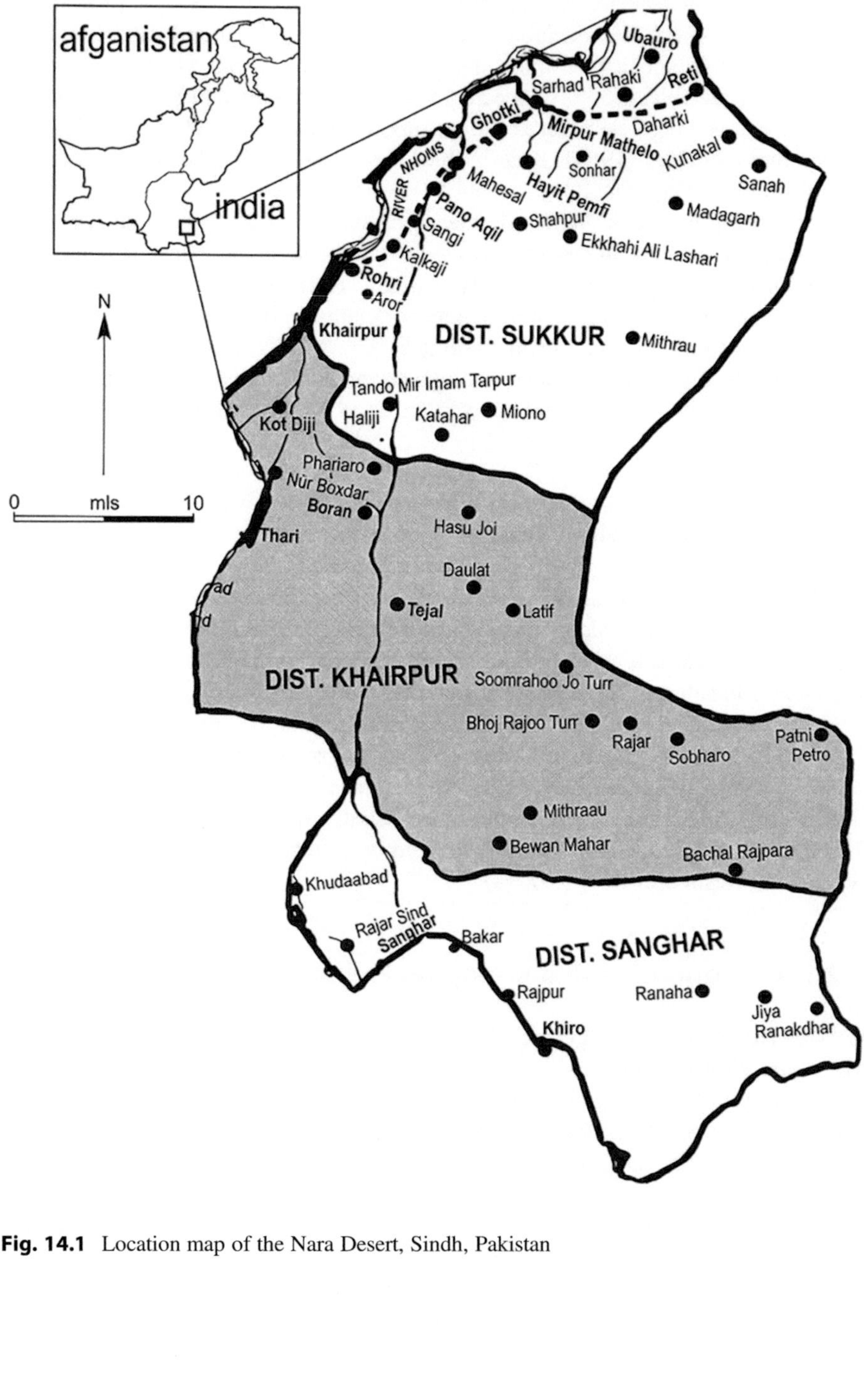

Fig. 14.1 Location map of the Nara Desert, Sindh, Pakistan

Fig. 14.2 Landscape of Nara Desert, Pakistan

moderately thick vegetation. The rainfall varies between 88 and 135 mm annually and mostly experienced during monsoon season. Water is also a significant problem and the groundwater source is found at very deep levels of 50–300 feet (Qureshi and Bhatti 2005b).

Nara Desert has a high floristic diversity. Considering the ecological importance, a detailed field visit was done to record the available species of plants, to determine the impact that endangers the ecosystem, and to suggest measures that can be taken to restore the deteriorating lands. The authors have already made elaborate descriptions on the flora and vegetation of this distinct landscape (Qureshi 2004, 2008a, b, 2011; Qureshi and Bhatti 2006, 2008a, b, c, 2009, 2014; Qureshi and Ahmad 2010; Qureshi et al. 2001a, b, 2009, 2010). The current communication provides the latest details of floral composition and status, which is important to the researchers, the planners and the policymakers in the aspects of conservation, sustainable use, and management of the ecosystems in arid environments.

14.2.2 Phytogeography

Phytogeographically, Ali and Qaiser (1986) categorize most of the study area to be Saharo–Sindian Region. A smaller area, particularly to the western rim, has floristic influences of the Indian Region, which points to transitional phytogeographic zone. The region consists mostly of xerophytic vegetation, which is adapted to arid and semi-arid climatic conditions. The hot arid deserts of Cholistan, Tharparkar, and Thal are also under the Saharo–Sindian phytogeographic zone and have similar ecological conditions and vegetation cover.

14.2.3 Flora and Vegetation

The floristic structure of the region is majorly Saharo–Sindian although the region has a small western part that has an Indian floristic influence. Vegetation in the area is largely xerophytic, and can withstand extreme heat, drought, and sandy soils. The floristic survey was a complete listing of 267 plant species (167 genera and 59 families) (Qureshi and Bhatti 2014). This biodiversity indicates the ecological abundance and adaptive mechanisms of the vegetative life that is present in this dry desert habitat. The chapter will discuss the forage palatability and inventory of an arid ecosystem in Sect. 14.3.

This part presents the methodology and the results of a research carried out between September 1998 and October 2001, which sought to put a systematic record of the forage resources that livestock has access to within a given arid to semi-arid area. The study was aimed at describing plant species, seasonality, palatability of the plants to various stock, and the foraging patterns. The seasonal availability of fields and their seasonal surveys are presented in Sect. 14.3.1.

A systematic survey of the whole study area was conducted during a period of 3 years to list all the possible forage species. This was a complete census of foliage of different types of plant life including grasses, herbs, shrubs, and trees. The main aim was to monitor and catalogue the yearly presence of these species and the time of the year, which they were most abundant and could be utilized by grazing animals.

14.3 Classification Based on Palatability and Preference by an Animal

The forage species reported on were examined in reference to two important criteria:

14.3.1 Differential Palatability of Portions of the Plant

Researchers found out what specific portions of each plant were eaten by the livestock (ex: the entire plant, or only leaves, young shoots, or floral parts).

14.3.2 Livestock Preference

The species were further grouped based on how they were thought to be preferred by the four primary types of livestock found in the region goats, sheep, cows, and camels. The plants were placed in a four-tier palatability scale depending on how many times the animals were observed to eat them:

Highly Palatable (HP): Species who are the primary selection, and the most desired by the most and the majority of grazing animals.

Moderately Palatable (MP): Species that are eaten on a regular basis but are not the best; they are a mediocre degree of preference.

Less Palatable (LP): Those species that are only consumed occasionally, usually when other more preferable food items are in short supply.

Rarely Palatable (RP): These are species that are never sampled and only eaten as a last resort when no other foodstuff can be found.

14.3.3 Data Collecting and Species Identification

The data was taken into the field to identify the issue with great scientific accuracy by collecting plant specimens and drying them using the standard methods before preserving and identifying them using existing literature and regional floras (Jafri 1966; Nasir and Ali 1970–1989; Ali and Nasir 1990–1991; Ali and Qaiser 1993–-1995; Batanouny 1981; Matthew 1981–1983; Shetty and Singh 1987; Boulos 1991; Bhandari 1978). Herbarium records were made of voucher specimens. In addition, to triangulate the field observations, the research used ethnobotanical knowledge. Author interviewed pastoralists, herders, and residents that take care of livestock in the desert ecosystem. This was an invaluable aid in giving knowledge of the real palatability and use of different plants that could not be discerned by the naked eye when merely in the field.

14.4 Inventory and Analysis of Forage Species

14.4.1 Forage Species Diversity

The study recorded a total of 151 plant species serving as livestock feed. These species belong to 128 genera and 43 families, indicating a significant diversity of forage resources in the region (Table 14.1).

14.4.2 Parts of Plants Consumed

An analysis of which plant parts were utilized revealed that the most common form of consumption was recorded as whole plants with 93 species (64.14%), being eaten entirely (Fig. 14.3). Leaves were the second most choice of the domesticated animals and eaten as the primary part (31 spp., 21.38%). However, shoots were utilized in 16 species (11.03%), which are often prized for their tenderness.

Table 14.1 Seasonal availability, palatability, and animal preference of fodder species from Nara Desert, Pakistan

S. No.	Plant species	Parts	Flowering period	Palatability	Animal preference
	Sedges				
1	*Cyperus arenarius* Retz.	Sh	September–November	HP	All
2	*C. conglomeratus* Rottb.	Sh	September–November	HP	All
3	*C. laevigatus* L.	Sh	September–November	MP	Goat, Cow
4	*C. rotundus* L.	Sh	June–September	HP	All
	Grasses				
5	*Aeluropus lagopoides* (L.) Trin. ex. Thw	Lvs, Sh	April–November	MP	Camel, Goat
6	*Aristida adscensionis* L.	Lvs, Sh	March–December	HP	Goat, Sheep
7	*A. funiculata* Trin. & Pupr.	Lvs, Sh	June–September	MP	Goat, Sheep
8	*A. mutabilis* Trin. & Rupr.	Lvs, Sh	March–December	MP	Sheep Goat
9	*Cenchrus biflorus* Roxb.	Lvs, Sh	January–April and September–November	MP	Cow, Goat, Sheep
10	*C. ciliaris* L.	Lvs, Sh	February–March	HP	Cow, Goat, Sheep
11	*Cymbopogon jwarancusa* (Jones) Schult.	Lvs, Sh	March–May and again in September	LP	Sheep Goat, Cow
12	*Dactyloctenium aegyptium* (L.) P. Beauv.	WP	July–October	HP	Cow, Goat, Sheep
13	*Desmostachya bippinata* (L.) Stapf.	Lvs, Sh	July–November	MP	Cow, Goat
14	*Dichanthium annulatum* (Forssk.) Stapf.	Lvs, Sh	March–April	HP	All
15	*Dinebra retroflexa* (Vahl.) Panzer.	WP	May–September	HP	Goat, Cow, Sheep
16	*Eragrostis minor* Host.	WP	June–September	HP	Goat, Cow, Sheep
17	*Lasiurus sindicus* Henr.	Lvs, Sh	March–May and September–October	HP	All
18	*Ochthochloa compressa* (Forssk.) Hilu.	Lvs, Sh	March–October	HP	All

(continued)

Table 14.1 (continued)

S. No.	Plant species	Parts	Flowering period	Palatability	Animal preference
19	*Panicum turgidum* Forssk.	Sh, Lvs	April–May	HP	All
20	*Saccharum spontaneum* L.	Lvs	July–September	RP	Cow
21	*Setaria pumila* (Pouir.) Roem. & Schult.	WP	June–August	HP	Goat, Cow
22	*Stipagrostis plumosa* (L.) Munro ex T. Anders.	WP	About round the year	HP	Goat, Cow
	Forbs				
23	*Blepharis sindica* Stocks ex. Anders	WP	July–September	HP	Goat, Cow
24	*Boerhavia rubicunda* Steud.	WP	July–August	HP	Goat, Sheep
25	*B. procumbens* Banks ex Roxb.	WP	September–August	HP	Goat, Sheep, Cow
26	*Achyranthes aspera* L.	WP	September–December	HP	All
27	*Alysicarpus monilifer* (L.) DC.	WP	October–November	HP	All
28	*A. scariosus* Grah. ex Thwaites	WP	November–March and August–September	HP	All
29	*Amaranthus graecizans* subsp. *thellungianus* (Nevski) Gusev.	WP	July–December	HP	Goat, Sheep, Cow
30	*A. virdis* L.	WP	August—September	HP	Goat, Sheep, Cow
31	*Amberboa ramosa* (Roxb.) Jafri.	WP	January–February	MP	Camel, cow
32	*Anticharis glandulosa* Aschers.	WP	October–May	LP	Camel, Goat, Sheep
33	*Arnebia hispidissima* (Sieber) DC.	WP	November–March	LP	Camel, cow
34	*Asphodelus tenuifolius* Cavan	WP	November–February	HP	All
35	*Cassia italica* (Mill.) F.W. Andr.	Lvs	October–December	LP	Cow, Goat
36	*Chenopodium album* L.	WP	January–April	HP	All
37	*Cistanche tubulosa* (Schrenk) Hook. f.	Sh	December–January	MP	Goat, Camel
38	*Cleome brachycarpa* Vahl.	WP	November–May	LP	Cow, Goat

(continued)

Table 14.1 (continued)

S. No.	Plant species	Parts	Flowering period	Palatability	Animal preference
39	*C. scaposa* D.C.	WP	August	HP	Cow, Goat
40	*C. viscosa* L.	WP	July–September	LP	Cow, Goat
41	*Commicarpus boisseiri* (Heimer) Cufod.	WP	September–October	HP	Goat, Sheep
42	*Convolvulus arvensis* L.	WP	Throughout the year	HP	Goat, Sheep, Cow
43	*C. glomeratus* Choisy.	WP	December–August	MP	Goat, Sheep, Cow
44	*C. prostratus* Forssk.	WP	November and January	HP	Goat, Sheep, Cow
45	*Corchorus aestuans* L.	WP	June–October	HP	Goat, Sheep, Cow
46	*C. depressus* (L.) Stocks.	WP	February–November	HP	Goat, Cow, Sheep
47	*C. tridens* L.	WP	July–November	HP	Goat, Cow, Sheep
48	*C. trilocularis* L.	WP	June–October	HP	Goat, Cow, Sheep
49	*Cressa cretica* L.	WP	Almost throughout the year	RP	Camel, Goat
50	*Ctenolapis cerasiformis* (Stocks) Hooker.f.	WP	October–November	RP	Goat
51	*Cucumis melo* var. *agrestis* Naudin.	WP	August–September	HP	All
52	*C. prophaterum* L.	WP	August–December	LP	Camel
53	*Cuscuta chinensis* Lam.	WP	February–April	HP	Goat, Sheep, Cow
54	*Digera muricata* (L.) Mart.	WP	September–October	HP	Goat, Sheep, Cow
55	*Euphorbia hirta* L.	WP	July–November	HP	Goat, Cow
56	*E. thymifolia* L.	WP	December–February	HP	Goat, Cow

(continued)

Table 14.1 (continued)

S. No.	Plant species	Parts	Flowering period	Palatability	Animal preference
57	*Fagonia bruguieri* DC. var. *rechingeri* Hadidi	WP	March–May	MP	Goat, Camel
58	*F. indica* var. *schweinfurthii* Hadidi	WP	Almost throughout year	MP	Goat, Camel
59	*Farsetia hamiltonii* Royale.	WP	August–November	HP	All
60	*Fumaria indica* (Hausskn.) H.N. Pugsley	WP	March–June	MP	Goat
61	*Gisekia pharancoides* L.	WP	Almost round the year	HP	Goat, Cow
62	*Glinus lotoides* L.	WP	February–April	RP	Camel
63	*Gynandropsis gynandra* (L.) Briq.	WP	June–July	HP	Cow, Goat, Sheep
64	*Heliotropium crispum* Desf.	WP	January–April	LP	Camel, Goat, Cow
65	*H. curassavicum* L.	WP	January	RP	Camel
66	*H. europeum* L.	WP	November–January	HP	Goat, Cow, Camel
67	*H. strigosum* Willd.	WP	July–September	MP	Goat, Cow
68	*H. subulatum* (Hochst. ex DC.) Vatke.	WP	October–December	MP	Goat, Cow
69	*Hibiscus lobatus* (J.A Murray) O. Ketze.	WP	September–October	HP	All
70	*Indigofera argentea* Burm. f.	Lvs, Sh	Throughout the year	HP	All
71	*I. cordifolia* Heyne ex Roth	WP	August–October	HP	All
72	*I. hochstetteri* Baker	WP	August–October	HP	All
73	*I. linifolia* (L.) Retz.	WP	July–October	HP	All
74	*Kickxia ramosissima* (Wall.) Janchen	WP	November–January	HP	Goat, Sheep
75	*Launaea procumbens* (Roxb) Ramayya & Rajagopal.	WP	January–May	HP	Goat, Camel, Sheep
76	*Leucas aspera* (Willd.) Link.	WP	June & again November	RP	Goat
77	*Limeum indicum* Stocks ex T. Anderson.	WP	March–September	HP	Goat, Cow
78	*Mollugo cerviana* (L.)Ser.	WP	February–April	HP	Goat, Sheep, Cow
79	*M. nudicaulis* Lamk.	WP	October–November	HP	Goat, Sheep, Cow

(continued)

Table 14.1 (continued)

S. No.	Plant species	Parts	Flowering period	Palatability	Animal preference
80	*M. pentaphylla* L.	WP	September–November	HP	Goat, Sheep, Cow
81	*Monsonia heliotropoides* (Cav.) Boiss.	WP	August–October	HP	Goat, Cow, Sheep
82	*Mukia maderaspatana* (L.) M. J. Roem.	WP	July–December	LP	All
83	*Neurada procumbens* L.	WP	September–December	HP	Goat, Cow
84	*Pavonia glechomaefolia* S. Abedin	WP	November–December	MP	All
85	*Phyla nodiflora* (L.) Greene.	WP	Almost round the year	LP	All
86	*Polygala erioptera* DC.	WP	Almost round the year	HP	Goat, Sheep
87	*Polygonum plebejum* R. Br.	WP	September–December	HP	Goat, Sheep, Cow
88	*Rhynchosia minima* (L.) DC.	WP	January–March	HP	All
89	*R. pulverulenta* Stocks	WP	January–April	HP	All
90	*R. schimperii* Hochst. ex Boiss.	Sh	September–March	LP	All
91	*Seetzenia lanata* (Willd.) Bullock	WP	February–May	LP	Camel
92	*Sesuvium sesuvioides* (Fenzl) Verdc.	WP	December	LP	Cow, Camel
93	*Solanum nigrum* var. *nigrum* L.	WP	Almost round the year	MP	All
94	*Tribulus longipetalus* Viv.	WP	August–November	HP	Goat, Sheep, Cow
95	*T. terrestris* L.	WP	July–September	HP	Goat, Sheep, Cow
96	*Trianthema triquetra* Rottl. and Willd.	WP	July–November	MP	camel
97	*Trichodesma indicum* (L.) R.Br.	WP	November–December	LP	Camel
98	*Zelya petandra* (L.) Jeffrey	WP	July–November	MP	Cow, Camel
99	*Zygophyllum simplex* L.	WP	May–August	MP	Camel
	Shrubs				
100	*Abutilon fruticosum* Guill. & Perr.	Lvs	October–November	LP	All

(continued)

Table 14.1 (continued)

S. No.	Plant species	Parts	Flowering period	Palatability	Animal preference
101	*Abutilon indicum* (L.) Sweet	Lvs	March–April	LP	All
102	*Acacia jacquemontii* Benth.	Lvs, Sh	February–May	HP	All
103	*Aerva javanica* var. *bovei* Webb.	Lvs	September–December	MP	Cow, Goat
104	*A. javanica* var. *javanica* (Burm.f.) Juss. ex.	Lvs	July–September	MP	Cow, Goat
105	*Alhagi maurorum* Medic.	Lvs, Sh	April–September	MP	All
106	*Arthrocnemum indicum* (Willd.) Moq.	Sh	October–November	MP	Camel
107	*Bergia aestivosa* Wight & Arn.	Lvs	November–December	MP	Goat, Sheep
108	*Cadaba farinosa* Forssk.	Lvs	October–December	MP	Goat, Camel
109	*Calligonium polygonoides* L.	Sh, Sh	March–April	HP	All
110	*Calotropis procera* (Willd.) R. Br.	Lvs, Sh	July–September	LP	Goat
111	*Capparis spinosa* L.	WP	April–May	HP	All
112	*Carthamus oxycantha* Bieb.	WP	May–August	MP	Camel, Cow
113	*Chrozophora plicata* (Vahl) A. Juss. Ex Spreng.	WP	March–July	MP	Goat, Camel
114	*Cocculus hirsutus* (L.) Diels	WP	February–March	HP	All
115	*Crotolaria burhia* Ham. ex Benth.	Lvs, Sh	September–October	RP	Goat
116	*Echinops echinatus* DC.	WP	August–October	MP	Camel
117	*Ephedra ciliata* Fisch. & Mey. ex C.A.Meyer.	WP	February–March	HP	All
118	*Grewia tenax* (Forssk.) Aschers & Schweinf.	Lvs	July–September	HP	All
119	*Haloxylon stocksii* (Boiss.) Benth. & Hook	Sh	November–December	MP	Camel
120	*Indigofera oblongifolia* Forssk.	Lvs, Sh	September–November	HP	All
121	*Leptadenia pyrotechnica* (Forssk.) Decne.	Lvs, Sh	August–January	MP	All
122	*Lycium edgeworthii* Dunal.	Lvs	October–January	LP	Camel, Goat
123	*Maerua crassifolia* Forssk.	Lvs, Fr	November–April	MP	Camel, Goat

(continued)

Table 14.1 (continued)

S. No.	Plant species	Parts	Flowering period	Palatability	Animal preference
124	*Malvastrum coromendelianum* (L.) Garcke	WP	August–January	LP	All
125	*Oxystelma esculentum* (L.f.) R. Br.	WP	December	HP	All
126	*Pluchea lanceolata* (DC.) Oliv. & Hiern.	Lvs	December	LP	Goat, Camel
127	*Prosopis juliflora* (Swartz) DC.	Fr	March–June	MP	Goat
128	*Salsola imbricata* Forssk.	Sh	August–October	RP	Camel
129	*Solanum surattense* Burm. f.	WP	March–April and July–September	LP	Camel
130	*Suaeda fruticosa* (L.) Forskssk.	Sh	November–July	MP	Camel
131	*Tephrosia falciformis* Romaswami.	Lvs, Sh		LP	Cow, Goat
132	*T. uniflora* var. *petrosa* (Blatter & Hallberg) Gillett & Ali	DP	June–August	LP	Goat, Cow
133	*T. villosa* (L.) Pers.	DP	January–September	LP	Goat, Cow
134	*Tamarix indica* willd.	Lvs	January–October	MP	Camel
135	*Tamarix passernioides* Del ex. Desv var. *macrocarpa* Ehrenb	Lvs	January–October	MP	Camel
136	*Zizyphus nummularia* (Burm. F.) Wt. & Arn.	Lvs, Sh	March–June	HP	All
	Trees				
137	*Acacia nilotica* (L.) Delile	Lvs, Sh	March–May and August–October	HP	All
138	*A. senegal* (L.) Willd.	Fr	August–October	HP	All
139	*Capparis decidua* (Forssk.) Edgew.	Sh	April & August–September	HP	Camel, Goat
140	*Cordia myxa* L.	Lvs	March–April	LP	Goat, Sheep
141	*Ficus religiosa* L.	Lvs	April–June	LP	Goat
142	*Moringa oleifera* Lamk.	Lvs	February–March	RP	Camel, Goat
143	*Prosopis cineraria* (L.) Druce	Lvs, Sh	April–August	HP	All
144	*P. juliflora* Swartz	Fr	August–October	MP	Goat, Sheep
145	*Salvadora oleoides* Decne.	Lvs	December–March	HP	Camel, Cow
146	*S. persica* L.	Lvs	March–June	MP	Camel

(continued)

Table 14.1 (continued)

S. No.	Plant species	Parts	Flowering period	Palatability	Animal preference
147	*Tamarix aphylla* (L.) Karst.	Lvs	June–October	MP	Camel
148	*Tecomella undulata* (Sm.) Seem.	Lvs	January–April	LP	Goat, Sheep
149	*Trichodesma indicum* (L.) R.Br.	WP	November–December	LP	Camel
150	*Zelya petandra* (L.) Jeffrey	WP	July–November	MP	Cow, Camel
151	*Zygophyllum simplex* L.	WP	May–August	MP	Camel

Legends: *HP* Highly palatable, *MP* Moderately palatable, *LP* Less palatable, *RP* Rarely palatable, *WP* Whole plant, *Lvs* Leaves, *Sh* Shoots, *Fr* Fruits, *DP* Dried plant

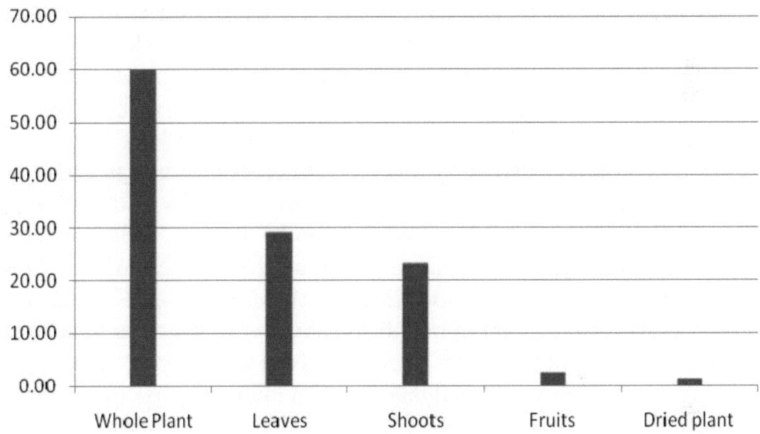

Fig. 14.3 Showing different parts of plants used as forage in Nara Desert, Pakistan

14.4.3 Grazing System and Fodder Management

The area has a continuous grazing system. The natural forage is found in mid-May until end of October. During any time not within this fruitful season, the animals still graze upon all the vegetation on the ground which is green or dry. Fodder conservation is practiced by the local communities to fight seasonal deficits. They harvest and hoard dried fodder in the summer season both in the wild and in the farms. *Chenopodium album, Cucumis melo* var. *agrestis, Cleome viscosa, Medicago polymorpha, Melilotus alba, M. indica, Trianthema portulacastrum, Tribulus terrestris* and numerous grasses and sedges such as *Aristida* spp. and *Cynodon dactylon, Dactyloctenium aegyptium, Ochthochloa compressa, Lasiurus scindicus*, etc., are some of the key stored species (Qureshi and Bhatti 2007a, b, c, d).

14.4.4 The Patterns of Seasonal Availability

Plant growth season is a time-period of rainfall and temperature, which occurs between April and October. Precipitation is also mainly experienced during the monsoon and this causes most of the annual plants to emerge and grow between September and October. The perennial species also remain constant throughout the year and play an important role of providing forage during lean months. The availability of species varied in a wide manner during the year. The greatest number of forage species were seen in the late summer and early in the fall with the highest diversity being recorded in September (69), then August (67), and lastly in October (64). In November (59), March (50), July (48), April (42), and December (39), the availability was moderate. There was a high shortage during January and February, where there were only 25 and 27 species of *M. crunogenus*, respectively (Fig. 14.4). On the basis of season classification, early spring produced a total of 28 species, summer had 105, autumn 43, and early winter 49. During these seasonal changes, the perennial species offered a stable but lower level of livestock forage. The finest forage which was composed of most palatable forbs and grasses/sedges were in greatest quantity in April and September. The abundance of the most palatable species however decreased drastically after November and lasted until February. This season is also associated with a drought season that brings about a severe shortage of feeds as livestock has to feed on less desirable, withered vegetation. Also, Hussain and Durrani (2009) reported similar trends in Harboi rangelands, which is known as the Kalat where the productivity is reduced in winter. *Carthamus oxyacantha, Cymbopogon jwarancusa, Echinops echinatus,* and *Fagonia indica* are desirable in the first wet season but not after maturity. On the other hand, *Tephrosia* spp. (drought-tolerant species) are grazed in dry state in case of extended dry seasons. The paper has also recorded that the palatability of particular species varies with season.

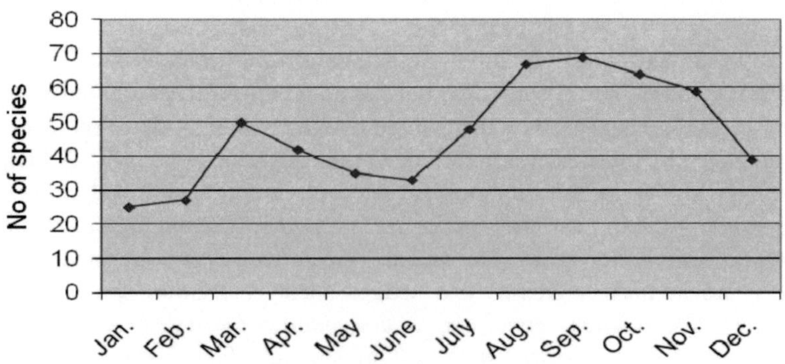

Fig. 14.4 Showing the availability of fodder species in different months

14.4.4.1 Peak Availability

The highest number of species was available between August and November, with a peak of 69 species in September.

14.4.4.2 Low Availability

The period from January to February represents a significant forage deficit, with only 25–27 species recorded. This is due to a combination of drought conditions and the natural life cycles of the plants.

14.4.4.3 Perennial Contribution

Perennial species played a critical role by providing a consistent, though reduced, forage base across all seasons.

14.4.4.4 Grazing Season

The effective grazing season aligns with plant growth, starting in April and ending in November.

14.4.5 Livestock Selectivity and Dietary Breadth

The study revealed distinct species selectivity among animal types (Table 14.2). Goats showed the broadest dietary range, consuming 124 species, including 44% forbs, 17% shrubs, and 11% grasses. Cows fed on 100 species, mainly 38% forbs and 13% shrubs. Camels browsed 85 species, with a higher preference for woody shrubs (19%) and trees (7%). Sheep grazed on 82 species, largely herbaceous forbs (30%) and grasses (9%). Notably, a core group of 41 species (27.15%) was consumed by all four types of livestock, indicating these are crucial, universally accepted forage resources. Most of these common species were utilized as whole plants.

14.4.6 Forage Resource Base and Palatability Classification

The study area is characterized by an arid to semi-arid climate, which has given rise to a plant community dominated by xerophytic species. These plants are uniquely

Table 14.2 Different life spans of species along with the percent palatability browsed by domesticated animals

	Browsed by animals			
Life span	Camel	Cow	Goat	Sheep
Species browsed	85	100	124	82
Sedges	2	3	3	2
Grasses	3	9	11	9
Forbs	24	38	44	30
Shrubs	19	13	17	10
Trees	7	3	6	4

Fig. 14.5 Showing the palatability of species in the Nara Desert

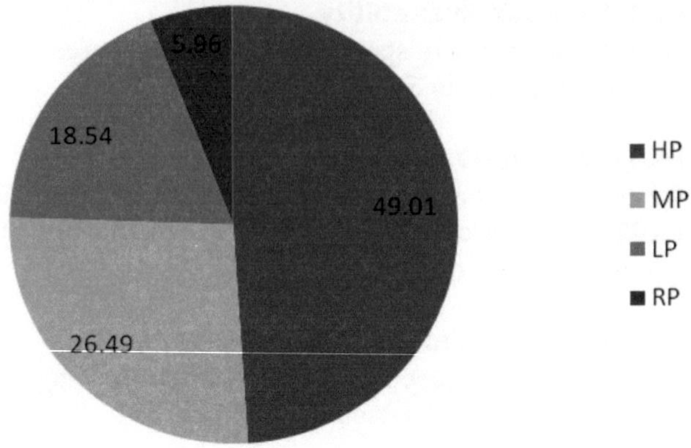

adapted to the region's environmental stresses, including high seasonal temperatures, low humidity, significant fluctuations in moisture availability, and a wide range of soil (edaphic) conditions (Qureshi and Bhatti 2005a).

Within this ecosystem, a total of 160 plant species were recorded. Of these, a significant majority i.e., 151 species, representing 128 genera and 43 families, were identified as being utilized as livestock feed. These forage species were systematically classified into a palatability hierarchy based on their desirability to grazing animals (Fig. 14.5). According to it, majority of species (74) were highly palatable (HP) contributing 49.01% of the total flora, followed by moderately palatable (MP) with 40 species (26.49%), less palatable (LP) with 28 species (18.54%) and rarely palatable (RP) with 9 species (5.96%).

14.4.7 Livestock Selectivity and Key Forage Species

The study revealed distinct dietary preferences among the different livestock types, reflecting their unique physiological adaptations and nutritional needs. A clear hierarchy of dietary breadth was observed. Goats demonstrated the widest dietary range, consuming 124 species. Cows fed on 100 species. Camels browsed on 85 species. Sheep utilized 82 species. Notably, a core group of 41 species (27.15%) served as a common forage base, being consumed by all four types of domesticated animals. The majority of these universally accepted species were utilized as whole plants. The most common and valuable forage species were cataloged according to the growth forms as below:

14.4.7.1 Trees and Shrubs

Prosopis cineraria, Acacia nilotica, A. jacquemontii, Capparis decidua, Calligonium polygonoides, Indigofera argentea, Grewia tenax and *Zizyphus nummularia.*

14.4.7.2 Grasses and Sedges

Aristida spp., *Cenchrus ciliaris, Cynodon dactylon, Cyperus arenarius, C. conglomeratus, C. rotundus, Dactyloctenium aegyptium, Eragrostis minor, Lasiurus sindicus, Ochthochloa compressa, Panicum turgidum, Setaria pumila* and *Stipagrostis plumosa.*

14.4.7.3 Forbs (Herbaceous Plants)

Asphodelus tenuifolius, Boerhavia procumbens, Cleome scaposa, Commicarpus boisseiri, Corchorus tridens, C. trilocularis, C. depressus, Digera muricata, Euphorbia hirta, E. thymifolia, various *Indigofera* species, *Kickxia ramosissima, Limeum indicum, Neurada procumbens, Polygonum plebejum, Tribulus longipetalus* and *T. terrestris.*

14.4.8 Ecological Impact of Palatability and Grazing Pressure

In contrast, species such as *Tephrosia* spp., *Dipterygium glaucum, Cassia italic, Crotalaria burhia*, and *Cymbopogon jwarancusa* were identified as the least palatable. Their low desirability means they are largely avoided by livestock, a factor that has allowed them to become dominant across large areas of the rangeland. This shift in plant community composition is directly linked to grazing pressure. Chronic overgrazing and the lopping of trees and shrubs have significantly reduced the populations of palatable and desired species. This degradation creates ecological opportunities for non-preferred species to establish and thrive, ultimately leading to the replacement of a diverse, nutritious forage base with a less productive one. This finding is consistent with a global body of research concluding that overgrazing systematically reduces palatable plant cover and diminishes overall species diversity (Khan 1996; Liu et al. 1996; Hickman et al. 1996; Makulbekova 1996; Hussain and Chughtai 1984; Hussain and Durrani 2007, 2008).

Goats mainly satisfy their nutritional requirements with the help of the forages such as grasses, browse (woody plants), weeds, forbs, small grains, hay, and silage (Fig. 14.6). Such types of plants usually have viable amounts of protein, energy, minerals, and vitamins. Goats are also fond of browsing on shrubs and bushes and not grazing on grasses (Wilson et al. 1995; Huston 1978; Grunwaldt et al. 1994). Their diet structure can be made up of approximately 70% browse, 25% grasses and 5% forbs seasonally (Vallentine 1990). On the contrary, sheep prefer grasses and forbs, and they browse more in the dry season. According to Migongo-Bake and Hensen (1987) more than 50% of sheep diets were made up of grasses in all seasons. Moreover, Kababia et al. (1992) established that goats fed on shrubs have an intake of between 79% and 86% of the total dry mass of the winter, spring, and summer.

The camels normally feed on a large selection of fodder plants such as thorny shrubs, halophytes, as well as aromatic species (Iqbal and Khan 2001). They are highly adapted to the harsh feeding environment of desert ecosystems and have a selective foraging behavior, so as to ensure high quality diet throughout the year. *Salvadora oleoides* (Figs. 14.7 and 14.8) was the favorite of the camel in Nara

Fig. 14.6 Goat and sheep herd in the Nara Desert, Pakistan

Fig. 14.7 Camel browsing on *Salvadora oleoides*

Desert. Most of these diets contain plants that are highly digestible (Rutagwenda et al. 1989) and high in crude protein (Kamoun and Steinmetz 1995; Woodward and Coppock 1995).

El-Keblawy (2003) found that due to the high pressure of grazing by camels in the United Arab Emirates, several palatable plants including *Crotalaria aegyptiaca,*

Fig. 14.8 Camel herd standing under shade of *Prosopis cineraria*

Indigofera articulata, Pennisetum divisum, Stipagrostis plumosa, and *Panicum turgidum* had significantly reduced. A rich species of these species was observed in sheltered enclosures but was never produced in overgrazed areas, which suggests the fact that they are important in camel diets in the arid rangelands.

Animals also have a seasonal effect on their feeding. Nyamangara and Ndlovu (1995) observed that seasonal changes in feed are observed whereby shrubs are noted to help cattle get the vital green until the herbaceous species that are annual grow up (Lichacar et al. 1996). Taste (e.g., sweet, sour, or bitter), odor, texture, and digestibility are all the elements of taste that determine palatability which is an important factor in dietary preference. It depends on a number of factors, such as the species of the animal, its age, physiological conditions (i.e., when it is pregnant), health, degree of hunger, and the properties of the plant itself. Such factors of the plant are the stage of growth, seasonal availability, phenology, morphology, chemical composition, relative abundance, and accessibility (Kababia et al. 1992; Grunwaldt et al. 1994; Nyamangara and Ndlovu 1995).

Black and Kenney (1984a, b) state that sheep prefer fresh forage to dried forage because of their taste, odor, and tactile differences. Moreover, fruits and flowers that are available seasonally are also a significant animal food (Pfister and Malechek 1986).

14.4.9 Ecological Future of Rangelands: A Delicate Balance

The long-term health and productivity of these rangelands are critically dependent on two interdependent factors: the composition and abundance of palatable plant species, and the management of the grazing animals that depend on them. A sustainable future requires a balance where the vegetation can recover from grazing pressure.

14.4.10 Documented Impact of Unmanaged Grazing

Evidence confirms that uncontrolled grazing leads to direct ecological degradation. Studies have consistently shown that key indicators of plant health such as fresh mass, plant height, and overall herbage cover are significantly reduced in overgrazed areas compared to protected pastures. This decline is a direct result of the prevalent, traditional grazing system in the area, which is largely unmanaged and unscientific. Livestock are often left to graze unsupervised, leading to:

14.4.10.1 Physical Damage

Animals trample palatable species as they walk, causing mechanical damage and soil compaction.

14.4.10.2 Competitive Disadvantage

Constant grazing pressure gives a competitive edge to less desirable, non-palatable plants. Species like *Cassia italica, Crotalaria burhia, Dipterygium glaucum*, and *Tephrosia* spp. are now dominating vast areas because they are avoided by livestock and face reduced competition.

14.4.11 Paradox of Valuable but Vulnerable Species

The rangeland is, in fact, home to a reservoir of highly nutritious and palatable grasses and sedges, including *Aristida* spp., *Cyperus* spp., *Eragrostis ciliaris, Dactyloctenium aegyptium, Lasiurus scindicus, Panicum turgidum,* and *Stipagrostis plumosa*. However, these very species are threatened by over-exploitation. Their populations are declining precisely because of their high palatability, as they are the first to be consumed and are often unable to recover.

A natural experiment underscores this point: in inaccessible areas where livestock cannot graze due to a lack of trails, these same palatable species thrive. This stark contrast provides compelling evidence that the current decline is not due to environmental unsuitability but is a direct consequence of management practices.

14.4.12 Compounding Pressures: Proximity and Fuelwood Harvesting

The degradation is most severe near human settlements. The perimeter of dwelling areas suffers from heavy grazing and harvesting pressure due to easy access. This pressure is compounded by the chopping of trees and shrubs for fuel, a practice that degrades the environment further by removing vital perennial plants that provide shade, stabilize soil, and serve as important browse, especially during dry seasons (Qureshi and Bhatti 2007a).

14.4.13 Hidden Dangers: Toxic Plants in the Forage Landscape

Beyond palatability, another critical factor in animal-plant relationships is toxicity (Mundy and Morris 1978). Several recorded species pose health risks to livestock. For instance, *Tribulus* species can cause photosensitization, a severe skin reaction to sunlight. *Solanum* species are known to cause poisoning (Muenscher 1960). *Chenopodium album* can cause kidney damage. This highlights the necessity for a more nuanced understanding of the rangeland ecosystem. Reporting merely on species availability is insufficient; a deeper understanding of animal-plant relationships, including toxicology, is essential for effective livestock and rangeland management (Hussain and Mustafa 1995).

14.4.14 Synthesis and Management Implication

The collective evidence leads to an inescapable conclusion: the degradation is not irreversible. The thriving palatable species in inaccessible zones demonstrate the inherent resilience of the ecosystem. Therefore, the implementation of scientific management strategies such as rotational grazing, the creation of seasonal resting areas, and community-based regulation of fuelwood harvesting is not just a recommendation but a prerequisite for maintaining productive pastures and ensuring the long-term sustainability of both the rangeland and the livelihoods it supports.

14.5 Conservation and Sustainable Management of the Nara Desert Rangelands

The fragile ecosystems of the Nara Desert, particularly its dune lands and slopes, are dominated by key perennial species such as the tussock grasses *Panicum turgidum* and *Lasiurus sindicus* along with *Cyperus* sedges. These foundational plants are subjected to intensive grazing pressure from goats, cows, and camels, often during critical growth phases. Research indicates that grazing during the early wet season has the most detrimental impact on species composition, as this is when plants are most vulnerable while investing energy into new growth. Consequently, these habitats are under immense ecological strain and require immediate conservation interventions.

14.6 From Resource Extraction to Sustainable Integration

The flora of this region is the cornerstone of local livelihoods, serving not only as fodder for livestock but also as a direct source of turf, fuel, food, and medicine for human communities. This multi-purpose use underscores the profound link between plant resources and human survival in this arid environment. To safeguard this vital resource base, it is imperative to initiate genetic conservation programs to collect and preserve the diversity of range grasses, legumes, and forbs. Protecting this genetic pool is the first step toward cultivating plant varieties capable of overcoming environmental stressors, producing high-yield forage, and fulfilling various conservation roles. The cornerstone of achieving this balance is the implementation of scientifically managed livestock grazing, which is essential for the sustainable use of arid land resources.

14.7 A Pathway to Enhanced Productivity and Community-Linked Recovery

Despite current degradation, the area possesses significant potential for enhanced productivity and increased carrying capacity, which would in turn improve overall biodiversity. Realizing this potential requires a dedicated, multifaceted approach:

14.7.1 Productivity and Nutrition Research

Further studies are urgently needed to quantify the biomass productivity of palatable species and analyze the nutritive and mineral content of preferred grasses and forbs. This data is crucial for developing effective rangeland health assessments and livestock nutrition models.

14.7.2 Re-vegetation Science

Research must focus on determining the most effective methods for re-vegetating depleted areas, using resilient, native species.

14.7.3 Community-Centered Implementation

Critically, none of these efforts will succeed without the active participation and buy-in of the local community. Conservation strategies must be co-developed with residents, integrating traditional knowledge and ensuring that management practices are both ecologically sound and socio-economically beneficial.

14.8 Conclusion

The current research will give in-depth report of the seasonal accessibility and palatability of native forage vegetable species in the Nara Desert, Pakistan. The study does not only add to the knowledge on the seasonal availability and palatability of indigenous forage plant species in the Nara Desert in the Sindh region, but it assists in the establishment of the floristic and pastoralist importance of the characteristics in the supporting of the livelihoods of the pastoralist populations in the arid conditions. The presence of 160 plant species including 151 as livestock forage not only marks the degree of floristic richness but also suggests the high degree of economic importance of such aspects to the pastoralist livelihoods. The wide range of palatability scores across different species and also across different times, is an indicator of the intelligent grazing behavior among the livestock with goats and cattle performing better than the others in adaptability and efficiency. In general, the superiority of highly and moderately palatable species does not only indicate the importance of such attributes in enhancing the nutritional condition of livestock at seasons when sufficient forage is easily accessible but it also denotes the restrictions and sufferings entailed to livestock at seasonal seasons when forage is limited. Additionally, there are strong ramifications of these seasonal restrictions in the livestock and pastoralist rangelands. Conclusively, in order to shift the state of resource depletion to sustainable productivity, a multifaceted approach must be adopted where the targeted scientific study and conservation of genetics are incorporated along with the practice of controlled grazing that is well based on mutual associations with the local population.

14.9 Recommendations

Seasonal Rangeland Management: Seasonal grazing plans have to be developed in order to address the vegetation shortage in some months particularly during winter. There should be controlled grazing or rotational grazing schemes to conserve high palatable species.

Nutritional and Mineral Profiling: Nutritional and mineral components analysis is required in highly and moderately palatable species of forages. This will assist in determining their contribution in enhancing the health of livestock. This will be of great assistance especially where there is limited availability of forage.

Conservation of Forage: It should be encouraged to conserve the excess forage at the seasons when forage thrives most by hay conservation or feed production in form of silage and which will be used during the dry winter months when feed is rare.

Promotion of Drought-Tolerant Species: It is therefore advisable that highly palatable and drought-resistant indigenous forage species be massively utilized as feeding material in the course of enrichment and restoration of the range land.

Livestock-Specific Fe: Considering the differing feeding patterns of the various kinds of livestock, both grazing and feeding recommendations about each species must be established with an objective of utilizing the maximum amount of forages and reducing the amount of grazing carried out on sensitive plants.

Development of Drought-resistant Species: Drought-resistant native forage species: in the recovery or improvement program of native pasture in these grasslands, preference should be given to palatable and drought-resistant native forage species.

Special Feeding Schemes: Due to the differences in feeding behavior exhibited by the diverse livestock species, species specific grazing, and supplementation processes are required to optimize the utilization of the forages and to mitigate the effect of grazing on sensitive plants.

Policy and Community Engagement: The policy makers should collaborate closely with the local community to reasonably integrate the traditional knowledge as practiced in the past and the recent scientific findings in order to achieve a sustainable exploitation of the rangelands in the Nara Desert.

Community Involvement and Public Policies: The policymakers and the locals should collaborate in ensuring a knowledge gap between the pastoralists and scientific knowledge is integrated to enable the Nara Desert to be managed in a sustainable manner.

References

Ali SI, Nasir YJ (1990–1991) Flora of Pakistan (fascicle series), Islamabad

Ali SI, Qaiser M (eds) (1993–1995) Flora of Pakistan (fascicle series), Islamabad

Ali SI, Qaiser M (1986) A phytogeographic analysis of the phanerogams of Pakistan and Kashmir. Proc R Soc Edinb 89B:89–101

Anon (1992) Pak. Swiss Range and Livestock Improvement Project

Batanouny KH (1981) Ecology and Flora of Qatar. Centre for scientific and applied Research, University of Qatar, Doha

Bhandari MM (1978) Flora of Indian Desert. Scientific Publishers, Jodhpur

Bhatti GR, Qureshi R, Shah SM (1998) Ethnobotany of Calotropis procera with especial reference to the people of Nara Desert. Scientific Sindh 05:13–22

Bhatti GR, Qureshi R, Shah M (2001a) Ethnobotany of Qadan Wari of Nara Desert. Pak J Bot 33 (especial issue):801–812

Bhatti GR, Qureshi R, Shah M (2001b) Ethnobotany of Qadan Wari of Nara Desert. Pak J Bot 33 (SI):801–812

Bhatti GR, Qureshi R, Shah M (2002) Ethnomedicinal observation of Cymbopogon jwarancusa (Jones) Schult. In Nara Desert (Sindh). In: Proceedings of the workshop on curriculum development in applied ethnobotany. WWF, pp 34–39

Black JL, Kenney PA (1984a) Factors affecting diet selection by sheep. Aust J Agric Res 35: 551–563

Black JL, Kenney PA (1984b) Factors affecting diet selection by sheep. II. Height and density of pasture. Aust J Agric Res 35:565–578

Boulos L (1991) Flora of Egypt. Al Hadara Publishing, Cairo
El-Keblawy A (2003) Effect of protection from grazing on species diversity, abundance and productivity in two regions of Abu-Dhabi Emirate, UAE. In: Alsharhan AS, Wood WW, Goudie AS, Fowler A, Abdellatif E (eds) Desertification in the third millennium. Swets Zeitlinger Publisher, Lisse, pp 217–226
Grunwaldt EG, Pedrani AR, Vich AI (1994) Goat grazing in arid piedmont of Argentina. Small Rumin Res 13:211–216
Hickman K, Hartnett D, Cochran R (1996) Effects of grazing systems and stocking rates on plant diversity in Kansas tall grass prairie. In: Proceedings of the Rangelands. In a sustainable biosphere. (Ed.): N.E. West. 5th International Congress 1995, Salt Lake City Utah, pp 228–229
Hussain F, Chughtai SR (1984) The effect of overgrazing on the development of herbaceous vegetation in Zangilora, Quetta, Baluchistan, Pakistan. Pak Area Study J 9(1):29–38
Hussain F, Durrani MJ (2007) Forage productivity of arid temperate Harboi rangeland, Kalat, Pakistan. Pak J Bot 39(5):1455–1470
Hussain F, Durrani MJ (2008) Mineral composition of some range grasses and shrubs from Harboi rangeland Kalat, Pakistan. Pak J Bot 40(6):2513–2523
Hussain F, Durrani MJ (2009) Seasonal availability, palatability and animal preferences of forage plants in Harboi arid range land, Kalat, Pakistan. Pak J Bot 41(2):539–554
Hussain F, Mustafa G (1995) Ecological studies on some pasture plants in relation to animal use found in Nasirabad Valley, Hunza, Pakistan. Pak J Pl Sci 1(2):263–272
Huston JE (1978) Symposium. Dairy goats. Forage utilization and nutrient requirements of the goats. J Dairy Sci 61:988–993
Iqbal A, Khan BB (2001) Feeding behaviour of camel. Pak J Agric Sci 38:58–63
Jafri SMH (1966) The Flora of Karachi. The Book Corporation, Karachi
Kababia D, Landan S, Perevolostsky A, Vecht Y, Eliasof L, Zeltzer S (1992) The feeding behaviour of milking goats in woody rangeland in the Judean Mountains. Hasssadch 72:1536–1540
Kamoun M, Steinmetz P (1995) Feeding behaviour, intake and digestion of the Camelus dromedarious at pasture. In: Tisserand J-L (ed). CIHEAM-IAMZ, Zaragoza, pp 51–57
Khan II (1996) Biodiversity depletion with respect to human and livestock population in Indian Desert. In: Proceedings of the Rangelands in a sustainable biosphere. (Ed.): N.E. West. 5th international Congress 1995, Salt Lake City Utah, pp 286–287
Lichacar S, Love RM, Rains DW, Laude HM (1996) The effect of some California soil on the early growth of Atriplex polycarpa (Terr.) Wats, and Atriplex repanda Phil. In: Proceedings of the Rangelands. In a sustainable biosphere. (Ed.): N.E. West. 5th International Congress 1995, Salt Lake City Utah, pp 300–301
Liu D, Yun J, Hao X, Meng S (1996) Grassland deterioration and control measures in arid and semi – arid region in China. In: Proceedings of the Rangelands. In a sustainable biosphere. (Ed.): N.E. West. 5th International Congress 1995, Salt Lake City Utah, pp 320–321
Makulbekova GB (1996) The ecological evaluation of the present condition of rangeland vegetation of Kazakhistan deserts. In: Proc. Rangelands in a sustainable biosphere. (Ed.): N.E. West. 5th International Congress 1995, Salt Lake City Utah, pp 338–339
Matthew KM (1981–1983) Flora of Tamilnadu Carnatic, vol 1–3. The Rapinat Herbarium, St. Joseph's College, Tiruchirapalli
Migongo-Bake W, Hensen RM (1987) Seasonal diets of camels, cattle, sheep and goats in a common range in Eastern Africa. J Range Manag 40:76–79
Muenscher WC (1960) Poisonous Plants. MacMillan Co., New York
Mundy BL, Morris DI (1978) Tasmanian plants toxic for animals. Publications Tasmanian Department of Agriculture, p 101
Nasir E, Ali SI (1970–1989) Flora of Pakistan (Fascicle series). Department of Botany, University of Karachi, Pakistan
Noor M (1978) Comparison of grazed and un-grazed vegetation of sub-alpine ecological zones at Sar. Pak J For 28:186–189

Nyamangara ME, Ndlovu LR (1995) Feeding behaviour, feed intake, chemicals and botanical composition of the diet of indigenous goats raised on natural vegetation in a semi-arid region of Zimbabwe. J Agric Sci 124:455–461

Omera RM, Hestera AJ, Gordona IJ, Swaineb MD, Raffique SM (2006) Seasonal changes in pasture biomass, production and offtake under the transhumance system in Northern Pakistan. J Arid Environ 67:641–660

Pfister JA, Malechek JC (1986) Dietary selection by goats and sheep in a deciduous woodland of north eastern Brazil. J Range Manag 39:24–28

Qureshi R (2004) Floristic and ethnobotanical study of Desert Nara region, Sindh. PhD thesis, Department of Botany, Shah Abdul Latif University, Khairpur, pp 1–454

Qureshi R (2008a) Preliminary floristic list of Chotiari Wetland Complex, Nawab Shah, Sindh, Pakistan. Pak J Bot 40(6):2281–2288

Qureshi R (2008b) Vegetation assessment of Sawan Wari of Nara desert, Pakistan. Pak J Bot 40(5): 1885–1896

Qureshi R (2011) Floral diversity, threats and conservation measures: a case study from Nara Desert, Pakistan. In: Guevara J, Sivaperuman C, Grünwaldt E (eds) Deserts: fauna, flora and environment. Nova Science Publishers, Inc., Hauppauge, pp 53–77. ISBN: 978-1-61470-038-8

Qureshi R (2012) The Flora of Nara Desert, Pakistan. Nova Science Publishers, Inc., Hauppauge, pp 1–317. Hardcover (ISBN: 978-1-62081-683-7)

Qureshi R (2018) Medicinal uses of trees and shrubs by the inhabitants of Nara Desert, Pakistan. In: Ozturk M, Hakeem KR (eds) Plant and human health, volume 1: ethnobotany and physiology. Springer International Publishing AG, Springer Nature, pp 391–407

Qureshi R (2020) Phytodiversity and the conservation status of the Nara Desert, Pakistan. In: Nangyal H, Khan MS (eds) Environmental pollution, biodiversity and sustainable development: issues and remediation. Apple Academic Press/CRC Press/Taylor & Francis Group, pp 51–76

Qureshi R, Ahmad M (2010) Some notes on the vegetation of Achhro Thar (White desert) of Nara region, Sindh, Pakistan. Pakistan J Bot 42(5):2985–2994

Qureshi R, Bhatti GR (2005a) Nara Desert, Pakistan: part I: soils, climate vegetation. Rangelands 27(5):27–31

Qureshi R, Bhatti GR (2005b) Nara Desert, Pakistan: part II: human life. Rangelands 27(5):32–35

Qureshi R, Bhatti GR (2006) Ethnobotanical observations of Achyranthes aspera Linn. And Aerva spp. with special reference to the people of Nara Desert. Hamdard Medicus XLIX(1):43–48

Qureshi R, Bhatti GR (2007a) Floristic list of monocot of Nara Desert Sindh, Pakistan. Hamdard Medicus XLIX(4):63–67

Qureshi R, Bhatti GR (2007b) Indigenous uses of Caesalpinia bonduc (L.) Roxb. With particular reference to the people of Nara Desert. Hamdard Medicus 50(2):112–118

Qureshi R, Bhatti GR (2007c) Nara Desert, Pakistan: part IV: destruction of natural habitats and its impact on plant diversity. Rangelands 29(1):30–33

Qureshi R, Bhatti GR (2007d) Nara Desert, Sindh, Pakistan: part III: range types and their plant resources. Rangelands 29(1):26–29

Qureshi R, Bhatti GR (2008a) Diversity of micro-habitat and its plant resources in Nara Desert, Pakistan. Pakistan J Bot 40(3):979–992

Qureshi R, Bhatti GR (2008b) Ethnobotany of plants used by the Thari people of Nara Desert, Pakistan. Fitoterapia 79:468–473

Qureshi R, Bhatti GR (2008c) Taxonomy of Scrophulariaceae from Nara Desert, Pakistan. Pakistan J Bot 40(3):973–978

Qureshi R, Bhatti GR (2009) Folklore uses Amaranthaceae family of Nara Desert, Sindh, Pakistan. Pakistan J Bot 41(4):1565–1572

Qureshi R, Bhatti GR (2014) Floral diversity of Nara Desert, Pakistan. In: Rafique M, Ibupoto KA, Hussain SA, Mahmood K (eds) Biodiversity of Pakistan. Pakistan Museum of Natural History, Islamabad, pp 413–431

Qureshi R, Bhatti GR, Shah M (2001a) Ethnomedicinal properties of Aloe barbadensis Mill. With particular references the people of Nara Desert. Hamdard Medicus XLIV(3):46–50

Qureshi R, Bhatti GR, Shah M (2001b) Ethnomedicinal properties of Aloe barbadensis Mill. with particular references the people of Nara Desert. Hamdard Medicus XLIV(3):46–50

Qureshi R, Bhatti GR, Chaudhary AK (2009) Phytodiversity under human induced degradation: a case study from Nara Desert, Pakistan. In: Proceedings of the international symposium on "compatible with climate change" by Shah Latif University Botanical Garden & Herbarium, Khairpur, pp 171–188

Qureshi R, Bhatti GR, Memon RA (2010) Ethnomedicinal uses of herbs from Nara Desert, Pakistan. Pak J Bot 42(2):839–885

Rutagwenda T, Kaske M, Engelhardt WV, Lechner-Doll M, Schultka W, Schwartz HJ (1989) Adaptation strategies of camels on a thorn bush savannah pasture: comparison with other domestic animals. Options Me'diterrane'ennes–Se'rie Se'minaires 2:69–73

Shetty BV, Singh V (1987) Flora of Rajasthan, vol I & II. Botanical Survey of India, Old Connaught Place Dehra Dun

Vallentine JF (1990) Grazing management. Academic

Wahid A (1990) Dietary composition and nutritional status of sheep and goats grazing in two rangeland types in Balochistan, Pakistan. PhD thesis, Oregon State University

Wilson AD, Leigh JH, Hindley NL, Mulham WE (1995) Comparison of the diets of goats and sheep on a Easuarina cristata – Heterodendrram oleifolium woodland community in Western New South Wales. Austr J Ex Agric Anim Husb 15:45–53

Woodward A, Coppock DL (1995) Role of plant defense in the utilization of native browse in Southern Ethiopia. Agrofor Syst 32:147–161

Impacts of Climate Change on Nutritional Quality of Fruit Crops

15

Muhammad Shafiq, Muhammad Zafar Iqbal, Mohammad Athar, Rahmatullah Qureshi, and Mirza Faisal Qaseem

Abstract

The changing climate has had an important effect on the development and growth of fruit-producing plant species all over the world, particularly in developing countries such as Pakistan, due to a variety of factors including rising global temperatures, rainfall patterns, and variations in the weather at the moment over the past decade. Fruit-bearing plant species are under severe threat from the planet's constant rising temperature as a result of various forms of human, industrial, and automotive activity, cutting down forests, fossil fuel burning, and overuse of conventional fertilizers to increase crop productivity. Scarcity of pure water, ecosystem disturbances, and soil degradation are all contributing factors to a decline in fruit crop land as compared to the previous few decades owing to climate change. Changes in climate circumstances also have an impact on the physical characteristics of soil, precipitation, the water process, soil quality and wellness, root structure and relationship with soil bulk density, porosity of the soil, and decreased macro and micronutrient accessibility to soil and plants. However, several fruit crops in Pakistan have been proven to be vulnerable to climate change activities. This book chapter aims to report on the effects of climate change on fruit crop growth performance. These data were gathered

M. Shafiq (✉) · M. Z. Iqbal
Department of Botany, University of Karachi, Karachi, Pakistan

M. Athar
California Department of Food and Agriculture, Sacramento, CA, USA

R. Qureshi
Department of Botany, Faculty of Sciences, Pir Mehr Ali Shah Arid Agriculture University, Rawalpindi, Pakistan
e-mail: rahmatullahq@uaar.edu.pk

M. F. Qaseem
Department of Botany, Pir Mehr Ali Shah Arid Agriculture University, Rawalpindi, Pakistan

R. Qureshi (ed.), *Plant Biodiversity*, https://doi.org/10.1007/978-981-92-1471-6_15

from several electronic websites and analyzed to determine the impact of global warming on commercial fruit crops grown around the world. To avoid a further decline of fruit crop production, rigorous enforcement against those guilty for disturbing the natural ecosystem of fruit crop cultivation regions in accordance with authorized environmental law is required immediately. Research, development, and use of new technology instruments would help maintain and promote the agricultural system while also improving fruit crop quality and output. It was found that in recent years, climate change has become a major driving force element affecting the growth and development of fruit crops, particularly in Pakistan, more than other industrialized countries. The studies also revealed some potential to aid fruit crop growers in adapting novel farming methods that might help in boosting the cultivation of high-quality fruit crops.

Keywords

Biotic and abiotic stress · Ecology · Environmental pollution · Global warming · Increasing temperature · Soil health, Toxicity and tolerance

Abbreviations

CC	Changing Climate
COBD	Conservation On Biological Diversity
GDP	Gross Domestic Product
WHO	World Health Organization

15.1 Introduction

Climate change, which is considered as long-term changes in temperature, precipitation, and the nature of the air, has emerged as one of the hottest issues of concern to the world of agriculture (Kotir 2011). Man-made activities such as deforestation, burning fossil fuels, industrialization, and overuse of agrochemicals have contributed to hastened global warming and have changed the natural ecosystems (Jetz et al. 2007; Grand et al. 2019). Nonetheless, climate change in the fruit crop is the issue that has attracted a lot of literature across the science field (Jones and Davis 2000; Sparks et al. 2000; Chuine and Beaubien 2001). Researchers have presented research works regarding the consequences of climate change and its implications on the special characteristics of the growth and development of fruit crops, variability in flowering, blossom, flowering date advance, changes in phenological characteristics and on respective flowering period (Vitasse et al. 2009; Dong et al. 2024). The situation has affected many economic sectors especially the agriculture, flowering, postharvest fruit, food quality, and livestock in the whole world due to the climate conditions their individual years have encountered (Mattos et al. 2013; Medda et al. 2022; Iftikhar et al. 2024).

The fruit crops are an essential element of the agricultural economies as they are characterized by their nutritional, economic, and ecological value. Climate variability is, however, threatening their productivity. Phenological events, in fact, have been recorded to undergo drastic changes in budburst, flowering and maturity of fruits in different regions in response to the impacts of climate change (Chuine and Beaubien 2001; Chmielewski et al. 2004). As an example, the long-term studies conducted in Europe and Japan show a shift in flowering and a reduction in developmental cycles in relation to an increase in temperatures (Doi and Katano 2008).

15.1.1 Climate Change: A Worldwide Menace to Foodstuff

The high frequency and intensity of high temperature can influence the population of species negatively, which causes the survival of the affected species to drive a response in the population of the species, as well as in geographical distribution of the species (Wang et al. 2023). Climate and land use change are two of the most rapidly growing global dangers to the natural ecosystem and wildlife (Jetz et al. 2007; Grand et al. 2019). The causal level of the climate change threat is a wide range of sustainable development challenges, in particular, the health, food insecurity, employment, gender equality, education, housing, and temperature increase, particularly in the most susceptible areas (Fig. 15.1), such as Pakistan and the African countries (Lundgren et al. 2013; Sellers 2016). It is assumed that the tropical areas are the first and most vulnerable to climate change and the temperature will increase by 0.4–1 °C, 0.8–3.2 °C, and 1.2–6 °C by the year 2020, 2050, and 2100

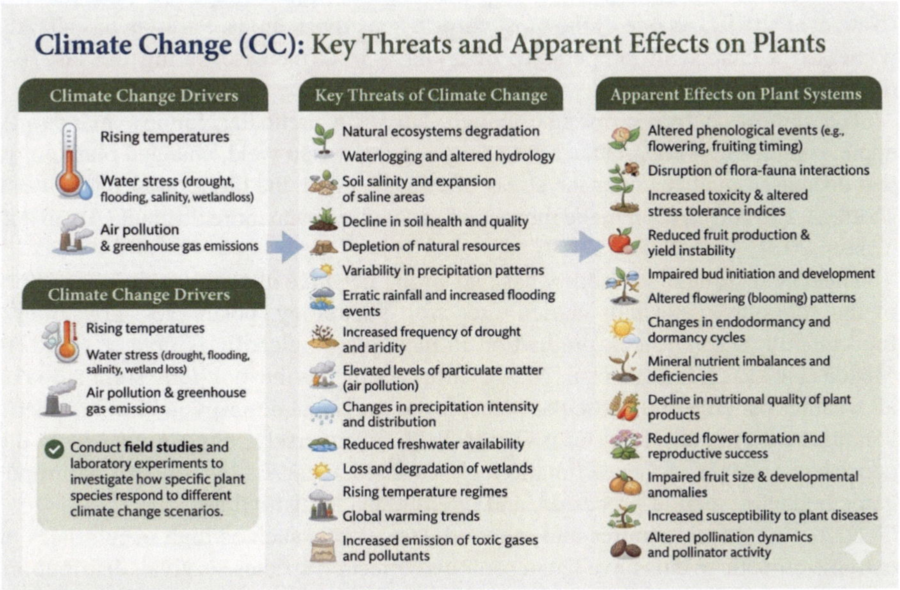

Fig. 15.1 Climate change, threats, and their effects on plant systems

relative to the 1961–1990 value (Nath et al. 2019). According to various professionals, the increase of temperature because of the recent climate change has influenced the life cycles and reproduction of majority of plant species.

The effects of changing climate on the quality of fruit crops are a topic of concern that has received widespread attention among the scientific community. The conceptual phenomenon has been understood by various scientists and summarized in Tables 15.1 and 15.2. The researchers have undertaken various studies about the effects of climate change and its implications on the growth and development of fruit crops, variability in flowering, blossom, flowering date advance, variations in phenological aspects, and also respective flowering duration (Sparks et al. 2000; Chuine and Beaubien 2001; Vitasse et al. 2009). Maatallah et al. (2024) measured physical variables in the leaves and the morphological/pomological variables in the fruits of six varieties of fig (*Ficus carica* L.) Sawoudi, Bayoudhi, Mlouki, Assal, Zidi, and Mozai that grew in the dry climate of the Gafsa oasis (in the center of Tunisia). These cultivars were distinguished by the peel colors that were different and included greenish, yellowish-brown, and dark purple. Moreover, the fruit varieties of these hybrids contained the most sugar levels in the peel and the meat and the highest solid soluble content values (22.23 and 20.83 degBx). Mlouki also bore the most fruit weight (66 g) when compared to the other cultivars used. The acidity of the fruits was also measured by the authors. Bayoudhi was the most valued (6.56 g MAE 100 mL) and Assal and Zidi the least.

The vulnerability is greater in tropical and subtropical areas such as Pakistan where the adaptive capacity is low, and agriculture is climate sensitive. The rise in temperature projected to occur by the end of the century, ranging between 0.4 and 6 °C, will have a great impact on crop distribution, productivity, and sustainability (Nath et al. 2019). As one of the most climate-sensitive nations, Pakistan has already witnessed a change in temperature of about 0.5 °C increase during the last five decades (Jaffery 2018).

The significant fruit crops like mango, citrus (in particular, kinnow mandarin), apple, peach, and pomegranate are undergoing decreased yield, changed phenology, and decreased quality under the stress of climate. Also, the higher pest rates, water shortage, and soil erosion make the task of production even more difficult (Awan and Yaseen 2017; Nawaz et al. 2019).

The recent climatic variability has had strong negative impacts on various sectors of the economy, especially agriculture, crop phenology, postharvest physiology, food quality, and livestock production in most places globally (Gupta et al. 2021; Medda et al. 2022; Iftikhar et al. 2024). Climate which is the long-term summation of all weather patterns in a particular place, is the statistical description of atmospheric conditions during the long-term (Kotir 2011). Conversely, climate change is the consistent changes in these climatic regimes that are caused by a mix of anthropogenic behavior, natural processes, and environmental disruptions.

High global temperatures and other adverse effects such as high occurrences of extreme climatic events have had a profound effect on species survival, distribution, and population dynamics. Intense thermal stress interferes with the functioning of physiological mechanisms, which limits the adaptability of species and changes the

Table 15.1 Climate change drivers and their impacts on commercial fruit crops

Factor/driver	Refined observations	References
Climatology	Long-term viticultural records (1952–1997) from Bordeaux, France, demonstrate that temperature variations significantly influence phenological stages such as budburst, flowering, and harvest. Recent decades show advancement of phenological events, shortened developmental intervals, and extended growing seasons, influencing varietal suitability and wine quality, with greater dependence on late-maturing cultivars like Cabernet Sauvignon	Jones and Davis (2000)
Temperature (Lychee physiology)	In eastern Australia, vegetative flush and flowering in lychee are strongly temperature-dependent. Vegetative growth occurs under warmer conditions (17–19 °C), while floral induction requires relatively cooler temperatures, highlighting the importance of thermal thresholds in reproductive development	Olesen et al. (2002)
Phenological shifts in trees	Rising temperatures between 1950 and 2004 in Beijing resulted in earlier and prolonged flowering periods in multiple tree species. Early-flowering species exhibited higher sensitivity to temperature increases, with flowering timing closely linked to pre-season thermal conditions	Lu et al. (2006)
Climate change and tropical fruits	Climatic variability has been associated with altered growth patterns, productivity, and quality attributes of tropical fruit crops, including mango, indicating their vulnerability to temperature and rainfall fluctuations	Bhriguvanshi (2010)
Climatic–biological interactions	Comparative studies of olive flowering in Italy and Spain revealed a strong relationship between pollen release patterns and spring temperature trends, confirming temperature as a key regulator of reproductive phenology	Orlandi et al. (2010)
Phenology as climate indicator	Long-term observations in Southern Africa indicate that flowering dates in apple and pear have advanced significantly ($\approx$1.6 days per year), primarily due to rising temperatures and reduced rainfall during critical seasonal transitions	Grab and Craparo (2011)
Temperature effects on crop duration	Climate records (1963–2009) from Australia show reduced duration of fruit development phases in several crops, including custard apple, litchi, macadamia, and mango. Warmer conditions accelerated crop maturation and altered seasonal growth patterns	Olesen (2011)
Climatic variability (coffee systems)	Variability in temperature, solar radiation, and humidity has been identified as a major factor influencing coffee yield fluctuations in Brazil.	Paes de Camargo (2010)

(continued)

Table 15.1 (continued)

Factor/driver	Refined observations	References
	Climate projections suggest further warming and increased rainfall variability, posing risks to crop stability	
Adaptation strategies	Long-term datasets indicate that increased temperature and reduced soil moisture accelerate fruit maturation in vineyards. Adaptive management practices, including irrigation optimization and crop load regulation, are critical to maintaining yield and quality	Webb et al. (2012)
Chilling and heat requirements	In subtropical fruit trees, the balance between winter chilling and heat accumulation is essential for proper flowering and fruit set, both of which are increasingly disrupted under climate change scenarios	Guo et al. (2014)
Global warming and fruit physiology	Elevated temperatures shorten vegetative and reproductive phases in fruit crops, influencing flowering duration and fruit firmness. High temperatures may also reduce enzymatic activity related to ripening, affecting fruit quality	Haokip et al. (2020)
Temperature and soil–plant interactions	Increasing temperatures contribute to depletion of soil organic carbon and micronutrients. In addition, altered thermal regimes affect flowering behavior, fruit set, and quality traits due to insufficient chilling and stress-induced physiological imbalances	Atreya and Kaphle (2020)
Phenology and dormancy dynamics	Climate change influences endodormancy and thermal requirements in temperate fruit crops such as apricot, potentially shifting cultivation zones geographically due to altered chilling accumulation patterns	Vanalli et al. (2021)
Warm climate effects on fruit quality	Fruit quality attributes, including soluble solids content and aroma in apple cultivars, are significantly affected by climatic variables and orchard management practices under warmer conditions	Pechan et al. (2023) and Iwanami et al. (2023)
Nutrition and fruit disorders	Climatic stress can disrupt nutrient balance and increase physiological disorders such as fruit cracking, mediated by stress-responsive genetic and biochemical pathways	Manzoor et al. (2024)
Climatic variability (Pakistan context)	Long-term analyses (1991–2020) reveal a negative association between climatic variables and apple production in Pakistan, indicating the sensitivity of temperate fruit crops to regional climate fluctuations	Chandio et al. (2024)

biogeographical distributions (Wang et al. 2023). At the same time, climate change and the shift in land use have become a key challenge to the biodiversity and ecosystem stability (Jetz et al. 2007; Grand et al. 2019). These transformations are

Table 15.2 Impacts of climatic variables on fruit crops and associated ecological processes

Climatic variable/ process	Refined scientific observation	References
Temperature and fruit set	Variations in temperature significantly regulate fruit set in sweet orange, influencing reproductive success and early fruit development under controlled conditions	Moss (1970)
High temperature and photosynthesis	Elevated temperatures under high light intensity reduce photosynthetic efficiency in longan and mango seedlings due to physiological limitations affecting carbon assimilation	Yamada et al. (1996)
Climate influence on grapevine	Climatic factors determine grapevine growth, berry composition, and wine quality, particularly affecting sugar accumulation and fermentation characteristics in viticultural systems	Jones and Davis (2000)
Climate change and agriculture	Modeling studies indicate that rising temperatures and climatic variability have measurable impacts on crop productivity and agricultural sustainability in Taiwan	Chang (2002)
Temperature and lychee phenology	Flowering and vegetative growth in lychee are closely linked to seasonal temperature patterns, with thermal conditions influencing shoot elongation and reproductive timing	Olesen et al. (2002)
Climate impacts on mangosteen	Production and yield stability of mangosteen in Thailand are sensitive to climatic fluctuations, particularly temperature and rainfall variability	Boonklong et al. (2006)
Phenology as climate indicator	Shifts in plant phenology provide strong evidence of ecosystem responses to climate change, linking species-level adjustments with broader ecosystem processes and carbon cycling	Cleland et al. (2007)
Phenological variability in temperate fruits	Climate change has altered the timing of developmental stages in apple and pear trees, leading to variability in growth cycles and production patterns	Guédon and Legave (2008)
Fruit ecology and wildlife interaction	Fruit development dynamics influence feeding behavior and survival strategies of wildlife species, highlighting ecological interdependence in forest ecosystems	Kagoro-Rugunda and Baranga (2009)
Climate effects on cherry trees	Long-term climatic changes have modified flowering timing and growth patterns in cherry trees and related species in Japan	Primack et al. (2009)
Climatic variability in coffee systems	Coffee production in Brazil is highly sensitive to temperature and rainfall variability, which directly affect physiological processes and yield stability	Paes de Camargo (2010)
Advancement of flowering dates	Observations from Southern Africa indicate earlier full bloom in apple and pear due to rising temperatures and altered precipitation regimes	Grab and Craparo (2011)
Winter chilling requirements	Declining winter chill accumulation under warming climates threatens the productivity of	Luedeling et al. (2011)

(continued)

Table 15.2 (continued)

Climatic variable/ process	Refined scientific observation	References
	temperate fruit and nut crops by disrupting dormancy release	
Temperature gradients and flowering patterns	Spatial variation in flowering timing of cherry blossoms in Japan reflects microclimatic differences, with urban and inland regions exhibiting earlier blooming compared to coastal areas	Ohashi et al. (2011)
Climate-driven phenological processes	Wildflower phenology is strongly influenced by climatic variability, affecting ecosystem functions such as energy exchange, water flux, and carbon dynamics	Richardson et al. (2013)
Long-term chilling decline impacts	Progressive reduction in winter chilling is projected to negatively affect fruit and nut production, particularly in regions such as California	Ramirez and Kallarackal (2015)
Seed dispersal and ecological interactions	Fruit consumption by animals, such as bears feeding on Prunus species, enhances seed dispersal and germination without damaging seed viability, indicating mutualistic ecological relationships	Koike et al. (2007)
Heat stress and human systems	Rising temperatures pose significant risks to human health and labor productivity, particularly in vulnerable regions such as Africa	Ncongwane et al. (2021)
Climate effects on fruit development	Climate variability affects pollination, fertilization, and fruit growth processes, thereby influencing overall fruit yield and quality	Maupilé et al. (2024)
Global warming and agriculture	Increasing climatic instability represents a major threat to agricultural productivity worldwide, with significant implications for food security and environmental sustainability	Mbewu et al. (2024)

not confined to the ecological level and impact the vital aspects of sustainable development such as food security, health of the population, employment, and social and economic stability, especially in such at-risk areas as Pakistan or some parts of Africa (Lundgren et al. 2013; Sellers 2016).

Global warming has been projected to increase steadily with a temperature range of 0.4–1 °C in 2020, 0.8–3.2 °C in 2050, and 1.2–6 °C in 2100 as compared to 1961–1990 temperature (Nath et al. 2019). These warming patterns have already brought about dramatic changes in the developmental process of plants such as phenology and reproductive cycles. In the case of increasing spring temperatures, long-term experiments, e.g., by Doi and Katano (2008) showed an improvement in the budburst timing in a series of tree species in Japan during a 50-year study, which was directly associated with an increase in temperature during spring. In the same way, the changes caused by climatic factors also have an impact on the nutritional value and health-related properties of tropical fruits, such as mango (Ramírez et al. 2010; Bhardwaj et al. 2023).

The empirical studies also support the fact that the agricultural productivity is extremely vulnerable to the changes in climate. An example is when Chang (2002) found that there was high yield variation among 60 cultivars in Taiwan in warmer conditions; this demonstrates how vulnerable agricultural systems are to temperate changes. Fruit crops are especially of great nutritional and economic value because they are highly enriched with necessary vitamins, minerals, and bioactive compounds (Pagadala et al. 2023). But their perennial character and physiological vulnerability make them very vulnerable to climatic disturbances, which results in lower productivity, diminished quality, and financial losses. Climate change is thus emerging to be a central issue of concern among farmers, horticulturists and researchers around the world.

15.2 Methodological Approach

The analysis is grounded on the information retrieved via various scientific databases and websites, such as ScienceDirect, Taylor and Francis, PubMed, MDPI, and Google Scholar among others. The chapter is logically structured in the prominent thematic parts: (i) introduction, (ii) effects on phenology and sensitive growth stages, (iii) life cycle responses to biotic and abiotic stresses, (iv) effects of cold season dynamics on nutrition and productivity, (v) habitat degradation and soil-related effects, (vi) climate change mitigation strategies, (vii) future research directions, and conclusions, which are supported by references, tables, and bibliographic analyses.

15.3 Climate Change Effects on Phenology on Bioactive Compounds in Fruits

Phenology, which is the study of periodic events in plant development is a sensitive indicator of climate change. Change in temperature and precipitation has a direct effect on flowering, fruiting, and seed development stages. Scientists have cited climate change to have led to physiological alterations such as shorter growth season, lower bud fertility, earlier bud break, blooming, fruiting, and maturity and earlier leaf opening or blooming (Ramirez and Kallarackal 2015; Chmielewski et al. 2004). The temperature of the air has significantly increased since the latter half of the 80 s, and this has had visible alterations in plant phenology across most parts of the globe to include phenological stages of naturally found plants, fruits trees, and field crops (Chmielewski and Rötzer 2001).

Another study showed CC affects the time lag between flowering and reduced maturity of fruits ultimately reduce the growth cycles (Haokip et al. 2020). Dong et al. (2024) reported that CC affects the distorted quality of fruits and changes in the degree of sugar accretion, acidity, and fruit size. Phenological responses are species-specific and are always correlated to climatic variables, which makes them reliable predictors of change in the environment (Gallinat et al. 2018; Rafique et al. 2024). In research of apple and pear in Europe, it was found that flowering date has changed to appear earlier since the late 1980s (Guédon and Legave 2008). Similarly, fig cultivars in Tunisia were found to be variant in the fruit biochemical characteristics in arid climatic conditions (Maatallah et al. 2024). The temperature limits are very crucial in the development of fruits. High amounts of rainfall and decreased accumulative temperature cause fruit acidity, and decreased diurnal temperature variation impacts sugar metabolism and ripening processes (Dong et al. 2024).

15.3.1 Phenological Changes of Fruit Crops

Doi and Katano (2008) quantified and compared the impacts of the climate change over the past half a century on the physiological leaf bloom phenology of four tree species in four sites in Japan. The obtained results demonstrated that the plant leaf budburst dates have also been changing over the last 50 years in four different Japanese locations leading to the conclusion of climate change modifying the budburst phenology of rising spring temperature at localities and even localities of the same latitude.

The study focused on growing seasons of a fruiting apple cultivar, golden delicious, and three varieties of pears trees that grew in different locations across France in relation to global warming (Guédon and Legave 2008). The results of a comparison of the temperature conditions conducted on the three sites as compared to the annual change of the monthly mean temperatures showed variability of monthly temperature mean with respect to the degree of longitude of location. During the past 40 years, it can be seen that the flowering of the apple and pear trees grown in the four regions (Three in France and one in Switzerland) have had a significant change. The findings indeed confirm the existence of severe abrupt shifts in the time cycle of different flowering dates at the conclusion of the 1980s to early and more prevalent dates (Guédon and Legave 2008).

The world has experienced the environmental change and this has indeed had a tremendous impact on the crop productivity of crops, primarily long-term bearing fruit crops like citrus. The study was a case study of the fruit quality of 12 citrus plantations (orange, *Citrus sinensis* L. cv. *bingtang*) in a subtropical region of Yunnan, China, in 2014–2022 (Dong et al. 2024). The results showed that substantial amount of rainfall (>220 mm) and decrease in cumulative temperature (<3150 °C) increases titratable acidity (>1.8%) in young fruits. Growth in size of the fruits (horizontally increasing diameter of 50–65 mm), excessive rainfall (300–400 mm), reduced cumulative temperature (less than 2400 °C) and a smaller range of diurnal temperatures (less than 10 °C) suppress the decrease in titratable acidity.

15.3.2 Effects of Temperature on Fruit Setting

Moss (1970) recorded the effect of temperature on the set of the fruit in sweet orange (*Citrus sinensis* L. Osbeck) cuttings. Excessive temperatures under high irradiance altered photosynthesis in longan (*Dimocarpus longan* Lour.) and mango (*Mangifera indica* L.) seedlings, and proved that reduction of net consumption rate is due to physiological constraint of high temperatures (Yamada et al. 1996).

Olesen et al. (2002) conducted a study on the effect of stem lengthening time on flowering in eastern Australian lychee (*Litchi chinensis* Sonn.). The greenery of the shoots was witnessed in the presence of an average temperature of over 17–19 °C during the early flush growth and lowering temperature in flowering. Flowering needs cool weather, in the primary areas of development the number of events per year with an average range of less than 20 °C was estimated (Olesen et al. 2002).

15.3.3 Phenological Indicators of Climate Change

Phenology is an effective measure of climate change (Grab and Craparo 2011). The effect of climatic change on the growth of crops in Southern Africa demonstrated that the average date of blooming of apples and pears has been affected by a 1.6 days per annum. A combination of warming areas and lowering autumn and early spring rainfall led to phase shifts (Grab and Craparo 2011). Climate warming due to human activities extended blooming of four tree species, *Prunus davidina*, *P. armeniaca*, *Robinia pseudoacacia*, and *Syringa oblata*, dramatically in accordance with the increasing temperatures in Beijing, China, between 1950 and 2004 (Lu et al. 2006). The early-flowering *P. davidina* species improved 2.9 days every decade between 1950 and 2004 as compared to other species, which improved 1.5–2.0 days per decade. *P. davidina*, flowering in early spring, was much more sensitive to minimum and maximum temperatures (2.88–2.96 days /°C). It was statistically found that in Beijing, the most vulnerable average temperature was above 30 days to the average blooming date which caused the plant blossoming to be affected (Lu et al. 2006).

Patterns of climate (1963–2009) in Alstonville in northern New South Wales, Australia, were examined (Olesen 2011). The average springtime development of cultivar custard apples in terms of days to reach maturity decreased by an estimated 13 days (20) between 1963/64 and 2008/09; forecasts of autumn litchi harvest development decreased by an estimated 8 days (7); forecasts of winter macadamia flush development by an estimated 17 days (12); and forecasts of summertime mango fruit development by an estimated 12–16 days (7–8) on the basis of cultivar (Olesen 2011).

15.4 Life Cycle Responses to Biotic and Abiotic Stress

According to the existing statistics of the literature, the temperature will reach 3 degrees Celsius in 2050 (Anum et al. 2014). The growth of plants is influenced by abiotic factors which are light, water, nutrients, and carbon dioxide. Plant and other living systems have their availability changed due to the changes (Ramirez and Kallarackal 2015). Fruit crops are especially vulnerable to long-term stressful environmental factors since these are perennial crops. The fruit crops are vulnerable to stress due to their long life cycles unlike crops with short life cycles who can adjust their life cycle, and finish it as fast as possible (Sarkar et al. 2021). Both abiotic (temperature, drought, salinity) and biotic (pests, diseases) factors are increased by climate change. Abiotic stress has less effects during photosynthesis because of high temperatures may be attributed due to change in the nutrient uptake and translocation (Rachappanavar et al. 2024a), as well as stress on water caused by the irregular rainfall (Mujtaba et al. 2024).

Warmer climate results in higher abundance of pests and pathogens that improve physiological processes and results in the survival and adaptation of insects (Atreya and Kaphle 2020). Radical climatic conditions like floods and droughts greatly interfere with systems of production of fruits (Aslam 2018). In Pakistan, 2022 floods have also demonstrated how dependent on climate extremes perennial crops are as these crops suffered massive destruction (Bhutta et al. 2022). Moreover, climate change impacts the dynamics of pollination that results in decreased fruit set and productivity (Maupilé et al. 2024).

15.4.1 Abiotic and Biotic Stressors

In addition to that the frequency of both biotic and abiotic stressors is increasing unpredictably. Both biotic and abiotic stress factors will always remain a threat to the fruit crops, and they may greatly interfere with the uptake and translocation of essential elements to result in low crop yield (Rachappanavar et al. 2024a). The duration of growth season is one of the highly important parameters in forestry, agricultural, and horticultural practices, and variation in the duration of it is highly influenced by the action of tempest-in-spring activities such as sprouting, leaf development, and flowering (Grab and Craparo 2011). Climate change, or rise in temperatures, precipitation, and climate variability, will result in a huge impact on the crop production and distribution within the region within the following decades with dire consequences on the food security (Rajatiya et al. 2018).

It was theorized that loss of rainfall will affect the tropical rainforests negatively and the West Africa is already undergoing a multidecadal trend of rainfall reduction (Polansky and Boesch 2013). United Nations Sustainable Development Goal 6 (SDG-6) includes the idea of clean water and sanitation that should be availed to all human beings, and the sustainable management of those resources. Global warming crisis made the scholars interested in the impact of flood disasters on the food safety and health industries in Pakistan (Devi 2022). The catastrophic flood in

Pakistan occurred, in June 2022, due to the melting of the glaciers and the significant precipitation in southern Punjab, in the larger part of Balochistan, and in the center of Sindh (Bhutta et al. 2022). It destroyed millions of farmlands and slaughtered many animals.

15.4.2 Stress and Quality of Fruits

The most severe issue related to climate change is the rise in temperature and it has been found that the extreme temperatures combined with the exposure to the increased levels of green house gases and ozone knows a direct and indirect impact on the quantity and quality of the fresh fruits and vegetable harvests (Moretti et al. 2010). The latter effect of temperature on fruit set was induced both with a fixed vapor pressure deficit (4–8 millibars) and with the percentage of the flowers that fruit being lower when the plant produced many flowers at the outset. There are fruit and nut trees which require to be frozen so that they can break the hibernation stage during the winter season and revert to normal growth in the spring period (Luedeling and Brown 2011).

The effects of heat stress were experienced on human health in Africa (Ncongwane et al. 2021). The growth of fruits plays a vital role in the propagation of floral plants and an essential source of food; however, adverse weather changes, lack of sufficient sunlight, and reduction in pollinators also lead to the fruit supply being highly sensitive to extreme weather conditions, inadequate sunlight exposure, and pollinators, rendering them susceptible to harsh weather conditions (Maupilé et al. 2024).

When evaluating the impact of the current sudden climate change on the ecosystem, plant developmental phases constitute the yearly recurrent cycle of phenology or plant life, which contributes significantly to the plant functioning, including the services it provides to society and the biophysical and chemical feedback of the climate system, and depends on the context of the environment and the current sudden climate change (Piao et al. 2019). Also, the impacts of the alterations in plant phenology are not only significant to the ecological carbon cycles but also to climatic feedbacks. The water has an essential role in the life on earth. Climate change and improper ways of utilizing water are increasing the strain on water availability in most of the developing countries, including Pakistan (Mujtaba et al. 2024).

15.5 Cold Season Implications to Fruits Crops, Nutrition, and Production

The climate change effect is not only affecting the quality and quantity of fruits produced in agricultural systems across the globe but is also having a fatal impact on the lives of the farmers and the well-being of the people (Stewart and Ahmed 2019). The findings showed how climate change affects five of the most common fruits globally, that is apples, bananas, grapes, oranges, and watermelon, as far as their nutritional value is concerned. The cold-season weather poses a severe drawback to the fruit production in the eastern parts of Canada, and the meteorological forecast predicts that there might be an increase of 2–6 °C of winter temperatures in 2050, at least (Rochette et al. 2004). With the research, probabilities would go up that a late frost would ruin flower buds in the south Ontario area, around the same in the Maritimes and the Ottawa valley-Southern Quebec areas and less in others.

The changes in temperatures have been significant in lowering the soil organic carbon and soil micronutrients (Atreya and Kaphle 2020). To survive, insect species may produce heat shock proteins, cryoprotectants, and osmolyte substances in their entire bodies, which enables them to be able to survive in the environment of Nepal. Warm climates promote early flowering in the subtropics and this can lead to low fruit set as a consequence of irregularities brought about by low night time temperatures and lack of chilling. This has a powerful impact on flower development and colouration of fruits and the degradation of their texture and taste (Atreya and Kaphle 2020).

Climate change impacts are also predicted on the development and endodormancy of apricot as well as a structure that delimits the environmental heat demand of apricot (*Prunus armeniaica*) in terms of south-to-north relocation of agriculturally based activities in France (Vanalli et al. 2021). Iwanami et al. (2023) examined meteorological and tree management parameters that affect the quantity of solution solids in the apple (*Malus x domestica* Borkh.) fruit and crop load plans when fruit are densely planted with high dissolved solids composition and scent.

Dormant breaking of the temperate fruit crops requires winter chilling. Climate warming lowers the hours of chilling; therefore, it alters flowering synchrony and fruit production. The reduced chilling has great impact on late or abnormal flowering, reduced fruit set and deterioration of quality and uniform sizes (Luedeling and Brown 2011).

Higher winter temperatures in places like Canada will increase the chances of late frost damage which will negatively impact the survival of buds and yield (Rochette et al. 2004). It also affects the quality of nutrition, as researchers report a shift in the sugar level, vitamins, and antioxidants in the fruits, including apple, grape, and citrus in different climatic conditions (Bhardwaj et al. 2023).

15.6 Habitat, Land, and Soil Degradation: Effects on Fruit Quality

Industrial pollution in Pakistan has strained the country a lot. As Ali (2000) notes, 11 or more ecosystems are facing the threat especially through loss and degradation of habitats. The process of land degradation is complicated, and it involves changes in soil structure and changes in output in agriculture (Andrade et al. 2015). The loss of soil causes the loss of subsoils which are valuable, nutrients, organic matters and porosity. The soil erosion does not only influence the physiological and microbial properties, but also it points out the physical property of the soil due to the association with the soil structure. Soil pollution deteriorates chemical performance of soil and thus invalidates the nutrition availability and reduces the microbial features are as well induced by the soil pollution and the soil compactability affects the soil porosity network. Due to being an agriculture oriented country, Pakistan is at risk due to global climatic changes, as it is ranked among those regions of the whole world that are estimated to have been worst affected by global climatic changes (Ullah 2017).

15.6.1 Health of Soils and Availability of Nutrients

Climatic changes also affect the physical characteristics of the soil, precipitation, the water process, the well-being and quality of the soil, the structure of roots and their connection to the soil bulk density, the soil porosity, and reduced accessibility of macro and micronutrients to the soil and plants (Andrade et al. 2015). Such processes of soil degradation have a direct impact on the bioactive compounds that are accumulated in the fruit crops because the mineral availability is fundamental to the biosynthesis of the secondary metabolites of medicinal properties (Rachappanavar et al. 2024b).

15.6.2 Effects of Fruit Crop Nutrition

Alteration of crop nutrition and enhances the cracking of fruits, including the key transcription factors in stress response (Manzoor et al. 2024). According to the study, the climatic factors were significantly found to negatively interfere with the cultivation of apple fruits in Pakistan between the years 1991 and 2020 (Chandio et al. 2024). At the end of the twenty-first century, a long term and stable reduction of chilling during winter had a negative economic impact on the fruits and nuts production of California, USA (Ramirez and Kallarackal 2015).

15.7 Case Studies: The Effects of Climate Change on Fruits Crops in the World

15.7.1 Pakistan: A Nation at Risk to Climate

Pakistan, a country which is blessed with diversity of ecological zones such as deserts, mountains, rivers, and rich floristic diversity (Nasir 1991) is becoming susceptible to climatic stresses. Its flora and fauna are being huge causative factors of both biotic and abiotic stressors, particularly due to increases in temperatures. The nation is one of the most climate-sensitive countries, and its impact on agriculture and rural sustainability is enormous (Awan and Yaseen 2017). Major fruit crops, including mango, citrus, apple, grape, peach, loquat, olive, and pomegranate, have also been experiencing significant losses in their yield and quality over the past decades, and that is due to climate change, as well as to environmental degradation, overuse of agrochemicals and insufficient use of modern agro-technology.

The climate change has disrupted major physiological and developmental activities of fruit crops such as flowering, pollination, fruit set, and maturation, which eventually decreases the yields and market quality. Pakistan, as well as Asia, is especially vulnerable to the increase of temperature and the loss of biodiversity (CE 2024). In Pakistan, the average annual temperature has risen by about 0.5 °C during the last 50 years (Jaffery 2018), exacerbating climatic pressure on agriculture.

The citrus crops, particularly, Kinnow mandarin grown in the plains of Punjab are very delicate to changes in the environment. Temperature and humidity fluctuations have negative impacts on biochemical and physiological characteristics, such as sugar-acid balance and firmness of the fruit. Also, climate change has worsened the occurrence of pests and diseases, augmented the irrigation needs, and augmented the cost of production. Vehari, Toba Tek Singh, and Sargodha are only some of the regions that have seen recurring heat waves, drought, and fogs which have led to decreased yield and poor quality of fruits (Nawaz et al. 2019).

Being aware of these issues, Pakistan developed its National Climate Change Policy in 2012 to advance adaptation and mitigation measures (COBD 2024). Current research activities are being directed toward learning the effects of climate change and establishing resilient systems of fruit production. The chapter is a synthesis of literature on the scientific literature available to assess the impacts of climatic change on fruit crops both on a regional and global basis.

The reasons behind the decline include overexploitation of chemical fertilizers, environmental degradation, failure to implement the recent eco-agro farming technologies, and ignorance and unavailability of up to date scientific equipment in the undeveloped and emerging nations than the developed countries. Among fruit crop growers, production is also influenced resulting in lower yield and changed growth pattern, phenology, blooming, pollination, flower development, fruit size, and fruiting.

15.7.2 Global Perspectives

The climate of Bordeaux, France influences the development of Grapevine, structure of grapes, and the production and quality of alcohol (Jones and Davis 2000). Paes de Camargo (2010) reported the effects of climatic variations and climatic change on Arabic coffee crop in Brazil. Boonklong et al. (2006) reported climate change that impacted on production of mangosteen in Thailand. Primack et al. (2009) conducted research to determine the effects that climate changes have on cherry trees and other species in Japan.

The wildflower phenology is highly reliant on the variations in the climate and regulates a few interactions between the vegetation and the climate by modifying the fluctuating seasons of the albedo, roughness of the surfaces length, the crown conductivity, and the water, energy, CO_2, and naturally occurring organic substance outflows (Richardson et al. 2013). The phenomenon of global warming is increasingly bringing a critical impact on both the human system and the natural system, and changes in climatic parameters pose a threat to the agricultural production process, and agriculture is one of the most affected fields (Mbewu et al. 2024).

15.8 Future Work Plan and Mitigation Strategies

15.8.1 Climate Change Impact Mitigation

The potential effects of climate change on fruit crops are the retarded increases in growth and propagation of plants, and reduced resistance to pollutant stress. The decisions of the fruit crops that have been adopted are some of the mitigation measures. Dispersal and counsel by government and nonprofit making agencies, stakeholders, policymakers and scientists in the field of fruit crop can lessen the impact of climatic change on production of fruits.

The dynamics in physiological stages were supported with adjustments in the climatic conditions and the techniques of biological network adaptation (Webb et al. 2012). Time series of wine fruit maturity trends were established to date back to up to 64 years of time series, meaning that two environmental conditions are putting a significant level of this rate of development in practice: increasing temperature and declines in the level of water content of soils. The crop production becomes less and changing the means of control have probably led to the untimely hardening. Some of the factors that contributed to the trend of younger mature were noted to be potential adaptation alternatives, which entail the water content in the soil and the yields of crops, through specific management measures (Webb et al. 2012).

15.8.2 Future Work Plan

WHO has been highly prolific on the climate change issue since over 40 years ago, coordinating global meetings, workshops, and conferences to limit the harm caused to the environment and to make sure that the effects of climate warming are such a nightmare to the generations to come (Iqbal 2020). Multiple steps can be taken to make sure that the crops of the fruits will not be damaged further and that the cooperation would be necessary. Climate change can result in periodic surveys and biological monitoring of the disturbed sites, which will serve as a point of reference where the changes occurring in the affected regions can be recorded. In the same context, mass research and publication of specific data on such resistant species ought to be done. Research requires to be done on ecology, productivity, and maintenance of the ecosystem.

Participation of the farmers requires the need to sensitize them on the effectiveness of the fluctuation of the climate on the quality of the fruits. The researchers are making an attempt of introducing the fruit crops that are conducive in the development in such adverse environment. Further studies need to be done to show the viability of technical solutions in preventing effects of increasing temperature on the fruits. Rambutan (*Nephelium lappaceum* L.) is a native tree of South East Asia belonging to Sapindaceae family. It is a medium-sized evergreen tree, whose height is 12–15 m and which improves the climate (Bhattacharjee et al. 2022).

15.9 Summary and the Revival of the Food Is Medicine Idea

The study has recognized and analyzed the significant contribution made by the change in climate toward the industrial production of fruits in Pakistan and other nations in the world. Different published research materials substantiated the high loss of production of fruit crop because of unplanned urbanization, loss of green spaces, industrialization, and more drastic changes of the current developing climatic conditions. They must constantly check the natural surroundings on one or more pollutants and resort to drastic measures in order to save the endangered species of the fruit crops. It takes much effort and effort so that the environment can be further festered as by-product of the climate change activities. More research on the damaged areas that can be replaced should explore the plantation that has greatly adaptive plant species that can endure changes in climatic conditions. The intelligent fertilizers and new nanotechnology can help to increase yield of the fruit crops as well as the establishment of equal ecological conditions. It was further known that the research will have interest to plant ecologists, hydrologists, farmers, horticulturists, water resource specialists, environmental scientists, and policymakers who want to resolve the issue of climate change as it applies to the high production of fruit crops.

15.9.1 The Revitalization of the Food Is Medicine Concept

The idea of Food is Medicine (FM), though not very new, has recently acquired a new scientific interest within a modern healthcare context. The idea is traditionally credited to Hippocrates, who stressed the curative power of food (Witkamp and van Norren 2018). Over the past few decades, FM has evolved to become less of a philosophical concept and has become more of an evidence-based practice in the mainstream of medical care. Recent dietary recommendations by most professional organizations also support plant-based dietary patterns with large consumption of whole grains, fruits, vegetables, and minimally processed foods and reduced intake of saturated fats and ultra-processed foods (Arnett et al. 2019; Visseren et al. 2021).

Nonetheless, the climate change is an existential threat to the FM paradigm since it jeopardizes the quality of nutrition and medicine of the foods on which it is based (Rachappanavar et al. 2024b; Chandio et al. 2024). It is not just an agricultural or even an environmental concern, but rather a basic public health necessity, the preservation of medicinal fruit crops under the conditions of climate change.

15.9.2 The Planetary Health, Recent Initiatives, and Policy Developments

Reactive to the chronic illnesses caused by a diet, mass policy programs have developed. It is also important to note that in 2022, the White House Initiative on Hunger, Nutrition, and Health has made more than 8 billion in commitments to enhance food access, fund nutrition research, and incorporate dietary interventions into healthcare systems (White House 2022). In 2012, Pakistan developed the National Climate Change Policy that addresses the problem of climate change adaptation (COBD 2024). Applications of FIM are not limited to the context of single health but to the idea of planetary health that can be described as the interdependence between the well-being of the human population and the environmental systems that remain sustainable (Whitmee et al. 2015). The EAT-Lancet Commission has suggested a radical change in diets toward consumption of plant-based foods as the solution to human and environmental health issues (Willett et al. 2019). Their results indicate that changes in diets would help to avert about 10–11 million deaths per year and have a profound impact on decreasing the environmental footprints. Overall, this chapter highlights the strategic role of medicinal fruits in the development of the systems-based sustainable strategy of health promotion and chronic disease management. Their ongoing exploration and use will play an important part in the creation of strong, clinically improved populace on an international scale, as long as an immediate measure is taken to counter the effects of global warming in these crucial food providers.

References

Ali S (2000) Impact of environmental degradation on biodiversity. Proc Pak Acad Sci 37(1):93–97

Andrade BO, Koch C, Boldrini II, Velez-Martin E, Hasenack H, Hermann JM, Kollmann J, Pillar VD, Overbeck GE (2015) Grassland degradation and restoration: a conceptual framework of stages and thresholds illustrated by southern Brazilian grasslands. Nat Conserv 13:95–104. https://doi.org/10.1016/j.ncon.2015.08.002

Anum T, Nazia T, Khuda B, Ashfaq M, Hassan S (2014) Food security in the context of climate change in Pakistan. Pak J Commer Soc Sci 8(2):540–550. https://hdl.handle.net/10419/188154

Arnett DK, Blumenthal RS, Albert MA, Buroker AB, Goldberger ZD, Hahn EJ, Himmelfarb CD, Khera A, Lloyd-Jones D, McEvoy JW, Michos ED, Miedema MD, Muñoz D, Smith SC Jr, Virani SS, Williams KA Sr, Yeboah J, Ziaeian B (2019) 2019 ACC/AHA guideline on the primary prevention of cardiovascular disease. Circulation 140(11):e596–e646. https://doi.org/10.1161/CIR.0000000000000678

Aslam M (2018) Flood management current state, challenges and prospects in Pakistan: a review. Mehran Univ Res J Eng Technol 37(2):297–314

Atreya PN, Kaphle M (2020) Visible evidence of climate change and its impact on fruit production in Nepal. Int J Agric Environ Food Sci 4(2):200–208. https://doi.org/10.31015/jaefs.2020.2.10

Awan AG, Yaseen G (2017) Global climate change and its impact on agriculture sector in Pakistan. Am J Trade Pol 4:41–48. https://doi.org/10.18034/ajtp.v4i3.425

Bhardwaj RL, Parashar A, Parewa HP, Vyas L (2023) An alarming decline in the nutritional quality of foods: the biggest challenge for future generations' health. Foods 13(6):877. https://doi.org/10.3390/foods13060877

Bhattacharjee P, Das S, Das SK, Chander S (2022) Rambutan (*Nephelium lappaceum* L.): a potential fruit for industrial use, serving nutraceutical and livelihood interests and enhancing climate resilience. S Afr J Bot 150:26–33. https://doi.org/10.1016/j.sajb.2022.06.064

Bhriguvanshi SR (2010) Impact of climate change on mango and tropical fruits. In: Singh HP, Singh JP, Lal SS (eds) Challenges of climate change-Indian horticulture. New Delhi: Westville Publishing House

Bhutta ZA, Bhutta SZ, Raza S, Sheikh AT (2022) Addressing the human costs and consequences of the Pakistan flood disaster. Lancet 400(10360):1287–1289. https://doi.org/10.1016/S0140-6736(22)01874-8

Boonklong B, Jaroensutasinee M, Jaroensutasin K (2006) Climate change affecting mangosteen production in Thailand. In: Proceedings of the 5th WSEAS international conference on environment, ecosystems and development, Venice

CE. Client Earth (C.E.). Client Earth in Asia. https://www.clientearth.asia/. 2008–2024. Visited on 22-11-2024.

Chandio AA, Gokmenoglu KK, Joyo MA, Jiang Y (2024) Modeling the climate change impacts on major fruits production: recent evidence from Pakistan. Sci Hortic 324:112618. https://doi.org/10.1016/j.scienta.2023.112618

Chang C (2002) The potential impact of climate change on Taiwan's agriculture. Agric Econ 27(1): 51–64. https://doi.org/10.1111/j.1574-0862.2002.tb00104.x

Cleland EE, Chuine I, Menzel A, Money HM, Schwartz MD (2007) Shifting plant phenology in response to global change. Trends Ecol Evol 22(7):357–365. https://doi.org/10.1016/j.tree.2007.04.003

Chmielewski F, Rötzer T (2001) Response of tree phenology to climate change across Europe. Agric For Meteorol 108(2):101–112. https://doi.org/10.1016/S0168-1923(01)00233-7

Chmielewski FM, Müller A, Bruns E (2004) Climate changes and trends in phenology of fruit trees and field crops in Germany, 1961–2000. Agric For Meteorol 121(1–2):69–78. https://doi.org/10.1016/S0168-1923(03)00161-8

Chuine I, Beaubien EG (2001) Phenology is a major determinant of tree species range. Ecol Lett 4: 500–510

COBD. COBD (conservation on biological diversity). Pakistan country profile. Biodiversity Facts; 2024. https://www.cbd.int/countries/profile?country=pk. Visited on 22-11-2024

Devi S (2022) Pakistan floods: impact on food security and health systems and its cost. Lancet 400(10355):799–800. https://doi.org/10.1016/S0140-6736(22)01732-9

Doi H, Katano I (2008) Phenological timings of leaf budburst with climate change in Japan. Agric For Meteorol 148(3):512–516. https://doi.org/10.1016/j.agrformet.2007.10.002

Dong ZH, Chen MX, Srivastava AK, Mahmood UH, Ishfaq M, Shi XJ, Zhang YQ, Moussa MG, Li XX, Hu CX, Zhang F (2024) Climate changes altered the citrus fruit quality: a 9-year case study in China. Sci Total Environ 923:171406. https://doi.org/10.1016/j.scitotenv.2024.171406

Gallinat AS, Primack RB, Willis CG, Nordt B, Stevens AD, Fahey RT, Whittemore AT, Du Y, Panchen ZA (2018) Patterns and predictors of fleshy fruit phenology at five international botanical gardens. Am J Bot 105(11):1824–1834

Guédon Y, Legave JM (2008) Analyzing the time-course variation of apple and pear tree dates of flowering stages in the global warming context. Ecol Model 219(1–2):189–199. https://doi.org/10.1016/j.ecolmodel.2008.08.010

Grab S, Craparo A (2011) Advance of apple and pear tree full bloom dates in response to climate change in the southwestern Cape, South Africa: 1973–2009. Agric For Meteorol 151(3): 406–413. https://doi.org/10.1016/j.agrformet.2010.11.001

Grand J, Wilsey C, Wu JX, Michel NL (2019) The future of North American grassland birds: incorporating persistent and emergent threats into full annual cycle conservation priorities. Conserv Sci Pract 1(4):e20. https://doi.org/10.1111/csp2.20

Guo L, Dai J, Ranjitkar S, Yu H, Xu J, Luedeling E (2014) Chilling and heat requirements for flowering in temperate fruit trees. Int J Biometeorol 58:1195–1206

Gupta DK, Keerthika A, Gupta CK, Shukla AK, Mohamed MBN, Jangid BL, Kumar C (2021) Horticulture based integrated farming systems, 1st edn. Imprint CRC Press, p 12. eBook First Published 2021. ISBN9781003245810

Haokip SW, Shankar K, Lalrinngheta J (2020) Climate change and its impact on fruit crops. J Pharmacogn Phytochem 9(1):435–438

Iftikhar S, Khattak HA, Saadat A, Ameer Z, Zakarya M (2024) Efficient fruit disease diagnosis on resource-constrained agriculture devices. J Saudi Soc Agric Sci. https://doi.org/10.1016/j.jssas.2024.07.002

Iqbal PP (2020) Effect of climate change on health in Pakistan. Proc Pak Acad Sci B Life Environ Sci 57(3):1–12. Pakistan Academy of Sciences. ISSN: 2518-4261 (print), ISSN 2518-427X (online), 2020

Iwanami H, Moriya-Tanaka Y, Hanada T, Baba T, Sakamoto D (2023) Meteorological and tree management factors related to soluble solids content of apple fruit and crop load management for producing high soluble solids content fruit in high density planted 'Fuji'. Sci Hortic 310: 111755. https://doi.org/10.1016/j.scienta.2022.111755

Jaffery R. The diplomat. The Pulse. Pakistan's biodiversity is disappearing, but no one seems to notice; 2018. https://thediplomat.com/2018/12/pakistans-biodiversity-is-disappearing-but-no-one-seems-to-notice/. Visited on 2211–2024

Jetz W, Wilcove DS, Dobson AP (2007) Projected impacts of climate and land use change on the global diversity of birds. PLoS Biol 5:1211–1219

Jones GV, Davis RE (2000) Climate influences on grapevine phenology, grape composition, and wine production and quality for Bordeaux, France. Am J Enol Vitic 51:249–261

Kagoro-Rugunda G, Baranga J (2009) The fruit phenology of Musanga leo-errerae and its importance for chimpanzee diet in Kalinzu Forest Uganda. Afr J Ecol 47(1):14–19. https://doi.org/10.1111/j.1365-2028.2007.00902.x

Koike S, Kasai S, Yamazaki K, Furubayashi K (2007) Fruit phenology of *Prunus jamasakura* and the feeding habit of the Asiatic black bear as a seed disperser. Ecol Res 23(2):385–392

Kotir JH (2011) Climate change and variability in Sub-Saharan Africa: a review of current and future trends and impacts on agriculture and food security. Environ Dev Sustain 13:587–605

Lu P, Yu Q, Liu J, Lee X (2006) Advance of tree flowering dates in response to urban climate change. Agric For Meteorol 138(1–4):120–131. https://doi.org/10.1016/j.agrformet.2006.04.002

Luedeling E, Brown PH (2011) A global analysis of the comparability of winter chill models for fruit and nut trees. Int J Biometeorol 55:411–421

Luedeling E, Girvetz EH, Semenov MA, Brown PH (2011) Climate change affects winter chill for temperate fruit and nut trees. PLoS One 6:e20155

Lundgren K, Kuklane K, Gao C, Holmér I (2013) Effects of heat stress on working populations when facing climate change. Ind Health 51:3–15

Maatallah S, Guizani M, Lahbib K, Montevecchi G, Santunione G, Hessini K, Dabbou S (2024) Physiological traits, fruit morphology and biochemical performance of six old fig genotypes grown in warm climates "Gafsa oasis" in Tunisia. J Agric Food Res 17:101253. https://doi.org/10.1016/j.jafr.2024.101253

Manzoor MA, Xu Y, Lv Z, Xu J, Shah IH, Sabir IA, Wang Y, Sun W, Liu X, Wang L, Liu R, Jiu S, Zhang C (2024) Horticulture crop under pressure: unraveling the impact of climate change on nutrition and fruit cracking. J Environ Manag 357:120759. https://doi.org/10.1016/j.jenvman.2024.120759

Mattos LM, Moretti CL, Jan S, Sargent SA, Lima CEP, Fontenelle MR (2013) Climate changes and potential impacts on quality of fruit and vegetable crops. In: Emerging technologies and management of crop stress tolerance, pp 467–486. https://doi.org/10.1016/B978-0-12-800876-8.00019-9

Maupilé L, Chaib J, Boualem A, Bendahmane A (2024) Parthenocarpy, a pollination-independent fruit set mechanism to ensure yield stability. Trends Plant Sci 29(11):1254–1265. https://doi.org/10.1016/j.tplants.2024.06.007

Mbewu AD, Elephant DE, Motsi H, Nyambo P (2024) Climate change effects on water footprint of crop production: a meta-analysis. Environ Chall 17:101033. https://doi.org/10.1016/j.envc.2024.101033

Medda S, Fadda A, Mulas M (2022) Influence of climate change on metabolism and biological characteristics in perennial woody fruit crops in the Mediterranean environment. Horticulturae 8(4):273. https://doi.org/10.3390/horticulturae8040273

Moretti CL, Mattos LM, Calbo AG, Sargent SA (2010) Climate changes and potential impacts on postharvest quality of fruit and vegetable crops: a review. Food Res Int 43(7):1824–1832. https://doi.org/10.1016/j.foodres.2009.10.013

Moss GI (1970) The influence of temperature on fruit set in cuttings of sweet orange (*Citrus sinensis* L. Osbeck). Hortic Res 10:97–107

Mujtaba G, Shah MUH, Hai A, Daud M, Hayat M (2024) A holistic approach to embracing the United Nation's Sustainable Development Goal (SDG-6) towards water security in Pakistan. J Water Process Eng 57:104691. https://doi.org/10.1016/j.jwpe.2023.104691

Nasir YJ (1991) Plant life of South Asia. In: Ali SI, Ghaffar A (eds) Proceeding of the International symposium on plant life of South Asia (24–27 Feb. 1990) Karachi, Pakistan. Shamim Printing Press, Karachi, p 229

Nath V, Kumar G, Pandey SD, Pandey S (2019) Impact of climate change on tropical fruit production systems and its mitigation strategies. In: Sheraz Mahdi S (ed) Climate change and agriculture in India: impact and adaptation. Springer, Cham. https://doi.org/10.1007/978-3-319-90086-5_11

Nawaz R, Abbasi NA, Hafiz IA, Khalid A, Ahmad T, Aftab M (2019) Impact of climate change on Kinnow Fruit Industry of Pakistan. Agrotechnology 8:186. https://doi.org/10.35248/2168-9881.19.8.186

Ncongwane KP, Botai JO, Sivakumar V, Botai CM (2021) A literature review of the impacts of heat stress on human health across Africa. Sustainability 13(9):5312. https://doi.org/10.3390/su13095312

Ohashi Y, Kawakami H, Shigeta Y, Ikeda H, Yamamoto N (2011) The phenology of cherry blossom (*Prunus yedoensis* "Somei-yoshino") and the geographic features contributing to its flowering. Int J Biometeorol 56(5):903–914. https://doi.org/10.1007/s00484-011-0496-4

Olesen T, Menzel CM, Wiltshire N, McConchie CA (2002) Flowering and shoot elongation of lychee in eastern Australia. Aust J Agric Res 53:977–983

Olesen T (2011) Late 20th century warming in a coastal horticultural region and its effects on tree phenology. N Z J Crop Hortic Sci 39(2):119–129. https://doi.org/10.1080/01140671.2010.550627

Orlandi F, Garcia-Mozo H, Galán C, Romano B, de la Guardia CD, Ruis L, del Mar Trigo M, Dominguez-Vilches E, Fornaciari M (2010) Olive flowering trends in a large Mediterranean area (Italy and Spain). Int J Biometeorol 54:151–163

Paes de Camargo MB (2010) The impact of climatic variability and climate change on Arabic coffee crop in Brazil. Bragantia 69:239–247

Pagadala T, Alam MA, Maxwell TM, Curran TJ (2023) Measuring flammability of crops, pastures, fruit trees, and weeds: a novel tool to fight wildfires in agricultural landscapes. Sci Total Environ 906:167489. https://doi.org/10.1016/j.scitotenv.2023.167489

Pechan PM, Bohle H, Obster F (2023) Reducing vulnerability of fruit orchards to climate change. Agric Syst 210:103713. https://doi.org/10.1016/j.agsy.2023.103713

Piao S, Liu Q, Chen A, Janssens IA, Fu Y, Dai J, Liu L, Lian X, Shen M, Zhu X (2019) Plant phenology and global climate change: current progresses and challenges. Glob Chang Biol 25(6):1922–1940

Polansky L, Boesch C (2013) Long-term changes in fruit phenology in a west African lowland tropical rain forest are not explained by rainfall. Biotropica 45(4):434–440

Primack RB, Higuchi H, Miller-Rushing AJ (2009) The impact of climate change on cherry trees and other species in Japan. Biol Conserv 142(9):1943–1949. https://doi.org/10.1016/j.biocon.2009.03.016

Rachappanavar V, Gupta SK, Jayaprakash GK, Abbas M (2024a) Silicon mediated heavy metal stress amelioration in fruit crops. Heliyon 10(18):e37425. https://doi.org/10.1016/j.heliyon.2024.e37425

Rachappanavar V, Kumar M, Negi N, Chowdhury S, Kapoor M, Singh S, Rustagi S, Rai AK, Shreaz S, Negi R, Yadav AN (2024b) Silicon derived benefits to combat biotic and abiotic stresses in fruit crops: current research and future challenges. Plant Physiol Biochem 211: 108680. https://doi.org/10.1016/j.plaphy.2024.108680

Rafique R, Ahmad T, Ahmed M, Khan MA (2024) Adapting the process based STICS model to simulate phenology and yield of table grapes- a high value fruit crop in a new emerging viticulture zone of South Asia. Sci Hortic 336:113419. https://doi.org/10.1016/j.scienta.2024.113419

Rajatiya J, Varu DK, Gohil P, Solanki M, Halepotara F, Gohil M, Mishra P, Solanki R (2018) Climate change: impact, mitigation and adaptation in fruit crops. Int J Pure Appl Biosci 6(1): 1161–1169. https://doi.org/10.18782/2320-7051.6161

Ramírez F, Davenport TL, Fischer G (2010) The number of leaves required for floral induction and translocation of the florigenic promoter in mango (*Mangifera indica* L.) in a tropical climate. Sci Hortic 123(4):443–453. https://doi.org/10.1016/j.scienta.2009.10.005

Ramirez F, Kallarackal J. Response of fruit trees to global climate change. SpringerBriefs in Plant Science. Ebook; 2015. https://link.springer.com/book/10.1007/978-3-319-14200-5

Richardson A, Keenan T, Migliavacca M, Ryu Y, Sonnentag O, Toomey M (2013) Climate change, phenology, and phenological control of vegetation feedbacks to the climate system. Agric For Meteorol 169:156–173

Rochette P, Bélanger G, Castonguay Y, Bootsma A, Mongrain D (2004) Climate change and winter damage to fruit trees in eastern Canada. Can J Plant Sci 84(4):1113–1125. https://doi.org/10.4141/P03-177

Sarkar T, Roy A, Choudhary SM, Sarkar SK (2021) Impact of climate change and adaptation strategies for fruit crops. In: Islam MN, van Amstel A (eds) India: Climate change impacts, mitigation and adaptation in developing countries. Springer Climate. Springer, Cham. https://doi.org/10.1007/978-3-030-67865-4_4

Sellers S (2016) Gender and climate change: a closer look at existing evidence. Global Gender and Climate Alliance, New York, NY

Sparks TH, Jeffree EP, Jeffree CE (2000) An examination of the relationship between flowering times and temperature at the National scale using long term phenological records from the U.K. Int J Biometeorol 44:82–87

Stewart AL, Ahmed S (2019) Effects of climate change on fruit nutrition. In: Fruit crops, pp 77–93. https://doi.org/10.1016/B978-0-12-818732-6.00007-1

Ullah S (2017) Climate change impact on agriculture of Pakistan – a leading agent to food security. Int J Environ Sci Natl Res 6(3):76–79

Vanalli C, Casagrandi R, Gatto M, Bevacqua D (2021) Shifts in the thermal niche of fruit trees under climate change: the case of peach cultivation in France. Agric For Meteorol 300:108327. https://doi.org/10.1016/j.agrformet.2021.108327

Visseren FLJ, Mach F, Smulders YM, Carballo D, Koskinas KC, Bäck M, Benetos A, Biffi A, Boavida JM, Capodanno D, Cosyns B, Crawford C, Davos CH, Desormais I, Di Angelantonio E, Franco OH, Halvorsen S, Hobbs FDR, Hollander M, Jankowska EA, Michal M, Sacco S, Sattar N, Tokgozoglu L, Tonstad S, Tsioufis KP, van Dis I, van Gelder IC, Wanner C, Williams B (2021) 2021 ESC guidelines on cardiovascular disease prevention in clinical practice. Eur Heart J 42(34):3227–3337. https://doi.org/10.1093/eurheartj/ehab484

Vitasse Y, Delzon S, Dufrêne E, Pontailler J, Louvet J, Kremer A, Michalet R (2009) Leaf phenology sensitivity to temperature in European trees: do within-species populations exhibit similar responses? Agric For Meteorol 149(5):735–744. https://doi.org/10.1016/j.agrformet.2008.10.019

Wang Y, Zhang X, Tian C, Guo X, Shu Q, Gu H, Feng P, Li F, Li B (2023) Effects of different diets on thermal tolerance in the silkworm, Bombyx mori. Ent Exp Appl 172(1):27–34. https://doi.org/10.1111/eea.13382

Webb LB, Whetton PH, Bhend J, Darbyshire R, Briggs PR, Barlow EW (2012) Earlier wine grape ripening driven by climatic warming and drying and management practices. Nat Clim Chang 2(4):259–264. https://doi.org/10.1038/nclimate1417

Whitmee S, Haines A, Beyrer C, Boltz F, Capon AG, de Souza Dias BF, Ezeh A, Frumkin H, Gong P, Head P, Horton R, Mace GM, Marten R, Myers SS, Nishtar S, Osofsky SA, Pattanayak SK, Pongsiri MJ, Romanelli C, Soucat A, Vega J, Yach D (2015) Safeguarding human health in the Anthropocene epoch: report of The Rockefeller Foundation–Lancet Commission on planetary health. Lancet 386(10007):1973–2028. https://doi.org/10.1016/S0140-6736(15)60901-1

White House. White House conference on hunger, nutrition, and health: national strategy. Washington, DC; 2022.

Willett W, Rockström J, Loken B, Springmann M, Lang T, Vermeulen S, Garnett T, Tilman D, DeClerck F, Wood A, Jonell M, Clark M, Gordon LJ, Fanzo J, Hawkes C, Zurayk R, Rivera JA, De Vries W, Majele Sibanda L, Afshin A, Chaudhary A, Herrero M, Agustina R, Branca F, Lartey A, Fan S, Crona B, Fox E, Bignet V, Troell M, Lindahl T, Singh S, Cornell SE, Srinath Reddy K, Narain S, Nishtar S, Murray CJL (2019) Food in the Anthropocene: the EAT–Lancet Commission on healthy diets from sustainable food systems. Lancet 393(10170):447–492. https://doi.org/10.1016/S0140-6736(18)31788-4

Witkamp RF, van Norren K (2018) Food is medicine: the evolution of a concept. Eur J Pharmacol 836:1–8

Yamada M, Fukumachi H, Hidaka T (1996) Photosynthesis in longan and mango as influenced by high temperatures under high irradiance. J Jpn Soc Hortic Sci 64:749–756

Ex Situ and In Situ Conservation Strategies for Medicinal Plants of Arid Climates

16

Mehmooda Munazir, Naima Shahzadi, Ammara Zafar, and Rahmatullah Qureshi

Abstract

Medicinal plants constitute an invaluable component of biodiversity, supporting traditional healthcare systems, local livelihoods, and ecosystem resilience, particularly in arid and semi-arid regions. Arid ecosystems, which cover nearly 41% of the Earth's terrestrial surface and sustain more than two billion people, harbor a rich diversity of medicinal plant species adapted to harsh environmental conditions. However, these valuable genetic resources are increasingly threatened by climate change, habitat degradation, overharvesting, overgrazing, deforestation, soil erosion, invasive species, urbanization, and weak conservation policies. This chapter examines the ecological, economic, and cultural significance of medicinal plants in arid ecosystems, with particular emphasis on Pakistan's diverse flora and ethnomedicinal heritage. Major threats to medicinal plant biodiversity and their implications for ecosystem stability and sustainable livelihoods are critically evaluated. The chapter further reviews contemporary conservation approaches, including *in situ* strategies such as protected areas, biodiversity hotspots, habitat restoration, and community-based conservation, alongside *ex situ* methods including botanical gardens, seed and gene banks, cryopreservation, *in vitro* conservation, and germplasm preservation. Emphasis is placed on integrating traditional ecological knowledge with modern conservation technologies, sustainable harvesting practices, policy interventions, and stakeholder participation. The chapter concludes that effective conservation of

M. Munazir (✉) · N. Shahzadi · A. Zafar
Department of Botany, Government College Women University Sialkot, Sialkot, Pakistan
e-mail: mehmooda.munazir@gcwus.edu.pk

R. Qureshi
Department of Botany, Faculty of Sciences, Pir Mehr Ali Shah Arid Agriculture University, Rawalpindi, Pakistan
e-mail: rahmatullahq@uaar.edu.pk

R. Qureshi (ed.), *Plant Biodiversity*, https://doi.org/10.1007/978-981-92-1471-6_16

medicinal plant biodiversity requires interdisciplinary collaboration, strengthened regulatory frameworks, public awareness, and adaptive management strategies to ensure the sustainable utilization and long-term preservation of these critical biological resources for future generations.

Keywords

In situ · Ex situ · Conservation · Medicinal plants · Arid ecosystem · Biodiversity · Climate change

Abbreviations

BSCs	Biological Soil Crusts
CBD	Convention on Biological Diversity
LN	Liquid Nitrogen
LOS	Low-Oxygen Storage
LPS	Low-Pressure Storage
MEAs	Multilateral Environmental Agreements
MOEFCC	Ministry of Environment, Forest, and Climate Change
NASA	National Aeronautics and Space Administration
PGRFA	Plant Genetic Resource for Food and Agriculture
SOC	Soil Organic Carbon
SOM	Soil Organic Matter
UNESCO	United Nations Educational, Scientific, and Cultural Organization
WRB	World Resource Base

16.1 Introduction

16.1.1 Importance of Biodiversity

As a key factor of terrestrial ecosystems, plant diversity is a comprehensive index of species composition and community structure, which is closely related to ecosystem stability. In addition, already affected by climate changes, it is highly susceptible to abiotic type of stresses, particularly in fragile arid environments. The loss of plant diversity can upset ecosystems, degrade resources, degrade environments, reduce products, and cause crises related to food.

16.1.2 Biodiversity in Arid Areas

Climate changes and anthropogenic activities have been identified as key drivers of vegetation cover dynamics with their impact on vegetation dynamics varying between regions. Moreover, an increase in human-related activities (e.g., large-scale agricultural land reclamation, poor water management) is known to be important. Multiple authors have discussed the impact of human-induced activities on the decline of biodiversity in arid environments. Land use change is also a major driver of this problem since land use land cover conversion severely decreases plant diversity. Decreasing land use changes are causing the groundwater sources in semi-arid areas to be depleted and degraded. Furthermore, dry ecosystems are strongly vulnerable to expected changes in precipitation patterns. Desertification can be defined as the long-term soil degradation in arid and sub-humid regions due to the effects of climate variability and anthropogenic activities. In hyper-arid deserts, the impact of human activity on the vegetation structure and on plant diversity is greater than the classic soil heterogeneity.

16.1.3 Impacts of a Decline in Biodiversity

Traditional knowledge associated with culturally important plants is under threat from recurring drought and invasive plant encroachment, which endangers biodiversity and retroactively threatens livelihoods and ceremonial and spiritual dependencies.

In drylands, the authors contend, vegetation loss and the increasing abundance of exotic or unwanted species are degrading the landscape, or both (Timis-Gansac et al. 2025).

16.2 Medicinal Plants of Arid Climate

16.2.1 Economic Value of Medicinal Plants

Pakistan possesses a rich and diverse resource base of medicinal plants that is distributed across almost all ecological zones of the country, ranging from alpine and temperate regions to arid deserts and coastal ecosystems. It is estimated that nearly 2,000 species of medicinal plants, including both higher and lower plants, occur in Pakistan. Despite this enormous biological wealth, many medicinal plant species are harvested unsustainably from the wild. In numerous cases, collectors sell medicinal herbs to intermediaries or traders offering the highest price, often leading to indiscriminate uprooting of entire plants. Such destructive harvesting practices severely reduce the natural regenerative capacity of these species and threaten the long-term sustainability of medicinal plant populations (Table 16.1).

Equally the tones of medicinal plants are used by the biochemists/pharmacologists/pharmaceutical companies to extract only some grams of secondary metabolites like alkaloids. This has led to consumption of medicinal plants. One can mention the examples of *Dioscorea, Paeonia, Podophyllum,* and *Saussurea* in this regard. In actual sense, there is no distinct medicinal plant sector which exists to develop relative policies to control their trade activities, facilitating development of new conservation practices as well as their sustainable use.

Table 16.1 Some of medicinal plants used for treating diseases from deserts of Pakistan (Hameed et al. 2011; Shaheen et al. 2014; Ghulam Yaseen et al. 2019)

Sr. no.	Plant species	Disease treated
1	*Bombax malabarium*	Stomach problem
2	*Nonea edgeworthii*	Cough
3	*Brasica compestris*	Jaundice, anti-lice, athlete foot
4	*Cassia fistula*	Constipation
5	*Capparis decidua*	Toothache, earache
6	*Spinacia oleracea*	Constipation
7	*Citrullus colocynthis*	Skin problem, digestive problems
8	*Cascuta reflexa*	Antidandruff
9	*Mentha longifolia*	Piles, pain, nausea
10	*Ficus religiosa*	Asthma, nausea, diarrhea
11	*Euclyptus camaldulensis*	Common cold, influenza
12	*Euphorbia hitra*	Cough
13	*Eclipta alba*	Asthma
14	*Zaleya pentandra*	Gastritis, stomach upsets
15	*Cyperus rotundus*	Blood disorders
16	*Aerva javanica*	Bladder and kidney stones
17	*Cressa cratica*	Urinary discharge
18	*Prospris cinereria*	Blood purification
19	*Zaleya pentandra*	Gastritis
20	*Dalbergia sissoo*	Nosebleed
21	*Acacia nilotica*	Eyes infection, sore throat, tooth ache, diarrhea
22	*Aloe vera*	Blood pressure, diabetes, tonsils, hair tonic
23	*Amaranthus albus*	Scorpio sting, body heat, ureter and kidney stone
24	*Calendula arvensis*	Induce sweating, heart disease, skin infections
25	*Calotropis procera*	Stomachache, toothache, malaria, dog bites
26	*Cucurbita maxima*	Nervous disorder, stomach ulcer, heart tonic, stomachache
27	*Cynodon dactylon*	Constipation, piles, redness of eyes, blood purifier, skin injuries and wounds
28	*Datura innoxia*	Mouth infections, gonorrhea, hydrophobia, earache
29	*Lantana camara*	Gastric problems, malaria, tetanus, rheumatism
30	*Ocimum basilicum*	Mouth sores, indigestion, cold, fever, ringworm, malarial fever

Most of the medicinal plants are utilized in the manufacturing of medicines, particularly in developing nations where they are still harvested wildly. The pressure is rising; the habitat is dwindling and therefore many plants are bound to fall under the pressure. Wild habitats are relieved by cultivation. It facilitates easier prevention of errors in identification and fight adulteration. Plant conservationists ought to be included and assist in developing programs on supply and conservation of medicinal plants, with the focus on the promotion of cultivation and processing (Shinwari and Shinwari 2010).

16.2.2 Threats to Survival of Medicinal Plants

16.2.2.1 Overharvesting

One of the biggest dangers to the survival of species is overharvesting. Cultivation of species which are known to be overharvested is a conservation measure that can result in the fulfillment of the growing market demand and preserve the wild population of the target species. This plan can be associated with the willingness of the international community to respect the right of local communities to exploit biological resources in order to sustain their livelihood.

16.2.2.2 Soil Erosion

Development of agricultural activities and overstocking leads to both water and wind erosion due to the decrease in vegetation cover. Not only in the Northern Areas is water erosion a specific issue, but it is being experienced in the agro-ecosystems of the barani lands e.g., Sulaiman Rodkohi and the Potwar plateau (traditional water harvesting system). Water erosion and the resulting soil erosion affect about 11 million hectares (Liu et al. 2019). Soil degradation leads to make the environment worse, arid lands among various continents are represented in Fig. 16.1.

16.2.2.3 Over-Exploitation of Plants

It is also because of the diversities in climatic and edaphic conditions that Pakistan is abundant in medicinal plants. Among the nearly 6000 species of vascular plants that have been identified to be growing in Pakistan, approximately 1000 species have been identified to have phytochemical constituents. About 350–400 herb species are traded in various drug markets and applied in major manufacturing units of yunani and homeopathic drugs. Plant-based drugs and products of diverse species have regularly been on the rise in demand over the past years. This has created mass harvesting and habitat destruction.

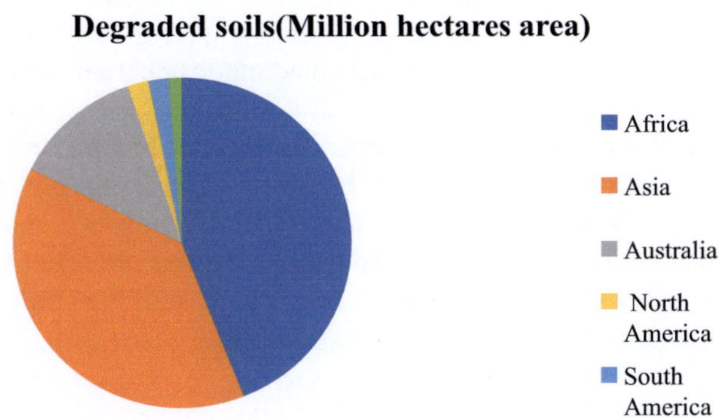

Fig. 16.1 Degraded soils in continents. (Naorem et al. 2023)

16.2.2.4 Introduced/Invasive Species

Introduced or alien species which are seen to be harmful to biodiversity can be introduced. This kind of bio-pollution has gained momentum in the last few years with the process of globalization which implies the accelerated and extensive exchange of goods between one location and another, promoting the dispersal of organisms in ship ballast water, in vessels, and evening merchandise. Many documented species extinctions have been caused by introduced species, particularly on islands, which rank second only to environmental degradation and ecological decline as a global cause of elimination (Shinwari and Shinwari 2010).

16.2.2.5 Deforestation

Taxation of the Himalayan forests is largely explained by the demographic overload and other similar impacts: more land is needed to be cultivated; more cattle are brought into the world; the remaining forest is exploited to satisfy the rising demands of fodder, fuelwood, and timber. Conversely, there have been hardly any social issues, including insights societal bonds, and values regarding natural reserves and how they are perceived by various social groups, matters of access and control to resources, and matters of power with regards to deforestation, discussed in the current article as potential driving forces to deforestation (Shinwari 2010).

16.2.2.6 Dependence of Local People and Afghan Refugees on Medicinal Plants

In Kohat, e.g., the Ghamkol Area, Peshawar Road, was a good Be*rberis lycium* and *Delphinium kohatense* (local endemic). Nowadays, the *Berberis lycium* and *Delphinium kohatense* have completely disappeared in the region (Shinwari et al. 2011). And due to these downward trends, a majority of the species, particularly those possessing medicinally properties, are raising red flags to join the "Threatened Species Club" (Shinwari 2010).

16.2.2.7 Standard Healthcare System and Extinction Scenario

Pharmaceutical use of 12% of the Pakistani flora and over 300 medicinal trade plants exist. Ten major Dawakhanas [Herbal manufacturers] of Pakistan are estimated to consume over 2million kg of 200 medicinal plants annually in 1990s while its consumption has grown more than a hundred-fold in the past two decades. The medicine of ancient Greece appears to have given rise to some of the applications of therapeutic flora in the Unani medicine for certain ailments.

16.2.2.8 Medicinal Plants Collection in Pakistan

Virtually medicinal flora in Pakistan is collected from indigenous locations. Regional harvesters know not the best acquisition methods. The medicinal plants that are collected in the respective locations are taken to the local and foreign trade markets via different dealer. Thus, the cost of the crude drugs rises up to over 100% along the trade chain (Shinwari 2010).

16.2.2.9 Anthropogenic Disturbances

The human activities can be directly or indirectly blamed as the cause of these disturbances. The anthropogenic disturbances that are major include human-caused forest fires, clearance of forests to extract timber, fodder, fuelwood, extraction of non-timber forest products, land use alterations, road construction, and so on.

16.2.2.10 Habitat Loss/Fragmentation

Habitat loss and habitat disruption is the main source of threat. A fraction of habitat occupied by a given species can merely be wiped out, either by urban and industrialization, or food and other natural resource production. One of the greatest causes of habitat destruction is land use change. Huge areas of forests are being invaded in order to satisfy need of ever-growing human population. Natural ecosystems convert to simplified agro-ecosystems leading to a great loss of biodiversity (Gairola et al. 2014).

16.3 Arid Ecosystems

Arid ecosystems are believed to be one of the most susceptible to the destabilizing impacts of the anthropogenic climatic change as most of the vegetative life has already been subjected to the higher extremes of climatic influence and resource constraints in the landscapes. When combined with a predicted warming of 1–5 °C in the global temperature in the next few decades, the severity and length of drought events are predicted to rise substantially in arid environments, probably leading to an acceleration in the degradation rates and desertification of arid landscapes (Millennium Ecosystem Assessment 2005). Without action, predicted rises in temperature and water stress related to near term anthropogenic climate change would go beyond the physiological tolerance of many arid zone species (Lewandrowski et al. 2021).

Drylands, which occupy approximately 41% of the terrestrial surface, can support more than 2 human beings (trisection of the human population in 2000) as well as demonstrate distinctive features in native vegetation, carbon turnover, weather condition, furthermore territory usage. These areas act as habitat to endemic and rare species and as a foraging area to the wild animals. During periods of sparse and erratic water supply, drylands are resilient; however, there is an increasing threat of human actions and climate change (Timis-Gansac et al. 2025).

16.3.1 Threats to Plants in Arid Zone

16.3.1.1 Scarcity of Water

Agricultural production is a primary constraint in arid places because of water shortage. Due to low precipitation and high evapotranspiration, arid areas have only little rainfall. This issue is worsened with the present condition of world climate change which causes severe climatic patterns and extended drought periods. The maximization of farming practice in production and water management enhances soil, water, and product quality. Due to limited water supply, the natural primary production of the soils in the arid regions is low, as well as its soil fertility (Naorem et al. 2023).

16.3.1.2 Salinity

Salinization of soil in arid areas has become a major environmental limitation in the arid areas posing a threat to the productivity of soils, agricultural sustainability, and food security. In case water-soluble salts get trapped in the soil, the soil becomes salty. It may arise naturally or due to ineffective anthropogenic activities, which in most cases is associated with agricultural practices. Salinization of soil is as a result of dry climates and low precipitations. This is caused by elevated rates of evaporation which enhance salt levels on the ground surface. Weak drainage or waterlogging does not allow the salts to be washed due to the absence of water transportation. Salt-rich water is being used to irrigate land and this raises the concentration of salt in soil. Intrusion of saltwater into ground water, inland transporting of salty air masses by coastal winds, submergence of seawater and subsequent saline drying, immersion and seawater, and mismanagement of fertilizer all cause the salinization of the soil. As of now, saline soils are found in no less than 100 countries, and cover an entire region of 932.2 Mha. This encompasses very large regions in Pakistan, India, United States, China, Argentina, Sudan, and most of the countries in Central and West Asia, and along the Mediterranean Coast.

16.3.1.3 Nutrient Imbalances

Jiao et al., pointed out that although arid soils are previously C, P, and N deficit, further depletion of arid soils and drought can decrease the accessibility of these nutrients. The loss of soil C and N by arid conditions is also caused by the reduction of the soil water supply and plant cover by arid conditions which directly or indirectly influence processes involving C and N and in particular the biological processes of photosynthesis, atmospheric fixation of N, and the task of microbes and soil enzymes. The rise in temperatures coupled with reduced botanical cover increases the process of soil drying and as a result, erosion occurs where particles, fertile compounds like clay are washed away. Even though available nutrients are not depleted by a low yearly rainfall and high rate of evaporation on arid sites, the processes may cause dry climate N reduction. Nitrogen and photosynthesis have a strong relationship; hence when the levels of N decline as a result of drought, plants will fail to generate as much food and counteract CO2 emissions.

16.3.1.4 Sandy Soils

Aggregation and soil texture have a strong influence on C storage, as both silt and clays size fractions can safeguard soil organic carbon (SOC) against decomposition. Soil organic and inorganic carbon present in arid soils are listed in Fig. 16.2. During the decomposition of organic matter, aggregates are created by sticking the organic material to the inorganic material like silt and clay. As per the World Resource Base (WRB), Arenosols are soils with over 700 g sand kg^{-1} (0.05–2.0 mm) and under 150 g clay kg^{-1} (less than 0.002 mm). The planet is covered with sandy types of

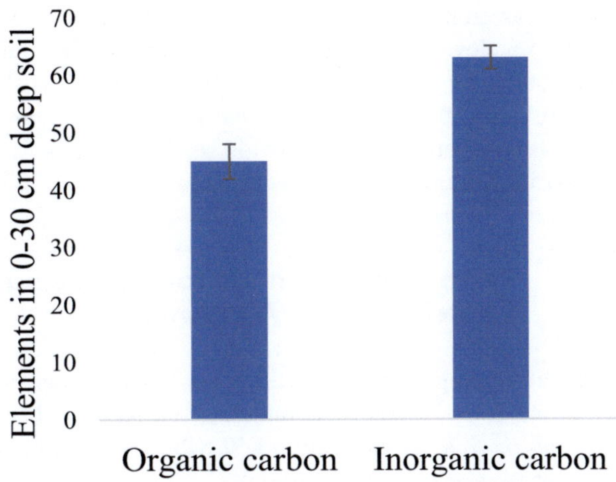

Fig. 16.2 Soil organic and inorganic carbon in arid soils. (Naorem et al. 2023)

soils. Like the name suggests, Arenosols are normally observed in a dry or semi-arid climate. Arenosols, that do not have a significant soil layering, are marked by ineffective development.

16.3.1.5 Soil Loss Through Erosion

Soil loss due to erosion or huge proportions of degraded soils are other features of the arid ecosystem. Most arid agroecosystems have very low prospects of increasing irrigation; therefore, alternative ways of optimizing land use must be sought. Arid regions are especially susceptible to topsoil weathering due to the absence of deep soil cover, low soil structural stability with a limited supply of SOC and a high human pressure. The prevention of soil erosion and maintenance of soil organic matter (SOM) under long-term conditions in arid regions have long been known to be dependent on vegetative cover. Even though proper protection of the soil surface is required in these areas, typical management approaches comprise intensive cultivation, overgrazing, increased feedstock requirements to generate bioenergy, and the requirements of livestock production (Naorem et al. 2023).

16.3.1.6 Climate Change

In 2008, NASA defined global warming as "the rise in the mean temperature of the ground level air and oceans due to increases in greenhouse gas concentrations," whereas it defined climate change as "a change in the climate of the Earth, or of a region of the Earth, that is persisting or may be expected to persist across a prolonged duration." In the MEA (2005), climate change is identified to be the most significant biodiversity decline factor and a grand obstacle for the future is also the conservation of biodiversity in the face of climate change.

- *Effects of Climate Change*

  In general, increased global temperatures are projected to cause

- 5–10%variation in rainfall
- 3–10% elevation in rainfall volumes during the rainfall season
- Fluctuations in stream flow in river basins of 5–10%
- Substantial drops of 15% in the mean temperature over the Arctic Ocean that leads to a permanent worldwide sea level rise between 4 and 7.5 m.
- 5–15% drop in crops production (Jovovic et al. 2020).

16.4 Restoration Methods for Improving Biodiversity Conservation in Arid Areas

- Intercropping and border cropping (also known as edge or margin planting), agroforestry systems combining moistened *Eucalyptus camaldulensis* and *Tamarix aphylla* cultivation with wheat, cluster beans (*Cyamous tetragocalobe*), or chickpeas (*Cicer arietinum*)
- Multi-crop “milpa” and “chichipera” cactus forest in Mexico
- Agroforestry systems based on olive in the Mediterranean area
- Trees are an important component of our environment because they conserve water, stabilize soil, and provide vital habitat for wildlife.
- Afforestation projects are important mechanisms for climate change mitigation, biodiversity conservation, and rehabilitation of arid zones. In addition, the diversity of native land animals in arid environments was strongly dependent on tree species.
- Planting other species of plants is another important technique. The tolerance aspect of barley provides it with the potential to cope with different biotic and abiotic stresses, and so it is among the selected cereals that are adapted to arid environments. *Prosopis flexuosa* DC (Leguminosae) is impactful in speedy restoration of vegetative cover in the Monte Desert of Argentina. Similarly, *Alhagi sparsifolia*, a hardy legume under dry conditions, adjusts root structure to tolerate extreme dryness, stabilize soil during windy conditions and reduce land degradation and desertification
- Biodiversity-pastoralist mixtures are good options for arid regions. For instance, managed grazing has been demonstrated to have good results in Australia. Systematic grazing and partial or no grazing are impactful for maintaining plant cover patterns, biotic diversity and ecological processes. Other approaches for economically viable and ecologically efficient cattle grazing have similarly been evaluated. Biological soil crusts (BSCs) are a critical element of arid ecosystems, serving to stabilize water land resources and to fix carbon and nitrogen. These surfaces include bryophytes, lichens, cyanobacteria, and associated food webs.
- Biocrust restoration, in which crust-forming organisms are inoculated, is an attractive rehabilitative technique.
- Within the Thar Desert, water conservation is now an important biodiversity conservation strategy in India (Timis-Gansac et al. 2025).

16.5 Need for Conservation

Biodiversity is the heterogeneity of living organisms on earth, including the wide variety of organisms, genetical and ecological diversity. However, the planet is now experiencing an unparalleled biodiversity crisis. Extinction of species due to different anthropogenic factors like habitat destruction, climate variation, pollution, and invasive species has accelerated the species extinction rate leading to depletion of ecosystem which jeopardizes the balance of life on earth. As a result of this crisis, conservation has become a worldwide issue. Two main approaches have been developed as main pillars of biodiversity conservation: in-situ conservation and ex-situ conservation. In-situ conservation means conservation and management of species and ecosystems in their natural habitats, whereas, ex-situ conservation includes the conservation of species in non-native environments, usually in specialized facilities and regulated environments. Both approaches are necessary parts of the wider conservation agenda and have individual merits in conservation protection.

16.6 In-situ Conservation

The central theme of in-situ conservation is the protection of nature through the protection of natural habitats, creation of national parks as natural reserve, protected areas, wildlife reserves, and other systems of conservation (Jovovic et al. 2020). By preserving habitats, in-situ conservation allows species to engage with their natural habitats, facilitating key ecological interactions and encouraging the coexistence of biological diversity. Moreover, in-situ conservation contributes to the maintenance of ecosystem services that are essential for human welfare, namely, clean water, air purification, and climate regulation. Conservation of intact habitats also aids in the ecosystem's resiliency, helping them to be more able to withstand environmental perturbations and climate change. In-situ preservation is at the core of biodiversity conservation, and yet it is also highly problematic. However, human footprint in addition to land conversion for agriculture, urbanization, and infrastructure development, is leading to habitat fragmentation and loss. In addition, the increasing threat of climate change serves to compound the pressure on natural habitats and some species may struggle to adapt to changing circumstances quickly enough. Hence the efficacy of in-situ conservation becomes diluted, and additional measures need to be taken for the survival of species.

16.6.1 Protected Areas

All places where the entire ecological system of living beings can be conserved are called protected areas. The in-situ conservation will be enhanced by investment in national conservation parks, ecological reserves, forest area reserves, wild sanctuaries, conservation of mangroves, Ramsar wetlands, traditional sites, and holy groves.

16.6.2 Wildlife Sanctuaries

A protected land that limits human activity to the minimal is what is referred to as a wildlife sanctuary. The Government and any other private body or individual may own such a protected area as long as the regulations are administered by the Government. In a wildlife refuge, no animals are allowed to be killed. Also, the trees should not be cut at all and specifically, they should not cut the forest to make room for farming. In India, there are a total of 531 wildlife sanctuaries reported that occupy an area of approximately 3.58% of the total geographical area of the country. Due to the serious threat of numerous wild species through over-exploitation, habitat loss, and alien species, wild nurseries further provide a viable pathway to in-situ protection of endemic, endangered, and demand medicinal plants.

16.6.3 National Parks

It will be the safeguarded region which is needed to protect as well as to propagate or develop wildlife in the safeguarded region or its environment irrespective of whether it is a sanctuary or not. In the territory of the national park, no human activity is allowed except that which might be authorized by the Chief Wildlife Warden of the state under conditions stipulated in Wildlife protection Act, 1972. As of 2015, 6555 national parks had been listed in the world. India has 103 national parks established covering 40,500 km^2 or 1.23% of the total geographical area of the country accessed in September 2015.

16.6.4 Mangrove Conservation Program

Under the Mangrove Conservation Program launched by MOEFCC, Government of India in 1987, 39 mangrove areas have been identified on the basis of their distinct ecosystems and biodiversity on the recommendations of National Committee on Wetlands, Mangroves, and Coral Reefs. There are 39 species of flora in 39 mangrove areas.

16.6.5 Conservation Areas of Medicinal Plants

In India the estimated number of medicinally used species in folk and recorded systems of medicine (Ayurveda, Siddha, Unani and Homoeopathy) is 6000–7000. An estimate of 960 medicinal plant species is traded with 178 species reaching volumes of greater than 100 metric tons of consumption per year. Through national and international agencies, some civil societies are engaged in in situ conservation of medicinal plant species. In India, there are now 108 MPCAs established in various states and these include: Karnataka (3), Kerala (9), Tamil Nadu (12), Maharashtra (13), Andhra Pradesh (8), Uttarakhand (7), Orissa (5), West Bengal (7), Madhya Pradesh (13), Arunachal Pradesh (7), Rajasthan (7), and Chhattisgarh (7).

16.6.6 Biodiversity Hotspots

A hotspot is any geographical region with at least 1500 endemic species of vascular plants. Biodiversity hotspots must meet two demanding requirements: they must contain at least 0.5 or 1500 endemic species of vascular plants and they must have lost more than 70% of their original vegetation. Out of the total 34 hotspots of biodiversity in the global map, 4 hotspots, including Himalayas (Western and Eastern Himalayas), Western Ghats and some parts of Sri Lanka, Northeast and Indo-Burma, and Nicobar Sundalands, are located in India.

16.6.7 World Heritage Sites

A World Heritage Site is a location, or a region, that is so important to mankind in the opinion of the United Nations Educational, Scientific and Cultural Organization (UNESCO) that it would be preserved by international agreement. This means that in order to become a World Heritage Site, a landmark must already be a landmark in itself, i.e., unique in some manner or another as a geographically and historically recognizable landmark.

Today we have 911 world heritage sites in 152 countries of which 39 of them are located in India in the states of Kerala, Karnataka, Tamil Nadu, and Maharashtra.

16.6.8 Sacred Grooves

The best example of in-situ conservation is the sacred groves. Sacred groves are areas of forest, ranging in size between 0.5 and 500 ha (some larger than 500 ha) that are safeguarded by religious groups, and possess some religious significance to the group that defends them. The mini forests with the greatest diversity are called sacred groves. Approximately 15,000 sacred groves in other areas of India have been reported. There are above 100 dreaded communities in Rajasthan. In sacred groves, hunting and wood-cutting are usually forbidden. There is also limited development in such patches. Much of sacred groove is connected to temples/monasteries/shrinery or cremation ground. The sacred grooves are common in most parts of India, particularly areas where the native people live. Other than that, sacred grooves hold a cultural value; various local cultural and religious celebrations are held in these patches. There is more than one ecosystem of sacred grooves, more than one food chain and food web. Sacred groves occur in Khasi and Jaintia Hills of

Meghalaya, Aravalli Hills of Rajasthan, Western Ghats of Karnataka and Maharashtra, Sarguja, Chanda and Bastar of Madhya Pradesh. The sacred groves of Meghalaya are the only remaining refuge of many threatened and rare plants (Dar et al. 2020).

16.6.9 Benefits of In Situ Conservation

The chapter concluded that in-situ conservation has a number of big advantages. By conserving natural habitats in this way, it is possible to maintain essential ecological processes like pollination, seed dispersal and nutrient cycling, which are critical for sustaining the productivity and resilience of ecosystems. Also, species within their natural environments can evolve to adapt to changing environmental conditions, which will enhance their probability of long-term survival and evolution to confront the new challenges. By protecting intact habitats, conservationists can protect entire ecological communities and many interactions that keep life going. The adaptive capacity of species in their natural habitats may increase their resistance to environmental changes and contribute to the overall healthy functioning and stability of ecosystems.

16.6.10 Limitations of the In-situ Conservation

While in-situ conservation has many advantages, several limitations were found. Human activities continue to destroy and fragment natural habitats. Interference with deforestation, urbanization, and infrastructure construction fragment wildlife corridors and migration routes, isolating populations and reducing gene flow. Further, climate change places an additional stress on protected areas, making it difficult for species to cope with challenges in their natural habitats. The limitations of in-situ conservation require a more active and adaptive approach to deal with the continuously increasing threats of human impact and climate change.

16.6.11 Challenges for In-situ Conservation of Wild Genetic Resources

Sectoralism and ignorance are two of the primary impediments to in-situ conservation of wild genetic resources. Efforts to protect areas are typically limited to ecosystem and species level and there is no special attention paid to conservation of crop genetic material. Protected areas agencies understand that the conservation of genetic resources is necessary; this is frequently viewed as an additional area of responsibility where no extra resources are typically present. Ministries of agriculture and their partners, who act on their behalf, are directly interested in conservation of wild relatives of crops, but they may not be aware of the significance of in-situ conservation. Partially, this is due to the fact that they are not necessarily the ones in charge of the right lands. Therefore, the cost of creating a protected zone can often be greater than the reward, especially when what is being sought is protection of a single wild relative of one crop species, or a single species. The second hindrance is the ignorance as to the extent and distribution of interpopulation genetic variation of wild relatives of crops. This is needed to answer questions such as, should in-situ conservation areas be created and, if so, what size should they be and how will they be used? This requires ecogeographical studies that establish the genetic variation of a species within a geographical ecological range of the species (Dar et al. 2020).

16.7 Ex-situ Conservation

Ex-situ conservation is literally translated to mean offsite conservation, i.e., conservation of the elements of biological diversity (seeds, clones, living young plants, pollen, or their DNA) outside of their native distributions in storage facilities known as gene banks. Such in-situ conserved species can undergo intraspecific processes of evolution and can be subject to natural selection, but such reserves are challenging to create, e.g., cost, size, and maintenance problems, political and social issues, natural calamities, etc. Ex-situ conservation seeks to develop and naturalize threatened species to keep them alive and sometimes yield a high yield of planting material to be used in the production of medicines. In the case of numerous plant species, including wild medicinal plants, not only can high potency be maintained in ex situ conditions away from the habitat, but the reproductive material may be selected and stored in seed banks to be replenished later. Germplasm collection is a prerequisite and the first step of ex-situ conservation.

16.7.1 Botanic Gardens

Ex-situ conservation depends on botanic gardens, and the gardens can preserve the eco systems to boost the survival rate of rare and endangered species. Botanic gardens: include a very wide range of plant species which are co-cultivated under comparable conditions, and which usually have taxonomically and ecologically diverse flora. But the botanical gardens are being threatened to lose germplasm due to disease, stress, and disaster and vast space and labor are required to maintain a minor portion of diversity. Over 1700 botanical gardens across the world contain botanic collections which are either conservation or instructional. Many botanical gardens require the protection of rare wild plants and endangered wild plants. According to a study 354,000 plants are grown in 1500 botanical gardens around the globe (Saaondo and Olalekan 2023). One of the largest of these conservation projects is the Millennium Seed Bank Project (MSBP) at the Royal Botanical Garden, Kew, England. In Calcutta, the Botanic Garden was established in 1787 on a total area of 110 ha and contained approximately 15,000 plants representing 2500 species (Dar et al. 2020).

16.7.2 Gene Banks

One of the forms of biorepositories is gene banks. Gene banks also depend on low temperatures to inhibit biochemical activities which cause cell death. Other banks spray liquid nitrogen (LN) at −196 °C. This deep-freezing method substitutes water within the cells by some other liquid that is glycerol that lowers the formation of ice crystals. There are many forms of material that could be stored in gene banks which can either use seeds in seed banks, frozen stem cuttings or laboratory-based test tube in vitro storage. Nowadays there are nearly 1500 gene banks containing 5.5 million accessions (FAO 1996). Some of these banks possess mainly genetic assets of simple staple crops as well as fodder, and usually do not have vegetable crops or forest genetic resources.

16.7.3 Seed Banks

Seed banks can finely conserve the ex-situ genetic variation of most plants compared with botanical gardens and thus endorsed to assist in saving biodiversity of wild growing plant forms. Genetic resources are stored in cold storage as seeds in 90% cases. The best seed bank is the Millennium Seed Bank Project at the Royal Botanic Gardens in Britain. The use of seed banks can help provide enhanced and rapid reach to plant material to assess its properties to obtain useful knowledge and evidences in safeguarding the surviving natural populations. The advantages of seed banking are quite significant, it is easy to store, does not require much space, relatively small amounts of labor and, therefore, can store large samples at a cost that is economically reasonable. Furthermore, genetic diversity can be conserved easily using seed: the samples are not large, have easy handling, do not need much storage, and can last for decades in most cases (Dar et al. 2020). The ability of seeds to tolerate very low temperature helped in making 1750 seed banks supporting conservation of 50,000 species (Walters and Pence 2021).

16.7.4 Pollen Banks

Pollen serves as an important basis of multi-allelic variation in a gene pool, and can thus potentially a productive propagule of gene banks. This allows the researcher more choice in planning his/her breeding program, because of the storage friendliness, consignment along with the ability to use it immediately. Techniques have been established to collect pollen, to desiccate it, to test its viability and to measure its remaining life including wide variety of species of interest, and this has demonstrated the fundamental importance of sufficient rapid pollen desiccation and many years of storage at very low temperatures to extend its lifespan. Propionic viability studies must be conducted in reconstituted medium after adequate rehydration and with valid stains, germination in controlled conditions, or natural pollination studies.

16.7.5 Storage Temperature

Many species of pollen can be stored at temperatures of 4 to −20 °C for limited duration of time. Dried and preserved pollen stored at 4–20 °C has a shelf life of 1–6 days to 1 year, and this is enough to use it in breeding programs. It is possible to store pollen at −80 °C or at LN (−196 °C) over the long term. When freeze-dried, the pollen may be stored in cryovials in LN or LN vapor in the cryobank. Containers and storage places should also be properly labeled to assist in sample recovery in future. Vials can be loaded in boxes or cryocanes afterwards and placed immediately into the vapor or liquid phase of liquid nitrogen (Dar et al. 2020).

16.7.6 In-vitro Germplasm Conservation

In-vitro techniques using ex-plant apical or auxiliary shoots, proliferating tissues as meristems and embryos are perfect for in-vitro preservation of germplasm of asexually or vegetatively propagated plants. This in-vitro method can additionally be applied to preserve the plants with stubborn and recalcitrant seeds and transgenic materials. In-vitro techniques are best method to store germplasm of endangered, rare, and threatened plant species (Chandran et al. 2023). In vitro conservation of genetic resources can be achieved by implementing three methods named below:

1. Cryopreservation (freeze preservation)
2. Low storage pressure and low storage oxygen
3. Cold storage

16.7.6.1 Cryopreservation

Cryopreservation is the maintenance of biological materials (seeds, plant embryos, shoot tips/meristems, and pollen) at extremes of low temperatures, typically that of liquid nitrogen at −196 °C. In this case, biochemical and the majority of physicochemical processes are suspended and constituents may be stored in the long duration of time. Cryopreservation of biological materials includes the following types of freezing:

- Over solid carbon dioxide (at −79 °C)
- Low temperature deep freezers (at −80 °C)
- In vapor phase nitrogen (at −150 °C)
- In liquid nitrogen (at −196 °C)

At the temperature of liquid nitrogen (−196 °C) the cells are in a totally dormant condition therefore can be stored over extended periods. The principle of the freeze preservation method is the conversion of the present water in the cells into solid state. Because of the salts and biomolecular components within the cells, the temperature at which the water in the cells is frozen is far lower (down to −68 °C) assessed relative to the freezing point of water (near 0 °C). To a great extent, the low temperature in the cells/tissues has prevented biochemical metabolism and biological degradations of the cells/tissues. Cryopreservation has actually been applied with success in preservation of the germplasm of many plant species such as wheat, cassava, rice, peanut, sugarcane, coconut, and strawberry, etc. It is possible to regenerate several plants using cells, meristematic tissue along with embryos in cryopreservation.

16.7.6.2 Low Pressure and Low Oxygen Storage

As alternatives to cryopreservation and cold storage, low-pressure storage (LPS) and low-oxygen storage (LOS) have been developed for use in germplasm conservation. Examples of plants (in various forms) that can be successfully cryopreserved cell suspension of Oryza *sativa*, *Glycine max, Zea mays, Nicotiana tabacum* and Capsicum annum, callus of Oryza *sativa*, *Capsicum annum* and *Saccharum* spp. protoplast *Zea mays* and *Nicotiana tabacum* Meristems *Solanum tuberosum* and *Cicer aietinum* zygotic embryos *Zea mays*, *Hordeum vulgare* and *Manihot esculenta* somatic embryos *Citrus sinensis*, *Daucus carota* and *Coffea arabica*.

16.7.6.2.1 Low-Pressure Storage (LPS)

When it is stored under low pressure, the pressure in the atmosphere encircling the plant tissue is minimized. This has the effect of partially decreasing the pressure of the gases around the germplasm. The resultant reduction in the partial pressure causes a reduction in the in vitro progression of plants. Low-pressure storage facilities can assist in short-term and long-term preservation of the plant material. The short-term storage is especially effective to prolong the durability of most plant parts, including fruits, cut flowers, vegetables, and plant cuttings. A low-pressure storage of the germplasm produced in culture is possible. Besides the preservation of germplasm, LPS inhibits the performance of pathogens and inhibits the sprouting of spores of plants within the culture system.

16.7.6.2.2 Low-Oxygen Storage (LOS)

The oxygen level is decreased in the low-oxygen storage, and the atmospheric pressure (260 mm Hg) is adjusted by adding non-reactive gases (in particular nitrogen). Below 50 mm Hg, the partial pressure of oxygen decreases plant growth. This happens by the reason; with decreased accessibility of oxygen the formation of CO_2 is low. Consequently, the photosynthetic performance is reduced, which then suppresses the plant tissue growth as well as size.

16.7.6.3 Cold Storage

Cold storage is simply the conservation of germplasm at low (nonfreezing) temperature (1–9 °C). Cold storage is more likely to slow down the growth of the plant material, and cryopreservation is more likely to arrest growth altogether. Therefore, cold storage is said to be a slow growth germplasm preservation method. The plant material (cells/tissues) will not experience cryogenic injuries. Long-term cold storage is easy and cheap and yields germplasm with acceptable survival probability. It has been tested and proven to store numerous in vitro grown shoots/plants of fruit tree species, e.g., grapes and strawberries. When a few drops of medium are added periodically (once in 2–3 months), virus-free strawberry plants could be stored at 10 °C over a period of about 6 years. A series of grape plants about 15 years in cold storage (9 °C) have been transferred to a new medium after a year (Dar et al. 2020).

16.7.7 The Benefits of Ex-situ Conservation

Ex-situ conservation is an important accompanying to in-situ conservation. This method provides a safety net for endangered species by keeping their captive populations outside their natural habitat. Ex-situ facilities such as botanical gardens, seed banks provide opportunities for research, genetic management, and public education, thus increasing awareness on the conservation of biodiversity (Saaondo and Olalekan 2023). The significance of ex-situ conservation, particularly with respect to endangered biodiversity species, speaks of its importance in the field of avoiding the extinction of species and involving the public. By keeping species in captivity and studying them, conservationists are gaining important information about their biology and behavior, which in turn will help them plan better for conservation.

16.7.8 Disadvantages of Ex-situ Conservation

Ex-situ conservation is challenging. Genetic diversity can be lost and that captive populations can suffer from genetic drift, which can undermine their long-term sustainability. Furthermore, the physiological and behavioral changes exhibited by captive animals can act to impede the success of reintroduction programs. The integration of ex-situ and in-situ conservation efforts can be optimally effective for reintroduction programs and will ensure that ex situ-reared species are able to adapt and thrive in their native environments (Naruka and Reddy 2023).

16.8 The Integration of Ex-situ and In-situ Conservation

The integration of ex-situ and in-situ conservation efforts can be optimally effective for reintroduction programs and will ensure that ex-situ-reared species are able to adapt and thrive in their native environments. Such cooperative breeding and reintroduction programs have proven to be successful with respect to improving the genetic well-being of captive populations and reintroduction attempts. In combination with habitat restoration programs seed banking is a powerful means for the protection of plant diversity and a valuable resource for ecosystem restoration, the successful fusion of in-situ and ex-situ conservation demonstrates the need for a holistic and collaborative approach to biodiversity conservation. By knowing when to apply these approaches, conservationists can ensure that they are as effective as possible and that species recovery efforts are more effective. The combination of ex-situ facilities with in-situ conservation activities can provide a more integrated and coordinated strategy for the preservation of endangered species and their habitats (Naruka and Reddy 2023).

16.9 Political Approaches to Conservation

Creating effective strategies for biodiversity conservation requires working in the messy space of politics where the demands of the ecological need to be translated into political reality. In this section, we explore the range of political approaches and processes that can drive collaboration, overcome obstacles, and further conservation:

- *Multi-stakeholder Engagement:* Engaging diverse contributors, involving government agencies, non-governmental organizations, indigenous association, indigenous groups, and private sectors, is crucial for fostering a collective commitment toward conservation. Including these stakeholders in decision-making processes adds legitimacy to policies, and ensures that a multiplicity of perspectives is addressed.
- *Science Communication:* Connecting science to politics in order to bridge the gap between scientific knowledge and political decisions is of the utmost importance. Efficient communication of scientific findings and ecological knowledge to policy makers and the public can lead to informed decision making and generation of support for conservation.
- *Economic Valuation of Biodiversity:* The articulation of economic benefits gained through biodiversity can be communicated to policy makers and industry. The quantification of ecosystem services such as clean water supply, pollination, and carbon sequestration can help gain support for conservation initiatives that ultimately can help human well-being.
- *Policy Integration and Mainstreaming:* Mainstreaming biodiversity conservation into other policy sectors—such as agriculture, energy, and infrastructure—can help to ensure that conservation objectives are integrated with wider development objectives. This mainstreaming approach allows conservation issues to be embedded into a wide variety of policy agendas.
- *Adaptive Management:* Conservation strategies need to be flexible as political landscapes change so rapidly. Adaptive management is thus the change in policy in response to monitoring and evaluation, as well as learning from successes and failures.
- *Incentive Mechanisms:* Incentive mechanisms that reward conservation efforts can encourage diverse stakeholders to actively participate in biodiversity protection. Rewards can range from payments for ecosystem services, to revenue sharing for ecotourism and green certifications.
- *Public Awareness and Advocacy:* Increasing public awareness of the importance of biodiversity will create a sense of responsibility and an incentive on citizens to support conservation policies. Grassroots movements and advocacy campaigns can shape political agendas and put conservation on local and national agendas.

- *Indigenous and Local Governance:* The use of participatory decision-making that recognizes the heritage-based knowledge of local and indigenous communities and gives them a sense of ownership in conservation efforts. "Conservation goals and local development goals do not necessarily have to be mutually exclusive, and participatory governance models can help to reconcile them."

16.9.1 International Collaboration and Diplomacy

International cooperation and diplomacy are critical in the world of biodiversity conservation, where borders are crossed, and nations work collectively toward the common goal of protecting Earth's biological diversity. In addition, international collaboration is not only needed to tackle complex conservation challenges but also to pool together expertise and resources to arrive at a comprehensive solution.

Diplomacy serves as a vital thread that weaves together countries, cultures, and agendas, allowing for mutual understanding, compromise, and collective action. Multilateral environmental agreements (MEAs) like the Convention on Biological Diversity (CBD) provide a framework for setting targets, sharing information and coordinating conservation efforts around the globe. International environmental conferences, such as the COP meetings, provide forums for diplomatic negotiations to reach consensus on the allocation of resources and protection measures. Differences in capacity and resources across countries can make it difficult for countries to collaborate effectively. Ultimately, diplomacy can be a powerful force for conservation success by promoting dialogue, facilitating knowledge sharing, and providing platforms for collective action. This diplomatic partnership propels nations toward a future where biodiversity flourishes and where the preservation of Earth's rich biological tapestry becomes a shared heritage for generations to come.

16.10 Future Directions and Recommendations

There are promising ways forward and suggestions that can inform the need to integrate political acumen into conservation of medicinal plants and its sustainable use.

- *Enhancing Interdisciplinary Collaboration:* Biodiversity conservation is a field that requires collaboration across various disciplines, such as ecology, political science, sociology, economics, and more. Harnessing interdisciplinary expertise to deepen understanding of the political and to enable innovative approaches to conservation problems.
- *Promoting Indigenous Knowledge and Community Engagement:* The contribution of regional knowledge and the participation of local communities in conservation are important factors for success. One of the future directions seems to be to focus on partnerships, which include equitable relationships, indigenous

wisdom, and local knowledge, to ensure that conservation efforts align with the cultural values of the communities affected.

- *Tapping into Technology and Data Science:* Technological advancements and data science present unparalleled opportunities for data collection, analysis, and visualization. The combination of geospatial technologies, remote sensing, and big data analytics will allow us to better understand the patterns of biodiversity and enable more informed decision-making.
- *Increasing Transparency and Accountability:* Political decision-making processes must be transparent to help build public trust and accountability. Future directions should focus on mechanisms that provide more transparency from open data to participatory governance models that engage stakeholders at every level.
- *Policy Coherence:* Advocacy for within- and cross-sectoral policy coherence and conservation practitioners must engage with policy makers to make sure that decisions made in other areas (such as agriculture, infrastructure, trade, energy, etc.) are compatible with biodiversity conservation goals.
- *Public Participation and Advocacy:* It is important for citizens to take ownership of biodiversity conservation. Increased public awareness campaigns and public education activities can help build public support for conservation measures and put political pressure on political entities to prioritize biodiversity.
- *Conflicting International Agreements:* International agreements relating to aspects of trade, development or human rights can come into conflict with biodiversity conservation objectives. These conflicts can undermine countries' ability to take conservation actions that are consistent with their international conservation commitments.
- *Cultivation Practices:* Cultivation of medicinally important plants helps in reducing the exploitation pressure exerted by human for their use in herbal medicine which will further improve the quality and identification of medicinally important plant species (Sharma and Reddy 2023).
- *Certification and Documentation:* Exploration of arid regions and ethnobotanical surveys help in listing of important plants and their uses. Documentation of medicinally important plants and its method to use helps in spreading awareness regarding uses (Hameed et al. 2011).
- *Sustainable Use Practice:* It is important for sustainable use of medicinal plant resources by implementation of sustainable solutions which creates balance between use and regeneration (Rajasekharan and Wani 2020).
- *Use Geographic Information System:* It helps in finding the gaps among ex-situ data collections and helps in in situ conservation (Kumar et al. 2025).
- *Hilfort as Site for In-situ Conservation:* Hillfort are known as being best sites for in situ conservation for longer periods of times for arid adapted medicinal plants by combining cultural resource management and biodiversity conservation (Labokas and Karpavičienė 2023).
- *Monitoring Genetic Erosion:* There is need to assess Plant Genetic Resource for Food and Agriculture (PGRFA) regularly to assess the species population and genetic erosion (Dar et al. 2020).

16.11 Conclusion

Biodiversity is where ecosystem balance, social well-being, and sustainable growth and development begin, but it is now threatened like never before by climate change, habitat loss, overexploitation, and invasive species. Arid areas are very susceptible because of weak ecosystems and increased human demands. In order to ensure ecological balance and cultural heritage, medicinal plants, conservation, sustainable regulation of natural environmental assets restoration are required. The in-situ and ex-situ conservation strategies contribute in a corresponding way to the conservation of the species diversity, the genetic continuity, and the ecosystem resilience. Nevertheless, ecological priorities should be combined with political, social, and economic aspects in order to ensure successful conservation. In the future, the need to build international collaboration, involve local populations, appreciate native knowledge, and embrace new technologies will play a vital role in overcoming existing challenges. Through cooperative, dynamic, and participatory approaches, humanity will be able to guarantee biodiversity to the current and future generations.

References

Chandran S, Raghu AV, Mohanan KV (2023) In vitro conservation of rare, endangered, and threatened plants. In: Conservation and sustainable utilization of bioresources. Springer, Singapore, pp 391–408

Dar JA, Kareem M, Zargar SM, Wani AA, Rasool S, Bhat KA (2020) Strategies for conservation of genetic resources. In: Rediscovery of genetic and genomic resources for future food security. Springer, Singapore, pp 315–334

FAO (1996) Report on the state of the World's plant genetic resources, Rome

Gairola SUMEET, Bhardwaj KN, Rana CS (2014) Threats to phytodiversity. Diversity of lower plants. ISBN No-9789382332411. IK International Publishing House Pvt. Ltd, New Delhi, pp 222–234

Ghulam Yaseen GY, Mushtaq Ahmad MA, Shehla Shinwari SS, Potter D, Muhammad Zafar MZ, Zhang GuoLin ZG et al (2019) Medicinal plant diversity used for livelihood of public health in deserts and arid regions of Sindh-Pakistan. Pak J Bot 51(2)

Hameed M, Ashraf M, Al-Quriany F, Nawaz T, Ahmad MSA, Younis A, Naz N (2011) Medicinal flora of the Cholistan desert: a review. Pak J Bot 43(2):39–50

Jovovic Z, Andjelkovic V, Przulj N, Mandic D (2020) Untapped genetic diversity of wild relatives for crop improvement. In: Rediscovery of genetic and genomic resources for future food security. Springer, Singapore, pp 25–65

Kumar V, Zadokar A, Kumar P, Sharma R, Sharma R, Siddiqui MW et al (2025) Advancing medicinal plant agriculture: integrating technology and precision agriculture for sustainability. PeerJ 13:e19058

Labokas J, Karpavičienė B (2023) On the prospects of in situ conservation of medicinal-and aromatic-plant genetic resources at ancient-hillfort sites: a case study from Lithuania. Plants 12(4):861

Lewandrowski W, Stevens JC, Webber BL, Dalziell EL, Trudgen MS, Bateman AM, Erickson TE (2021) Global change impacts on arid zone ecosystems: seedling establishment processes are threatened by temperature and water stress. Ecol Evol 11(12):8071–8084

Liu H, Gale SW, Cheuk ML, Fischer GA (2019) Conservation impacts of commercial cultivation of endangered and overharvested plants. Conserv Biol 33(2):288–299

MEA (2005) Ecosystems and human well-being: biodiversity synthesis. World Resources Institute, Washington, DC

Naruka KPS, Reddy PB (2023) Assessing the efficacy of in-situ vs. ex-situ biodiversity conservation measures. In: Recent advances in biodiversity research, vol 67

Naorem A, Jayaraman S, Dang YP, Dalal RC, Sinha NK, Rao CS, Patra AK (2023) Soil constraints in an arid environment—challenges, prospects, and implications. Agronomy 13(1):220

Rajasekharan PE, Wani SH (2020) Future of threatened medicinal plants in the era of anthropocene and climate change. In: Conservation and utilization of threatened medicinal plants. Springer, Cham, pp 533–548

Saaondo T, Olalekan O (2023) Ex situ and in situ conservation of plant genetic resources in Nigeria. ScienceOpen Preprints

Shaheen H, Qureshi R, Akram A, Gulfraz M (2014) Inventory of medicinal flora from Thal desert, Punjab, Pakistan. Afr J Tradit Complement Altern Med 11(3):282–290

Sharma P, Reddy PB (2023) Comparative analysis of software solutions for precise biodiversity index assessment. In: Recent advances in biodiversity research, vol 85

Shinwari MI, Shinwari MI (2010) Botanical diversity in Pakistan; past, present and future. World Environ 1:85–104

Shinwari ZK (2010) Medicinal plants research in Pakistan. J Med Plants Res 4(3):161–176

Shinwari S, Qureshi R, Baydoun E (2011) Ethnobotanical study of Kohat Pass (Pakistan). Pak J Bot 43(SI):135–139

Timis-Gansac V, Dinca L, Constandache C, Murariu G, Cheregi G, Timofte CSC (2025) Conservation biodiversity in arid areas: a review. Sustainability 17(6):2422

Walters C, Pence VC (2021) The unique role of seed banking and cryobiotechnologies in plant conservation. Plants People Planet 3(1):83–91

Community-Led Plant Biodiversity Conservation in Pakistan: Case Studies and Lesson Learnt

17

Mehmooda Munazir, Naima Shahzadi, Ammara Zafar, and Rahmatullah Qureshi

Abstract

Pakistan possesses remarkable plant biodiversity, encompassing more than 6,000 vascular plant species distributed across diverse ecosystems ranging from coastal zones and deserts to alpine regions. This rich biological heritage plays a critical role in maintaining ecosystem stability, supporting livelihoods, and sustaining traditional healthcare systems. However, plant biodiversity is increasingly threatened by habitat degradation, deforestation, overgrazing, unsustainable resource utilization, invasive species, climate change, and the erosion of indigenous knowledge. This chapter examines the status, threats, and conservation of plant biodiversity in Pakistan through selected case studies from diverse ecological regions, including deserts, forests, mountain ecosystems, and rangelands. Particular emphasis is placed on the role of local communities, traditional ecological knowledge, and community-led conservation initiatives in safeguarding biodiversity and promoting sustainable resource management. The chapter highlights successful conservation approaches, including participatory forest management, habitat restoration, biodiversity monitoring, and the documentation and transmission of ethnobotanical knowledge. Both *in situ* and *ex situ* conservation strategies, such as protected areas, botanical gardens, seed and gene banks, and germplasm conservation programs, are critically reviewed. The chapter underscores the importance of integrating scientific research, traditional knowledge systems, policy support, and stakeholder engagement to strengthen

M. Munazir (✉) · N. Shahzadi · A. Zafar
Department of Botany, Government College Women University Sialkot, Sialkot, Pakistan
e-mail: mehmooda.munazir@gcwus.edu.pk

R. Qureshi
Department of Botany, Faculty of Sciences, Pir Mehr Ali Shah Arid Agriculture University, Rawalpindi, Pakistan
e-mail: rahmatullahq@uaar.edu.pk

R. Qureshi (ed.), *Plant Biodiversity*, https://doi.org/10.1007/978-981-92-1471-6_17

biodiversity conservation efforts. It concludes that community-centered conservation, supported by effective governance and institutional collaboration, is essential for preserving Pakistan's plant biodiversity and ensuring ecosystem resilience and sustainable development in the face of growing environmental challenges.

Keywords

Plant biodiversity · Conservation strategies · Traditional knowledge · Ecological restoration · Pakistan flora · Policy framework · Case studies

Abbreviations

BAP	Biodiversity Action Plan
DMR	Direct Matrix Ranking
FL	Fidelity Level
ICF	Informant Consensus Factor
IKS	Indigenous Knowledge Systems
ILO	International Labor Office
IUCN	International Union for Conservation of Nature
IVI	Importance Value Index
LOC	Line of Control
NCS	National Conservation Strategy
NGOs	Non-Governmental Organizations
PR	Preference Ranking
SPCS	Sarhad Provincial Conservation Strategy
TDF	Tropical Dry Forests
TK	Traditional Knowledge
WWF	World Wildlife Fund

17.1 Introduction

17.1.1 Geographical Extent of Pakistan

Pakistan comprises an area of more than 882,000 km^2, with 24° and 37° north and 61° and 75° east. Pakistan makes some extensions located 1700 km north-side of the Arabian Sea coastline, as well as mouths of the Indus River to its headwater's mountains High mountains of the Himalaya and Hindu Kush and Karakorum ranges have many of them over 8000 m. Pakistan has a coast line of approximately 1046 km and water territory of 22,820 square km and economic zone about 196,600 km^2 (Anwar et al. 2008)

17.1.2 Biogeographical Regions, Biomes, and Vegetative Zones

Pakistan possesses three biogeography worlds that contain lives separated and are classified as having distinct biota, and four of the seven regions in the planet. It has been found that there are ten biomes (desert, temperate grassland, tropical seasonal forest, and mountain) (Cox and Moore 1993). Roughly, two-thirds of the country is mountainous; climatic and species variation in the country are brought about by changes in altitude; short distances. Pakistan has a rich landscape of terrestrial biomes in 12 major vegetative zones. About two percent (2%) of the general population has the condition (Ahmed et al. 2006).

17.1.3 Significance of Forest

Forests play many important roles; they provide timber and fuel wood, medicine, and biodiversity (Shahbaz and Suleri 2009). They also control microclimate, shelter wildlife and food quality as well as availability. Nevertheless, the biodiversity of the tropical dry forests is fast degrading as a result of the joint impact of climate change and humans. Habitat degradation has been compounded by global warming due to too much emission of greenhouses gases and thus drastic measures have to be taken in adherence to the international sustainability plans. The plant development in these forests depends heavily upon soil, temperature, availability of water and nutrients, and distribution of carbon, nitrogen and phosphorus imbalances in various ecosystems become disrupted upon long periods of droughts and warmer temperatures. Even reserves forests are undergoing change in structure, range changes of species and loss of carbon storage. The dominant plant species are a key to vegetation recovery and community stability since they have adaptive features helping to ensure ecological stability under the conditions of stress. Deforestation, mining, agriculture, urbanization, overexploitation of resources, and other human activities damage habitats directly by destroying them but also indirectly by causing the introduction of invasive species that kill the natural regeneration process. These mixed aspects endanger threatened species, slow the biodiversity restoration, and have the possibility of converting dry forests into savannas. The conservation of such ecosystems requires combined conservation measures, sustainable use of resources, and effective policy measures to be able to conserve the biodiversity as well as ensuring that future generations are able to serve the goals of the ecosystems (Shahzad et al. 2024). This chapter aims to identify the ecological, cultural, and economic importance of plant biodiversity in Pakistan, discuss the contribution that local communities can make in ensuring its conservation, report practical case studies of community biodiversity conservation, document the challenges and achievements of such initiatives, outline people to use and consider lessons and good practices, and to recommend ways of integrating scientific and traditional knowledge to reinforce plant biodiversity conservation (Shahzad et al. 2024).

17.2 Importance of Biodiversity

Biodiversity contributes to maintaining a healthy environment and the existence of the humans is precarious without it. Biodiversity's services to human beings, free of charge, are estimated in billions of dollars annually, for clean air and water, pollination of crops, formation and protection of soil, control of crop pests, production of food, fuel, fibers, and drugs. Biodiversity is one factor in reducing the risk of such disasters as floods, slides and avalanches, soil erosion, desertification, and siltation of water bodies. In this sense, biodiversity helps save billions of rupees annually that would otherwise be required to combat such disasters (Ahmad et al. 2015).

17.3 Floral Diversity in Pakistan

Pakistan has a variety of flora due to its variation in altitudes and geography. Over 6000 vascular plant species are found in this area (Stewart 1972) of which 5600 species are known so far in the flora of Pakistan which belong to 22 families and nearly 150 genera (Nasir and Ali 1972). They also have at least 189 lower plants, including at least 153 Sino-Himalayan pteridophytes (ferns and allies) and 36 Euro-Siberian. However, algae, liverworts, mosses, and lichens remain comparatively less explored. In Pakistan, approximately 87 genera and 3383 species of fungi have been documented (Mirza and Qureshi 1978). Approximately 80% of all endemic flowering plants are confined to western and northern mountains of Pakistan. There are 5 families with more than 20 recorded endemics, which are Leguminosae (57), Asteraceae (50), Apiaceae (34), Poaceae (32), and Brassicaceae (21). Endemic species are most numerous in the genus Astragalus.

17.4 Role of Traditional Knowledge in Conservation of Plant Biodiversity

Indigenous people and the local communities across the region have long resisted the commercialization of their cultural heritage including the use of traditional art, crafts, symbols, designs as well as the patenting medicinal plant knowledge and the copyrighting or the publication of traditional stories and songs. Other communities also desire to assure a fair use of their traditional knowledge (TK) under conditions established by the traditions of the community, or to have benefit sharing when their traditional knowledge (TK) was used.

The last couple of years, we are hearing about the indigenous knowledge in different fields. It encompasses practice such as intercropping, natural pest management, crop diversification, and seed selection in agriculture; fish breeding and in biology; traditional medicine in human health care; soil and water conservation, irrigation in natural resource management; and preservation of oral traditions and native languages within education system.

Indigenous knowledge contributions to these sectors have not been recognized until recently and there is now an uptake in interest by academicians, as well as, policymakers. Several governmental and non-governmental organizations and international agencies including World Bank, International Labor Office (ILO), UNESCO, and FAO have become to recognize the contribution of indigenous knowledge systems (IKS) in promoting sustainable development. This growing awareness is increasing evident in the policies and programs adopted by many nations around the world.

17.5 Threats to Traditional Knowledge

The traditional, as well as folklore medicine, tends to be passed by generations in the domestic and local remedies. Even the developed world is resorting to use of herbal drugs and herbal medicines. This is rich heritage in predominantly the combined family system of northern Pakistan, which is being kept mostly by the older women. The forthcoming deterioration of cultural norms is going to decrease this knowledge day by day. This does not necessarily guarantee how elders are going to pass on traditional knowledge to the younger. There has also come the long overdue requirement of this source of knowledge to be documented so that future generations can have access to the knowledge that it holds. Though naturally some indigenous knowledge is lost because practices are changed or not utilized throughout extended periods of time, the existing rate can be explained by modernization and homogenization of cultures, the existing educational systems that assume that macro-level problems could only be solved with the help of the global knowledge base, and the gradual development of institutions that cater the needs of grass roots innovations. Under high population growth, in-migration, and governmental relocation plans (large-scale development projects), standards of living are being compromised. Short-term economic benefits are preferred over environmental-sensitive local practices as the poverty levels increase. Replacement of agricultural and forestry patterns with the mono cropping patterns translates to a loss of biodiversity and consequently a loss in indigenous knowledge. Forest clearance brings the extinction of certain valuable, but little-known medicinal plants and as a consequence the wisdom relating to the plants also decreases (Shah 2008).

The loss of traditional knowledge is also linked to a growing communication gap, as both children and adults spend less time engaging with their local communities. Since the sharing of native knowledge mostly is oral, it will be subjected to modification, especially where the inhabitants relocate to other areas, or where the lifestyle of those living tends to be less similar to that of the earlier ancestry. The intellectual property laws have attracted international focus over the past years in a bid to ensure the preservation and promotion of TK. The 1992 Convention on Biological Diversity did recognize the role that TK can play with respect to protecting species, ecosystems, and landscapes and, as such, did incorporate laws with respect to it access and use. It has been found that the developmental activities that neglect or ignore locally generated technologies; local knowledge systems, and those which do not develop the policy formulations in the desirous manner concerning local environment are often not capable of producing the expected results. Though it has this been proved that the TK of the Indigenous People is very helpful and productive, the world on the other hand seems to have ignored it. Some of these threats are the environment, urbanization, globalization, global warming, climate change, etc. The rural areas host 80% of the population of Pakistan which has a very wealth history of use of plants. Folk medicine is practiced by more than 95% of the people in the Northern part of Pakistan when it comes to primary health care. The extensive reliance on traditional medicines can be attributed to people's strong connection with the local plant species, inherited knowledge of their uses, the affordability and the easy availability of medicinal plants, limited access to the modern allopathic treatments, and generally low economic status of these communities (Ahmad et al. 2015).

17.6 Major Threats to Biodiversity

Pakistan is a country with a great biodiversity; there are severe threats to it. The wasting of biodiversity resource goes through overgrazing, over harvesting, water-logging and salinization, deforestation, land conversion, soil erosion, the desertification and alien invasive species, and chemical contamination. The prime cause of recent extinction rate is disintegration and destruction of natural habitat, which is an extreme problem in all biomes (Watson and Watson 1995; UNEP 1995). In Pakistan, the green resources like woody biomass is dwindling at an average of 4.6 per annum (GOP 1992; Hosier 1993). The cover of the mangrove forest has reduced by half over the last 20 years. Decreased 2600 Km^2 in 1970s to 1300 Km^2 in mid 1990s. They are now rapidly depleting of this natural wealth (Saifullah 1997).

Since the coverage includes forests, shrubs, and planted trees on the farmlands, it is uncovered that it only takes 4.2 million hectares or 4.8% of the country. This has been taken without considering the plantation and shrub forests, and this leaves it at 2.4 million hectares (2.7%). Woody biomass is depleting at the rate of 4–6% annually and the use is expected to rise according to percentage ratio to population (3% per year); thus, the biomass can be lost in 10 years. The list of the internationally threatened or near-threatened mammals in Pakistan by the International Union on Conservation of Nature (IUCN 1998) comprised 37 species and 14 subspecies. It also hosts other 25 endangered species of birds, and 10 species of reptiles which are internationally endangered. The character and makeup of the species currently in existence is the outcome of no less than 3.5 billion years of evolution, which encompassed speciation, migration, and extinction, and, in more recent times too, human activity. It is estimated that there are 7 to 20 million species that are alive (IUCN Red List 1997). The information presented in taxonomic-literature shows that only 1.75 million present species have been characterized. In the western Himalaya, floristic diversities are in a deplorable state. The various risks that the plants in the region are exposed to have been identified.

17.6.1 Risk of Habitat Destruction of Medicinal Plants

Nearly everywhere on this planet, man has a powerful influence on natural habitats of flora and fauna, not just in high alpine areas, but on both coast pillars as well as in the deserts, and in the rain forests. The woods are felled to provide farmland or meet the demand of the industry in wood, and in so doing the forest in which our home ultimately must lie, are essentially altered or banished. Naturally, such shifts affect medicinal plants, as well. There are numerous ways that habitat destruction can have impacts on the medicinal plants. The shortest impact is caused due to the direct extinction of species in a particular area through destruction of the ecosystem. Clearly evident is the emerging expansion in the so-called, slash-and-burn clearings. Without its natural cover of vegetation, the soil is easily subjected to uncontrolled erosion, more so when it is not re-grown back into forest. Then crops that were indigenous to a given region are displaced permanently and along with it the herbal medicines. Harmful harvesting may have the same effect on a habitat.

17.6.2 Commercial Exploitation and Overharvesting of Medicinal Species

Historically, there had been use of medicinal herbs on the basis of collection and local use. Along with the increasing urbanization and expansion of the trade relations that have been following the urbanization, the medicinal plants have joined trade in a bigger scale. Medicinal plants have transformed into a product and received an isolated value that no longer related to its original use. Increased volumes of trade can also be attributed in part to the fact that people (in industrialized countries, in particular) are yearning to have a natural healing in the form of natural agents. However, this commercial boom has led to the excessive harvesting of wild plant species, motivated by short-term financial gains while neglecting ecological stability. Consequently, overexploitation combined with other environmental pressure has pushed many valuable plants like *Podophyllum hexandrum*, *Valeriana jatamansi*, *Geranium wallichianum*, *Bergenia ciliata*, *Primula denticulata*, and *Saussurea lappa*, toward endangerment or near extinction in their native habitats.

17.6.3 Artificial Genetic Modification and Their Impact on Medicinal Plants

In nature, mutations (in other words, genetic changes) always take place; they become the foundation of the evolution process and new species development. New species evolve when mutation or a combination of mutations allows the emergence of mutants and non-mutants, then the formation of groups, populations, or subpopulations. It is a value-free and continual evolutionary process. Mutation is another way of genetic changes initiated by human beings. Unlike the natural process, artificial mutations are normally aimed at something. The specific induced genetic characters on medicinal plants in terms of cultivation and development of different varieties are usually meant to increase ratio of some compounds. The transformations and the following adjustment to new conditions, which have been artificially established, as opposed to nature, occur incredibly rapidly, sometimes only after a couple of years. The intended outcomes are met early on at the cost of the disruption of the balance of the vegetation.

17.7 Case Study and Lessons Learned from the Cholistan Desert Study

17.7.1 Description

In this case study, we provide a case study on how various fencing regimes have impacted the community structure of perennial plant species in the Cholistan Desert, Pakistan. The experiment was carried out in three large locations, and it gave me powerful information about biodiversity management in dry areas (Khan et al. 2025).

17.7.2 Study Sites and Methodology

- The study was conducted in the Cholistan Desert, South of Punjab, Pakistan, an area with sandy deserts, no drinkable water, and minimal rains.
- The study was done in three main locations: Derawar Fort (Biodiversity Park IUB), Mauj Garh Fort (Livestock Farm), and Lal Suhanra National Park (Sheikh Mohammed Bin Zayed Al-Nahyan Conservation and Breeding Center).
- Each location was further broken down into 3 sub-locations, namely sand dune, inter dune and clayey land. Eleven $10 \times 10\ m^2$ quadrats were placed randomly in every sub-site bringing a total of 33 quadrats at each site.
- The researchers used two seasons (Fall, November–December) 2021 and Spring (February–March) 2022, to observe seasonal changes.
- Notable phytosociological parameters were measured which included diversity index, importance value index, relative frequency, relative cover, relative density, cover, and relative frequency.

17.7.3 Key Findings

17.7.3.1 Total Species Richness

In all three locations, 31 families and 57 species were listed at the end of the study period.

17.7.3.2 Dominant Species

Species such as *Ochthochloa compressa*, *Suaeda fruticosa,* and *Lasiurus scindicus* repeatedly had high relative density, relative frequency, relative cover and importance value.

17.7.3.3 Moderate and Low Value Species

Generally, *Cenchrus ciliaris* and *Cymbopogon jwarancusa* had moderate values of phytosociological parameters, whereas *Ziziphus nummularia*, *Salvadora oleiodes*, *Aristida hystricula*, *Calotropis procera*, and *Heliotropium crispum* had very low values of phytosociological parameters.

17.7.3.4 Diversity Index

The mean of the diversity index of all three sites was observed to be 0.8009822.

17.7.4 Site-Specific Observations

17.7.4.1 Derawar Fort

Ochthochloa compressa exhibited a very high relative density and IVI in the sand dune regions with *Lasiurus scindicus* and *Calligonum polygonoides* moderately high. Relative density of *Tamarix aphylla* was high in inter-dune areas whereas Cenchrus ciliaris exhibited high relative cover. Relative cover and IVI of *Suaeda fruticosa* were high in clayey land.

17.7.4.2 Mauj Garh Fort

Cenchrus ciliaris and *Ochthochloa compressa* often had high relative density and IVI when growing in sand dunes. *Lasiurus scinduis* was very dominant in inter-dune regions. *Haloxylon salicarnicum* was found to be very dense and cover in clayey land.

17.7.4.3 Lal Suhanra National Park

Lasiurus scindius was found to have very high relative density and IVI in the sand dune areas. *Cymbopogon jwarancusa* was very dominant in inter-dune areas. *Suaeda fruticosa* demonstrated very high relative density, frequency, and cover in the clayey land. Anthropogenic Factors are the physically analysis during the phytosociology study for Fall and Spring season 2021–2022 of the three sites Derawar Fort (Biodiversity park IUB) showed in Fig. 17.1 Mauj Garh Fort (Livestock farm) in Fig. 17.2 Lal Suhanra National Park (Sheikh Mohammed Bin Zayed Al-Nahyan Conservation and Breeding Center) in Fig. 17.3.

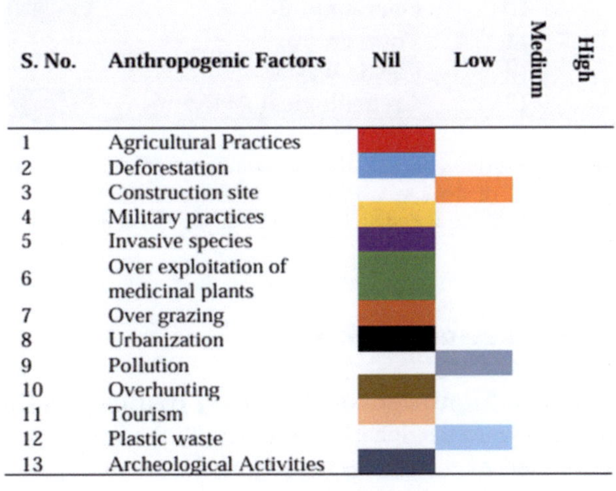

Fig. 17.1 Anthropogenic Factors during phytosociology study for Fall and Spring season 2021–2022 of the site Derawar Fort (Biodiversity Park IUB). With sub sites Sand dune, Inter dune and Clayey land (Khan et al. 2025)

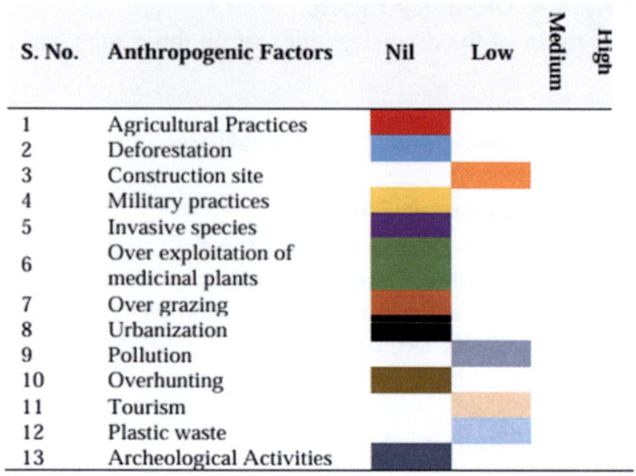

Fig. 17.2 Anthropogenic Factors in course of phytosociology study for the season 2021–2022 of the site Mauj Garh Fall and Spring season. Fort (Livestock farm). Sand dune, Inter with sub sites Sand dune, Inter. dune and Clayey land (Khan et al. 2025)

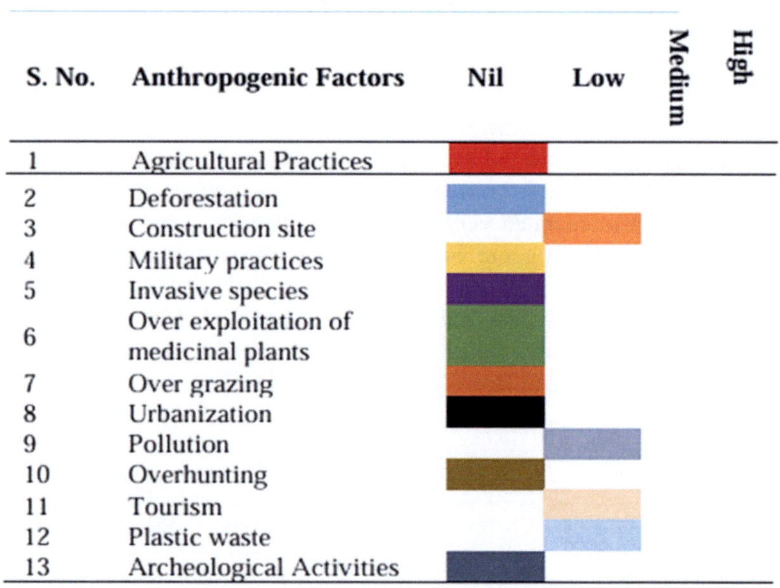

Fig. 17.3 Anthropogenic Factors during phytosociology study for Fall and Spring season 2021–2022 of the site Lal Suhanra National Park (Sheikh Mohammed Bin Zayed Al-Nahyan Conservation and Breeding Center). With sub sites Sand dune, Inter dune and Clayey land (Khan et al. 2025)

17.7.5 Lessons Learned

17.7.5.1 Significance of Fencing Regimes in Conservation

In the research, emphasis was made on the importance of the sites used, which are controlled by varied fencing regime in conserving numerous native desert species. This is in correlation with the hypothesis that fencing practices may be beneficial to conserve valuable plant species by reducing the number of human activities and impacts of predators.

17.7.5.2 Biodiversity Evaluation in Arid Situations

Phytosociological assessment techniques such as density, frequency, cover, and importance value index were found to be useful in measuring structures of plant communities in the Cholistan Desert. It is a holistic methodology of ecological life of drylands.

17.7.5.3 Dominant Species as Indicators

It is observed that some species such as *Ochthochloa compressa*, *Suaeda fruticosa*, and *Lasiurus scindicus* showed a consistent dominance in the various sites and parameters indicating their adaptability and ecological importance in the Cholistan desert. They may be used as an indicator species of ecosystem health or conservation objectives.

17.7.5.4 Site-Specific Management Requirements

The pattern of species dominance and species diversity across sand dune, inter dune, and clayey land sub-sites in each of the major locations underscores the importance of site-specific conservation and management plans. This micro-habitat difference may not be well addressed using a one-size-fits-all approach.

17.7.5.5 Handling of Anthropogenic and Biotic Factors

Implicitly in the study, it is indicated that the conservation process in these sites is crucial since the native species are otherwise experiencing various biotic, abiotic, and anthropogenic factors. This highlights the larger problem of resource harvesting and climate change affecting desert ecology.

To conclude, this study on the Cholistan Desert offers a strong illustration of how careful phytosociological observations can guide conservation practices in desert conditions. It emphasizes the importance of isolated spaces and fencing to conserve the species of native plants and outlines the necessity of customized management strategies, depending on the specific nature of the micro-habitat of the desert ecosystem (Khan et al. 2025).

17.8 Case Study: Comparative Ethnobotanical Study of Cholistan and Pothwar

17.8.1 Description

This study is comparative ethnobotanical research based on therapeutic flora and traditional medicinal use in two different parts of Punjab in Pakistan; the Cholistan desert (a dry region) and the Pothwar Plateau (a semi-arid region). The objective of the study was to record and compare the traditional knowledge of the use of medicinal plants that is rapidly declining because of the absence of qualitative and quantitative studies (Malik et al. 2015).

17.8.2 Areas of Study and Their Nature

17.8.2.1 Cholistan Desert

It is a barren land measuring about the size of 26,000 km^2 with a subtropical dry climate with scorching hot temperatures in the summer, monsoon rains, occasional long dry spells, and hot winds during summer. It is mostly pastoral-based and the locals, who are mostly pastoral nomads, have been living a nomadic life over centuries. The desert has been subdivided into Lesser Cholistan (northern, 7770 km^2) and Greater Cholistan (southern, 8130 km^2).

17.8.2.2 Pothwar Plateau

It is a semi-arid area covering 23,160 km^2 in the North-western part of the Punjab that consists of many topographical features like ridges, troughs, and basins. Pothwar has mostly rain-fed agriculture, and the population is mostly Muslim, with minorities of Christians, Hindus, and Sikhs.

17.8.3 Methodology

17.8.3.1 Data Collection

The research used qualitative research approach which included unstructured interviews, informal meetings, open-ended discussions, and group discussions with local people and traditional healers. Information was also cross checked with reported literature.

17.8.3.2 Participants

136 local individuals were interviewed (traditional healers: Hakeems, and domestic women aged 20–80 years). A verbal consent was obtained and no less than three traditional healers were consulted to ensure that the plant was used on certain diseases.

17.8.3.3 Plant Identification and Documentation

The identities of plant specimens were identified through field identification with the assistance of traditional healers. Data were documented about the parts of plants used, how they were collected, processed, and prepared into a drug, properties, how to take it, dosage and diseases treated.

17.8.4 Key Findings

17.8.4.1 Medicinal Plant Diversity

In Cholistan, the study identified 67 species of plants belonging to 29 families and which are used in the treatment of 123 diseases in humans. Eighty-six species of plants (38 families) were recorded in Pothwar.

17.8.4.2 Shared Species

The floristic similarity between the two study areas was less than 10.5%, suggesting marked differences in species composition, ecological conditions, and the traditional utilization of plant resources by local communities. Such remarkable species are *Calotropis procera subsp. hamiltonii*, *Citrullus colocynthis*, *Cyperus rotundus*, *Acacia nilotica*, *Boerhavia procumbens*, *Ziziphus nummularia*, *Solanum surattense*, *Withania somnifera*, and *Peganum harmala*.

17.8.4.3 Dominant Families

Poaceae, Zygophyllaceae, and Capparidaceae in Cholistan were the most diverse with medicinal species. Asteraceae ranked highest in Pothwar in number of medicinal plants.

17.8.4.4 Frequent Disorders Treated

In Cholistan, 14 plants were applied in gastrointestinal disease, 16 in skin diseases, 10 in respiratory illnesses, and 10 in musculoskeletal and joint illnesses. Additional applications were seen in male and female sexual disorders, urinary tract infection, diabetes, fever, liver diseases, jaundice, renal stones, and CNS disorders. In Pothwar, the number of species applied in jaundices and liver disorders reached 22, in antidiabetic use, 22, in skin problems, digestive disorders, etc.

17.8.4.5 Parts of the Plant Used

Whole plant (35 species), leaves (17 species), roots (17 species), and fruits (16 species) were used widely in Cholistan. Whether it is leaves (39 species) or whole plant (17 species), Pothwar was the most common to use.

17.8.4.6 Conventional Administration

Herbal preparations would be typically administered using water, cow/buffalo milk, goat milk, sheep milk, lassi, or arq (distillate) to facilitate absorption, reduce side effects, and also as nutritional support.

17.8.4.7 Dosage and Standardization

According to the study, there were no standardized criteria on herbal remedies, and the doses were varied according to age, physical status, illness condition, and severity. Different plant parts (%) utilized for medicinal purposes against different diseases explained in Fig. 17.4 and gender and age years of traditional healers explained in Table 17.1.

17.8.5 Lessons Learned and Recommendations

The research pinpoints some important lessons learned and offers some recommendations on the future of traditional medicine in Pakistan.

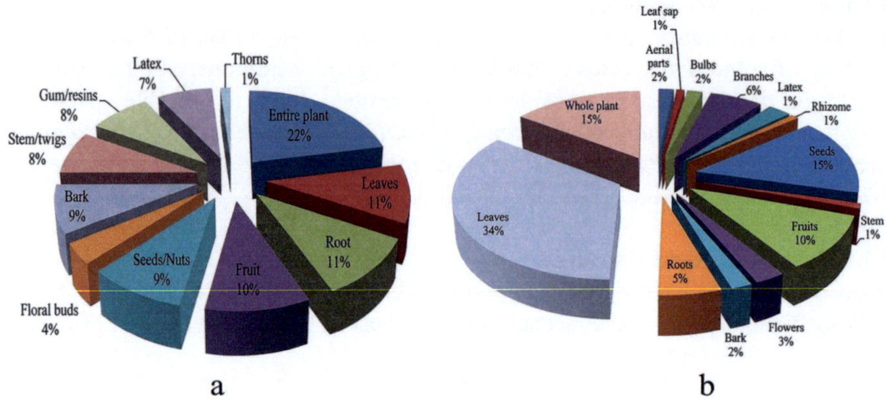

Fig. 17.4 Different plant parts (%) utilized for medicinal purposes against different diseases in (**a**) Cholistan Desert and (**b**) Pothwar Plateau (Malik et al. 2015)

Table 17.1 Gender and age years of traditional healers interviewed in the above study (Malik et al. 2015)

Gender	Age-Years (20–40)	Age-Years (40–60)	Age-Years (60–80)	Total
Female	7	16	5	28
Male	53	38	17	108

17.8.5.1 Importance of Traditional Knowledge

In both Cholistan and Pothwar, the local people have a long-standing traditional knowledge of medicinal plants, being a major provider of healthcare, particularly to individuals unable to access modern healthcare. This folk wisdom is customarily transmitted orally which makes it very important to document it.

17.8.5.2 Indigenous Knowledge Is Threatened by the Factors Listed Below

The risk of losing this wealth of traditional knowledge is very high in the near future because it has not been documented properly or preserved in written form. Some of the causes of this loss are urbanization, population increase, destruction of habitats, excessive exploitation of plant species, non-existence of the law and a lack of awareness. The research highlights that the ancient endemic remedies are waning because there is shortage in qualitative and quantitative research.

17.8.5.3 Significance of Documentation and Scientific Verification

The study highlights the urgency to record this unrecorded knowledge particularly in less-explored regions such as the Cholistan desert. Such records are special and critical to the preservation of medical plant species and to the identification of their pharmacological applications in human medicine. Research and developmental activities are advised to be directed in the scientific discovery of plant potential and enhancement of traditional herbal therapy by pharmaco-chemical research as a method of determining their efficacy.

17.8.5.4 Preference of Traditional Medicine

In Cholistan, especially the local population tend to favor traditional medicine because it is cheap and seems to have an effect, with less side effects than allopathic therapy. To a great number, it forms part of their social life and culture.

17.8.5.5 Conservation/Sustainable Use

The study underscores the fact that, the Cholistan desert, although very sparsely vegetated, is an important source of medicinal plants. But environmental degradation, long droughts, grazing pressure, and fuel cutting threaten this resource.

The value of ethnobotanical research in biodiversity conservation is increasingly being recognized. The harvesting of these plants should be carried out sustainably to ensure their conservation. Therefore, local communities should be sensitized to the importance of these plant resources and encouraged to adopt sustainable harvesting and extraction practices to enhance economic returns while conserving biodiversity.

Finally, this ethnobotanical work, covering the whole spectrum of traditional medicine practices, is an important step in the documentation of the rich, but at risk, traditional medicine of Pakistan, in the Cholistan and Pothwar areas. It not only offers important information on the medicinal applications of the local flora but it also emphasizes the emerging need to systematically record, scientifically substantiate and ecologically conserve this important heritage to leave in the hands of the future generations (Malik et al. 2015).

17.9 Case Study and Lessons Learned from Plant Conservation in Neelum Valley

17.9.1 Description

This case study was about the conservation of plant resources and medicinal significance of Sharda, Neelum Valley, Azad Jammu and Kashmir in Pakistan. It brings to the fore the dangers to these resources and also suggests ways to conserve them (Ahmad et al. 2012).

17.9.2 Study Area and Focus

The study site was Sharda located in the Neelum Valley of Azad Jammu and Kashmir (AJ and K), which has temperate climate and various landforms. The study concerned in recording the traditional knowledge of rural populations concerning medicinal plants, determining the state of their conservation and the influences on the same plants.

17.9.3 Methodology

The field surveys were conducted between May 2009 and August 2010 and included the collection of plant specimens and the acquisition of vegetation data through quadrat sampling. One hundred and thirty-six local people (136 males, 136 females, and 136 herbalists) were interviewed to collect data on the use of medicinal plants. The changes in plant availability, exploitation, habitat, conservation, and threats were observed. Conservation status of the plant species was determined through a comparison between the ecological amplitude and distribution of the plant species and the IUCN (2001, 2008) criteria.

17.9.4 Most Important Results Regarding Plant Status and Usage

A total of 39 vascular plant species, comprising 36 genera and 28 families, were identified. Conservation status showed 5 spp. as being critically endangered (12.82%), 6 species as endangered (17.95%), 10 species as vulnerable (25.65%), 11 species as secure or least concerned (28.21%), and 6 species as rarely distributed (15.38%). Leaves were the most commonly used parts of a plant (28.85%), then whole plants (15.38%), bark (13.46%), roots (11.54%), wood (9.62%), fruit (7.69%), flowers and seeds (5.77% each), and stem (3.85% of the total). The prevalent herbs in the study area were 46.15%, trees 38.46%, and shrubs 15.38%. Conservation status and medicinal use of some important plants of Sharda, Neelum valley, AJ and K explained in Table 17.2.

17.9.5 Lessons Learned and Recommendations

17.9.5.1 Plant Resources Threats

The analysis revealed that the Neelum Valley plant communities are greatly perturbed by various factors. These are fuel deforestation, excessive use of medicinal resources to treat the disease, population growth, increased tourism, and simple ignorance regarding conservation. Random logging contributes to the disappearance not only of forests but also of indigenous species, endangering them with extinction. *Cedrus deodora*, *Pinus wallichiana* and *Abies pindrow* all are forest resources that are facing immense pressure due to fuel wood gathering. Several economically significant medicinal plants, including *Saussurea lappa*, *Jurinea dolomea*, *Aconitum heterophyllum*, and *Berberis lyceum*, are widely exploited and are even nearing extinction because of high exploitation rates.

17.9.5.2 Significance of Local Communities

Local communities (especially poor villagers) gather medicinal herbs to use them either personally or for income sometimes as a secondary occupation. This shows that they rely on these resources and how they are exploited in the process.

Table 17.2 Conservation status and medicinal use of some important plants of Sharda, Neelum valley, AJ and K (Ahmad et al. 2012)

Botanical name	Local name	Part (s) used	Medicinal uses
List of endangered species			
Saussurea lappa (Falc.) Lipsch.	Kuth	Root	Acts as a tonic, stimulant, pain reliever, respiratory cleanser, and diuretic
Jurinea dolomiaea Boiss	Google Toof	Root	Used as a cordial and for treating colic
Berberis lycium Royle	Sumbal	Bark	Functions as an aperient, digestive aid, antipyretic, and eye tonic
Betula utilis D. Don	Borzai	Fruit	Possesses antiseptic, carminative, and antibacterial properties
Dioscorea bulbifera Decne	Contres	Stem, Leaf	Used as a diuretic and to relieve colic
Geranium wallichianum D. Don	Ratanjog	Root	Known for its astringent and coagulating properties
Morchella esculenta L.	Gucchi	Whole Plant	Used as a tonic and for treating joint pain and insomnia
Abies pindrow Royle	Rewar	Leaf, Seed	Possesses antiseptic, stimulant, and expectorant properties
Cedrus deodara Roxb. ex Lamb	Deodar	Bark, Wood, Leaf	Acts as a cardiac stimulant, laxative, antiseptic, and diuretic
Pinus wallichiana A.B. Jackson	Biyar	Fruit, Wood	Used as an anthelmintic
Bistorta amplexicaulis (D. Don) Greene	Masloon	Root, Leaf	Used as a nerve tonic and sedative
Aconitum heterophyllum Wall. ex Royle	Patrees	Root	Possesses analgesic, anti-inflammatory, and febrifuge properties
Bergenia ciliata (Haw.) Sternb.	Butpawah	Root	Used as a laxative, analgesic, astringent, and diuretic
Viola canescens Wall.	Banfsha	Leaf, Flower	Exhibits antibacterial, analgesic, expectorant, and febrifuge properties
List of vulnerable species			
Actaea spicata L.	Khapakh	Root, Leaf, Flower	Functions as a spasmolytic, cytostatic, emetic, and purgative
Plantago ovata Forssk.	Ispghol	Seed	Exhibits demulcent, laxative, astringent, refrigerant, and tonic effects
Eucalyptus camaldulensis Hook.f.	Lond	Leaf, Wood	Used as a sedative, anesthetic, antiseptic, and expectorant
Acacia nilotica Willd.	Kikar	Leaf, Bark	Used as an astringent, stimulant, and tonic
Ajuga bracteosa Wall. ex Benth.	Jane Adam	Whole Plant	Exhibits aromatic, astringent, and tonic effects
Plectranthus rugosus Wall. ex Benth.	Peemar	Whole Plant	Used as a bitter, cooling, and diuretic herb

(continued)

Table 17.2 (continued)

Botanical name	Local name	Part (s) used	Medicinal uses
Quercus incana Roxb.	Reen	Bark	Acts as an astringent and relieves colic
Quercus baloot (Desf.) A. DC.	Chor	Bark, Wood	Used as an astringent
Dryopteris ramosa L.	Longer	Stem, Leaf	Exhibits diuretic, aid digestion, antibiotic, astringent, and antipyretic effects
Dryopteris stewartii L.	Daid	Whole Plant	Acts as a diuretic and antimicrobial
Viburnum cotinifolium D. Don	Guch	Leaf, Fruit	Functions as a sedative, astringent, laxative, and blood purifier
Impatiens glandulifera Royle	Bun Til	Aerial Parts	Used as an emetic, cathartic, and diuretic
List of rare species			
Taraxacum officinale L.	Hand	Root, Leaf	Serves as an alterative, diuretic, and antiseptic agent
Achillea millefolium L.	Khati Booti	Leaf	Known for its demulcent, tonic, and stimulating properties
Cynoglossum lanceolatum Forssk.	Landi	Whole Plant	Used as an analgesic, antispasmodic, astringent, emollient, and narcotic
Diospyros lotus L.	Amlok	Fruit	Used as a febrifuge, astringent, and sedative
Ailanthus excelsa Roxb. ex Willd.	Darawa	Bark, Root, Leaf	Acts as an astringent
Morus alba L.	Safaid Toot	Root, Leaf, Fruit	Exhibits antibacterial, astringent, and ophthalmic properties
Olea ferruginea Royle	Rons Patar	Whole Plant	Serves as a laxative, expectorant, and tonic
List of secure species			
Salix tetrasperma Roxb.	Bans	Leaf, Bark, Wood	Acts as an anodyne and febrifuge
Galium boreale L.	Guch	Whole Plant	Used as a diuretic and emetic
Juglans regia L.	Akhrot	Leaf, Bark, Fruit	Used as an astringent, tonic, and anthelmintic
Cannabis sativa L.	Bhang	Leaf, Flower	Acts as a carminative, astringent, sedative, and narcotic
Indigofera heterantha Wall. ex Brandis	Zand	Whole Plant	Known for its anthelmintic properties

17.9.6 Conservation Strategies

To prevent the extinction of these plant species the study recommends implementing specific and direct conservation strategies.

17.9.6.1 Documentation and Preservation of Indigenous Knowledge

It is significant to document and preserve traditional knowledge properly.

17.9.6.2 Community Training

Community should be provided with adequate education on conservation and sustainable use of flora with medicinal value.

17.9.6.3 Minimization of Deterioration of Use

Trees should be decreased in terms of overgrazing and cutting.

17.9.6.4 Monitoring and Infrastructure

Semi permanent monitoring programs shall be prepared and alternative source of fuel like natural gas shall be adopted.

17.9.6.5 Forest Management and Healthcare

Forest management with good biodiversity influence and provision of good health facilities to the local communities should be supported.

17.9.6.6 Awareness Programs

Awareness campaigns; this should be initiated at the grass root level, which will promote conservation activities.

Overall, the Sharda, Neelum Valley case study suggests that there is an urgent problem when it comes to the excess utilization and exploitation of medicinal plants resources and habitat of a given plant. The most important lesson that has been imbued is that concerted conservation practice is not just a good thing, but is a necessity and must incorporate the local population as a stakeholder, must encourage sustainable exploitation of the resources, must be sensitive of the socio-economic factors underpinning the same so that further degradation of biodiversity is avoided (Ahmad et al. 2012).

17.10 Case Study and Lesson Learned from District Kohat, Pakistan

17.10.1 Description

This case study contains the ethnobotanical study of the semi-arid subtropical area of Pakistan with reference to the District Kohat. The objective of the study was to record the traditional use of plants, the conservation status, and the importance of medicinal flora. This ethnobotanical survey was conducted in District Kohat in the eastern belt of Khyber Pakhtunkhwa, Pakistan as the main case study area. It is an arid area with a rich ecosystem composed of different plant species that sustain a rich pool of traditional knowledge and biodiversity. The region lies in the intersection of different ecological zones and a wide range of plants are nourished by distinctive soil compositions and microclimates (Zamin et al. 2024).

17.10.2 Climatic and Geographical Features

Kohat District covers an area of approximately 297,300 hectares and is bounded by Jamrud (Peshawar) to the north, Bannu to the south, the Indus River to the east, and Hangu to the west. The district lies at an average elevation of about 558 m above sea level. The soils range from clay to sandy loam, are generally alkaline in nature, and consist predominantly of sand and silt particles. The climate is characterized by hot summers and mild winters, with temperatures ranging from 40–50 °C during June and 6–18 °C during January. Rainfall occurs throughout the year, with the highest precipitation typically recorded during the months of August and March.

17.10.3 Ethnobotanical Significance

Semi-arid subtropical areas of Pakistan such as Kohat can provide variety of exceptional ethnobotanical research opportunities. Conventional knowledge here involves many facets of livelihood, including medicines, food, building equipment, rituals, and cosmetics among others. The research was to support the conventional applications of native plants and evaluate their level of conservation in District Kohat.

17.10.4 Research Methodology

This was done through an ethnobotanical approach which included semi-structured questionnaires and corner meetings to collect primary information about plant species. Quantitative methods included Pearson correlation tests and ethnobotanical indices such as informant consensus factor (Fic), fidelity level (FL), preference ranking (PR), and direct matrix ranking (DMR) to determine the importance and conservation status of recorded medicinal flora. Between March 2017 and October 2022, fieldwork and household surveys were carried out in five particular precincts, i.e., Jarma, Seni Gumbat, Lachi, Dara Adam Khel, and Usterzai were surveyed. Route of administration of the reported MPs showed in Fig. 17.5 and novel uses of the reported plant species in the semi-arid region showed in Fig. 17.6.

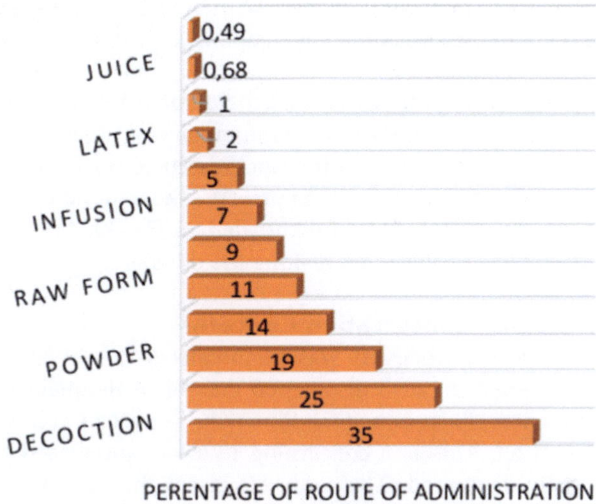

Fig. 17.5 Route of administration of the reported MPs (Zamin et al. 2024)

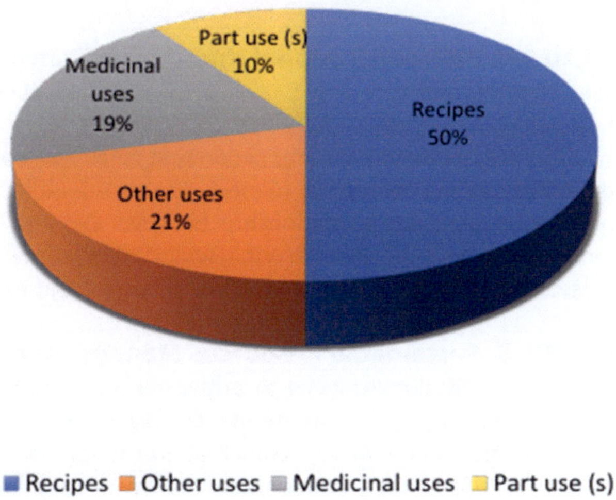

Fig. 17.6 Novel uses of the reported plant species in the semi-arid region (Zamin et al. 2024)

17.10.5 Lessons Learned

The study produced a number of valuable findings and lessons about the uses and conservation of plants in the area:

17.10.5.1 Richness of Traditional Knowledge

The analysis revealed 103 plant species of which 36 were new recipes, 15 other uses, 14 medicinal uses, and 7 specific part uses. Herbs formed the most prevalent type of vegetation and the most common parts used were leaves and young twigs. This points out the abundance of traditional know-how held by local populations in reference to indigenous plants and their various uses. The results are in support of the successful use of natural vegetation as folk remedies by local people.

17.10.5.2 Significance of Ethnobotanical Research

The case study highlights the importance of novel use of particular plant species in customary practices. It points out the criticality of documentation of the plant species and traditional knowledge to preserve the cultural tradition and medical flora, so as to prevent the loss of useful knowledge. This kind of research is useful in preserving the cultural heritage and making the development sustainable.

17.10.5.3 Conservation Concerns

An important discovery was the threatened conservation of numerous plant species. The researchers also discovered that 44 of the plant species were on the brink of extinction, 30 species at medium risk, 17 species at lower risk, and 12 species not at risk at all. This is a concerning figure of threatened plant species that should be rapidly conserved. The study expresses the concern of the possible extinction of medicinal species of plants through unsustainability, including unscientific use and overexploitation.

17.10.5.4 Intergenerational Knowledge Transfer Gap

The study found that there is a worrying disconnect in the passing down of traditional knowledge among the older and younger generations. This is in line with the results that the young generation cares less about traditional medicine, yet there has been a rebirth of interest due to scientific inventions. The researchers observed a significant relationship between age and traditional knowledge, with individual who were older having wider knowledge. This shows the necessity to protect the culture in order to save the diversity and identity.

17.10.5.5 Sustainable Resources Management

The study indicates the need to utilize resources in a sustainable manner and the urgency to decrease the environmental degradation. It points out that a number of factors related to the large number of consumptions of some species render them helpful in other ways rather than medicinal including fire wood, cash income, fencing, building construction, and fodder. This utility is an advantage but it adds to the stress on such plant populations.

In short, the case study in District Kohat demonstrated a rich and threatened ethnobotanical heritage. The presented lessons learned are related to the paramount importance of ethnobotanical research in recording traditional knowledge, the urgency of conservation efforts, the generation gap in knowledge, and sustainable practices to preserve precious plant resources and related cultural practices (Zamin et al. 2024).

17.11 Case Study: Phytodiversity and Conservation Status of Nara Desert, Pakistan

17.11.1 Geographical and Climatic Context

17.11.1.1 Location and Topography

Nara Desert is located in Sindh Province, Pakistan between 26° to 28° North and 68° to 70° East. It is a divers landscape with sandy hills, sharp inclined and extensive flat areas at low altitude. It is a plain of undulating sandy soil with numerous sand dunes, and as we approach the East, the number and size of these dunes increase. The altitude of the desert is 50–115 m above sea level (Qureshi 2020).

17.11.1.2 Climate

This is a hot sandy desert with severe temperatures of between 20 °C and 45 °C. It is hyper-arid to arid, has long dry seasons, heavy wind speeds, low moisture levels, high evapotranspiration, and very low rainfall, which normally averages 88 to 135 mm per year, mostly in the monsoon months. Water resources are very limited and the groundwater is located 50 to 300 feet beneath the surface.

17.11.2 Biodiversity and Phytogeography

17.11.2.1 Flora Composition

It took more than 10 years of exploration (September 1998 to December 2008) to list 267 species of plants in the Nara Desert. This contains 3 fern, 1 gymnosperm and 64 monocots, spread over 167 genera, and 59 families.

17.11.2.2 Dominant Life Forms

Floristic composition shows a predominance of herbs (42.70%), grasses (17.60%), shrubs (13.11%), subshrubs (10.86%), sedges (5.24%), and trees (4.87%).

17.11.2.3 Phytogeographical Genesis

A vast preponderance of the flora is of Saharo-Sindian descent, the smaller amount in the westward region being of Indian character. The vegetation is xerophytic, and accustomed to dry climate throughout the whole territory.

17.11.3 Human Dependence and Threats to the Ecosystem

17.11.3.1 Resource Provision

Nara Desert is a vulnerable ecosystem that sustains the existence of people and livestock by supplying them with basic resources in the form of fuel, food, fodder, shelter, as well as ethno-medicine. The people in the area also depend on indigenous plants, as a source of various day-to-day items such as traditional medicine, forage, construction, and craft items.

17.11.3.2 Climatic Threats

The region is under severe environmental stress with increasing cases of drought, desertification, deforestation, and soil wash that inflict huge burden on the ecology of the area. Some of the causes of desertification include low amounts of rainfall, low moisture levels and high temperatures.

17.11.3.3 Anthropogenic Threats

Humans are the greatest cause of degradation:

17.11.3.3.1 Overgrazing

Transient cattle overgraze pastures, particularly those in inter-dunal valleys to the extent of destroying desirable forage species, in some cases entirely.

17.11.3.3.2 Deforestation

Overcutting and lopping of trees especially *Prosopis cineraria* to make coal is fast reducing their population.

17.11.3.3.3 Gas Exploratory Activities

Gas companies construction of roads, well drilling and camping on which clear centuries old climax vegetation and place extreme stress on natural habitat.

17.11.3.3.4 Hunting Activities

Hundreds of plant species are broken and destroyed by the construction of access roads during the hunting of the game particularly the Houbara Bustard.

17.11.3.3.5 Excessive Exploitation of Species

During this study, the researchers noted that the following field trips experienced an extinction of annual species such as *Mollugo nudicaulis*, *Tephrosia falciformis*, *Neurada procumbense,* and *Schweinfurthia papilionacea* due to dry spells and overexploitation. Extensive use of the land is continually posing a threat to many important plant species.

17.11.3.4 Conservation Status

Of the 267 species, 102 (38.20) were common, 85 (31.84) were rare and 49 (18.35) were very common. Most importantly, 17 species (6.37%) were determined to be threatened and 14 species vulnerable. Summary of conservation ststus of the taxa showed in Fig. 17.7.

17.11.3.5 Lessons Learned and Recommendations

17.11.3.5.1 Sense of Urgency to Rehabilitate

The high rate of degradation of valuable plant species requires immediate rehabilitation to curb overgrazing, deforestation and habitat destruction.

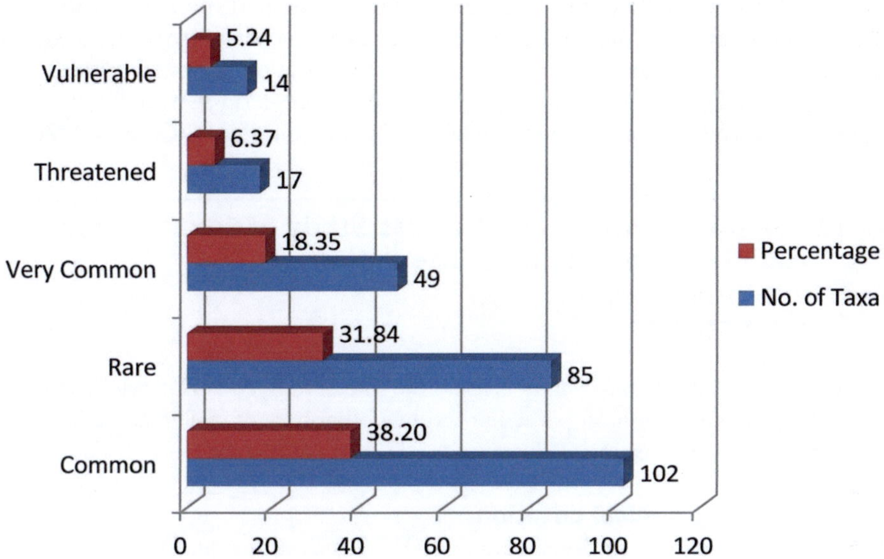

Fig. 17.7 Summary of conservation status of the taxa (Qureshi 2020)

17.11.3.5.2 Reforest Native Tree Species

Native trees are an important part of the ecosystem. The plant is named *Prosopis cineraria* as the lifeline of the inhabitants as it has various applications and it needs special attention in planting. Another useful and fast-growing species is *Acacia senegal*.

17.11.3.5.3 Sand Dune Stabilization

Native trees such as *Acacia senegal*, *Albizzia lebbeck*, *Cordia rothii*, *Dalbergia sissoo*, *Prosopis cineraria*, *Salvadora oleoides*, *Zizyphus jujuba,* and *Z. mauritiana* are naturally adapted to arid conditions and are drought tolerant and can thus be used to stabilize sand dunes. *Calligonum polygonoides* and *Zizyphus nummularia*, as well as *Lasiurus sindicus* and *Panicum turgidum*, are also important shrubs to prevent soil erosion.

17.11.3.5.4 Germplasm and Seed Bank Protection

Since livestock forms a large base of the local population, and since the desert flora provides livestock feed, inter-dunal valleys require protection. Such regions could be important germplasm and seed banks to conserve genetic diversity of range and forage grasses, legumes, and other forbs.

17.11.3.5.5 Sustainable Utilization Practices

Sustainable utilization of the plant resources is necessary to reduce or eliminate the loss of annual species by dry spells and overexploitation.

In summary, the Nara Desert case study shows clearly how an ecologically problematic area can seem when a combination of climate changes and unproductive human activities is introduced. The overall findings support the urgency to adopt combined conservation measures, with priority given to rehabilitation of degraded areas with reforestation of native species, stabilization of sand dunes, and safeguarding of genetic diversity as a guarantee of long-term sustainability of this vulnerable ecosystem and the livelihoods to which it contributes (Qureshi 2020).

17.12 Lesson Learned from the Case Study of Soona Valley on Plant Biodiversity Conservation

17.12.1 Description

A case study described in the study is about the role of indigenous knowledge toward biodiversity conservation, in particular the tree ethnobotany in Soona Valley, District Bhimber Azad Kashmir, Pakistan.

17.12.1.1 Geographical Location

The research was done in Soona Valley (SV), Tehsil Samahni, District Bhimber, Azad Kashmir, Pakistan. This region lies in the Line of Control (LOC) and acts as borderland between India and Pakistan in Kashmir territorial region (Ishtiaq Ch et al. 2013).

17.12.1.2 Valuable Asset to Conservation

It brings out the issue that the indigenous knowledge is a valuable resource to the knowledge of conservation of plant biodiversity particularly, woody plants. The use of woody plants to serve many purposes in the community, such as wooden materials, fodder, construction, and ethnomedicines, can be found in the knowledge of the local community. This traditional knowledge will act as a basis of new drug discovery and help in targeting the conservation efforts of the endangered species.

17.12.1.3 Extensive Ethnobotanical Profile

The study was able to create an extensive ethnobotanical profile of the indigenous information on 58 woody species in 24 families of which 30 of the species are identified as key significant species. This proves how highly local knowledge of plant utility is, not just in terms of the most frequently used parts (fruit, root, stem, flower, bark) but also in the way to prepare ethnomedicines (decoctions, extractions).

17.12.2 Threats to Plant Biodiversity

17.12.2.1 Anthropogenic Pressures

The study identifies some main threats that encounter woody plant species of which the major factor regards human activities. Among them is house building, fuel/

fodder consumption, urbanization, agricultural expansion, and the human activity of timber mafias. These threats are aggravated by rapid growth of population.

17.12.2.2 Conflict-Induced Damage

Exceptional threat cited is that biodiversity of the Soona valley suffers bad damage due to shelling across the border and firing between India and Pakistan due to its location on the Line of Control (LOC).

17.12.2.3 Specific Trees

Specific tree species such as *Terminalia belerica*, *Butea monosperma*, *Terminalia arjuna*, *Phyllanthus embllica*, *Tribulus terristris*, and *Cordia obliqa* to name a few are found to be endangered due to these forces.

17.12.3 Strategies for Effective Conservation

The study gives a firm recommendation to incorporate communities in conservation and forestation programs thus employing methods such as farm forestry. Educational programs in schools, colleges, and in the streets as well as creating of publicity materials are also very helpful.

17.12.3.1 Policy and Monitoring

It is necessary to apply a rigorous monitoring methodology to forests and to penalize abuses. The study also advocates the control of timber mafia by engaging government officials and the masses in timber mafia committees.

17.12.3.2 Alternate Resources

As an alternative practice, woody plants should be less exploited by generating alternative sources of fuel wood such as Kerosene oil and LPG at subsidized prices and appropriate stoves and furnaces may also be used that minimizes the trucks of woody plants.

17.12.3.3 Strategic Plantings

Plantations with fast-growing species of plants such as *Albizzia lebbeck* and *Eucalyptus globulus* can be established in wastelands and barren lands so as to relieve the pressure on the native plants and to also improve the fuel woods needs of the population. Frequency of most endangered tree species in Soona Valley explained in Table 17.3.

This study highlights that effective conservation of plant biodiversity in regions such as Soona Valley is a multi-dimensional approach that combines traditional expertise, human-driven and conflict-related pressures, and the development of strong conservation policies that simultaneously involve the community in conservation actions and state protection of such activities.

Table 17.3 Frequency of most Endangered endangered Tree tree Species species in Soona soona Valley valley (Ishtiaq Ch et al. 2013)

Botanical names (scientific names)	Frequency of most endangered tree species (%)
Terminalia belarica	25.50
Butea monosperma	20.50
Terminalia arjuna	15.00
Phyllanthus embllica	10.40
Tribulus terristris	8.25
Cordia obliqa	6.60
Ficus bengalensis	5.75
Launea coromandaliana	4.40
Phoenix dactylifera	3.60

17.13 Case Study for Biodiversity Conservation in Tropical Dry Forests of Sialkot

17.13.1 Location of TDF

The case study in the study is a study conducted in the plant biodiversity evaluated in the Sialkot in the tropical dry forests (also known as the TDF) of Pakistan. This is an area of focus because of the rich flora and fauna that offer different ecological services to the human societies in the area. The study highlights the fact that assessment of the biodiversity is important in order to establish the sustainable forest management policies. The main aim of the study was examining faith in vegetation and the impacts of environmental and anthropogenic factors on forest biodiversity in five endangered forest areas in District Sialkot, Ghalotian, Kishan Garh, Daburgi Chanda Singh, Pir Kot, and Ghulab Garh. The research was also work toward:

17.13.1.1 Evaluation of Species Diversity and Richness

Quantitatively and qualitatively utilize ecological approaches to determine diversity and richness of flora in the wild.

17.13.1.2 Find Priority Areas and Vulnerable Species

Identify priority areas to conserve plants in the prospect of future warming, vulnerable species, and propose urgent approaches to urgently conserve them. Summary of the floristic diversity of the Sialkot Range Forest, District Sialkot, Punjab, Pakistan, explained in Table 17.4.

17.13.2 Importance of the Study Area

The tropical dry forests of the studied research areas in Sialkot are considered extremely important for their ecosystems and the livelihoods of local communities.

Table 17.4 Summary of the floristic diversity of the Sialkot Range Forest, District Sialkot, Punjab, Pakistan (Shahzad et al. 2024)

Plants groups	Genera	Families	Species
Angiosperms	130	55	162
Bryophytes	1	1	1
Pteridophytes	7	6	7
Total	138	62	170

Table 17.5 Phytosociological analysis of 5 selected sites of Sialkot Range Forest, District Sialkot, Punjab, Pakistan (Shahzad et al. 2024)

Site names	Location (Lat, Long)	Perennial	Annual	Trees	Shrubs	Herbs	Total species
Ghalotian Forest	32°27′88″ N, 74°34′74″ E	11	20	2	4	25	31
Kishan Garh Forest	32°49′68″ N, 74°24′90″ E	15	28	8	4	31	43
Daburgi Chanda Singh Forest	32°49′81″ N, 74°22′71″ E	10	16	5	5	16	26
Pir Kot Forest	32°50′21″ N, 74°23′76″ E	9	12	6	3	12	21
Gulab Garh Forest	32°48′56″ N, 74°23′10″ E	21	28	11	8	30	49
Total		66	104	32	24	114	170

Despite their significance, the flora of Punjab, particularly District Sialkot, remains largely under-researched, highlighting a need for greater access and resources for study. The research reinforces the importance of vegetational assessment and understanding the significant environmental drivers influencing plant species diversity in TDF areas (Shahzad et al. 2024). Phytosociological analysis of 5 selected sites of Sialkot Range Forest, District Sialkot, Punjab, Pakistan explained in Table 17.5.

17.13.3 Priorities and Strategies for Plant Biodiversity Conservation.

This study highlights the imperativeness of conservation of TDFs in the region of Sialkot. Proposed key conservation strategies are:

17.13.3.1 Deterring Illegal Utilization of Natural Resources

This is vital in ensuring that forests are not exploited unlawfully to the extent of eliminating forest biodiversity.

17.13.3.2 Restorations and Conservation

Strategies will be based on restoring and conserving plants in the special areas of planting and protecting rare species at the local level.

17.13.3.3 Dispelling Habitat Fragmentation

To improve the probability of species dispersal and variability to changing circumstances, it is desirable that efforts should be directed toward reducing habitat fragmentation and deforestation.

17.13.3.4 Collaboration and Awareness

Forest and irrigation departments, community organization and non-governmental organization have a crucial role in protecting wild flora and also in enforcing effective management methodologies. It is also necessary to raise the awareness of local people in relation to the need of preserving nature

17.13.3.5 The Policy and Research

The results add value to the policies in the country in that importance should be attached to these ecosystems and provide more research needed to raise awareness on the threats facing the diversity of plants.

Finally, a case study in Sialkot, Pakistan, contributes many details about the importance of a complicated set of environmental and human interactions on the biodiversity of TDF. It underlines the need of extensive evaluation of biodiversity and emphasizes the particular conservation measures that can be used to avert adverse effects of climate change and anthropogenic pressures.

17.14 Case Study for Plant Biodiversity Conservation Strategies in Montane Ecosystem

The success in the conservation of both plant biodiversity and that of montane ecosystems such as the Himalaya depends not only on applying contemporary science, but also on synergistic combination of traditional knowledge and policies. The significant measures include an integration of the combination of ecological researches and ethnobotanical research, as well as the implementation of certain mechanisms of conservation.

17.14.1 Integrated Ecological and Ethnobotanical Approaches

17.14.1.1 Merging Data Sets

Conservation of montane ecosystems requires not only trans-disciplinary studies, but also the merging of ecology and social information. It is a combination of the quantitative ecological (phytosociological) methods with ethnobotanical practices with the view of assessing and characterizing the ecosystem services supplied and maintained by plant communities. This combined analysis is useful in determining anthropogenic stress on both community and species diversity level of plants.

17.14.1.2 Evaluation of Anthropogenic Activities

The state of ecosystems strongly relates with the plant biodiversity levels, the process of classification of vegetation covers a critical number one step toward managing and preserving an ecosystem. Phytosociological and ethnobotanical data can be combined to reveal species on the most critical anthropogenic pressure (importance value indices, IVIs, and use values, UVs), respectively. This can be used to define over utilized species that are at great risk (*Juglans regia* and *Paeonia emodi*).

17.14.1.3 Finding Indicator Species

Ecological indicator species can be found based off large-scale vegetation description and statistical analysis, whereas traditional indicator species are found by ethnobotanical surveys. A species with high values of fidelity that have limited areas or fragmented habitats are said to be of high conservation importance. Six of the more important plant species used for traditional medicine in the western Himalayas showed in Fig. 17.8.

17.14.2 Direct Conservation Measures

17.14.2.1 In Situ and Ex Situ Conservation Strategies

This is a conservation strategy, which involves maintaining the species within its natural environment (habitat). Ex situ, or out-of-situ conservation: It is a

Fig. 17.8 Important medicinal plants of the western Himalayas: (**a**) *Podophyllum hexandrum*, (**b**) *Rheum austral*, (**c**) *Primula denticulate*, (**d**) *Polygonatum verticillatum*, (**e**) *Fritillaria roylei*, (**f**) *Paeonia emodi* (Khan et al. 2013)

conservation strategy in which the species are not maintained in their natural habitat but rather in botanical gardens. Much is to be done to restore plant biodiversity and also to counter ecosystem degradation through causing plantation, help create areas that are to be preserved and parks as such.

17.14.2.2 Prioritization and Domestication of Rare Species

To prioritize the species requiring conservation and the rare species that are to be potentially subjected to domestication is an important process. It is stressed that IUCN Red List criteria should be used in order to identify and prioritize species in need of conservation, mostly endemics and rare species.

17.14.2.3 Ensuring Conservation

The threats to the montane ecosystems include the growing pressure of grazing, harvesting to use as food or other purposes, and effects of climate change that are causing extinction to the species (Shah 2008). There is an urgent need to implement sustainable strategies regarding the utilization of resources especially in less economically developed nations where indigenous people commonly depend on vegetative wealth.

17.14.3 Policy and Knowledge Integration

17.14.3.1 Policy Implementation

Policy instruments are required including policy instruments on the new government and corporate policies that are required to have systematic incorporation of the ecosystem services within the economic decision-making. These include incorporation to natural resource management decisions of scientific information about biodiversity and threats to plants.

17.14.3.2 Traditional Ecological Knowledge

Traditional ecological knowledge, particularly on the older generation, can help in exploring the use of plants and this may guide conservation methods. This knowledge, however, is on risk of being lost through urbanization and industrialization. Participatory exchange of knowledge between vegetation scientist, locals and conservation managers is required.

17.14.3.3 Red Data Book

The lack of a complete Red Data Book of plant species endangered in the Himalayas raises a noteworthy concern of the absence of detailed documentation, a crucial ingredient of the successful conservation (Khan et al. 2013). The map of the Naran Valley, western Himalaya, depicting various vegetation zones and conservation status of endemic species are provided in Fig. 17.9.

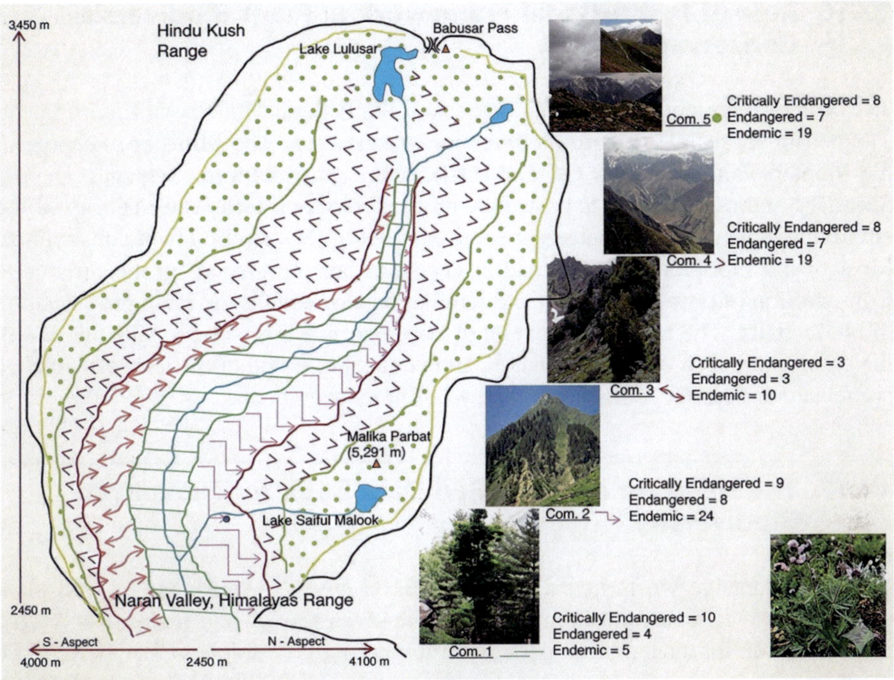

Fig. 17.9 Map of the Naran Valley, western Himalaya with vegetation zones and conservation status of endemic species (Khan et al. 2011a, b, 2012a, b; Khan 2012)

17.15 Community Involvement in the Conservation Strategies for Plant Biodiversity in Pakistan

17.15.1 Key Challenges for the Biodiversity Conservation in Pakistan

The issues that are considered most serious to the conservation of biodiversity in Pakistan are as follows (Anwar and Shank 2000):

- The necessity of concomitant policy reforms and institutional reforms and institution strengthening.
- Mainstreaming of biodiversity conservation programs in the sectoral programs.
- Enhanced knowledge of the entire biodiversity and efficient tools such as securing their sustainable use.
- Creating community-wide systems of biodiversity control.
- Designing and making permanent procedures to track important components of biodiversity; and improved application of currently existing plans.

17.16 Role of Institutional Framework in Plant Biodiversity Conservation

The center of attraction of national interests is the Federal Ministry of Environment. The consumer scale equals to biodiversity conservation. The Minister in charge of the Ministry is the Director General (DG). A Secretary with the help of a Deputy Secretary and a Section Officer deals with biodiversity issues. The majority of the environmental conscious matters are concentrated in DG. environment conventions. Most of the biodiversity conservation decisions are dominated by the provinces. Conservation of provincial forest and wildlife departments is the main proclamation of biodiversity. The Inspector General of Forests office is the one to take the lead in the co-ordination of policies, research, and education. Some of the issues of interest are related to forestry, rangelands, and wildlife management.

17.17 The Need for an IUCN Red Data Book for Himalayan Biodiversity

No comprehensive work, i.e., a Red Data Book covering those endangered plant species of the Himalayan region, is available, even though the region has unique endemic flora. Published work on a small representative sample of IUCN Red List plant species is severely restricted and disjointed (Ali 2008; Alam and Ali 2009, 2010; Ali and Qaiser 2011) therefore it is not possible to find many comparators to assess endangered and critical plant species on their national level (Ali 2008). New initiatives are, however, developing and partial descriptions concerning a very low number of species can be observed in some recent publications (Alam and Ali 2010; Ali and Qaiser 2011). It is perhaps not inappropriate to compare the available case studies of the Himalayan Range by other researchers who present studies of areas of the Himalaya that are closer to those reported by Khan (2012) in terms of the possible applicability of relating endemism with ecosystem services (Samant and Dhar 1997; Aumeeruddy 2003; Kala 2007; Qureshi et al. 2007) and the case study of Khan (2012) could be the pioneering force to establish.

17.18 Role of NGO'S

WWF and IUCN are the most prominent international NGOs that are engaged in the sphere of biodiversity conservation. There are also other local NGOs involved in different areas of environmental protection and conservations in the form of biodiversity. The majority operates very closely with local community-based organizations in the conservation of biodiversity. NGOs especially prevail in mountainous region of northern Pakistan, where the Aga Khan prevails. Community mobilization and community organization were introduced by Rural Support Program (AKRSP) (Anwar and Shank 2000).

17.19 Development and Implementation of Pakistan's Biodiversity Action Plan

In 1992, Pakistan became one of the first nations to sign the CBD and the Pakistani Parliament ratified. He joined the government as a cabinet in 1994. Pakistan launched the mechanism of fulfilling its obligation under CBD. This led to development of the Biodiversity Action Plan in 1993. There were three processes that were considered as country study by AWS and these processes were known as BAP (CBD 1992). It gives a brief summary of the conditions and patterns of biodiversity, strategic goals and objectives. It specifies the objectives, and describes a plan of action.

Ministry of Environment, which is also the national focal point of CBD has overall mandate to practice BAP. The BAP suggests the establishment of a National Biodiversity Steering Committee which is to be headed by the Minister of the Environment and comprises of the members of the relevant federal ministries and departments, provinces, and NGOs to coordinate the implementation process. The BAP also allows Provincial Steering Committees since most of the activities will be carried out by provinces. These committees headed by the ministry of concerned provinces of forest, fisheries, and wildlife would comprise of the membership of the respective departments, non-governmental organizations, and community organizations. These fora will guarantee co-ordination at every tier in order to conserve biodiversity in Pakistan (GOP 2000).

17.20 Action Plan for Biodiversity Conservation

The Government of Pakistan devised and adopted National Conservation Strategy (NCS) in 1992, which is recognized by the World Bank as a National Environmental Action Plan. The NCS has14 programs and lots of them are about diversity. But the operations did not have biologically particular measures for loss or preservation in general terms. Eighth Five Year Plan (1993–1998) reflects the requirement for adopting provincial conservation plans to effectuate NCS recommendations. Provincial conservation plans have been developed in the Sarhad and Balochistan with preparation of other two northern Areas is in progress. There'll be a minimum of two district-based conversances established, they said. Local attention to biodiversity is also far more explicit in such strategies (GOP 2000).

The Sarhad Provincial Conservation Strategy (SPCS) is greater emphases on biological diversity by including a dedicated chapter on biodiversity, parks and protected areas has been included. Nonetheless, it fails to meet the need of the CBD thoroughly. A comparable hierarchically approach to conservation planning is also observed at the local level with district conservation strategies formulated in Chitral and Abbottabad. These strategies were formulated after the development of the Biodiversity Action Plan (BAP) and in this respect, they will specifically cover the recommendations of the BAP within the local context (GOP 2000).

The environment ministry is a national focal point of the CBD, and it is responsible for implementing process, i.e., BAP. The establishment of a Biodiversity steering Committee at the federal level and chaired by the Minister of environment should be proposed. The BAP calls for the formation of Provincial steering Committees since most implementation activities will occur at the provincial level representatives of relevant ministries and departments of the federal government, provinces and NGOs. These committees will be chaired by the respective provincial ministers of forests, fisheries, and wildlife and will comprise representations of the relevant departments, the NGOs and community organizations. It will offer an effective co-ordination among all the stakeholders aimed at the conservation of the biodiversity across the country.

17.21 Customized Approaches for the Biodiversity Conservation in Pakistan

Pakistan has completed quite a lot of work already to create biodiversity specific documents. Similar to the NCS and the BAP, it also means the replacement of calorie counting. The implementation stage, however, has not been so successful or seen very minimal success on its part. The results of these documents should be well propagated at different levels. These documents should be presented in a suitable way, e.g., executive summaries and briefings to top level government officials, policy makers and NGOs. Ultimately, biodiversity should be presented as an economic priority to all, rather than merely as an incentive package.

Upon successfully identifying and addressing biodiversity issues in the right manner, all the departments in the public aspect need to be involved in managing it. To preserve biodiversity in a certain or both of the ways, the enlistment of biodiversity must take place. Biodiversity concerns/issues pertinent to each department can. They will be separated and the solutions to individuals proposed. Lastly, the departments must come up with its own annual plan of action incorporating its issues of biodiversity This might turn into an activity that is scheduled on a yearly basis. High political figures of the country should be made aware of problems surrounding biodiversity. Diversity is also associated with food security, social security, and the stability of the nation in general.

17.22 Conclusion

Plant biodiversity in Pakistan represents an incomparable natural heritage, which promotes ecological stability, economic growth, and cultural identity. The rich flora, a variety of ecosystems, and cultural heritage of the country are under the increasing pressures of the habitat degradation, overexploitation, and indigenous knowledge loss.

The biodiversity of plants is a natural heritage and the foundation of ecological stability, cultural identity and human existence in Pakistan. Cases provided in this chapter—in deserts, in the montane ecosystems—help bring to light the multitude of conservation challenges and opportunities, and, when put together, demonstrate how biodiversity protection cannot achieve success without the active participation of the local communities.

The Cholistan Desert research reported the role of fencing regimes and phytosociological measurements in informing conservation in arid landscapes. The success of strong species like Ochthochloa compressa and Lasiurus scindicus reiterated the relevance of indicator species in the monitoring of the ecosystem. In the same vein, the comparative ethnobotanical study Cholistan and Pothwar presented the richness of traditional knowledge and its primary importance in the medical field, and cautioned that, unless it is documented in writing, the knowledge will be lost soon.

The conservation of numerous medicinal species such as Saussurea lappa and Aconitum heterophyllum in the Neelum Valley in Azad Jammu and Kashmir highlighted the level of human stress and vulnerability of montane eco systems. The case has shown that the exploitation can be reduced by training, documentation, provision of alternative fuel and awareness programs. In relocating to District Kohat, the ethnobotanical survey both indicated the richness of traditional plant use and documented disturbing deficiencies in intergenerational knowledge transmission and a high percentage of endangered species. This demands immediate conservation measures and the need to enliven cultural knowledge systems.

The Nara Desert research highlighted the accumulated effects of overgrazing, deforestation, gas exploration and hunting on the delicate desert ecosystems. The results support rehabilitation by means of reforestation using native drought-resistant species, stabilization of sand dunes, and seed banking in order to protect the ecological and economic life of desert societies. Putnam Indigenous knowledge of 58 woody species in Soona Valley illustrated how local people rely on biodiversity to fulfill various demands-medicine, timber, fuel, and food. Nevertheless, the dangers of anthropogenic forces and even inter-country strife point to the necessity of community involvement, policy implementation, and plotting plantations to ease pressure on native plants.

The evaluation of the floristic diversity in the Sialkot tropical dry forests (TDFs) revealed a high number of anthropogenic and climatic risks, which supports the urgency of the approaches related to habitat recovery, discouragement of unlawful exploitation, and mitigation of fragmentation. These results necessitate closer interaction between the communities, government departments and the NGOs in the preservation of TDF ecosystems. The example of the montane ecosystems in the western Himalayas further proved that successful conservation should include a combination of phytosociological data and ethnobotanical information, a focus on endangered species such as Juglans regia and Paeonia emodi and compiling a complete Red Data Book of the Himalayas. This example demonstrates that a combination of science and traditional ecological knowledge can be used in policy and conservation planning both locally and nationally. Case studies explained make it clear that conservation cannot be solely based on contemporary science and policy; instead, it should combine the wisdom of local communities, effective institutional structures, and specific research. Biodiversity Action Plan (BAP), provincial conservation initiatives, and initiatives organized by non-governmental organizations have provided a framework but have yet to be implemented with greater co-ordination, accountability, and political will.

Collectively, these case studies pass a single message; community stewardship is the key to success in biodiversity conservation in Pakistan. In deserts, valleys, forests, and mountains, ethnobotanical knowledge has sustained healthcare, livelihoods, and cultural resilience in local communities. However, climate change, modernization, war, and unplanned exploitation is eroding this treasure trove of knowledge and biodiversity.

Moving on, Pakistan needs to embrace a multi-pronged strategy that will integrate community empowerment, scientific research, policy reforms, and awareness campaigns. The most important steps to take are strengthening of participatory forest management, documenting indigenous knowledge, ensuring alternative livelihoods, and the enforcement of policies that are biodiversity friendly. The conservation of species is not the sole meaning of conservation, but an ecosystem-wide assurance of services, cultural heritage, and resiliency into the future.

After all, preservation of plant biodiversity in Pakistan is a collective burden. With the ability of communities, policymakers, scientists, and institutions to collaborate in true partnership, Pakistan can not only be able to preserve its rich floral diversity, but will also be able to serve as a model in community-led conservation in the Global South. The experiences of Cholistan, Pothwar, Neelum, Kohat, Nara, Soona, Sialkot, and the Himalayas lead to a concise direction: conservation of biodiversity should be locally based, nationally sponsored, and internationally appreciated. It is possible that by reconciling the ideals of scientific innovation with the wisdom of the culture and connecting the goals of conservation with the socio-economic ones, Pakistan can ensure the success of its botanical wealth to the generations to come, and enhance the resilience to environmental and developmental stress.

References

Ahmad KS, Rahmatullah Qureshi RQ, Mansoor Hameed MH, Farooq Ahmad FA, Tahira Nawaz TN (2012) Conservation assessment and medicinal importance of some plants resources from Sharda, Neelum Valley, Azad Jammu and Kashmir, Pakistan

Ahmad M, Bano A, Zafar M, Sultana S, Rashid S (2015) Interdependence of biodiversity, applied ethnobotany, and conservation in higher ecosystems of northern Pakistan under fast climatic changes. In: Climate change impacts on high-altitude ecosystems. Springer, pp 455–489

Ahmed M, Husain T, Sheikh AH, Hussain SS, Siddiqui MF (2006) Phytosociology and structure of Himalayan forests from different climatic zones of Pakistan. Pak J Bot 38(2):361–383

Alam J, Ali SI (2009) Conservation status of Astragalus gilgitensis Ali (Fabaceae): a critically endangered species in the Gilgit District, Pakistan. Phyton 48:211–225

Alam J, Ali SI (2010) Contribution to the red list of the Plants of Pakistan. Pak J Bot 42(5): 2967–2971

Ali H, Qaiser M (2011) Contribution to the red list of Pakistan: a case study of Silene longisepala (Caryophyllaceae). Oryx 45(4):522–527

Ali SI (2008) Significance of flora with special reference to Pakistan. Pak J Bot 40(3):967–971

Anwar M, Jasra AW, Ahmad I (2008) Biodiversity conservation status in Pakistan-a review. Pak J For 58(1):39–48

Anwar M, Shank C (2000) Developing and implementing biodiversity strategies and action plans: lessons from Pakistan. In: Biodiversity planning in Asia. IUCN Regional Biodiversity Program, Colombo, Sri Lanka

Aumeeruddy Y (2003) Applied Ethnobotany; case studies from the Himalayan region. Godalming: WWF; 2003. People and plants working paper, 12:3–4

CBD (1992) Convention of biological diversity. United Nations Environment Program

Cox CB, Moore PD (1993) Biogeography: an ecological and evolutionary approach. Blackwell Scientific Publications, p 326

GOP (1992) Forestry sector master plan. Government of Pakistan

GOP (2000) Biodiversity action plan of Pakistan. Ministry of Environment/IUCN/WWF, Islamabad

Hosier R (1993) Forest energy in Pakistan: the evidence for sustainability. Pakistan household. Energy Strategy Study (HESS), Islamabad, GOP, Energy Wing. Energy Sector Management Assistance Program

IUCN (1997) IUCN red list categories. IUCN Species Survival Commission

IUCN, Gland IUCN (1998) Guidelines for re–introductions. Prepared by the IUCN/SSC Reintroductions Specialist Group–IUCN, Gland, Switzerland and Cambridge, UK

IUCN (2001) IUCN red list categories and criteria: version 3.1 IUCN. Species Survival Commission 2:30. IUCN, Gland, Switzerland and Cambridge, UK

IUCN (2008) IUCN red list of threatened species. www.iucnredlist.org cited on March 02, 2008.

Ishtiaq Ch DM, Maqbool M, Hussain T, Shah A (2013) Role of indigenous knowledge in biodiversity conservation of an area: a case study on tree ethnobotany of Soona Valley, district Bhimber Azad Kashmir, Pakistan. Pak J Bot 45:157–164

Kala CP (2007) Local preferences of ethnobotanical species in the Indian Himalaya: implications for environmental conservation. Curr Sci 93:1828–1834

Khan SM (2012) Plant communities and vegetation ecosystem services in the Naran Valley, Western Himalaya. University of Leicester, PhD Thesis

Khan SM, Page SE, Ahmad H, Harper DM (2013) Sustainable utilization and conservation of plant biodiversity in montane ecosystems: the western Himalayas as a case study. Ann Bot 112(3): 479–501. https://doi.org/10.1093/aob/mct125

Khan MI, Habib-Ur-Rehman, Ahmed M, Asghar F, Abdul Latif, Ejaz ul Haq, Nadeem M, Zulqarnain SM, Naz N, Akhtar H (2025) Effect of different fencing regimes on community structure of perennial plant species in Cholistan Desert, Pakistan. Indus J Biosci Res 3(1): 394–428

Khan MS, Khan SM, Abdullah, Liu J, Wu ZY, Hussain J, Zeb SA (2025) Ecological assessment of *Iris hookeriana* across subalpine and alpine regions of the Hindu-Himalayas. Front For Glob Change, Sec Forest Management Volume 8 -2025. https://doi.org/10.3389/ffgc.2025.1539025

Khan SM, Harper DM, Page S, Ahmad H (2011a) Species and community diversity of vascular flora along environmental gradient in Naran Valley: a multivariate approach through indicator species analysis. Pak J Bot 43:2337–2346

Khan SM, Harper DM, Page S, Ahmad H (2011b) Residual value analyses of the medicinal flora of the Western Himalayas: the Naran Valley, Pakistan. Pak J Bot 43(SI):97–104

Khan SM, Ahmad H, Page S, Harper DM (2012a) Anthropogenic influences on the natural ecosystem of the Naran Valley in the western Himalayas. Pak J Bot 44(SI):231–238

Khan SM, Page S, Ahmad H, Shaheen H, Harper DM (2012b) Vegetation dynamics in the Western Himalayas, diversity indices and climate change. J Sci Technol Dev 31:232–243

Malik S, Ahmad S, Sadiq A, Alam K, Wariss HM, Ahmad I et al (2015) A comparative ethnobotanical study of Cholistan (an arid area) and Pothwar (a semi-arid area) of Pakistan for traditional medicines. J Ethnobiol Ethnomed 11:31. https://doi.org/10.1186/s13002-015-0018-2

Mirza JH, Qureshi MSA (1978) Fungi of Pakistan. Dept. of Plant Pathology. University of Agri culture, Faisalabad, Pakistan, p 311

Nasir E, Ali SI (1972) Flora of West Pakistan: an annotated catalogue of the vascular plants of West Pakistan and Kashmir: Fakhri

Qureshi R (2020) Phytodiversity and conservation status of the Nara Desert, Pakistan. In: Environmental pollution, biodiversity, and sustainable development. Apple Academic Press, pp 52–76

Qureshi RA, Ghufran MA, Sultana KN, Ashraf M, Khan AG (2007) Ethnobotanical studies of medicinal plants of Gilgit District and surrounding areas. Ethnobot Res Appl 5:115–122

Saifullah SM (1997) Management of the Indus Delta mangroves. In: Coastal zone management imperative for maritime developing nations. Dordrecht: Springer Netherlands, pp 333–346

Samant SS, Dhar U (1997) Diversity, endemism and economic potential of wild edible plants of Indian Himalaya. Int J Sustain Dev 4(3):179–191

Shah A (2008) Loss of biodiversity and extinctions. Global Issues. http://www.globalissues.org/article/171/loss-of-biodiversity-and-extinctions

Shahbaz B, Suleri AQ (2009) The political economy of forest management in Pakistan. In: Cronin R, Pandya A (eds) Exploiting natural resources: Growth, instability, and conflict in the Middle East and Asia. Washington, DC: The Henry L. Stimson Center, pp 21–31

Shahzad K, Ali WS, Muhammad S, Dai J, Zeb U, Zhu M (2024) Assessment of plant biodiversity in tropical dry forests of Sialkot, Pakistan; insight into environmental, anthropogenic influence and conservation strategies. Front For Glob Change 7. https://doi.org/10.3389/ffgc.2024.1362117

Stewart RR (1972) An annotated catalogue of the vascular plants of West Pakistan and Kashmir

Watson RT, Watson R (1995) Global biodiversity assessment

Zamin M, Adnan M, Begum S, Ullah I (2024) Novel plant uses and their conservation status in a semiarid subtropical region of Pakistan. Ethnobot Res Appl 29. https://doi.org/10.32859/era.29.13.1-49

Biodiversity in Action: Case Studies of Resilient Solutions for Food Security

18

Muhammad Majeed, Naila Parveen Abbasi, Ishrat Fatima, Muddaser Shah, Muhammad Noor ul Hasan, Muzzammil Hussain, Saba Nazir, Samreen Fatima, Muhammad Anas, Waseem Ahmed Khattak, Rahmatullah Qureshi, Syed Asad Abbas, and Ömer Kiliç

M. Majeed (✉)
Department of Botany, University of Gujrat, Gujrat, Punjab, Pakistan
e-mail: m.majeed@uog.edu.pk; https://www.researchgate.net/profile/Muhammad-Majeed-18; https://scholar.google.com/citations?user=wkuePhsAAAAJ&hl=en

N. P. Abbasi
Department of Geography, Shah Abdul Latif University Khairpur Mirs, Khairpur, Sindh, Pakistan
e-mail: naila.abbassi@salu.edu.pk

I. Fatima
Department of Biological Sciences, University of Veterinary and Animal Sciences, Lahore, Pakistan

M. Shah
Department of Botany, Bacha Khan University, Charsadda, Khyber Pakhtunkhwa, Pakistan
e-mail: muddasershah@awkum.edu.pk

M. N. u. Hasan
Department of Entomology, Faculty of Agricultural Sciences and Technology, University of Layyah, Layyah, Punjab, Pakistan

M. Hussain
Department of Wildlife & Ecology, Faculty of Life Sciences, University of Okara, Okara, Punjab, Pakistan

S. Nazir
Dr. Muhammad Ajmal Khan Institute of Sustainable Halophyte Utilization, University of Karachi, Karachi, Sindh, Pakistan

S. Fatima
Department of Biological Sciences, University of Baltistan, Skardu, Gilgit Baltistan, Pakistan

M. Anas · W. A. Khattak
Department of Plant Sciences, Faculty of Biological Sciences, Quaid-I-Azam University, Islamabad, Pakistan
e-mail: Muhammadanas@bs.qau.edu.pk; waseemkhattack@bs.qau.edu.pk

R. Qureshi (ed.), *Plant Biodiversity*, https://doi.org/10.1007/978-981-92-1471-6_18

Abstract

Biodiversity is fundamental to ecosystem stability, agricultural productivity, climate resilience, and global food security. However, accelerating biodiversity loss driven by climate change, habitat degradation, land-use change, pollution, invasive species, and overexploitation poses significant threats to sustainable food systems worldwide. This chapter examines the interrelationships among biodiversity, food security, and climate change, highlighting the critical role of agrobiodiversity in enhancing agricultural resilience, resource-use efficiency, and adaptive capacity to environmental stresses. Particular emphasis is placed on the conservation of crop genetic resources, traditional landraces, and diversified farming systems that support sustainable agricultural transitions. The chapter reviews biodiversity-based approaches, including agroecological practices, polycultures, integrated pest management, agroforestry, conservation agriculture, and nature-based solutions, which improve ecosystem services, soil health, carbon sequestration, water conservation, and nutritional security. Emerging innovations such as biomimicry in urban agriculture, climate-resilient crop improvement, and the integration of traditional ecological knowledge with modern technologies are also discussed. Furthermore, the socioeconomic and policy dimensions influencing the adoption of biodiversity-centered agricultural systems are critically evaluated. The chapter concludes that conserving and integrating biodiversity into food production systems is essential for achieving climate adaptation, sustainable agriculture, and long-term food security. Effective implementation requires multi-stakeholder collaboration, supportive policies, community participation, and context-specific strategies that align environmental sustainability with socioeconomic development and the Sustainable Development Goals.

Keyword

Agrobiodiversity · Agricultural resilience · Biomimicry · Food security · Nature-based solution (Nbs) · Urban agriculture

R. Qureshi
Department of Botany, Faculty of Sciences, Pir Mehr Ali Shah Arid Agriculture University, Rawalpindi, Punjab, Pakistan
e-mail: rahmatullahq@uaar.edu.pk

S. A. Abbas
Department of Forestry, Kohat University of Science and Technology, Kohat, Khyber Pakhtunkhwa, Pakistan

Ö. Kiliç
Pharmacy Faculty, Adıyaman Üniversitesi, Adıyaman, Turkey

Abbreviations

BGI	Blue-Green Infrastructure
CC	Climate Change
CIDSE	Cooperation Internationale pour le Développement and la Solidarité
EbA	Ecosystem-Based Adaptation
FSN	Food Security and Nutrition
INKOTA	Information, Koordination, Tagungen
NbS	Nature-Based Solutions
NCS	Natural Climate Solutions
REDD	Reduced Emissions from Deforestation and Degradation
SFP	Sahara Forest Project
UA	Urban Agriculture

18.1 Introduction

Biodiversity is defined as the diversity of biological organisms found worldwide. Biodiversity includes the diversity of genes within a single species, the diversity of various species, and finally, the diversity of ecosystems in different environments. There have been living organisms on Earth for roughly 3.5 billion years, but the number of species on Earth today has ranged from ten million to up to 100 million; however, only 1.75 million species have actually been named to date (García-Vega and Newbold 2020).

Anthropogenic pressures on biodiversity include overexploitation of species, land degradation, nitrogen deposition, pollution, biological invasions, diversion of land and water resources, habitat fragmentation, urbanization, and industrial activity. However, among these, the most critical and overarching threat is how climate change (CC) interacts synergistically with pre-existing stressors. While direct threats such as habitat fragmentation, invasive species, pollution, and overexploitation are well recognized, the underlying drivers are far more complex and interconnected. Overpopulation, overconsumption, political instability, and socio-economic factors exacerbate the problem. Notably, the literature often overlooks the roles of social and economic frameworks, policy coherence, good governance, legislative enforcement, and corruption—all of which significantly influence the severity of biodiversity threats (Chakraborty and Chakraborty 2021).

Climate represents the average weather conditions in a given area and is based on long-term statistical data on various weather conditions. Conversely, CC represents the changes in the average weather conditions caused by human activities and natural factors such as ozone depletion and the effects of greenhouse gases. The changes in the climate can be attributed to several factors, including changes in solar irradiance, long-term changes in the Earth's orbit such as eccentricity, axial tilt, and the precession of the equinoxes. In addition, climate changes can be caused by natural and human activities that affect the environment. Although climate changes are caused by changes in external factors such as natural and human activities, the changes in the climate in the future are mainly predicted based on the effects of increasing greenhouse gases and human activities. Therefore, the twentieth century recorded the highest rate of climate change in the past thousand years. The average temperatures during this period increased by 0.6 °C. However, the increase in temperature in the future is expected to be higher than that recorded in the twentieth century. The increase in the coming years is estimated to be in the range of 0.1–2 °C per decade (Neale et al. 2021).

Biodiversity, agricultural yield, and food security are expected to experience significant changes due to changes in global climate patterns. One of the changes expected to happen due to changes in climate patterns is the movement of plant species to higher altitudes and latitudes, away from the climatic conditions to which they have adapted over the years. There is now substantial evidence showing that changes in the locations of plant species are already in motion (Muluneh 2021). For example, it has been noted that there have been changes in elevation in places such as Alaska, Scandinavia, the Alps, and the Mediterranean basin. As species move to a higher latitude and altitude, the populations on the leading side of the species' movement are most probably to increase. This change has resulted in the decrease of the number of the population and increase in extinction levels on the trailing side. In species with geographical boundaries affected by climatic conditions, increased temperatures can contribute to a more successful reproduction and survival at the northern and high altitude extremes of their range. Conversely, at the equatorial and lower altitudinal limits, a species may be outcompeted, primarily due to factors like water availability and higher temperatures (Chakraborty and Chakraborty 2021). However, the effects of changing climatic conditions are expected to vary across continental and oceanic ecosystems, which could lead to limited range expansion and reductions (Arif et al. 2021; Muhammad et al. 2024b; Rasool et al. 2025).

While sustainable food production is key to advancing sustainable development, it currently poses a threat to the success of sustainable development initiatives and the 2030 Agenda for Sustainable Development. With this, the agricultural industry is encouraged to enhance its sustainability efforts in adapting to the effects of climate change without compromising the success of global food security initiatives. Thus, it is essential to investigate the effects of climate change on biodiversity and food security from a global perspective, as climate change is a global challenge that requires united action (Caron et al. 2018). Therefore, this chapter seeks to identify, evaluate, and integrate scientific evidence related to the relationships among climate change, biodiversity, and food insecurity.

Consequently, the main objectives of the study are to (i) inspect the effects of climatic change on biodiversity as well as species' range moves, (ii) examine an interactive connection among biodiversity and food, (iii) investigate threats to biodiversity and their effects, (iv) describe the adaptation and mitigation procedures for a fluctuating climate. The graphical presentation of key objectives is reflected in Fig. 18.1.

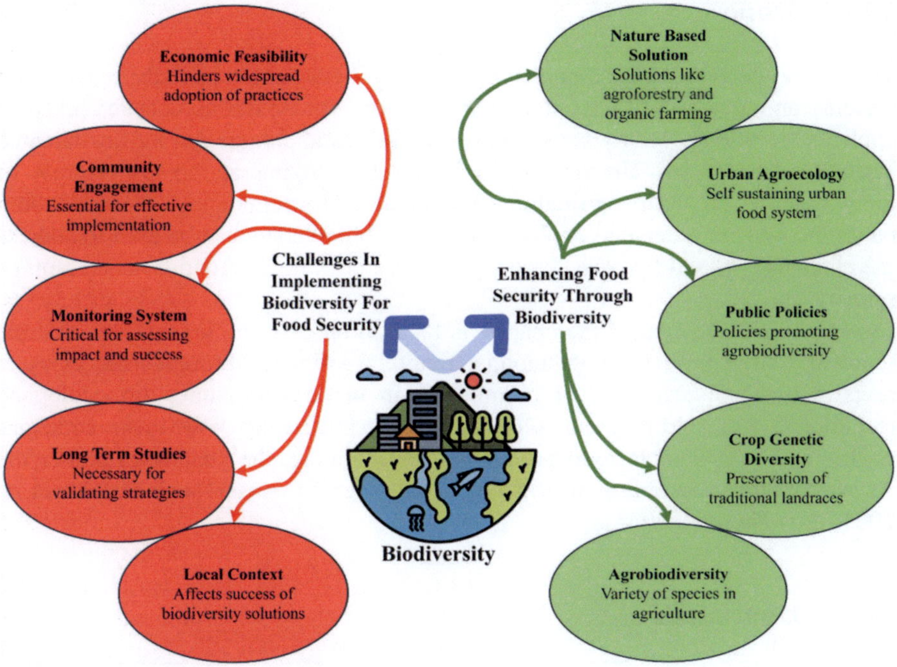

Fig. 18.1 Important of biodiversity in human wellbeing

18.1.1 Significance

The chapter addresses one of the most urgent global challenges, ensuring food availability amid the threats of climate change and biodiversity loss. By clarifying the links between biodiversity, agricultural methods, and food systems, the chapter highlights the essential role of biodiversity in strengthening agricultural resilience (Muluneh 2021). However, the significance of this chapter lies in its in-depth discussion and analysis of the contribution that biodiversity makes toward food security. Biodiversity is vital for key ecosystem services like soil fertility, pollination, nutrient cycling, and pest management, which are all critical for successful farming. In addition, it is vital for supporting genetic diversity, which is critical for developing crops and animals that will be able to withstand environmental challenges, pests, and diseases. This chapter also highlights innovative ways to incorporate biodiversity into agricultural practices. By presenting case studies on biomimicry in urban farming, agroecological principles, and nature-based solutions (NbS), it provides practical insights on how to use biodiversity to create resilient food systems. These approaches demonstrate the feasibility of developing sustainable and climate-resilient agricultural systems, especially in areas vulnerable to climate-related challenges. It serves as an important resource for fostering transformative progress toward sustainable agricultural practices, environmental protection, and global food security (Arshad et al. 2022).

18.1.2 Scope

The chapter delves into the complex relationships between biodiversity, agricultural systems, and food security. It encompasses both the theoretical aspects and practical applications of biodiversity conservation, as well as its integration into sustainable agricultural practices. The discussion begins by looking at the critical role of biodiversity in supporting agricultural ecosystems. This foundational understanding paves the way for a detailed exploration of how biodiversity can be used to address the challenges posed by climate change, environmental degradation, and resource scarcity.

A significant section of the chapter is focused on case studies that demonstrate innovative strategies for strengthening agricultural resilience through biodiversity. It reviews practical examples of biomimicry in urban agriculture, agroecological techniques, and nature-based solutions (NbS), illustrating how these strategies contribute to sustainable food production. Furthermore, the chapter examines the socio-economic and policy dimensions of biodiversity conservation (Bashir et al. 2023; Barwant et al. 2024).

18.1.3 Limitations

Although this chapter gives valuable insights on the complex link between biodiversity and food security, it also has some limitations. One of the limitations of this chapter is that, although it gives useful information on the complex link between biodiversity and food security, it is limited by the use of case studies to explain how biodiversity can contribute to food security. While case studies provide practical insights, their context-specific nature may limit their relevance to different regions or agricultural systems. The diverse ecological, social, and economic conditions around the world suggest that solutions that work well in one context may not be applicable elsewhere. Additionally, some regions or ecosystems might be underrepresented in the literature, leading to potential gaps in the chapter's coverage. The chapter highlights the importance of integrating biodiversity into agricultural practices, but it may overlook some specific challenges that come with implementation. Issues like socioeconomic disparities, insufficient policy support, and a lack of awareness among stakeholders are mentioned but not explored in depth. This gap could limit the practical application of the suggested solutions, especially in low-income or resource-limited settings (Bhattacharya et al. 2023).

18.2 Agrobiodiversity and Genetic Diversity

Agro-biodiversity and genetic diversity are demanding constituents of sustainable agriculture that impact on food security and ecosystem resilience. In agricultural systems Agro-biodiversity mentions to the diversity of life forms, while genetic diversity concerns to the variation within species that allows adaptation to environmental changes. For the development of robust agricultural practices, the interaction between these two forms of diversity is essential. A wide range of species includes crops, livestock, and microorganisms known as agro-biodiversity includes, adds to food security by increasing nutritional diversity and resilience against pests and diseases. The loss of agro-biodiversity, determined by monocultures and industrial agriculture, threatens global food security and increases susceptibility to climate change (Khan et al. 2022b).

Genetic diversity is necessary in plant species for the adaptation and evolution, facilitating them to survive environmental stresses (Haq et al. 2022). To enhance genetic variability, techniques such as mutation breeding and CRISPR are employed, which is essential for improving crop traits like yield and resistance. Effective conservation of agro-biodiversity and genetic resources can be effectively conserved through in-situ and ex-situ methods, including biotechnological approaches like cryopreservation. By creating diverse ecosystems that support various species and improve soil health, Agroforestry practices can enhance agro-biodiversity. Although the focus on agro-biodiversity and genetic diversity is fundamental for sustainable agriculture, although that some argue that the dependence on modern biotechnological methods may minimize traditional practices that have traditionally maintained these diversities (Dhakad et al. 2022). For whole agricultural sustainability, balancing the innovation with traditional knowledge is essential.

18.3 The Role of Genetic Diversity in Agricultural Resilience

Increasing the diversity of genes in crops is a significant factor for climate resilience from a number of agricultural viewpoints, as climate change poses a significant threat to agricultural production. It is the characteristic of plant species that they are differentiated for their tolerance to stress. The significance of this tolerance can be realized in the regions of the country where there is a presence of stress (Dhakad et al. 2022). Certain crops, such as millets, pearl millet, and sorghum, have developed resistance to drought for long periods in arid and semi-arid areas such as Rajasthan and some areas of Maharashtra, where water scarcity is a main limiting factor in agricultural development, which is of principal importance. In low-moisture conditions, the genetic diversity within these species tolerates them to maintain productivity and survive. The genetic diversity in plants also provides resistance against a range of pests and diseases. For example, in Southern India certain varieties of finger millet reveal resistance to blast disease, a major threat to millet crops. This resistance is a direct result of the genetic diversity intrinsic in these traditional varieties. It is commonly known as millets for their ability to grow well in nutrient-poor, marginal soils where other crops would fail. This is mostly important in these areas where soil fertility is often low. In millet, the genetic adaptability allows them to consume nutrients well and grow in less-than-ideal soil conditions (Patil 2020).

In plant species, the intrinsic genetic diversity, makes them compatible to adapt the challenges posed by climate change, which faces various climatic challenges.

Millets have shown a significant capacity to adapt to varying temperatures. Millets like foxtail and barnyard display resilience in the regions where temperature fluctuations between seasons are important, adjusting their growth cycles to the usual temperatures. Due to climate change with changing rainfall patterns, millets' have ability to grow under low water availability becomes increasingly important. The cultivation of millets ensures food security and agricultural sustainability in regions where rainfall is occasional. In plants the genetic diversity not only adds to their adaptability to environmental stresses but also to their nutritional profile. Millets, unlike many other crops, can maintain their nutritional quality even under stress conditions. For example, pearl millet and finger millet still produce grains with high nutritional value in drought conditions, providing primary nutrients like iron and calcium to the population (Das and Rakshit 2016).

Utilization of genetic diversity for biofortification is a favorable way to increase the nutritional value of millets. Breeding programs are focusing on development of plant varieties by utilizing the genetic variability present within these species with increased micronutrient content. There are different challenges in the conservation and utilization of genetic diversity. The loss of genetic diversity, also known as genetic decline, is one of the major challenges. This is mainly because of the exchange of traditional varieties with high-yielding and commercially viable varieties. The change in the pattern of agricultural regions has led to the reduced cultivation of diverse plant varieties, especially those that were commonly found in semi-arid and arid regions. The lack of access to available genetic resources is also one of the significant challenges. Provision and rational property issues often impact the resources availability for breeding and research purposes. The efforts to utilize the existing genetic diversity in breeding programs are hampering by this. The conservation of the remaining biodiversity in agro-ecosystems is depending on agriculture which is also one of the greatest threats to biodiversity. The substitution of local varieties with improved or exotic varieties and species is one of the leading reasons of genetic degradation of crops and other agro-biodiversity (Nonić and Šijačić-Nikolić 2021). Thus, sustainable farming practices that will be approved to use and preserve biodiversity could eventually improve the worth of the environment and reduce the agricultural expansion at the expense of natural forests (Fig. 18.2).

18.4 Conservation Techniques

There is growing interest in techniques that facilitate harmonization in agricultural production with biodiversity conservation and sustainable utilization in controlling the adverse impacts of extensive agriculture. When compared with wild systems, biodiversity in global agriculture is unmarked. If we are to conserve domesticated biodiversity in plant species and achieve global sustainable development goals, biodiversity goals have to be marked. Plant scientists have recognized gene editing techniques in biotechnology as game-changing techniques with the aim of attaining the objective of increasing genetic diversity in crops. Such exhibits are not new, with

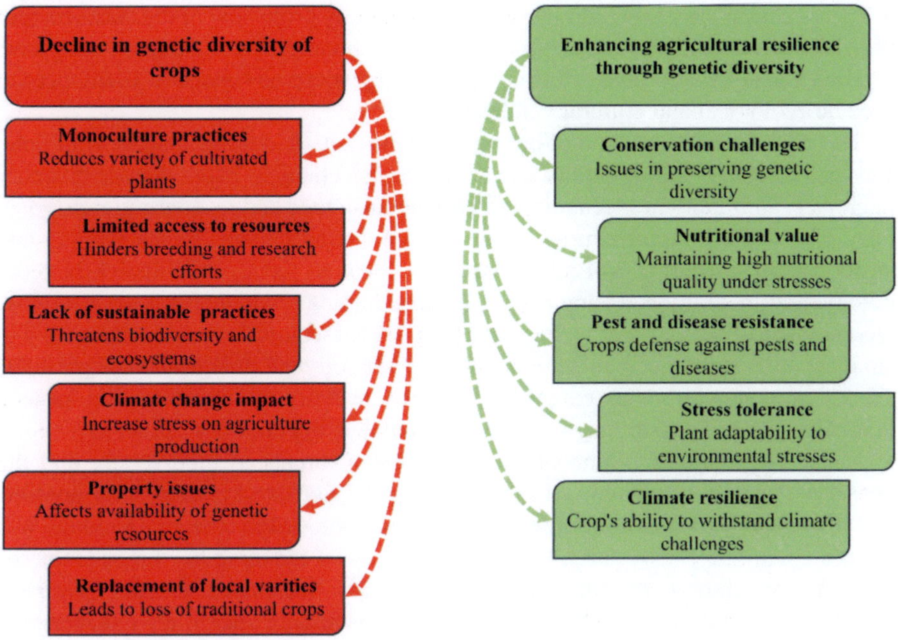

Fig. 18.2 Decline in crop genetic diversity and its impact on agricultural resilience

scientists in the history of crop breeding having discovered numerous innovations in similar measures, with the most recent being the prospect that recombinant DNA would unveil paradigm-changing techniques. To target a particular location in an organism's genome, gene-editing techniques utilize site-specific nucleases (ZFNs, TALENs, CRISPR-Cas9), which are guided either by protein engineering or RNA sequence complementarity (Chandrasegaran and Carroll 2016). Such gene editing techniques have been recognized as effective tools in enhancing crop biodiversity with the rapid advancement in genetic sequencing techniques.

By re-starting crop adaptation, gene editing also offers possibility to mostly reversal the breeding process in ways that enhance genetic diversity. Crop adaptation depends upon a combination of artificial selection by humans and suddenly occurring genetic mutations. As an example, grains in wild rice explode to scatter the seed far and wide. A mutation occurred so that the grains were no longer scattered during the domestication of rice which was advantageous to the early agricultural societies and thus was chosen to be cultivated. Even today, when rice wild relatives have acquired positive changes such as adapting to different growth environments, their grains continue to burst. Through gene editing, methods like "De novo adaptation" generate mutations known to be markers of adaptation in crop wild relatives or neglected crops. As the goal of assertion is useful characteristics that are maintained by these plants like nutrient content or stress tolerance, and the technique also facilitates desired domesticated characteristics, like non-shattering/scattering grains, this has been shown to have the technique in action (Kreiner et al. 2021). Whereas de novo adaptation via editing of key genes is potentially promising in the enhancement of genetic diversity, in order to put the process to practical use in crops, which have reached the cultivation phase, follow-up breeding would be needed to adapt them to specific ecological and cultural conditions.

The need for an expansion of crop genetic diversity is only increasing with time, and therefore, it is appropriate to advocate for such changes. More genetic diversity in crops is an important mechanism of climate insurance and response to nutritional demands in many varied settings of agriculture, including subsistence farming and an industrial monocrop. Not all of these modification techniques involve biotechnology, and in addition, easing access to biotechnology, whether it is gene editing, recombinant DNA, or other technologies yet to be developed, should not be used to supplant other techniques. In addition, dealing with climate change and hunger in the short run will require an extensive array of techniques for increasing genetic diversity, some of which may involve judiciously using crop biotechnology.

Modern agriculture is one of the most severe threats to biodiversity and the conservation of the remaining biodiversity in the agro-ecosystems. It is possible that innovative biodiversity-based agriculture may be sustainable and yielding, and therefore help the conservation of the wild species by reducing the negative impact of the modern mono-cropping agricultural practice. Sustainable farming practices, which harness and preserve biodiversity, could eventually lead to the enhancement of the quality of the environment and the reduction of the undesirable effect of agriculture on biodiversity and agricultural activities encroaching natural forests (Duru et al. 2015). Therefore, there is a strong argument in favor of assisting conservation that is compatible with support requirements of the affected people. Integrated agro forestry systems have been promoted as an appropriate direction toward a better livelihood since it is also advancing in biodiversity in working landscapes by the incorporation of more species into agriculture.

The traditional agro-forest systems have been complex systems that have been developed with the aim of imitating natural systems in which farmers and dwellers have sought to utilize biodiversity of forest resources or adopt numerous tree species in different productive niches in their farms. The first traditional agro-forest system is practiced at the tree canopy in parkland systems where crops have been grown in home-gardens, agro-forests. The home-gardens not only have high biodiversity but also have an ancient form of land utilization that is highly sustainable and meets numerous requirements. Such practices have been effective in biodiversity since they provide appropriate habitat for both plant and animal species in farm areas, reduced pressure on existing forests, and conservation of tree species on farms using in-situ. It has been argued that it preserves the value of biological interactions and biodiversity than some of the new agroforestry technologies using in-situ conservation of tree species on farms, reduced pressure on existing forests, and traditional systems. Agroforestry systems have high biodiversity in both flora and fauna. For example, an example of small-scale farmers may grow numerous crops, for example, cassava, peanuts, sunflower seeds, beans, potatoes, and soybeans in association with trees (Duru et al. 2015). For example, in a field, high abundance of fauna and flora species plays an important role in adapting to environmental changes.

Genetic resources can now be conserved, managed, and maintained in traditional agro-ecosystem. Under such traditional farming systems, crop land races are a common form of genetic resources that are conserved, for example, farmers save livestock and retain diversity by sharing genetic resources via stock purchase and sharing systems. Crop land races conservation is also practiced by local farmers. The local farmers also practice crop land races in conserving. Moreover, the conventional agro-ecosystems have also helped the conservationist and plant breeders in collecting samples from the fields of the local farmers, which are then conserved in gene banks (off-site) for later use. However, there is an increasing awareness of the need to conserve this type of conservation in an in-situ (on-site) manner. in order to the successfully conserve the genetic resource that enables them to continuously adapt to the environment.

18.5 Biomimicry in Urban Agriculture

Biomimicry in urban agriculture uses natural systems to improve food production's sustainability and resilience. Urban agriculture, by replicating ecological processes, can address issues including food security, environmental degradation, and urbanization pressures. This method encourages not only efficient resource usage but also community involvement and ecological balance. The sections below discuss essential aspects of biomimicry in urban agriculture. Biomimicry can enhance urban food systems by incorporating conventional ecological knowledge and natural ecosystems, for establishing sustainable food systems in cities prone to external supply disruptions. Biomimicry promotes sustainable agriculture practices that mimic the resilience of natural ecosystems. Case studies show that biomimetic designs can improve agricultural efficacy while reducing ecological impacts, hence promoting resource conservation (Stojanovic 2019). The use of biophilic design ideas in peri-urban regions can help to transition to more sustainable agricultural interfaces, hence improving urban metabolism and foodscapes. Mathematical models can assess the biomimicry of urban architecture, boosting walkability and efficient food distribution networks. While biomimicry offers creative solutions for urban agriculture, there are still difficulties to its general adoption, notably in terms of technical integration and socioeconomic aspects that may impede its implementation (Fig. 18.3).

18.6 Leveraging Ecological Processes in Urban Farming

A regenerative city is significantly more than just a sustainable one. The regenerative approach perceives the urban green space as a productive terrain that yields food and provides biodiversity. The so-called urban wastelands are positively understood in this point of view. During recent years, urban agriculture (UA) has become a burning issue in the context of sustainable development. Accordingly, the urban fabric combined with local food production is vital due to the environmental reasons.

Fig. 18.3 Cycle of biomimicry in urban agriculture

The major issues are the decrease in food miles and the necessity of produced food, which places an amount of strain on the environment. Meanwhile, UA encourages regeneration and restoration. A city farm is a section of farmland that is situated inside a city. In a similar vein, farms are set up on unoccupied land, just like community gardens. They differ in that the farm area is utilized mostly for the tillage of edible plants, whereas relaxation and leisure activities are either nonexistent or only infrequently occur to a far lesser extent (Mohan et al. 2020). Examples of space that are related to the idea of city farms include the vegetable gardens that were developed in both Europe and the United States of America as a result of the two world wars. However, as late as the second half of the twentieth century, agricultural land that was developed in cities started being referred to as city farms. A farm was developed in the suburbs of London in 1972, and another was developed in London in 1984.

The primary goal of urban farming is to provide fresh produce. When it comes to war, political unrest, or food product price speculation, local food production is very crucial. Thus, one factor influencing regional food security is urban agriculture. As a result, some cities implement agricultural areas top-down, for example, at the level of the national or local government. In this instance, urban agriculture is regarded as one of the components that make up the local food system, which includes the area associated with food production, distribution, and consumption as well as the organization of food production and distribution. In several countries, city farms are also set up by informal (like cooperatives) and nongovernmental (like NGOs) entities to safeguard regional food security. Bottom-up farms could potentially be a component of regional food systems (Dobele et al. 2022).

18.7 Case Studies of Biomimetic Solutions in Urban Agriculture

Cuban city farms are examples of those set up to guarantee the local food security. The US government's 1962 embargo and the economic crises associated with the collapse of the Eastern Bloc in the 1990s provide historical context for the phenomena of urban agriculture in Cuba. The food shortage and economic difficult circumstances that were brought about by the loss of economic trading partners and economic isolation forced Cubans to develop an alternative economy that is based on the internal market rather than relying on external markets for export and import. The Cuban government decided to integrate bottom-up garden projects into systematic and organized planning at the central government level in the 1990s following the economic opportunities that urban gardening provided. The Cubans were able to develop small farms for local food production in both urban and suburban areas following the changes. As early as 1998, Havana had established the Department of Urban Agriculture (Spanish Grupo Nacional de Agricultura Urbana) (Koont 2016).

The AgroCité in Paris serves as an illustration of a city farm in Europe that is included into the local food chain. This farm was created between 2010 and 2014 as part of the R-Urban project by the previously mentioned architectural firm, Atelier d'Architecture Autogérée (AAA). The project's main focus was on ways that bottom-up efforts in urban and ecological agriculture could collaborate. The philosopher André Gorz is used by the architects in their explanation of the R-Urban concept: "produce what we consume and consume what we produce." This goal led to the R-Urban project, which examined the potential for closed production-consumer cycles in the city, recycling methods, neighborhood self-organization, and common housing. Lastly, two projects were completed in Paris: the recycling education facility RecyLab and the AgroCité farm (Nowysz et al. 2022).

The objective of the vast Sahara Forest Project (SFP) in Qatar, Tunisia, and Jordan is to demonstrate the possibility of applying biomimicry in various closed-loop model problems. This includes the harvesting of fresh water in a passive manner, transitioning to a solar economy, regeneration of land, carbon sequestration in soil, closure of the nutrient cycle, and utilization of large populations. It aims to use restorative techniques to reduce the strain on natural resources, reverse the trend of desertification, and plant vegetation in arid areas (Verbrugghe et al. 2023). Architects and engineers took inspiration from the Namib Desert beetle's capacity to withstand fog because the climate does not permit the use of fresh water for crops. Although the Namibian Desert receives only 1 cm of precipitation annually, fog events from the Atlantic Ocean occur there in the mornings. This creature developed to effectively capture fog with its back, which alternates between hydrophilic and hydrophobic regions where droplets are transferred to the beetle's mouth and where they gather (Hamdy 2022). In order to provide the most suitable conditions for growth and to carry out agricultural activities throughout the year, this biological technique was incorporated into saltwater-cooled greenhouses. The temperature of the greenhouse's interior is reduced by the flow of humid air, which is created by fans blowing the desert's air onto seawater to facilitate the process of evaporation.

The source of freshwater for irrigation is created by condensing moist air through pipes that are cooled by seawater. In a CSP plant, solar power generates electricity that powers turbines, generating the necessary electricity for the operation of the greenhouse. Cucumbers have been grown in a pilot plant in Qatar, utilizing half the amount of freshwater that would be required in a normal environment. The majority of the characteristics of an ecosystem are met by SFP, including nutrient-rich crops, diversity, energy benefits from solar power, and a regenerative nature (Rani et al. 2024; Rasool et al. 2024).

The Gate Residence (CGR) in Cairo incorporates bioclimatic design, biomimicry, and vernacular architecture. Utilizing endemic plan species, geothermal energy, the solar cycle, prevailing wind directions, and other factors to make the design as passive as possible, the project combines traditional wind-catcher techniques from Egypt, Iran, and the Gulf region with a significant use of renewable energy systems to save 50% of energy demand. Its goal is to turn the area into a forest and the city into an ecosystem. Traditional wind catchers known as "malqaf" were used in Egyptian buildings as early as 1300 BC. It is a shaft that rises far above the structure and faces the predominant wind through a unidirectional aperture. It captures the wind that blows over the structure and directs it into the house's areas via internal air ducts. Similar to termite mounds and the EDH, the "maqlaf" and a higher tower on the other side of the building for hot air escape function together in bigger structures, such as mosques. The vernacular "maqlaf" is ignored by "modern" buildings from the 1960s onwards, which require massive cooling and energy-intensive systems. The CGR uses the "malqaf" in the shape of nine enormous chimneys, or "megatrees," which direct airflow in three different directions: upward via a temperature gradient aided by wind, downward by direct wind entrance, and upward via a solar-assisted temperature gradient (Verbrugghe et al. 2023).

18.8 Challenges and Opportunities in Biomimetic Integration

Applications of biomimicry in a variety of fields, including technology and material science, have shown promise in generating creative answers to today's problems. However, there are a number of obstacles to biomimetic applications in the built environment, such as the lack of biological expertise among planners and architects and the absence of a suitable, shared method for converting biomimetic ideas into urban-relevant strategies. The current absence of an integrated methodology and the dearth of finished examples that offer measurable proof of the advantages of regenerative built environments are the primary obstacles to regenerative development implementation. Regenerative design is defined as "building capacity not things," in which structures are not viewed as objects but rather as systems that interact with one another, the natural world, and their human occupants. Prominent experts on regenerative design contend that a change from a built environment that eventually damages ecosystems to one that revitalizes local environments and renews the ability of ecosystems to flourish will necessitate a fundamental rethinking of both the architectural design and, as Hunt notes, the competitive economic landscape of the built environment today (Borham et al. 2024).

According to some academics, the use of biomimicry in the built environment faces a number of difficulties, despite the fact that the strategy of using biomimetic techniques is promising to achieve regenerative built environments. First of all, the term "biomimicry" lacks a clear and consistent meaning, making it difficult to comprehend the implications of using its abstract principles. Secondly, there is a lack of relevant approaches to help integrate it into urban and architectural design. Thirdly, a major issue that deters architects and urban planners from attempting to use biomimicry in their designs is the lack of widespread access to biological knowledge. Because of this, some academics have proposed early on in the design process to include a biologist on the design team (Ahamed et al. 2022). Last but not least, there are additional knowledge hurdles that prevent designers from applying biological principles to designs and technologies that would be used in the built environment, even if they were aware of them. Even though more people are becoming aware of biomimicry's potential, it is still not widely used (Jamil et al. 2022; Khoja et al. 2022; Nasar et al. 2024b; Shafqat et al. 2024).

18.9 Agroecological Approaches

By combining ecological and socioeconomic principles, agroecological techniques offer a revolutionary framework for sustainable agriculture that improves the resilience and production of food systems. These approaches address important global issues like food security, climate change, and biodiversity loss by emphasizing biodiversity, environmental services, and social fairness. The fundamental ideas, methods, and effects of agroecology are described in the sections that follow. Agroecology merges ecological principles with the socio-economic realities of agriculture, offering a transformative path toward sustainability. It enhances food system productivity, resilience, and equity by conserving biodiversity, supporting ecosystem services, and upholding social justice. Through agricultural diversification such as crop rotation, intercropping, and integrated pest management (IPM), agroecology reduces reliance on synthetic inputs, instead harnessing natural processes for nutrient cycling, pest control, and soil fertility. It also strengthens the livelihoods of smallholder farmers by promoting resource-efficient, locally adapted practices. Agroforestry further contributes by sequestering carbon, preserving soil health, and boosting climate resilience. Collectively, these approaches foster sustainable food production, protect the environment, and support local economies. Yet, realizing agroecology's potential requires supportive policies, capacity development, increased research funding, and educational initiatives (Reddy 2017). Even though agroecology has many benefits, obstacles including low knowledge and a lack of funding prevent it from being widely used. To reach its greatest potential in changing agricultural practices toward sustainability, these obstacles must be removed.

18.10 Core Principles of Agroecology for Sustainable Farming

Since the 2000s, agroecology has broadened its scope to include the entire food system, moving beyond the field, farm, and agroecosystem scales. Over the years, the range of issues covered expanded, as did the many agroecological expressions (research, practice, and social movements). This was mirrored in the growing number and variety of agroecological principles. There are different sets of agroecological principles in the scientific literature. They are applied at different scales (field, farm, landscape, or food system) or to different aspects of food systems (production, governance, etc.). They include both normative elements that state values (for example, food systems ought to be equitable) and causative elements that describe relationships (for example, more biodiverse food production systems are likely to be more resilient), as in science. In addition to certain more general socio-economic, cultural, and political ideas, agroecology is now linked to a set of guidelines for the ecological and agricultural management of agri-food systems. These latter ideas have just lately surfaced in the literature as a result of social movements' use of agroecology (Wezel et al. 2020).

Many contend that merely adopting certain practices and altering certain technologies is insufficient to discourse food security and nutrition (FSN) and make so-called industrial agricultural systems sustainable; instead, agroecological principles must be applied and farming systems must be redesigned. More precisely, some of these principles speak to the enhancement of ecological services and processes, such as those related to soil, water, air, and biodiversity. Among these are (i) biomass recycling; (ii) functional biodiversity enhancement; (iii) providing favorable soil conditions for plant growth; (iv) minimizing losses; (v) species and genetic resource diversification within the agroecosystem; and (vi) enhancing advantageous biological interactions and synergies. The authors added two more specific animal production principles to agroecological methods involving animals: (i) implementing management strategies that aim to improve animal health, and (ii) increasing variety within animal production systems to strengthen their resilience concepts of agroecology, particularly as they relate to grass-based farming systems (Migliorini and Wezel 2017). The socioeconomic principles of agroecology are financial independence, access to and autonomy in the markets, diversity of knowledge and experience, equality in society, democracy in the management of the community, and the production of community knowledge.

Agroecological principles were also created by Cooperation Internationale pour le Développement and la Solidarité (CIDSE) in collaboration with other civil society organizations. They divided the many tenets into four groups: political, economic, sociocultural, and environmental. Some of these ideas are in line with the goals and demands of numerous civil society organizations, who are working to promote family and smallholder farming as well as sustainable livelihoods in the Global South with equitable production and market conditions. Similar to this, the Information, Koordination, Tagungen (INKOTA) network identified eleven co-equal components to fully utilize agroecology, emphasizing components pertaining to participation, rights, control over livelihoods, and decision-making voice (Wezel et al. 2020).

The ten elements of agroecology, namely diversity, information co-creation, synergies, efficiency, recycling, resilience, human and social values, cultural and food traditions, responsible governance, and circular and solidarity economy, were first presented in a paper by the FAO. The aim of the HLPE report was to integrate these various points of view on agroecology principles by compiling and summarizing the many publications that reflect an increasingly large number of principles, existing statements of principles, and elements. This led to the presentation of 13 principles that include both normative and causative elements. The principles also make it possible to differentiate between biodiversity and economic diversification, which are mixed in the single element of diversity, and to be clear about maintaining soil and animal health, whereas these elements are included in the elaboration on many elements. The elements presented by the FAO are also different from one another, whereas the resulting principles are presented as action elements that include both normative (ensuring animal health and welfare) and causative (increasing participation in decision-making supports decentralized governance and hence facilitates local adaptive management) elements. Resilience and efficiency are quantifiable system elements or outcomes, whereas responsible governance, circular economy, and solidarity economy are associated with governance and improvement of food systems (Barrios et al. 2020).

In agriculture, improvements in one type of efficiency, such as yield per land, have often gone hand-in-hand with decreases in other types of efficiency, such as yield per fossil fuel input or biodiversity loss. The concept of efficiency is broad and connects outputs to inputs in such a way that numerous types of efficiencies are possible. One important aspect of the consolidated principles is that, despite their generic formulation, they are implemented locally in reality, leading to a variety of agro ecological practices that are appropriate for the local environment. The importance of co-creation of knowledge, which includes the equitable participation of various stakeholders, such as local knowledge of farmers in developing countries, is also strongly emphasized in the FAO elements as well as in the consolidated principles. It is also a basic tenet of transdisciplinary in an agricultural context.

18.11 Practices Enhancing Soil Health and Biodiversity Conservation

Particularly rich and diversified microbiomes and fauna are evolving in soils, which are living habitats. The biological functioning that results directly affects a number of ecosystem services in addition to soil fertility. Therefore, microbial communities participate in geochemical cycles where the various phases are catalyzed by microbial enzymes. Since these activities have an impact on plant development and environmental quality, they must be modified. Generally speaking, agroecosystem stability and productivity are impacted by biodiversity (Fig. 18.4). Therefore, when planning cropping systems and assessing their effects, it is crucial to consider soil

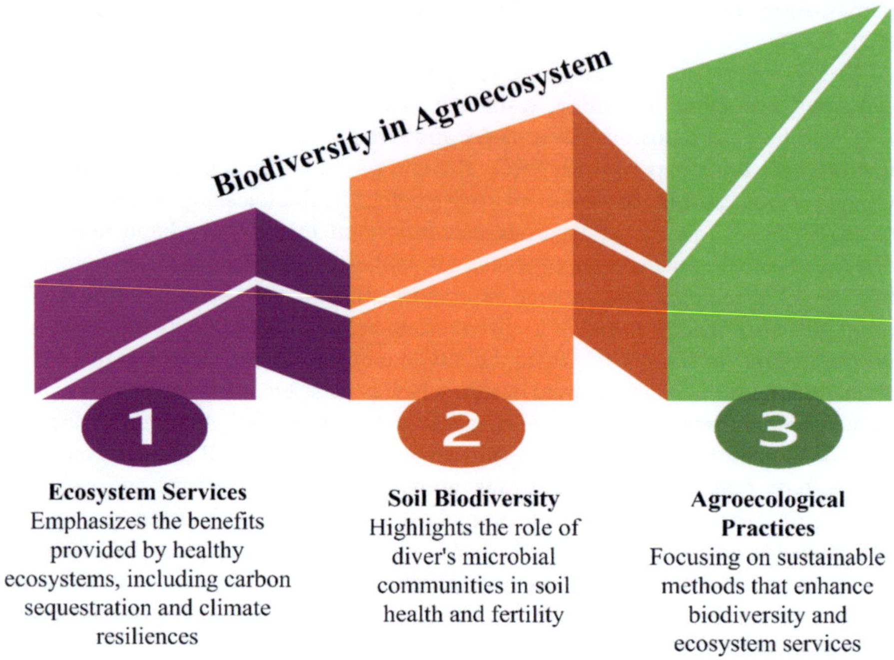

Fig. 18.4 Key components of biodiversity in the agroecosystems

biodiversity and biological functioning. In addition to using fossil fuels, inputs, and natural resources economically, agriculture must inevitably sustain high and consistent yields while incorporating biodiversity into the production process itself. More and improved production is implied. These issues necessitate the creation of novel farming systems that provide the anticipated ecosystem services while using less inputs and showing greater environmental respect. By making agroecosystems resilient to disturbances and after disturbances have passed, biodiversity does in fact contribute to their productivity and the ecological insurance that goes along with their stability. The fundamental topic of biodiversity conservation highlights how agroecology, through techniques like agroforestry, cover crops, and reduced chemical inputs, helps preserve native species, pollinators, and beneficial organisms. Crop diversification, polyculture, and integrated pest management are examples of agroecological concepts that enhance food security by boosting nutrition and stability. Agroecology promotes soil health, carbon sequestration, and greenhouse gas reductions, leading to farming systems that are climate resilient (Kazemi et al. 2018).

Agroecological methods are essential for reducing the effects of climate change. The utilization of agroforestry systems, decreased reliance on artificial fertilizers and pesticides, and soil carbon storage are some of the tactics that help lower greenhouse gas emissions. Furthermore, the varied cropping systems that agroecology promotes improve soil health and increase resistance to stressors brought on by the climate. Agroecology shows itself to be a ray of hope among the alarming decline in biodiversity. Agroecology actively supports biodiversity through promoting a variety of crop types, mixed agricultural systems, and habitat preservation. In addition to improving ecosystem health, lowering chemical inputs and encouraging natural pest management methods also greatly aid in the preservation of biodiversity worldwide. Effective water management is made possible by the implementation of strategies like drip irrigation, rainwater collection, and soil conservation. Enhancing soil structure and water-retention capabilities in agroecological systems reduces water demand while preserving agricultural output (Migliorini and Wezel 2017). A variety of techniques, including rotational grazing, cover crops, and conservation tillage, have been embraced by farmers. These methods improve soil water retention and reduce water runoff, making crops and pastures more resilient to prolonged dry spells. As a result, there is now greater resistance to drought and less susceptibility to production hazards associated to water.

The corn-soybean-wheat rotation is a traditional agroecological technique that has gained widespread use in the US. During the course of several seasons, corn, soybeans, and wheat are grown in succession on the same field. By adjusting nutrient requirements and decreasing insect accumulation, this technique improves soil fertility and produces resilient and sustainable crop yields. Farmers in sub-Saharan Africa now have the ability to adapt to shifting climate patterns thanks to the International Maize and Wheat Improvement Center's (CIMMYT) creation of drought-tolerant maize varieties. Farmers can escape drought spells and obtain more consistent crops thanks to these upgraded maize varieties' shorter growth cycles (Prasanna 2016).

Cover crops and residue mulching are two strategies used by Brazilian farmers that practice no-till agriculture to enhance soil structure and lessen erosion. By improving soil health, decreasing soil evaporation, and increasing water infiltration, these techniques allow for adaptation to changing rainfall patterns. Parts of Africa have used the "push-pull" intercropping strategy. It entails planting sorghum and maize intercropped with desmodium, a legume that fixes nitrogen, and forage grasses. Desmodium adds nitrogen to the soil and keeps stem borers away from sorghum and maize. This agroecological method improves soil fertility and pest management. Holistic planned grazing in Australia entails the regulated movement of cattle between pastures. This method improves biodiversity, lessens soil erosion, and supports the health of grasslands (O'Grady et al. 2024). By guaranteeing appropriate land management and sustained pasture utilization, it adjusts to shifting climatic circumstances.

18.12 Socioeconomic Impacts of Agroecological Farming

When compared to industrial and conventional agriculture, agroecology may yield greater financial benefits. It is now crucial to assess agroecology's socio-economic viability, which is a key factor in its upscaling, since a broader adoption of the practice can hasten the achievement of sustainability goals like those established by the European Green Deal, the Sustainable Development Goals, and, more recently, the UN Food System Summit through the Coalition for Food Systems Transformation through Agroecology. Depending on the practice under evaluation, the findings of agroecology's socioeconomic success differ considerably. Over 50% of the results are favorable for the two management strategies that are most frequently assessed, namely intercropping and agroforestry. Compared to 26% and 36% of bad outcomes, respectively, these activities are linked to 53% of positive outcomes. In terms of tillage management, the results are overwhelmingly favorable (58%), followed by cover crops and mulching (58%), and pest and disease control (67%). Positive and negative results for cropping system diversification are nearly equal (35% vs. 33%, respectively). This category shows a high proportion of neutral and inconclusive results (32% for both combined). Negative consequences are more prevalent than positive ones in the case of organic fertilization (51% negative vs 43% positive). Both the package of practices and the management of landscape aspects have a high percentage of neutral and inconclusive results (44% for the former practice combined and 53% for the package of practices) (Mouratiadou et al. 2024).

By establishing the ideal conditions to optimize ecosystem services and biodiversity conservation at the lowest possible cost to farmers, agro-ecological practices that improve ecosystem functionality (such as offering pollinator nesting sites or improving soil functionality) have positive socioeconomic effects. In general, agro-ecological methods are more frequently linked to favorable socioeconomic results than to unfavorable or neutral ones. In such cases, the most widely studied ones such as intercropping and agroforestry. In the majority of reviewed socio-economic metrics—particularly financial capital indicators—positive effects on income, revenues, productivity, and efficiency are commonly reported, with a smaller but still favorable impact on production costs. Compared to conventional methods, agro-ecological techniques are often associated with higher labor requirements and costs. However, they can also generate higher returns on investment. The three core principles of conservation agriculture (CA)—minimum soil disturbance, permanent soil cover, and crop rotation—are applied to achieve sustainable and profitable farming (Khan et al. 2022a; Khan et al. 2024; Nasar et al. 2024a). Rather than being site-specific prescriptions, these principles serve as adaptive goals for effectively scaling CA across diverse production environments. Under certain conditions, CA has been shown to maintain or increase crop yields, enhance soil fertility, and reduce soil erosion relative to conventional agriculture.

Transitioning from conventional tillage to conservation tillage systems (e.g., no-till, strip-till, ridge-till, mulch-till) can substantially reduce diesel fuel energy consumption. This reduction results from fewer tillage operations, adapted irrigation practices, and lower labor inputs (Farooq and Siddique 2015). Fewer field operations save time and money by lowering labor, fuel, and machinery maintenance costs, while also reducing soil compaction—thereby improving plant growth conditions. Additionally, the use of cover crops decreases the need for costly chemical inputs.

18.13 Nature-Based Solutions (NbS)

Nature-based solutions (NbS) utilizes nature for coping with environmental challenges and also for socio-economic benefits. NbS helps in enhancing sustainability, resilience, and productivity in cities and agriculture. NbS is a group of broad approaches that encompass a vast array of terms and approaches related to nature, for solving societal challenges (Fig. 18.5). Other terms that are associated with NbS include ecosystem-based adaptation (EbA), agroforestry, agro-ecology, green and blue infrastructure, natural climate solutions (NCS), reduced emissions from deforestation and degradation (REDD), blue carbon, and natural capital. According to IUCN these approaches categorize into five main NbS approaches: Curative (e.g., ecological engineering, ecological restoration); Issue-specific (e.g., ecosystem-based adaptation, ecosystem-based disaster risk reduction); Infrastructure-related (e.g., green infrastructure); Management (e.g., Integrated coastal zone management); and Protection (e.g., Area-based conservation) (Seddon et al. 2021).

NbS can significantly enhance quality of urban environments by incorporating green infrastructure GI (green roofs and solar energy systems). Research in Torino, Italy, recognized 2450 m^2 of appropriate roof area for GI, which could be utilized to

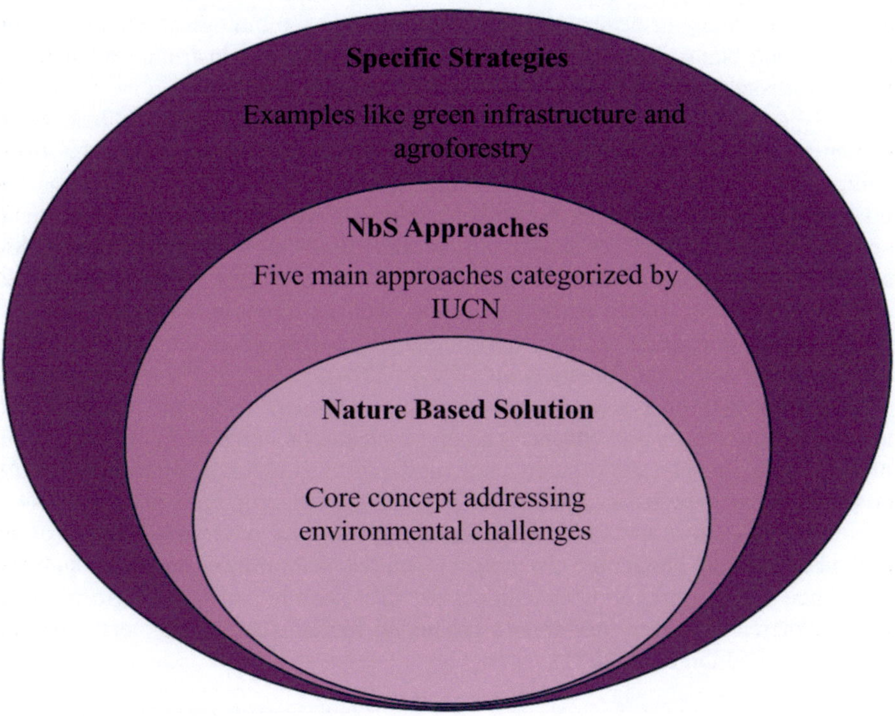

Fig. 18.5 Nature based solutions framework

contribute urban sustainability thereby uplifting quality of life. Coastal NbS can influence to minimize negative climate impacts. These strategies include habitat creation to hybrid structures, facilitating ecological economic assistances. Appropriate execution can ensure lasting practicability. Advanced ecosystem services and enhanced yield agriculture, can be acquired by Agricultural NbS. Economic assessment of NbS in field of agricultural is needed to incorporate them as sustainable development and conservation strategies. In order to maximize NbS advantages concerns such as operation costs and maintenance necessities must be addressed.

18.13.1 Enhancing Agricultural Productivity Through NbS

The collaborative efforts in the recent past have been aimed at improving the quality of the soil, both in terms of chemical-physical (such as increasing the cover crops, reducing tillage intensity and frequency, increasing the use of perennials, and developing site-specific soil and crop management to increase soil organic carbon, etc.) and biological (such as using microorganisms to improve soil quality, etc.) components. These concepts and implementations are aimed at: (1) manipulating the potential use of microorganisms to manage soil degradation, thus improving the hydraulic characteristics of the soil, such as infiltration, water retention, and reducing hydrophobicity; (2) developing protocols for the nationwide monitoring of soil health; (3) developing better indicators for assessing the biological, chemical, and physical well-being of the soil; (4) developing sensors and instruments for faster on-site assessments.

NbS can reproduce natural processes and meet the demands of land-water management concepts, which require the simultaneous improvement of water availability and quality, as well as agricultural productivity. Natural ecosystems are preserved, and this leads to the enhancement of natural processes in artificial ecosystems. Acquiring NbS in agricultural landscapes, serve to restore soil health, soil moisture, carbon mitigation (through soil and forestry), downstream water quality protections, biodiversity benefits as well as agricultural production and supply chains to acquire net-zero environmental impacts while achieving food and water security and meet climate goals (Katyal 2024).

Adoption of these systems help to minimize sediment and nutrient runoff into adjacent watercourses and enhance carbon sequestration and storage. Conservation of agriculture enhance soil organic matter and improved sediment retention; though, crop productivity varies on practices employed, climate, crop type, and biophysical conditions. Agroforestry systems conserve diversity of wild foods and provide pollinator habitat, altogether can help to combat malnutrition and micronutrient deficiencies. Incorporation trees with pasture Silvopasture (subset of Agroforestry), aims to increase pasture quality and producing fodder serves to protect soils and vegetation (Seddon et al. 2021).

18.13.2 Economic and Ecological Benefits of NbS Implementation

NbS have been accepted as global experiences to achieve agriculture-derived resilience. Such as, trees in croplands or agroforestry a can serve multiple economic, cultural and ecological assistances. Agroforestry's differentiated cropping systems copies natural forests form an important part of indigenous food production systems around the world and are also being used as a modern agricultural top management practice in non-traditional contexts. These approaches appear to be hardy, productive, pest resistant, nutrient-conserving, and biodiverse, providing multiple economic, cultural and ecological benefits. For instance, agroforestry can provide fuelwood, cultivated foods, timber, and medicinal plants to serve local communities, while also backup biodiversity. These systems minimize sediment and nutrient runoff into adjacent watercourses and enhance carbon sequestration and storage. Agroforestry systems and Silvopasture conserve biodiversity and producing fodder while also protecting soils and vegetation (Raymond et al. 2017).

Another type of NbS in agricultural landscapes, preserve agriculture by conservation tillage, crop rotations, and cover crops has gained traction in many parts of the world. The ideologies of conservation agriculture have been part of traditional agricultural systems for generations in some regions of the world such as United States, Brazil, Argentina, Australia, and Canada which ultimately has been cited as benefits for farmers. Adoption of blue-green infrastructure system in Augustenborg has been known to flood risk reduction (minor and severe flood events), hence the positive effects of the on socioeconomic development and environmental impacts, as well as the enhancement of biodiversity (50%) by creating an environment conducive to the survival of local flora and fauna. The present application of blue-green infrastructure (BGI) is to mitigate the adverse effects of urbanization (flooding) and to counter the expected effects of climate change (flooding and droughts).BGI is serves positive environmental and social functions. It serves as apart from stormwater management (sustainable urban drainage systems), providing greenways and ecological networks, which are important components in the concept of green infrastructure (Keita and Kourouma 2024). Blue-green infrastructure provides services as purification of water, flood control, water storage, and heat control. In past nature-based solutions for stormwater control were installed in Augustenborg area in southern Sweden to adapt to flooding and achieve sustainable urban development, involve residents in development of their neighborhood and enhance biodiversity for sustainable environment (Tassadduq et al. 2022; Ullah et al. 2022).

Though, embracing of NbS in agriculture is providing promising benefits to livelihoods but several studies suggest reservations of framers to adopt due high initial costs, labor cost, or customs and preferences. Collective economic, attitudinal and farm structural features effect to a range of situations from riparian buffer zones in the EU to coastal zone supervision in Southeast Asia.

18.14 Discussion

Global climate change is subjected to natural factors and human activities that impact biodiversity, agricultural production, and food security. The most affected are the less adapted and endemic species that are prone to extinction. Therefore, global climate change addresses issues such as extinction of species, food security globally, and primary health. Though climate change impacts biodiversity and food security, little is known about the extent of the problem. Thus, the objectives of this chapter are to recognize, appraise, and synthesize the relationship that exists between climate change and biodiversity and food security. Data and outputs obtained from previous publications were used. The global climate changes led to the migration of species. Thus, the extinction rates of 1103 species under migration scenarios provide 21–23% with unlimited migration and 38–52% with no migration at all (Neale et al. 2021).

Short-term environmental changes induce plastic phenotype, which is beneficial to the species in dealing with the effects of climate change. Moreover, communities that depend on rain-fed agricultural activities are adversely affected by climate change. It is a known fact that there is an optimum tolerance level beyond which growth is affected. It is estimated that agricultural crops in Africa are expected to lose over 30% of their yield by 2050. The best way to combat food insecurity is to protect biodiversity instead of cultivating more land for agricultural purposes, creating new fish stocks (a costly measure) (Amir et al. 2020). Thus, modifying food waste, rewarding those who are food insecure, conserving biodiversity, using genetic resources, and using ecological knowledge can combat biodiversity loss, thus ensuring food security in the context of climate change. Food security can be ensured through strong policies, using high-yielding varieties of crops, developing climate resilient irrigation structures, and agricultural systems. Highly recommended strategies to cope global climate change include restoration of degraded land, land use changes, switching to bio-energy, sustainable forest management, and community-based biodiversity conservation (García-Vega and Newbold 2020).

Maintenance of bio-diverse ecosystems is essential for world's sustainable food production, provision for food industries and related services. Human food form both wild and cultivated (vegetables, fruits, nuts, meat) and adjuncts to food (food colorants, flavoring, and preservatives) can be obtained by preserving products of biodiversity in wide array. Among the estimated 300,000 flowering species, approximately 40% of these are thought to be edible to man, actual numbers may be far higher perhaps 150 domesticated food species. Over 75% of the human food supply is, of 12 species of plant (bananas, beans, cassava, maize, millet, potatoes, rice, sorghum, soybean sugar cane, sweet potatoes, and wheat). As the annual production of cereals, roots, and tubers in the world was approximately 2.07 billion tons, 0.64 billion tones in 1996–1998, respectively (Muluneh 2021). Biodiversity conservation not only holds promise of food security but it also offers services needed to sustain the agricultural system (watershed services, soil fertility, pollination, seed dispersal, nutrient cycling, natural pest and disease control). The avian and pollinator diversity and counts ultimately decrease in intensive agricultural systems (weeding away wild insects, predators, and other pests) as a result of the homogenization of insect populations. Adoption of conventional agriculture not only reduce species' diversity and ecosystem function but also boasts synanthropic species (well adapted to anthropogenic landscapes) (Dhakad et al. 2022).

Biodiversity can be arranged into different interrelated levels, which include genetic diversity, species and subspecies diversity, diversity of functional qualities, diversity between populations or communities of species, ecosystem or habitat diversity, diversity between large zones of the landscape, and finally, global diversity. The examination also considers food security and biodiversity. Enough food (hypothetically and in terms of average calorific units) is being produced to support the global human population could be produced as an alternative production regime supporting biodiversity. The relationship between global and regional biodiversity is somewhat more complicated, in that regions of the world where food is available in sufficient amounts do not automatically mean that food is available in all regions of the globe. There is not necessarily an amount of biodiversity that each region should have, but rather each ecosystem must develop desirable qualities such as stability, resilience, recovery, and production (Patil 2020).

Due to widespread global environmental impacts, there is an urgent need for modification of traditional farming systems in order to achieve food security and nutrition that provides social and economic equity, build, and protect ecosystems. This has led to the development of several alternatives, e.g., nutrient cycling, pest control, or pollination, in order to achieve resilient and productive farms. The main aim of biodiversity conservation is to minimize external inputs with ecosystem services, or explore complementarity or positive interactions (Dawson et al. 2019). The implementation of alternatives is recognized as foundations for global changes in agricultural practices, which is in line with the UN 2030 Agenda for Sustainable Development and IPBES.

Ecologically inspired approaches have considerable taxonomic level attractiveness, at the field scale, for biodiversity conservation. For instance, polyculture crop can display greater yield resilience compared to mono-culture plantings. Crop yield resilience is greater to climatic agitations in polyculture (genetically diverse) forage systems; stocking multiple livestock species may also display greater resilience. There is limited evidence for crop species in polyculture having merits; increased crop type diversity in rotation may display greater yield resilience. Breeding crop and forage species that display greater resilience to perturbations and are less impacted by climatic agitations, with high plasticity, is also an approach. For instance, crops like maize display local adaptation to extreme climates, similar to a resource (Petrie and Bates 2017). Agroecosystem broadly, enhancing resilience of crops by improving soil condition, such as by agroforestry or adding organic matter; or by maintaining wild bee diversity to provide a resilient pollination service (Sinha et al. 2023; Tariq et al. 2023).

Crop resilience can be achieved at the farm scale, by diverse crop rotations. Farmer show integrated way to manage for resilience. This entails flexibility in crop selection and/or forage and/or livestock selection and husbandry, and the adoption of new technologies in reaction to changing conditions. For instance, the selection of rice varieties, the expansion of irrigation systems and the adoption of technologies such as micro-irrigation are adopted by farmers to result in drought-resistant rice in India. Traditional farming approach such as raised bed cultivation systems adopted for seasonally flooded savannas may confer resilience, in Latin American. Nevertheless, restoring traditional farming ought not be seen as a remedy as method of acquiring resilience, and it does not necessarily help to cope with climate and other environmental changes. In the same way, there are less evidences about biodiverse landscapes in and around farms enhance agroecosystem resilience.

Resilience in food production systems such as crop diversification in the EU should be ensured at the regional and global levels in a coordinated way. Educating farmers and developing capacity and social networks might be more effective in helping farmers to organize themselves in addressing issues of resilience. Globally there is a need for these institutions to adapt and transform complex and interacting consequences for acquiring resilience food production, impacts on resources, costs and collective actions (Borham et al. 2024). Development of by developing links between biodiversity and agroecosystem resilience can be break through development in Ecological and agronomic research.

18.15 Future Prospects

Aim to reduce future biodiversity loss and food scarcity can be achieved by assessing appropriate approaches and strategies to address global climate change. Global climate change and its multidimensional effects on biodiversity and food security need to be understood in terms of scientific evidence. This study also focuses on the need to understand and address the tracing, quantification, and examination of the current and predicted effects of climate change on biodiversity and global food security. The scientific evidence and information on global issues like climate change, biodiversity loss, and food security, along with their mutual relationship and interlinkages, are significant for understanding and developing solutions for these issues. Similarly, this paper also aims to create better understanding and mutual communication between scientific communities, researchers, and international and national policymakers regarding issues like climate change, biodiversity loss, and food security. Moreover, it also aims to create awareness and understanding regarding the impacts of climate change on biodiversity and food security, along with associated strategies for adapting to climate change. In this context, it is imperative that climate change, biodiversity loss, and food security need to be addressed by the collective wisdom and intelligence of all nations and communities worldwide. For this paper, data and scientific information from the aforementioned publications have been used by re-reading and re-interpreting it in terms of the impact of global climate change on biodiversity and food security. For this purpose, relevant and available published articles have been used up to 2021 for the preparation of this paper. Quantitative data regarding numerical value for climate change's impact on biodiversity and food security have been adopted from the aforementioned literature. In addition, previously developed climate model scenarios and emission scenarios were also used (Muhammad et al. 2024a).

Researching and releasing high-yielding varieties that are resistant to biotic and abiotic stresses and are suitable for various agro-ecologies are recommended as important adaptation and mitigation strategies against the effects of global climate changes. Therefore, varieties that are resistant to biotic and abiotic stresses under changing climatic conditions should be made available at the farm level with adequate financial and policy support. In addition to this, the adoption of production technologies that improve water productivity through irrigation practices is also important. In this regard, further research on the ecological requirements of species, ecosystem-based climate change mitigation measures, species adaptation pathways, and the symbiotic relationships between plants and microbes is important. In this context, further research on the development of technologies such as bio-fertilizers, wild edible plants, and microbial inoculants that are less dependent on scarce resources such as arable lands and water is important. In this regard, the development of climate-resilient irrigation structures or systems and the development and use of bioenergy and alternative energy resources are important. In this context, the development and improvement of natural resource management and natural resource utilization practices, agro-biodiversity conservation, and climate-smart agriculture are important. In addition to this, the application of biodiversity conservation practices and the implementation of land restoration programs at the farm level are important. Therefore, in the context of the above-mentioned issues and concerns, it is suggested that the scope of the future research and development should be focused on the exploration and innovation of climate-resilient technologies and the development of alternative and carbon-neutral energy resources and the development and utilization of bioenergy and wild food resources.

18.16 Conclusion

Biodiversity across genetic, species, and ecosystem levels forms the foundation of agricultural resilience. It underpins essential ecosystem services such as nutrient cycling, pest regulation, and pollination, which together sustain productivity in the face of climate change, resource depletion, and habitat loss. Practical strategies that harness biodiversity include urban agriculture inspired by biomimetic principles, agroecological practices (polycultures, integrated pest management, carbon sequestration), and nature-based solutions (NbS) for soil and water conservation. While these approaches offer a balanced pathway toward productivity, environmental sustainability, and social well-being, their widespread adoption is hindered by socioeconomic inequities, weak policy support, and a lack of long-term empirical data. Context-specific interventions tailored to local environmental, cultural, and economic conditions are therefore essential. Climate change exacerbates vulnerabilities by altering species distributions and ecosystem functions. Proactive measures such as breeding stress-resistant crop varieties, restoring degraded ecosystems, and integrating traditional ecological knowledge with modern innovations are urgently needed. Empowering smallholder farmers and local communities through education, capacity-building, and equitable resource access is critical to success. By embedding biodiversity into global food systems, we can advance multiple Sustainable Development Goals (SDGs), including Zero Hunger, Climate Action, and Life on Land. Realizing this transformative pathway demands cooperative efforts at local, national, and international levels to drive research, foster innovation, and scale effective practices. Ultimately, biodiversity is not merely a conservation target—it is the vital foundation upon which resilient, sustainable, and equitable food systems are built for future generations.

References

Ahamed MK, Wang H, Hazell PJ (2022) From biology to biomimicry: using nature to build better structures–a review. Constr Build Mater 320:126195. https://www.sciencedirect.com/science/article/pii/S095006182103926X

Amir S, Saqib Z, Khan MI, Ali A, Khan MA, Bokhari SA (2020) Determinants of farmers' adaptation to climate change in rain-fed agriculture of Pakistan. Arab J Geosci 13:1–19. https://doi.org/10.1007/s12517-020-06019-w

Arif U, Bhatti KH, Ajaib M, Wagay NA, Majeed M, Zeb J, Hameed A, Kiani J (2021) Ethnobotanical indigenous knowledge of Tehsil Charhoi, District Kotli, Azad Jammu and Kashmir, Pakistan. Ethnobot Res Appl 22:10.32859/ERA.22.50.1-24

Arshad F, Waheed M, Harun N, Fatima K, Khan BA, Fatima K, Abbas Z, Jabeen S, Majeed M (2022) Indigenous farmer's perception about fodder and foraging species of semi-arid lowlands of Pakistan: a case study of District Kasur, Pakistan. Taiwania 67(4):510–523. https://doi.org/10.6165/tai.2022.67.510

Barrios E, Gemmill-Herren B, Bicksler A, Siliprandi E, Brathwaite R, Moller S et al (2020) The 10 elements of agroecology: enabling transitions towards sustainable agriculture and food systems through visual narratives. Ecosyst People 16(1):230–247. https://www.tandfonline.com/doi/full/10.1080/26395916.2020.1808705

Barwant MM, Rasool A, Mannam VK, Hussain A, Abdullah M, nisa Memon Z et al (2024) Nanotechnology applications in livestock farming and animal health. In: Harnessing nanoomics and nanozymes for sustainable agriculture. IGI Global, pp 413–442. https://doi.org/10.4018/979-8-3693-1890-4.ch020

Bashir SM, Altaf M, Hussain T, Umair M, Majeed M, Mangrio WM, Khan AM, Gulshan AB, Hamed MH, Ashraf S, Amjad MS, Bussmann RW, Abbasi AM, Casini R, Alataway A, Dewidar AZ, Al-Yafrsi M, Amin MH, Elansary HO (2023) Vernacular taxonomy, cultural and ethnopharmacological applications of avian and mammalian species in the vicinity of Ayubia National Park, Himalayan Region. Biology 12(4). https://doi.org/10.3390/biology12040609

Bhattacharya R, Bhattacharya R, Majeed M, Bhandari S, Aziz R, Sinha D et al (2023) Environmental rehabilitation of industrial waste dumping site. In: Biohydrometallurgical processes. CRC Press, pp 44–89. https://doi.org/10.1201/9781003451457-4

Borham O, Croxford B, Wilson D (2024) Biomimetic strategies for sustainable resilient cities: review across scales and city systems. Biomimetics 9(9):514. https://doi.org/10.3390/biomimetics9090514

Caron P, Ferrero y de Loma-Osorio G, Nabarro D, Hainzelin E, Guillou M, Andersen I et al (2018) Food systems for sustainable development: proposals for a profound four-part transformation. Agron Sustain Dev 38:1–12. https://doi.org/10.1007/s13593-018-0519-1

Chakraborty SK, Chakraborty SK (2021) Land-use changes: floodplains, dams, and reservoirs–integrated river basins management. In: Riverine ecology volume 2: biodiversity conservation, conflicts and resolution, pp 531–607. https://doi.org/10.1007/978-3-030-53941-2_6

Chandrasegaran S, Carroll D (2016) Origins of programmable nucleases for genome engineering. J Mol Biol 428(5):963–989. https://www.sciencedirect.com/science/article/pii/S0022283615006063

Das IK, Rakshit S (2016) Millets, their importance, and production constraints. In: Biotic stress resistance in millets. Academic Press, pp 3–19. https://doi.org/10.1016/B978-0-12-804549-7.00001-9

Dawson IK, Park SE, Attwood SJ, Jamnadass R, Powell W, Sunderland T, Carsan S (2019) Contributions of biodiversity to the sustainable intensification of food production. Glob Food Sec 21:23–37. https://www.sciencedirect.com/science/article/pii/S2211912418301019

Dhakad AK, Mohanapuria P, Ranjan R, Vaishnav V, Yadav SK (2022) Molecular approaches in conservation and restoration of agrobiodiversity. In: Agro-biodiversity and agri-ecosystem management. Springer, Singapore, pp 169–216. https://doi.org/10.1007/978-981-19-0928-3_10

Dobele M, Zvirbule A, Dobele A, Muska A (2022) Factors affecting the development of urban agriculture in Latvia. Res Rural Dev 37

Duru M, Therond O, Martin G, Martin-Clouaire R, Magne MA, Justes E et al (2015) How to implement biodiversity-based agriculture to enhance ecosystem services: a review. Agron Sustain Dev 35:1259–1281. https://doi.org/10.1007/s13593-015-0306-1

Farooq M, Siddique KH (2015) Conservation agriculture: concepts, brief history. In: And impacts on agricultural systems. Springer, pp 3–17. https://link.springer.com/chapter/10.1007/978-3-319-11620-4_1

García-Vega D, Newbold T (2020) Assessing the effects of land use on biodiversity in the world's drylands and Mediterranean environments. Biodivers Conserv 29(2):393–408. https://doi.org/10.1007/s10531-019-01888-4

Hamdy EA (2022) Achieving thermal comfort by simulating the ecosystems of the desert environment. In: Advances in architecture, engineering and technology. Springer, Cham, pp 227–242. https://link.springer.com/chapter/10.1007/978-3-030-86913-7_15

Haq SM, Yaqoob U, Majeed M, Amjad MS, Hassan M, Ahmad R et al (2022) Quantitative ethnoveterinary study on plant resource utilization by indigenous communities in high-altitude regions. Front Vet Sci 9:944046

Jamil MD, Waheed M, Akhtar S, Bangash N, Chaudhari SK, Majeed M, Hussain M, Ali K, Jones DA (2022) Invasive plants diversity, ecological status, and distribution pattern in relation to edaphic factors in different habitat types of district Mandi Bahauddin, Punjab, Pakistan. Sustainability 14(20). https://doi.org/10.3390/su142013312

Katyal N (2024) Understanding about nature-based approaches for sustainable consumption, production and development. In: Nature-based solutions in achieving sustainable development goals: harmonizing nature and progress. Springer, Cham, pp 317–367. https://link.springer.com/chapter/10.1007/978-3-031-76128-7_11

Kazemi H, Klug H, Kamkar B (2018) New services and roles of biodiversity in modern agroecosystems: a review. Ecol Indic 93:1126–1135. https://www.sciencedirect.com/science/article/pii/S1470160X18304588

Keita K, Kourouma S (2024) Blue-green infrastructure for urban resilience and sustainability in developing countries. In: Blue-green infrastructure for sustainable urban settlements: implications for developing countries under climate change. Springer, Cham, pp 31–55. https://link.springer.com/chapter/10.1007/978-3-031-62293-9_2

Khan AM, Altaf M, Hussain T, Hamed MH, Safdar U, Ayub A et al (2024) Ethnopharmacological uses of fauna among the people of Central Punjab, Pakistan. Front Vet Sci 11:1351693. https://doi.org/10.3389/fvets.2024.1351693

Khan AM, Li Q, Saqib Z, Khan N, Habib T, Khalid N, Majeed M, Tariq A (2022a) MaxEnt modelling and impact of climate change on habitat suitability variations of economically important chilgoza pine (Pinus gerardiana Wall.) in South Asia. Forests 13(5). https://doi.org/10.3390/f13050715

Khan SAKU, Rahman MM, Islam MM (2022b) Agro-biodiversity across the food chain. In: Biodiversity, functional ecosystems and sustainable food production. Springer, Cham, pp 1–40. https://doi.org/10.1007/978-3-031-07434-9_1

Khoja AA, Haq SM, Majeed M, Hassan M, Waheed M, Yaqoob U, Bussmann RW, Alataway A, Dewidar AZ, Al-Yafrsi M, Elansary HO, Yessoufou K, Zaman W (2022) Diversity, ecological and traditional knowledge of pteridophytes in the Western Himalayas. Diversity 14(8). https://doi.org/10.3390/d14080628

Koont S (2016) Sustainable urban agriculture in Cuba. University Press of Florida. https://books.google.com.pk/books?hl=en&lr=&id=3YDSEAAAQBAJ&oi=fnd&pg=PA1881&dq=The+US+government%27s+1962+embargo+and+the+economic+crises+associated+with+the+collapse+of+the+Eastern+Bloc+in+the+1990s+provide+historical+context+for+the+phenomena+of+urban+agriculture+in+Cuba.+&ots=qSNmbR8_04&sig=hND_YAv2mX49JzhQBz97kSWF_yg&redir_esc=y

Kreiner JM, Caballero A, Wright SI, Stinchcombe JR (2021) Phenotypic patterns and genomic underpinnings of contemporary agricultural adaptation across the range of Amaranthus tuberculatus. In: Population genomics of herbicide resistance & contemporary agricultural adaptation in Amaranthus tuberculatus, vol 126

Migliorini P, Wezel A (2017) Converging and diverging principles and practices of organic agriculture regulations and agroecology. A review. Agron Sustain Dev 37:1–18. https://doi.org/10.1007/s13593-017-0472-4

Mohan SV, Hemalatha M, Amulya K, Velvizhi G, Chiranjeevi P, Sarkar O et al (2020) Decentralized urban farming through keyhole garden: a case study with circular economy and regenerative perspective. Mater Circ Econ 2:1–12. https://doi.org/10.1007/s42824-020-00011-1

Mouratiadou I, Wezel A, Kamilia K, Marchetti A, Paracchini ML, Bàrberi P (2024) The socio-economic performance of agroecology. A review. Agron Sustain Dev 44(2):19. https://doi.org/10.1007/s13593-024-00945-9

Muhammad M, Basit A, Arooj A, Dixit G, Majeed M, Sinha D et al (2024a) Roles of gibberellins in plant defense against biotic and abiotic stress. In: Plant growth regulators to manage biotic and abiotic stress in agroecosystems. CRC Press, pp 183–210

Muhammad M, Basit A, Majeed M, Shah AA, Ullah I, Mohamed HI et al (2024b) Bacterial pigments and their applications. In: Bacterial secondary metabolites. Elsevier, pp 277–298. https://doi.org/10.1016/B978-0-323-95251-4.00019-3

Muluneh MG (2021) Impact of climate change on biodiversity and food security: a global perspective—a review article. Agric Food Secur 10(1):1–25. https://doi.org/10.1186/s40066-021-00318-5

Nasar S, Muhammad M, Siddiqi EH, Majeed M, Ameen N, Ullah S et al (2024a) OMICS-based knowledge for achieving food and nutritional security. In: OMICs-based techniques for global food security, pp 67–90. https://doi.org/10.1002/9781394209156.ch4

Nasar S, Rasool A, Ikram W, Shujat S, Majeed M, Muhammad M, Saleem S (2024b) Nano-osmics: a new approach to plant breeding. In: Sustainable agriculture: nanotechnology and biotechnology for crop production and protection, vol 129. https://doi.org/10.1515/9783111234694-008

Neale RE, Barnes PW, Robson TM, Neale PJ, Williamson CE, Zepp RG et al (2021) Environmental effects of stratospheric ozone depletion, UV radiation, and interactions with climate change: UNEP Environmental Effects Assessment Panel, Update 2020. Photochem Photobiol Sci 20(1): 1–67. https://doi.org/10.1007/s43630-020-00001-x

Nonić M, Šijačić-Nikolić M (2021) Genetic diversity: sources, threats, and conservation. In: Life on land, pp 421–435. https://link.springer.com/referenceworkentry/10.1007/978-3-319-95981-8_53

Nowysz A, Mazur Ł, Vaverková MD, Koda E, Winkler J (2022) Urban agriculture as an alternative source of food and water security in today's sustainable cities. Int J Environ Res Public Health 19(23):15597. https://doi.org/10.3390/ijerph192315597

O'Grady AP, Mendham DS, Mokany K, Smith GS, Stewart SB, Harrison MT (2024) Grazing systems and natural capital: influence of grazing management on natural capital in extensive livestock production systems. Nat Based Sol 100181. https://www.sciencedirect.com/science/article/pii/S2772411524000727

Patil DA (2020) Agrobiodiversity and advances in the development of millets in changing environment. In: Sustainable agriculture in the era of climate change, pp 643–673. https://doi.org/10.1007/978-3-030-45669-6_27

Petrie CA, Bates J (2017) 'Multi-cropping', intercropping and adaptation to variable environments in Indus South Asia. J World Prehist 30:81–130. https://doi.org/10.1007/s10963-017-9101-z

Prasanna BM (2016) Developing and deploying abiotic stress-tolerant maize varieties in the tropics: challenges and opportunities. In: Molecular breeding for sustainable crop improvement, vol 2, pp 61–77. https://link.springer.com/chapter/10.1007/978-3-319-27090-6_3

Rani J, Rasool A, Sher H, Fatima R, Hussain J, Majeed M et al (2024) Precision farming with nanoscale sensors: monitoring soil, water, and crop health. In: Harnessing nanoomics and nanozymes for sustainable agriculture. IGI Global, pp 78–99. https://doi.org/10.4018/979-8-3693-1890-4.ch005

Rasool A, Feroz E, Qayyum Khan Z, Qayyum Khan M, Majeed M, Kousar S (2025) Chapter 19 Sustainable agriculture: a nanoosmic-based management. In: Rajput V, Singh A, Ghazaryan K, Minkina T, Al-Tawaha A (eds) Sustainable agriculture: nanotechnology, biotechnology, management and food security. De Gruyter, Berlin, Boston, pp 411–430. https://doi.org/10.1515/9783111369754-019

Rasool A, Tariq M, Asif M, Shah GM, Fatima R, Majeed M et al (2024) Nanotechnology for improved crop resilience in challenging environments. In: Harnessing nanoomics and nanozymes for sustainable agriculture. IGI Global, pp 149–175. https://doi.org/10.4018/979-8-3693-1890-4.ch008

Raymond CM, Frantzeskaki N, Kabisch N, Berry P, Breil M, Nita MR et al (2017) A framework for assessing and implementing the co-benefits of nature-based solutions in urban areas. Environ Sci Pol 77:15–24. https://www.sciencedirect.com/science/article/pii/S1462901117306317

Reddy PP (2017) Agro-ecological approaches to pest management for sustainable agriculture. Springer, Singapore, pp 1–339. https://link.springer.com/book/10.1007/978-981-10-4325-3

Seddon N, Smith A, Smith P, Key I, Chausson A, Girardin C et al (2021) Getting the message right on nature-based solutions to climate change. Glob Chang Biol 27(8):1518–1546. https://doi.org/10.1111/gcb.15513

Shafqat A, Rasool A, Fatima R, Majeed M, Rahman KU, Narejo AH et al (2024) Nanomaterials for sustainable pest and disease management in agriculture. In: Harnessing nanoomics and nanozymes for sustainable agriculture, pp 100–125. https://doi.org/10.4018/979-8-3693-1890-4.ch006

Sinha D, Maurya AK, Abdi G, Majeed M, Agarwal R, Mukherjee R, Ganguly S, Aziz R, Bhatia M, Majgaonkar A, Seal S, Das M, Banerjee S, Chowdhury S, Adeyemi SB, Chen JT (2023) Integrated genomic selection for accelerating breeding programs of climate-smart cereals. In: Genes, vol 14(7). NLM (Medline). https://doi.org/10.3390/genes14071484

Stojanovic M (2019) Biomimicry in agriculture: is the ecological system-design model the future agricultural paradigm? J Agric Environ Ethics 32(5):789–804. https://doi.org/10.1007/s10806-017-9702-7

Tariq A, Mumtaz F, Majeed M, Zeng X (2023) Spatio-temporal assessment of land use land cover based on trajectories and cellular automata Markov modelling and its impact on land surface temperature of Lahore district Pakistan. Environ Monit Assess 195(1). https://doi.org/10.1007/s10661-022-10738-w

Tassadduq SS, Akhtar S, Waheed M, Bangash N, Nayab DE, Majeed M, Abbasi S, Muhammad M, Alataway A, Dewidar AZ, Elansary HO, Yessoufou K (2022) Ecological distribution patterns of wild grasses and abiotic factors. Sustainability 14(18). https://doi.org/10.3390/su141811117

Ullah I, Aslam B, Shah SHIA, Tariq A, Qin S, Majeed M, Havenith HB (2022) An integrated approach of machine learning, remote sensing, and GIS data for the landslide susceptibility mapping. Land 11(8). https://doi.org/10.3390/land11081265

Verbrugghe N, Rubinacci E, Khan AZ (2023) Biomimicry in architecture: a review of definitions, case studies, and design methods. Biomimetics 8(1):107. https://doi.org/10.3390/biomimetics8010107

Wezel A, Herren BG, Kerr RB, Barrios E, Gonçalves ALR, Sinclair F (2020) Agroecological principles and elements and their implications for transitioning to sustainable food systems. A review. Agron Sustain Dev 40:1–13. https://doi.org/10.1007/S13593-020-00646-Z

Biodiversity Under Climate Stress: Pathways to Resilient and Sustainable Food Systems

19

Jameel M. Al-Khayri, Aftab Umar, Ambar Amin, Mehrab Khalil, Rahmatullah Qureshi, Muhammad Majeed, and Ozodbek S. Abduraimov

Abstract

Incorporating plant biodiversity in Climate-Smart Agriculture (CSA) is crucial in planning for climate change and food security. Plant and animal species are basic elements of any ecosystem that sustain agricultural production through services such as nutrient cycling, pest control, and soil health. Some of the practices include agroforestry, intercropping, and conservation tillage, which can help CSA impact climate change challenges favorably without undermining

J. M. Al-Khayri (✉)
Department of Agricultural Biotechnology, College of Agriculture and Food Sciences, King Faisal University, Al-Ahsa, Saudi Arabia

Department of Botany, University of Agriculture, Faisalabad, Punjab, Pakistan
e-mail: jkhayri@kfu.edu.sa

A. Umar
Department of Agricultural Biotechnology, College of Agriculture and Food Sciences, King Faisal University, Al-Ahsa, Saudi Arabia

A. Amin
Department of Biotechnology, Lahore College for Women University, Lahore, Pakistan

M. Khalil
Department of Zoology, Wildlife and Fisheries, University of Agriculture, Faisalabad, Punjab, Pakistan

R. Qureshi
Department of Botany, Faculty of Sciences, Pir Mehr Ali Shah Arid Agriculture University, Rawalpindi, Pakistan
e-mail: rahmatullahq@uaar.edu.pk

M. Majeed (✉)
Department of Botany, University of Gujarat, Ahmedabad, Punjab, Pakistan

O. S. Abduraimov
Institute of Botany, Academy of Sciences, Republic of Uzbekistan, Tashkent, Uzbekistan

R. Qureshi (ed.), *Plant Biodiversity*, https://doi.org/10.1007/978-981-92-1471-6_19

ecological stability. However, the actual achievement of such strategies to promote the conservation of biodiversity is not without challenges such as financial constraints, lack of technology, and poor policy coherence. Monocultures and high-input farming systems have in the past emphasized production to the detriment of environmental considerations and have thus caused habitat destruction and species simplification. This chapter calls for policies, incentives, and community support in enhancing the practice of biodiversity in agriculture. Measures such as agri-environmental schemes, biodiversity credit trading, and payments for ecosystem services are presented as powerful means of encouraging farmers. Furthermore, the use of traditional knowledge in the formulation of biodiversity management strategies and participation ensures that the strategies developed fit for culture and relevant to the local situation. There is now a great appreciation of technology in the assessment of biodiversity in various ecosystems as well as in the formulation of adaptive management measures through the use of remote sensing and GIS. The research also underlines the importance of international cooperation and policy harmonization for the further development of the biodiversity-related CSA practices. The convention on biological diversity (CBD) and other national and global policies must therefore harmonize to ensure that there are proper policies that can foster conservation of biodiversity. This chapter has discussed how economic, ecological, and social advantages make the incorporation of biological diversification into CSA a sound method for attaining sustainable agriculture. Technically, these systems cut on production costs and increase output, especially in the constraining conditions of resource based farming systems. Conservation wise, they help protect species, habitats, and ecosystem services that are vital for the overall sustainability of the environment.

Keywords

Plant biodiversity · Climate-smart agriculture · Agroecology · Ecosystem services · Sustainable food systems

Abbreviations

AES	Agri-environmental Schemes
CBD	Convention on Biological Diversity
CSA	Climate-Smart Agriculture
GHG	Greenhouse Gases
GIS	Geographical Information Systems
R&D	Research and Development
SDGs	Sustainable Development Goals

Table 19.1 Ecosystem services provided by plant biodiversity in agricultural systems

Ecosystem service	Mechanism	Benefits
Nutrient cycling	Diverse root systems enhance nutrient uptake and recycling	Reduced fertilizer requirement, improved soil fertility
Pest regulation	Habitat for natural predators, chemical diversity	Reduced pesticide use, crop protection
Pollination	Flowering plants attract and support pollinators	Increased crop yield and quality
Soil formation	Root systems stabilize soil, organic matter accumulation	Erosion control, soil structure improvement
Water regulation	Diverse canopies intercept rainfall, root systems enhance infiltration	Reduced runoff, improved water retention
Carbon sequestration	Biomass accumulation, soil organic carbon storage	Climate change mitigation
Microclimate regulation	Canopy cover moderates temperature extremes	Reduced heat stress on crops

19.1 Introduction

Agriculture is one of the most important sectors that feed the people, support income, and offer vital ecosystem services. But this important sector is becoming more vulnerable to climate change, environmental pollution, and species decline. Current agricultural production systems and practices, commonly known as monoculture and chemically based, have focused on increasing yields rather than the sustainability of production. They have resulted in soil fatigue, water deficits, loss of ecosystem, and heightened risk of climate change-related shocks (Bhattacharya et al. 2024; Majeed et al. 2024a). Incorporating plant diversity into agricultural systems can provide solutions to these problems while creating the necessary buffer against future shocks. Table 19.1 describes a summary of key ecosystem services provided by plant biodiversity in agricultural landscapes, highlighting the mechanisms and benefits associated with each service (Hassan et al. 2022a; Majeed et al. 2022a).

Plant biodiversity is the number of plant species that can be found in a particular area, either in an agricultural or natural habitat. Given that these diverse organisms are essential to the health of agroecosystems and their ability to perform important services like nutrient cycling, pest management, water retention, and pollination (Arif et al. 2021; Muhammad et al. 2024b; Qureshi 2025; Rasool et al. 2025). In conventional farming systems, including those applied by indigenous people, different types of cropping, including intercropping, agroforestry, and crop cycling, have been vital to sustenance. These approaches have in the past reduced risks of pests, diseases, and climate changes through the development of diversified socio-ecological systems that balance the utilization and replenishment of resources.

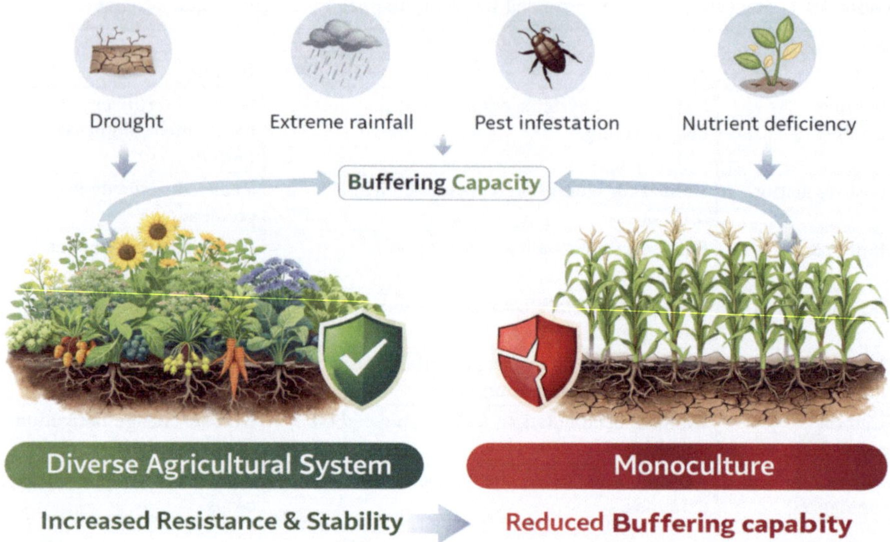

Fig. 19.1 Resilience of diverse agricultural systems versus monocultures under various stress conditions

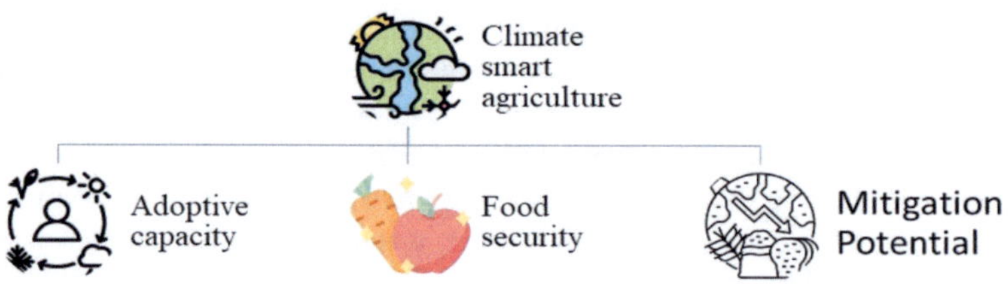

Fig. 19.2 Climate-smart adaptation and its impacts

Nevertheless, the positive role of biological diversity has been progressively abandoned with the development of monocultures aimed at achieving higher productivity and competitiveness in the market. These uniform systems are especially susceptible to pest and diseases incidence and to climatic variations (Fig. 19.1). For instance, one pest can cause severe reduction in monoculture crops because they all possess almost the same genes. Likewise, over-dependency on synthetic fertilizers and pesticides has reduced the quality of soil health and water, thus destructing the stability of agricultural systems (Haq et al. 2022b).

Climate change adds on to these challenges through factors such as, unpredictable rainfall, drought, and increased instances of disasters. Newly defined as climate-smart agriculture (CSA), this approach aims to address these increasing pressures through an integrated approach of improving productivity, increasing resilience and reducing the negative impacts on the environment (Fig. 19.2). Thus, embedding plant diversity into CSA can help farmers and government develop systems that are

Table 19.2 Key principles and practices of climate-smart agriculture

Pillar	Objective	Practices	Expected outcomes	References
Productivity	Sustainable increase in agricultural productivity	Improved crop varieties, efficient water management, integrated nutrient management	Enhanced food security, increased farm income	Haq et al. (2022a)
Adaptation	Building resilience to climate shocks	Crop diversification, agroforestry, conservation agriculture, drought-resistant varieties	Reduced vulnerability, stable yields under stress	Majeed et al. (2023a)
Mitigation	Reducing greenhouse gas emissions	Carbon sequestration, reduced tillage, agroforestry, methane reduction strategies	Lower carbon footprint, enhanced carbon storage	Tariq et al. (2023)

resilient to climate change and are sustainable in the future (Haq et al. 2022b; Majeed et al. 2024a, b). Biodiverse systems can control pests, build soil structure, and increase water penetration. Also, these systems offer other services like carbon storage as part of measures that help in combating climate change. Integrating biodiversity in to CSA is the fusion of age-old wisdom and recent developments (Arshad et al. 2022; Naveed et al. 2024c). Table 19.2 outlines the key principles and practices of climate-smart agriculture, categorized by the three pillars such as productivity, adaptation, and mitigation (Haq et al. 2022a; Majeed et al. 2023a).

In conventional farming systems, including those applied by indigenous people, different types of cropping, including intercropping, agroforestry, and crop cycling, have been vital to sustenance. These approaches have in the past reduced risks of pests, diseases, and climate changes through the development of diversified social-ecological systems that balance the utilization and replenishment of resources. Nevertheless, the positive role of biological diversity has been progressively abandoned with the development of monocultures aimed at achieving higher productivity and competitiveness in the market. These uniform systems are especially susceptible to pest and diseases incidence and to climatic variations. For instance, one pest can cause severe reduction in monoculture crops because they all possess almost the same genes. Likewise, over-dependency on synthetic fertilizers and pesticides has reduced the quality of soil health and water, thus destructing the stability of agricultural systems. Climate change adds on to these challenges through factors such as, unpredictable rainfall, drought, and increased instances of disasters. Newly defined as climate-smart agriculture (CSA), this approach aims to address these increasing pressures through an integrated approach of improving productivity, increasing resilience, and reducing the negative impacts on the environment. Thus, embedding plant diversity into CSA can help farmers and government develop systems that are resilient to climate change and are sustainable in the future. Biodiverse systems can control pests, build soil structure, and increase water penetration. Also, these systems offer other services like carbon storage as part of measures that help in

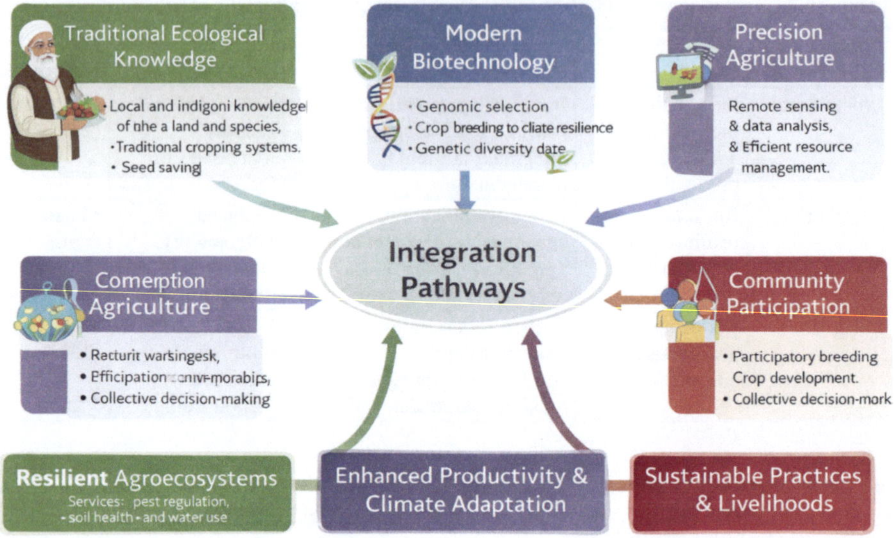

Fig. 19.3 Diagrammatic pathways between traditional ecological knowledge, modern biotechnology, precision agriculture, and community participation for sustainable agricultural

combating climate change. Integrating biodiversity in to CSA is the fusion of age-old wisdom and recent developments (Arshad et al. 2022; Naveed et al. 2024c).

Traditional farming systems like, crop diversification to match with the ecological conditions has been observed to maintain the production system. The above-mentioned methods and techniques can be complimented with new technologies such as precision agriculture, remote sensing, and genomic selection. For instance, genomic selection makes it feasible to develop crop varieties that can perform well in environments characterized by climate change such as drought stress, pest, and diseases. These innovations therefore augment conventional approaches to the management of biological diversity to produce systems that are both indigenous and technologically informed (Hussain et al. 2023; Rasool et al. 2024). A conceptual framework integrating traditional knowledge with modern technologies in biodiversity-based agriculture is shown in Fig. 19.3 (Rasool et al. 2025; Muhammad et al. 2023).

Biodiversity integration is not limited to the farm level but also has positive impacts on the whole landscape. Complex agriculture systems are more resilient to climate variability than the mono-cropped systems. For instance, intercropping of trees with crops preserves water, maintains the soil integrity, and supports useful fauna like bees and other insects that prey on pests. These systems not only enhance agricultural durability but also assist with the objectives of other forms of conservation. Research has also indicated that landscapes with high plant species' diversities are more efficient at withstanding climate changes and give more predictable production output (Aziz et al. 2023a; Naveed et al. 2024b).

19.2 Plant Biodiversity in Climate-Smart Agriculture

The inclusion of plant diversification in climate-smart agriculture is important at the global, regional, and local levels as it solves the problem of sustainable food production, environmental protection and climate change. Globally, plant genetic resources contribute to the stability of the agricultural production systems by reducing their vulnerability to climate change factors such as droughts and extreme weather conditions. Diverse small scale agroecosystems have a great potential for carbon sequestration and thus contribute to reduction on greenhouse gases (Hassan et al. 2022b; Hussain et al. 2024). The carbon sequestration is potential across different agricultural systems, demonstrating the significant contribution of biodiversity-based systems to climate change mitigation (Majeed et al. 2022c; Waheed et al. 2022) as shown in Table 19.3.

Through the support of sustainable farming, the integration of biodiversity is consistent with other global objectives, including the United Nations Sustainable Development Goals (SDGs) and the Convention on Biological Diversity (CBD). For instance, target 2.4 of the SDGs aims at increasing adaptation to sustainable and resilient practices to enhance agricultural productivity for food security and healthy ecosystems. These global frameworks stress the role of biological diversity in establishing sustainable agriculture systems that will support climate change and conserve the earth's assets (Jamil et al. 2022; Khoja et al. 2022; Nasar et al. 2024b).

The use of CSA for the preservation of biological diversity has been most effective in responding to the specific regional and climatic characteristics. In the tropics like the African Sahel, Agroforestry has been used to halt soil degradation

Table 19.3 Carbon sequestration potential of different agricultural systems

Agricultural system	Carbon sequestration rate (t C/ha/year)	Soil organic carbon (t/ha)	Aboveground biomass carbon (t/ha)	References
Conventional monoculture	0.5–1.2	25–35	5–10	Majeed et al. (2022c)
Conservation agriculture	1.5–2.5	35–50	10–20	Waheed et al. (2022)
Agroforestry systems	2.5–4.0	50–80	30–60	Hassan et al. (2022b)
Silvopastoral systems	3.0–5.0	60–90	40–80	Hussain et al. (2024)
Diverse cropping systems	1.8–3.0	40–65	15–35	Aziz et al. (2023b)

Table 19.4 Biodiversity-based farming practices across different climatic regions

Climatic region	Key practices	Biodiversity components	Adaptation benefits	References
Tropical (Sahel, Africa)	Agroforestry, parkland systems	Acacia spp., native grasses	Soil fertility restoration, drought resistance	Khan et al. (2022)
Tropical (Southeast Asia)	Rice-fish systems, home gardens	Multiple rice varieties, fruit trees, aquatic species	Flood resilience, diversified income	Nasar et al. (2024a)
Mediterranean	Silvopastoral systems, intercropping	Olive, carob, cereal-legume mixtures	Water conservation, heat tolerance	Majeed et al. (2021c)
Temperate (Europe)	Crop rotation, cover cropping	Legumes, brassicas, native wildflowers	Soil health, pest suppression	Tariq et al. (2023)
Temperate (North America)	Prairie strips, conservation buffers	Native perennial grasses, forbs	Erosion control, pollinator habitat	Aziz et al. (2023b)
Arid/semi-arid	Dryland agroforestry, water harvesting	Drought-tolerant trees, millet, sorghum	Water efficiency, temperature moderation	Waheed et al. (2023)

and enhance soil fertility. These systems combine trees and shrubs with crops, improving water holding capacity, decreasing soil erosion, and offering shade to heat-sensitive crops. In the same way, in the temperate climates, crop succession and cover crops have been used in order to preserve the soil and to minimize the use of chemical fertilizers and pesticides (Khan et al. 2022, 2024; Nasar et al. 2024a). These regional practices show how the approaches to support biological diversification can be suited to the particular territories, agricultural land use, and climate conditions. Table 19.4 provides a comparative analysis of biodiversity-based farming practices across different climatic regions, highlighting region-specific adaptations and outcomes (Majeed et al. 2021c; Tariq et al. 2023).

At the local level, the small farmer also gains a lot from practicing plant diversification in his or her farming system. Such diverse farming practices as intercropping and traditional crop varieties help the farmer cut the cost of synthetic inputs while at the same time enhancing crop resistance to diseases, pests and drought. Also, micro-level interventions such as seed banks, and participatory breeding promote farmers' conservation and use of plant genetic resources. Such practices not only enhance yield in agriculture but also enhance the bond between people and the farm and the farm land (Majeed et al. 2020; Waheed et al. 2022). Biodiversity-based farming has socio-economic benefits to smallholder farmers, showing income diversification, input cost reduction, and yield stability compared to conventional monoculture (Majeed et al. 2021b; Rasool et al. 2024). Such kind of farming system socially is important to mobilize communities to engage in the conservation of biodiversity and at the same time enhancing their standards of living (Aziz et al. 2023b; Naveed et al. 2024a). All these attributes, therefore, support the importance of diversification in the maintenance of agricultural systems at all levels and thus the need for the conservation of biodiversity.

19.3 Understanding Climate-Smart Agriculture

Climate-smart agriculture (CSA) is a new concept which seeks to solve the twin problems of climate change and food insecurity. It evolved into a model which not only provides measures to deal with climate change in relation to agriculture but also provides strategies for climate change mitigation in regard to greenhouse gas GHG emissions. At its core, CSA strives for a "triple win": improving the yield of crops, enhancing the adaptation to climate change, and reducing or avoiding the emissions of GHG through new practices, strategies, and sources of funding (Haq et al. 2022a; Majeed et al. 2024c). This dynamic model captures an evolution in agricultural development that brings together the challenge of food security with the challenge of climate change. The summary of CSA interventions across three pillars is provided in Table 19.5.

One of the greatest strengths of CSA is its emphasis on the double bottom line of productivity and sustainability. Agriculture is under pressure from the rising population and changing diet, together with the effects of climate change. CSA provides means to increase crop production productivity and to reduce food system vulnerability by promoting adaptation measures. These strategies include enhancement of water productivity, crop and livestock system diversification, and release of climate adaptive seed varieties (Majeed et al. 2021a; Ullah et al. 2022). It is not only beneficial for increasing the agricultural yields but also assist the farmers to cope with the climate.

A special aspect of CSA is its emphasis on the mainstreaming of mitigation actions into agricultural development. Globally, agriculture is a major source of GHG emissions, and through the practices such as sustainable land management, decreased deforestation, and carbon sequestration, CSA seeks to solve it. Activities like agroforestry, conservation tillage, and organic farming follow the mitigation principles of CSA, which is evidence that farming can move from being a part of the

Table 19.5 Climate-smart agriculture interventions and indicators across the triple pillars

Pillar	Interventions	Indicators	Monitoring tools	References
Productivity	Improved varieties, efficient irrigation, integrated soil fertility	Yield (t/ha), income ($/ha), food security index	Field surveys, remote sensing	Majeed et al. (2022b)
Adaptation	Crop diversification, agroforestry, water harvesting	Yield stability, drought tolerance, pest resistance	Phenological monitoring, climate data	Ullah et al. (2022)
Mitigation	Reduced tillage, cover cropping, agroforestry	GHG emissions (t CO_2 eq/ha), carbon stock (t/ha)	Eddy covariance, soil sampling	Tariq et al. (2023)

Table 19.6 Greenhouse gas emission reduction potential of CSA practices

CSA practice	GHG reduction (t CO_2 eq/ha/year)	Primary mechanisms	Co-benefits	References
No-till farming	0.5–1.5	Reduced soil carbon oxidation	Soil health improvement	Muhammad et al. (2023)
Cover cropping	0.8–2.0	Carbon sequestration, reduced N_2O	Erosion control, nutrient retention	Naveed et al. (2024c)
Agroforestry	2.0–5.0	Biomass carbon storage	Biodiversity habitat, timber products	Hassan et al. (2022a)
Integrated nutrient management	0.3–1.0	Reduced N_2O from fertilizer	Improved nutrient use efficiency	Majeed et al. (2022c)
Rice intensification	1.0–3.0	Reduced methane emissions	Water savings, yield increase	Haq et al. (2022b)

problem to being part of the solution (Majeed et al. 2021c; Sinha et al. 2023; Tariq et al. 2023). Also, the use of technologies in farming such as precision agriculture and renewable energy sources on farms confirms that CSA is capable of integrating climate change measures with agricultural output. The potential for reducing greenhouse gas emissions from various CSA techniques in comparison to traditional agriculture is measured in Table 19.6.

The CSA framework is hinged on the use of context-based strategies since farming systems are different in different regions and, therefore, practices must be suited to the local environment, resources, and social economic settings. For instance, in the Sub-Saharan Africa region where smallholder agriculture is the norm, CSA embraces low-cost and resource conserving approaches that can easily be adopted at scale to address local challenges. As with developed countries, CSA promotes the use of modern technologies and effective policies for the implementation of large-scale mitigation and adaptation (Majeed et al. 2022b, 2023c).

19.4 Plant Biodiversity: The Foundation of Resilient Agriculture

Plant genetic resources form the basis of sustainable agriculture and food production coupled with other related products that help farm systems to cope with biophysical and socio-economic stresses. Biodiversity provides resilience and productivity in agricultural systems through supporting biological processes including nutrient cycling, pest management and water use. Different plant species are important in providing some level of insurance to agriculture and other farming systems especially in the sense of whether the systems can bounce back after having suffered some form of stress such as droughts, pests, and diseases (Hussain et al. 2023; Rasool et al. 2024). This role is crucial in the current environment of climate change

Table 19.7 Components of plant biodiversity and their functional roles in agroecosystems

Biodiversity Component	Examples	Functional Role	Ecosystem Service	References
Crop genetic diversity	Landraces, heirloom varieties	Stress tolerance, yield stability	Food security, adaptation	Majeed et al. (2020)
Associated wild Flora	Native grasses, forbs	Pollinator habitat, natural enemy refuge	Pollination, pest control	Jamil et al. (2022)
Cover crops	Clover, rye, vetch	Soil cover, nitrogen fixation	Soil health, nutrient cycling	Waheed et al. (2022)
Trees in agroforestry	Acacia, Gliricidia, Faidherbia	Shade, nutrient cycling, carbon storage	Microclimate regulation, climate mitigation	Hassan et al. (2022b)
Hedgerows and buffers	Native shrubs, perennial grasses	Habitat connectivity, windbreak	Biodiversity corridor, erosion control	Aziz et al. (2023b)
Soil microbiome	Mycorrhizae, rhizobia	Nutrient acquisition, disease suppression	Soil fertility, plant health	Muhammad et al. (2024b)

that has produced disturbances in weather and climate risks that put immense pressure on world food production. A brief summary of plant biodiversity components and their functional roles in agricultural systems is provided in Table 19.7 (Khan et al. 2022; Khoja et al. 2022).

Due to high level of species diversity, a complex agricultural is typically more adaptable. For instance, a mix of nutrient-efficient and drought-resistant plant species can support yield maintenance in the face of climate uncertainty. Through genetic resources from wild relatives and traditional land races, biodiversity also contributes to the breeding of stress-tolerant crop varieties that improve food security (Hassan et al. 2022a; Shafqat et al. 2024; Rani et al. 2024). This study demonstrates that while incorporating other plant species into tropical agroforestry systems aids in pollination and carbon storage, yields remain unaffected.

Plant diversity is essential to the fertility and health of the soil; hence, crop diversification increases the soil's microbial activity and nutrients. Mixed and intercropping are two examples (legumes improve soil structure and nitrogen fixing). According to the research, having a variety of plant species improves soil regeneration over the long term by increasing carbon buildup and nutrient content (Hassan et al. 2022b; Hussain et al. 2024). Data on improvements in soil health under various biodiversity-based farming systems in comparison to monoculture are shown in Table 19.8 (Waheed et al. 2023; Majeed et al. 2022c).

Table 19.8 Soil health parameters under different farming systems

Farming system	Soil organic matter (%)	Soil microbial biomass (μg C/g)	Aggregate stability (%)	Available N (kg/ha)	References
Conventional monoculture	1.2–1.8	150–250	25–35	80–120	Waheed et al. (2023)
Crop rotation	1.8–2.5	250–400	35–50	120–180	Majeed et al. (2022c)
Cover cropping	2.0–3.0	300–500	40–60	140–200	(Hassan et al. 2022b)
Agroforestry	2.5–4.0	400–700	50–75	160–240	Hussain et al. (2024)
Diverse perennial systems	3.0–5.0	500–900	60–85	180–280	Aziz et al. (2023b)

19.5 Practical Strategies for Biodiversity Integration

Diversified cropping systems, such as agroforestry, stand out as a transformative approach for integrating biodiversity into agriculture. A multiple-layered system that integrates trees, shrubs, and crops in one land use area is referred to as Agroforestry. This process leverages the arrangements of complementary relationships established between the plant species, which can actively support environmentally robust conditions such as climate change, or pest invasion (Majeed et al. 2021c; Sinha et al. 2023; Tariq et al. 2023). For instance, trees in agroforestry systems offer shades needed to monitor heat stress on crop as well as to reduce evaporation rates for water thereby covering water needs. Also, they extend extensive root systems that fix the soil and pull nutrients from deeper parts that help other crops with less root penetration. Apart from ecological benefits, multiple products from trees, food crops, and other crops make agroforestry lucrative and secure income source for farmers. Table 19.9 presents a typology of agroforestry systems with their characteristics, suitable climates, and key biodiversity and productivity outcomes (Khan et al. 2022; Nasar et al. 2024a).

In addition to agroforestry, poly-cropping and intercropping are also other important sources of land diversification. For instance, planting legumes alongside cereals not only enhances nitrogen fixation in the soil but also reduces competition for nutrients due to their differing root architectures. These integrated systems enhance efficient utilization of resources with a general enhancement of the biosphere and the farming practices (Muhammad et al. 2024a).

Another strategy for integrating biodiversity is returning to native plant stocks adapted to climate resilience. Because native plants are susceptible to local pests,

Table 19.9 Typology of agroforestry systems for biodiversity integration

Agroforestry system	Structure	Suitable climate	Biodiversity benefits	Productivity outcomes	References
Silvoarable	Trees + crops	Temperate, Mediterranean	Habitat for birds and insects, soil biota	20–40% yield increase in crops	Khan et al. (2022)
Silvopastoral	Trees + pasture + livestock	Tropical, temperate	Enhanced forage diversity, wildlife corridor	15–30% increase in livestock productivity	Nasar et al. (2024a)
Homegardens	Multi-story tree-crop mixtures	Tropical, subtropical	High species richness (50–200 species)	Year-round food production	Majeed et al. (2022a)
Riparian buffers	Trees + shrubs along watercourses	All regions	Aquatic habitat, filter strips	Water quality improvement	Aziz et al. (2023b)
Shade systems	Overstory trees + understory crops	Tropical	Pollinator habitat, carbon storage	10–25% yield premium for shade-grown crops	Hassan et al. (2022b)
Windbreaks	Linear tree/ shrub rows	Arid, temperate	Wildlife corridor, shelter for crops	10–30% yield increase in protected crops	Tariq et al. (2023)

illnesses, and harsh weather, they always do better than foreigners. Additionally, this reduces the usage of chemicals that deteriorate ecosystems, such as fertilizers and pesticides. Nonetheless, native species are prized as historical and cultural assets connected to customary diet and indigenous knowledge (Majeed et al. 2022a; Muhammad et al. 2023; Qureshi et al. 2024a, b). Examples of native crop species revival projects in various regions are included in Table 19.10, emphasizing the nutritional advantages and adaptability characteristics of these species (Majeed et al. 2021b; Waheed et al. 2023).

19.6 Learning from Success: Case Studies of Biodiversity

Examining successful international biodiversity initiatives provides a framework for sustainable farming methods. Incorporating biodiversity into farming can yield social, economic, and ecological benefits, as demonstrated by global examples of sustainable agriculture and perspectives from farmers and communities (Majeed et al. 2021c; Sinha et al. 2023; Tariq et al. 2023). A review of successful case studies

Table 19.10 Revival of native crop species for climate-resilient agriculture

Crop species	Region of revival	Climate adaptation traits	Nutritional benefits	Socio-economic outcomes	References
Millet (pearl, finger)	Sub-Saharan Africa, India	Drought tolerance, heat resistance	High iron, zinc, fiber	Improved food security, women's empowerment	Majeed et al. (2021b)
Sorghum	Africa, South Asia	Water-use efficiency, heat tolerance	Gluten-free, antioxidants	Income diversification	Waheed et al. (2023)
Quinoa	Andean region, North America	Frost tolerance, salt tolerance	Complete protein, minerals	Export opportunities	Arshad et al. (2022)
Amaranth	Central America, Asia	Heat tolerance, C4 photosynthesis	High protein, lysine content	Value-added products	Aziz et al. (2023a)
Teff	Ethiopia	Drought tolerance, low input requirement	High iron, calcium	Traditional food preservation	Nasar et al. (2024b)
Fonio	West Africa	Poor soil tolerance, short growing cycle	High methionine, cysteine	Income for smallholders	Muhammad et al. (2024b)

from around the globe is shown in Table 19.11, emphasizing important results and lessons gained (Bashir et al. 2023; Jamil et al. 2022; Khoja et al. 2022).

19.7 Challenges and Future Priorities

Incorporating diversification into agriculture is a multifaceted yet crucial endeavor for guaranteeing sustainable food production and ecological resilience. Contemporary agricultural methods, frequently defined by monocultures and high-input utilization, have substantially contributed to the decline of biodiversity. Monocultures streamline landscapes, diminishing the variety of habitats essential for sustaining pollinators, natural pest regulators, and other advantageous creatures (Majeed et al. 2022a; Muhammad et al. 2023). This loss is most apparent in regions where non-crop ecosystems have been removed or deteriorated. Table 19.12 presents an analysis of barriers to biodiversity adoption in agriculture, categorized by type with proposed solutions (Ullah et al. 2022; Naveed et al. 2024a).

Table 19.11 Case studies of biodiversity integration in agriculture

Case study location	Practice	Biodiversity outcomes	Productivity outcomes	Socio-economic outcomes	Key lessons	References
Organic farms, Europe	Reduced inputs, hedgerows, buffer zones	30–50% increase in species richness	Comparable to conventional yields	Premium prices, ecosystem service payments	Long-term commitment needed	Bashir et al. (2023)
Small-scale farms, Sweden	Wildflower strips, diversified cropping	Higher butterfly and bee abundance	Stable yields	Farmer satisfaction, aesthetic value	Landscape-scale coordination	Jamil et al. (2022)
LandCare, Australia	Native vegetation restoration	Improved habitat connectivity	Reduced erosion, improved soil	Increased property value	Community-driven approach	Khoja et al. (2022)
GRPI, Ethiopia/ Nepal	Traditional crop valuation, participatory breeding	Conservation of landraces	15–25% yield increase	Market access, income diversification	Participatory approach essential	Majeed et al. (2020)
Coffee agroforestry, Latin America	Shade-grown coffee	High bird and pollinator diversity	10–20% yield premium	Certification income, carbon credits	Market linkages critical	Hassan et al. (2022b)
Rice-fish systems, Southeast Asia	Integrated rice-fish culture	Aquatic biodiversity conservation	10–30% yield increase	Income diversification, protein source	Traditional knowledge integration	Nasar et al. (2024a)

Table 19.12 Barriers to biodiversity adoption in agriculture and proposed solutions

Barrier category	Specific barriers	Impact level	Proposed solutions	References
Economic	High transition costs, delayed returns, lack of credit	High	Subsidies, low-interest loans, payment for ecosystem services	Ullah et al. (2022)
Policy	Inconsistent policies, lack of incentives, monoculture subsidies	High	Policy reform, biodiversity incentives, removal of harmful subsidies	Majeed et al. (2023a)
Technical	Knowledge gaps, lack of extension services, site-specific complexity	Medium	Farmer training, participatory research, digital extension tools	Naveed et al. (2024a)
Social	Cultural barriers, risk aversion, lack of farmer organization	Medium	Peer learning networks, farmer field schools, women's groups	Aziz et al. (2023b)
Institutional	Weak governance, fragmented institutions, poor coordination	High	Institutional strengthening, multi-stakeholder platforms	Bashir et al. (2023)
Climate	Increased uncertainty, extreme events, shifting growing conditions	High	Climate-resilient practices, risk insurance, early warning systems	Tariq et al. (2023)

19.8 Conclusion

The incorporation of plant diversity into climate-smart agriculture constitutes a revolutionary strategy for tackling the interconnected issues of global warming, food security, and biodiversity depletion. Ecological services essential to agriculture, such pollination, pest control, and soil fertility, are supported by plant biodiversity. By incorporating biodiversity into agricultural practices, more robust systems that withstand environmental shocks while maintaining sustainability and productivity can be established. In conclusion, including plant diversity into climate-smart agriculture is a practical approach to creating resilient and sustainable food systems as well as an ecological necessity. Despite current obstacles, a clear route forward is outlined by the incorporation of new practices, supportive laws, community involvement, and technology advancements. By promoting biodiversity, agriculture may transform into a system that supports both humanity and the environment, ensuring a secure future for the next generation.

References

Arif U, Bhatti KH, Ajaib M, Wagay NA, Majeed M, Zeb J, Hameed A, Kiani J (2021) Ethnobotanical indigenous knowledge of Tehsil Charhoi, district Kotli, Azad Jammu and Kashmir, Pakistan. Ethnobot Res Appl 22:1–24

Arshad F, Waheed M, Harun N, Fatima K, Khan BA, Fatima K, Abbas Z, Jabeen S, Majeed M (2022) Indigenous farmer's perception about fodder and foraging species of semi-arid lowlands of Pakistan: a case study of district Kasur, Pakistan. Taiwania 67(4):510–523

Aziz A, Anwar MM, Majeed M, Albanai JA, Almohamad H, Abdo HG (2023a) Quantifying the impacts of urbanization on urban green, evidences from Maga City, Lahore Pakistan. Discov Sustain 4(1):1–11

Aziz A, Anwar MM, Majeed M, Fatima S, Mehdi SS, Mangrio WM, Elbouzidi A, Abdullah M, Shaukat S, Zahid N, Mahmoud EA, Casini R, Yessoufou K, Elansary HO (2023b) Quantifying landscape and social amenities as ecosystem services in rapidly changing peri-urban landscape. Land 12(2):477

Bashir SM, Altaf M, Hussain T, Umair M, Majeed M, Mangrio WM, Khan AM, Gulshan AB, Hamed MH, Ashraf S, Amjad MS, Bussmann RW, Abbasi AM, Casini R, Alataway A, Dewidar AZ, Al-Yafrsi M, Amin MH, Elansary HO (2023) Vernacular taxonomy, cultural and ethnopharmacological applications of avian and mammalian species in the vicinity of Ayubia National Park, Himalayan Region. Biology 12(4):609

Bhattacharya R, Bhattacharya R, Majeed M, Bhandari S, Aziz R, Sinha D, Niyogi SS (2024) Environmental rehabilitation of industrial waste dumping site. In: Biohydrometallurgical processes. CRC Press, Boca Raton, pp 44–89

Haq SM, Tariq A, Li Q, Yaqoob U, Majeed M, Hassan M, Fatima S, Kumar M, Bussmann RW, Moazzam MFU, Aslam M (2022a) Influence of edaphic properties in determining forest community patterns of the Zabarwan Mountain range in the Kashmir Himalayas. Forests 13(8):1214

Haq SM, Yaqoob U, Majeed M, Amjad MS, Hassan M, Ahmad R, Morales-de la Nuez A (2022b) Quantitative ethnoveterinary study on plant resource utilization by indigenous communities in high-altitude regions. Front Vet Sci 9:944046

Hassan M, Haq SM, Ahmad R, Majeed M, Sahito HA, Shirani M, Mubeen I, Aziz MA, Pieroni A, Bussmann RW, Alataway A, Dewidar AZ, Al-Yafrsi M, Elansary HO, Yessoufou K (2022a) Traditional use of wild and domestic fauna among different ethnic groups in the Western Himalayas—a cross cultural analysis. Animals 12(17):2276

Hassan M, Haq SM, Majeed M, Umair M, Sahito HA, Shirani M, Waheed M, Aziz R, Ahmad R, Bussmann RW, Alataway A, Dewidar AZ, El-Abedin TKZ, Al-Yafrsi M, Elansary HO, Yessoufou K (2022b) Traditional food and medicine: ethno-traditional usage of fish fauna across the valley of Kashmir: a Western Himalayan region. Diversity 14(6):455

Hussain S, Raza A, Abdo HG, Mubeen M, Tariq A, Nasim W, Majeed M, Almohamad H, Al Dughairi AA (2023) Relation of land surface temperature with different vegetation indices using multi-temporal remote sensing data in Sahiwal region, Pakistan. Geosci Lett 10(1):33

Hussain F, Salam IU, Memon ZN, Abdullah M, Abbas G, Akbar M, Moda HM (2024) Occurrence of fungal microbial contamination in drinking water of megacity of Karachi (Pakistan) and their physico-chemical control. Heliyon 10(7):e28926

Jamil MD, Waheed M, Akhtar S, Bangash N, Chaudhari SK, Majeed M, Hussain M, Ali K, Jones DA (2022) Invasive plants diversity, ecological status, and distribution pattern in relation to edaphic factors in different habitat types of district Mandi Bahauddin, Punjab, Pakistan. Sustainability 14(20):13312

Khan AM, Li Q, Saqib Z, Khan N, Habib T, Khalid N, Majeed M, Tariq A (2022) MaxEnt modelling and impact of climate change on habitat suitability variations of economically important Chilgoza Pine (*Pinus gerardiana* Wall.) in South Asia. Forests 13(5):715

Khan AM, Altaf M, Hussain T, Hamed MH, Safdar U, Ayub A, Mahmoud EA (2024) Ethnopharmacological uses of fauna among the people of Central Punjab, Pakistan. Front Vet Sci 11:1351693

Khoja AA, Haq SM, Majeed M, Hassan M, Waheed M, Yaqoob U, Bussmann RW, Alataway A, Dewidar AZ, Al-Yafrsi M, Elansary HO, Yessoufou K, Zaman W (2022) Diversity, ecological and traditional knowledge of pteridophytes in the Western Himalayas. Diversity 14(8):628

Majeed M, Bhatti KH, Amjad MS, Abbasi M, Bussmann RW, Nawaz F, Rashid A, Mehmood A, Malik MM, Khan WM, Ali S (2020) Ethno-veterinary uses of Poaceae in Punjab, Pakistan. PLoS One 15(11):e0241705

Majeed M, Bhatti KH, Amjad MS (2021a) Impact of climatic variations on the flowering phenology of plant species in Jhelum District, Punjab, Pakistan. Appl Ecol Environ Res 19(5):3343–3376

Majeed M, Bhatti KH, Pieroni A, Sõukand R, Bussmann RW, Khan AM, Chaudhari SK, Aziz MA, Amjad MS (2021b) Gathered wild food plants among diverse religious groups in Jhelum District, Punjab, Pakistan. Foods 10(3):594

Majeed M, Tariq A, Anwar MM, Khan AM, Arshad F, Mumtaz F, Farhan M, Zhang L, Zafar A, Aziz M, Abbasi S, Rahman G, Hussain S, Waheed M, Fatima K, Shaukat S (2021c) Monitoring of land use–land cover change and potential causal factors of climate change in Jhelum District, Punjab, Pakistan, through GIS and multi-temporal satellite data. Land 10(10):1026

Majeed M, Khan AM, Habib T, Anwar MM, Sahito HA, Khan N, Ali K (2022a) Vegetation analysis and environmental indicators of an arid tropical forest ecosystem of Pakistan. Ecol Indic 142:109291

Majeed M, Lu L, Haq SM, Waheed M, Sahito HA, Fatima S, Aziz R, Bussmann RW, Tariq A, Ullah I, Aslam M (2022b) Spatiotemporal distribution patterns of climbers along an abiotic gradient in Jhelum District, Punjab, Pakistan. Forests 13(8):1244

Majeed M, Tariq A, Haq SM, Waheed M, Anwar MM, Li Q, Aslam M, Abbasi S, Mousa BG, Jamil A (2022c) A detailed ecological exploration of the distribution patterns of wild Poaceae from the Jhelum District (Punjab), Pakistan. Sustainability 14(7):3786

Majeed M, Lu L, Anwar MM, Tariq A, Qin S, El-Hefnawy ME, El-Sharnouby M, Li Q, Alasmari A (2023a) Prediction of flash flood susceptibility using integrating analytic hierarchy process (AHP) and frequency ratio (FR) algorithms. Front Environ Sci 10:1037547

Majeed M, Muhammad M, Hussain T, Ali M, Naz T, Nawaz S, Abbas G (2023b) Nanopriming-based management of biotic stresses for sustainable agriculture. In: Nanopriming approach to sustainable agriculture. IGI Global, Hershey, pp 263–289

Majeed M, Muhammad M, Nawaz S, Naz T, Iqbal MM, Zahid N, Mehboob M (2023c) Nanopriming for crop management for sustainable agriculture: an overview. In: Nanopriming approach to sustainable agriculture. IGI Global, Hershey, pp 110–141

Majeed M, Gulshan AB, Muhammad M, Hussain M, Aziz R, Abdullah M, Tariq M (2024a) An investigation into the reactions of biochar in heavy metal polluted soils. In: Biochar-assisted remediation of contaminated soils under changing climate. Elsevier, Amsterdam, pp 247–276

Majeed M, Noor W, Rasool A, Saleem M, Bashir I, Fatima R, Rafique A (2024b) Nanotechnology for soil remediation and land restoration. In: Harnessing NanoOmics and nanozymes for sustainable agriculture. IGI Global, Hershey, pp 60–77

Muhammad M, Waheed A, Wahab A, Majeed M, Nazim M, Liu YH, Li WJ (2023) Soil salinity and drought tolerance: an evaluation of plant growth, productivity, microbial diversity, and amelioration strategies. Plant Stress 10:100319

Muhammad M, Basit A, Arooj A, Dixit G, Majeed M, Sinha D, Li WJ (2024a) Roles of gibberellins in plant defense against biotic and abiotic stress. In: Plant growth regulators to manage biotic and abiotic stress in agroecosystems. CRC Press, Boca Raton, pp 183–210

Muhammad M, Basit A, Majeed M, Shah AA, Ullah I, Mohamed HI, Ghanaim AM (2024b) Bacterial pigments and their applications. In: Bacterial secondary metabolites. Elsevier, Amsterdam, pp 277–298

Nasar S, Muhammad M, Siddiqi EH, Majeed M, Ameen N, Ullah S, Abdullah M (2024a) OMICS-based knowledge for achieving food and nutritional security. In: OMICs-based techniques for global food security. Wiley, Hoboken, pp 67–90

Nasar S, Rasool A, Ikram W, Shujat S, Majeed M, Muhammad M, Saleem S (2024b) Nano-osmics: a new approach to plant breeding. In: Sustainable agriculture: nanotechnology and biotechnology for crop production and protection. De Gruyter, Berlin, pp 129–148

Naveed M, Majeed M, Jabeen K, Hanif N, Naveed R, Saleem S, Khan N (2024a) Plant disease diagnosis with artificial intelligence (AI). In: Microbial data intelligence and computational techniques for sustainable computing. Springer, Singapore, pp 217–234

Naveed M, Memon ZN, Abdullah M, Makhdoom SI, Azeem A, Mehmood S, Majeed M (2024b) Use cases and future aspects of intelligent techniques in microbial data analysis. In: Microbial data intelligence and computational techniques for sustainable computing. Springer, Singapore, pp 259–280

Naveed M, Zahid N, Fatima I, Saleem A, Majeed M, Abid A, Qasim A (2024c) Sustainable AI-driven applications for plant care and treatment. In: Microbial data intelligence and computational techniques for sustainable computing. Springer, Singapore, pp 235–258

Qureshi R (2025) Plant diversity and useful ecosystem services of the Cholistan Desert, Pakistan. J Native Alien Plant Stud 21:1–21

Qureshi R, Arif IA, Al-Faraj Asiri TK, Alsheikh AA, Baig MB (2024a) Role of plants in restoring degraded ecosystems to improve biodiversity in alula region—Saudi Arabia. In: Fahad S, Saud S, Nawaz T, Gu L, Ahmad M, Zhou R (eds) Environment, climate, plant and vegetation growth. Springer International Publishing AG, Springer Nature, pp 205–228

Qureshi R, Baig MB, Belgacem AO (2024b) Climate change and innovative ecological interventions through climate-smart agriculture in the Cholistan Desert of Pakistan. In: Behnassi M, Al-Shaikh aa, Baig MB, Qureshi RH, Faraj TKA (eds) Climate-smart and resilient food systems and security. Springer International Publishing AG, Springer Nature, pp 215–257

Rani J, Rasool A, Sher H, Fatima R, Hussain J, Majeed M, Hussain K (2024) Precision farming with nanoscale sensors: monitoring soil, water, and crop health. In: Harnessing NanoOmics and nanozymes for sustainable agriculture. IGI Global, Hershey, pp 78–99

Rasool A, Tariq M, Asif M, Shah GM, Fatima R, Majeed M, Ozodbek A (2024) Nanotechnology for improved crop resilience in challenging environments. In: Harnessing NanoOmics and nanozymes for sustainable agriculture. IGI Global, Hershey, pp 149–175

Rasool A, Feroz E, Qayyum Khan Z, Qayyum Khan M, Majeed M, Kousar S (2025) Sustainable agriculture: a nanoosmic-based management. In: Rajput V, Singh A, Ghazaryan K, Minkina T, Al-Tawaha A (eds) Sustainable agriculture: nanotechnology, biotechnology, management and food security. De Gruyter, Berlin, pp 411–430

Shafqat A, Rasool A, Fatima R, Majeed M, Rahman KU, Narejo AH, Farid A (2024) Nanomaterials for sustainable pest and disease management in agriculture. In: Harnessing NanoOmics and nanozymes for sustainable agriculture. IGI Global, Hershey, pp 100–125

Sinha D, Maurya AK, Abdi G, Majeed M, Agarwal R, Mukherjee R, Ganguly S, Aziz R, Bhatia M, Majgaonkar A, Seal S, Das M, Banerjee S, Chowdhury S, Adeyemi SB, Chen JT (2023) Integrated genomic selection for accelerating breeding programs of climate-smart cereals. Genes 14(7):1484

Tariq A, Mumtaz F, Majeed M, Zeng X (2023) Spatio-temporal assessment of land use land cover based on trajectories and cellular automata Markov modelling and its impact on land surface temperature of Lahore district Pakistan. Environ Monit Assess 195(1):1–18

Ullah I, Aslam B, Shah SHIA, Tariq A, Qin S, Majeed M, Havenith HB (2022) An integrated approach of machine learning, remote sensing, and GIS data for the landslide susceptibility mapping. Land 11(8):1265

Waheed M, Arshad F, Majeed M, Fatima S, Mukhtar N, Aziz R, Mangrio WM, Almohamad H, Al Dughairi AA, Al-Mutiry M, Abdo HG (2022) Community structure and distribution pattern of woody vegetation in response to soil properties in semi-arid lowland district Kasur, Punjab, Pakistan. Land 11(12):2145

Waheed M, Arshad F, Majeed M, Haq SM, Aziz R, Bussmann RW, Ali K, Subhan F, Jones DA, Zaitouny A (2023) Potential distribution of a noxious weed (*Solanum viarum* Dunal), current status, and future invasion risk based on MaxEnt modeling. Geol Ecol Landsc 9(1):183–198

Traditional Food Plant Resources of the Mountain Ranges in the Indo-Pak Subcontinent

20

Rahmatullah Qureshi, Chudary Sadam Hussain, and Noor-Ul-Ain Zafar

Abstract

The mountainous range of the subcontinent of Indo-Pak, and specifically, the Himalaya, Karakoram, Hindu Kush, and the neighboring hill systems is a significant hub of agrobiodiversity and traditional ecological knowledge. These mountains harbored 190 wild edible plant species distributed in 78 families, reflecting a rich agrobiodiversity and traditional ecological knowledge hotspot. Rosaceae, Asteraceae, Apiaceae, and Lamiaceae were the most important families contributing wild food plant species, indicating broad botanical knowledge. Leaves (35.48%) and fruits (29.03%) are the most used plant parts, followed by aerial parts, seeds, and roots. Geographically, Khyber Pakhtunkhwa accounts for nearly half of all records, followed by Azad Jammu & Kashmir and Punjab, with the Western Himalayas and Hindu Kush showing transboundary species sharing. Cooked vegetables (49.48%) and raw fruits (30.73%) are the primary food categories, with herbal teas (7.81%) serving both nutritional and medicinal purposes. Among the 190 species, 87 (45.8%) show cross-regional consensus (e.g., *Chenopodium album, Mentha longifolia, Oxalis corniculata, Portulaca oleracea*), while 103 species (54.2%) are single reference, single territory records, highlighting localized knowledge and risk of loss. New territorial records (*Apteranthes tuberculata, Diplazium esculentum, Notholirion thomsonianum*) and novel preparations (e.g., *Cannabis sativa* seeds with sugar) extend previous national inventories. These wild food resources are vital for food security, dietary diversity, and climate adaptation in one of the world's most

R. Qureshi (✉)
Department of Botany, Faculty of Sciences, Pir Mehr Ali Shah Arid Agriculture University, Rawalpindi, Pakistan
e-mail: rahmatullahq@uaar.edu.pk

C. S. Hussain · N.-Ul-A. Zafar
Department of Botany, Pir Mehr Ali Shah Arid Agriculture University, Rawalpindi, Pakistan

R. Qureshi (ed.), *Plant Biodiversity*, https://doi.org/10.1007/978-981-92-1471-6_20

vulnerable mountain regions. Conservation priorities include in situ and ex situ protection, systematic documentation in under-surveyed areas (Waziristan, Chitral, Upper Kohistan), and integration into national food and climate policies.

Keywords

Wild edible plants · Traditional foods · Indo-Pak subcontinent · Indigenous knowledge · Sustainable livelihoods · Biodiversity conservation

Abbreviations

FAO	Food and Agriculture Organization
IPBES	Intergovernmental Science-Policy Platform on Biodiversity and Ecosystem Services
SDGs	Sustainable Development Goals
UN	United Nations
WEPs	Wild edible plants

20.1 Introduction

20.1.1 Global Context of Food Insecurity and Wild Edibles

Malnutrition is one of the most severe problems of the world, which endangers the well-being and health of hundreds of millions of individuals. Afshin et al. (2019) argue that low-quality diets are the cause of 20% of untimely deaths caused by chronic diseases and contribute to the increasing healthcare expenses. By 2050, the global population is projected to reach 9.77 billion, which is likely to cause increased diet-related health issues (Desa 2019; WHO 2021). According to the Food and Agriculture Organization (FAO 2024), in 2023, an estimated 282 million individuals in 59 countries were acutely food-insecure, with an estimated 582 million individuals projected to be chronically undernourished by 2030. The low and middle-income countries are especially at risk of this crisis.

Wild plants are also a significant part of the food supply in the world and about 1 billion people around the world eat wild foods on a daily basis (Lulekal et al. 2011). The edible weed is one of the most useful types of wild edible plants (WEPs), largely used and distributed throughout history. The most prominent ones are *Taraxacum officinale, Sonchus arvensis, S. oleraceus,* and *Chenopodium album* (Åhlberg 2019). Weeds and other WEPs are important sources of food, especially during times of famine, and have been shown experimentally to have bioactive compounds that have therapeutic effects against a number of diseases. On a case-by-case basis, the context-specific management of edible weeds as alternative crops can be taken into account (Menalled and Ebel 2025). The growing interest in edible weeds has become an exciting area of research where agroecosystem sustainability can be improved, and the effects can be beneficial on the agroecosystem, food system, and population health levels (Borsari and Vidrine 2024).

20.1.2 Wild Edible Plants in the Indo-Pak Subcontinent

Wild edible plant gathering is mainly done by poor and rural communities, and it is a normal survival mechanism which is mostly adopted especially in the winter seasons. In evolutionary terms of human-environment interaction, WEPs can be seen as the entire spectrum of uncultivated plants that humans use (Shirsat and Koche 2024). The plants yield edible portions that are naturally occurring in local habitats like forests, farmlands, and uncultivated lands. Different foods such as fresh produce, cooked foods, condiments, decoctions, and sources of minerals and energy are prepared using different parts of the plant, such as the leaves, seeds, fruits, roots, bark, and tubers (Vinceti et al. 2012).

Wild edible vegetation forms a vital part of the subsistence mechanisms of historic indigenous agricultural/hunting communities. WEPs also play a vital role as alternative food sources in marginalized communities in cases where there is a lack of cultivated food production (Majeed et al. 2025). Their use as famine foods not only satisfies hunger, but also has partial nutritional value. WEPs, as backup resources, and in certain cases the most crucial source of survival, are particularly crucial in regions where drought and famine are probable outcomes (Ong and Kim 2017).

20.1.3 Indigenous Knowledge and Cultural Significance

Dependence on wild food plants is also very elevated, especially in tribal areas, as they are usually the poorest areas. The low dependence on commercial foodstuffs, home-prepared foods, and the consumption of foods helps to maintain a close relationship between the human populations and wild vegetation resources (Kumar et al. 2014). Indigenous knowledge of wild food species is strongly connected to culture and has been orally passed down through the generations. The past coexistence between humans and plants has served to control this and has helped to avert the extinction of biodiversity. Moreover, the scientific interest in the possible health benefits of wild fruits, especially their preventive effects against multiple diseases and disorders, is growing (Anwar et al. 2024).

Medicinal qualities of wild fruits are determined by the chemical and nutritional makeup of the fruits. The issue of household food security has played a crucial role in the lives of most rural people. In the past, wild edible plants have been used as important sources of solutions to nutritional, economic, and health-related issues (Mishra et al. 2021). The indigenous societies in these areas have used medicinal and aromatic plants as subsistence and cultural activities. Despite being scientifically recognized as an important source of supplementary and alternative foods in rural livelihoods, wild edible vegetables have received relatively little research attention in comparison with domesticated crops though they may be more widely used than domesticated foods (Abbasi et al. 2015; Mishra et al. 2021).

20.1.4 Study Rationale and Objectives

The majority of research that is available has focused on medicinally useful plants whereas wild edible plants have not been explored extensively. The presence of healers and the medicinal plants have been among the main forms of healthcare in most of the developing countries in the pre-colonial food systems (Bodeker 1995; Patwardhan and Partwardhan 2005). Globally, over 90% of the plant species are wild-harvested in the herbal medicine sector. More than three-out-of-four herbal preparations in the Himalaya are based on plants that are vulnerable to overuse (Abbasi et al. 2015; Shinwari et al. 2024). A great part of these medicinal plants occurs in the subalpine and alpine areas of the Himalaya and thus are very susceptible to environmental pressures and unregulated extraction (Awan et al. 2023; Hussain et al. 2023; N. Khan et al. 2023). Low food supply in cultivations leads to an increased use of wild-harvested resources by consumers. The present research has been conducted to record the diversity of wild food plants utilized by tribal people as well as to analyze folk knowledge of wild food plants and their role in the indigenous food systems.

20.1.5 Household Food Security and Wild Edible Plants

Medicinal plants have been discovered to be a significant source of healthcare in Asia, with more than 80 percent of the population being served by them due to availability, low cost and acceptance by the culture. These are traditionally viewed as scientifically simplistic, cost-effective, and preferred by marginalized communities, and also lead to household food security (Zahara et al. 2019). The use of hundreds of plant species in family food systems is an idea practiced in traditional medicine of some South and East Asian countries. In the region, preservation and promotion of the cultural and spiritual aspects of household food security are widely supported by the community (Asprilla-Perea and Díaz-Puente 2019).

Besides, it has been articulated that indigenous knowledge systems need to be included into the wider interventions to deliver food to all since it is cost-effective and proven to be valuable. In fact, nearly 80 percent of the citizens of developing countries are using plant-based foods as a primary source of their healthcare, as such products are natural, non-narcotic, free of side effects, and both preventive and curative (M. S. Iqbal et al. 2021). According to the world health organization (WHO), herbal medicines are entirely labeled foodstuffs with substances obtained either in the raw form or in the form of plant preparations as a result of aerial or underground plant parts (WHO 2021). In modern times, 25 percent or higher of pharmaceutical drugs are plant-derived, and most synthetic drugs are patterned after plant-derived compounds. Pakistan, India, and China, which make a significant portion of biodiversity on the planet and contain many rare species, are some of the most valuable sources of raw plant products used as food and medicine (Khan et al. 2013; Majid et al. 2020). The United Nations estimates that about 33% of foods in highly industrialized countries are of plant origin (Nakai 2018).

20.1.6 Himalayan Wild Fruits as a Powerful Food Security

Availability of healthy nutrition is key to staying healthy and promoting economic growth. The issue of adequate food and nutrition is a big problem in the communities within the Himalayan region. Nonetheless, the socioeconomic status and nutritional security can be improved by continuing to use natural resources such as pastures, forests, and water systems in an ecologically sustainable manner. The inhabitants of the Himalayas rely on wild plants and fruits as sources of food, nutrition, and income (Bashir et al. 2023; Sinha et al. 2023).

Himalayas have a variety of wild fruits that can grow in adverse climatic conditions without cultivation, leading to a variety of secondary metabolites. According to our findings, the nutritional values of the forest products produced by the Himalayas differ significantly from the nutritional values of mass-produced foods (Rasul et al. 2018). Wild fruits are one of the sources of free food and can be found on the menu of about one out of six human beings in the world. Literature has reported a variety of studies where Himalayan wild fruit has been reported to contain biologically active compounds, antioxidants, minerals, and vitamins, such as *Berberis asiatica, Ficus palmata*, and *Fragaria indica* (Ahmed et al. 2023).

Despite this potential, limited research has been done on the health-promoting properties, chemical compositions, and nutraceutical profiles of the Himalayan wild fruits. The species require a lot of information on their botanical traits, potential yield, farmers' preferences, and consumer demands in order to rank them for development. Increasing production of these native fruits locally will entail information about the existing utilization patterns, production, cultivar improvement, nutrient profiling, after-harvest preservation techniques, and processing plans (Haq et al. 2022).

20.1.7 Wild Edible Plants as a Food Security Asset

The Food and Agriculture Organization (FAO) defines wild edible plants as plants that grow naturally in self-sustaining communities in wild or semi-natural ecosystems and can survive without direct human intervention. These resources have historically been part of human populations, and the species consumed as well as the techniques of consumption have adapted to a variety of environments and cultural practices (Jamil et al. 2022). Exploration of wild edible plant resources has a long history, as researchers acknowledge that these products can reinforce global food security by adding value to diets during scarcity periods and supplying unique nutrients (Bonthoux and Chollet 2024).

Long-term research in Africa has shown that wild fruits benefit nutritional intake and can replace expensive exotic fruits. Therefore, they are considered a valuable resource for helping rural families cope with the impact of climate change (Akinola et al. 2020). Major socioeconomic changes, rural-to-urban migration, and urban population expansion, however, have resulted in the loss of traditional knowledge and practices related to the collection of edible wild plants in places like the

Bamenda highlands over the last century. Habitat loss brought about by agricultural and pastoral development continues to affect rural populations, perpetuating poverty and food insecurity. Simultaneously, the world has rediscovered an interest in wild food and medicinal species. Wild edible plants that have been retained in rural areas are more likely to be documented in ethnobotanical studies. The traditional utilization of wild edible plants at a given location can be captured by interviewing local people, observing contemporary behavior, and reviewing records from the last several decades (Fongnzossie et al. 2020).

20.1.8 Folk Collection and Uses of Wild Edible Plants

The collection of wild plants, along with hunting, became a vital nutritional source for early humans, and it remained important even after the introduction of agriculture and animal domestication (Majeed et al. 2021). This connection between human beings and plants forms the foundation of the discipline of ethnobotany, which aims to record the various species of wild plants used by local communities for food, trade, and medicine (Cheek et al. 2021). Wild edible plants remain an important part of the local population in many regions, even as modern agricultural technology spreads. Economic crises, long-term conflict, and persistent political and cultural turmoil have contributed to the continued significance of wild plants as alternative food sources (León-Lobos et al. 2022).

One of the earliest figures to focus on the connection between nutrition and health was the physician Al-Razi, a major figure in Arab-Islamic medicine. His popular book, dedicated to the virtues of food and the prevention of its evil consequences, puts forth the idea that one can find medicine in one's food. For this reason, Al-Razi is considered among the pioneers of what is today known as nutraceutical science or health nutrition (Al-Fatimi 2021). The nutritional value and antioxidant effects of a range of wild edible plants have been the primary objectives of phytochemical research, along with their role in human health and nutrition. WEPs contain primary metabolites (carbohydrates, lipids, proteins, minerals, vitamins) essential for survival, as well as secondary metabolites. Wild food plants are characterized by higher levels of inorganic metabolites compared to domesticated species (Sharma et al. 2024).

Hunger and malnutrition remain burning concerns worldwide, with an estimated two billion people suffering from micronutrient deficiencies. These inadequacies increase vulnerability to illness and make economic growth very difficult. Climate change also contributes to food insecurity, especially in resource-constrained areas like sub-Saharan Africa, where crop production is expected to decrease drastically. This vulnerability was highlighted by the food price crisis of 2006–2008, which affected some 12 million people in the region (Mittal 2009). Resilience, defined as the ability to resist disruption and to transform, has become a core concept in policies intended to reduce the negative outcomes of climate change on food systems. However, in Africa, limited technological innovation, poor infrastructure, and extreme climate conditions contribute to a lack of adaptive capacity, increasing the continent's exposure to challenges (Savary et al. 2020). Although various studies have explored agricultural adaptability in relation to socioeconomic and governance institutions, very few have investigated the contributions of WEPs and other indigenous practices as components of local welfare systems. This gap emphasizes the necessity of recognizing wild edible plants as alternative sources of food and essential actors for improving food security (Djagba et al. 2019).

20.1.9 Indigenous Knowledge and Wild Edible Plants in Pakistan

The tribal belt of Pakistan has a rich and very diverse ecosystem of plains, woodlands, and pastures, which together nourish a great variety of wild fruits and vegetables. In addition to nutrition, most wild fruits are medically important, and some species have been studied globally to discover their pharmacological effects (Abdullah et al. 2021; Gayathiri et al. 2024). In recent years, some of these plants have been used as functional foods because they are health-promoting, offering nutritional benefits and potential protection against disease. Their wider significance in fostering economic growth, alleviating poverty, increasing dietary diversity, improving food security, and fostering agricultural diversification has been highlighted in many studies. In traditional societies where access to stable and diverse food sources is limited, wild food plants are essential for maintaining nutritional health (Mgalula 2024). Their significance is particularly clear during crises, such as the COVID-19 pandemic and persistent geopolitical and tribal tensions that have increased food insecurity and hunger (Abdullah et al. 2021; Gbashi et al. 2021).

Pakistan, the world's fifth-most populous country and classified as lower-middle-income, ranks as the eleventh poorest-nourished nation globally, with nearly 60% of the population at risk of starvation (Hameed et al. 2021). Geographic isolation, arid weather, limited access to urban markets, and conflicts have combined to cause high food insecurity in the tribal belt. The conflict zone on the border of Pakistan and Afghanistan is the Hindu Kush mountains region. Poverty and malnutrition in the area are further exacerbated by frequent natural disasters, high population growth, limited food supply, and subsistence livelihood methods (Pain 2002). Sustainable management and use of wild food plants have the potential to reduce malnutrition and food insecurity, representing essential assets for low-income groups. Nevertheless, there is a shortage of scientific records regarding native fruit and vegetable species used in localized food systems in Pakistan, especially in the Hindu Kush and adjacent Afghan regions (Abdullah et al. 2021).

20.1.10 Ecological Status of Wild Food Plants

Interest in the exploration of plant resources for food and medicine has remained high. Availability of fruits and leaves from different tree varieties is very important, particularly in developing nations, where they have played a significant role in ensuring food security and better human health. Regular consumption of wild fruits can lower the risk of chronic diseases such as diabetes, obesity, cancer, and cardiovascular disorders (Mozaffarian 2016). Medicinal plants are the foundation of conventional healthcare systems in most areas. South Asia has an estimated population of 500 million individuals who use wild plants to prevent and treat diseases (Stakman 1957). With the side effects of modern medical facilities, plant-based remedies have gained further attention.

Medicinal plants are in high demand in Ethiopia, where they are used due to cultural beliefs, a religious approach based on herbal treatment, and comparatively affordable treatment (Hassan et al. 2024; Khoja et al. 2022; Tassadduq et al. 2022). Traditionally, humans have relied on hunting and gathering plants to sustain their basic needs, using plants and animals as sources of food, medicine, cosmetics, and cultural practices.

20.2 Methodology

20.2.1 Review Methodology

Various search engines and electronic databases were used to conduct a systematic review of original research articles. The mountain ranges that were the target of the literature search were the Indo-Pak Subcontinent mountain ranges. Relevant studies were identified using the search terms such as "wild edibles", "wild edible plant", "wild edible herbs", "wild edible fruits", "wild edible vegetables", "forage/feed", etc. Following the retrieval of the related metadata, all record titles and abstracts were filtered to ensure that only relevant records were selected and to eliminate duplicate records. Then, key articles were assessed in full text in order to obtain detailed information.

20.2.2 Data Source and Compilation

A list of edible wild plant species was elaborated, prepared, and provided in Table 20.1. Data were recorded for each species, containing the botanical name, folk names (local/common names), family, collection time/season, plant part used, type of food preparation, mode of consumption, location(s) where reportedly used, reference citation, and source(s).

20.2.3 Analytical Approach

The following steps of analysis were undertaken in order to assess patterns of use and to distinguish between widely accepted or generally accepted (consensus) uses and local or new reports.

Table 20.1 List of plant species along with local names, family, period of procurement, food category, and method of preparation from the reported areas

Sr. no.	Botanical name	Folk name	Family	Gathering period	Used part	Food Category	Mode of consumption	Reported area	Reference
1.	*Abies pindrow* Royle	Tung/ fir	Pinaceae		Bark	Herbal Tea	The inner soft part of the bark is boiled and prepared as tea	Neelum Valley (Azad Kashmir), Western Himalayas	Assad et al. (2021)
2.	*Achillea millefolium* L.	Dand jari/ Yarrow	Asteraceae		Young stem and Leaves	Vegetable	Aerial parts are being utilized as a vegetable	Neelum Valley (Azad Kashmir)	Naz et al. (2021)
3.	*Adiantum capillus veneris* L.	Sumbal	Adiantaceae		Leaves	Decoction	leaves extract used by children as infusion	Neelum Valley (Azad Kashmir)	Ansari and Ekhlasi-Kazaj (2012), Ilyas et al. (2013)
4.	*Aegopodium burtii* E.Nasir	Kamasla	Apiaceae		Whole plant	fodder/ forage	Fodder, milk curding	Hazara, Chail Sar to Ganja Kandao, KPK	Ilyas et al. (2013)
5.	*Aerva javanica* (Burm. f.) Juss. Ex Schult.	Shorakai / Java aerva, kapok bush	Amaranthaceae	late September	Leaves	fodder/ forage	Leafy shoots are used in curry. It provides grazing for livestock.	Hazara KPK	Ahmad et al. (2019)
6.	*Ajuga integrifolia* Buch.-Ham.	Jaan adam	Lamiaceae		Leaves and Stem	Herbal Tea	Aerial parts or leaves extract are being utilized as a herbal drink	Neelum Valley (Azad Kashmir)	Ullah et al. (2021)
7.	*Allium carolinianum* DC.	Khokhai/ rich piyaz	Amaryllidaceae	June-September	Whole plant	Vegetable	The leaves and bulbs are eaten with bread.	Gilgit baltistan, Upper Kohistan (KP), Pakistan	Ahmad and Pieroni (2016)
8.	*Allium griffithianum* Boiss	Ogakai/ Pyaizokai	Amaryllidaceae	Jul–Aug	whole plant	Vegetable	Fresh leaves are cooked and used as a vegetable	Swat, Lakki Marwat	Ullah et al. (2024), Yahya et al. (2025)

(continued)

Table 20.1 (continued)

Sr. no.	Botanical name	Folk name	Family	Gathering period	Used part	Food Category	Mode of consumption	Reported area	Reference
9.	*Allium humile* Kunth.	Duna/ Pahari Pyaz/ Small Alpine onion	Amaryllidaceae	May–July	Leaves	Vegetable	Leaves are fried and kept in the air to make them crispy Vegetable	Western Himalaya Kashmir Valley (India)	Lanzes and Bhushan (2022), Malik et al. (2021)
10.	*Allium victorialis* L.	Jungle rohan	Amaryllidaceae	April–June	Leaves	Vegetable	Leaves are used as vegetable	Western Himalaya Kashmir Valley (India)	Song et al. (2019)
11.	*Alnus nitida* Endl.	Geeray	Betulaceae	October–June	Stem/ leaves	Vegetable	Leaf poultice applied on painful sites to relieve pain. The decoction of leaves used to treat sore feet. The plant is used as fuel wood and soil binder.	KPK, Gilgit Baltistan, AJK	Ilyas et al. (2013)
12.	*Amaranthus caudatus* L.	Lesa/ Ganhar/ Chalwairay	Amaranthaceae	May–September	Leaves & Seeds	Vegetable	Seeds and leaves are fried in ghee and used as vegetable	Western Himalaya Kashmir Valley (India)	Verma (2020), Priyadarshini et al. (2024)
13.	*Amaranthus crassipes* Schltdl.	Jangli Chulai	Amaranthaceae	April–December	Leaves	Vegetable	Fresh leaves are fried, and during the winter season the dried biomass is also used as a vegetable	Hindu kush Mountains	Jamshed et al. (2018)
14.	*Amaranthus dubius* Mart.	Leesi	Amaranthaceae	May–August	Seeds	Vegetable	Seeds are fried in ghee and used as a vegetable	Western Himalaya Kashmir Valley (India)	Malik et al. (2021)

15.	*Amaranthus spinosus* L.	Ghinyar, chalvary / thorny pigweed	Amaranthaceae	October–June	Leaves, Shoots	Vegetable	Fresh leaves are cooked and used as vegetable	D. I. Khan, Bannu and Kohat, KPK	Ahmad et al. (2019), Muhammad et al. (2022)
16.	*Amaranthus viridis* L.	Leesa/ Ganhar/ Chalwaye	Amaranthaceae	May–August	Leaves & Seeds	Vegetable	Seeds and leaves are fried in ghee and used as a vegetable, dried seed powder are used for vision improvement	Western Himalaya Kashmir Valley (India)	Ilyas et al. (2013), Kumar et al. (2022)
17.	*Angelica archangelica* L.	Choud	Apiaceae	July–August	Roots	Condiment	Dried roots are used as flavoring agent	Western Himalaya Kashmir Valley (India)	Kaur and Bhatti (2021)
18.	*Angelica glauca* Edgew	Choor/Da Ghar Zankai	Apiaceae	July–August	Roots	Condiment	Dried roots are used as flavoring agent	Western Himalaya Kashmir Valley (India)/Qalagai Hills, Kabal	Ilyas et al. (2013), Sood et al. (2019)
19.	*Apium graveolens* L.	Surkhai / celery	Apiaceae	October–April	Leaves	Vegetable	fresh leaves consumed raw as well as cooked	Entire province, KPK	Ahmad et al. (2019)
20.	*Apium leptophyllum* L.	celery	Apiaceae		Leaves	Vegetable	Leaves and seeds infusion is used as emmenagogue and cholagogue	KPK, AJK, Punjab, Gilgit Baltistan	Ahmad et al. (2019)
21.	*Apteranthes tuberculata* (N.E. Br.) Meve & Liede	chungee, chunga, pawuni	Apocynaceae	january–june	Aerial Part	Vegetable	Fresh leaves are cooked and used as a vegetable	Waziristan KPK	Khan et al. (2025), Ullah et al. (2025)
22.	*Arenaria serpyllifolia* L.	Nyman/ Thyme leave sandwart	Caryophyllaceae		Leaves	Vegetable	Leaves are cooked and used as vegetable	Neelum Valley (Azad Kashmir)	Ullah et al. (2019)

(continued)

Table 20.1 (continued)

Sr. no.	Botanical name	Folk name	Family	Gathering period	Used part	Food Category	Mode of consumption	Reported area	Reference
23.	*Arnebia benthamii* (Wall)	Gaozaban	Boraginaceae		Roots	Herbal Tea	Roots are dried and boiled in water with the addition of milk and sugar, and a good source of herbal Tea	Neelum Valley (Azad Kashmir)	Ganie et al. (2014)
24.	*Asparagus gracilis* Royle	Tendorli	Asparagaceae	March–April	Aerial Part	Vegetable	Fresh aerial parts are cooked and used as a vegetable	Swat, lower Dir	Ahmad et al. (2015), Talha et al. (2025)
25.	*Asparagus officinalis* L.	Asparagus	Asparagaceae	June–July	Aerial Part	Vegetable	Leaves are cooked and used as vegetable	Hindu kush Mountains, Kurran Valley KPK	Hussain et al. (2023), Mishwani et al. (2024)
26.	*Bauhinia variegata* L.	Kachnar	Fabaceae	January to April	Flowers	Vegetable	Floral buds are cooked along with minced meet as delicious dish	Punjab, KPK, AJK	Iqbal et al. (2019)
27.	*Berberis calliobotrys* Bien. ex	Sarmay	Berberidaceae	October	Fruits	Fruit	Directly consumed, Eaten Raw	Balochistan, KPK and Punjab	Ahmad and Pieroni (2016)
28.	*Berberis kashmirana* Ahrendt	Sumbal	Berberidaceae		Fruits	Fruits	Taken as fruit	Neelum Valley (Azad Kashmir)	Khan et al. (2012)
29.	*Berberis lyceum* Royles	Sumbal/ kawary	Berberidaceae		Fruits and leaves	Vegetable	leaves are used as vegetable and fruits are consumed	Neelum Valley (Azad Kashmir)	Qudsia et al. (2019)

30.	*Bergenia stracheyi* (Hook.f. & Thomson) Engl.	Batpywa	Saxifragaceae		Roots	Herbal Tea	Roots are dried and boiled in water with the addition of milk and sugar, and a good source of herbal Tea	Neelum Valley (Azad Kashmir)	Siddiq et al. (2012)
31.	*Bergera koenigii* L.	Kari Patta	Rutaceae	Mar–Oct	Leaves	Vegetable	Leaves are used as condiments.	KPK, Sindh	Jabeen et al. (2024)
32.	*Betula utilis* D.Don	Purz/Bhurj	Betulaceae		Bark	Herbal Tea	The inner soft part of the bark is boiled and prepared as tea	Neelum Valley (Azad Kashmir), Western Himalayas	Safdar et al. (2017)
33.	*Bistorta affinis* (D. Don) Greene	Mashroom	Polygonaceae		Rhizome	Herbal Tea	Roots are dried and boiled in water with the addition of milk and sugar, and a good source of herbal Tea	Neelum Valley (Azad Kashmir), Deosai GB	Sultana et al. (2022)
34.	*Brassica oleraceavar.* acephala	Kale/Haaq	Brassicaceae	March–June	Leaves	Vegetable	Leaves are often steamed, boiled, or added to soups and stews.	Western Himalaya Kashmir AJK	Malik et al. (2021)
35.	*Caltha alba* Cambess	Makar pat	Ranunculaceae	March–April	Leaves	Vegetable	Fresh aerial parts are cooked and used as a vegetable	Swat	Shah et al. (2025), Sher et al. (2020)
36.	*Cannabis sativa* L.	Bhangu/ bhang	Cannabaceae		Seeds	Vegetable	Seeds are roasted and then consumed with sugar.	Western Himalaya Kashmir Valley (India), AJK Pakistan, North-West Punjab	Pellati et al. (2018)

(continued)

Table 20.1 (continued)

Sr. no.	Botanical name	Folk name	Family	Gathering period	Used part	Food Category	Mode of consumption	Reported area	Reference
37.	*Capsella bursa-pastoris* (L.) Medik	Mirchi/ Shamsho	Brassicaceae		Leaves	Vegetable	Leaves are cooked and used as vegetable	Neelum Valley (Azad Kashmir), Western Himalayas	Dar et al. (2021)
38.	*Capparis decidua* (Forssk) Edgew	Karri, dela	Capparaceae	March–April	Aerial Part	Fruit	Ripened fruits are eaten as raw while un-ripened fruit as used in making pickle as well as cooked as vegetable	Swat, Kashmir	Zareef et al. (2023)
39.	*Caragana ambigua* Stocks	Zaray	Fabaceae	June	Flowers	Vegetable	Directly consumed, Eaten Raw	Balochistan, Waziristan KPK	Ahmad and Pieroni (2016)
40.	*Caralluma tuberculata* N. E. Br.	Pamanai, pamankey / caralluma	Apocynaceae	January to June	Aerial parts	Vegetable	Stem is used fresh or cooked against diabetes, rheumatism, and joint pains.	Punjab, KPK, AJK	Ahmad et al. (2019)
41.	*Carthamus oxyacantha* M. Bieb	Kariza	Asteraceae	July–September	Seeds	fodder	Seeds are cooked in bread and mix the butter oil	Hindu kush Mountains, Punjab and Uttar Pradesh (India)	Ahmad et al. (2007), Khan et al. (2026)
42.	*Celtis australis* L.	Tagha	Ulmaceae	September–October	Fruits	Fruits	Taken as fruit	Waziristan KPK	Khan et al. (2021), Ilyas et al. (2013)
43.	*Celtis caucasica* Willd.	Togha	Cannabaceae	March–June	Fruits	Fruits	Taken as fruit	Hindu kush Mountains, Kurran Valley KPK	Batool et al. (2025)

44.	*Chenopodium album* L.	bathoo/ Sarmay	Amaranthaceae	March–June	Leaves	Vegetable	Leaves are boiled in water and added with cow ghee, used as a vegetable, used to relieve stomach pain and constipation	Western Himalaya Kashmir Valley (India)/ Qalagai Hills, Kabal Valley, Swat	Ilyas et al. (2013), Poonia and Upadhayay (2015)
45.	*Chenopodium murale* L.	Thor suma, bathu / goosefoot	Amaranthaceae		Leaves	Vegetable	Fresh and cooked leaves are used as digestive, tonic, and anthelmintic.	North Waziristan, Upper Kohistan (KP), Pakistan	Ahmad et al. (2019)
46.	*Cichorium intybus* L.	Kaasni/Han	Asteraceae	April–July	Leaves	Vegetable	Leaves are cooked and used as vegetable	Neelum Valley (Azad Kashmir)	Ilyas et al. (2013), Khalid et al. (2018)
47.	*Cirsium arvense* (L.)Scop.	Da khwarak azghai	Asteraceae	May–August	Stems	Vegetable	The ectoderm of the semi-mature stem is removed. The remaining white fleshy part is eaten raw.	Punjab	Ahmad and Pieroni (2016), Ali et al. (2016), Khan et al. (2013)
48.	*Convolvulus arvensis* L.	Ero, Kheroon	Convolvulaceae		Leaves	Vegetable	Fresh leaves are cooked and used as a vegetable	Upper Kohistan (KP), Pakistan	Amin et al. (2023), Ashfaq et al. (2020)
49.	*Cotoneaster microphyllus* WalL. ex Lindl	Mamanna/ Kharava	Rosaceae		Fruits	Fruits	Taken as fruit	DadarNalah, Waziristan KPK	Ali et al. (2021), Javid et al. (2025), Khan et al. (2013)
50.	*Coriandrum sativum* L.	Dhania	Apiaceae	May	Leaves, seeds	Vegetable	Leaves are used as aroma while seeds spices	Thakht-e-Sulaiman Hills, North-West Pakistan, Punjab, KPK, AJK, GB, Sindh	Ahmad and Pieroni (2016)

(continued)

Table 20.1 (continued)

Sr. no.	Botanical name	Folk name	Family	Gathering period	Used part	Food Category	Mode of consumption	Reported area	Reference
51.	*Cotoneaster minutus* Klotz	Sharavo	Rosaceae	September	Fruit	Fruit	Directly consumed, Eaten Raw	Thakht e-Sulaiman Hills, North-West Pakistan	Ahmad and Pieroni (2016)
52.	*Cotoneaster nummalrius* Fisch. & C.A. Mey	Sherawa	Rosaceae	April–June	Fruits	Fruits	Taken as fruit	Waziristan KPK	Nasab et al. (2025)
53.	*Cotoneaster pruinosus* Klotz	Pushthawergai	Rosaceae	August	Fruit	Fruit	Directly consumed, Eaten Raw	Thakht e-Sulaiman Hills, North-West Pakistan	Ahmad and Pieroni (2016)
54.	*Crataegus oxyacantha* L	Ghowanza/ Hawthorn	Rosaceae		Fruits	Fruits	Taken as fruit	Waziristan KPK	Ali et al. (2017), Saeedand Ahmad (2016)
55.	*Cuminum cyminum* L.	Zeera	Apiaceae	April–May	Seeds	Condiment	Seeds are used as aroma ana flavouring in differents foods	Hindu kush Mountains	Mughal (2022)
56.	*Cynoglossum lanceolatum* Forssk	Jezgai	Boraginaceae	April–July	Leaves	Vegetable	Young leaves are boiled and used as a pickle	Thakht e-Sulaiman Hills, North-West Pakistan	Ahmad and Pieroni (2016)
57.	*Descurainia Sophia* (L.) Webb ex Prantl	Jenjarr	Brassicaceae	March–June	Leaves	Vegetable	Leaves are cooked and used as vegetable	Hindu kush Mountains, Kurran Valley KPK	Khan et al. (2016)
58.	*Digera muricata* L.	Lulur	Aristolochiaceae	June–October	Leaves	Vegetable	Leaves are cooked and used as vegetable	Hindu kush Mountains, Nara desert Sindh	Khan et al. (2013), Qureshi et al. (2010)

59.	*Diplazium esculentum* (Retz.) Sw	Ladora/ Kasrod	Athyriaceae	March–April	Leaves	Vegetable	Fresh leaves are cooked and used as a vegetable	Swat, Kashmir	Haq et al. (2022)
60.	*Diplazium maximum* (Retz)	Langru/ Kasrod	Dryopteridaceae		Leaves	Vegetable	Fresh leaves are being used as a vegetable and also dried for the winter season.	Western Himalaya Kashmir Valley (India)	Sareen et al. (2021)
61.	*Dryopteris filixmas* (L.) Schott	Kunji	Dryopteridaceae		Aerial Part	Vegetable	Aerial parts are fried in oil with toma toes, red chilies, and ginger and used as vegetable	Neelum Valley (Azad Kashmir)	Khan et al. (2024)
62.	*Dryopteris ramose* (C. Hope) C. Chr.	Kunjay / hand fern	Dryopteridaceae	June–October	Circinate leaves	Vegetable	leaves are cooked and eaten as a vegetable	Hazara and Malakand, KPK, AJK, Punjab	Ahmad et al. (2019)
63.	*Elaeagnus angustifolia* L.	Sinjur	Elaeagnaceae		Fruits	Fruits	Taken as fruit	chitral, Waziristan KPK	Aziz et al. (2017), Azmat et al. (2020)
64.	*Elaeagnus umbellata* Thunb	Gohi	Elaeagnaceae		Fruits	Fruits	Taken as fruit	Upper Kohistan (KP), Pakistan	Amin et al. (2023), Sabir and Riaz (2005)
65.	*Eryngium biebersteini*anum Nevski ex Bobrov	Mangaz/Ali kanda	Apiaceae	July–September	Shoot	fodder/ forage	Fodder for Sheep	Swat KPK	Ilyas et al. (2013)
66.	*Fagopyrum esculentum* Moench	Katha	Polygonaceae		Aerial Part	Vegetable	Leaves are fried in oil with tomatoes, red chilies, and ginger and used as vegetable	Neelum Valley (Azad Kashmir)	Debiage et al. (2016), Hussain et al. (2023)

(continued)

Table 20.1 (continued)

Sr. no.	Botanical name	Folk name	Family	Gathering period	Used part	Food Category	Mode of consumption	Reported area	Reference
67.	*Ficus palmata* Forssk.	Injar	Moraceae	April (Leaves), July (Fruit)	Leaves, Fruit	Vegetable	The fruits are directly consumed.The young fresh leaves are boiled in water to soften completely and eaten with bread	Thakht e-Sulaiman Hills, North-West Pakistan	Ahmad and Pieroni (2016)
68.	*Fragaria nubicola* (Hook.f.) LindL.ex Lacaita	Ghra Strawberry	Rosaceae	May–August	Fruits	Fruits	Fresh fruits are consumed	Waziristan KPK	Khalid et al. (2023), Qarni et al. (2022), Roshan et al. (2019)
69.	*Geranium pratense* L.	Gugchuk	Geraniaceae	July–August	Roots	Herbal Tea	Roots are dried and boiled in water and a good source of kashmiri Tea	Western Himalaya Kashmir Valley (India), AJK Pakistan	Khoja et al. (2022)
70.	*Geranium wallichianum* Oliv.	Ratanjog	Geraniaceae	July–August	Roots	Herbal Tea	Roots are dried and boiled in water and a good source of kashmiri Tea	Western Himalaya, Kashmir Valley (India), temperate and sub-alpine zones of Pakistan	Ashraf et al. (2025), Khoja et al. (2022)
71.	*Grewia tenax* (Forssk.) Fiori	Pasthawnai	Malvaceae	August	Fruit	Fruit	Directly consumed, Eaten Raw	Thakht e-Sulaiman Hills, North-West Pakistan	Ahmad and Pieroni (2016)

72.	*Grewia villosa* Willd.	Injarai	Malvaceae	July	Fruit	Fruit	Directly consumed, Eaten Raw	Punjab (Kala-Chitta Hills, Campbell Pur District)	Ahmad and Pieroni (2016)
73.	*Hedera nepalensis* K.Koch	Palool Zeelai	Araliaceae	October–April	Leaves		The powder of leaves is given to reduce blood sugar.	Malakand, KPK	Ilyas et al. (2013)
74.	*Heracleum candicans* Wall.	hogweed	Apiaceae	May–June	Leaves	Vegetable	Leaves are boiled in water and then fried and serve as vegetable	Western Himalaya Kashmir Valley (India) Northwestern Hindukush and Himalayas	Gillani et al. (2024, Butola et al. (2010)
75.	*Hypericum perforatum* L.	Esperke	Hypericaceae		Aerial Part	Vegetable	Fresh leaves are cooked and used as a vegetable, and dry leaves are used in herbal tea	Upper Kohistan (KP), Pakistan	Amin et al. (2023), Saddiqe et al. (2010)
76.	*Impatiens edgeworthii* Hook. f.	Khanroye, Benthal	Balsaminaceae		Leaves & Seeds	Vegetable	Fresh leaves are cooked and used as a vegetable, and seeds are consumed as raw	Upper Kohistan (KP), Pakistan	Amin et al. (2023), Mushtaq et al. (2013)
77.	*Impatiens glandulifera* Royle	Tilfid	Balsaminaceae,		Fruits	Fruits	Ripe fruits are eaten	Western Himalaya Kashmir Valley (India)	Coakley and Petti (2021)
78.	*Iris hookeriana* Foster	Sheeziay	Iridaceae	May–June	Leaves	Vegetable	Fresh leaves are cooked and used as a vegetable	Waziristan KPK	Khalid et al. (2023), Khan et al. (2025)

(continued)

Table 20.1 (continued)

Sr. no.	Botanical name	Folk name	Family	Gathering period	Used part	Food Category	Mode of consumption	Reported area	Reference
79.	*Juglans regia* L.	Ghuz	Juglandaceae	September–October	Nut, Bark, Wood	Nut/Fruit	Edible nut is used as brain and heart tonic. The bark is used for teeth cleaning. Wood is used for high quality furniture. Fruit epicarp is used as hair dye.	KPK	Ilyas et al. (2013)
80.	*Lactuca dissecta* D.Don	Paywerka	Asteraceae	April–August	Leaves	Vegetable	Directly consumed, Eaten Raw	Thakht e-Sulaiman Hills, North-West Pakistan	Ahmad and Pieroni (2016)
81.	*Lantana camara* L.	Bhangi Booti/ Panj Phali	Verbenaceae	Apr–Jul	Fruit/ Vegetable	Fruits/ Leaves	Fresh leaves are eaten raw and also make herbal tea	KPK, Punjab, AJK	Jabeen et al. (2024)
82.	*Lathyrus aphaca* L.	Jangli mater, kukarmany	Fabaceae	March–June	Aerial Part	Vegetable	Leaves are cooked and used as vegetable	Hindu kush Mountains	Ahmad et al. (2019), Tanveer et al. (2012)
83.	*Launaea procumbens* (Roxb.) Ramayya & Rajagopal	Sondrashi	Asteraceae	April–August	Leaves	Vegetable	Directly consumed, Eaten Raw	Punjab, Sindh, Khyber Pakhtunkhwa, AJK	Ahmad and Pieroni (2016)
84.	*Leontopodium himalayanum* DC.	Taliban ka qaleel	Asteraceae		Aerial Part	Raw	Aerial parts are tasted as bubble gum	Upper Kohistan (KP), Pakistan	Amin et al. (2023)

85.	*Lepidium apetalum* Willd.	Bashky, burchan / garden cress	Brassicaceae	April–June	Leaves	Vegetable	Leaves decoction is used against liver and spleen disorders.	Punjab, Sindh, Khyber Pakhtunkhwa, Balochistan	Ahmad et al. (2019)
86.	*Lepidium draba* L.	Bashka, Ghurghwast	Brassicaceae	March–May	Leaves	Vegetable	Fresh leaves are cooked and used as a vegetable	Waziristan KPK	Abbas et al. (2020), Afzal et al. (2016), Khalid et al. (2023)
87.	*Limonium cabulicum* (Boiss.)	Botyarai	Plumbaginaceae	June–August	Leaves	Herbal Tea	The fresh leaves are boiled in water containing sugar. This is then taken as tea	Swat, KPK	Ahmad and Pieroni (2016)
88.	*Lithospermum officinale* L.	Noraki Saba, Bashka	Boraginaceae		Aerial Part	Vegetable	Fresh aerial parts are cooked and used as a vegetable	Waziristan KPK	Ahmad et al. (2019)
89.	*Malcolmia cabulica* (Boiss.) Hook. f. & Thomson	Sag	Brassicaceae	March–May	Aerial shoots	Not specified	No medicinal uses were recorded.	Balochistan, Khyber Pakhtunkhwa, and Punjab	Ahmad et al. (2019)
90.	*Malva neglecta* Wall.	Sochal	Malvaceae	March–April	Leaves	Vegetable	Leaves are fried and used as vegetable	Western Himalaya Kashmir Valley (India)	Abdullah et al. (2021), Ilyas et al. (2013), Khan et al. (2024), Khalid et al. (2023)
91.	*Malva sylvestris* L.	Khwazamary, samchal / common mallow	Malvaceae	April–July	Leaves	Vegetable	Fresh leaves decoction is orally given to cure constipation. Fresh leaves are also used against toothache	KPK, AJK	Ahmad et al. (2019)

(continued)

Table 20.1 (continued)

Sr. no.	Botanical name	Folk name	Family	Gathering period	Used part	Food Category	Mode of consumption	Reported area	Reference
92.	*Marsilea quadrifolia* L.	Devasthal	Marsilleaceae	May–July	Leaves	Vegetable	Leaves are cooked and used as vegetable	Hindu kush Mountains, Maidan Valley Dir KPK	Abdullah et al. (2021), Gul et al. (2016)
93.	*Medicago polymorpha* L.	Speshtlarai, Speshtai	Fabacaeae	March–June	Aerial Part	Vegetable	Leaves are cooked and used as vegetable	Hindu kush Mountains, North Waziristan KPK	Ahmad et al. (2019), Khalid et al. (2023), Shah et al., (2025), Yousfi et al. (2015)
94.	*Mentha longifolia* L.	Vaylanai / horse mint	Lamiaceae	May–June	Shoots	Vegetable	Fresh leaves tea is made for cough and cold. Also used as carminative and against vomiting and indigestion.	Punjab, KPK, AJK, Balochistan	Ahmad et al. (2019), Ilyas et al. (2013), Lal et al. (2024)
95.	*Mentha royleana* Wall. ex Benth.,	Speer, Vilanay	Lamiaceae	March–October	Leaves	Vegetable	Eaten as raw/ salad	KPK, AJK, Punjab, Sindh	Ahmad et al. (2019)
96.	*Mirabilus jalapa* L.	Gul-e-Badah	Nyctaginaceae	May–August	Leaves/ root		Poultice is applied to abscesses. The root powder is given as aphrodisiac and used in tonsillitis. Planted as ornamental.	KPK, Balochistan, Punjab	Ilyas et al. (2013)
97.	*Myrsine Africana* L.	Manrogayah	Primulaceae	July–September	Fruits	Fruits	Taken as fruit	Hindu kush Mountains	Gul et al. (2017), Khan and Abdullah (2020)

98.	*Nannorrhops ritchieana* (Griff.) Aitch	Mazari palm	Arecaceae	September–October	Fruits	Fruits	Taken as fruit	Hindu kush Mountains, Waziristan and Kurram districts KPK, Jacobabad Sindh	Ali et al. (2020)
99.	*Nasturtium officinale* R.Br.	Aabi Salad/ Tarmera/ Talib shah/ water cress	Brassicaceae	March–November	Leaves	Vegetable	Leaves are cooked and used as vegetable	Hindu kush Mountains, Punjab, KPK	Amin et al. (2023), Khan et al. (2023)
100.	*Nepeta erecta* Benth.	Peshoo Botay	Lamiaceae	June–August	Shoot	Herbal Tea	The boiled water is given to ease coughs, reduce temperature, headache and scalp irritation.	Punjab, KPK	Ilyas et al. (2013)
101.	*Notholirion thomsonianum* (Royle) Stapf	Shyajey	Liliaceae	April–May	Leaves	Vegetable	Young fresh leaves are boiled in water for ½ hour. The extra water is poured out, and then the leaves are moved to another container containing ghee with fried onions & red chilies. Kept on the fire until water evaporates & ghee appears on top. This is then eaten with bread.	KPK, AJK	Ahmad and Pieroni (2016)

(continued)

Table 20.1 (continued)

Sr. no.	Botanical name	Folk name	Family	Gathering period	Used part	Food Category	Mode of consumption	Reported area	Reference
102.	*Olea ferruginea* WalL. ex Aitch	Kowo/ Shwawan	Oleaceae	September	Fruits	Fruits	Taken as fruit	Hindukush range, Upper Kohistan (KP), Pakistan	Ahmad and Pieroni (2016), Amin et al. (2023), Nasrullah et al. (2015)
103.	*Opuntia monacantha* (Willd.) Haw	Zuqam/ Nagphana	Cactaceae	September–October	Fruits	Fruits	Taken as fruit	Teri Khel (Mianwali), Chabb (Attock)	Majeed et al. (2022)
104.	*Oxalis corniculata* L.	Khattibooti	Oxalidaceae	March–May	Leaves	Vegetable	Fresh leaves are cooked and used as a vegetable	Waziristan KPK	Abid (2010), Iqbal et al. (2019)
105.	*Ocimum basilicum* L.	Niazbo	Lamiaceae		Leaves	Vegetable	Fresh shoot is spread over every kind of cooked dish as flavoring agent	Western Himalaya Kashmir Valley (India), KPK, AJK, Punjab	Feng et al. (2024), Iqbal et al. (2019)
106.	*Oxyria digyna* (L.) Hill	Huli	Polygonaceae		Aerial Part	Vegetable	Fresh leaves are cooked and used as a vegetable	Eastern Hindukush, Upper Kohistan (KP), Pakistan	Amin et al. (2023), Ullah and Badshah (2023)
107.	*Papaver rhoeas* L.	Gul-e-lala, Lal poshta	Papaveraceae	March–May	leaves	Vegetable	Leaves are cooked and used as vegetable	Hindu kush Mountains, Kashmir Himalayas, India	Abbasi et al. (2015), Dar et al. (2010)
108.	*Parrotiopsis jacquemontiana* (Decne.) Rehder	Pashoo, Pashoot, Passat	Hamamelidaceae		Fruits	Fruits	Taken as fruit	Upper Kohistan (KP), Pakistan	Ali and Juyal (2018)
109.	*Pedicularis oederi* Vahl.	Khana por	Orobanchaceae		Leaves	Vegetable	Fresh leaves are cooked and used as a vegetable	Gilgit baltistan, Upper Kohistan (KP), Pakistan	Hameed et al. (2021)

110.	*Periploca aphylla* Decne.	Barrarr	Apocynaceae	May	Fruit and young stems	Fruit	The fresh green fruits and young stems are chewed, and the sweet-tasting sap is swallowed.	Punjab, KPK, AJK, Balochistan	Ahmad et al. (2011), Ahmad and Pieroni (2016), Khalid et al. (2023)
111.	*Periploca hydaspidis* Falc.	Khwaza waley	Apocynaceae	May	Fruit	Fruit	Directly consumed, Eaten Raw	KPK, AJK	Ahmad and Pieroni (2016)
112.	*Physalis angulata* L.	Bhamolay Shrub	Solanaceae	Dec–Apr	Fruit	Fruit	Ripen fruits are eaten raw	KPK	Jabeen et al. (2024)
113.	*Physalis divaricata* D.Don	Band malkhovj	Solanaceae	May	Fruit	Fruit	Directly consumed, Eaten Raw	Balochistan, KPK	Ahmad and Pieroni (2016)
114.	*Phytolacca acinose* Roxb.	Jharkha	Phytolaccaceae		Leaves	Vegetable	Leaves are being utilized as a vegetable	Western Himalaya Kashmir Valley (India), Lesser Himalayas	Bailey (2021)
115.	*Pinus gerardiana* WalL. ex D. Don	Chonge, Tholash, Shentah	Pinaceae		Fruits	Fruits	Taken as fruit	Lalku Valley Swat, Upper Kohistan (KP), Pakistan	Shah et al. (2025)
116.	*Pinus wallichiana* A. B. Jacks.	Nashtar	Pinaceae	October	Fruit/ seeds/Sap/ Honeydew	Fruit	Directly consumed, Eaten Raw; honeydew eaten as a sweet treat or used like honey	KPK, Northern areas GB	Ahmad and Pieroni (2016)
117.	*Pistacia khinjuk* Stocks	Khakow	Anacardiaceae		Fruits	Fruits	Taken as fruit	Upper Kohistan (KP), Pakistan	Sher et al. (2020)
118.	*Plantago lanceolata* L.	Ghuai Jabai	Plantaginaceae	March–April	Leaves	Vegetable	Fresh leaves are cooked and used as a vegetable	Swat	Ilyas et al. (2013)

(continued)

Table 20.1 (continued)

Sr. no.	Botanical name	Folk name	Family	Gathering period	Used part	Food Category	Mode of consumption	Reported area	Reference
119.	*Plantago major* L.	Jangli ispaghol	Plantaginaceae,		Leaves	Vegetable	Fresh leaves are cooked and used as a vegetable	Waziristan KPK	Hassan et al. (2019), Iqbal et al. (2019)
120.	*Plantago ovata* Forssk	Isphagol,Sath	Plantaginaceae	June–October	Seeds	Raw	Fresh and dry seeds are consumed	Waziristan KPK	Khalid et al. (2023), Yousaf et al. (2024)
121.	*Podophyllum emodi* WalL. ex Royle	Ghara marchaka	Berberidaceae		Fruits	Fruits	Taken as fruit	Waziristan KPK, Jammu and Kashmir	Ahmad et al. (2015), Sher et al. (2010)
122.	*Podophyllum hexandrum* Royle	Shargoye, Khakori	Berberidaceae		Fruits	Fruits	Taken as fruit	Gilgit, Chitral, Kashmir, Upper Kohistan (KP), Pakistan	Choudhary et al. (2021)
123.	*Polygonum aviculare* L.	Anjbar	Polygonaceae	March–November	Leaves	Vegetable	Leaves are cooked and used as vegetable	Hindu kush Mountains, Okara Punjab	Arshad et al. (2024)
124.	*Polygonum plebeium* R.Br.	Chemti Sag, Dubia Sag and Anjaban	Polygonaceae		Aerial Part	Vegetable	Fresh aerial parts are cooked and used as a vegetable	Waziristan KPK	Saleem et al. (2020)
125.	*Portulaca oleracea* L.	Pechel	Portulacaceae		Aerial Part	Vegetable	Fresh leaves are fried, and during the winter season the dried biomass is also used as a vegetable	Upper Kohistan (KP), Pakistan	Amin et al. (2023), Naeem et al. (2022)

126.	*Portulaca quadrifida* L.	Warkhorai, lorank / chicken weed	Portulacaceae	August	Leaves, aerial shoots	Vegetable	Slightly warmed leaves are applied topically on swelling joints and as erysipelas. Fresh leaves are used as diuretic in dysuria and as vermifuge.	KPK,North Waziristan	Ahmad et al. (2019)
127.	*Portulaca grandiflora* L.	Gul Dopheri	Portulacaceae	Jun–Sep	Aerial parts/ leaves	Vegetable	Boiled leaves are fried in oil with tomato, onion and red chilies.	KPK, Northern region, Punjab	Khan and Abdullah (2020)
128.	*Primula elliptica* Royle	Lozum lelo	Primulaceae		Aerial Part	Condiment	The dried powder of the aerial parts is used for flavoring food.	Western Himalayas, Upper Kohistan (KP), Pakistan	Abbas et al. (2024), Khan et al. (2021)
129.	*Prunus cornuta* (Wall)	badara	Rosaceae		Fruits	Fruits	Ripe fruits are eaten	Western Himalaya Kashmir Valley (India)	Zanit and Mochi (2021)
130.	*Prunus jacquemontii* Hook.f.	Afghan Bush	Rosaceae		Fruits	Fruits	Taken as fruit	Waziristan KPK, AJK	Ali (2023)
131.	*Pteridium aquilinum* L.	Kawsay/ Kakran	Dennstaedtiaceae		Leaves	Vegetable	Fresh leaves are cooked and used as a vegetable	Waziristan KPK	Vetter (2010)
132.	*Pteridium revolutum* (Blume)	Jungle kunji	Dennstaediaceae	April–May	Stem	Vegetable	Stem is fresh as well as dried and boiled in water along with species like fennel and used as vegetable	Western Himalaya Kashmir Valley (India)	Bagri and Dhingra (2023)

(continued)

Table 20.1 (continued)

Sr. no.	Botanical name	Folk name	Family	Gathering period	Used part	Food Category	Mode of consumption	Reported area	Reference
133.	*Punica granatum* L.	Dango, Daroye, Daroo, Jangliannar	Lythraceae	July–August	Fruits	Fruits	Taken as fruit	Balochistan (Khuzdar, Kalat, and Quetta), Khyber Pakhtunkhwa (Dir, Chitral, Kurram, swat), and Punjab	Amin et al. (2023), Shah et al. (2024)
134.	*Pyracantha* M. Roam	Khrasharavo	Rosaceae	September	Fruit	Fruit/ Herbal tea	Direct consumed, Eaten Raw	Punjab, KPK	Ahmad and Pieroni (2016)
135.	*Pyrus pashia* Buch. & Ham.	Shegal	Rosaceae		Fruits	Fruits	Ripe fruits are eaten	Western Himalaya Kashmir Valley (India)	Ali et al. (2016)
136.	*Quercus baloot* Griff	Chairay	Fagaceae	December–Feb	seeds	Fruits	Taken as fruit	Hindu kush Mountains	Batool et al. (2025), Khurram et al. (2012)
137.	*Quercus semecarpifolia* Sm.	Khasru	Fagaceae		Fruits	Fruits	Fresh fruits are consumed	Upper Kohistan (KP), Pakistan	Quyoom et al. (2024)
138.	*Rheum australe* D. Don	Rewand Cheeni	Polygonaceae		Stem	Raw	After peeling off the fresh stem, the inner part is eaten	Nanda Devi National Park (NDNP), Uttarakhand, India	Rana et al. (2010)
139.	*Ribes alpestre* Wall. ex Decne.	Sheenkorai	Grossulariaceae	July	Fruit	Fruit	Directly consumed, Eaten Raw	Punjab, KPK, AJK	Ahmad and Pieroni (2016)

140.	*Rosa moschata* Herrm.	Khorach	Rosaceae	August	Fruit	Fruit	Directly consumed, Eaten Raw	Punjab, KPK	Ali et al. (2014), Ahmad and Pieroni (2016), Khan and Abdullah (2020)
141.	*Rubus caesius* L.	Gounch	Rosaceae	July–August	Fruits	Fruits	Fresh and raw consumed as fruit.	Western Himalaya Kashmir Valley (India)	Grochowski et al. (2020)
142.	*Rubus fruticosus* L.	Karwara, Ach, Akhara and Baganrra	Rosaceae		Fruits	Raw	Fresh fruits are consumed	North Waziristan, Upper Kohistan (KP), Pakistan	Khalid et al. (2023), Zia-Ul-Haq et al. (2014)
143.	*Rubus niveus* Thunb	Garachey	Rosaceae		Fruits	Raw	Fresh fruits are consumed	Azad Jammu and Kashmir, Upper Kohistan (KP), Pakistan	Ahmad et al. (2015), Sultana et al. (2019)
144.	*Rubus sanctus* Schreb	Bana, Karwara	Rosaceae	June–July	Fruits	Fruits	Fresh fruits are consumed	Swat	Ali et al. (2023)
145.	*Rubus ulmifolius* Schott	Gharangavo	Rosaceae	August	Fruit	Fruit	Directly consumed, Eaten Raw	Swat, KPK	Ahmad and Pieroni (2016)
146.	*Rumex chalepensis* Mill	Spin Zamda	Polygonaceae		Leaves	Vegetable	Fresh leaves are cooked and used as a vegetable	Waziristan KPK	Hasan et al. (1995)
147.	*Rumex dentatus* L.	jangli palak	Polygonaceae		Leaves	Vegetable	Leaves are consumed by the local community	Western Himalaya Kashmir Valley (India)	Choudhary et al. (2021), Ilyas et al. (2013)

(continued)

Table 20.1 (continued)

Sr. no.	Botanical name	Folk name	Family	Gathering period	Used part	Food Category	Mode of consumption	Reported area	Reference
148.	*Rumex hastatus* D. Don	Tarooky, takely	Polygonaceae	March–June	Leaves	Vegetable	Fresh leaves are fried, and during the winter season the dried biomass is also used as a vegetable	Hindu kush Mountains, Hazara KPK	Ahmad et al. (2019), Ahmad et al. (2015)
149.	*Salvadora oleoides* Decne	Waan	Salvadoraceae	Feb–Jun	Fruit	Fruit	Ripen fruits are eaten raw	Punjab, KPK, Balochistan,AJK	Jabeen et al. (2024)
150.	*Salvadora persica* L.	Pilu	Salvadoraceae	Feb–Jun	Fruit	Fruit	Ripen fruits are eaten raw	KPK. AJK, Punjab	Jabeen et al. (2024)
151.	*Salvia moorcroftiana* Wall. Ex	Dersai	Lamiaceae	May–June	Stem	Vegetable	The external green part of the semi-matured stem is removed, and the remaining white juicy part is eaten raw.	KPK, Kashmir	Ahmad and Pieroni (2016), Gani et al. (2019)
152.	*Salvia mukerjeei* Bennet & Raizada	Matter jarrai / glass-wort	Lamiaceae	March to September	Leaves	Vegetable	Roots decoction is used in cold and cough	Swat, Mansehra, and Hazara, KPK	Ahmad et al. (2019)
153.	*Sambucus nigra* L.	Khesh gen and Gandula	Adoxaceae,		Fruits	Fruits	Taken as fruit	swat, Waziristan KPK	Khan et al. (2023)
154.	*Scorzonera raddeana* Fisch. & C.A. Mey	Mutayer/ Arghuch/ Dagham/ Asguch	Asteraceae		Leaves	Vegetable	Fresh leaves are cooked and used as a vegetable	Waziristan KPK, Balochistan	Abid et al. (2022), Kousar et al. (2023)

155.	*Sideroxylon mascatense* (A. DC.) T.D. Penn	Gurgura, Jangli falsa	Sapotaceae	June–July	Fruits	Fruits	Taken as fruit	Hindu kush Mountains, North-Western Punjab	Ahmad and Pieroni (2016), Ali et al. (2022), Hussain et al. (2023), Malik et al. (2024)
156.	*Silene conoidea* L.	Mangotay	Caryophylaceae	March–May	Fruits	Fruits	Taken as fruit	Hindu kush Mountain	Khan and Abdullah (2020)
157.	*Silene vulgaris* (Moench)	Bibdughas	Caryophyllaceae		Leaves	Vegetable	Leaves are boiled, cooked, and being utilized as a vegetable	Western Himalaya Kashmir Valley (India)	Kumar et al. (2014)
158.	*Solanum americanum* Mill.	Kachmacho, malgabai / black nightshade	Solanaceae		Leaves	Vegetable	Fresh leaves are used as poultice for rheumatic and joints pains. Leaves decoction is used against enlarged spleen and in heart problems. Cooked leaves are used as anticonstipation.	KPK,AJK	Ahmad et al. (2019), Ullah et al. (2018)
159.	*Solanum villosum* Mill	Kachmach	Solanaceae,		Fruits and leaves	Raw	Fruits and leaves are consumed as raw	Waziristan KPK	Ali et al. (2022), Zahara et al. (2019)
160.	*Solena amplexicaulis* (Lam.) Gandhi	Karkundai/ Bankakra	Cucurbitaceae	August–September	Unripe fruit	Fruits	Strengthen body	Theni District (Western Ghats), Southern India	Jeyaprakash et al. (2011)

(continued)

Table 20.1 (continued)

Sr. no.	Botanical name	Folk name	Family	Gathering period	Used part	Food Category	Mode of consumption	Reported area	Reference
161.	*Spiraea canescens* D. Don	Srawany	Rosaceae	May	Fruit, stem, Bark	Herbal Tea	The fruits were directly consumed. The green stem cortex is boiled in water with sugar and then consumed as tea.	Chitral, Hazara KPK	Ahmad and Pieroni (2016)
162.	*Stellaria media* L.	Olalai / Chickweed	Caryophyllaceae	March–April	Aerial Part	Vegetable	Fresh leaves are cooked and used as a vegetable	Swat, Punjab	Aleem et al. (2023), Shujaat Ali et al., Gul et al. (2025), Ullah et al. (2019)
163.	*Tanacetum artemisioides* DC.	Tansy	Asteraceae		Leaves	Vegetable	Fresh leaves are cooked and used as a vegetable	Waziristan KPK	Bukhari et al. (2007)
164.	*Taraxacum campylodes* G. E. Han glund	Huand / dandelion	Asteraceae	June–July	Leaves	Vegetable	Rhizome decoction is used against jaundice and as diuretic. Leaves tea is used during hepatitis and kidney problems.	Gilgit-Baltistan, Khyber Pakhtunkhwa, Azad Jammu and Kashmir	Ahmad et al. (2019)
165.	*Taraxacum officinale* (Weber ex Wiggers)	Hand	Asteraceae		Leaves	Cooked	Jaundice, liver disorder, stomach pain	Neelum Valley (Azad Kashmir), Kurram Valley, KPK	Ahmad et al. (2017)

166.	*Teucrium royleanum* Wall. ex Benth	Aalba	Lamiaceae		Roots	Decoction	Roots are ground in water, and the filter water is taken as an herbal drink	Neelum Valley (Azad Kashmir)	Ahmad et al. (2011)
167.	*Teucrium stocksianum* Boiss	Kastorai	Lamiaceae	March–April	Fruits	Fruits	Fresh and dry fruits are consumed	Waziristan KPK	Khalid et al. (2023), Shah et al. (2012)
168.	*Thymus linearis* Benth.	Pannay	Lamiaceae		Aerial Part	Herbal Tea	Aerial parts are boiled in water, with the addition of milk and sugar, taken as an herbal tea	Neelum Valley (Azad Kashmir), Kurram Valley, KPK	Hussain et al. (2023), Shirazi (2018)
169.	*Torilis leptophylla* (L.) Rchb.f.	Gajar Booti	Apiaceae		Aerial Part	Vegetable/ fodder	Fresh leaves are cooked and used as a vegetable	Lower Dir, Upper Kohistan (KP), Pakistan	Nasreen et al. (2023)
170.	*Tragopogon gracilis* D. Don	Shabey	Asteraceae	July	Aerial parts	Vegetable	Consumed as raw or cooked form	KPK, balochistan	Ahmad and Pieroni (2016)
171.	*Trianthema portulacastrum* L.	Pandraosh/ It-sit	Aizoaceae		Leaves	Vegetable	Fresh leaves are cooked and used as a vegetable	Waziristan KPK	Khan et al. (2013)
172.	*Trifolium repens* L.	Zangali shautal	Fabaceae	March–August	Shoot	Fodder	Fodder, honey bee species	KPK, AJK, GB, Punjab	Ahmad et al. (2021), Hussain et al. (2023), Ilyas et al. (2013), Varin et al. (2010)
173.	*Trigonella gracilis* Benth.	Shpestaray / fenugreek	Fabaceae		Leaves	Not specified	No medicinal uses were recorded.	KPK, North-West Pakistan	Ahmad et al. (2019)

(continued)

Table 20.1 (continued)

Sr. no.	Botanical name	Folk name	Family	Gathering period	Used part	Food Category	Mode of consumption	Reported area	Reference
174.	*Trillium govanianum* Wall.	kalijiri	Melanthiaceae	April–May	Leaves	Vegetable	Leaves are cooked and used as vegetable	Western Himalaya Kashmir Valley (India)	Vidyarthi et al. (2013) Priyadarshini et al. (2024)
175.	*Tulipa lehmanniana* Merckl.	Sondai	Liliaceae	April–June	Bulb	Vegetable/ Fodder	Bulbs are eaten as a vegetable; Leaves used as fodder for goats	Balochistan, KPK, Punjab, AJK	Ahmad and Pieroni (2016)
176.	*Typha latifolia* L.	Kunder	Typhaceae	April–August	Rhizome, Fruit	Vegetable	A delicious and nutritious sweet under the local name of `Boori` is made from fruit while pounded roots are cooked as vegetables	KPK, Punjab, Sindh	Ahmad et al. (2019)
177.	*Urtica dioica* L.	Zahoye / Keyori/ Sezonky	Urticaceae		Aerial Part	Vegetable	Fresh aerial parts are cooked and used as a vegetable	Upper Kohistan (KP), Koh-e-Safaid Range, Pakistan	Hussain et al. (2018), Mohammad Ilyas et al. (2013)
178.	*Veronica anagallis* L.	Obo Saba	Plantaginaceae		Leaves	Vegetable	Fresh leaves are cooked and used as a vegetable	Waziristan KPK, Kashmir	Majeed et al. (2021)
179.	*Viburnum cotinifolium* D. Don	Aoza/ Margharawa	Adoxaceae	JuneAugust	leaves	Fruits	Fresh fruits are consumed	Upper Kohistan (KP), Waziristan, Pakistan	Amin et al. (2023), Khalid et al. (2023), Muhaisen et al. (2002)

180.	*Viburnum grandiflorum* Wall. ex-DC	Guchh, Ghaz mewa	Adoxaceae		Fruits	Fruits	Taken as fruit	Neelum Valley (Azad Kashmir), Abbotabad, KPK	Ali et al. (2025), Shah et al. (2024)
181.	*Vicia faba* L.	Pervatha, chilow / beans	Fabaceae	May–September	Leaves, Shoots	Not specified	No medicinal uses were recorded.	Gilgit baltistan, Upper Kohistan (KP), Pakistan	Ahmad et al. (2019)
182.	*Vicia sativa* L.	Jangli mattar, Raari	Fabaceae	April–May	Fruits	Fruits	Fresh fruits are consumed	Swat	Choudhary et al. (2021), Majeed et al. (2021)
183.	*Viscum cruciatum* Sieber ex Boiss.	Da shwawna lewanai	Santalaceae	May	Leaves	Herbal Tea	The leaves are used as tea	AJK, KPK	Ahmad and Pieroni (2016)
184.	*Vitis flexuosa* Thunb.	Malavo	Vitaceae	June	Fruit	Fruit	Directly consumed, Eaten Raw	KPK, Gilgit Baltistan	Ahmad and Pieroni (2016)
185.	*Withania coagulans* (Stocks) Dunal	Shapyanga	Solanaceae	September–October	Fruits	Fruits	Taken as fruit	Waziristan KPK	Iqbal et al. (2023), Khalid et al. (2023)
186.	*Xanthium strumarium* L.	Ghat Jishkay	Balsaminaceae	June–October	Leaves	Vegetable	The juice is given in chronic mild fever.	Punjab, KPK	Ilyas et al. (2013)
187.	*Ziziphus jujuba* Mill.	SeoBair, JangliBaiR, Bara Baira	Rhamnaceae	August–Sept	Fruits	Fruits	Taken as fruit	Hindu kush Mountains, North Punjab, KPK, AJK, Gilgit Baltistan	Ahmad and Pieroni (2016), Malik et al. (2024), Muhammad et al. (2022)
188.	*Ziziphus mauritiana* Lam.	Beri	Rhamnaceae	Feb–Jun	Fruit	Fruit	Ripened fruits are eaten raw. Fruits are dried to made herbal tea	Punjab, Sindh, KPK	(Melnyk (1996), Jabeen et al. (2024)

(continued)

Table 20.1 (continued)

Sr. no.	Botanical name	Folk name	Family	Gathering period	Used part	Food Category	Mode of consumption	Reported area	Reference
189.	*Ziziphus nummularia* (Burm.f.) Wight & Arn	Bar	Rhamnaceae	October–Nov	Fruits	Fruits	Taken as fruit	Hindu kush Mountains, Kurran Valley KPK	Batool et al. (2025), Uddin et al. (2021)
190.	*Zygophyllum creticum* (L.) Christenh.& Byng	Dhamman	Zygophyllaceae	Mar–Nov	Whole Plant	Vegetable/ herbal tea	Plant is dried to used as a phaki and herbal tea	Sindh	Jabeen et al. (2024)

20.2.4 Recognition of Multi-Reference Species

Those species reported by two or more independent sources were considered to have cross-regional consensus. Such species were deemed to be consistently observed within various areas of study.

20.2.5 Geographical Spread Analysis

The number of different areas of occurrence of each species (e.g., KPK, AJK, Punjab, Western Himalaya) was recorded. The most frequently utilized species was marked as the widely distributed species which appeared in three or more territories.

20.2.6 Detection of Unique or Novel Uses

Species that were only reported by a single reference and were limited to one territory were defined as locally unique or possibly novel. A special effort was made to record previously unrecorded uses in major national ethnobotanical reviews.

20.2.7 Use Pattern Categorization

Based on a literature review, the following three patterns of plant uses were developed:

20.2.7.1 Consensus Uses
Same or very similar usages found reported by more than one author in various geographic locations.

20.2.7.2 Diversity of Application
The identical part of the plant is applied, but the method of application varies according to region.

20.2.7.3 Territory-Specific Uses
Only documented in a single area, possibly as a consequence of local ecology or cultural localism.

20.3 Results

20.3.1 Plant Diversity and Family Distribution

A total of 190 plant species belonging to 78 families have been reported as the wild food resources from the mountain ranges of the Indo-Pak Subcontinent (Table 20.1). The pictorial view of each plant species is provided in Plate 20.1. Rosaceae was the

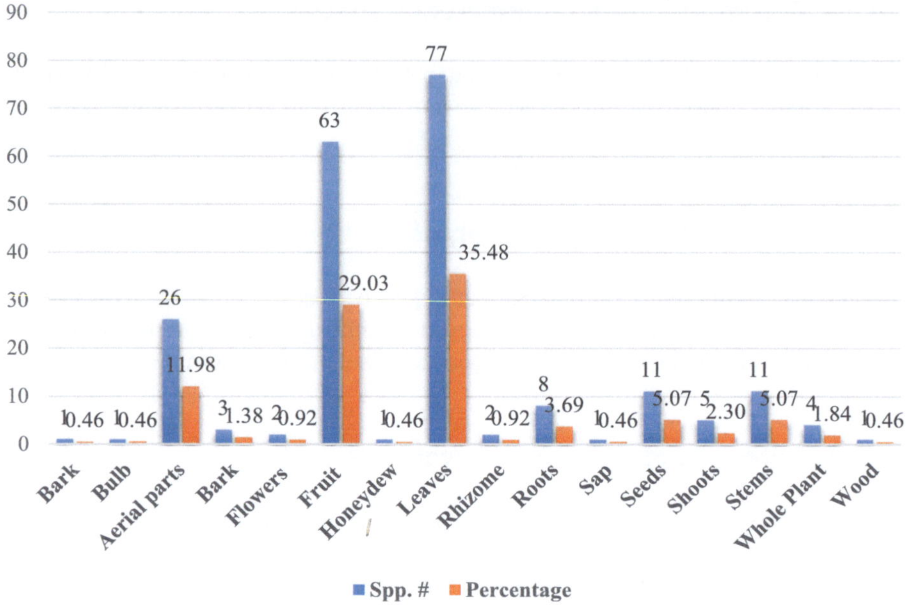

Plate 20.1 Pictorial view of the recorded plant species from the mountains of the Indo-Pak subcontinent.

largest family that contributed 17 species (8.95%), followed by Asteraceae (12 spp., 6.32%), Apiaceae & Lamiaceae (10 spp., 5.26 each), Polygonaceae (9 spp., 4.74%), Amaranthaceae & Fabaceae (8 spp., 4.21 each), Brassicaceae (7 spp., 3.68%), Berberidaceae & Solanaceae (5 spp., 2.63% each), the rest of the 60 families were represented by contributed 1–4 species (Table 20.2). This distribution shows that although there are some families that control the wild food resource, overall there is a large number of plant families that contribute to the overall diversity, which underscores the existence of a wide ethnobotanical knowledge base from various growing localities of the Indo-Pak subcontinent.

20.3.2 Plant Parts Used

Leaves were the most commonly used plant part, with 77 species (35.48%) being used as a food source, as reported in various literature (Fig. 20.1). The 2nd2nd most widespread part was fruit, employed in 63 plant species (29.03%). The aerial parts (shoots and above-ground parts) were used in 26 species (11.98%), followed by seeds & stems (11 spp., 5.07% each), roots (8 spp., 3.69%), shoots (5 spp., 2.30%), whole plants (4 spp., 1.84%), and bark (3 spp., 1.38%). Other components like bark, flowers, bulbs, rhizomes, nuts, and sap were used less as a proportion, ranging from 1 to 2 species (Fig. 20.1). The leafiness and fruitiness imply that local harvesting patterns consider renewable, readily available, and nutrient-dense plant components, a sustainable trend when done in a proper way.

Table 20.2 Contribution of families in terms of species diversity of the wild plant resources recorded for the area

S. no.	Family	Spp. #	Percentage
1	Adiantaceae	1	0.53
2	Adoxaceae	3	1.58
3	Aizoaceae	1	0.53
4	Amaranthaceae	8	4.21
5	Amaryllidaceae	4	2.11
6	Anacardiaceae	1	0.53
7	Apiaceae	10	5.26
8	Apocynaceae	4	2.11
9	Araliaceae	1	0.53
10	Arecaceae	1	0.53
11	Aristolochiaceae	1	0.53
12	Asparagaceae	2	1.05
13	Asteraceae	12	6.32
14	Athyriaceae	1	0.53
15	Balsaminaceae	3	1.58
16	Berberidaceae	5	2.63
17	Betulaceae	2	1.05
18	Boraginaceae	3	1.58
19	Brassicaceae	7	3.68
20	Cactaceae	1	0.53
21	Cannabaceae	2	1.05
22	Capparaceae	1	0.53
23	Caryophyllaceae	4	2.11
24	Convolvulaceae	1	0.53
25	Cucurbitaceae	1	0.53
26	Dennstaedtiaceae	2	1.05
27	Dryopteridaceae	3	1.58
28	Elaeagnaceae	2	1.05
29	Fabaceae	8	4.21
30	Fagaceae	2	1.05
31	Geraniaceae	2	1.05
32	Grossulariaceae	1	0.53
33	Hamamelidaceae	1	0.53
34	Hypericaceae	1	0.53
35	Iridaceae	1	0.53
36	Juglandaceae	1	0.53
37	Lamiaceae	10	5.26
38	Liliaceae	2	1.05
39	Lythraceae	1	0.53
40	Malvaceae	4	2.11
41	Marsilleaceae	1	0.53
42	Melanthiaceae	1	0.53

(continued)

Table 20.2 (continued)

S. no.	Family	Spp. #	Percentage
43	Moraceae	1	0.53
44	Nyctaginaceae	1	0.53
45	Oleaceae	1	0.53
46	Orobanchaceae	1	0.53
47	Oxalidaceae	1	0.53
48	Papaveraceae	1	0.53
49	Phytolaccaceae	1	0.53
50	Pinaceae	3	1.58
51	Plantaginaceae	4	2.11
52	Plumbaginaceae	1	0.53
53	Polygonaceae	9	4.74
54	Portulacaceae	3	1.58
55	Primulaceae	2	1.05
56	Ranunculaceae	1	0.53
57	Rhamnaceae	3	1.58
58	Rosaceae	17	8.95
59	Rutaceae	1	0.53
60	Salvadoraceae	2	1.05
61	Santalaceae	1	0.53
62	Sapotaceae	1	0.53
63	Saxifragaceae	1	0.53
64	Solanaceae	5	2.63
65	Typhaceae	1	0.53
66	Ulmaceae	1	0.53
67	Urticaceae	1	0.53
68	Verbenaceae	1	0.53
69	Vitaceae	1	0.53
70	Zygophyllaceae	1	0.53
		190	100.00

20.3.3 Geographic Distribution

Wild food plants were reported from a wide variety of locations across the Indo-Pak subcontinent. Table 20.3 reflects the localities of the reported areas of the wild food plants. The majority of species were reported to be found in a variety of locations, with Khyber Pakhtunkhwa (KPK) being the most commonly reported location, mentioned in 93 records (48.95% of all mentions). Azad Jammu and Kashmir (AJK) ranked 2nd with 29 records (15.26%), and Punjab 3rd3rd with 23 records (12.11%). Records (both on the Indian side and on the AJK side) in the Western Himalayas were 18 (9.47%), whereas the Hindu Kush Mountains had 13 records (6.84%). There were 12 records in Upper Kohistan (part of KPK) (6.32%), 9 records in Gilgit Baltistan (4.74%), 8 records in Balochistan, and 7 records in Swat and

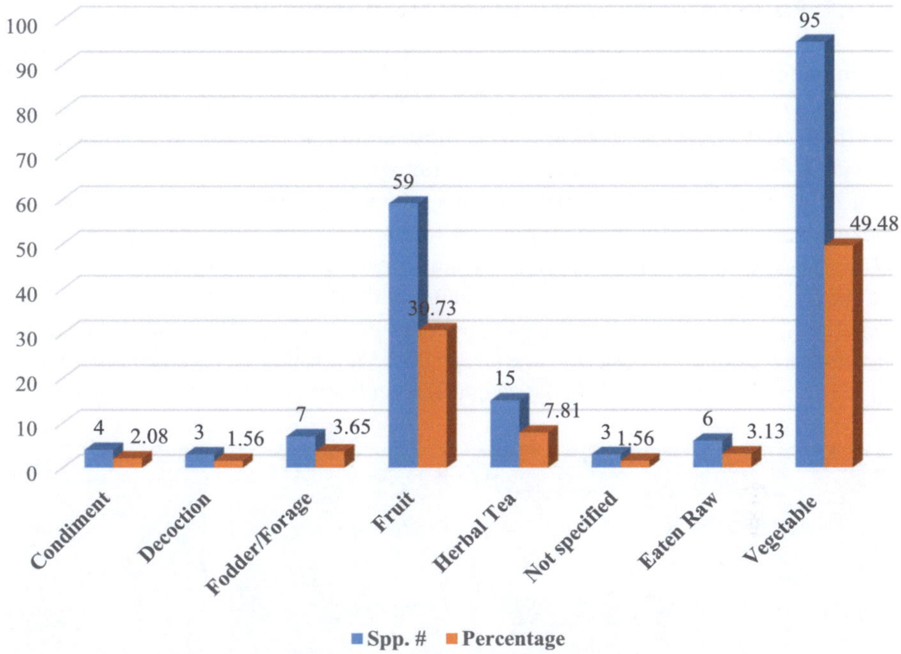

Fig. 20.1 Showing number of species percentage used as wild food plant resources

Table 20.3 Geographical distribution of collected information on wild plant resources

Geographic region	Number of species	Percentage
Khyber Pakhtunkhwa (KPK)	93	48.95%
Azad Jammu and Kashmir (AJK)	29	15.26%
Punjab	23	12.11%
Western Himalayas (Indian side + AJK)	18	9.47%
Hindu Kush Mountains	13	6.84%
Upper Kohistan (KPK)	12	6.32%
Gilgit Baltistan	9	4.74%
Balochistan	8	4.21%
Swat	7	3.68%
Waziristan	7	3.68%
Other regions (Sindh, Neelum Valley, Thakht-e-Sulaiman Hills, Chitral)	Lower	<3% each

Waziristan (both in KPK). Other regions, such as Sindh, Neelum Valley, Thakht-e-Sulaiman Hills, and Chitral, had fewer records. High density of records in KPK, AJK, and the Western Himalayas suggests that the areas are centers of wild food plant diversity and traditional knowledge, probably because of mountainous topography, climatic diversity, and cultural dependence on wild foods.

20.3.4 Food Categories and Preparation Methods

Based on the compiled data, wild edible plants were utilized in eight food categories such as condiment, decoction, eaten raw, fodder/forage, fruit, herbal tea, not specified, and vegetable cooking. The most common category was recorded as vegetable cooking, with 95 plant species contributing 49.48% of the total recorded flora (Fig. 20.2). For this purpose, leaves and young shoots are boiled, fried, or cooked with spices and ghee. Fruits are frequently used parts, distributed across 59 species (30.73%), which are directly consumed raw, fresh, or dried. Herbal teas are the 3rd most important use of the plants, with 15 species (7.81%). These were prepared by boiling roots, bark, leaves, or aerial parts, often with sugar. Some of the species are reported as fodder/forage (7 spp., 3.65%), followed by eaten raw (6 spp., 3.13%), condiment (4 spp., 2.08%), decoction, and not specified (3 spp., 1.5% each).

Fig. 20.2 Food categories and method of preparation

Fig. 20.2 (continued)

20.3.5 Multi-Reference Species (Consensus)

Out of 190 species, eighty-seven (45.8%) were referenced at least twice, that indicates their food applications are well known in various regions of study. These species constitute the core wild food heritage of northern Pakistan and the western Himalayas. Four species, such as *Chenopodium album, Mentha longifolia, Oxalis*

Fig. 20.2 (continued)

corniculata, and *Portulaca oleracea*, were reported from at least two localities with the same pattern of their use. *Mentha longifolia* was reported from Punjab, KPK, AJK, Balochistan as a condiment, and its pattern of use for making chatni and as herbal tea, which is used as a carminative and for coughs as well as colds. The remaining three species were reported from two localities with the same kind of use.

Fig. 20.2 (continued)

Chenopodium album was reported from Western Himalaya (India), Qalagai Hills (Swat), and its leaves are cooked as vegetables and used in digestive problems. *Oxalis corniculata* was reported from Waziristan, KPK and Punjab, and its fresh leaves are cooked as vegetables. *Portulaca oleracea* is reported from Upper Kohistan and many KPK areas and used as a potherb. These species probably are the most credible and culturally pertinent wild foods in the region. They are used regularly in various linguistic and ecological areas and imply the presence of strong traditional knowledge.

20.3.6 Geographical Distribution and Widespread Uses

There were 58 species (30.5%) which were reported in three or more different territories. Most common species (at least 4 territories) are *Mentha longifolia* (Punjab, KPK, AJK, Balochistan), *Chenopodium album* (Western Himalaya,

ontopodium himalayanum DC.

Lepidium draba L.

Lithospermum officinale L.

Malva neglecta Wallr.

Marsilea quadrifolia L.

Medicago polymorpha L.

Mentha longifolia L.

Myrsine Africana L.

Nannorrhops ritchieana (Griff.) Aitch

Nasturtium officinale R.

Olea ferruginea WalL. ex Aitch

Opuntia monacantha (Willd.) Haw

Oxalis corniculata L

Oxyria digyna (L.) Hill

Papaver rhoeas L.

Parrotiopsis jacquemontiana

Fig. 20.2 (continued)

Qalagai Hills, Swat), *Amaranthus viridis* (Western Himalaya, various areas of KPK), and *Oxalis corniculata* (Waziristan, Punjab, KPK). The most geographically abundant types of food are leafy vegetables and herbal teas, presumably due to the fact that the same or related species can be found in altitudinal gradients as well as the preparation methods of those foods can be easily transferred.

Fig. 20.2 (continued)

20.3.7 Unique/New Uses (Single Reference, Single Territory)

Out of 190 plant species, 103 species (54.2%) were reported by only one reference and from a single territory. Some of these are potentially new ethnobotanical records of either Pakistan or the area. Some of the territory-specific species are provided in Table 20.4. According to it, *Cannabis sativa* seeds are roasted with sugar and a sweet preparation was made from Western Himalaya (India) and AJK. *Notholirion*

Fig. 20.2 (continued)

thomsonianum was recorded in a multi-step cooking technique unique to KPK/AJK. Young stems/fruits of *Periploca aphylla* are chewed to extract sweet sap. The raw stem of *Rheum australe* is consumed as a snack in Chitral and Upper Kohistan.

20.3.8 Pattern of Use Consensus Versus Novelty

It has been observed that the majority of plant species (103, 54.2%) are only known to have unique or single reference species from a single territory (Table 20.5). This high percentage indicates that the knowledge of wild food plants is still very localized and that the areas have been under-surveyed. It is also an indication that there is a great chance of recording new or original uses before they are lost.

Two or more reference citations also validate the traditional knowledge in which 87 species (45.8%) are reported. Widespread species showing $\geq$3 territories are frequently recorded (Table 20.5).

Fig. 20.2 (continued)

Table 20.4 Novel and territory-specific unique uses of plant species

Species	Unique reported area	Unique use
Aegopodium burtii	Hazara, Chail Sar to Ganja Kandao	Fodder and milk curding
Apteranthes tuberculata	Waziristan KPK	Fresh leaves cooked as vegetable
Cannabis sativa	Western Himalaya (India), AJK, North-West Punjab	Seeds roasted with sugar
Notholirion thomsonianum	KPK, AJK	Young leaves boiled, then fried with ghee, onions, chilies
Periploca aphylla	Punjab, KPK, AJK, Balochistan	Green fruits and young stems chewed for sweet sap
Rheum australe	Gharam Chasma (Chitral), Upper Kohistan	Fresh peeled stem inner part eaten raw

Table 20.5 Pattern of uses of plant species recorded from the report areas

Category	Number of species	Percentage	Implication
Consensus (≥2 references)	87	45.8%	Robust, cross-validated traditional knowledge
Widespread (≥3 territories)	58	30.5%	Ecologically adaptable species with broad cultural acceptance
Unique (single reference + single territory)	103	54.2%	High local endemism of knowledge; potential for novel documentation

20.3.9 Comparison with Previous Literature

The current inventory offers in comparison to previous national-scale ethnobotanical lists (e.g., Rehman et al. 2024; Ahmad and Pieroni 2016; Ilyas et al. 2013), species records of new territory like *Apteranthes tuberculata* (Waziristan), *Diplazium esculentum* (Swat), and *Notholirion thomsonianum* (KPK/AJK). New preparations of food that had not been previously reported, such as *Cannabis sativa* seeds with sugar, *Rheum australe* crude stem, were recorded. Long geographical ranges of species of *Mentha longifolia* and *Chenopodium album* that are known are now being established in Balochistan and Upper Kohistan.

In almost half of the species, consensus uses are in place which has created an effective core of wild food plants that are shared in the northern part of Pakistan and the western Himalayas. Distinct/new applications are numerous (more than 50% of species), and further field work, particularly in remote or less studied regions like Waziristan, Upper Kohistan and Chitral is important. The leafy vegetables and herbal teas are mostly found geographically and thus their flexibility and cultural universality. The inventory assures that the knowledge of wild food plants is not only diverse but also localized and numerous distinctive uses are in danger of being lost unless the knowledge is further documented in ethnobotanical literature.

20.4 Discussion

20.4.1 Plant Diversity and Bio-cultural Richness

The present investigation indicated that 190 wild edible plant species belonging to 78 families of the mountain ranges of the Indo-Pak subcontinent are being used as wild edibles (Table 20.1). This rich taxonomic diversity highlights the region as an important agrobiodiversity and traditional ecological knowledge hotspot. The prevalence of the family Rosaceae (17 spp., 8.95%), Asteraceae (12 spp.), and Apiaceae/ Lamiaceae (10 spp. each) corresponds to the global trends where these three families are often reported as being rich sources of wild edibles (Abbasi et al. 2015; Shrestha and Dhillion 2006). The family Rosaceae is outstanding due to its edible fruits (e.g., Rubus, Prunus, Crataegus), frequently gathered and eaten directly, and Asteraceae and Apiaceae are sources of large amounts of leafy vegetables and condiments.

It is worth mentioning that 60 families with only 1–4 species each (with a significant proportion of the total diversity) are present in the area, which means that the knowledge of wild food in the area is not limited to several dominant plant groups but is quite evenly distributed in the botanical spectrum. This trend is typical of the indigenous food systems which have been developed over centuries of empirical trial and error as well as cultural transmission (Kumar et al. 2014). Families with few species in them also contribute different flavors, textures, and nutritional value, thus increasing the diversity of the diet. Furthermore, such extensive family-scale distribution implies that the local communities have a deep knowledge of their plants and can use a wide variety of taxa as food (Majeed et al. 2021).

The current list has territorial new records, such as *Apteranthes tuberculata* (Waziristan), *Diplazium esculentum* (Swat), and *Notholirion thomsonianum* (KPK/AJK) when compared to previous national-scale lists (Ahmad and Pieroni 2016; Ilyas et al. 2013). These contributions underscore the idea that, despite the fairly well-researched area, there are still systematic records of new species of wild food and applications yet to be documented. The discovery supports the idea that the ethnobotanical fieldwork in under-surveyed or conflict-stricken regions (e.g. Waziristan, Upper Kohistan, Chitral) is critical to document fading knowledge (Pain 2002; Abdullah et al. 2021).

20.4.2 Plant Part Use

The two most common uses of plant parts are leaves (77 species, 35.48%) and fruits (63 species, 29.03%), which comprise almost two thirds of the entire list of known uses in plants. This preponderance is an ecological expediency and wisdom of nutrition. Leaves can be the most readily available, renewable, and fast-growing sections of a plant, thus able to be harvested multiple times without killing the particular plant (Vinceti et al. 2012). They contain a lot of vitamins (A, C, K), minerals (iron, calcium), and dietary fiber, and are usually eaten as cooked vegetables, a cooking technique in which antinutritional factors are minimized and palatability is increased (Borelli et al. 2020).

On the other hand, fruits are typically harvested when they are already ripe and eaten raw, which are easy to obtain, antioxidant, and micronutrient-rich, and are often lacking in the staple-based diets (Ahmed et al. 2023; Bhattarai et al. 2023; Devkota et al. 2023). The dominance of fruit consumption (almost one-third of all species) indicates the significance of wild fruits as the fast food of nature in the remote and mountainous communities, where they might have limited access to grown fruits due to the elevation, climate, or market isolation (Rasul et al. 2018).

It is notable that the proportion of underground parts (roots, rhizomes, bulbs) and bark (3.69% and 1.38% respectively) is relatively low. These sections tend to be less renewable and harvesting of them in most cases harms or kills the plant. The minimal dependence on these parts implies that the local harvesters are increasingly adherent to the principles of sustainable harvesting, without any formal regulations. This is in line with traditional ecological knowledge that values long-term sustainability of resources more than short-term profitability (Berkes 2017).

20.4.3 Hotspots of Geography and Distribution of Knowledge

The geographic analysis showed that Khyber Pakhtunkhwa (KPK) has almost half of all records (48.95%), Azad Jammu and Kashmir (15.26%), and Punjab (12.11%). This disproportionate distribution is not just an artificial by-product of research intensity, but a real ecological and cultural fact. A tremendous variety of habitats, including the snow-covered peaks of the Hindu Kush and the semi-arid plains of Dera Ismail Khan, are included in KPK, and this variety of habitat supports an equally high number of wild edible plants (Khan et al. 2013). In addition, numerous ethnic groups (Pashtun, Kohistani, Gujjar) that live in KPK still use wild foods in their everyday food and traditional medicine, thus maintaining and passing the knowledge between generations (Ahmad and Pieroni 2016).

The transboundary nature of the wild food traditions can be highlighted by the high prevalence of the records of the Western Himalayas (Indian and Pakistani sides) and the Hindu Kush (6.84%). *Chenopodium album, Mentha longifolia* and *Oxalis corniculata* are just a few examples of species that cross political boundaries and this means that biocultural knowledge transcends the influence of modern geopolitics (Mir et al. 2021). This result suggests significant implications to the conservation strategies applicable to the region: cross-border collaboration is necessary to save the genetic resources, as well as knowledge systems that support them (Haq et al. 2022).

On the other hand, there were few records in regions like Sindh, Neelum Valley and Thakht e Sulaiman Hills. It is still unclear whether this is a manifestation of reduced real diversity, undersampling or fast loss of knowledge as a result of conflict and displacement. Since these regions have been highly affected by sociopolitical changes, conducting special ethnobotanical surveys in these regions is a priority until the local knowledge is lost forever (Pain 2002).

20.4.4 Preparation Techniques of Food: Vegetables, Fruits, and Herbal Teas

The leading status of cooked vegetables (95 species, 49.48%) as the most popular type of food corresponds to the popularity of cooking the leafy greens as a daily meal, and in many cases, it is served with bread (roti) and meat in some instances. This tendency is characteristic of agricultural communities in the mountains where fresh vegetables are not available in the winter season, and wild greens are the source of necessary vitamins and minerals (Borelli et al. 2020). Cooking (boiling, ghee, and spices) not only enhances the flavor but also enhances the bioavailability of fat-soluble vitamins and nutrients, including 8 carotenes (van der Meer et al. 2022).

The most common foods consumed are fruits (59 spp., 30.73), as they are an easy source of energy and micronutrients consumed raw. Wild fruits are also dried and preserved in lean seasons in most Himalayan communities, which helps to achieve year-round food security (Rasul et al. 2018). The third category, which is most important, is the herbal tea (15 species, 7.81%), which is made of roots (e.g., *Bergenia stracheyi, Geranium wallichianum*), bark (e.g., *Betula utilis, Abies pindrow*), or leaves (e.g., *Thymus linearis, Nepeta erecta*). These teas are not just drinks; they are prophylactic and curative treatments of coughs, fevers, digestive problems, and elevation-related illnesses, thus filling the food-medicine void (Hassan et al. 2024; Khoja et al. 2022; Tassadduq et al. 2022).

The fact that the percentage of condiments (2.08), decoctions (1.5), and unspecified purposes (1.5) is quite low does not make them less significant. Herbs like *Angelica glauca* roots and *Cuminum cyminum* seeds are culturally essential to add flavor to conventional foods; decoctions are a concentrated form of herbal treatment commonly used in more serious disease. The not specified sub-group (1.5) shows lapses in the original literature, which should be addressed in future studies.

20.4.5 Consensus and Cultural Heritage of Wild Foods

Out of the 190 species, 87 (45.8%) were mentioned more than twice, creating a pool of culturally core wild foods which are commonly identified in most linguistic and ecological areas. There were four species reported that had two or more localities with the same use pattern: *Chenopodium album, Mentha longifolia, Oxalis corniculata*, and *Portulaca oleracea. Their unanimity in the treatment of digestive issues and cough/cold, and their use as a general vegetable*, suggests a high level of trans-regional agreement on their edibility and effectiveness. These species are especially likely to be used to promote interventions on a wild-food basis since they are already trusted and known to local communities (Bonthoux and Chollet 2024)

The most common is *Mentha longifolia*, which is reported in Punjab, KPK, AJK and Balochistan. It is almost universally used as a condiment (chutney) and as an herbal tea to treat carminative complaints, cough, and cold. The discovery confirms the medicinal status of mint in the official pharmacopoeias and the folk medicine

(Iqbal et al. 2021). The good agreement with the *Chenopodium album* (Western Himalaya, Qalagai Hills) as a cooked vegetable to aid in digestive health is based on previous research that records its laxative and stomachic effects (Poonia and Upadhayay 2015).

These species bottlenecks have the potential to be domestication and value chain entry points since they are already widely accepted and need little behavioral modification. Any intervention of such kind, however, should be done in a way that does not violate traditional rights, but rather is subject to community consent (FPIC) to prevent bio-piracy and equitable benefit sharing.

20.4.6 Geographically Widespread: Signs of Portable Knowledge

It has been observed that thirty (35%) of the total number of species ($N = 58$) were reported in three or more territories and this showed that some of the wild food practices have high transferability across altitudinal and cultural gradients. The geographically most widespread ones are *Mentha longifolia* ($\geq$4 territories), *Chenopodium album* ($\geq$3), *Amaranthus viridis* ($\geq$3), and *Oxalis corniculata* ($\geq$3). They are usually leafy greens or herbal teas—foods that are not complicated to prepare, that use common household tools, and are based on the use of plant species that inhabit an extensive variety of environments (lowland plains to montane slopes). The widespread *nature* of their uses signifies that they are climate-resilient and socially tested. The migration of communities (e.g., to a new location because of a conflict or climate change) often brings with it information about how to prepare these familiar wild foods, thus preserving dietary diversity and nutritional security in new locations (Fongnzossie et al. 2020). This fact confirms the idea that the traditional food knowledge is a dynamic, adaptive resource, but not a cultural relic.

20.4.7 Unique and Potentially Novel Uses: A Window into Local Innovation

It is interesting to note that 103 species (54.2% of the total number of species) appeared in a single territory in only one reference. Although some of these might be the result of undersampling, most probably are real local innovations or idiosyncratic biocultural adaptations. According to Manzoor et al. (2023) seeds of *Cannabis sativa* are roasted with sugar to make a sweet preparation (Western Himalaya, AJK). This application has not been frequently documented in previous ethnobotanical collections and could indicate a certain cultural tradition that is associated with religious or festive events (Pellati et al. 2018). Young leaves of *Notholirion thomsonianum* are boiled, drained, and cooked with ghee, onions, and chilies, found only in KPK/AJK. This exquisite technique implies a great value of this species in the culture. Fruits and young shoots of *Periploca aphylla* are chewed to get sweet sap (Punjab, KPK, AJK, Balochistan). Such use is documented in various territories, but is otherwise quite uncommon; it might be a folk energy-giving snack.

Raw stem of *Rheum australe* is eaten as a snack in Chitral and Upper Kohistan. It is a local use that has not been well recorded, but is worth exploring since the potential of this as a refreshing, low-calorie food is considerable.

These special records demonstrate that a big percentage of wild food knowledge is local and unpublished. They also cite that there is a very high risk of knowledge loss since such specialized applications are often concentrated in the hands of a small number of older people within a society and might not be passed to the younger generations (Majeed et al. 2021). Ethnobotanical records should thus prioritize such under-documented territories and reach out to the knowledge bearers on the ground by using participatory approaches (FPIC, community protocols) to make sure that the special uses are not lost to extinction.

20.4.8 Comparison with Past National Inventories

The current inventory includes a comparison to the previous national-scale compilations (e.g., Ahmad and Pieroni 2016; Ilyas et al. 2013; Rehman et al. 2024). New territorial records of *Apteranthes tuberculata* (Waziristan), *Diplazium esculentum* (Swat), and *Notholirion thomsonianum* (KPK/AJK) have been reported. Besides, new preparation techniques like *Cannabis sativa* seeds, sugar, and raw stem of *Rheum austral* have been reported for the first time.

Broad amplitude or widespread geographical distances of species such as *Mentha longifolia* (since proven in Balochistan) and *Chenopodium album* (since found in Upper Kohistan) are indicative of wider dissemination of information and cross-cultural use.

These additions indicate that the wild food plant knowledge base is not exhaustive. In a relatively well-studied area, the field surveys that are conducted in continuation reveal new data. This particularly applies to remote, conflict-ridden or high-altitude regions, which have been traditionally under-surveilled because of the logistical challenges (Pain 2002). The results are very much in favor of a systematic multiyear ethnobotanical program of all administrative units of the Indo-Pak mountain arch.

20.4.9 Food Security, Nutrition, and Climate Adaptation Implications

The great diversity of wild edible plants reported here can be directly related to the aims of Sustainable Development Goal 2 (Zero Hunger) and Target 2.4 of the Kunming Montreal Global Biodiversity Framework (2022). Wild foods tend to be high in micronutrients not found in staple crops, and seasonally available in periods of hunger in-between cultivated foods (Borelli et al. 2020). In mountainous areas of Pakistan, where malnutrition is chronically high (Hameed et al. 2021), the sustainable utilization of these plants may be used to a considerable extent to enhance the quality of the diet and its resilience.

Besides, most wild edibles have features that allow them to withstand changes in climate Since most of them are usually drought resistant, are accustomed to unfertile soils, and can be cultivated without intensive inputs (Majeed et al. 2025). Wild food plants would become a buffer as growing conditions are changing due to climate change (Intergovernmental Panel on Climate 2022) and offer alternative nutrition sources to staple crops when they fail. This importance is particularly needed in the Hindu Kush Himalaya region, which is getting warmer more rapidly than the rest of the world (Intergovernmental Panel on Climate 2022).

20.4.10 Threats and Conservation Priority

In spite of this value, wild food plants and traditional knowledge associated with them are under imminent threat. The following consequences can happen if proper steps are not taken for their protection:

- Loss of traditional knowledge: Urbanization, formal education, and the effects of the media have been increasing the lack of knowledge of young generations about wild food traditions (Shirsat and Koche 2024).
- Overharvesting: Due to high demand on some medicinal and edible species (e.g., *Podophyllum hexandrum*, *Trillium govanianum*), there has been a decrease in populations (Awan et al. 2023).
- Change of land use: Agricultural intensification, development of infrastructure, and deforestation decrease the habitats of wild plants (Jamil et al. 2022).
- Climate variability: Dependence on changes in temperature and precipitation influence wild species flowering, fruiting and distributing (Majeed et al. 2021).

The following are some proposed priorities which should be taken into account for conservation such as:

(i) *In situ* conservation: Protecting wild species through community participation would act as wild plant reserves and employing wild food species in agroforestry.
(ii) *Ex situ* conservation: Preservation of priority species in gene banks by storing seeds.
(iii) Documentation: Systematic ethnobotanical survey of under-surveyed regions (Waziristan, Chitral, Upper Kohistan) using the data.
(iv) Awareness and education: Educating school curricula and extensions to promote nutritional and cultural value of wild foods.
(v) Policy integration: Acknowledging wild food plants in country food security and climate change adaptation policies (e.g., National Climate Change Policy, 2012 (Pakistan 2012), in Pakistan; Kunming Montreal Framework by the CBD).

20.4.11 Limitations and Future Research Directions

The current review has a number of limitations:

- Geographic discontinuities: There are geographic gaps (e.g., Balochistan, Sindh) that might be under represented because of lack of ethnobotanical studies and not because of lack of diversity.
- Time gaps: A significant number of the studies mentioned are not older than 10–15 years; the knowledge might have been forgotten by that time.
- Absence of quantitative nutritional data: In most species only qualitative use data exist, chemical analyses are desperately required.
- No measure of sustainability of harvest: Not many studies examine the sustainability of current levels of collection.

Based on these limitations, the following recommendations are proposed to carry out detailed studies on:

- Anti-nutritional and quantitative nutritional profiling of promising species.
- Action research with communities to collaboratively develop sustainable harvesting guidelines.
- Longitudinal research to monitor the knowledge and practice changes over time.

20.5 Conclusion

The Indo-Pak subcontinent mountain ranges hold a particularly rich and diverse wild food plant heritage, of at least 190 species in 78 families. The most frequent parts are leaves and fruits, and cooked vegetables make the food category most dominant. Although there is a common core of species (e.g., *Mentha longifolia, Chenopodium album*) that has cross-regional consensus, most of the reported uses are very localized (54.2% single reference, single territory) and so reflect local innovation as well as the danger of knowledge loss. Geographic hotspots, like KPK, AJK, and the Western Himalayas, need further documentation efforts, and under-surveilled areas, such as Waziristan and Chitral, need immediate attention. The checklist adds new territorial records to earlier national lists and new methods of preparation, and proves that even those areas that have been studied extensively possess surprises. The resources of wild foods are crucial to food security, dietary diversity, and climate adaptability of one of the most vulnerable mountainous areas in the world. Their conservation needs a combined strategy which includes ethnobotanical documentation, sustainable use practices, policy support as well as the active participation of the indigenous holders of knowledge.

References

Abbas W, Hussain W, Hussain W, Badshah L, Hussain K, Pieroni A (2020) Traditional wild vegetables gathered by four religious groups in Kurram District, Khyber Pakhtunkhwa, North-West Pakistan. Genet Resour Crop Evol 67(6):1521–1536

Abbas H, Ali H, Khan SW, Khadim S (2024) Ethnobotanical applications of medicinal plants in the alpine flora of Marukh and Asogaha Nallah, Haramosh Valley, District Gilgit, Pakistan. Ethnobot Res Appl 27(45):1–24

Abbasi AM, Shah MH, Khan MA (2015) Wild edible vegetables of lesser. Springer, Himalayas

Abdullah A, Khan SM, Pieroni A, Haq A, Haq ZU, Ahmad Z et al (2021) A comprehensive appraisal of the wild food plants and food system of tribal cultures in the Hindu Kush Mountain Range; a way forward for balancing human nutrition and food security. Sustainability 13(9): 5258

Abid R (2010) Floral morphs and mode of pollination in Oxalis corniculata L. from Pakistan. Pak J Bot 42(6):4027–4033

Abid R, Mehdi SR, Riaz S, Kanwal D, Imran M, Ather A (2022) Cypsela morphology and its taxonomic significance within the genus Scorzonera L.(Cichoreae-Asteraceae) from Pakistan and Kashmir. Pak J Bot 54(6):2219–2231

Afshin A, Sur PJ, Fay KA, Cornaby L, Ferrara G, Salama JS, Mullany EC, Abate KH, Abbafati C, Abebe Z (2019) Health effects of dietary risks in 195 countries, 1990–2017: a systematic analysis for the Global Burden of Disease Study 2017. Lancet 393(10184):1958–1972

Afzal M, Ali MI, Munir MA, Zeeshan MA, Sharif MN, Aslam M (2016) Genetic association among morphological traits of Lepidium draba. Bullet Biol Allied Sci Res 2016(1):1–1

Åhlberg M (2019) Totuus syötävistä luonnonkasveista eli miksi uskallan syödä lähiluonnon kasveista kestävästi keräämääni ruokaa: OSA I: Tieteellisiä perusteita käytännönläheisesti

Ahmad K, Pieroni A (2016) Folk knowledge of wild food plants among the tribal communities of Thakht-e-Sulaiman Hills, North-West Pakistan. J Ethnobiol Ethnomed 12(1):17

Ahmad M, Waheed I, Khalil-ur-Rehman M, Niaz U, Hassan SS (2007) A review on Carthamus oxycantha. Pak J Pharm 20(1):37

Ahmad I, Ibrar M, Barkatullah, Ali N (2011) Ethnobotanical study of tehsil kabal, swat district, KPK, Pakistan. J Bot 2011(1):368572

Ahmad L, Semotiuk A, Zafar M, Ahmad M, Sultana S, Liu Q-R et al (2015) Ethnopharmacological documentation of medicinal plants used for hypertension among the local communities of DIR Lower, Pakistan. J Ethnopharmacol 175:138–146. https://doi.org/10.1016/j.jep.2015.09.014

Ahmad M, Masood S, Sultana S, Hadda TB, Bader A, Zafar M (2015) Antioxidant and nutraceutical value of wild medicinal Rubus berries. Pak J Pharm Sci 28(1):241–247

Ahmad KS, Hamid A, Nawaz F et al (2017) Ethnopharmacological studies of indigenous plants in Kel village, Neelum Valley, Azad Kashmir, Pakistan. J Ethnobiol Ethnomed 13:68. https://doi.org/10.1186/s13002-017-0196-1

Ahmad K, Weckerle CS, Nazir A (2019) Ethnobotanical investigation of wild vegetables used among local communities in Northwest Pakistan. Acta Soc Bot Pol 88(1)

Ahmad L, Riaz M, Jan HA, Semotiuk AJ, Ahmad I, Khan I, Bussmann RW (2021) An ethnobotanical survey of wild food plants used by the local communities of Kumrat Valley in District Upper Dir, Pakistan. Ethnobot Res Appl 22:1–13

Ahmed M, Bose I, Goksen G, Roy S (2023) Himalayan sources of anthocyanins and its multifunctional applications: a review. Foods 12(11):2203

Akinola R, Pereira LM, Mabhaudhi T, De Bruin F-M, Rusch L (2020) A review of indigenous food crops in Africa and the implications for more sustainable and healthy food systems. Sustainability 12(8):3493

Aleem A, Aslam B, Alim MB, Hussain A, Faisal MN, Majeed W (2023) Phytochemical analysis and gastroprotective effect of Stellaria media (L.) Vill. methanolic extract on piroxicam-induced gastric ulcer in Wistar rats. Pak J Pharm Sci 36:1425

Al-Fatimi M (2021) Wild edible plants traditionally collected and used in southern Yemen. J Ethnobiol Ethnomed 17(1):49

Ali S (2023) A study on different plant species of the Rosaceae family and their ethnobotanical uses among the local communities at Swat district, Pakistan. Ethnobot Res Appl 25:1–16

Ali N, Alam H, Khan A, Ahmed G, Shah WA, Nabi M, Junaid M (2014) Antispasmodic and antidiarrhoeal activity of the fruit of Rosa moschata (J). BMC Complement Altern Med 14 (1):485

Ali M, Juyal D (2018) A review on pharmacognostical and phytochemical evaluation of Pyrus pashia Buch-Ham ex D. Don. Pharm Innov 7(5, Part C):186

Ali MI, Mahmood Z, Ahmad M, Afzal M, Mun MA, Sharif MN, Shakeel A (2016) Genetic variability in Cirsium arvense under different environmental conditions. Bullet Biol Allied Sci Res 2016(1):3–3

Ali M, Muhammad S, Shah MR, Khan A, Rashid U, Farooq U, Latif A (2017) Neurologically potent molecules from Crataegus oxyacantha; isolation, anticholinesterase inhibition, and molecular docking. Front Pharmacol 8:327

Ali M, Hussain K, Ullah M, Ali U, Khan SU, Bussmann RW, Shao JW (2020) Nannorrhops ritchieana (Griff.) Aitch.(Arecaceae)–a traditional multipurpose plant species of Pakistan. Ethnobot Res Appl 19(35):1–10

Ali M, Ullah H, Bari WU, Ul Islam N, Zahoor M, Ullah R, Bari A (2021) Phytochemical isolation and biological screening of Cotoneaster microphyllus. Int J Food Prop 24(1):1318–1334

Ali JS, Riaz N, Mannan A, Latif M, Zia M (2022) Antioxidant, antimicrobial, enzyme inhibition, and cytotoxicity guided investigation of Sideroxylon mascatense (A. DC.) TD Penn. leaves extracts. Nat Prod Res 36(16):4227–4230

Ali S, Qureshi R, Munazir M, Shah SA, Hussain W, Bussmann RW (2023) Patterns of wild edible plants consumption of District Swat, Pakistan: emphasizing the importance of plants in uplifting the quality of life of local residents. J Ethnobiol Ethnomed 22:12

Ali MZ, Saboor A, Ullah R, Haq QSU, Rahman FU, Ullah F, Ullah R (2025) Evaluation of Viburnum Grandiflorum crude extract as an antibacterial agent. Rev J Neurol Med Sci Rev 3 (1):223–232

Amin M, Aziz MA, Pieroni A, Nazir A, Al-Ghamdi AA, Kangal A, Abbasi AM (2023) Edible wild plant species used by different linguistic groups of Kohistan Upper Khyber Pakhtunkhwa (KP), Pakistan. J Ethnobiol Ethnomed 19(1):6

Amin M, Aziz MA, Manduzai AK, Pieroni A, Alkahtani J, AbdelGawwad MR, Abbasi AM (2024) Standard medical ethnobotany of Kohistan, North Pakistan. J Ethnobiol Ethnomed 20(1):64

Ansari R, Ekhlasi-Kazaj K (2012) Adiantum capillus-veneris. L: phytochemical constituents, traditional uses and pharmacological properties: a review. J Adv Sci Res 3(04):15–20

Anwar T, Qureshi H, Shahzadi S, Siddiqi EH, Ali HM, Abdelhamid MM, Nazim M (2024) Exploring the benefits of wild plants in dietary nutrition: investigating perspectives, choices, health impacts and sustainable practices. BMC Complementary Med Ther 24(1):86

Arshad M, Hanif U, Maqsood M, Sohail S, Ghaffar F, Aziz R (2024) Microscopic determination of adulterations through powder microscopy in medicinal plants Kasni and Anjbar (Cichorium intybus L. and Polygonum aviculare L.) collected from the local market of Okara City. GU J Phytosci:1–11

Ashfaq S, Ahmad M, Zafar M, Sultana S, Nazish M, Khan AN (2020) Systematics of medicinally important weeds of genus Convolvulus: Convolvulaceae. Planta Daninha 38:e020216203

Ashraf Z, Sajjad Y, Khan SA, Akhtar G, Ismail AM, Shalaby TA et al (2025) Growth dilatory effects of PEG and Sucrose on Geranium wallichianum: an in vitro approach for conservation. Phyton-Int J Exp Botan 94(3):987–1006

Asprilla-Perea J, Díaz-Puente JM (2019) Importance of wild foods to household food security in tropical forest areas. Food Secur 11(1):15–22

Assad R, Reshi ZA, Mir SH, Rashid I, Shouche Y, Dhotre D (2021) Bioprospecting appraisal of Himalayan pindrow fir for pharmacological applications. In: Phytomedicine. Academic Press, pp 461–482

Awan MS, Dar MEUI, Hussain K, Sabir S, Iqbal T, Mehmood A, Habib T (2023) Ethnomedicinal utilization and conservation status of highland flora from Western Himalayas of Azad Jammu and Kashmir, Pakistan. Ethnobot Res Appl 26:1–20

Aziz MA, Khan AH, Adnan M, Izatullah I (2017) Traditional uses of medicinal plants reported by the indigenous communities and local herbal practitioners of Bajaur Agency, Federally Administrated Tribal Areas, Pakistan. J Ethnopharmacol 198:268–281

Azmat MA, Khan AA, Khan IA, Buerkert A, Wiehle M (2020) Morphology, biochemistry, and management of Russian olive (Elaeagnus angustifolia L.) accessions in Gilgit-Baltistan, northern Pakistan. J Agric Rural Dev Trop Subtrop 121(2):151–160

Bagri AS, Dhingra GK (2023) Diversity and ethno-botanical uses of Pteridophytes in the district Tehri Garhwal, India. Life sciences research and development, p 105

Bailly C (2021) Medicinal properties and anti-inflammatory components of Phytolacca (Shanglu). Digit Chin Med 4(3):159–169

Bashir SM, Altaf M, Hussain T, Umair M, Majeed M, Mangrio WM et al (2023) Vernacular taxonomy, cultural and ethnopharmacological applications of avian and mammalian species in the vicinity of Ayubia National Park, Himalayan Region. Biology 12(4):609

Batool R, Nazar A, Adnan M, Khursheed Z, Mohsin F, Hussain W (2025) Cross culture comparison in ethno-pharmacological uses of plants between two geographical regions of Northwest Pakistan. Ethnobot Res Appl 30:1–21

Berkes F (2017) Sacred ecology. Routledge

Bhattarai K, Dhakal D, Pandey I, Subba B, Sharma KR (2023) Phytochemistry, biological activities, and chemical profiling of Berberis asiatica

Bodeker G (1995) Traditional health systems: policy, biodiversity, and global interdependence. J Altern Complement Med 1(3):231–243

Bonthoux S, Chollet S (2024) Wilding cities for biodiversity and people: a transdisciplinary framework. Biol Rev 99(4):1458–1480

Borelli T, Hunter D, Padulosi S, Amaya N, Meldrum G, de Oliveira Beltrame DM, Samarasinghe G, Wasike VW, Güner B, Tan A (2020) Local solutions for sustainable food systems: the contribution of orphan crops and wild edible species. Agronomy 10(2):231

Borsari B, Vidrine MF (2024) Edible weeds in agricultural landscapes as a research emphasis for sustainability. In: An agenda for sustainable development research. Springer, pp 21–37

Bukhari IA, Khan RA, Gilani A-UH, Shah AJ, Hussain J, Ahmad VU (2007) The analgesic, anti-inflammatory and calcium antagonist potential ofTanacetum artemisioides. Arch Pharm Res 30 (3):303–312. https://doi.org/10.1007/BF02977610

Butola JS, Vashistha RK, Malik AR, Samant SS (2010) Assessment of inter-population variability in Heracleum candicans wall with emphasis on seed characteristics and germination behavior. J Med Plants Res 4(15):1523–1534

Cheek M, Hatt S, Onana JM (2021) The endemic plant species of Bali Ngemba Forest Reserve, Bamenda Highlands Cameroon, with a new Endangered cloud-forest tree species Vepris onanae (Rutaceae). BioRxiv, 2021.2010. 2006.463416

Choudhary AK, Yadav DS, Sood P, Rahi S, Arya K, Thakur SK, Singh R (2021) Post-emergence herbicides for effective weed management, enhanced wheat productivity, profitability and quality in North-Western Himalayas: a 'Participatory-Mode'Technology development and dissemination. Sustainability 13(10):5425

Coakley S, Petti C (2021) Impacts of the invasive Impatiens glandulifera: lessons learned from one of Europe's top invasive species. Biology 10(7):619

Dar GH, Rashid T, Naqshi A, Khuroo AA, Malik AH (2010) Two new species of Papaver L. (Papaveraceae) from Kashmir Himalaya, India. Pak J Bot Special (42):57–62

Dar MA, Mir RH, Mohi-ud-din R, Mir PA, Masoodi MH, Akbar S et al (2021) Capsella bursa-pastoris (L.) medic: an insight into its pharmacology, expository traditional uses and extensive phytochemistry. Curr Tradit Med 7(2):168–179

Debiage RR, Regildo MÃ, Porto PNP, Yoshihara E, de Mello Peixoto ECT (2016) Fagopyrum esculentum Moench: a crop with many purposes in agriculture and human nutrition. Afr J Agric Res 11(12):983–989

Desa U (2019) United Nations Department of economic and social affairs. Population Division. World Population Prospects

Devkota HP, Belwal T, Bhatt ID (2023) Nutraceuticals from Himalayan fruits and berries: opportunities and challenges. Himalayan Fruits and Berries, pp 497–505

Djagba JF, Zwart SJ, Houssou CS, Tenté BH, Kiepe P (2019) Ecological sustainability and environmental risks of agricultural intensification in inland valleys in Benin. Environ Dev Sustain 21(4):1869–1890

FAO. (2024). The State of Food Security and Nutrition in the World 2024 – Financing to end hunger, food insecurity and malnutrition in all its forms. Rome. https://openknowledge.fao.org/items/ebe19244-9611-443c-a2a6-25cec697b361.

Feng Q, Yang W, Ma X, Peng Z, Wang G (2024) Investigation on the anti-α-glucosidase mechanism of Aspergillus triazolata A from Oxalis corniculate L. Int J Biol Macromol 279:135457

Fongnzossie EF, Nyangono CFB, Biwole AB, Ebai PNB, Ndifongwa NB, Motove J, Dibong SD (2020) Wild edible plants and mushrooms of the Bamenda Highlands in Cameroon: ethnobotanical assessment and potentials for enhancing food security. J Ethnobiol Ethnomed 16(1):12

Framework B (2022) Decision adopted by the conference of the parties to the convention on biological diversity. Paper presented at the Conference of the Parties to the Convention on Biological Diversity (15th: 2021-2022. Kunming, China; Montréal, Canada and Nairobi) (DHLAUTH) 939522

Gani R, Bhat ZA, Dar MA, Shah Z (2019) Pharmacognostic and phytochemical characteristics of the aerial part of Salvia moorcroftiana Wall. Ex Benth. Growing wild in Kashmir Valley, India. Pharmaceutical. Methods 10(1):3–47

Ganie SA, Dar TA, Hamid R, Zargar O, Abeer SU, Masood A, Zargar MA (2014) In vitro antioxidant and cytotoxic activities of Arnebia benthamii (Wall ex. G. Don): a critically endangered medicinal plant of Kashmir valley. Oxidative Med Cell Longev 2014(1):792574

Gayathiri E, Gobinath R, Jayanthij J, Prakash P, Ragunathan M (2024) Sustainable biodiversity conservation in tribal area. In: Sustainability of natural resources. CRC Press, pp 24–40

Gbashi S, Adebo O, Adebiyi JA, Targuma S, Tebele S, Areo OM et al (2021) Food safety, food security and genetically modified organisms in Africa: a current perspective. Biotechnol Genet Eng Rev 37(1):30–63

Gillani SW, Ahmad M, Zafar M, Haq SM, Waheed M, Manzoor M et al (2024) An insight into indigenous ethnobotanical knowledge of medicinal and aromatic plants from Kashmir Himalayan region. Ethnobot Res Appl 28:1–21

Grochowski DM, Strawa JW, Granica S, Tomczyk M (2020) Secondary metabolites of Rubus caesius (Rosaceae). Biochem Syst Ecol 92:104111

Gul A, Alam J, Ahmad H, Shah GM, Hussain M, Dogan Y, Rahman K (2016) Traditional, medicinal and food uses of Pteridophytes of district Mansehra (Pakistan) and their some adjacent areas. Int J Biosci 9(5):116–133

Gul H, Ahmad M, Zafar M, Sheeraz Ahmad M, Abid A, Hira S, Gulfraz M (2017) The in vitro and in vivo biological activities of the leaf of Cape Myrtle, Myrsine africana L. Phytother Res 31 (9):1305–1309

Gul H, Anwar I, Ibrar KU, Hussain A, Fazal M (2025) Ethnobotanical study of indigenous medicinal plants of district Swabi, Khyber Pakhtunkhwa, Pakistan. Pak J Weed Sci Res 31(2):90

Hameed A, Zafar M, Ahmad M, Sultana S, Bahadur S, Anjum F et al (2021) Chemo-taxonomic and biological potential of highly therapeutic plant Pedicularis groenlandica Retz. using multiple microscopic techniques. Microsc Res Tech 84(12):2890–2905

Haq F, Irfan M, Fraser-Jenkins C (2022) Multivariate statistical analysis of the pteridophytic diversity of district Battagram, Khyber Pakhtunkhwa, Pakistan. Acta Ecol Sin 42(4):322–331

Hasan A, Ahmed I, Jay M, Voirin B (1995) Flavonoid glycosides and an anthraquinone from Rumex chalepensis. Phytochemistry 39(5):1211–1213

Hassan N, Din MU, Shuaib M, Ul-Hassan F, Zhu Y, Chen Y, Iqbal A (2019) Quantitative analysis of medicinal plants consumption in the highest mountainous region of Bahrain Valley, Northern Pakistan. Ukrainian J Ecol 9(1):35–49

Hassan M, Mir TA, Jan M, Amjad MS, Aziz MA, Pieroni A et al (2024) Foraging for the future: traditional culinary uses of wild plants in the Western Himalayas–Kashmir Valley (India). J Ethnobiol Ethnomed 20(1):66

Hussain W, Badshah L, Ullah M, Ali M, Ali A, Hussain F (2018) Quantitative study of medicinal plants used by the communities residing in Koh-e-Safaid Range, northern Pakistani-Afghan borders. J Ethnobiol Ethnomed 14(1):30

Hussain K, Dar M, Shaheen H, Sabir S, Habib T, Mehmood A et al (2023) A comprehensive checklist of orchids diversity in Western Himalayas of Azad Kashmir, Pakistan. Int J Forest Sci 4:218–228

Hussain S, Ullah F, Shah A, Ullah I, Mehmood S, Gul I et al (2023) Quantitative ethnomedicinal studies of wild edible fruits used by the indigenous people of the Surghar Range, Pakistan. Ethnobot Res Appl 26:1–17

Ilyas M, Qureshi R, Shinwari ZK, Arshad M, Mirza SN (2013) Some ethnoecological aspects of the plants of qalagai hills, Kabal valley, Swat, Pakistan. Int J Agric Biol (Pakistan) 15(5):801

Intergovernmental Panel on Climate C (2022) Climate change 2022: impacts, adaptation and vulnerability. Working Group II Contribution to the Sixth Assessment Report of the Intergovernmental Panel on Climate Change. Retrieved from Cambridge

Iqbal J, Shah A, Sarvat R, Adnan M, Parveen N, Nuzhat R (2019) Documentation of folk herbal uses of medicinally important wild vegetables used by the tribal communities of Sargodha Region, Pakistan. Planta Daninha 37:e019189207

Iqbal MS, Ahmad KS, Ali MA, Akbar M, Mehmood A, Nawaz F et al (2021) An ethnobotanical study of wetland flora of Head Maralla Punjab Pakistan. PLoS One 16(10):e0258167

Iqbal U, Rehman FU, Aslam MU, Gul MF, Farooq U, Ameer A et al (2023) Survival tactics of an endangered species Withania coagulans (Stocks) Dunal to arid environments. Environ Monit Assess 195(11):1363

Jabeen S, Arshad F, Harun N, Waheed M, Alamri S, Haq SM et al (2024) Folk knowledge and perceptions about the use of wild fruits and vegetables–cross-cultural knowledge in the Pipli Pahar reserved forest of Okara, Pakistan. Plants 13(6):832

Jamil MD, Waheed M, Akhtar S, Bangash N, Chaudhari SK, Majeed M et al (2022) Invasive plants diversity, ecological status, and distribution pattern in relation to edaphic factors in different habitat types of district Mandi Bahauddin, Punjab, Pakistan. Sustainability 14(20):13312

Jamshed S, Ullah A, Khan R, Hussain A (2018) Describing one generic and two new species record to the flora of Pakistan using LM and SEM methods. Microsc Res Tech 81(5):469–473. https://doi.org/10.1002/jemt.23000

Javid W, Wani AA, Najar RA, Lateef S (2025) Phenotypic characterization of Cotoneaster germplasm in the Kashmir Himalayas. Genet Resour Crop Evol 72(1):217–233

Jeyaprakash K, Ayyanar M, Geetha KN, Sekar T (2011) Traditional uses of medicinal plants among the tribal people in Theni District (Western Ghats), Southern India. Asian Pacific J Trop Biomed 1(1):S20–S25

Kaur A, Bhatti R (2021) Understanding the phytochemistry and molecular insights to the pharmacology of Angelica archangelica L.(garden angelica) and its bioactive components. Phytother Res 35(11):5961–5979

Khalid A, Shahid S, Khan SA, Kanwal S, Yaqoob A, Rasool ZG, Rizwan K (2018) Antioxidant activity and hepatoprotective effect of Cichorium intybus (Kasni) seed extract against carbon tetrachloride-induced liver toxicity in rats. Trop J Pharm Res 17(8):1531–1538

Khalid N, Badshah L, Shah AA, Ullah A, Khan N, Aziz MA et al (2023) Wild food plants gathered by four cultural groups in North Waziristan, Pakistan. Genet Resour Crop Evol 70(4):1243–1276

Khan SM, Abdullah A (2020) Contribution of the wild food plants in the food system of Tribal Belt of Pakistan; The Pak-Afghan Border Region. Sustainability 13(9):5258
Khan SM, Abdullah A (2020) Contribution of the wild food plants in the food system of tribal belt of Pakistan. The Pak-Afghan Border Region
Khan SM, Page S, Ahmad H, Harper D (2012) Anthropogenic influences on the natural ecosystem of the Naran Valley in the western Himalayas. Pak J Bot 44:231–238
Khan SM, Page SE, Ahmad H, Harper DM (2013) Sustainable utilization and conservation of plant biodiversity in montane ecosystems: the western Himalayas as a case study. Ann Bot 112(3): 479–501
Khan MT, Khan I, Khan MI, Hussain Z, Ayub S, Khan N, Khan I (2016) Ethnobotanical study of wild flora in the remote areas of Nothern Pakistan. Wulfania J 23(10):149–165
Khan A, Ihsan M, Hazrat A, Nisar M, Laiq M, Bibi M, Nawaz MA (2021) Evaluation of genetic diversity in wild Celtis australis L. genotypes using multivariate analysis. Sarhad J Agric 37 (4):1418–1425
Khan N, Ullah R, Okla MK, Abdel-Maksoud MA, Saleh IA, Abu-Harirah HA et al (2023) Environmental and anthropogenic drivers of watercress (Nasturtium officinale) communities in char-lands and water channels across the Swat River Basin: implication for conservation planning. Front Plant Sci 14:1225030
Khan S, Masoodi TH, Ajaz-Ul-Islam M, Pala NA, Mugloo JA, Raja A et al (2024) Wild edible plants sustaining food security among indigenous communities in Kashmir Himalayas, India. Indian J Tradit Knowl 23(10):977–987
Khan I, Bibi S, Shehzad J, Riaz Z, Khan MS, Ghorbanpour M, Mustafa G (2025) Apteranthes tuberculata's antidiabetic potential: exploring phytochemicals, screening antioxidant activity, and validating DPP-4 inhibition using in vitro and in Silico approaches. Food Sci Nutrit 13(7): e70494
Khan MA, Gamrat R, Khan MT, Fawad M (2026) Ecology, biology and importance of wild safflower (Carthamus oxyacantha (M. Bieb)). Pak J Bot 58(2):563–570
Khoja AA, Haq SM, Majeed M, Hassan M, Waheed M, Yaqoob U et al (2022) Diversity, ecological and traditional knowledge of pteridophytes in the western Himalayas. Diversity 14(8):628
Khurram M, Hameed A, Khan MA, Amin MU, Hassan M, Ullah N et al (2012) Antibacterial potentials of Quercus baloot Griff. J Med Plants Res 6(7):1244–1249
Kousar S, Khan N, Abbas Z, Harun N, Ishaq K, Khan T, Bussmann R (2023) Documenting the wild edible plant's resources used by Pushtoons, Balochis and Brahvis as survival strategy in the arid Province of Balochistan, NW Pakistan
Kumar P, Rana PK, Himshikha, Singhal VK, Gupta RC (2014) Cytogeography and phenomenon of cytomixis in Silene vulgaris from cold regions of Northwest Himalayas (India). Plant Syst Evol 300(5):831–842
Kumar A, Katiyar A, Gautam V, Singh R, Dubey A (2022) A comprehensive review on anti-cancer properties of Amaranthus viridis. Prevent 25:27
Lal M, Gogoi A, Tamang R, Dutta P, Perveen K, Alshaikh NA, Begum T (2024) Exploring the bioactive potential of Mentha longifolia from Northeast India: an inclusive study on phytochemical composition and biological activities. J Essent Oil Bear Plants 27(4):1102–1120
Lanzes T, Bhushan A (2022) Indigenous vegetables resources and their utilization under North Western Himalayan Region. Int J Eng Appl Sci Technol 7(2):374–380
León-Lobos P, Díaz-Forestier J, Díaz R, Celis-Diez JL, Diazgranados M, Ulian T (2022) Patterns of traditional and modern uses of wild edible native plants of Chile: challenges and future perspectives. Plants 11(6):744
Lulekal E, Asfaw Z, Kelbessa E, Van Damme P (2011) Wild edible plants in Ethiopia: a review on their potential to combat food insecurity. Afrika Focus 24(2):71–122
Majeed M, Bhatti KH, Pieroni A, Sõukand R, Bussmann RW, Khan AM et al (2021) Gathered wild food plants among diverse religious groups in Jhelum District, Punjab, Pakistan. Foods 10(3): 594
Majeed S, Zafar M, Ahmad M, Zafar S, Ghufran A, Ayoub M, Nabila (2022) Morpho-palynological and anatomical studies in desert cacti (Opuntia dillenii and Opuntia monacantha) using light and scanning electron microscopy. Microsc Res Tech 85(8):2801–2812

Majeed M, Sahito HA, Mangrio WM, Muhammad M, Hussain M, Aziz R et al (2025) Wild edible plants as alternate food sources and agents of food security. In: Exploring traditional wild edible plants. CRC Press, pp 125–143
Majid A, Ahmad H, Saqib Z, Rahman IU, Khan U, Alam J et al (2020) Exploring threatened traditional knowledge; ethnomedicinal studies of rare endemic flora from Lesser Himalayan region of Pakistan. Rev Bras 29:785–792
Malik G, Singh D, Malik AA, Gulzar A (2021) Underutilized vegetable species of Indian Himalayas. In: Neglected and underutilized crops-towards nutritional security and sustainability. Springer, pp 113–132
Malik A, Adnan M, Ben Hasher FF, Abbasi AM, Gatasheh MK, Ullah Z et al (2024) Wild food plant knowledge in multicultural peri-urban area of North-Western Punjab, Pakistan. J Ethnobiol Ethnomed 20(1):99
Manzoor M, Ahmad M, Zafar M, Gillani SW, Shaheen H, Pieroni A et al (2023) The local medicinal plant knowledge in Kashmir Western Himalaya: a way to foster ecological transition via community-centred health seeking strategies. J Ethnobiol Ethnomed 19(1):56
Melnyk M (1996) Indigenous enterprise for the domestication of trees and the commercialization of their fruits. In: Domestication and commercialization of non-timber forest products, p 97
Menalled F, Ebel R (2025) Edible weeds as crops. In: Agroecology of edible weeds and noncrop plants. Elsevier, pp 75–102
Mgalula ME (2024) An ethnobotanical study of wild edible fruits in miombo woodlands of Tabora region in Western Tanzania. J Ethnobiol Ethnomed 20(1):23
Mir TA, Jan M, Khare RK (2021) Ethnomedicinal application of plants in Doodhganga forest range of district Budgam, Jammu and Kashmir, India. Eur J Integr Med 46:101366
Mishra A, Swamy SL, Thakur TK, Bhat R, Bijalwan A, Kumar A (2021) Use of wild edible plants: can they meet the dietary and nutritional needs of indigenous communities in Central India. Foods 10(7):1453
Mishwani BK, Shah M, Anwar N (2024) Evaluation of conservation status of important Flora at Talash Valley District Dir (Lower) Pakistan. Available at SSRN 4962910
Mittal A (2009) The 2008 food price crisis: rethinking food security policies. UN, New York
Mohammad Ilyas MI, Rahmatullah Qureshi RQ, Shinwari Z, Muhammad Arshad MA, Mirza S, Zia-ul-Haq Z-U-H (2013) Some ethnoecological aspects of the plants of Qalagai hills, Kabal valley, swat, Pakistan
Mozaffarian D (2016) Dietary and policy priorities for cardiovascular disease, diabetes, and obesity: a comprehensive review. Circulation 133(2):187–225
Mughal SS (2022) A review on potential antioxidant effects of Cumin (Cuminum cyminum), phytochemical profile and its uses. Authorea Preprints
Muhaisen HM, Ilyas M, Mushfiq M, Parveen M, Basudan OA (2002) Flavonoid from Viburnum cotinifolium. J Chem Res 2002(10):480–481
Muhammad N, Luo Z, Yang M, Liu Z, Liu M (2022) The nutritional, medicinal, and drought-resistance properties of Ziziphus Mill. make it an important candidate for alleviating food insecurity in arid regions—A case of Pakistan. Horticulturae 8(10):867
Mushtaq M, Anwer N, Waqar MA, Latif S, Shahid SA, Azam A (2013) Antioxidant, antimicrobial potential and phytochemical attributes of impatiens edgeworthii. Asian J Chem 25(17):9799
Naeem MY, Jabran K, Özden M, Bakhsh A (2022) Assessment of morphological and biochemical characteristics of common purslane (Portulaca oleracea L.) accessions. Pak J Agric Sci 59 (6):1017
Nakai J (2018) Food and Agriculture Organization of the United Nations and the sustainable development goals. Sustain Dev 22:1–450
Nasab FK, Zeraatkar A, Mehdiyeva NP, Salimov R, Bussmann RW, Paniagua-Zambrana NY, Maisaia I (2025) Cotoneaster multiflorus Bunge, Cotoneaster nummularius Fisch. In: Mey CA, Cotoneaster racemiflora (Desf.) C. Koch, Cotoneaster uniflorus Bunge (Rosaceae) (eds) Ethnobotany of the Caucasus. Springer Nature Switzerland, Cham, pp 1839–1847
Nasreen N, Semmar N, Farman M, Lacaille-Dubois M-A, Ahmed NS (2023) Phytochemical, geographical, and pharmacological retrospect of genus Torilis. Curr Top Med Chem 23 (24):2300–2331

Nasrullah K, Fayaz A, Kishwar A, Shahid S (2015) Composition, structure and regeneration dynamics of Olea ferruginea Royle forests from Hindukush range of Pakistan. J Mt Sci 12 (3):647–658

Naz S, Farooq U, Ma H, Sarwar R, Riaz N (2021) Three new acrylic acid derivatives from Achillea mellifolium as potential thymidine phosphorylase inhibitor: molecular docking and MD simulation studies. J Biomol Struct Dyn 39(18):7138–7149

Ong HG, Kim Y-D (2017) The role of wild edible plants in household food security among transitioning hunter-gatherers: evidence from The Philippines. Food Secur 9(1):11–24

Pain A (2002) Understanding and monitoring livelihoods under conditions of chronic conflict: lessons from Afghanistan. Overseas Development Institute

Pakistan GO (2012) National climate change policy. Ministry of Climate Change, Islamabad

Patwardhan B, Partwardhan A (2005) Traditional medicine: modern approach for affordable global health. World Health Organization Switzerland

Pellati F, Borgonetti V, Brighenti V, Biagi M, Benvenuti S, Corsi L (2018) Cannabis sativa L. and nonpsychoactive cannabinoids: their chemistry and role against oxidative stress, inflammation, and cancer. Biomed Res Int 2018(1):1691428

Poonia A, Upadhayay A (2015) Chenopodium album Linn: review of nutritive value and biological properties. J Food Sci Technol 52(7):3977–3985

Priyadarshini S, Tudu S, Dash SS, Biswal AK, Sahu SC (2024) Wild edible plants: diversity, use pattern and livelihood linkage in eastern India. Genet Resour Crop Evol 71(6):3111–3133

Qarni A, Muhammad K, Wahab A, Ali A, Khizar C, Ullah I et al (2022) Molecular characterization of wild and cultivated strawberry (Fragaria× ananassa) through DNA barcode markers. Genet Res 2022:9249561

Qudsia NH, Sultana K, Shah SA (2019) Berberidaceae from moist temperate forests, northern areas, Pakistan. J Bioresour Manag 6(4):1

Qureshi R, Bhatti GR, Memon RA (2010) Ethnomedicinal uses of herbs from northern part of Nara desert, Pakistan. Pak J Bot 42(2):839–851

Quyoom I, Bhat BA, Najar ZH, Tanveer S (2024) Winter diet composition of Himalayan serow Capricornis sumatraensis thar in Bani Wildlife Sanctuary: implications for the conservation of Quercus semecarpifolia forest. Biologia 79(3):853–861

Rana C, Sharma A, Kumar N, Dangwal L, Tiwari J (2010) Ethnopharmacology of some important medicinal plants of Nanda Devi National Park (NDNP) Uttarakhand, India. Nat Sci 8:9–14

Rasul G, Hussain A, Mahapatra B, Dangol N (2018) Food and nutrition security in the Hindu Kush Himalayan region. J Sci Food Agric 98(2):429–438

Rehman S, Iqbal Z, Qureshi R, Shah GM, Afzal A, Rahman KU et al (2024) Quantitative ethnomedicinal survey of wild edible fruits used by the indigenous community in North Waziristan, Khyber Pakhtunkhwa, Pakistan. Ethnobot Res Appl 28:1–20

Roshan R, Ahmed S, Ul Hassan MM (2019) Fragaria nubicola (Rosaceae): a review of medicinal uses, phytochemistry and pharmacology. J Pharmacogn Phytochem 8(4):3390–3393

Sabir SM, Riaz K (2005) Morphological, biochemical and elemental analysis of Elaeagnus umbellata, a multipurpose wild shrub from Pakistan. J Appl Hortic 7(2):113–116

Saddiqe Z, Naeem I, Maimoona A (2010) A review of the antibacterial activity of Hypericum perforatum L. J Ethnopharmacol 131(3):511–521

Saeed F, Ahmad M (2016) Sambucus nigra Linn: in vivo studies on rabbits. Int J Pharm Pharm Sci 8 (7):340–345

Safdar I, Bibi Y, Hussain M, Iqbal M, Saira H, Shaheen S, Mehboob H (2017) Review on current status of Betula utilis: an important medicinal plant from Himalaya. J Bot Sci 6:1–7

Saleem U, Khalid S, Zaib S, Anwar F, Ahmad B, Ullah I, Ayaz M (2020) Phytochemical analysis and wound healing studies on ethnomedicinally important plant Malva neglecta Wallr. J Ethnopharmacol 249:112401

Sareen B, Bhattacharya A, Srivatsan V (2021) Nutritional characterization and chemical composition of Diplazium maximum (D. Don) C. Chr. J Food Sci Technol 58(3):844–854

Savary S, Akter S, Almekinders C, Harris J, Korsten L, Rötter R et al (2020) Mapping disruption and resilience mechanisms in food systems. Food Secur 12(4):695–717

Shah SMM, Ullah F, Shah SMH, Zahoor M, Sadiq A (2012) Analysis of chemical constituents and antinociceptive potential of essential oil of Teucrium Stocksianum Bioss collected from the North West of Pakistan. BMC Complement Altern Med 12(1):244

Shah SHA, Shah GM, Ali N, Alam J, Rehman S, Alia A, Basit A (2024) Ethnomedicinal survey of medicinal plants traditionally used in Sakhra Valley district Swat, Pakistan. Ethnobot Res Appl 29:1–28

Shah SHA, Shah GM, Rehman S, Ali N, Alam J, Waheed F et al (2025) Ethnobotanical study of wild edible plants traditionally used by the inhabitants of Lalku Valley, District Swat, Pakistan. Ethnobot Res Appl 30:1–22

Sharma K, Gupta S, Srivatsan V, Yadav SK (2024) Documentation of wild edible plants (WEPs) consumption in North-Western Himalayas: the untapped genetic resources for ensuring nutritional security. Indian J Plant Genet Resour 37(03):404–424

Sher H, Ahmad A, Eleyemeni M, Fazl-i-Hadi S, Hazrat Sher HS (2010) Impact of the nomadic grazing on medicinal plants diversity in Miandam, Swat-Pakistan. Am Eurasian J Sustain Agric 4:152

Sher H, Inamuddin I, Khan Z, Bussmann RW, Rahman IU (2020) Medicinal plant diversity of Hindubaig Mountain, Lalku Valley, District Swat, Pakistan. Ethnobot Res Appl 20:1–13

Shinwari AK, Ahmad M, Alsahli AA, Israr A, Shinwari ZK, Shinwari S et al (2024) Himalayan medicinal plants wealth, threats and conservation for securing the future of biodiversity. Pak J Bot 56(6):2327–2350

Shirazi M (2018) In vivo biological investigation of methanolic extract of Thymus linearis whole plant. Am J Ethnomed 5(1-2):1–5

Shirsat R, Koche D (2024) Wild edible plants (WEPs): A review on their importance, possible threats, and conservation. Contemp Res Perspect Biol Sci; Son, PL, Ed.; BP International, China 2:97–115

Shrestha PM, Dhillion SS (2006) Diversity and traditional knowledge concerning wild food species in a locally managed forest in Nepal. Agrofor Syst 66(1):55–63. https://doi.org/10.1007/s10457-005-6642-4

Siddiq F, Fatima I, Malik A, Afza N, Iqbal L, Lateef M, Khan SW (2012) Biologically active bergenin derivatives from Bergenia stracheyi. Chem Biodivers 9(1):91–98

Sinha D, Maurya AK, Abdi G, Majeed M, Agarwal R, Mukherjee R et al (2023) Integrated genomic selection for accelerating breeding programs of climate-smart cereals. Genes 14(7):1484

Song JM, Yi JS, Cheong EJ (2019) Growing three wild herbaceous plants, Allium victorialis var. platyphyllum, Ligularia stenocephala, and Panax ginseng in the Understory of Larix kaempferi Plantation. J Korean Soc Forest Sci 108(4):522–530

Sood M, Aziem S, Iqbal K, Hussain A, Mahato S (2019) Economic potential of Angelica glauca Edgew: an endangered plant of western Himalaya. G-J Environ Sci Technol 6(5):28–30

Stakman E (1957) Problems in preventing plant disease epidemics. Am J Bot:259–267

Sultana K, Manzoor M, Farooq MS (2019) Relative abundance and diversity of Rosaceae from protected parks in northern areas of Pakistan. J Bioresour Manag 6(2):4

Sultana K, Ahmad B, Ahmad Z, Rauf A, Al-Awthan YS, Bahattab O, Ramadan MF (2022) Effect of Bistorta affinis (D. Don) and Malcolmia cabulica (Boiss) essential oils on human and foodborne pathogens, insects and termites. J Pure Appl Microbiol 5:14–21

Talha M, Shah SM, Ullah M, Pirzada AS, Hussain S (2025) Molecular docking and antileishmanial potential of isolated compounds from Asparagus gracilis. PHYTONutrients 4:91–96

Tanveer A, Tabassum F, Abbas RN, Tahir M, Javaid MM, Aziz A, Siddiqui MH (2012) Seed germination ecology of Lathyrus aphaca L. and Vicia sativa L. in comparison with Triticum aestivum L. Pak J Weed Sci Res 18(3):293

Tassadduq SS, Akhtar S, Waheed M, Bangash N, Nayab D-E, Majeed M et al (2022) Ecological distribution patterns of wild grasses and abiotic factors. Sustainability 14(18):11117

Uddin N, Ali N, Muhammad N, Zehra SA, Nisar M, Khan MKU (2021) Morphological attributes and total seed protein revealed diversity in Ziziphus nummularia (Burm. F.) Wight & Arn. Populations from malakand division, Pakistan. Pak J Bot 53(5):1727–1735

Ullah H, Badshah L (2023) Nutritional and mineral analysis of the ultimate wild food plants of Lotkuh, Chitral, The Eastern Hindukush Pakistan. Heliyon 9(3):e14449

Ullah Z, Rashid A, Shah M, Ali M, Gul H, Khan A (2018) Ethnobotanical studies of medicinal plants of Malakand district. FUUAST J Biol 8(1):161–167

Ullah F, Papini A, Shah SN, Zaman W, Sohail A, Iqbal M (2019) Seed micromorphology and its taxonomic evidence in subfamily Alsinoideae (Caryophyllaceae). Microsc Res Tech 82(3):250–259

Ullah MA, Gul FZ, Khan T, Bajwa MN, Drouet S, Tungmunnithum D, Abbasi BH (2021) Differential induction of antioxidant and anti-inflammatory phytochemicals in agitated micro-shoot cultures of Ajuga integrifolia Buch. Ham. ex D. Don with biotic elicitors. AMB Express 11(1):137

Ullah T, Khan SM, Abdullah A, Sulaiman N, Ullah A, Khan MS, Zeb SA, Pieroni A (2024) Bio-cultural diversity for food security: traditional wild food plants and their Folk Cuisine in Lakki Marwat, Northwestern Pakistan. Diversity 16(11):684

Ullah A, Badshah L, Uddin N, Islam M, Akhtar F, Tahira (2025) Pattern of geographical distribution and diversity of vascular plants: a case study of South Waziristan, Pakistan. Arct Antarct Alp Res 57(1):2458736

van der Meer L et al (2022) Bioavailability of nutrients from wild leafy vegetables. Food Chem 380: 132–145

Varin S, Cliquet JB, Personeni E, Avice JC, Lemauviel-Lavenant S (2010) How does sulphur availability modify N acquisition of white clover (Trifolium repens L.)? J Exp Bot 61(1):225–234

Verma NJKC (2020) A review on nutrition value of Amaranth (Amaranthus caudatus L.): the crop of future. J Pharmacogn Phytochem 9(4):317–319

Vetter J (2010) Toxicological and medicinal aspects of the most frequent fern species, Pteridium aquilinum (L.) Kuhn. In: Working with ferns: issues and applications. Springer New York, New York, pp 361–375

Vidyarthi S, Samant SS, Sharma P (2013) Dwindling status of Trillium govanianum Wall. ex D. Don-A case study from Kullu district of Himachal Pradesh, India. J Med Plants Res 7(8):392–397

Vinceti B, Eyzaguirre P, Johns T (2012) The nutritional role of forest plant foods for rural communities. In: Human health and forests. Routledge, pp 63–96

WHO (World Health Organization) (2021). The State of Food Security and Nutrition in the World 2021: Transforming Food Systems for Food Security, Improved Nutrition, and Affordable Healthy Diets for All (vol 2021). Food & Agriculture Organnization

Yahya M, Khan MA, Ullah H, Sultana K, Ullah A, Ahsan H, Ali D (2025) Phytochemical screening and antioxidant potential of methanolic extract of aerial parts of Allium griffithianum Boiss (Amaryllidaceae). Chem Select 10(23):e01189

Yousaf MM, Shah MJ, Hussain M, Bashir A, Zafar MB, Hayat M, Majeed U (2024) Effects of different combination of fertilizer on growth and yield in Plantago ovata (Forsk.) Isapghol. Pak J Agric Res 37(3):190

Yousfi N, Saïdi I, Slama I, Abdelly C (2015) Phenology, leaf gas exchange, growth and seed yield in Medicago polymorpha L. populations affected by water deficit and subsequent recovery. Flora-Morphol Distrib Funct Ecol Plants 214:50–60

Zahara K, Ahmad N, Bibi Y, Bibi F, Sadaf HM, Sardar N (2019) An insight to therapeutic potential and phytochemical profile of Solanum villosum (L). Med Drug Discov 2:100007

Zanit SB, Mochi SA (2021) Trees for agroforestry in Dhanore Village in Rajouri district of Jammu & Kashmir, India. East Himalayan Soc Spermatophyte Taxon 15(2):149–163

Zareef H, Gul MT, Qureshi R, Aati H, Munazir M (2023) Application of ethnobotanical indices to document the use of plants in traditional medicines in Rawalpindi district, Punjab-Pakistan. Ethnobot Res Appl 25:1–29

Zia-Ul-Haq M, Riaz M, De Feo V, Jaafar HZ, Moga M (2014) Rubus fruticosus L.: constituents, biological activities and health related uses. Molecules 19(8):10998–11029

Impact of Climate Change on Food Security and Biodiversity of Pakistan

21

Muhammad Athar, Muhammad Zafar Iqbal, Ozair Chaudhry, Mohammad Shafiq, Nyla Mansha, and Rahmatullah Qureshi

Abstract

Climate change has emerged as one of the most significant threats to biodiversity conservation and food security, particularly in developing countries such as Pakistan, where agriculture and natural resources are central to livelihoods and economic stability. Rising temperatures, erratic precipitation patterns, prolonged droughts, floods, glacier melt, land degradation, deforestation, and unsustainable resource exploitation are accelerating biodiversity loss and undermining agricultural productivity. These environmental changes are threatening diverse ecosystems and vulnerable species while adversely affecting major food crops, including wheat, rice, maize, cotton, and sugarcane, resulting in reduced yields, disrupted food supply chains, and heightened risks of food insecurity. Additional pressures from pollution, soil degradation, water scarcity, invasive species, and habitat fragmentation further exacerbate ecological and agricultural

M. Athar
California Department of Food and Agriculture, Sacramento, CA, USA

M. Z. Iqbal
Department of Botany, University of Karachi, Karachi, Pakistan

O. Chaudhry
DECAC, Environment and Climate Advisory, Planning & Economic Development, Durham, Regional Municipality, Rossland, ON, Canada

M. Shafiq (✉)
Department of Botany, University of Karachi, Karachi, Pakistan

N. Mansha
The Design Build Inc., Sacramento, CA, USA

R. Qureshi
Department of Botany, Faculty of Sciences, Pir Mehr Ali Shah Arid Agriculture University, Rawalpindi, Pakistan
e-mail: rahmatullahq@uaar.edu.pk

R. Qureshi (ed.), *Plant Biodiversity*, https://doi.org/10.1007/978-981-92-1471-6_21

vulnerabilities. This chapter examines the complex interactions between climate change, biodiversity, and food security in Pakistan, highlighting the ecological, socioeconomic, and policy dimensions of these challenges. It evaluates existing conservation and climate adaptation frameworks and explores strategies such as climate-resilient agriculture, ecosystem restoration, sustainable natural resource management, biodiversity conservation, technological innovation, and community-based adaptation. The chapter emphasizes that safeguarding biodiversity and ensuring food security require integrated, multi-stakeholder approaches involving government institutions, scientific communities, local populations, and the private sector. Strengthening policy implementation, public awareness, and climate adaptation measures is essential for building resilient ecosystems, sustainable food systems, and long-term environmental sustainability in Pakistan.

Keywords

Climate change · Food security · Pakistan agriculture · Sustainable development · Crop productivity · Environmental pollution · Water scarcity

Abbreviations

ARDL	Autoregressive distributed lag
BD	Biological diversity
Br	Brassinosteroids
FOP	Flora of Pakistan
GHG	Greenhouse Gas
MeJA	Methyl jasmonates
NBSAP	National Biodiversity Strategy and Action Plan
NGOs	Nongovernmental organizations
PGR	Plant growth regulators
RCPs	Representative concentration pathways
Tr	Triazoles
WHO	World Health Organization

21.1 Introduction

Pakistan is losing its native biodiversity at an increased rate due to the changing climatic conditions. Climate change variability has become a major factor in the decline in biodiversity and decreased yields (Siddique and García Molinos 2024; Ebrahimi et al. 2025). The same trends are reported in the developing world, especially in the African nations, which show that such ecosystems are highly susceptible to climate change (Lundgren et al. 2013; Sellers 2016; Ncongwane et al. 2021). The climatic variation in Pakistan is also significant, as the minimum temperatures in the mountainous parts of the country are as low as -0C and the maximum temperatures are 52 degC in the plain areas of the country like Sibi (Ali 1978; FOP 2016). An increased average annual temperature has become evident over the past decades, which has had its negative impact on agriculture and rural livelihood (Awan and Yaseen 2017; Jaffery 2018).

Pakistan is the 31st ranked country according to the Climate Change Performance Index in 2025 (CCPI 2025). The policy of the National Food Security of 2018 was formulated to boost agricultural productivity and climate resilience (Malik 2022). With its agrarian base and food systems input, global climate change is becoming more of a threat to ecosystems and wildlife in Pakistan (Jetz et al. 2007; Grand et al. 2019). Global warming is caused by increases in temperature and changing patterns of distribution and survival of species (Wang et al. 2023). The mega floods of 2022 were examples of such effects, which led to mass displacement, fatalities, and devastating agricultural, livestock, and infrastructure damage. On top of that, unsustainable agricultural methods and resource mining, such as oil and gas exploration, have corrupted native grasslands (Naeth et al. 2020), and overgrazing has disturbed alpine ecosystems (Bao et al. 2015).

Climate change is a major threat to the food production systems and biodiversity both on regional and global levels (Naz et al. 2024; Wheeler and Von Braun 2013). These shocks disrupt the stability of the food system, especially by disrupting food supply. Greener evaluations also bring out climate-related hazards, including heatwaves, droughts, and floods, and consequences to agriculture, energy, urban systems, and the well-being of society, including health, poverty, and migration (WGB and ADB 2021).

Pakistan is a developing South Asian state that is very vulnerable to such effects. The environmental conditions are shifting and climate-sensitive species can no longer survive in the changing conditions, and the process of deforestation and forest degradation is growing. It has reported a decrease of about 1.27 percent per year in the coniferous forests, most of which is due to decreased precipitation (Ahmad et al. 2012). In addition to this, regional studies show that the composition of forests is highly controlled by climatic and edaphic factors, as the vegetation dynamics in the Tanawal area of the Lesser Himalayas have been modulated by higher temperatures (33.67 degC in 2015–2016), which may be explained by higher temperatures (Farooq et al. 2019). The issue of environmental degradation and loss of forests is still a major issue in the ecological scenario in Pakistan (Lin and Ullah 2024). Comprehensively, climatic change is having an intense impact on the economic well-being and livelihoods of the developing nations. This chapter tries to critically evaluate its effects on biodiversity and food security in Pakistan. The evidence synthesized is supposed to educate policymakers, environmental managers, researchers and other interested parties to establish appropriate adaptation and mitigation measures.

21.2 Literature Search Strategy

To compile this chapter, an extensive literature review was carried out on articles published between 1978 and 2025. Keywords such as global warming, greenhouse gases, drought, soil health, pollution, heavy metals, ecological impacts, toxicity, and tolerance were used to retrieve relevant publications on the subject of biodiversity and food security on major databases, such as Google Scholar, Web of Science, and Scopus. First of all, nearly 650 records in English were identified. Following the relevance, duplication, and alignment with the scope of the study, 146 peer-reviewed articles were included. The chapters of books were omitted to avoid variations in citation limits. The references were organized in the APA citation style.

21.3 Climate Change Context

There are 17 global priorities identified by the United Nations Sustainable Development Goals (UNSDGs), Goal 13 of which highlights the need to act urgently to combat climate change and the numerous effects it has been having on all continents. The increased emissions of greenhouse gases have amplified climate-associated risks, such as wildfires, floods, droughts, and extreme weather, which are posing a major threat to ecosystems and human health. Sustainable development can only be achieved through global cooperation as promoted by the UNSDG framework (UN 2025).

Goal 15 pays attention to the terrestrial ecosystems, and its aim is to stop the decline of biodiversity and encourage sustainable land use (SDG 2023). The agricultural development, especially in the growing of oil palm, has been one of the largest contributors to deforestation, with a significant amount of forest loss between 2000 and 2018. In Pakistan, the damaging effect of events caused by climate, such as floods, heatwaves, and droughts, especially in Sindh, has had significant destructive effects on infrastructure, livelihoods, and agricultural systems (Ajani and van der Geest 2021).

Although the trends on poverty reduction have been remarkable, it is estimated that some 575 million individuals might continue to languish in abject poverty in case present trends are maintained. The elimination of extreme poverty, which is living below US$2.15 per day (2017 purchasing power parity), is one of the priorities of the 2030 Agenda of Sustainable Development. Despite the tremendous changes realized throughout the world, there is a need to work hard to ensure that poverty is eradicated in all its manifestations.

The Sustainable Development Goals (SDGs) also focus more on eradicating hunger and enhancing food security by the year 2030 (Table 21.1). Access to safe water and sanitation resources, safe, affordable, and sustainable energy resources should also be ensured. The structure also promotes building of inclusive, safe, resilient, and sustainable cities, responsible consumption and production patterns.

Climate change and its extensive effects need urgent and concerted efforts to combat them. Moreover, conservation and sustainable use of marine ecosystems,

Table 21.1 Sustainable development goals—SDG-(United Nations 2025)

Geest	Geest	Geest
1	No poverty	With current trends, 575 million people will continue to live in abject poverty. One of the key aims of the 2030 Agenda of Sustainable Development is to eradicate extreme poverty in the world by 2030. Extreme poverty, which is traditionally regarded as living under the poverty threshold of 2017 purchasing power parity of less than 2.15 US dollars per person per day, has decreased significantly in the last few decades, but some significant obstacles emerge.
2	Zero hunger	This is a goal to eradicate hunger through means of access to safe and nutritious foods and sustainable agriculture by 2030.
3	Good health and well-being	Secure healthy living and well-being of all at any age.
4	Quality education	Deliver quality education that is inclusive and equitable, and encourages the learning life-long
5	Gender equality	Empower and achieve gender equality
6	Clean water and sanitation	Be sure the water and sanitation of all are available and under a sustainable management
7	Affordable and clean energy	Secure affordable, reliable, sustainable, and modern energy
8	Decent work and economic growth	Enhance growth in the economy, productive and decent work toward all
9	Industry, innovation, and infrastructure	Develop resilient infrastructure, enhance inclusive industrialization, and make innovations
10	Reduced inequality	Make countries and countries less unequal
11	Sustainable cities and communities	Make cities inclusive, safe, resilient and sustainable
12	Responsible consumption and production	Secure sustainable consumption and production pattern
13	Climate action	Clean and sustainable use of oceans, seas, and renewable resources
14	Life below water	Conserve and sustainably use the oceans, seas and marine resources
15	Life on land	Safeguard, renew and advance sustainable utilization of land ecosystem and biodiversity
16	Peace, justice, and strong institutions	It is proposed to encourage peaceful and inclusive societies and access to justice and accountable institutions
17	Partnerships on the goals	Enhance global partnerships to help in the achievement of sustainable development goals

protection and restoration of terrestrial ecosystems including forests are also major priorities. The fight against desertification, the reversal of land degradation, and avoiding biodiversity loss are still the focus areas to achieve long-term sustainability of the environment.

21.4 Biological Diversity

Biological diversity, often abbreviated as biodiversity, refers to the variability of living organisms in the broader sense, that is, plants, animals, and microorganisms, and ecological complexes of which they form a part. It is important for the ability to stabilize and sustain resilience and stability in ecosystems. Many ecosystem services related to nutrient cycling, pollination, climate, and soil fertility are also based on biodiversity and are critical to human well-being and sustainable development (Molina-Venegas et al. 2021; Studocu 2024).

Nevertheless, there are continued climatic changes that have been threatening both the regional biodiversity and global biodiversity; some of them include rising temperatures, changes in precipitation, frequency of droughts, and spread in aridity. These environmental pressures have negative effects on the distribution of species, dynamics of the ecosystems, and agricultural productivity that contributes to low yields of crops and livestock activity. These factors combined are contributing to the loss of biodiversity, and this has been widely recorded in the literature (Bennett 1990).

Pakistan is home to great biological richness due to the variety in the ecological zones of the country, which include alpine areas, deserts, coastal ecosystems, etc. (Anwar et al. 2022). Although the policies including the National Biodiversity Strategy and Action Plan (PNBSAP 2017) and the National Climate Change Policy aim to safeguard biodiversity in the country, it is still very susceptible to the effects of climate change (COBD 2024). Although desert biomes are intrinsically sparse in terms of species, they are so delicate and vulnerable to environmental perturbations (Agwan 1999). Due to the effect of pollution and industrial release into the environment, the vegetation and species diversity have notably reduced in urban-industrial areas, including Karachi (Shafiq and Iqbal 1988; Iqbal et al. 1999; Shafiq et al. 1992; Shafiq et al. 2018).

The issue of biodiversity loss is closely interconnected with land degradation, deforestation, and climate change on the global level. These problems are further aggravated by the fact that in Pakistan there is rapid urbanization, overgrazing, and unsustainable use of resources. Forest ecosystems are in extreme distress, and the tree cover is being reduced, which endangers habitats and the existence of species. It is reported that further deforestation is threatening major species, such as markhor, squirrels, and woodpeckers, so the ecological balance is being destabilized (Zia 2022).

Migration of avifauna has also been interrupted due to climate change. Another group of birds known to demonstrate the decreasing tendency of visitation to Pakistan is several migratory birds such as houbara bustard, pintails, mallards, shovelers, coots, and Siberian cranes, which have been greatly impacted by the lack of fresh water and habitat degradation on migration paths (Shamsulhaq 2002; Iqbal et al. 2018a, b). The anthropogenic pressures have degraded the coastal ecosystem, especially the mangroves in the coastal belt of Sindh and Balochistan, as well as in Port Qasim, Kemari, and other areas (Ahmed and Rana 1999).

The same applies to forest ecosystems in northern areas, such as Sathan Gali (Mansehra) and neighboring districts of Azar Jammu and Kashmir, where the lack of rehabilitation interventions, combined with overgrazing, has caused considerable ecological disturbances, thereby causing the loss of vegetation cover and impeding the functioning of the ecosystem (Khan et al. 2019a, b). Also, other abiotic limiting factors like waterlogging and salinity caused by soil have cropped up as significant problems that are posing serious threats to conservation of biodiversity (Baig and Al-Subaiee 2009).

Habitat fragmentation, overexploiting plant resources, and the introduction of invasive species are a few examples of anthropogenic activities that have been accelerating the process of biodiversity loss. As an example, the reef plants that have been at risk include *Asparagus adscendens* and Viola, which are unsustainable collections in Margalla Hills National Park (Shinwari and Khan 2000). Ethnobotanical research in places like Nowshera and Timergara (Khyber Pakhtunkhwa) found that biotic pressure on local flora was strong because of grazing and local use activities (Ali et al. 2019; Shuaib et al. 2019).

On the whole, there is a growing threat to biodiversity in Pakistan due to the intricate combination of climatic and anthropogenic factors. The proposed solutions to these issues involve a combination of conservation mechanisms, such as the creation of conservation areas, sustainable land use, restoration of the damaged ecosystems, and the enactment of the environmental laws. Research, community participation, and policy execution should be strengthened in order to protect biodiversity and provide long-term ecological and food security.

Pakistan National Conservation Strategy also identified 14 priority programs such as biodiversity conservation, creation of new national parks and endangered species protection (Table 21.2). In spite of such efforts, all over the world, there has been a decrease in the population of wildlife especially the grassland birds owing to degradation of habitat and other ecological disturbances.

The destruction of habitats, overgrazing, disease outbreaks, and man-made pressures (hunting, trapping) have been identified as causes of biodiversity loss in the western Himalayan part of Pakistan, especially District Battagram. Secured lands like the Deosai National Park are important in saving major species, e.g., the Himalayan brown bear (Ursus arctos isabellinus), endangered in the Pakistani mammalian RedList.

Markhor, snow leopard (Panthera uncia), Asian white-backed vulture, Indus River dolphin, golden mahseer and Marco Polo sheep are several of the iconic species in Pakistan that are facing the risk of extinction. Recent estimates show that a variety of taxa are endangered and in a decline, such as around 20 mammalians, 25 avian, 6 reptilians, 5 fish, and 8 marine mollusk species (Harmain 2024).

The increase in temperature and the irregularity of precipitation in the Hindu Kush Himalayan region are among the greatest threats to the biodiversity. Climate change is being credited with increasingly causing habitat shifts and ecological instability, including effects on the survival and distribution of species. An example is the snow leopard species in the Bagrot and Haramosh valleys in the Trans-Himalayan region, which are experiencing pressure because of the changing climatic

Table 21.2 The impact and identification of climate change threats on biodiversity

Description and impact	References
A total of 14 core program including in the National Conservation Strategy of Pakistan that defines primary priority programs, such as biodiversity conservation, creation of the protected areas, and the protection of the endangered species	Siddiqui et al. (2007)
There is a steady decline in the population of grassland birds around the world especially in steep and delicate areas	Knopf (1994), With et al. (2008)
Habitat degradation, overgrazing, disease pressures, as well as illegal hunting and trapping, have been cited as the causes of biodiversity loss in district Battagram (Western Himalaya, Pakistan)	Haq (2012)
Deosai National Park is an important conservation zone of Himalayan brown bear (Ursus arctos isabellinus) that is endangered in Pakistan	UNESCO (2016)
Many of the iconic species, such as markhor, snow leopard, Asian white-backed vulture, Indus River dolphin, golden mahseer, and Marco Polo sheep, are experiencing a drastic drop in population and are on the verge of extinction	ITN (2022)
Many of the species are now facing the risk of extinction such as 20 mammals, 25 birds, 6 reptiles, 5 fish species, and 8 marine mollusks	Burca (2023)
In the Hindu Kush Himalayan area, the increase in temperature and the erratic weather conditions are significant factors of biodiversity loss	Ali et al. (2021), Chettri et al. (2023)
Risks affecting biodiversity related to climate are also being associated with economic vulnerabilities, such as financial instability and business risks	(Khan 2001; Adamolekun (2024)
Population of snow leopard (*Panthera uncia*) in the Bagrot and Haramosh valleys of the Trans Himalayan region is reducing because of changes in habitats caused by climatic conditions	Hani et al. (2024)
Honey bees that pollinate a range of plants are very important to food security and ecological stability, but their numbers have been endangered by various environmental and biological stresses	Munir et al. (2024)
Climate change is also causing loss of habitat and is therefore likely to cause changes in the distribution pattern of the species like Egyptian vulture in Punjab in Pakistan	William et al. (2024)
Pakistan is faced with a rising threat of the negative impacts of climate change on its forest ecosystems	Imran (2025)
Continuous destruction of the environment has been increasing the level of extinction among various species of plants and animals across the world	Segala et al. (2025).
Exceptional populations of native weasels in the mountainous areas of the Pakistani peninsula, into central and East Asia and the Siberian islands are experiencing an ongoing reduction in the number of mature animals	WOP (2025)

conditions. On the same note, predictive research indicates that the habitat changes caused by climate change will probably affect population distributions of certain species like the Egyptian vulture in Punjab.

Honey bees and other pollinators are important in biodiversity conservation and food security, but due to a complex of biological, chemical, and environmental stresses, these insect groups are at risk. The forest ecosystems of Pakistan are also at great risk from climate change, which leads to ecological degradation and loss of species in the bigger ecological context.

These pressures have resulted in an ecological crisis which has caused the decline and even extinction of many plant and animal species. The native fauna, like weasels spread throughout the mountainous zones of Pakistan and beyond into Central Asia and Siberia, also faces a decrease in population. The solution to these problems lies in intensified conservation, protection of habitats, and management strategies that are climate-resistant.

21.5 Biological Diversity, Protection, and Restoration

Biodiversity conservation enhances both the natural and agroecosystems and ecosystem stability (Van Schalkwyk et al. 2019; Douglas et al. 2020). Nonetheless, agricultural land development, coniferous encroachment, and lack of prey have all been extensively cited as the major causes of biodiversity loss (Jaksié et al. 1992; White et al. 1996; Anderson and Inouye 2001; Ryall and Fahrig 2006). Recovery of these types of ecosystems, and especially sagebrush communities, is complicated and slow (Knapp 1996).

Ecological restoration aims at the restoration of damaged ecosystems by reintroduction of indigenous species and ecological processes. There is a need to strike a balance between biodiversity conservation and sustainable economic use (Lashari et al. 2020). Besides that, the pollinators, such as butterflies and bees, contribute greatly to the biodiversity of grasslands (Leone et al. 2022). Extensive studies indicate that climate change ecosystem resistance, stability, and tolerance are determined by the interplay of biotic and abiotic factors. The restoration measures and grassland restoration techniques have been well documented all over the world, especially in North America (Gann et al. 2019).

21.6 Wetlands and Degradation of Habitat

The destruction of habitats and loss of natural resources has played a significant role in the loss of biodiversity in the past few decades. It has been calculated that over half of the wetlands in the world during the past 200 years have been lost, and this has resulted in a diminished ecological and biogeochemical activity (VanMeter and Basu 2015). The same patterns have also been witnessed in Pakistan, whereby the populations of waterfowl were reduced in places like Taunsa Barrage as a result of habitat destruction (Azam and Hasan 1999).

Forced changes in natural habitats and industrial contamination have impacted the natural habitats and endangered biodiversity through anthropogenic stressors (Zahid 2023). The presence of toxic pollutants in the soil changes physicochemical processes, decreases microbial life, and eliminates the supply of nutrients, which eventually influence vegetation growth and the operation of the ecosystem.

21.7 Native Species Decadence, Pollution, and Overexploitation

Cases of depletion of the native species diversity have been reported in various parts of Pakistan because of overexploitation, fuelwood harvesting, and grazing strains. Research conducted in Sarban Hills, Bhimber Hills, and Nikyal Valley reported decreases in vegetation diversity as well as in structural composition (Amjad et al. 2014; Ijaiz et al. 2018; Nafeesa et al. 2021).

Forest ecosystems have also been affected by climatic variability, especially decreased precipitation. In the case of Punjab, coniferous forests have undergone ecological pressure, thereby leading to the extinction of a number of butterfly species in the last few decades (Saadat 2012). Such tendencies mean a greater ecological disturbance and biodiversity depletion.

21.8 Degradation of Ecosystems, Degradation of the Land, and Alien Species

Climate change and habitat fragmentation, as well as degradation of the land, also become the major causes of biodiversity loss. Previous research findings have described the high levels of vegetation changes and fragmentation's impact on ecosystems (Avon et al. 2010; Davies et al. 2011; Dhar et al. 2019; Grant et al. 2020; Elsinger et al. 2022; Cullingham et al. 2023).

The other significant cause of the lowering of biodiversity is invasive species, which interfere with native ecological balance and compete with others (de Freitas et al. 2020). According to global evaluations, ecological degradation is rapidly accelerating the loss of species and augmenting the dangers of extinction at an appalling pace (Ceballos et al. 1999; Luck 2007).

21.9 Food Security and Food Production

There has been a growing pressure on food production systems with changes in climatic and environmental conditions like rising temperatures, water shortages, drought, salinity, agrochemicals and pollution. All these contribute to a decrease in the productivity of crops and pose a risk to the availability of food, especially in low-income nations.

Edaphic conditions, salinity of soil, pH and availability of nutrients determine the agricultural productivity in Pakistan. Biotic and abiotic stresses have increased the problem of food security in various regions (Muluneh 2021). The effects of climate change on livestock, deforestation, and desertification have become well-publicized (Ahmad and Farooq 2010; Anjum et al. 2010; Arshad and Shafqat 2012; Ahmed et al. 2015; Abdela and Jilo 2016; Abrar and Maryiam 2023; Lodhi et al. 2024). Agricultural productivity is also further affected by the agro-climatic groupings and thermal limits (Chaudhary and Rasul 2004; Zahid and Rasul 2011). Moreover, in other areas like the Indus Delta, increased risks to the production of sustainable crops have been precipitation and the fluctuation of temperatures (Rasul et al. 2012).

Food security has several aspects, among them being availability, accessibility, affordability, safety, nutritional quality, and livelihood sustainability (Videnović et al. 2013; Ateeq-Ur-Rehman et al. 2018; Farrukh et al. 2020; Nasrullah et al. 2022; Ashraf et al. 2023). Fading productivity of land, especially for staple foods like barley, poses a great challenge to long-term food sustainability.

21.10 Heat Stress and Increasing Temperatures

The increase in temperatures and heat stress is potentially dangerous to crop productivity. The rise in temperature levels adversely impacts the growth, development, and yield of crops (Ali and Erenstein 2016). It is estimated that a rise in temperature of 1.5degC-3degC can cause major losses in wheat yields of areas like Swat (Hussain and Mudasser 2007).

21.11 Water Scarcity and Flooding

Water is a resource over which human life and agricultural productivity depend. Extreme weather caused by the climate, including floods and droughts, also shook the food systems and health of the population (Devi 2022). Heavy rain and glacial melt led to the 2022 floods in Pakistan that resulted in major devastation of crops, livestock, and arable land (Bhutta et al. 2022). It is projected that the risk of floods will grow in future as glacial melting persists (Aslam 2018).

21.12 Arid and Drought and Salinity Effects

The key limitations to agricultural productivity include aridity, drought, and salinity. These aspects restrict the growth of plants, decrease their productivity, and deteriorate the quality of the soil. Droughts play a great role in the growth of crops, and the use of management methods like overgrazing and use of herbicides can also lead to low abundance of plants (Vaness and Wilson 2007).

The change in precipitation and temperature patterns has resulted in the increased abiotic stress on crops due to climate change (Mahmood et al. 2012). The adverse effects of these stresses on wheat and maize production have been reported in Pakistan (Anum et al. 2014; Rizwanullah et al. 2023; Khan and Wang 2024). Wheat, rice, cotton, sugarcane, maize, and pulses are major crops that have been affected by climatic variability. High temperatures, especially, have been reported to be the cause of low productivity of maize and decreased yields in arid regions (Shakoor et al. 2011).

21.13 Pollution

It is estimated that environmental degradation will have a significant impact on the economy of Pakistan and the potential GDP will decline to 18–20 percent by 2050 as the risks posed by climate extremes continue to rise (URAAN 2025). The various sources of pollution, which are greenhouse gas emissions, hydrogen sulfide, ammonia, particulate matter, combustion of fossil fuels, *and* heavy metals (lead, cadmium, copper, chromium, mercury, boron, zinc, and arsenic), are significant causes of decreasing productivity in agriculture.

Unreasonable and uneven application of fertilizers also contributes to the further degradation of the environment, which has an adverse effect on the health of *the* soil and yields. Moreover, the unregulated release of municipal and industrial effluents, including pesticides, insecticides, organic and nitrogenous substances, suspended solids, detergents, phenols and sulfuric compounds etc. has caused massive contamination of air, water, and soil systems. Previous surveys had reported that several ecosystems in Pakistan face great danger (Ali 2000). The process of land degradation remains complicated and multifaceted and keeps jeopardizing the sustainability of the natural resources (Andrade et al. 2015). Also, other processes like soil erosion and cycling of nutrients are very sensitive to the variability of precipitation and the disruption of vegetation cover (Smith et al. 1984).

21.14 Engineering Sector Role in Mitigating Climatic Change

The engineering fields are essential in reducing climatic changes and improving environmental sustainability in the regions and globally. The major interventions are *the* layout and implementation of renewable energy systems, efficient waste and water management technologies, recycling infrastructure, and energy-saving technologies that do not impact the ecology much.

The method of urban and environmental engineering, which includes climate-resilience infrastructure, flood management methods, establishing urban green areas, and introducing green roofs, helps to mitigate urban heat stress and elevate environmental quality. The engineering division in Pakistan is also taking action in these directions by advancing sustainable technologies and infrastructure construction to decrease environmental degradation and encourage climate resiliency.

Policy-based research has equally given practical measures in the reduction of carbon emissions and *the* degradation of the environment (Chien et al. 2021). Moreover, some frameworks have been put forward to be used in engineering practices to achieve net-zero greenhouse gas emissions (Younger et al. 2022).

The system of water resources is one of the most urgent issues, and drought and floods have an increased impact on groundwater systems. Extreme temperature changes and the growing frequency of extreme heat events in the world are the enemies of the population, infrastructure, and the stability of society. The reasons why these challenges have triggered an urgent focus on the need to have sustainable, equitable, and resilient cooling and adaptation strategies to protect communities and reduce the impact of further climate effects (Sayin and Peters 2025).

Climatic stressors were observed to lower the yields of wheat by about 6–9% in arid, semi-arid and sub-humid Pakistan. The long-term patterns of production (1960–2009) show that the output of wheat relies heavily on climatic variables and a rise in temperature of 1 °C will lead to a reduction in yield by 5–7%. There is variability in the region, especially in Punjab, where wheat productivity varies differently to the varying climatic conditions. In general, the key crops in Pakistan such as wheat, rice, cotton, sugarcane, and maize are very susceptible to climate change (Table 21.3). Climate-smart agricultural practices and technologies have been proposed as an important tool for reducing these effects, especially for crops like cotton. Farms that depend on rain are particularly susceptible, with such communities being the most endangered by the prevailing and forecasted climatic conditions.

The productivity of crops is greatly dependent on water supply and some reported effects have been between a 12 percent decrease and a 24 percent growth as far as the moisture content is concerned. Exogenous use of plant growth regulators (PGRs) in rice during high day and night temperatures has also been observed to have a negative impact on the morphology of plants (e.g., a decline in the leaf area in cultivars IR-64 and Huanghuazhan). Examination of climatic data (1989–2015) of major crops in Pakistan, such as wheat, rice, maize, and sugarcane shows that there are especially adverse impacts of an increase in maximum temperatures on the production of wheat. These tendencies have raised the level of discussion on the national food security. The socioeconomic evaluation of selected districts in Punjab (2014) indicated that 73–85 percent of agricultural households suffered economically due to the changing weather conditions. Estimated projections of temperatures in Sindh show that a 1 °C rise will further lower the production of rice in the region in the next few years. Review of literature on simulation studies comparing present (1980–2010) climate and future climates in Punjab is indicative that increased temperatures may reduce rice production by about 7.3 percent. Likewise, statistical results on the changes in rain patterns (1981–2017) suggest that the changes in precipitation decrease the establishment of plants at the level of transplanting. There are economic estimates of huge production losses, of which climate change leads to a decrease of 222.30 thousand metric tons of rice, which translates to a loss of about 27.79 billion PKR in one season in Punjab. In addition to crops, the livestock sector, in terms of production systems as well as supply chains, is also impacted negatively by climate variability.

Table 21.3 The impact of climate change on food production

Description and impact	References
Climatic stressors have been found to decrease wheat production by around 6–9 percent in arid, semi-arid and sub humid agro- ecological zones in Pakistan	Sultana and Ali (2006)
The long-term trends (1960–2009) have shown that the trend in production of wheat in Pakistan heavily depends on climate related variables	Janjua et al. (2010)
An increment of 1 °C in temperature is linked with a reduction of approximately 5–7% in the productivity of wheat	Aggarwal and Sivakumar (2011)
The spatial variability is also observed in Punjab in which the production of wheat responds differently to the changing climatic conditions	Ashfaq et al. (2011)
There is a wide fluctuation between crop productivity and water availability with a range of 12 to 24 percentage change depending on the level of moisture	Zhu et al. (2013)
The use of plant growth regulators (PGRs) in rice varieties (IR-64 and Huanghuazhan) under both high day and night temperatures affects the plant morphology and decrease the leaf area which has final effects on the yield characteristics	Fahad et al. (2016)
Temperature trends (1989–2015) indicate that increasing maximum temperatures will have a negative impact on large scale crops, especially wheat	Ali et al. (2017)
The issue of climate variability has become a key issue that has caused food insecurity	Ullah (2017)
Self- reports based on household level surveys in 5 districts of Punjab (2014) indicated that 73–85 percent of farmers reported economic losses under new climatic conditions	Naqvi et al. 2017.
The projected rise of the temperatures by 1degC will have severe impacts on the productivity of rice in Sindh province	Joyo et al. (2018)
Embracing of climate-sensitive agricultural methods has been cited to be a viable plan of curbing the risks of climatic changes on cotton production	Imran et al. (2018)
The comparison of the two simulation models of the baseline (1980–2010) and future show the possibility of 7.3% decline of the rice yields in Punjab because of the increasing temperature level	Shabbir et al. (2020)
All major staple crops in Pakistan among them are wheat, rice, cotton, sugarcane and maize, which are susceptible to climatic variability and extremes	Syed et al. (2021)
Statistical (1981–2017) analyses indicate that the variability of rainfall has adverse impacts on establishment of rice especially by decreasing the number of plants during the transplantation of rice	Abbas and Mayo (2021)
The reduction of yield in rice due to climatic conditions in Punjab has caused economic losses of 222.30 thousand metric tons which would translate to about 27.79 billion PKR in one season	Janjua et al. (2021)
The effects of climate change are not limited to the livestock value chain since it affects production, processing, and consumption of livestock worldwide	Godde et al. (2021)
Farming systems using rainfed in Pakistan are also highly susceptible to climate variability, and related communities of farmers are at high risk	Mahmood et al. (2021)

21.15 Future Proposal of Work and Recommendations

Both the anthropogenic and natural driver actions are increasing the rate at which biodiversity is declining and food security is under threat. Planning is therefore necessary to reduce disturbances of the ecological system and preserve wildlife. Creation and proper management of the protected areas can lead to a substantial increase in the stability and resiliency of the ecosystems. Meanwhile, the community-based education and awareness programs are to be reinforced to motivate the population to engage in the conservation activities. Academic and institutional levels should develop short-, medium-, and long-term research strategies that are sponsored by both the national and international funding mechanisms. Long-term research projects can help in the attainment of quantifiable conservation and productivity objectives. It is especially critical to strengthen the collaboration between the developed and developing nations in conducting research to share the solutions to climate-related issues. Knowledge exchange, technical collaboration, and financial assistance are available through international forums like the Paris Agreement, the UNFCCC, and subsequent global climate forums (i.e., COP meetings).

Periodic ecological evaluations with the use of periodic surveys and biomonitoring of the disturbed habitats are necessary in order to produce baseline information and track the changes of the environment. Degraded ecosystems are advised to be restored scientifically to recover their ecological functions (Halle 2007; Vaughn et al. 2010). These strategies are aiming at speeding up the restoration of the ecosystems that have been affected by structural and functional disturbances.

There is a need to carry out more research to overcome climate-induced food production limitations. Among the priority areas are the breeding of stress- and metal-tolerant crops, implementation of climate-sensitive agriculture, and incorporation of new technologies in farm systems like nanotechnology in farming, which are sustainable. The recent research also highlights the need to focus on the need to combat food insecurity, pollution, and malnutrition using integrated solutions (Khan et al. 2024). The management of the biodiversity under legal and policy frameworks, especially in the protected areas, also needs to be strengthened (Daran et al. 2023).

Other interventions involve incorporation of agricultural insurance programs, water conservation, degraded land restoration, forest management, and implementation of sustainable land-use activities to alleviate heat stress. More recommendations should be adopted based on scholarly studies, conferences, and policy discussions to deal with environmental degradation.

21.16 Conclusion

Climate change is a multi-layered and broad issue that needs joint planning and management to attain the low-carbon developmental directions. Besides minimization of greenhouse gas emissions, it is important to control human activities that lead to environmental degradation, loss of biodiversity and low agricultural productivity. The solution to these problems requires a multi-sectoral solution involving governmental bodies, the scientific community, and civil society. Effective implementation requires strengthening of the governance systems, especially environmental laws and urban infrastructure. More control and surveillance of industrial and municipal waste, both in rural and urban regions, should be done to limit environmental pollution. Some of the policy interventions that can help in mitigating the environmental effects include encouraging the use of electric vehicles, imposing industrial emission regulations, and enhancing the effectiveness of the public health check systems. Moreover, the uncontrolled growth of urban areas, a high concentration of agrochemicals, soil erosion, salinization, and unpredictable climatic changes are still endangering the biodiversity and food production systems. Environmental management needs a great deal of determination, policy implementation and cooperation. Through the incorporation of conservation measures, agriculture that is resistant to climate changes, and good governance, one can reduce the subsequent destruction of the environment and guarantee ecological and food security in the long term.

References

Abbas S, Mayo ZA (2021) Impact of temperature and rainfall on rice production in Punjab, Pakistan. Environ Dev Sustain 23:1706–1728. https://doi.org/10.1007/s10668-020-00647-8

Abdela N, Jilo K (2016) Impact of climate change on livestock health: a review. Glob Vet 16(5): 419–424

Abrar MA, Maryiam M (2023) Climate change impact on food security in Pakistan. Pak J Multidisc Res 4(1):131–146

Adamolekun G (2024) Firm biodiversity risk, climate vulnerabilities, and bankruptcy risk. J Int Financial Markets Instit Money 97:102075. https://doi.org/10.1016/j.intfin.2024.102075

Aggarwal P, Sivakumar MV (2011) Global climate change and food security in South Asia: an adaptation and mitigation framework. In: Climate change and food security in South Asia. Springer, pp 253–275

Agwan AR (1999) Islam and the issue of biodiversity. J Islamic Sci 15(1–2):9–24

Ahmad M, Farooq U (2010) The state of food security in Pakistan: future challenges and coping strategies. Pak Dev Rev 49(4):903–923

Ahmad SS, Abbasi Q, Jabeen R, Shah MT (2012) Decline of conifer forest cover in Pakistan: a GIS approach. Pak J Bot 44(2):511–514

Ahmed K, Shahbaz M, Qasim A, Long W (2015) The linkages between deforestation, energy and growth for environmental degradation in Pakistan. Ecol Indic 49(1):95–103

Ahmed N, Rana MI (1999) Valuation of non-market resources of Arabian Sea: a case study of aquarium products. An evidence of drift woods. In: Seminar on aquatic biodiversity of Pakistan, 20–22 March 1999. p 16. B.C.C. & T. Press, University of Karachi, Pakistan; p 100

Ali A, Erenstein O (2016) Assessing farmer use of climate change adaptation practices and impacts on food security and poverty in Pakistan. Clim Risk Manag 16:183–194. https://doi.org/10.1016/j.crm.2016.12.001

Ali S, Shah SZ, Khan MS, Khan WM, Khan Z, Hassan N, Zeb U (2019) Floristic list, ecological features and biological spectrum of district Nowshera, Khyber Pakhtunkhwa, Pakistan. Acta Ecol Sin 39(2):133–141. https://doi.org/10.1016/j.chnaes.2018.08.007

Ali S, Liu Y, Ishaq M, Shah T, Abdullah, Ilyas A, Din IU (2017) Climate change and its impact on the yield of major food crops: evidence from Pakistan. Foods 6(6):39. https://doi.org/10.3390/foods6060039

Ali S, Saeed A, Kiani RS, Muhammad S, Khan F, Babar R, Khan A, Iqbal MS, Goheer MA (2021) Future climatic changes, extreme events, related uncertainties, and policy recommendations in the Hindu Kush sub-regions of Pakistan. Theor Appl Climatol 143(1):193–209. https://doi.org/10.1007/s00704-020-03399-7

Ali S (2000) Impact of environmental degradation on biodiversity. Proc Pak Acad Sci 37(1):93–97

Ali SI (1978) The Flora of Pakistan: some general and analytical remarks. Notes R Bot Garden Edinburgh 36:427–439

Amjad MS, Arshad M, Chaudhari SK (2014) Structural diversity, its components and regenerating capacity of lesser Himalayan forests vegetation of Nikyal valley District Kotli (A.K), Pakistan. Asian Pac J Trop Med 7:S454–S460

Anderson JE, Inouye RS (2001) Landscape scale changes in plant species abundance and biodiversity of a sagebrush steppe over 45 years. Ecol Monogr 71:531–556

Andrade BO, Koch C, Boldrini II, Velez-Martin E, Hasenack H, Hermann JM, Kollmann J, Pillar VD, Overbeck GE (2015) Grassland degradation and restoration: a conceptual framework of stages and thresholds illustrated by southern Brazilian grasslands. Nat Conserv 13:95–104. https://doi.org/10.1016/j.ncon.2015.08.002

Ajani A, van der Geest K (2021) Climate change in rural Pakistan: evidence and experiences from a people centered perspective. Sustain Sci 16:1999–2011

Anjum SA, Wang LC, Xue L, Saleem MF, Wang GX, Zou CM (2010) Desertification in Pakistan: causes, impacts and management. J Food Agric Environ 8(1):1203–1208

Anum T, Nazia T, Khuda B, Ashfaq M, Hassan S (2014) Food security in the context of climate change in Pakistan. Pak J Commer Soc Sci 8(2):540–550. https://hdl.handle.net/10419/188154S

Anwar M, Rais M, Baig MB, Behnassi M (2022) Impacts of climate change on biodiversity in Pakistan: current challenges and policy recommendations. In: Behnassi M, Gupta H, Barjees Baig M, Noorka IR (eds) The food security, biodiversity, and climate nexus. Springer, Cham. https://doi.org/10.1007/978-3-031-12586-7_6

Arshad S, Shafqat A (2012) Food security indicators, distribution and techniques for agriculture sustainability in Pakistan. Int J Appl Sci Technol 2(5):137–147

Ashfaq M, Zulfiqar F, Sarwar I, Quddus MA, Baig IA (2011) Impact of climate change on wheat productivity in mixed cropping system of Punjab. Soil Environ 30(2):110–114

Ashraf W, Rehman A, Rabbani M, Shaukat W, Wang J (2023) Aflatoxins posing threat to food safety and security in Pakistan: call for a one health approach. Food Chem Toxicol 180:114006. https://doi.org/10.1016/j.fct.2023.114006

Aslam M (2018) Flood management current state, challenges and prospects in Pakistan: a review. Mehran Univ Res J Eng Technol 37(2):297–314

Ateeq-Ur-Rehman M, Siddiqui BN, Hashmi N, Masud K, Adeel M, Khan MRA (2018) Climate change impact on rural livelihoods of small landholder: a case of Rajanpur, Pakistan. Int J Appl Agric Sci 4(2):28. https://doi.org/10.11648/j.ijaas.20180402.11

Avon C, Berges L, Dumas Y, Dupouey JL (2010) Does the effect of forest roads extend a few meters or more into the adjacent forest? A study on understory plant diversity in managed oak stands. For Ecol Manage 259:1546–1555. https://doi.org/10.1016/j.foreco.2010.01.031

Awan AG, Yaseen G (2017) Global climate change and its impact on agriculture sector in Pakistan. Am J Trade Pol 4:41–48. https://doi.org/10.18034/ajtp.v4i3.425

Azam MM, Hasan (1999) Aquatic biodiversity of Taunsa barrage, wildlife sanctuary. In: Seminar on aquatic biodiversity of Pakistan, 20–22 March 1999. p 18. B.C.C. & T. Press, University of Karachi, Karachi, Pakistan, p 100

Baig MB, Al-Subaiee FS (2009) Biodiversity in Pakistan: key issues. Biodiversity 10(4):20–29. https://doi.org/10.1080/14888386.2009.9712858

Bao G, Suetsugu K, Wang H, Yao X, Liu L, Ou J, Li C (2015) Effects of the hemiparasitic plant, *Pedicularis kansuensis*, on plant community structure in a degraded grassland. Ecol Res 30: 507–515

Bennett A (1990) Introduction. In: Biological diversity and developing countries, issues and options. ODA, London, pp 6–10

Bhutta ZA, Bhutta SZ, Raza S, Sheikh AT (2022) Addressing the human costs and consequences of the Pakistan flood disaster. Lancet 400(10360):1287–1289. https://doi.org/10.1016/S0140-6736(22)01874-8

Burca JD (2023) Constructive voices. Biodiversity. December 11. https://constructive-voices.com/pakistan-biodiversity/. Visited on 22/11/2024

CCPI (2025) Climate change performance index (CCPI). Available from https://ccpi.org/ country/pak/. Visited on 21-07-2025

Ceballos G, Pacheco J, List R (1999) Influence of prairie dogs (*Cynomys ludovicianus*) on habitat heterogeneity and mammalian diversity in Mexico. J Arid Environ 41(2):61–172. https://doi.org/10.1006/jare.1998.0479

Chaudhary QZ, Rasul G (2004) Agro-climatic classification of Pakistan. Science Vision, Vol. 9 No. 1–2 (Jul-Dec 2003) & No. 3–4 (Jan-Jun 2004), 59–66)

Chettri N, Adhikari B, Chaudhary S, Wangchuk K (2023) Changing discourses in the third pole: a systematic review of climate change impact on biodiversity in the Hindu Kush Himalaya. Ecol Indic 155:111046. https://doi.org/10.1016/j.ecolind.2023.111046

Chien F, Ajaz T, Andlib Z, Chau KY, Ahmad P, Sharif A (2021) The role of technology innovation, renewable energy and globalization in reducing environmental degradation in Pakistan: a step towards sustainable environment. Renew Energy 177:308–317. https://doi.org/10.1016/j.renene.2021.05.101

Conservation On Biological Diversity (COBD) (2024) Pakistan country profile. Biodiversity Facts. https://www.cbd.int/countries/profile?country=pk. Visited on 22-11-2024

Cullingham C, Stephens T, Swan K, Wilson S, Janes J, Matchett M, Griebel R, Moehrenschlager A (2023) Genetic analysis reveals hidden threats and new motivation for conservation translocation of black-tailed prairie dogs at the northern limit of their range. Glob Ecol Conserv 46: e02591. https://doi.org/10.1016/j.gecco.2023.e02591

Daran ED, Ajlani RA, Zia-ud-Din M, Elhajraoui FE (2023) Management of biodiversity in Pakistan protected areas and its legal implications. J Wildlife Biodiv 7(supplementary issue):1–13. https://doi.org/10.5281/zenodo.10014389

Davies KW, Boyd CS, Beck JL, Bates JD, Svejcar TJ, Gregg MA (2011) Saving the sagebrush sea: an ecosystem conservation plan for big sagebrush plant communities. Biol Conserv 144:2573–2584

de Freitas TC, Guarino EDG, Gomes GC, Molina AR, da Luz Real IM, Beltrame R (2020) The effect of seed ingestion by a native, generalist bird on the germination of worldwide potentially invasive trees species *Pittosporum undulatum* and *Schinus terebinthifolia*. Acta Oecol 108: 103639. https://doi.org/10.1016/j.actao.2020.103639

Devi S (2022) Pakistan floods: impact on food security and health systems and its cost. Lancet 400(10355):799–800. https://doi.org/10.1016/S0140-6736(22)01732-9

Dhar A, Comeau PG, Vassov R (2019) Effects of cover soil stockpiling on plant community development following reclamation of oil sands sites in Alberta. Restor Ecol 27:352–360. https://doi.org/10.1111/rec.12858

Douglas MR, Anthonysamy WJ, Mussmann SM, Davis MA, Louis W, Douglas ME (2020) Multi-targeted management of upland game birds at the agroecosystem interface in midwestern North America. PLOS ONE *15*(4):e0230735. https://doi.org/10.1371/journal.pone.0230735

Ebrahimi E, Ahmadzadeh F, Abdoli A, Araújo MB, Naimi B (2025) Refined gap analysis for biodiversity conservation under climate change. Biol Conserv 305:111054. https://doi.org/10.1016/j.biocon.2025.111054

Elsinger ME, Dhar A, Naeth MA (2022) Recovery of plains rough fescue grasslands on reclaimed well sites. J Nat Conserv 66:126122. https://doi.org/10.1016/j.jnc.2021.126122

Flora of Pakistan (F.O.P) (2016) Pakistan plant data base in "Flora of Pakistan". http://www.tropicos.org/Project/Pakistan

Fahad S, Hussain S, Saud S, Hassan S, Ihsan Z, Shah AN, Wu C, Yousaf M, Nasim W (2016) Exogenously applied plant growth regulators enhance the morphophysiological growth and yield of rice under high temperature. Front Plant Sci. https://doi.org/10.3389/fpls.2016.012500

Farooq M, Hussain M, Saqib Z, Khan KR, Shah AH, Shah GM, Gul S, Anjum W, Sthanadar AA (2019) Spatial distribution and correlation of important edaphic and climatic gradients in Tanawal area of Himalayas, Pakistan. Acta Ecol Sin 39(4):267–272. https://doi.org/10.1016/j.chnaes.2018.09.014

Farrukh MU, Bashir MK, Hassan S, Adil SA, Kragt ME (2020) Mapping the food security studies in India, Pakistan and Bangladesh: review of research priorities and gaps. Glob Food Sec 26: 100370. https://doi.org/10.1016/j.gfs.2020.100370

Gann GD, McDonald T, Walder B, Aronson J, Nelson CR, Jonson J, Hallett JG, Eisenberg C, Guariguata MR, Liu J, Hua F, Echeverría C, Gonzales E, Shaw N, Decleer K, Dixon KW (2019) International principles and standards for the practice of ecological restoration. 2nd ed. Restor Ecol 27(S1):–S46. https://doi.org/10.1111/rec.13035

Godde C, Mason-D'Croz D, Mayberry D, Thornton P, Herrero M (2021) Impacts of climate change on the livestock food supply chain; a review of the evidence. Glob Food Secur 28:100488. https://doi.org/10.1016/j.gfs.2020.100488

Grand J, Wilsey C, Wu JX, Michel NL (2019) The future of north American grassland birds: incorporating persistent and emergent threats into full annual cycle conservation priorities. Conserv Sci Pract 1(4):e20. https://doi.org/10.1111/csp2.20

Grant TA, Shaffer TL, Flanders B (2020) Resiliency of native prairies to invasion by Kentucky bluegrass, smooth brome, and woody vegetation. Rangeland Ecol Manage 73:321–328. https://doi.org/10.1016/j.rama.2019.10.013

Halle S (2007) Present state and future perspectives of restoration ecology—Introduction. Restor Ecol 15(2):304–306. https://doi.org/10.1111/j.1526-100X.2007.00213.x

Hani UE, Haq SM, Shabbir R, Waheed M, Jabeen A, Hussain M, Najar ZH, Alhimaidi AR, Amran RA, Bussmann RW (2024) Geospatial assessment of climate and human pressure on snow leopard habitat in the Trans-Himalayan region of Pakistan. Glob Ecol Conserv 53:e03024. https://doi.org/10.1016/j.gecco.2024.e03024

Haq FJALS (2012) The critically endangered flora and fauna of district. Battagram Pakistan. Adv Life Sci 2:118–123

Harmain M (2024) Explorer Pakistan. Endangered species in Pakistan: preserving our wildlife. https://explorepk.net/endangered-species-in-pakistan/. Visited on 22/11/2024

Hussain SS, Mudasser M (2007) Prospects for wheat production under changing climate in mountain areas of Pakistan–an econometric analysis. Agric Syst 94:494–501

Ijaiz F, Iqbal J, Alam R, Kausar IU, Rahman A, Afzal N, Ali M, Shah AM, Shah I, Ahmed, Calixto ES (2018) Decline in timber and fuel wood species diversity driven by anthropogenic disturbance: a multivariate approach in Sarban Hills (Abbottabad), Pakistan. Ukarine J Ecol 8:11–16

Imran M, Ali A, Ashfaq M, Hassan S, Culas R, Ma C (2018) Impact of Climate Smart Agriculture (CSA) practices on cotton production and livelihood of farmers in Punjab, Pakistan. Sustainability 10(6):2101

Imran N (2025) From hegemony to nature-based solutions: evolution of Pakistan's forestry sector. Trees For People 20:100892. https://doi.org/10.1016/j.tfp.2025.100892

Iqbal M, Shaukat SS, Kazmi A (1999) Diversity of fish communities in Pakistan's coastal waters (Northern Arabian Sea). Abs. 14. Pp 28. Seminar on Aquatic Biodiversity of Pakistan. B.C.C. & T. Press, University of Karachi, p 100

Iqbal M, Khan SM, Khan MA, Ahmad Z, Ahmad H (2018a) A novel approach to phytosociological classification of weeds flora of an agro-ecological system through cluster, two way cluster and indicator species analyses. Ecol Indic 84:590–606

Iqbal MZ, Shafiq M, Kabir M, Farooqi ZR (2018b) A review: impact of anthropogenic activities on biodiversity in Pakistan: a review. Biosci Res 15(1):481–489

International The News (ITN) (2022, February 15th) 50 exotic animal species face extinction in Pakistan. https://www.thenews.com.pk/print/933700-50-exotic-animal-species-face-extinction-in-pakistan. Visited on 22/11/2024

Jaffery R (2018) The diplomat. The pulse. Pakistan's biodiversity is disappearing, but no one seems to notice. https://thediplomat.com/2018/12/pakistans-biodiversity-is-disappearing-but-no-one-seems-to-notice/. Visited on 22-11-2024

Jaksié FM, Jiménez JE, Castro SA, Feinsinger P (1992) Numerical and functional response of predators to a long term decline in mammalian prey at a semi arid Neotropical site. Oecologia 89:90–101. pmid:28313400

Janjua AA, Aslam M, Sultana N, Batool Z (2021) Identification of climate induced optimal rice yield and vulnerable districts rankings of the Punjab, Pakistan. Sci Rep 11(1):1–15. https://doi.org/10.1038/s41598-021-02691-4

Janjua PZ, Samad G, Khan NU, Nasir M (2010) Impact of climate change on wheat production: a case study of Pakistan [with comments]. Pak Dev Rev 49:799–822

Jetz W, Wilcove DS, Dobson AP (2007) Projected impacts of climate and land-use change on the global diversity of birds. PLoS Biol 5:1211–1219

Joyo M, Ram N, Magsi H (2018) Risk assessment of climate variability on rice productivity in Sindh province of Pakistan. Pak J Agric Agric Eng Vet Sci 34(1):68–77

Khan A, Murad W, Ali S, Shah SS, Halim SA, Khalid A, Kashtoh H, Khan A, Al-Harrasi A (2024) Contribution of mushroom farming to mitigating food scarcity: current status, challenges and potential future prospects in Pakistan. Heliyon:e40362. https://doi.org/10.1016/j.heliyon.2024.e40362

Khan A, Wang C (2024) Exploring the consequence of ecological and agronomic determinants on wheat production instabilities in Khyber Pakhtunkhwa, Pakistan: perspectives from dynamic autoregressive distributed lag analysis. Agric Water Manag 300:108889. https://doi.org/10.1016/j.agwat.2024.108889

Khan AM, Qureshi R, Saqib Z (2019a) Multivariate analyses of the vegetation of the western Himalayan forests of Muzaffarabad district, Azad Jammu and Kashmir, Pakistan. Ecol Indicat 104:723–736. https://doi.org/10.1016/j.ecolind.2019.05.048

Khan KR, Iqbal Z, Alam J, Farooq M, Shah AH, Hussain M, Shah GM, Ali D (2019b) Comparative study on floristic diversity of protected and unprotected Forests of Sathan Gali, District Mansehra, KP, Pakistan. Acta Ecol Sin 39(4):273–279. https://doi.org/10.1016/j.chnaes.2018.09.016

Khan SR (2001) Economic policies and measures in biodiversity conservation: implementation status. https://www.jstor.org/stable/resrep00638.11

Knapp PA (1996) Cheatgrass (*Bromus tectorum* L) dominance in the Great Basin Desert. Glob Environ Chang 6:37–52

Knopf FL (1994) Avian assemblages on altered grasslands. Stud Avian Biol 15:247–257

Lashari AH, Li W, Hassan M et al (2020) Biodiversity conservation in China–Pakistan Economic Corridor region with strategic environmental assessment. Environ Sci Pollut Res 27:33500–33502. https://doi.org/10.1007/s11356-020-09955-y

Leone JB, Pennarola NP, Larson JL, Oberhauser K, Larson DL (2022) Divergent responses of butterflies and bees to burning and grazing management in tallgrass prairies. Ecol Evol 12(12): e9532. https://doi.org/10.1002/ece3.9532

Lin B, Ullah S (2024) Evaluating forest depletion and structural change effects on environmental sustainability in Pakistan: through the lens of the load capacity factor. J Environ Manag 353: 120174

Lodhi S, Ayyubi MS, Hayat S, Iqbal Z (2024) Unravelling the effects of climate change on agriculture of Pakistan: an exploratory analysis. Qlantic J Soc Sci 5(2):142–158

Luck GW (2007) A review of the relationships between human population density and biodiversity. Biol Rev 82:607–645. https://doi.org/10.1111/j.1469-185X.2007.00028.x

Lundgren K, Kuklane K, Gao C, Holmér I (2013) Effects of heat stress on working populations when facing climate change. Ind Health 51:3–15

Mahmood N, Ahmad B, Hassan S, Bakhsh K (2012) Impact of temperature and precipitation on rice productivity in rice-wheat cropping system of Punjab province. J Anim Plant Sci 22(4):993–997

Mahmood N, Arshad M, Mehmood Y, Shahzad MF, Kächele H (2021) Farmers' perceptions and role of institutional arrangements in climate change adaptation: insights from rainfed Pakistan. Climate Risk Manage 32. https://doi.org/10.1016/j.crm.2021.100288

Malik S (2022) Climate change and food insecurity in Pakistan. South Asian Voices. 10th August, 2022. https://southasianvoices.org/climate-change-and-food-insecurity-in-pakistan/. Visited on 22-11-2024

Molina-Venegas R, Rodríguez MÁ, Mabberley DJ (2021) A global database of plant services for humankind. PLoS One 16(6):e0253069. https://doi.org/10.1371/journal.pone.0253069

Muluneh MG (2021) Impact of climate change on biodiversity and food security: a global perspective—a review article. Agric Food Secur 10:36. https://doi.org/10.1186/s40066-021-00318-5

Munir M, Munir AR, Khalid T (2024) Biodiversity and challenges of honey bee population in Pakistan. Sci Lett 12(1):27–42. https://doi.org/10.47262/SL/12.1.132023950

Naeth MA, Locky AD, Wilkinson SR, Nannt MN, Bryks CL, Low CH (2020) Pipeline impacts and recovery of dry mixed grass prairie soil and plant communities. Rangeland Ecol Manage 73: 619–628. https://doi.org/10.1016/j.rama.2020.06.003

Nafeesa Z, Haq SM, Bashir F, Gaus G, Mazher AM, Rasool A, Rashid N (2021) Observations on the floristic, life-form, leaf-size spectra and habitat diversity of vegetation in the Bhimber hills of Kashmir Himalayas. Acta Ecol Sin 41(3):228–234. https://doi.org/10.1016/j.chnaes.2021.03.0

Naqvi SAA, Ashfaq M, Adil SA, Ahmed A (2017) Current agricultural production system of Punjab is vulnerable to climate change: impact assessment. J Agric Res 55(1):125–135

Nasrullah M, Liang L, Rizwanullah M (2022) Estimating the efficiency gap of maize yield across different irrigation method in Pakistan. Irrig Drain:1–16. https://doi.org/10.1002/ird.2750

Naz S, Iqbal Z, Begum S (2024) Climate change's impact on food security in Pakistan: challenges and mitigation strategies. Pak Soc Sci Rev 8(2):426–438. https://doi.org/10.35484/pssr.2024(8-II-S)37

Ncongwane KP, Botai JO, Sivakumar V, Botai CM (2021) A literature review of the impacts of heat stress on human health across Africa. Sustainability 13(9):5312. https://doi.org/10.3390/su13095312

PNBSAP (2017) Pakistan National Biodiversity Strategy and Action Plan for achieving Aichi Biodiversity Targets and Sustainable Development Goals Government of Pakistan 2017–2030, p 111

Rasul G, Afzal M, Zahid M, Bukhari SA (2012, December) Climate change in Pakistan. Focused on Sindh. Pakistan Meteorological Department P.O. Box No. 1214, Sector H-8/2 Islamabad-Pakistan. Technical Report No. PMD-25/2012, p 55

Rizwanullah M, Yang A, Nasrullah M, Zhou X, Rahim A (2023) Resilience in maize production for food security: evaluating the role of climate-related abiotic stress in Pakistan. Heliyon 9(11): e22140. https://doi.org/10.1016/j.heliyon.2023.e22140

Ryall KL, Fahrig L (2006) Response of predators to loss and fragmentation of prey habitat: a review of theory. Ecology 87:1086–1093. pmid: 16761585

Saadat HB (2012) Climate change and biodiversity (A Pakistan case study): impact of climate change on biodiversity of butterflies in Northern coniferous forests of Punjab, Pakistan Paperback – November 19. LAP Academic Publishing. p 112

Sayin L, Peters T (2025) Briefing: how do we adapt to the new hot reality?. Proceedings of the Institution of Civil Engineers - Civil Engineering. https://doi.org/10.1680/jcien.25.00080

SDG (2023) The Sustainable Development Goals Report 2023: Special Edition. Available from https://unstats.un.org/sdgs/report/2023/?_gl=1*5wxoqh*_ga*MTkxMTA1 Mzkx NS4 xNzUyNzIwMTkw*_ga_TK9BQL5X7Z*czE3NTI3MjMwMjgkbzIkZzAkd DE3 NT I3 MjMwMjgkajYwJGwwJGgw. Accessed on 17-07-2025

Segala FV, Guido G, Stroffolini G, Masini L, Cattaneo P, Moro L, Motta L, Gobbi F, Nicastri E, Vita S, Iatta R, Otranto D, Locantore P, Occa E, Putoto G, Saracino A, Di Gennaro F (2025) Insights into the ecological and climate crisis: emerging infections threatening human health. Acta Trop 262:107531. https://doi.org/10.1016/j.actatropica.2025.107531

Sellers S. Gender and climate change: a closer look at existing evidence; Global gender and climate alliance: New York, 2016

Shabbir G, Khaliq T, Ahmad A, Saqib M (2020) Assessing the climate change impacts and adaptation strategies for rice production in Punjab, Pakistan. Environ Sci Pollut Res 27(18): 22568–22578

Shafiq M, Iqbal MZ, Habib I (1992) Phytosociological studies around the industrial areas of Landhi, Karachi. New Agric 3(2):179–188

Shafiq M, Iqbal MZ (1988) Phytosociological studies around the industrial areas of Korangi, Karachi. Pak J Sci Ind Res 31(8):569–573

Shafiq M, Mehboob S, Iqbal MZ, Kabir M, Farooqi ZR (2018) Influence of lead on seedling growth of wheat (*Triticum aestivum* L.). Civ Environ Res 10(7):42–48. https://doi.org/10.7176/CER

Shakoor U, Saboor A, Ali I, Mohsin AQ (2011) Impact of climate change on agriculture. Empirical evidence from arid region. Pak J Agric Sci 48(4):327–333

Shamsulhaq M (2002, Sunday, Jan. 20) Environmental damage reduces bird's migration. The daily Dawn, Karachi, p 5

Shinwari MI, Khan MA (2000) Folk use of medicinal herbs of Margalla Hills National Park, Islamabad. J Ethnopharmacol 69(1):45–56. https://doi.org/10.1016/S0378-8741(99)00135-X

Shuaib M, Ahmed S, Ali K, Ilyas M, Hussain F, Urooj Z, Shah SS, Kumar T, Shah M, Khan I, Hussain F (2019) Ethnobotanical and ecological assessment of plant resources at District Dir, Tehsil Timergara, Khyber Pakhtunkhwa, Pakistan. Acta Ecologica Sinica 39(1):109–115. https://doi.org/10.1016/j.chnaes.2018.04.006

Siddique MT, García Molinos J (2024) Risk from future climate change to Pakistan's protected area network: a composite analysis for hotspot identification. Sci Total Environ 916:169948. https://doi.org/10.1016/j.scitotenv.2024.169948

Siddiqui PJ, Farooq S, Shafique S, Burhan Z, Farooqi Z (2007) Conservation and management of biodiversity in Pakistan through the establishment of marine protected areas. Ocean Coast Manag 51(5):377–382. https://doi.org/10.1016/j.ocecoaman.2008.01.006

Smith SJ, Williams JR, Menzel RG, Coleman GA (1984) Prediction of sediment yield from southern plains grasslands with the modified universal soil loss equation. J Range Manag 37: 95–297

Studocu (2024) Conservations issues and challenges in Pakistan. https://www.studocu.com/row/document/comsats-university-islamabad/introduction-to-biosciences/conservation-of-biodiversity-in-pakistan/58608061. Visited on 22-11-2024

Sultana H, Ali N (2006) Vulnerability of wheat production in different climatic zones of Pakistan under climate change scenarios using CSM-CERES-Wheat Model. In: Second International Young Scientists' Global Change Conference, Beijing, pp 7–9

Syed A, Raza T, Bhatti TT, Eash NS (2021) Climate impacts on the agricultural sector of Pakistan: risks and solutions. Environ Chall 6:100433. 10.1016/j.envc.2021.100433

Ullah S (2017) Climate change impact on agriculture of Pakistan- a leading agent to food security. Int J Environ Sci Nat Resour 6(3):76–79

United Nations (UN) (2025) Sustainable development goals. The 17 goals. Available from: https://www.un.org/sustainabledevelopment/goal-of-the-month/. Accessed on 17-07-2025

UNESCO (2016) UNESCO world heritage convention. Deosai National Park. https://whc.unesco.org/en/tentativelists/6115/. Visited on 22-11-2024. Date of Submission:12/04/2016, Criteria: (ix)(x), Category: Natural, Submitted by: Government of Pakistan, Director General of Archeology, State, Province or Region:Gilgit - Baltistan – Coordinates: N35.018930 E 75.414393, Ref.: 6115

URAAN (2025) URAAN Pakistan book. Climate change. Available from https://uraanpakistan.pk/climate-change/. Visited on 25-07-2025

Van Schalkwyk J, Pryke JS, Samways MJ (2019) Contribution of common vs. rare species to species diversity patterns in conservation corridors. Ecol Indic 104:279–288. https://doi.org/10.1016/j.ecolind.2019.05.014

Vaness BM, Wilson SD (2007) Impact and management of crested wheatgrass (*Agropyron cristatum*) in the northern Great Plains. Can J Plant Sci 87:1023–1028

VanMeter KJ, Basu NB (2015) Signatures of human impact: size distributions and spatial organization of wetlands in the prairie pothole landscape. Ecol Appl 25:451–465

Vaughn KJ, Porensky LM, Wilkerson ML, Balachowski J, Peffer E, Riginos C, Young TP (2010) Restoration ecology. Nat Educ Knowl 3(10):66

Videnović Ž, Dumanović Ž, Simić M, Srdić J, Babić M, Dragičević V (2013) Genetic potential and maize production in Serbia. Genetika 45(3):667–677. https://doi.org/10.2298/GENSR1303667V

Wang Y, Zhang X, Tian C, Guo X, Shu Q, Gu H, Feng P, Li F, Li B (2023) Effects of different diets on thermal tolerance in the silkworm, Bombyx mori. Entomologia Experimentalis et Applicata 172(1):27–34. https://doi.org/10.1111/eea.13382

WBG and ADB. World Bank Group (2021) Climate risk country profile: Pakistan. The World Bank Group and the Asian Development Bank, p 32

Wheeler T, Von Braun J (2013) Climate change impacts on global food security. Science 341(6145):508–513

White PJ, Vanderbilt White CA, Ralls K (1996) Functional and numerical responses of kit foxes to a short term decline in mammalian prey. J Mammal 77:370–376

William G, Saqib Z, Naeem N, Kamran A, Masih A, Rafique A (2024) Predicting climate driven habitat shifts for the Egyptian vulture in Punjab, Pakistan. J Nat Conserv 83:126774. https://doi.org/10.1016/j.jnc.2024.126774

With KA, King AW, Jensen WE (2008) Remaining large grasslands may not be sufficient to prevent grassland bird declines. Biol Conserv 141:3152–3167

Wildlife of Pakistan (WOP) (2025) Mountain Weasel. Available from https://wildlife.com.pk/endangered/mountain-weasel/. Visited on 20-07-2025

Younger S, Parry D, Meigh D (2022) Preparing for the future: the impact of climate change on the civil engineering profession. Proc Instit Civil Eng 175(2):87–93. https://doi.org/10.1680/jcien.21.00174

Zahid M, Rasul G (2011) Changing trends of thermal extremes in Pakistan. Clim Chang. https://doi.org/10.1007/s10584-011-0390-4

Zahid R (2023) Assessing Pakistan's efforts in protecting the environment. https://iips.com.pk/assessing-pakistans-efforts-in-protecting-the-environment/. Visited on 22-11-2024

Zhu T, Ringler C, Iqbal MM, Sulser TB, Goheer MA (2013) Climate change impacts and adaptation options for water and food in Pakistan: scenario analysis using an integrated global water and food projections model. Water Int 38(5):651–669. https://doi.org/10.1080/02508060.2013.830682

Zia A (2022) Threats to biodiversity in Pakistan. ENews Pakistan. https://earthjournalism.net/stories/threats-to-biodiversity-in-pakistan. Visited on 18-11-2024

Integrating Biodiversity into Agri-Food Systems: A Transition Roadmap for Sustainable Agriculture in Morocco

22

Moha Haddouch and Rahmatullah Qureshi

Abstract

The agri-food systems of Morocco have been the basis of national food security, rural living, and trade competitiveness, but they are becoming limited by biodiversity loss, climate change, water scarcity, and new sustainability demands in the international market. This chapter outlines a policy-focused transition roadmap to mainstream biodiversity into agri-food systems of Morocco with specific reference to the modern irrigated export agriculture and the traditional agroecosystems with high biodiversity. The analysis is based on the Kunming-Montreal Global Biodiversity Framework (KMGBF 2020–2030) and is aligned to the main national strategies, such as Generation green 2020–2030, Natural biodiversity Strategy and Action Plan (NBSAP), the National Climate Plan, and the Strategy of the forests of Morocco 2020–2030. The analysis is grounded in a review of documents, consultation with stakeholders, surveys, SWOT, and risk analysis. Empirical data on the Souss-Massa region can be used to depict divergent directions toward export-oriented intensive regimes, as well as the traditional agro-sylvo-pastoral regimes. The suggested roadmap builds three strategic axes: (i) decarbonization and resilience of modern agriculture; (ii) the ability of traditional systems to be sustainable with the help of biodiversity conservation and models of the social and solidarity economy; and (iii) facilitates institutional, regulatory, and financial frameworks. The chapter ends by making policy-relevant recommendations to operationalize biodiversity mainstreaming in agri-food value chains on the regional and national levels.

M. Haddouch
Independent Consultant, SECADERD, Hay Ibn Rochd Wifak Temara, Temara, Morocco

R. Qureshi (✉)
Department of Botany, Faculty of Sciences, Pir Mehr Ali Shah Arid Agriculture University, Rawalpindi, Pakistan
e-mail: rahmatullahq@uaar.edu.pk

R. Qureshi (ed.), *Plant Biodiversity*, https://doi.org/10.1007/978-981-92-1471-6_22

Keywords

Agrobiodiversity · Agroecology · Ecosystem services · Climate adaptation · Morocco · Policy integration

Abbreviations

ABR	Arganeraie Biosphere Reserve
AMPs	Aromatic and Medicinal Plants
CSR	Corporate Social Responsibility
F2F	Farm to fork
KMGBF	Kunming-Montreal Global Biodiversity Framework
NBSAP	National Biodiversity Strategy and Action Plan
NSDS	National Sustainable Development Strategy
NZNP	Net-zero Nature-positive
SIBEs	Sites of Biological and Ecological Interests
SSE	Social and Solidarity Economy

22.1 Introduction

Morocco is the second largest biodiversity hot spot in the Mediterranean region, which has about 31,000 species of plants and animals, which include almost 930 endemic species of plants. The nine million hectares of forest cover in the country was one of the densest forest systems in the region that show remains of species of world significance such as Atlas cedar (*Cedrus atlantica*) and argan (*Argania spinosa*). There are five major ecosystems in Morocco that include forests and steppes, Saharan ecosystems, coastal and marine, wetlands, and cave systems. These natural resources are organized in 17 biodiversity corridors, 10 national parks, 4 biosphere reserves, and 24 Ramsar sites. Almost 70% of national surface water resources are produced in mountainous areas, which are also inhabited by nearly 30% of the population (Haddouch 2015).

The agri-food economy of Morocco is mainly centered on modern-day agricultural perimeters, such as Souss-Massa, Haouz, Doukkala, Tadla, Gharb, Loukkos, Berkane, and Draa-Tafilalet. The available statistics show that food exports in 2018 were about USD 6.1 billion, with exports of fruits and vegetables taking close to 40% of the total (Ministry of Agriculture). The European Union is the main market, which takes approximately 65% of the Moroccan agri-food exports. As a result, European preferences and regulations are becoming more influential in shaping the Moroccan value chains. The recent shifts in the ethical and sustainability-based markets, accelerated during the COVID-19 crisis, stimulated increased environmental and social demands, especially within the context of the EU Green Deal and the Farm-to-Fork Strategy. Within this context, it has become a pressing issue to reverse biodiversity depletion in a changing climate and develop better-functioning adaptive capacity.

The present chapter is thus a proposal to mainstream biodiversity into the Moroccan agri-food systems, as per KMGBF, the Generation Green Strategy, the Forests of Morocco 2020–2030 Strategy, and national adaptation framework. Research conducted by Expertise France in the Souss-Massa area is used to draw empirical evidence and field experience (MTEDD 2023).

22.2 Methodological Approach

The overall aim of the research is to make a workable roadmap of improving biodiversity mainstreaming in agricultural sectors, with a view to accomplishing pertinent KMGBF national targets chiefly by:

(i) the spatial planning tools that are sensitive to biodiversity;
(ii) regulatory and economic tools...
(iii) mechanisms of capacity-building.

The methodology was involved to integrate the sectoral priority evaluation with stakeholders, the high-impact value chains, the alignment with the national strategies and KMGBF objectives, and the systematic participatory process in the development and validation of the roadmap. The need for capacity building was evaluated using surveys, interviews, and focus group discussions where specific modalities and recommendations for training were developed.

22.3 Agri-Food Systems of the Souss-Massa Region: Threats and Opportunities

The study was intended to evaluate the obstacles and facilitating factors to successful implementation of biodiversity in agricultural and agri-food systems in the Souss-Massa Region, in the context of the goals of the Kunming-Montreal Global Biodiversity Framework (KMGBF). The evaluation helped to highlight sector-specific strengths, weaknesses, opportunities, and gaps, which were, in turn, used to develop a strategic roadmap on biodiversity mainstreaming at regional scale.

The methodological framework was a combination of the revised and thorough documentary analysis and a workshop of the participatory inception that took place in Agadir, including institutional actors, professional organizations, agricultural enterprises, women's cooperatives, and representatives of civil societies. These were supplemented by quantitative and qualitative surveys, SWOT analysis, and assessment of sustainability risk with regard to the key agricultural sector groups identified in the region.

22.3.1 Regional Resources and Agricultural Biodiversity

The documentary review also validates that the Souss-Massa Region has significant resources on how to incorporate biodiversity into agricultural and agri-food systems, especially in reference to KMGBF Targets 7, 10, and 14. It covers a very diverse biogeographic terrain, including the Souss Plain, the Anti-Atlas and High Atlas mountain ranges, the Haha plateaus to the north, and the Tata oases to the southeast. Climates are semi-arid to arid, with ecosystems consisting of highly adapted species of flora and fauna and a high degree of endemism characteristic of arid and mountainous regions.

This ecological variability finds reflection in a biodiversity of market economic systems conditioned by the agroecological situation and supported by several-centuries-old traditional knowledge, especially in mountainous and oasis regions. The ecological and farm-based biodiversity and the ecological services they offer to the region, including soil fertility, pollination, water management and climate regulation, are fundamental to the operations of the regional agri-food systems.

To describe this diversity, the study was based on an agroecological zoning approach, which considers representativeness of agricultural production systems and related ecosystem services at the regional level. There were three primary areas of interest:

22.3.1.1 Natural Ecosystem Zones

This zone, which corresponds more or less to the main zones of the Arganeraie Biosphere Reserve (ABR), is concentrated in mountainous regions. These areas enjoy permanent protection, allowing the conservation of natural biodiversity, ecological surveillance, research work and reconstruction of ecosystems, with much of this founded on areas designated as Sites of Biological and Ecological Interest (SIBEs).

22.3.1.2 Traditional Agricultural Areas

It occurs in ABR buffer zones and oasis systems, where agriculture can still be maintained in harmony with ecological sustainability. These regions are defined by agrobiodiversity, stable production frameworks, and local environments that are based on quality-labeled merchandise.

22.3.1.3 Modern Agricultural Zones

This zone is predominantly situated in the plains, where the production systems are based on the choice of, and frequently imported, genetic material and are oriented toward intensive and export-oriented agriculture. These zones emphasize two major production systems: intensive export-oriented agriculture that is concentrated in the Souss and Massa plains and the diversified food producing systems of the foothills, plateaus, mountain and oases. Animal production (more specifically goat breeding) and beekeeping were examined independently because of their different ways of interaction, both positive and negative, with biodiversity.

22.3.2 Ongoing Initiatives and Stakeholder Engagement

The research found circa 15 projects of different sizes carried out or in progress in the Souss-Massa Region that directly or indirectly apply to biodiversity integration in agriculture. These initiatives have great prospects of synergy, knowledge enrichment, and best practice demonstration of biodiversity-friendly agriculture systems through the leadership of institutional actors and civil society organizations.

There was extensive stakeholder involvement. Besides the first-hand interviews with the institutional representatives, professionals, and thematic experts, as well as the focus group discussions on the traditional agricultural systems, the study examined the answers of 34 participants in the form of detailed online questionnaires, which represented the institutional bodies, inter-professional organizations, the agricultural enterprises, and women cooperatives.

Survey analysis was conducted, centered on three fundamental dimensions, which are economic viability, environmental sustainability and biodiversity integration, and socio-cultural sustainability.

22.3.3 Economic Performance and Constraints

The surveys used in drawing economic indicators indicate the high dynamism of the agricultural sector in the Souss-Massa Region. It shows that high value-chain integration amounts to around 55%, while product processing stands at around 75%. Dates, vegetables and argan oil have a high local demand. Early vegetables and aromatic and medicinal plants (AMPs) have a long-term demand worldwide. Over two-thirds of the participants affirmed the economic significance of the two food security sectors (cereals, honey, goat breeding, argan, market gardening) and export-oriented sectors (citrus, tomatoes, argan products). The level of product specialization is greater in export sectors, with only 20% of the surveyed respondents reporting serious post-harvest losses, which were primarily related to specific food products, including argan fruits.

Employment creation comes right after sector's economic performance, with food producing and export sectors registering significant job creation. In addition, there was high adherence to certification programs—especially food safety and quality standards, which portrayed a positive institutional and professional setup to integrate biodiversity. Nevertheless, there are still economic restraints. The main bottleneck was seen in limited production capacity, much of which has been caused by the rising frequency and severity of drought. This ratifies climate change as the greatest overall threat to biodiversity and agricultural productivity in the area.

22.3.4 Environmental and Socio-cultural Dimensions

Climate change was reported by most respondents (71%) as the major cause of agricultural ecosystem degradation in mountainous regions, and the pollution threat mostly affects intensive export-oriented systems in the plains, with new effects on food crops also emerging. In the socio-cultural context, the traditional agricultural systems are important in conserving biodiversity because of the transfer of intergenerational knowledge and use of local varieties and practices. Local products play an important role in food security, nutrition and culture. Women, youth, and vulnerable social groups in agriculture are more incorporated in both traditional and solidarity-based systems, but their involvement in decision making is nonetheless minimal, especially in modern-day agriculture.

Compliance with human rights is significantly maintained in export-based systems because of certification demands, and the working conditions in subsistence industries are frequently more insecure. On the other hand, chemical exposure poses higher health risks to the population than in conventional food producing sectors.

22.3.5 Major Lapses and Industry-Particular Problems

A SWOT was performed and shown in Table 22.1, which includes a sustainability risk study on 4 big sector groups including: (i) intensive export crops, (ii) extensive food crops, (iii) animal production (goat breeding) and (iv) beekeeping. Examples of cross-cutting gaps are:

- Poor awareness and understanding of the advantages of biodiversity among farmers and decision-makers;
- Lack of economic and policy incentives to promote practices that are friendly to biodiversity;
- Lack of access to local genetic materials and technical advisory support;
- Lacunae in applied research specific to the regional agroecological context;
- Lack of co-ordination between the stakeholders in agricultural development and biodiversity conservation.

Integration through biodiversity has been limited in the contemporary export-based farming by dependence on monocultures, synthetic inputs, mechanization, and a large water and soil footprint, but is still enforced by certification. However, corporate social responsibility (CSR) activities embraced by a few giant organizations offer access points to mainstreaming biodiversity, especially in the context of European Green Deal demands. Community-based and landscape-level strategies are needed to conserve and restore biodiversity in subsistence and domestic cash-crop land uses, and the forestry development organizations are central to this need. The production systems of animals need technological and organizational advances to cope with the shortage of forage and enhance sustainability. The beekeeping industry is currently under threat due to climate change and the extinction of the Saharan yellow bee that requires specific actions regarding conservation and restoring the habitat.

In general, the discussion shows that the Souss-Massa Region has the ecological, institutional, as well as socio-economic prerequisites to work out coherent strategic

Table 22.1 Sector-wise analysis of biodiversity values, assets, gaps, opportunities, and risks

Sector group	Biodiversity values	Strengths and assets	Gaps and weaknesses	Opportunities	Risks and threats
Productions for export (early fruits, red fruits, citrus)	Sustainable management of sectors Economic competitiveness	Professional commitment European green Deal Commitment to water policy RB framework plans	Limited regulatory capacities Insufficient coordination	Program PAGE Ecolabel MFX Generation green strategy (SGG) and SFM	Extension of modern agriculture Overexploitation of water resources Agrochemical toxicity
Food and cash crop production (cereals, arboriculture, PAM, Argan products, etc.)	Food security Building resilience	Local know-how Dynamic NGOs	Lack of knowledge and limited awareness Lack of incentives Limited access to resources	TREEA program Agroecology project Label FAO SIPAM SGG, SFM APA Nagoya	Overexploitation of biological resources
Production animals	Sustainable livestock management Role in peasant treasury Food security	ANOC program	Insufficient research and development	Inter-professional program contract SGG and SFM	Overgrazing Diseases and parasites Droughts and climate change
Beekeeping	Ecological pollination service Protection of local breeds of bees Food security	Establishment of an SDOQ and organic certification system	Delicate community management Poorly coherent policies	SGG and SFM	Invasive black races Diseases and pests Collapse syndrome Droughts and climate change

Legends: *SGG* Generation Green Strategy (Stratégie Génération Verte), *SFM* Sustainable Forest Management (or related framework), *PAM* Protected Agricultural Modules (or similar, based on context), *SDOQ* System of Quality and Origin Differentiation (Système de Différenciation et de la Qualité), *APA Nagoya* Access and Benefit-Sharing under the Nagoya Protocol

planning of an agri-food roadmap with integrated biodiversity. The diagnostic evaluation established the priority issues and leverage points that already have been used to inform the current and planned projects. Such results form the empirical foundation to scale up the roadmap at the national level, which can advance the policies of agricultural development based on sustainability, inclusivity, and integration to support the national efforts, like the PAGE program.

22.4 Alignment to National Strategy, Revised NBSAP of Morocco, and Kunming-Montreal Global Biodiversity Framework (KMGBF 2030)

To ensure the successful implementation of biodiversity integration into the Moroccan agri-food systems, it is necessary that the proposed roadmap be fully consistent with national development strategies, the revised National Biodiversity Strategy and Action Plan (NBSAP), and the goals of the Kunming-Montreal Global Biodiversity Framework (KMGBF). This kind of alignment makes sure that there is coherence of the policies, that there is institutional ownership, and also that there is mainstreaming of biodiversity into agricultural planning, implementation and monitoring.

22.4.1 National Sustainable Development Strategy (SNDD)

The roadmap aligns with the ideals of the National Sustainable Development Strategy (SNDD) of Morocco, which fosters equitable economic development, sustainable natural resource development, and balanced development of the territory. The roadmap enables biodiversity-friendly agricultural practices, which contribute to intergenerational equity, production system resilience, and sustainable ecosystem use with consideration of area specificities and local socio-economic demands.

22.4.2 National Climate Plan/Territorial Plan to Fight Global Warming

The roadmap will lead to the goals of the National Climate Plan (2020–2030) of Morocco, which will form the basis of low-carbon, climate-resilient development by reducing greenhouse gas emissions, enhancing energy use efficiency, and increasing access to renewable energy.

At the regional level, the roadmap is in support of the Souss-Massa Territorial Plan to Combat Global Warming (PTRC), organized around two complementary pillars:

- Mitigation, by means of the minimization of greenhouse gas emissions;
- Adaptation, such as sustainable management of water resources, coastal defense, climate-resilient agriculture and enhanced adaptive capacity of locals.

The roadmap has the greatest role to play in the adaptation pillar, especially in the promotion of resilient agricultural systems, agroecological practices, local knowledge, and agrobiodiversity—important factors toward coping with droughts and extreme climatic events.

22.4.3 Generation Green Strategy (2020–2030)

Introduced in 2020, the Generation Green Strategy is the main policy document concerning agricultural development in Morocco. It is organized in terms of two strategic orientations:

- Human capital and social inclusion, with the development of a new agricultural middle class, support to youth entrepreneurship, mobilization of collective lands, and reinforcement of professional organizations;
- Competitive and sustainable agricultural development, with the goal of doubling agricultural GDP and agricultural exports by 2030 and enhancing the performance of value chains and innovation ability as well as resilience.

The roadmap is also very much aligned with Generation Green, as it integrates conservation of biodiversity and improvement of ecosystem services in development of value chains, natural resource management and territorial planning. This correlation reinforces the sustainability aspect of the strategy and also makes sure that ecological integrity is not affected by agricultural growth.

22.4.4 Forest Strategy (2020–2030)

The Forests of Morocco 2020–2030 Strategy is aimed at conserving the forest resources, encouraging participatory management, and restoring ecosystem resilience. To achieve these goals, the roadmap will incorporate the drivers of deforestation, agroforestry systems, silvo-pastoral systems, and ecological connectivity of agricultural landscapes and forest ecosystems—which are critical in conserving biodiversity on the landscape level.

22.4.5 KMGBF Targets and National Biodiversity Strategy and Action Plan (NBSAP)

The roadmap is being formulated in parallel with the revised NBSAP of Morocco, in line with the KMGBF (2020–2030). It represents the four pillars of the Convention on Biological Diversity:

- Biodiversity conservation and restoration;
- Careful utilization of biological resources;
- Distribution of benefits derived on genetic resource;
- Knowledge generation and capacity building.

The roadmap specifically correlates with the following targets of KMGBF:

- Target 6 (Invasive Alien Species): Prevention, early detection, and management by reducing the risks and impacts;
- Target 7 (Pollution Reduction): Minimal pollution of the agricultural inputs, especially pesticides and loss of nutrients;
- Target 10 (Sustainable Agriculture): Implementing practices that are biodiversity-friendly, agroecology, and sustainable intensification, which is adjusted to both modern intensive systems and traditional extensive systems;
- Target 14 (Policy Integration): Policy development for the importance of biodiversity into agricultural policies, planning tools and governance of value-chains.

Each of the actions suggested in the roadmap clearly is associated with one or more of these KMGBF targets, including alignment with international biodiversity commitments.

22.5 Roadmap for the Effective Integration of Biodiversity into Morocco's Agri-Food Systems

22.5.1 General Objectives of the Roadmap

The general objectives of the roadmap will escort a transformative approach to the agricultural development in Morocco whereby biodiversity can be systematically engrained in the agri-food systems. First, the roadmap is focused on strengthening sustainable agricultural practices that would not only increase the agrobiodiversity but also use natural ecological processes and reduce the use of inputs that contaminate the ecosystems, i.e., the chemical fertilizers and pesticides. Second, it is improving the provision of ecosystem services and actively restoring natural and agricultural biodiversity through certain actions like the reintroduction of locally endemic species of flora and fauna, restoration of degraded habitats, and the establishment of in situ conservation sites on agricultural landscapes. Third, the roadmap is aimed at empowering local communities by ensuring that biodiversity conservation is economically viable and socially desirable, and by ensuring that livelihoods and food security are not compromised. These are rather aligned with the Kunming-Montreal Global Biodiversity Framework (KMGBF 2020–2030) and reflect the ideas of the Lima Action Plan on Biosphere Reserves which are centered around the concept of the integrated approach to biodiversity conservation, sustainable use and community engagement, research and education, political advocacy and international cooperation. All these are goals that bring sanity to the operationalization of biodiversity mainstreaming and to the development of resilient, equitable and sustainable agricultural systems across Morocco.

22.5.2 Values, Strengths, Gaps, Opportunities, and Risks

The opportunities and gaps synthesis of integration of the biodiversity systems into the agri-food systems of Morocco illustrates the multidimensional character which is interrelated and should be considered when implementing effective interventions. Biodiversity has intrinsic advantages as it helps to make the agricultural value chain robust and sustainable, economically competitive in the expanding and challenging domestic and international markets and provides food security and cultural identity in the conventional agro-ecosystems. The assets it possesses include traditional knowledge, already well-established, including Agdal systems, transhumance of bee hives, good engagement of certification schemes, consistency with international projects like the European Green Deal, new water resource management frameworks, well-organized governance systems in biosphere reserves, and active inclusivity of the civil society. However, the gaps remain extremely big and include the challenges of law enforcement, the absence of awareness and incentives toward biodiversity-friendly activities, access to genetic resources and technical advising services, applied research, and a loose cross-sectoral coordination. These gaps have in common with good opportunities, like the PAGE program, Green Trail Morocco, eco-labels, the Generation Green and forest strategies, the TREEA program and the soon-to-be-adopted Nagoya Protocol, which can make biodiversity integrate better. At the same time, the threats to the integrity of the ecosystem may be seen as the development of intensive agriculture, abusing water and biological resources, and overgrazing. It is on this backdrop that the strategic vision of the roadmap is to enable an agri-food system that actively defends the biodiversity, restores the dwindling ecosystems, maximally utilizes water, adds worth to local products, provides modest incomes for farmers and sustainable employment in country areas, and aligns environmental stewardship with social and economic sustainability.

22.5.3 Strategic Axes of Biodiversity Integration of Moroccan Agri-Food Systems

The agri-food systems of Morocco biodiversity-integration roadmap is designed as three strategic axes (Table 22.2), complementary and reinforcing to each other, with the purpose of informing policies, practice, and investment in the modern and traditional agricultural sectors as well as sustainable rural development.

Table 22.2 Strategic axis, goals, actions, responsibilities, risks, and mitigation measures

Strategies axis	Goals (corresponding target in the CMB)	Actions	Responsible (partners)	Risks	Mitigation measures risks
A. Strengthening the resilience of modern agricultural sectors intended mainly for export, to contribute to achieving carbon neutrality	Adopt circular economy principles from 2026 in line with the EU F2F strategy (business Target 16)	Implemented by all exporters sustainable water resources management plans, as part of their certification plans	Businesses (DRA, ABH, MFX, CB)	Low adoption due to limited access	Training and awareness
		Establish plans for sustainable waste management by all exporters, as part of their certification plans	Businesses (DRA, MFX, CB)	Low adoption due to limited access	Training and awareness
		Comply with at least EU requirements in packaging material by all exporters	Businesses (DRA, MFX)	Low adoption due to limited access	Training and awareness
	Minimize the consequences of climate change and halve GHG emissions by 2035 to achieve carbon neutrality in 2050 (Target 8)	Perform energy audits and improve energy efficiency and effectiveness	Businesses (DRA, ADEREE, MTEDD, MEF)	Low adoption due to limited access	(i) Raise awareness about the Morocco PME subsidy and others (ii) Implication instance de coordination PTRC
		Carry out the certification of carbon balance sheets in response to the requirements of the MACF (carbon border adjustment mechanism), the European green Deal from 2026	Businesses (certifying bodies, MFX)	Low adoption due to limited access	(i) Training and awareness (ii) Implication instance de coordination PTRC

	Reduce the use of pesticides by at least 50% (Target 7)	Implementation of a plan for the use of bio pesticides, bio-fertilizers and IPM (integrated disease management) from 2030 in accordance with the F2F strategy	(i) Businesses (ii) DRA, INRA, ONSSA, MTEDD, ONCA, MEF	Low adoption for limited access	Training and public incentives
B. Sustainability of traditional agricultural sectors through the maintenance and restoration of biodiversity, within the framework of social and solidarity economy models	Ensure that at least 30% of key biodiversity areas are effectively conserved (Target 3), including the creation of a network of ecological corridors by 2035 (Target 3).	Creation of new protection zones and corridors with the involvement of the ODF	ANEF (ONG, CT)	Low social acceptance	Raising awareness among rights holders within the ODF
	Encourage biodiversity offsets and revitalize the practice of Agdal (target 19)	Forestry CMD program over 100,000 ha in 2030 supported by the ODF	ANEF (ONG, CT, Wilaya)	Low acceptance	Strong involvement of rights holders
	Minimize the consequences of climate change (Target 8) and halve AFOLU emissions by 2030)	Support nature-based solutions (terraces, domestication, soil conservation, etc.)	WEAR (ANDZOA, DPA)	Low adherence	(i) Initiate pilot experiments in NZNP green pathways (ii) Implication instance de coordination PTRC
	Ensure that at least 30% of degraded ecosystems are subject to active restoration and rehabilitation measures by 2035 (Target 2)	Restoration of wetland habitats and honey, almond and carob orchards: (30,000 ha per year by 2030)	DRA and ANEF	Lack of institutional coordination	Implication instance de coordination PTRC
	At least 10% of agricultural land, are reserved for environmentally friendly practices by 2030 (Target 10)	Support for organic farming, agroecology and agroforestry and agritourism	(i) MAPMDREF (ii) (Program beneficiaries TREEA)	Weak institutional coordination	Involvement of EF to ensure coordination

(continued)

Table 22.2 (continued)

Strategies axis	Goals (corresponding target in the CMB)	Actions	Responsible (partners)	Risks	Mitigation measures risks
	Contribute to the resilience of oasis agricultural systems (Target 10)	Hydro-agricultural developments to secure the rational irrigation of traditional oases	ABH	Water insecurity	Equitable sharing of water resources
	Protect the yellow bee and strengthen pollination services (Targets 11)	Pollination charter, transhumance plan, yellow queen bee nurseries and measurement quarantine against black bees	FIMAP (ADA, DRA and other inter-professional organizations)	Lack of coordination	Training and awareness
	Conservation and protection of local seed systems, plants and breeds, on site and from the situation for the 13 current flagship local products and the future identified local products (Target 10)	Constitution of a regional gene bank of local ecotypes, local domestic animal breeds and a national observatory of the yellow bee	INRA (DRA)	Lack of sufficient resources	Advocacy
		Encourage conservation in-situ local ecotypes through financial support	ADA (INRA, DRA)	Low conservation interest	Awareness
		Review of the political and regulatory framework for promotion of the social and solidarity economy	ANDZOA-DRA-ANEF	Insufficient consistency with TREEA	Involvement of civil society for better collaboration
		Regional mapping of STAR metrics	MTEDD	Unavailability of experts	Online coaching
		Establish a regional plan for creating green jobs	MTEDD	Low employability	Raising awareness for the creation of green jobs
		Take financial incentives to encourage good practices agroecological through an FDA support project	MTEDD	Complexity of grant award procedures	Consultation with the profession

		Ban on neonicotinoid pesticides and other bee pests (Target 7)	ON (Interprofessions, professional chambers)	Weak means of control	Regulatory review
		Develop a resource mobilization plan to achieve capacity building	MTEDD	Complexity of procedures for allocating budget lines	Agreement with the profession and the Ministry of Finance
		Conduct a search for solutions to secure the rational irrigation of traditional oases	(i) ABH (ii) (MAPM)	Water insecurity	Equitable sharing of water resources
		Include the practice of Agdal in the forest code	ANEF (ANDZOA, DRA, interprofessions)	Delayed revision of the forest code	Advocacy
		Integration of biosphere reserves into law 22-07 on protected areas	MTEDD (ANEF, ANDZOA, MAB committee)	Delayed revision of the law 22-07	Advocacy
		Regulation of the use of treated wastewater to define standards for the reuse of agricultural effluents—implementation from 2028	(i) MTEDD (ii) (ABH, DRA, ONSSA)	Delay of adoption	Advocacy
		Adoption of the Nagoya protocol law from 2026 (Target 13)	MTEDD (ANEF, ANDZOA, interprofessions)	Delay adoption of bill	Register an ABS Nagoya component for advocacy
		Develop and carry out a project to develop a water management method in partnership with AWS, relating to groundwater exploitation contracts and a Souss water fund (PSE)	(i) MTEDD (ii) (ABH, AWS, companies)	Weak inter-institutional coordination	EF-AWS and ABH partnership agreement

(continued)

Table 22.2 (continued)

Strategies axis	Goals (corresponding target in the CMB)	Actions	Responsible (partners)	Risks	Mitigation measures risks
		Mobilize funds for the diversification of local products	(DRA, ANDZOA, ANEF, INRA, inter-professional organizations)	Weak institutional coordination	Partnership agreement
C2. Institutional measures to strengthen coordination and partnerships (Target 14)	Development of ELSA spatial planning maps by 2026	(i) MATNUHPV (ii) UNDP	Limited access to data	Partnership	
	Expand the remit of the regional coordinating body responsible for CCs to include biodiversity and the NSSD	MTEDD (CC and biodiversity central commissions)	Insufficient human and material resources/limited coordination	Regular monitoring and evaluation	
	Develop a global framework for dialogue and partnership (water, agriculture/ agroecology, forest management, civil society)	CRSM (AREP, Interprofessions, NGO, ABH, DRA, ANEF, etc.)	Insufficient coordination	Follow up and independent assessment of progress made	
	Revise the RB and NBSAP management framework plans to integrate the proposed roadmap to align with the new global BR strategy planned for 2025 and the CMBKM	ANEF	Non-binding nature of law 22-07 on protected areas	Revise law 22-07 to integrate RB	
	Establish a national policy on carbon credits	(i) MTEDD (ii) ANDZOA (iii) FAO	Lack of regulatory tools	Setting out appropriate regulations	
	Setting a monitoring and evaluation framework	MAPMDREF	Lack of coordination	Partnership	

Legends: *ABH* Agence de Bassin Hydraulique (Hydraulic Basin Agency), *ABS* Access and Benefit-Sharing, *ADA* Agence pour le Développement Agricole (Agricultural Development Agency), *ADEREE* Agence Nationale pour le Développement des Energies Renouvelables et de l'Efficacité Energétique (National Agency for Renewable Energy and Energy Efficiency), *AFOLU* Agriculture, Forestry and Other Land Use, *Agdal* Traditional Moroccan community-based system of rotational grazing and resource management, *ANEF* Agence Nationale des Eaux et Forêts (National Water and Forests Agency), *ANDZOA* Agence Nationale pour le Développement des Zones Oasiennes et de l'Arganier (National Agency for Oasis and Argan Zones Development), *APA* Autorité de Protection des Accès (Access Protection Authority), *AREP* Association pour la Recherche et l'Echange des Pratiques (Association for Research and Exchange of Practices), *AWS* Alliance for Water Stewardship, *BR* Biosphere Reserve, *CB* Certification Body, *CC* Climate Change, *CMB* Kunming-Montreal Global Biodiversity Framework (often abbreviated as GBF; here CMB likely refers to Cadre Mondial de la Biodiversité), *CMD* Carbon Management and Development (or specific program name), *CRSM* Conseil Régional du Secteur Maritime (Regional Maritime Sector Council), *CT* Collectivité Territoriale (Territorial Community), *DPA* Direction Provinciale de l'Agriculture (Provincial Agriculture Directorate), *DRA* Direction Régionale de l'Agriculture (Regional Agriculture Directorate), *EF* Établissement Forestier (Forest Establishment), *ELSA* Ecological Landscape Spatial Analysis (or specific mapping tool), *EU* European Union, *F2F* Farm-to-Fork (EU Strategy), *FAO* Food and Agriculture Organization, *FDA* Fonds de Développement Agricole (Agricultural Development Fund), *FIMAP* Fédération Interprofessionnelle de la Filière Apicole (Interprofessional Federation of the Beekeeping Sector), *GHG* Greenhouse Gas, *INRA* Institut National de la Recherche Agronomique (National Institute of Agricultural Research), *IPM* Integrated Pest (or Disease) Management, *MAB* Man and the Biosphere (UNESCO program), *MACF* Carbon Border Adjustment Mechanism (EU), *MAPM* Ministère de l'Agriculture et de la Pêche Maritime (Ministry of Agriculture and Maritime Fisheries), *MAPMDREF* Ministère de l'Agriculture, de la Pêche Maritime, du Développement Rural et des Eaux et Forêts (Ministry of Agriculture, Maritime Fisheries, Rural Development and Water and Forests), *MATNUHPV* Ministère de l'Aménagement du Territoire National, de l'Urbanisme, de l'Habitat et de la Politique de la Ville (Ministry of National Land Use Planning, Urbanism, Housing and City Policy), *MEF* Ministère de l'Économie et des Finances (Ministry of Economy and Finance), *MFX* Maroc Fruit Export (or similar export promotion agency), *MTEDD* Ministère de la Transition Énergétique et du Développement Durable (Ministry of Energy Transition and Sustainable Development), *NBSAP* National Biodiversity Strategy and Action Plan, *NGO* Non-Governmental Organization, *NSSD* National Strategy for Sustainable Development, *NZNP* New Zealand National Park (or specific initiative), *ODF* Office de Développement Forestier (Forest Development Office), *ON* Office National, *ONCA* Office National du Conseil Agricole (National Office of Agricultural Advisory), *ONG* Organisation Non-Gouvernementale (NGO), *ONSSA* Office National de Sécurité Sanitaire des Produits Alimentaires (National Office for Food Safety), *PAGE* Programme d'Appui à la Gestion de l'Environnement (Environmental Management Support Program), *PTRC* Projet de Transition vers une Résilience Climatique (Climate Resilience Transition Project), *PME* Petite et Moyenne Entreprise (Small and Medium Enterprise), *PSE* Payment for Ecosystem Services, *RB* River Basin (or Réserve de Biosphère—Biosphere Reserve), *SDOQ* Système de Différenciation et de la Qualité (System of Differentiation and Quality), *SFM* Sustainable Forest Management, *SGG* Stratégie Génération Verte (Green Generation Strategy—Morocco's agricultural strategy), *SIPAM* Systèmes Ingénieux du Patrimoine Agricole Mondial (Globally Important Agricultural Heritage Systems—FAO program), *STAR* Spatial Tool for the Assessment of Resilience (or specific metric), *TREEA* Transformative Resilience and Ecosystem Economics for Agriculture (or specific project), *UNDP* United Nations Development Programme, *WEAR* Water, Energy, Agriculture and Resilience (or specific entity), *Wilaya* Administrative region (governorate) in Morocco

22.5.3.1 Improving the Resilience of the Modern Export-Oriented Agriculture

Export-oriented irrigated agriculture in the region of Souss-Massa is primarily focused on high-value crops and is concentrated in the plains of Souss and Massa, the main agricultural products are primeurs, citrus fruits, tomatoes, and argan products. These are highly productive systems which are likely to suffer water deficiency, climatic changes and biodiversity loss. The previous strategic axis will be capable of addressing those problems by bringing an innovation toward Net-Zero, Nature-Positive (NZNP) trajectories. The interventions predominately include the adoption of agroecological methods, reduction of chemical fertilizers and pesticides, soil protection, precision of water and adoption of nature based methods, such as hedges, inter-cultivation and habitat corridors. This axis will see the contemporary sectors of production meet KMGBF Target 10 on sustainable land management by enhancing soil and water management and reinstating ecological processes within the sceneries of production, and deliver toward the long-term decarbonization targets of Morocco. The European Green Deal compliance is the requirement that massive farms and export cooperatives become sustainability-certified and introduce environmental reporting as a part of their European Green Deal compliance and make their supply chains biodiversity-friendly, which, in turn, would have a positive ecological and economic impact.

22.5.3.2 Traditional Agriculture, Biodiversity, and Social and Solidarity Economic Models

They are very agrobiodiverse agro-sylvo-pastoral systems, particularly in the Anti-Atlas, the High Atlas and the oases, with centuries of local experience, such as Agdal rotational grazing systems, and transhumance of bees, such as the Inzerki apiary. The second strategic axis is concerned with the preservation and recovery of such systems containing biodiversity and their sustainability in terms of the economy, using Social and Solidarity Economy (SSE) models. Part of the relevant actions includes helping the cooperatives, farmer associations, and agricultural social enterprises, which pool their resources to produce, process, and market their products and make them economically viable and value-added. The training programs will raise capacity in the location so that the integration of the conservation of biodiversity into the agricultural decision-making and participatory governing procedures will be able to assist the women, youth, and the minority to contribute actively to the management of the resources. Other restoration initiatives, such as local crop varieties, cull cattle breed conservation, and preserving pollinator habitat, are beneficial as a side effect in that they have an improvement in ecological integrity, food security, and cultural heritage. This axis defines the way that the conventional agriculture can be profitable and simultaneously able to sustain some significant ecosystem services.

22.5.3.3 Making the Enabling Environment to Biodiversity Integration

The third axis could be used to endorse the roadmap, as it establishes the institutional, regulatory and financial environment, which would be required in the successful achievement of biodiversity in agri-food value chains. They include built-in spatial planning to give space to conservation, sustainable agriculture and restorative ecology, lawful and regulatory practices to encourage practices that are biodiversity-friendly, and multi-stakeholder dialogue between farmers, cooperatives, agribusiness, scientists, civil society and policy-makers. Moreover, this axis has put its emphasis on mobilization of both the public and private funds, including green investments, ecosystem services payments, innovations and capacity-building investments. All this gives an enabling environment where biodiversity is not only surviving but it is being exploited well to achieve food security, human wellbeing, economic diversification and climate resiliency. This, in its case of Souss-Massa, can be aligned with the works of Generation Green, TREEA and the National PAGE program, keeping it in pace with the national plans and international obligations, including the obligations of the Targets of the CBD KMGBF.

The roadmap is a combination of these three strategic pillars that aims at capturing the whole spectrum of the agribusiness nature of the biodiversity challenges and opportunities in Morocco. In modern export agriculture, resiliency and capacity to address the impact of climate change increase; traditional systems are ecologically and culturally sound and achieve socio-economic sustainability; and institutional frameworks provide the governance, financial and policy support necessary to survive in the long-term. Combined, these axes will provide a synergistic route to a sustainable, nature-affirming, and inclusive agricultural sector, that is to say, biodiversity is both a pillar of the Moroccan agri-food security objectives, rural development, and climate adaptation, and their driver.

22.6 Strategic Plan for Biodiversity Integration

The roadmap is organized around three interlinked strategic pillars.

22.6.1 Strengthening the Resilience of Modern Export-Oriented Agriculture

Agricultural systems in the modern world have to shift to Net-Zero, Nature-Positive (NZNP) directions. This includes turning to agroecology, climate-smarter solutions, nature-based solutions, and less use of chemicals in accordance with KMGBF Target 10. To meet the requirements of the EU Farm-to-Fork (F2F) strategy, corporations are advised to embrace the concept of a circular economy, which includes longer lifespan of a product, reuse, repair, and recycling of materials among others. This transition that will be adhered to in 2026 will require an evaluation of environmental impacts, adjustments of value chains, multi-stakeholder cooperation, and disclosure. The long-term targets are the halving of pesticide application by 2030; this will be accompanied by a reduction of greenhouse gas emissions by 2035 and full carbon neutrality by 2050. Through nature-positive solutions, contemporary agriculture can help promote climate mitigation, biodiversity recovery, and sustainable economic development at the same time.

22.6.2 Sustainable Biodiversity in Traditional Agriculture

Conservation and restoration of biodiversity need specific efforts in traditional agroecosystems that have a high number of endemic species and local knowledge. SSE models such as cooperatives, social enterprises and social associations help create equitable management of natural resources and make it economically viable. Some of the major actions involve the protection of 30% of biodiversity-important regions, the restoration of 5000 ha of wetlands, the restoration of bee-producing habitats, the creation of ecological lanes, the promotion of pollination pacts, and the protection of local seeds and breeds. These measures are supplemented by multilateral mechanisms of Nagoya Protocol and ITPGRFA that guarantee equitable sharing of benefits and create green workplaces and oasis resilience.

22.6.3 Enabling Environment: Building Environment to Facilitate Biodiversity Inclusion

Axis C serves the two above-mentioned axes by institutionalizing the concept of biodiversity in policies, regulations, and governance. There are measures such as formal recognition of biosphere reserves, incorporation of Agdal practices into forest codes, restrictions on the use of pesticides, and regulatory reforms in compliance with the Nagoya Protocol. The complementary measures include a biodiversity-sensitive budget, green job skills measurement, rational oasis irrigation planning studies, social and solidarity economy policy promotion, and financial incentive systems. There are institutional arrangements used to strengthen the coordination and the regional collaboration, bilateral and multilateral projects on sustainable water management, and biotechnological innovation of products in the local market. The combination of these measures enables the mainstreaming of biodiversity in all agricultural and agri-food systems in the long run.

22.7 Policy Implications and Recommendations

The roadmap should be implemented in a multi-level program which involves a mixture of:

- Enhancing sustainability in contemporary agriculture to achieve export and climatic goals.
- Protecting and preserving biodiversity in conventional systems and supporting social inclusion and SSE models.
- Building sound institutional structures, financial rewards and policy coherence.

In such a way, Morocco will be able to reach agri-food systems that are resilient, biodiversity-rich, secure in food security, economic growth, and environmental stability to meet national and international commitments, such as KMGBF, Paris Agreement, and SDGs.

22.8 Conclusion

Souss-Massa is the high biodiversity region with a more diversified agricultural system that is increasingly facing the pressure of the environment, climate variability, and food security issues. The chapter has also traced the road map toward adoption of biodiversity in the agri-food systems that can give the vision within the strategy that needs to transform the region into sustainable, resilient, and equitable agricultural development. The strategic synthesis of the strengths, weaknesses, opportunities, and threats of the sector in the study identifies the areas of leverage that are to be employed in improving agrobiodiversity, increasing ecosystem services, and improving the livelihoods of the local people. The three pillars of the roadmap, i.e., the strengthening of the sustainability of the present-day export-oriented agriculture, the preservation of the traditional forms of agriculture with the models of the biodiversity and social and solidarity economy, and the creation of the favorable institutional and regulatory environment, are complementary to reaching the reconciliation between the economic growth and the sustainability of the environment and social equity. The application of strategies such as agroecology, solution-oriented nature, the circular economy, and the position of local products as the key force behind the biodiversity-oriented agriculture is presented. The roadmap will require political determination on the implementation, stakeholders' coordination, capacity building, and mobilization of financial resources. The proposed SMART goals and action plan present useful guidelines on how the biodiversity integration would be achieved in the agri-food chains of the region. In conclusion, this roadmap is a unique opportunity for the Souss-Massa region to transform it into the model of innovative and sustainable agricultural development. By utilizing the ecological and cultural resources available in the region, bridging the existing gaps, and utilizing the emerging opportunities, the region could achieve resilient food systems, biodiversity conservation, and contribute to the national and global biodiversity and sustainable development goals.

References

Adam A (2020) Elements for a rural geohistory of Mediterranean beekeeping (Corsica and Southwest Morocco). Thesis Paul Valery University Montpellier 3

Agency for Agricultural Development (2020) The Generation Green Strategy

Ahmed B (2023) COP15 at the CBD: contribution to the deciphering of the Kunming-Montreal agreement on the global biodiversity framework for post-2020 – what prospects for Morocco? CNDB, Morocco

Alain K (2024) Ecological taxation how to implement a tax bonus-malus mechanism? CIRAD

ANEF, "Forests of Morocco 2020–2030" Strategy

AZURA (2022) rapport RSE 2021–2022

BEI (2014) EIB study on the Social and Solidarity Economy in the Mediterranean: https://www.eib.org/attachments/country/femip_study_femise_economie_sociale_solidaire_fr.pdf

BETAF (2022) This profile of the Souss Massa region

CBD (2022) Kunming Montreal Global Biodiversity Framework including the theory of change and targets 7, 10 and 14

CDB (2022) Décisions de la quinzième CdP. In: Fifteenth meeting of the conference of the parties to the convention on biological diversity (Part Two). https://www.cbd.int/meetings/COP-15
CESE (2015) Report on the Social and Solidarity Economy. https://www.cese.ma/media/2020/10/Rapport-L%E2%80%99Economie-Sociale-et-Solidaire-un-levier-pour-une-croissance-inclusive%E2%80%8B.pdf
FAO (2020) FAO strategy on mainstreaming biodiversity across agricultural sectors. Rome. https://doi.org/10.4060/ca7722en
Faouzi H (2015) Agdal in the dynamics of the agrarian systems of the argan plantations of Haha (Western High Atlas, Morocco). The Agdal in the Dynamics of Farming Systems of the Haha (Western High Atlas, Morocco). https://doi.org/10.4000/etudescaribeennes.5569
FRB (2021) Indicators-and-tools-to-measure-the-impact-of-human-activity-on-biodiversity. Foundation for Biodiversity Research, 51 pages
GEF- UNEP (2018) Mainstreaming biodiversity conservation and sustainable use into production landscapes
Haddouch (2015) A circular economy approach to agro-biodiversity conservation in the Souss-Massa Draa region of Morocco. Responsible Natural Resource Economies in Africa
IUCN (2020) General guidance for using the IUCN global standard for nature-based solutions, 1st edn. IUCN, Gland
Ministry of the Environment of the Kingdom of Morocco -UNDP-GEF: National Strategy and Action Plan for the Biological Diversity of Morocco, 2016–2020
MTEDD (2023) Initiative pour l'intégration de la biodiversité dans les systèmes agroalimentaires pour la mise en œuvre du Cadre mondial de la biodiversité Kunming-Montréal – Région de Souss Massa
PAGE (2022) Morocco's transition to a green economy: current status and inventory. MTEDD
Payment contract model for ecosystem services applied to the honey sector, 2021, Souss Massa Regional Directorate of Agriculture
Sous Massa Regional Council (2018) Summary of the territorial plan to combat global warming in the sous Massa region Souss Mass
Study for the development of a bee pollination charter in the Souss Massa region, 2018, Souss Massa Regional Directorate of Agriculture
UNEP (2022) Guidance on integrating human rights in National Biodiversity Strategy and action plans (NBSAPs)
World Bank 2017 and 2021. Program for results for strengthening agri-food value chains - Environmental and Social Systems Assessment (ESA)
WWF (2021) Study on the identification of drivers of biodiversity loss and key sectors impacting biodiversity in Tunisia. BIODEV 30
WWF (2022a) Practical steps to reform and repurpose biodiversity harmful subsidies in the Post-2020 Global Biodiversity Framework WWF. Discussion Paper, 11 p, March 2022
WWF (2022b) Integrating Biodiversity into Development, Guyana National Report, BIODEV30-Agro-industrial sub-sector selection report for the project

MIX
Papier aus verantwortungsvollen Quellen
Paper from responsible sources
FSC® C105338

If you have any concerns about our products,
you can contact us on
ProductSafety@springernature.com

In case Publisher is established outside the EU,
the EU authorized representative is:
Springer Nature Customer Service Center GmbH
Europaplatz 3, 69115 Heidelberg, Germany

Printed by Libri Plureos GmbH
in Hamburg, Germany